36th Congress of the European Brewery Convention (EBC 2017)

Ljubljana, Slovenia
14 - 18 May 2017

ISBN: 978-1-5108-5391-1

Printed from e-media with permission by:

Curran Associates, Inc.
57 Morehouse Lane
Red Hook, NY 12571

Printed by Curran Associates, Inc. (2018)

European Brewery Convention
c/o The Brewers of Europe
Rue Caroly 23-25
1050 Brussels
Belgium

Phone: 32 2 551 1810
Fax: 32 2 660 9402

info@brewersofeurope.org

Additional copies of this publication are available from:

Curran Associates, Inc.
57 Morehouse Lane
Red Hook, NY 12571 USA
Phone: 845-758-0400
Fax: 845-758-2633
Email: curran@proceedings.com
Web: www.proceedings.com

SCIENTIFIC PROGRAMME

Monday, 15 May 2017		
	Opening Session – **Tiago Brandão, HR Director, Unicer Bebidas, S.A., Porto, PT, and EBC President**	
08:30 – 09:00	Opening: Opening address by ZPPP chairman **Matej Oset**, President of the Association of Slovene Brewers 1	
09:00 – 09:30	L1: Key-note 1: **Paul Bruijn** (Heineken, Zoeterwoude, The Netherlands): The PEF project at The Brewers of Europe. An Update 7	
09:30 – 10:00	L2: Key-note 2: **Larry Nelson** (Brewers' Guardian, Reigate, UK): The continued craft brewing phenomena: Industry competitive responses 22	
10:00 – 10:30	Refreshment Break	
	Plenary Session – **Tiago Brandão, HR Director, Unicer Bebidas, S.A., Porto, PT, and EBC President**	
10:30 – 11:00	L3: Key note 4: **Henrike Vorwerk** (VLB, Berlin, Germany): Joint Euromalt/Gushing Project: Conclusions N/A	
11:00 – 11:30	L4: **Julien Billard**, et al. (IFBM-Qualtech, Nancy, France): Identification of factors at the origin of the phenomenon of beer gushing and development of a premature detection technique of the gushing risk 30	
11:30 – 12:00	L5: **Takahiro Yamazaki** (Kirin, Yokohama, Japan): Development of novel hop-derived bitter acid oxides with body fat-reducing effect 42	
12:00 – 12:30	L6: Key note 3: **Peter Kraemer** (AB-Inbev, US): A better future to beer market 56	
12:30 – 14:00	Lunch	
	Myriam Shingleton, VP Group Development, Carlsberg Breweries A/S, Obernai, FR Yeast & Fermentation I (Parallel I)	**Paul Bruijn, Senior Global Lead Safety & Environment, Heineken Supply Chain, Zoeterwoude, NL** Environmental (Parallel II)
14:00 – 14:30	L7: **Krogerus**, et al. (VTT, Espoo, Finland): Construction of novel lager yeast strains through a fertile allotetraploid intermediate. 72	L8: **Blomenhofer**, et al. (Ziemann-Holvrieka, Ludwigsburg, Germany): An energy self-sufficient brewery – the flagship project becomes a reality. 86
14:30 –15:00	L9: **Aritomo**, et al. (Sapporo, Yaizu, Japan): High-throughput mechanical measurement of brewing yeast cells by atomic force microscopy 99	L10: **Ritala**, et al. (VTT, Espoo, Finland): Reuse of second steeping water improves eco-efficiency of malting. 115
15:00 – 15:30	L11: **Gosselin**, (Université Libre de Bruxelles (ULB), Brussels, Belgium): Shelf-life and consistency of active dry yeast for breweries. 124	L12: **Offenbacher**, (BDI BioEnergy, Grambach, Austria): Spent grain to energy – biogas plant Göss. 139
15:30 – 16:00	L13: **Noba**, et al. (Asahi, Moriya-shi, Japan): Elucidation of the formation mechanism of 2-mercapto-3-methyl-1-butanol (2M3MB) in beer. 145	L14: **Folz**, (Pentair, The Netherlands): The role of a beer membrane filter in an environmental friendly brewery. 163
16:00 – 16:30	Refreshment Break	
17:00 – 17:30	TOUR at the UNION BREWERY and MUSEUM OF BREWING, Ljubljana	
17:30 – 23:00	RECEPTION at the UNION BREWERY, Ljubljana	

Linhart Hall   Kosovel Hall

Tuesday, 16 May 2017		
	Clemens Forster, Supply Chain Support Manager, Brau Union Österreich AG, Linz, AT Analysis I (Parallel I)	**Carsten Zufall, Gen. Corporativo de Technol. e Innov., Cervecería Polar, Caracas, VE** Process Development (Parallel II)
08:30 – 09:00	L16: **Oomuro** (Asahi, Moriya-shi, Japan): 179 Genetic analysis of bottom-fermenting brewer's yeast imparting good kire (crispness) to beer ...	L17: **Sparacio** (Simatec, Vaie, Italy): Scaling & logistics rethought: multi-function mash-tun, kettle & whirlpool. 201
09:00 – 09:30	L18: **Lafontaine**, et al. 212 (Oregon State University, Corvallis, USA): Understanding the impact hopping rate has on the aroma quality and intensity of dry hopped beers.	L19: **Jörg**, et al. (Centec, Maintal, Germany): Improved dealcoholization process of beer and its market potential. 238
09:30 – 10:00	L20: **Peterherr** (Anton Paar, Graz, Austria): Alcolyzer beer ME – Results of alcohol analysis on beer samples with an alcohol content lower than 0.4% v/v. 268	L21: **Michel**, et al. (GEA Brewery Systems, Kitzingen, Germany): The use of hops in brewhouse and cold block. 249
10:00 – 10:30	Refreshment Break	
	Martin Ketterer, Chief Technical Officer, Brauerei Schützengarten, St. Gallen, CH Analysis II (Parallel I)	**Sandra Stelma, Head of Science, Diageo, Dublin, IE** Product Development (Parallel II)
10:30 – 11:00	L22: **De Clippeleer**, et al. N/A (KU Leuven, Ghent, Belgium): Towards a new comprehensive methodology for the evaluation of free and bound aldehydes in malt, wort and beer.	L23: **Kunov-Kruse**, et al. (SpecShell, Valby, Denmark): Perspectives of real-time carbohydrate profiling. N/A
11:00 – 11:30	L24: **Steinhaus** (Leibniz Institute, Freising, Germany): Influence of variety, provenance, and processing on the concentrations of 4-mercapto-4-methyl-2-pentanone (4MMP) in hops. N/A	L25: **Ito**, et al. (Asahi, Moriya-shi, Japan): Development of a non-alcoholic beer with a novel idea undertaken by casting off preconceptions about beer. 279
11:30 – 12:00	L26: **Watson**, et al. (UGent, Belgium): Approach for quantitative analysis of hordein peptides in gluten-free malt beers with High-Resolution-Orbitrap-MS.	L27: **Steffen**, et al. (University College Cork, Ireland): Polyols as natural sweeteners from lactic acid bacteria in oat wort-based beverages.
12:00 – 12:30	L28: **Olšovská** (RIBM, Prague, Czech Rep.): Journey into history - analysis of 100-year-old beer. 299	L29: **Matsche** (Barth-Haas, Nuremberg, Germany): Influence of yeast fermentation on dry hop aroma and flavour. 312
12:30 – 14:00	Lunch	
	Anneli Ritala, Principal Scientist, VTT Technical Research Centre of Finland Ltd., Espoo, FI Sensory (Parallel I)	**Monica Mandrutiu, Brewing Lead Europe & Innovation Brewer, Asahi Breweries Europe Ltd., RO** Yeast & Fermentation II (Parallel II)
14:00 – 14:30	L30: **Gadzov** (FlavorActiv, Chinnor, UK): Off-flavours due to microbiological problems in speciality beers in Europe. 322	L31: **Meier-Dörnberg**, et al. (TUM WZW, Freising, Germany): Genetic and phenotypic characterization of different top-fermenting Saccharomyces cerevisiae ale yeast isolates. 332
14:30 –15:00	L32: **Baldus**, et al. (TU Berlin, Germany): Chemical-analytical characterization of potential prooxidative effects of malt and beer-derived antioxidants. N/A	L33: **Schmitt**, et al. (IFBM-Qualtech, Nancy, France): EBC wort turbidity: nature and impact of malting and brewing process. 344
15:00 – 15:30	L34: **Oladokun**, et al. (University of Nottingham, UK): New insights into bitterness perception of beer. 366	L35: **Preiss**, et al. (Guelph University, Ont., Canada): Characterization of Kveik, traditional Saccharomyces yeasts domesticated by Norwegian homebrewers. 380
15:30 – 16:00	Refreshment Break	Posters
16:00 – 17:00	Poster presenters to be available for Q & A	
	Free Evening	

SCIENTIFIC PROGRAMME

Tuesday, 16 May, 17:00 – 20:00

Public lectures (held only in Slovenian)

Congress Venue

Public lectures about precious beer accompained with the degustation of Slovenian beers: the aim of these lectures is to bring the science and love for beer to the general Slovenian public.
The main message: The beer Congress is in our town, Ljubljana in Slovenia, not any Congress but one of the most important World Beer Congresses ... What do we know about beer in general, what do we know about Slovenian beer and about our green gold: hops? The red line in the lectures will be hops and Slovenian beers, guests will be able to learn the basics: how beer is made, where the hops joins the beer, what can Slovenia offer in the hops world, the influence of hops on the beer as a product – the varieties, basic techniques for tasting beer, ending with the final lecture about Beer & Health and clear message for responsible drinking. Tasting of the Slovenian beer samples, gift for each participant.

Price: € 10,00

Tickets are available from 20. April 2017, can be purchased on-site at the Congress Venue or via internet sales.
No pre-registration required.
Tickets will be available one month before the lectures will take place.

Max: 220 Pax

Združenje Slovenskih Pivovarn GIZ

The Brewers of Europe

Zgodba o slovenskem hmelju in pivu

Wednesday, 17 May 2017		
	Anne Pietercelie, Brewing Science and Fermentation Technology, Institut Meurice – HELB, Brussels, BE Beer Quality (Parallel I)	**Deniz Bilge, Snr. Consultant & Deputy Head, Research Institute for Beer & Beverage Production, VLB Berlin, DE** Brewhouse (Parallel II)
08:30 – 09:00	L36: **Voigt**, et al. (Univ. Appl. Sc., Trier, Germany): A new analytical approach to malt aroma compounds. N/A	L37: **Harmegnies**, et al. (Meura, Péruwelz, Belgium): MEURASTREAM - A global concept of wort production with reduced consumptions of energy and process water. 391
09:00 – 09:30	L38: **De Roos**, et al. (VUB, Brussels, Belgium): The role of acetic acid bacteria (AAB) in lambic beer production processes is underestimated. N/A	L39: **Becher**, et al. (Ziemann-Holvrieka, Ludwigsburg, Germany): The transformation of old tradition – a novel brewhouse concept. 403
09:30 – 10:00	L40: **Neiens** (Leibniz Institute, Freising, Germany): Aroma-active compounds in novel German flavor hops. N/A	L41 **Zarnkow**, et al. (TUM WZW BLQ, Freising, Germany): Influence of the grist fractions on the final beer quality. 417
10:00 – 10:30	Refreshment Break	
	Roland Folz, Hygienic Process Solutions, Pentair Process Technologies, Venlo, NL Filtration & Packaging (Parallel I)	**Stefan Hanke, Leiter Versuchsbrauerei, Bitburger Braugruppe, Bitburg, DE** Raw Materials (Parallel II)
10:30 – 11:00	L42: **Zacharias**, et al. (Krones, Neutraubling, Germany): Solving the challenges of natural compressible filter aids for processing precoats on candle filters. 549	L43: **Decloedt**, et al. (UGent, Belgium): Industrial feasibility of the production of gluten-free barley malt beers on a day-to-day basis using different technologies.
11:00 – 11:30	L44: **Jastřembská** (Univ. Pardubice, Czech Rep.): Membrane separation technologies in brewing. 434	L45: **Lagemann**, et al. (Bitburger Braugruppe, Bitburg, Germany): Behavior of plant protection substances during hop growing and beer production. N/A
11:30 – 12:00	L46: **Osterroth**, et al. (TUM, Freising, Germany): Energy consumption of beverage bottling plants. 445	L47: **Boivin**, et al. (IFBM-Qualtech, Nancy, France): EBC wort turbidity: Nature and impact of malting and brewing process 460
12:00 – 12:30	L48: **Werner**, et al. (TUM WZW, Freising, Germany): Using pulsatile target jets to clean spent grain loaded filter cloths. 483	L49: **Maye** (Hopsteiner, New York, USA): Dry hopping and its effect on beer bitterness, the IBU test, foam, and Ph. 492
12:30 – 14:00	Lunch	
	Benet Fité Luis, Director de Calidad, I+D+I y Medio Ambiente, Madrid, ES, and EBC Vice-President Generalist Session	
14:00 – 14:30	L50: **Sirrine** (Michigan State University, Suttons Bay, USA): Hop production in re-emerging U.S. growing regions: How craft beer is reshaping the raw material supply sector. N/A	
14:30 – 15:00	L51: **Hutzler**, et al. (TUM WZW BLQ, Freising, Germany): Advanced Yeast Hunting. 513	
15:00 – 15:30	L52: **Biendl** (Hopsteiner, Mainburg, Germany): How to analyze dry hopped beer? 528	
15:30 – 16:00	Refreshment Break	
16:00 – 17:15	**Tiago Brandão, John Brauer** CLOSING CEREMONY: 1) Summary of papers and impressions 2) Announcement of best paper and best poster 3) Acknowledgements to session chairs and LOC/Interplan 4) Announcement of congress 2019 5) Organisational announcements re. farewell party	
19:30 – 23:00	Farewell Party/Buffet Dinner at the National Gallery	

Linhart Hall   Kosovel Hall

POSTERSESSION ON TUESDAY, 16 MAY 2017 AT 16:00 PM

	Abstract Title	First Author	Institute, City, Country
	Analysis and methodology (raw materials, product, research)		
P001	Development of a mycotoxin screening sensor system usable in industrial grain sorting – OptiScreen – N/A	Frank-Jürgen, Methner	Technische Universität Berlin, Berlin, Germany
P002	An improved beer color independent enzymatic method for α- and β-glucan determination in beverages N/A	Thomas, Kunz	Technische Universität Berlin, Berlin, Germany
P003	Improved HS Trap GC-MS analysis of hop aroma compounds in dry hopped beer 568	Annika, Lagemann	Bitburger Braugruppe GmbH, Bitburg, Germany
P004	Improved PCR detection of beer spoiling bacteria – optimized enrichment with NBB and innovative lateral flow detection 569	Agnes Johanna, Huber	Döhler GmbH, Darmstadt, Germany
P005	Key-role of humulinones, polyfunctional thiols and and terpenols in Belgian dry-hopped beers N/A	Carlos, Silva Ferreira	Université catholique de Louvain, Louvain-la-Neuve, Belgium
P006	Volatile fingerprinting of coriander and orange peel in view of improved flavour-consistency of specialty beers and distilled spirits 570	Filip, Van Opstaele	KU Leuven, Ghent, Belgium
P007	A new RT-PCR-based method for the direct monitoring of spoilage bacteria in beer samples N/A	Daniel, Gerhards	Pall GeneDisc Technologies, Bruz, France
P008	LC-MS/MS analysis of hop flavonoids in dry-hopped beers 571	Christina, Schmidt	Hopsteiner - HHV, Mainburg, Germany
P009	Chemical studies on bitter acid oxides derived from stored hops (Humulus lupulus L.) 572	Yoshimasa, Taniguchi	Kirin Company Limited, Kanazawa-ku, Yokohama-shi, Japan
P010	State of the art solutions for beer analysis N/A	Uwe, Oppermann	Shimadzu Europa GmbH, Duisburg, Germany
P011	Development of a carbohydrate mass balance during malting and mashing 573	Niels Albert, Langenaeken	KU Leuven, Heverlee, Belgium
P012	Application of a central composite design for optimization of an HS-SPME-GC-MS method for determination of volatiles in beer 574	Jakub, Nešpor	University of Chemistry and Technology, Prague, Czech Republic
P013	Comparative study of accelerated shelf-life tests on physical ageing of beverages containing weighted flavour emulsions 575	Sören, Rossmann	Institute of Food Technology ILT.NRW, Lemgo, Germany
P014	Sensory and analytical changes in beer-mix beverages during shelf life 576	Helmut, Klein	Brau Union Österreich AG, Linz, Austria
P015	The language of hops – How to implement and apply a common description of hop flavours 577	Elisabeth, Wiesen	Joh. Barth & Sohn, Nürnberg, Germany
P016	First identification of cysteinylated and glutathionylated precursors of 3-mercaptohexan-1-ol in Barley. What is the influence of malting process? 578	Aurélie, Roland	Nyseos, Montpellier, France
P017	Investigation of mycotoxins in different beers with the Mycotoxin Screening System using several cleanup techniques 579	Carola, Schultz	Shimadzu Europa GmbH, Duisburg, Germany
P018	Evaluation of a rapid haze stability measurement for beer N/A	Kevin van Koerten	NIZO, Ede, Netherlands

P039	Contribution of compositional profile of iso-alpha acid to the quality of bitterness N/A	Toru, Kishimoto	ASAHI BREWERIES, LTD., Moriya-shi, Ibaraki Japan
P040	Effect of retronasal aroma on kire of beer N/A	Seiko, Miyashita	ASAHI BREWERIES, LTD., Moriya, Japan
Beer, society, health and nutrition			
P041	Occurrence of biogenic amines in craft corn beers and commercial barley beers 596	Luis Alberto, Villarreal-Sanchez	Metropolitan Autonomous University, Mexico, Mexico
P042	Impact of dry hopping techniques on xanthohumol content of beer 597	Armindo, Melo	Pharmacy Faculty of University of Porto, Porto Portugal
P043	Evaluation of α-acids content of beer produced by dry hopping techniques 598	Julio Cesar, Machado Jr	Pharmacy Faculty University of Porto, Porto, Portugal
P044	Antagonistic effects of tetrahydro-iso-alpha acids on breast cancer cells N/A	Dominique, Gallo	Institut Meurice - Commission Communautaire Française (Cocof), Brussels, Belgium
P045	Antiproliferative activity of hop-derived compounds against human colon cancer Caco-2 cells N/A	Miguel A., Faria	LAQV-REQUIMTE, Porto, Portugal
P046	Brewing yeasts secrete trophic, anti-inflammatory and antioxidant biomolecules in their surrounding medium: why not in beer? N/A	Monique, Dillemans	Meurice R&D - Institut Meurice, Brussels, Belgium
Consumer studies			
P048	Beer taxation in Europe – A consumer friendly framework? N/A	Jan, Lichota	Universidad Nacional de Educación a Distancia, Madrid, Spain
P049	Uncovering patterns of consumers' preference for beer: a case study with craft beer (CB). N/A	Gianluca, Donadini	Università Cattolica del Sacro Cuore, Piacenza, Italy
Environmental issues and sustainability			
P050	Further use of PVPP regeneration caustic for CIP applications – A significant step to enhance an environmental friendly beer production 599	Philipp, Zeuschner	VLB Berlin, Berlin, Germany
P051	Valorization of brewer's spent yeast 600	Isabel, Ferreira	REQUIMTE, University of Porto, Faculty of Pharmacy, Porto, Portugal
P052	Study on the implementation of a biocatalytic power-to-gas reactor in a brewery 601	Timo, Broeker	Institute of Food Technology ILT.NRW, Lemgo, Germany
P053	Wet oxidation of spent grains from breweries for the production of fermentable sugars 602	Jan, Schneider	Institute of Food Technology ILT.NRW, Lemgo, Germany
Filtration			
P054	Effect of wort filtration type on final blond beer: industrial case study 603	Anita, Van Landschoot	Ghent University, Ghent, Belgium
P055	Faster wort and beer filtration, from lab to large brewery scale 604	Wilbert, Heijne	DSM, Delft, Netherlands
P056	Studying fouling in crossflow microfiltration of rough beer 605	Àron, Varga	Szent István University, Budapest, Hungary
P057	Filtration and stabilization optimization and its effect on quality relevant parameters such as EAP 606	Thomas, Zimmermann	BASF South East Asia Pte Ltd., Vice Director Regional Technical & Application Management, Singapore, Singapore

Hop breeding, agronomy and processing			
P058	Effect of harvest time on health promoting activities of hop polyphenols 607	Takako, Inui	Suntory Beer Ltd., Osaka, Japan
P059	Are different hop pellet properties for dry hopping and hop dosing in the brewhouse justified? 608	Adrianw, Forster	HVG eG., Wolnzach, Ġermany
P060	Identification of geranic acid contributing to varietal aroma in SORACHI ACE and synergy with other hop-derived flavor compounds 609	Atsushi, Tanigawa	SAPPORO BREWERIES LTD, Yaizu Shizuoka, Japan
P061	Recent hop varieties from Slovenian breeding program 610	Andreja, Čerenak	Slovenian Institute of Hop Research and Brewing, Žalec, Slovenia
P062	Speeding up hop breeding through in vitro seed germination and precocious sex determination 611	Claudia, Liberatore	UNIPR, Parma, Italy
P063	Dry Hopping – a new look at techniques, utilisation and economics 612	Ray, Marriott	Totally Natural Solutions Ltd, East Peckham, United Kingdom
P064	Description and classification of two new huell aroma breeding varieties 613	Andreas, Gahr	Hopfenveredlung St. Johann GmbH, Train – St. Johann, Germany
Malting: malt production and quality			
P065	Selective steering of roasting processes to reduce prooxidative effects of roasted malt products N/A	Thomas, Kunz	Technische Universität Berlin, Berlin, Germany
P066	Varietal comparison of cytolysis of barley during malting 614	Xiang S, Yin	Rahr Malting Co., Shakopee, United States
P067	A study on advanced DON reduction technologies in malting 615	Patricia,Aron	Rahr Malting Co., Shakopee, United States
P068	Characterisation of key aroma compounds in two special malts by means of aroma extract dilution analysis 616	Veronika, Mall	German Research Center for Food Chemistry, Leibniz Insititute, Freising, Germany
Management issues			
P069	Comparative analysis of brewhouse efficiency, energy and materials cost in three types of microbreweries 617	Panagiotis, Tataridis	Technological Educational Institution of Athens Aegaleo, Athens, Greece
P070	Management of research and developent against extreme changes of market N/A	Yoshinori, Ito	Asahi Breweries, Moriya, Japan
P071	WS Brew – Weihenstephan Standards for brewing processes 618	Raik, Bär	Technical University of Munich / Chair of Food Packaging Technology, Freising, Germany
New process development			
P072	Development of a continous fermentation procedure for the main fermentation by immobilisation of the yeast in a crossflow reactor N/A	Burghard, Meyer	VLB-Berlin, Berlin, Germany
P073	Investigation of a new crystallizer with included post-treatment for ion depletion of spring water for the brewery industry 619	Lars, Erlbeck	Mannheim University of Applied Sciences, Mannheim, Germany
P074	Development of a membrane assisted bioreactor for the continuous fermentation of wort into beer N/A	Brecht, De Causmaecker	KU Leuven, Faculty of Engineering Technology, Technology Campus Ghent, Ghent, Belgium
P075	Strategies to improve lactic acid production from brewing-relevant lactic acid bacteria strains in wort substrates N/A	Konstantin, Bellut	University College Cork, Cork, Ireland

P076	Flash pasteurization: How long is the optimal heat-holding time/temperature for achieving the best beer quality? 620	Roland, Feilner	Krones AG, Neutraubling, Germany
P077	On the way to the "Smart Brewery" How to better profit from automation and IT technology. N/A	Emilian, Axinia	COPA-DATA GmbH, Salzburg, Austria
P078	Monitoring the CO_2 purity level during fermentation as a parameter of quality monitoring and control 621	Martin, Jörg	Centec GmbH, Maintal, Germany
P079	"Asahi Super Dry" development strategies for beer dispensers challengingly contributing to the increase in its brand value N/A	Junichi, Kitano	Asahi Breweries, Ltd., Moriya-Shi,Ibaraki, Japan
P080	Hop citrus character – why are all the molecules floral? 622	Ray, Marriott	Totally Natural Solutions Ltd, East Peckham, United Kingdom
P081	Mini-oak chips: a promising tool for development of flavour enriched beer 623	Vojtěch, Hanko	University of Chemistry and Technology, Prague, Czech Republic
P082	Insight into the American IPA. A deconstruction of America's popular beer style	Thomas, Shellhammer	Oregon State University, Corvallis, United States
Novel raw materials for beer production			
P083	Can brewing with unmalted quinoa produce good beers? 624	Edyta, Kordialik-Bogacka	Lodz University of Technology, Lodz, Poland
P084	Sensory profile of a gluten free beer made 100% with corn malt 625	María Angelica, Romero-Medina	Universidad Autónoma Metropolitana, Mexico City, Mexico
P085	Comparative brewing performances of quinoa, amaranth, einkorn, millet, buckwheat and chia to produce beer 626	Anne, Pietercelie	Institut Meurice, Brussels, Belgium
P086	CO_2 from Anaerobic Digestion sources a novel and safe source for the food and beverage industry? 627	Frank, Verkoelen	Haffmans BV, Venlo, Netherlands
P087	Characterization of free and esterified phenolic compounds in blue corn malt beers N/A	Alex M. Daniela, Flores-Calderon	Metropolitan Autonomous University, Mexico City, Mexico
Wort production			
P088	Analysis of significant factors in the formation of sugars during the mash process in beer production and its modeling with neural networks 628	Héctor, Rivera	Tecnológico Nacional de México Instituto Tecnológico de Celaya, Celaya, Mexico
P089	Evaluation of poly phenols and bitter substances derived from hops and their yield by producing hopped wort concentrates for the production of beer and beer type malt beverages 629	Jean, Titze	Anhalt University of Applied Sciences, Koethen, Germany
P090	Monitoring of an onion-like off-flavor component and its precursor in large-scale brewing 630	Taku, Irie	Asahi Breweries, Ltd., Moriya, Japan
Yeast and fermentation			
P091	Characterization of ale yeasts – a study of technological, analytical, biological and sensory attributes 631	Fabian, Webersinke	Brau Union Österreich AG, Linz, Austria
P092	Genetic and phenotypic changes in brewing yeast Saccharomyces pastorianus during long term serial re-pitching 632	Neža, Čadež	University of Ljubljana, Ljubljana, Slovenia
P093	The use of selected strains of Lachancea thermotolerans for brewing of sour beer 633	John, Sheppard	North Carolina State Brewery, Raleigh, United States

HISTORY OF EBC 17

70 YEARS OF EBC. THE EARLY YEARS.

Before World War II there was no international organisation looking after brewers, brewing and beer. In those days, the dichotomy of brewing practices – the Anglo-Saxon, Irish top fermented beers on one hand, the Bohemian-Czech, German-Austrian and Danish bottom fermentation systems on the other. However, even in those days, there was one organisation which could be called an early form of an international brewing association and that was the London-based Institute of Brewing. The institute was beginning to extend its contacts to the brewing world into the wider realm of the British Empire by the 1930's. On the European continent obviously no such unity could be expected, given the hostile political climate prevalent at that time.

Still, it needs to be stressed that Germany was leading the field in brewing science with names still familiar to us today, the VLB, or Weihenstephan. An important name from those days was Prof. Max Delbrück, chair of brewing science in Berlin, lending his name to the Delbrück point, now familiar to most post-graduate science students across the globe. Valuable research was also performed in Copenhagen at the Carlsberg Laboratories and it was here even before World War I that Søren Sørensen described the concept of pH value, pertaining to the relative strengths of acids and bases as measured by the concentration of hydrogen ions in solution.

It is at this point that I wish to pay special tribute to Mr. Philippe Kreiss, Director General of the Brasseries de la Meuse in eastern France. Born in German occupied Alsace in 1887, he was convinced of the need for European brewers to forge alliances for better raw material procurement, and to exchange ideas and protocols leading to an improved product quality. He had a keen interest and respect for scientific work connected with barley, hops and fermentations but realised that most of the work was done in relative isolation, as there was virtually no connectivity among the various brewing researchers in Europe at that time. Immediately after the war Philippe Kreiss sprung into action, contacting breweries and research institutions.

Mr. Philippe Kreiss, Director General of the Brasseries de la Meuse (formerly G. Schneider-Domeck in Schiltigheim, Alsace) and first EBC President (1947 – 1963)

The first exploratory meeting took place in Paris under the name Centre Continental de Brasserie. As a result of these discussions, and owing to the most helpful attitude of the former CBK (as the Dutch brewers' association was known at the time), the first congress could be announced. The Dutch generosity extended to the registration fees as they offered to paid for everyone of the 188 delegates at the first congress in 1947, taking place in the small seaside resort of Scheveningen. At this first EBC congress, the name European Brewery Convention was adopted and Philippe Kreiss became the first president of the association.

It was a diverse assemblage of academics, agronomists and brewers that got together in Scheveningen and over the years a pattern would establish itself. There were the eminent professors and deans from universities offering brewing degrees attending the congress, alongside other scientists who had chosen to be in the employment of brewing companies. I some cases these brewing companies had become large and powerful leading to the establishment of research institutes as early forerunners of R&D departments. Worth noting here were the departments at Guinness in Ireland, the Whitbread and Bass laboratories in the UK, and the central research station of Munich-based breweries. I have already mentioned the very fine contributions of the Carlsberg Laboratories.

I know that it is an impossible task to highlight the achievements of some individuals in furthering our knowledge of the intrinsics of beer whilst disregarding possibly equally deserving contributions. For what it is worth, here is my personal check-list of some key presentations given during the first 20 years of the EBC, 1947 to 1967.

- **1949 (EBC Lucerne). B.D. Hartong:** Properties and composition of chill haze. In this paper, Dr. Hartong describes the mechanisms responsible for the formation of cold-induced turbidity in beer. This had an enormous impact of the science of filtration (pre-complexing of haze removed at cold temperatures), as well as being able to understand how hazes would form under certain conditions in bottle and, by inference, to subsequent consumer reaction.
- **1951 (EBC Brighton). A. Lund & R.B. Gilliland:** Investigation into the flocculation phenomenon of yeast. This was an early attempt by this Danish-Scottish research group at explaining the electrostatic interactions between yeast cells at the end of fermentation. This led to a better understanding of beer maturation, especially the kinetics of vicinal diketone control in beer.
- **1953 (EBC Nice). J. de Clerck:** Investigation into the relationship between malt-derived proteins and colloidal stability in beer. Prof. De Clerck's formative work resulted in a better understanding of stability in beer and laid the foundations for avoidance or reduction strategies for haze in beer.
- **1955 (EBC Baden-Baden). M. Meilgaard:** Hop utilisation in the brewing process and how to improve it; (together with P. Kolbach's work on the effect of pH on hop utilisation). Here we witness the beginning of optimising raw materials from a cost basis whilst taking care not to compromise product quality.

Prof. Dr. Hugo Theorell, Swedish biochemist and recipient of the 1955 Nobel prize for Physiology & Medicine. Chairman of the organising committee for EBC 1965 in Stockholm and key note speaker.

- **1957 (EBC Copenhagen). E. Urion, L. & M.S. Chapon:** Systematic studies on the oxidation of beer. This study, carried out at the IFBM laboratories, made an enormous contribution to the then infant science of flavour stability which, with increasing volumes of beer being filled as bottles and cans, was becoming more and more important to brewers exporting their products further afield.
- **1959 (EBC Rome). J.S. Hough & A.D. Rudin:** Continuous fermentations. The post-war boom taking root in Europe in the late 1950 and early 1960's had meant taking a long and hard look at capacity constraints. There was a sizeable fraternity in those days believing in the concept of continuous fermentations. Putting this into practice, with consistent results on product quality, has proved – for the most part – an elusive and difficult goal for brewers.
- **1961 (EBC Vienna). A. Hartl & W. Kleber:** Separation and identification of substances making up hop bitter components. It had been known before that other, non-alpha-acid bitter components contribute to the sensation of bitterness in beer. These researchers used an early generation chromatographic technique (adsorption to polyamide 9 – perlon) to elucidate these compounds.
- **1963 (EBC Brussels). A.M. MacLeod:** Gibberellic acid in the germination of barley. Based on the work by Prof. Anna MacLeod, most malting plants nowadays use trace amounts of a plant growth substance (GA3) to achieve a high level of germination during the process.
- **1965 (EBC Stockholm). T.-M. Enari, et al:** Identification of different ß-amylases in barley and malt. By identifying different ß-amylases and elucidating their behaviour, a better understanding of the biosynthesis of saccharifying enzymes was gained; additionally, maltsters and brewers now had better tools to control enzyme formation thereby optimising the production of fermentable extract.
- **1967 (EBC Madrid). M. Verzele:** Progress in the analysis of hop derived substances. This research group based in Ghent successfully optimised a paper-based chromatography procedure for the analysis of bitterness inducing hop fractions, before the advent of chromatography columns.

In 1963, at the age of 76, Monsieur Kreiss passed away. The brewing world was in mourning. His successor was none other than Dr. Jean De Clerck, professor at the brewing department of the Université catholique de Louvain in Belgium. With Jean De Clerck at the helm, the EBC entered the late sixties and early seventies against the backdrop of impressively rising production and consumption figures for beer in Europe.

At the president's table (Lucerne 1949): Prof. Jean De Clerck, Mr. Philippe Kreiss and Dr. F. Mendlik (Secretary General of the EBC).

John M. Brauer
(EBC Executive Officer), Brussels April 2017

LOCAL ORGANISING COMMITTEE 2017

Matej Oset is the President of the Slovene Brewers' Association and Supply Chain Director of Pivovarna Laško Union, d.o.o., a member of the Heineken group of companies. Born 1968 in Celje, Matej studied at the University of Ljubljana. After graduating with an MBA from Bled Business School (Slovenia), he held the job of CTO (Chief Technical Officer) of Pivovarna Laško Group for 6 years. He was an EBC council member from 1996 to 2003 and played a key role in motivating for Slovenia to gain full membership of The Brewers of Europe in 2012.

Alenka Lesjak is the Secretary General of the Association of Slovene Brewers. Born 1968 in Celje she holds a B.Sc. (Chemical Sciences, biochemical engineering) from the University of Ljubljana, studied Economics at Faculty of Economics and Business. Worked in the breweries more than twenty years, starting in the technology, former Head of quality control, has an extensive knowledge for overall brewing and ecology and sustainable projects. Former head of beer product development in connection to the customer researches, organising educational events from the brewing field for the past few years before joining the Association of Slovene Brewers in 2016.

WORDS OF WELCOME

WELCOME BY THE EBC PRESIDENT

On behalf of the European Brewery Convention it is a pleasure to welcome you to the 36th EBC congress in Ljubljana.

This EBC congress is set to take place 14 to 18 May 2017. Our congress will be held at the Cankarjev Dom conferencing facilities in direct vicinty of the historical part of Ljubljana.

We are happy to report that the venue chosen provides an excellent setting for our technical, technological and scientific lectures and presentations.

As in previous congresses, there will be a good balance between the technical lectures and social programme. Based on requests from various sources throughout the past few years have led us to reduce the overall number of speakers slightly; however, the number of posters has stayed unchanged.

With my experience of the EBC congress 2015 in Porto in mind, I very much support the initiatve of the Local Organising Committee to offer a "broad-band" technical tour on Thursday, 18 May, to all delegates. This means combining a visit to the IHPS, the Slovenian hop research and brewing institute, to the Laško brewery all in one day. This approach will ensure maximum exposure of key components of the Slovenian brewing sector to delegates of the EBC congress.

No visit of Ljubljana would be complete without a visit to the city's flagship brewery, the Union. The brewery combines state-of-the-art technology housed in a historic building. On the same site, the brewery housed a wonderful museum, displaying "breweriana" and other beer-related items, such as drinking vessels and early advertising of Union beers.

With my best wishes for a superb congress,

Tiago M. Brandão

HR Director Unicer Bebidas SA and EBC President

CONGRESS VENUE

Cankarjev Dom

Cultural and Congress Centre
Prešernova c. 10
Ljubljana - Slovenia

Phone +386 (0)1 24 17 100
E-mail info@cd-cc.si
www.cd-cc.si/en

The Cankarjev Dom is located in the heart of Ljublijana and can be reached within only 25 minutes from the airport.

INTERVIEW

SUPPORTING THE RENAISSANCE IN EUROPEAN BREWING

The beer landscape in Europe is changing, fast. Our communications director, Jan de Grave, spoke to Tiago M. Brandão, EBC President and HR Director of Unicer Bebidas S.A. and Pierre-Olivier Bergeron, Secretary General of The Brewers of Europe, to understand how their respective organisations view and are adapting to these changes.

Tiago M. Brandão

Pierre-Olivier Bergeron

Jan de Grave:
Can you explain the relationship between The Brewers of Europe and EBC, for those attendees who are unfamiliar with the workings of both associations?

Pierre-Olivier Bergeron:
The Brewers of Europe is the voice of the Europe's 7,500 breweries in Brussels, uniting 29 national beer associations. The EBC is since 2007 The Brewers of Europe's scientific and technological arm, with a focus on technical aspects of beer brewing knowledge.

We are operating in a very dynamic and exciting sector, and have therefore been developing a roadmap for the future. You could say that the European brewing landscape is experiencing something of a renaissance.

Jan de Grave:
A European brewing renaissance – can you elaborate?

Pierre-Olivier Bergeron:
What we're seeing is an increasing number of people who are interested in beer and how it is made. In some parts of Europe, also where wine making has traditionally dominated the local market, we've seen a number of small, local breweries emerging and reaching out to a new generation of consumers. We're also seeing a re-invigoration of medium-sized breweries who are exporting more of their products at a European and global level. To add to the picture, alongside these new breweries and re-energised medium breweries, the large breweries are continuing to invest in the sector and to sell their beer at a global level.

Jan de Grave:
What does this mean for The Brewers of Europe and the EBC?

Pierre-Olivier Bergeron:
It represents a unique and exciting opportunity for the brewing sector in Europe. The question is: how can we connect with these new breweries, and create a real community of European brewers and suppliers, irrespective of their location or size? How can we best transfer knowledge, expertise and experience to serve all brewers across Europe?

Tiago M. Brandão:
The toolkit on the calculation of the nutritional values of beer developed by The Brewers of Europe and EBC might be a good example of the added value we can bring to those new brewers. This toolkit is aimed at helping brewers to determine the nutritional values of their beers by drawing brewers' attention to available methods and advice that would help them to analyse or calculate the nutritional values of their beers.

Jan de Grave:
How will you bring the sector together to create this community?

Pierre-Olivier Bergeron:
For some time now, we've been considering hosting a major event for the entire European beer and brewing sector. And having seen the success of the EBC over its remarkable 36 editions, we were very keen to work with the EBC team to create this event. I'm delighted to announce that our idea will be realised, and that we'll be hosting the first edition of this new event in 2018, with EBC team closely involved.

The plan is for the event to be hosted in a different European country each year (much the same as the EBC has been). We feel this will help act as a global showcase for the variety, breadth and quality of European beer.

Jan de Grave:
How is the EBC involved in these plans?

Tiago M. Brandão:
We support the vision of a truly representative event and see the need for a showcase that caters for the needs of all brewers, something for those just starting in the sector and for those more established professionals.

Since the inaugural EBC Congress was held in 1947, we have created a landmark event in the brewing calendar, that offers an unrivalled opportunity to share knowledge and has played its part in developing methods for quality control, procedures in brewing, and quality management institutions throughout the world. The change is that the 37th EBC Congress to be held in Antwerp will form part of a bigger event. Actually, everything you would expect from the EBC Congress, but more!

Meanwhile, we will continue to listen to the needs of our members wishing to host technical events. For example, there have been a lot of recent advances in the science of hops. We are planning an EBC symposium on this topic, to be hosted with the support of the German and Bavarian Brewers' Federations. Hops plays a key role in the production of beer – this fact is shared by all brewers, no matter what size! So, watch our for more information regarding this symposium in 2018.

It is evident that the EBC will support the new format even in years when there is no EBC congress. Our proven track-record of putting on brewing technical meetings will help to make those events more successful. In this particular case we are looking at filtration seminar/symposium to highlight the recent developments in this important arena.

Jan de Grave:
What can we expect from the 2018 new format event?

Pierre-Olivier Bergeron:
The brewing renaissance that we are experiencing means that we aim to host an event that welcomes everyone.

It will feature several different 'tracks' to cater for different interests, which will run concurrently, including the technical and commercial side of brewing, growth strategies and market trends and regulation.

Jan de Grave:
What are the benefits for attendees?

Pierre-Olivier Bergeron:
We all have so much to learn from each other. For example, you can see how the experience and knowledge held by The Brewers of Europe and the large brewers could be very helpful to smaller brewers facing real challenges like how to scale up businesses while retaining quality and consistency. We can offer scientific and technical support, how to approach marketing to grow the customer base, or how to navigate an ever increasing set of regulations, including areas such as labelling. Plus, we will make sure to recruit the most important brewing stakeholders and suppliers, so that networking, procurement and even trading opportunities may occur.

Plans are well underway. It's really important to us that the event adds value to the whole sector, and particularly smaller brewers, so we're ensuring they are involved in the event design and development. We have, so far, entered an implementation stage and we have set up a programme team which is chaired by the former President of the Belgian Brewers Association, Mr. Theo Vervloet.

We will be increasing our advertising activities in the coming months, so make sure to look out for more information.

LIST OF EXHIBITORS AND SPONSORS

Actemium	Belgium, Aalter	20
Anton Paar GmbH	Austria, Graz	01
APPLITEK NV	Belgium, Nazareth	26
Bucher Unipektin AG - Competence Center Filtration	Switzerland, St. Gallen	19
CARA TECHNOLOGY LTD/AROXA	United Kingdom, Leatherhead, Surrey	12
DSM Food Specialties B.V.	The Netherlands, Delft	SPONSOR
EJDFoodSci - PhD school in brewing	Poland, Krakow	05
FERMENTIS	France, Marcq-en-Baroeul cedex	24
FlavorActiV	United Kingdom, Oxfordshire	02
GEA Brewery Systems GmbH	Germany, Kitzingen	SPONSOR
Hamilton Bonaduz AG	Switzerland, Bonaduz	25
I.F.B.M.-QUALTECH Groupe	France, Vandoeuvre-les-Nancy	03
IHPS and Society of slovene hop growers	Germany, München	19A
Institute of Brewing & Distilling	United Kingdom, London	22
Institute of Hop research and Brewing SIHRB	Slovenia, Žalec	
KD VeSave A/S	Denmark, Vejle	11
KHS GmbH	Germany, Dortmund	16
LEUVEN INSTITUTE FOR BEER RESEARCH	Belgium, Leuven (Heverlee)	04
MEURA	Belgium, Peruwelz	23
National Collection of Yeast Cultures	United Kingdom, Norwich	27
PACKO PUMPS	Belgium, Diksmuide	07
Pall Food and Beverage	Germany, Bad Kreuznach	13
Pentair Food & Beverage	The Netherlands, Venlo	18
RITZENHOFF AG / STEKLARNA HRASTNIK	Germany, Marsberg / Slovenia, Hrastnik	29
SHIMADZU EUROPA GmbH	Germany, Duisburg	09
Siemens AG	Germany, Nürnberg	28
Skalar Analytical B.V.	The Netherlands, Breda	21
SPX Flow Technology Santorso s.r.l.	Italy, Santorso (VI)	14
Stabifix Brauerei-Technik KG	Germany, Gräfelfing	SPONSOR
stayconcrete GmbH	Luxembourg, Contern	30
TENSID-CHEMIE GmbH	Germany, Muggensturm	10
Thermo Fisher Scientific	Finland, Vantaa	17
VELP Scientifica	Italy, Usmate	32
VLB Berlin	Germany, Berlin	06

WE THANK OUR SPONSORS

ACTEMIUM
Anton Paar
FlavorActiV™
BUCHER filtrox
cara TECHNOLOGY
HRASTNIK1860
European Joint Doctorate Food Science – International PhD school in malting & brewing
Fermentis – LESAFFRE FOR BEVERAGES
KD VeSave
GEA – engineering for a better world
HAMILTON
LAŠKO – Pivovarna Union – PART OF THE HEINEKEN COMPANY
IBD – Institute of Brewing & Distilling
iFBM-Qualtech GROUPE
KHS – Filling and Packaging – Worldwide
KU LEUVEN – INSTITUTE FOR BEER RESEARCH
MEURA
ncyc – www.ncyc.co.uk
Packo Pumps
PALL
PENTAIR
RITZENHOFF Cristal
SHIMADZU – Excellence in Science
SIEMENS – Ingenuity for life
Skalar
SPXFLOW – APV – Seital Separation Technology
stabifix
stayconcrete
TC TENSID CHEMIE – HYGIENE CONCEPTS TECHNOLOGY
ThermoFisher SCIENTIFIC
AppliTek
VELP SCIENTIFICA
VLB BERLIN
DSM – BRIGHT SCIENCE. BRIGHTER LIVING.

WELCOME TO SLOVENIA

ZDRUŽENJE SLOVENSKIH PIVOVARN GIZ

May 15th 2017

IN THE HEART OF EUROPE

36. EBC CONGRESS,
Ljubljana, May 15th 2017

IN THE HEART OF SLOVENIA

LJUBLJANA - the green capital of Europe 2016

36. EBC CONGRESS,
Ljubljana, May 15th 2017

EBC 2017 HOST IS SLOVENE BREWERS ASSOCIATION

CELEBRATING TWO ANNIVERSARIES - 70 YEARS OF EBC & 20 YEARS OF SLOVENE BREWERS ASSOCIATION

ZDRUŽENJE SLOVENSKIH PIVOVARN

36. EBC CONGRESS,
Ljubljana, May 15th 2017

Slovenian economy is slowly recovering

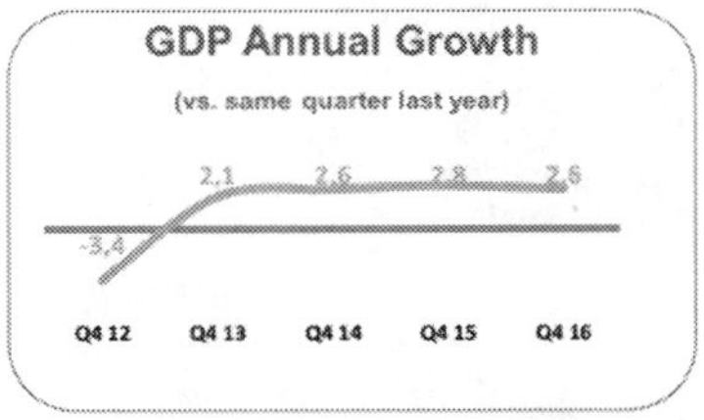

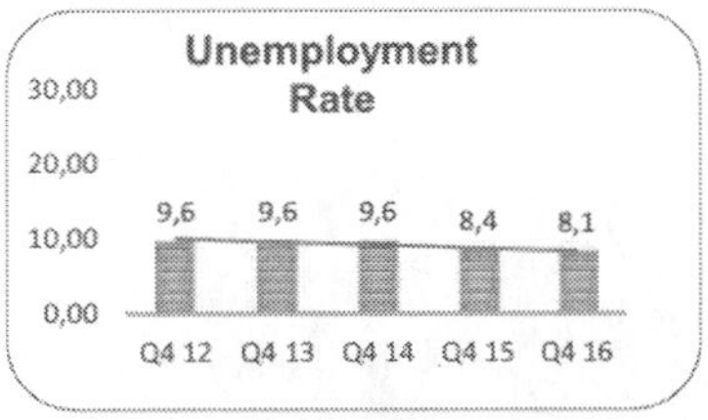

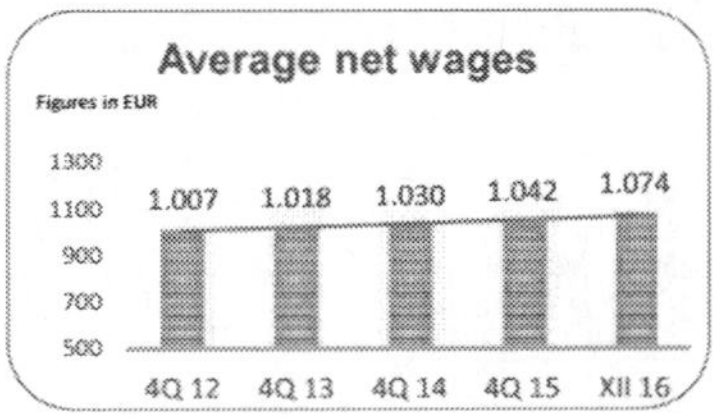

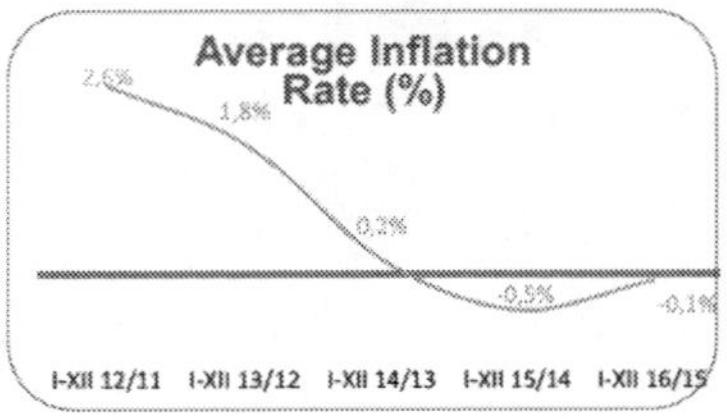

ZDRUŽENJE SLOVENSKIH PIVOVARN

Source: Statistical Office of the Republic of Slovenia.

36. EBC CONGRESS,
Ljubljana, May 15th 2017

Slovene Beer Market

Private labels and Excise Duty influence market the most

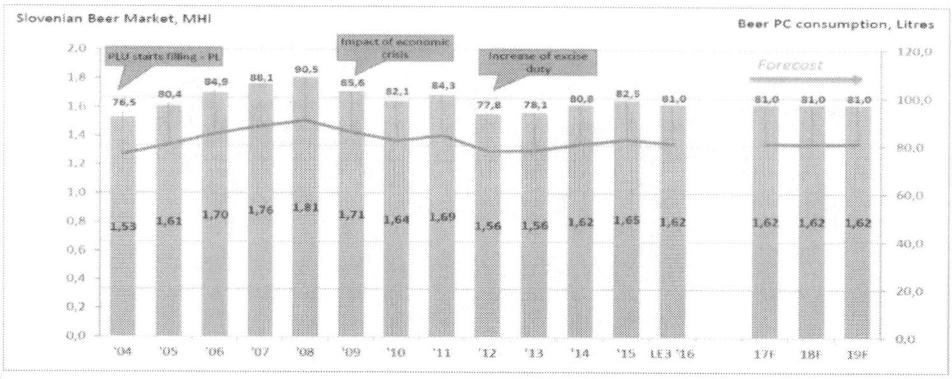

Source: Statistical Office of the Republic of Slovenia.

36. EBC CONGRESS,
Ljubljana, May 15th 2017

Slovene Beer market facts

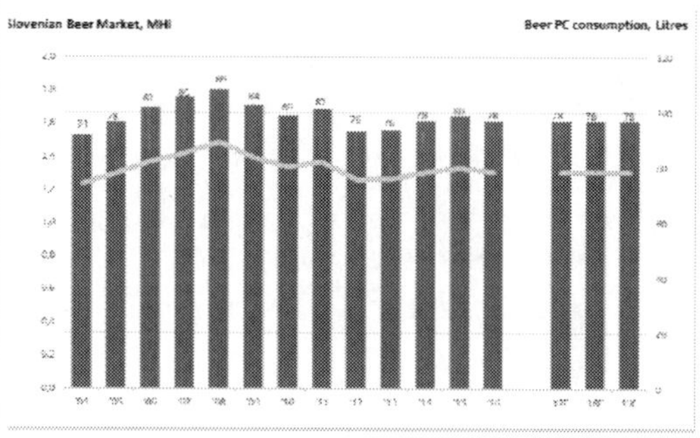

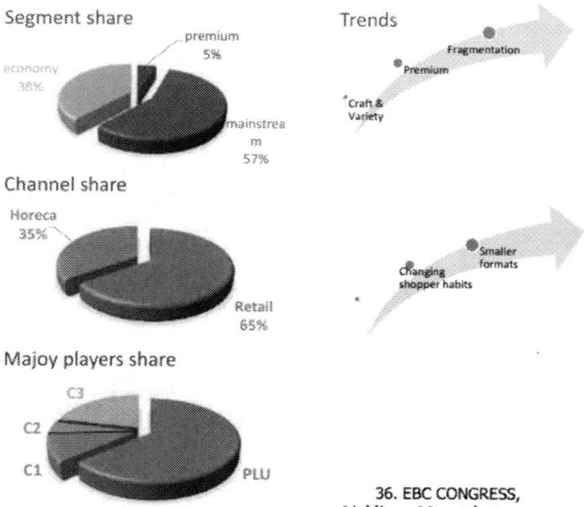

tistical Office of the Republic of Slovenia.
, internal data

36. EBC CONGRESS,
Ljubljana, May 15th 2017

Excise Duty on Beer

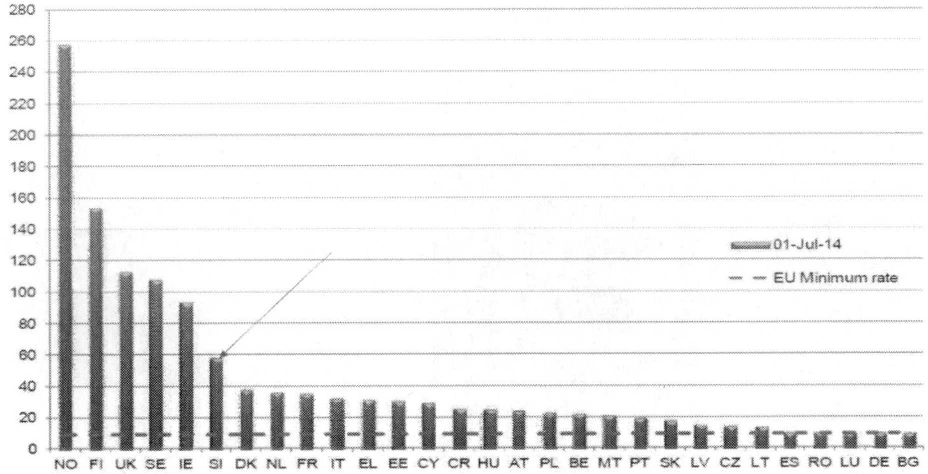

36. EBC CONGRESS,
Ljubljana, May 15th 2017

tistical Office of the Republic of Slovenia.
s. internal data

Slovene Beer market facts

Current situation

- Stagnating beer consumption 78 l/pc
- Increasing Fragmentation; 12 (sub)brands on average
- High excise duty – 6th place in EU
- Mostly lager market
- High share of private label
- Can over glass

But the future looks bright

- ✓ Bust of innovations on the market
- ✓ Growth of micro breweries/brands
- ✓ Awarenes of health benefits
- ✓ Collaboration among major brewers and and craftbrewers.

ZDRUŽENJE SLOVENSKIH PIVOVARN

36. EBC CONGRESS,
Ljubljana, May 15th 2017

Slovene Beer market trends

Like most of Europe

- Premiumisation vs economisation
- Wellness and moderate drinking
- Consumer demand for variety
- Increasing presence of craft
- Use of new technologies in marketing and sales

And also

- Leveraging from both scales
- Highest number of breweries per capita
- Vineyards

ZDRUŽENJE SLOVENSKIH PIVOVARN GIZ

36. EBC CONGRESS,
Ljubljana, May 15th 2017

THANK YOU FOR YOUR ATTENTION

ZDRUŽENJE SLOVENSKIH PIVOVARN GIZ

May 15th 2017

The Brewers of Europe

Building the Single Market for Green Products

Paul Bruijn

Senior Global Lead Safety and Environment
Global Production, HEINEKEN Supply Chain
Coordinator The Brewers of Europe BEER PEF PILOT

The Brewers of Europe

WHY?

=

Confusion, mistrust
Free-riders win
Costs

- More than 400 environmental labels in the world
- Only for GHGs, 80 leading reporting methods and initiatives
- Issues:
  - What is green?
  - How do I prove that my product or company is green?
  - If I choose one approach, will it be accepted by everyone?
  - Do I have to prove I'm green in different ways to different clients?
  - Will consumers and business partners understand my claim?

SINGLE MARKET FOR GREEN PRODUCTS

The Brewers of Europe

➢ **Objective**

To improve the availability of clear, reliable and comparable information on the environmental performance of products and organisations

➢ **How?**

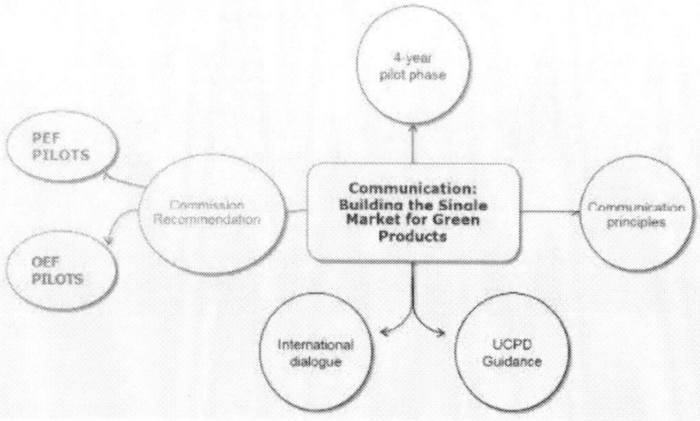

PEF PILOTS: goals

The Brewers of Europe

- ➢ To set up and validate the process of the development of Product Environmental Footprint Category Rules – PEFCRs
- ➢ To test different compliance and verification systems
- ➢ To test different B2B and B2C communication vehicles for Product Environmental Footprint information

- ➢ The pilot will inform future policy and regulation as it relates to product environmental impact assessments, verification systems, declaration activities, as well as communication methods.

THE BREWERS OF EUROPE: motivation to join the pilots

- To influence methodological development concerning Product Environmental Footprinting (PEF), which longer term **will** lead to policy
- To be sure that future PEF methodologies developed are acceptable and practical to use for the entire beer industry. (Both big and small brewers)
- To position the brewing sector and The Brewers of Europe as a proactive, ambitious, environmental concerned industry and association, which are willing to invest in a methodological development, in order to ensure, that our consumer in the future will get an objective and correct information about the environmental impacts from our products.

PEF GOVERNANCE

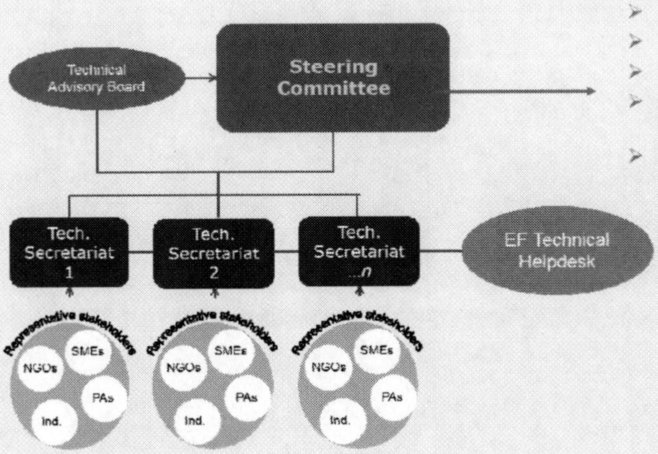

- MS representatives
- Commission
- Representative from pilots
- Representatives of main stakeholder groups
- Approvals, monitoring and conflict resolution

THE BEER PEF PILOT

TS MEMBERS

CONSULTANTS

BEER PEFCR: content

- Clear definition of scope
- Definition of representative product(s)
- Benchmark(s)
- Identification of the most relevant life cycle stages, impacts, and processes (for which primary data shall be produced)
- Calculation rules
- Additional environmental information

The Brewers of Europe

BEER PEFCR: content

- Clear definition of scope
- Definition of representative product(s)
- Benchmark(s)
- Identification of the most relevant life cycle stages, impacts, and processes (for which primary data shall be produced)
- Calculation rules
- Additional environmental information

The Brewers of Europe

SYSTEM BOUNDARIES

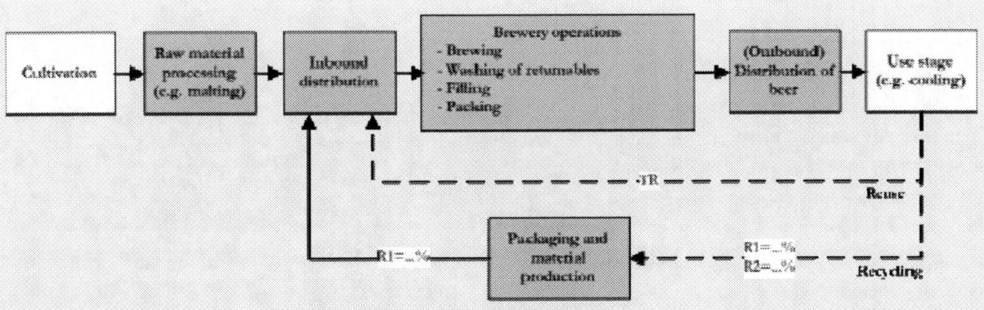

The Brewers of Europe

BEER PEFCR: content

- Clear definition of scope
- **Definition of representative product(s)**
- Benchmark(s)
- Identification of the most relevant life cycle stages, impacts, and processes (for which primary data shall be produced)
- Calculation rules
- Additional environmental information

The Brewers of Europe

BEER = BEER

- **CPA/NACE code = C11.0.5 - Manufacture of beer**

This class includes also low alcohol or non-alcoholic beer.

- **Unit of analysis**

What: A refreshing beer consumed in a social setting.
Where: Consumed in the European Union
How much: One hectolitre of beer (1hl; equal to 100l).
How long: Until at least 6 months after the expiry date.
How well: Served at the advised serving temperature.

If the beer cannot be preserved within 1 month after production, the default losses of 2% must be increased to 7%.

For communication purposes the results can be translated to stock keeping units (SKUs).

The Brewers of Europe

BEER PEFCR: content

- Clear definition of scope
- Definition of representative product(s)
- **Benchmark(s)**
- Identification of the most relevant life cycle stages, impacts, and processes (for which primary data shall be produced)
- Calculation rules
- Additional environmental information

The Brewers of Europe

BENCHMARK

- For the determination of the representative product a virtual beer is created:
  1. Beer types sold in Europe
  2. Recipes of the beer types

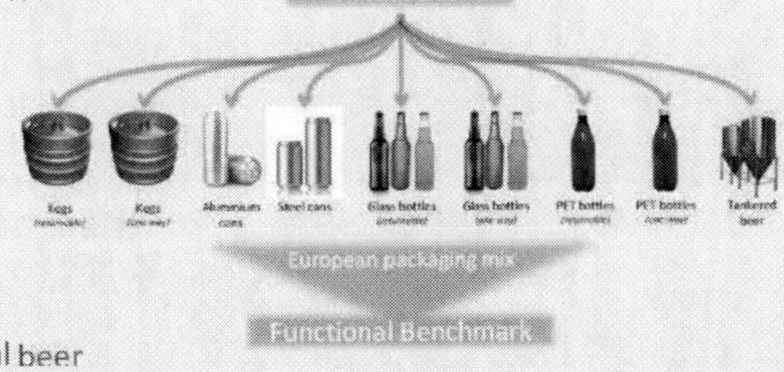

- Bill of materials for virtual beer
- The virtual beer is packed in the European average packaging mix, to come to 1 functional benchmark

The Brewers of Europe

BENCHMARK BEER : data and assumptions

Additives	-
Miscellaneous	
Ascorbic acid/ascorbate	0.004
Benzoic acid/benzoates	0.02
Caramel	0.7
Citric acid	0.002
Gumarabic	0.001
Sulphites	0.03
Lactic acid	0.035
Lysozyme	0.008
Propane-1,2-diol alginate	0.01
Sorbic acid/sorbates	0.02
Sweeteners	
Acesulfame K	0.04
Aspartame	0.06
Aspartame/acesulfame k salt	0.04
Neohesperidine DC	0.00
Neotame	0.00
Saccharin	0.01
Steviol	0.01
Sucralose	0.03

Processing aids	-
Yeast	0.163
Enzymes	
Cytolytic enzyme blends	0.001
Amylolytic blends	0.004
Proteases	0.003
Fermentation blends	0.005
Acetolactatedecarboxylase	0.002
Proline-specific endoprotease	0.002
Filter aids	
Diatomaceous earth (calcined)	0.066
Diatomaceous earth (uncalcined)	0
Diatomaceous earth (flux calcined)	0.01
Perlite	0.019
Filter sheets	0.003
Cartridge filters	0.001
Cellulose fibres	0.005
PVP	0.002
PVPP	0.002
Silica hydrogel	0.012
Silica xerogel	0.001
Bentonite	0.001
Antifoam	
dimethylpolysiloxane	0.1
Fining agents	
Tannic acid	0.001
Isinglass	0.006
Polysaccharide auxiliary finings	0.001
Polysilicate auxiliary finings	0.001
Silica sol	0.006
Carrageenan	0.002

Other inputs in the brewing process	-
Brewing salts and pH regulators	0.05
Calcium chloride	
Calcium sulphate	
calcium hydroxide	
citric acid	
hydrochloric acid	
phosphoric acid	
potassium hydroxide	
sodium carbonate	
sulphuric acid	
Yeast foods	
Yeast foods	0.0001
Zinc chloride/sulphate	0.00007
Cleansers	
Nitric acid	0.205
Peracetic acid	0.002
Phosphoric acid	0.205
Quaternary ammonium compounds	0.0001
Sodium hydroxide	2.367
Sodium hypochlorite	0.00001
Sulphamic acid	0.00001
Sulphuric acid	0.0001

The Brewers of Europe

BEER PEFCR: content

- Clear definition of scope
- Definition of representative product(s)
- Benchmark(s)
- **Identification of the most relevant life cycle stages, impacts, and processes (for which primary data shall be produced)**
- Calculation rules
- Additional environmental information

HOTSPOTS

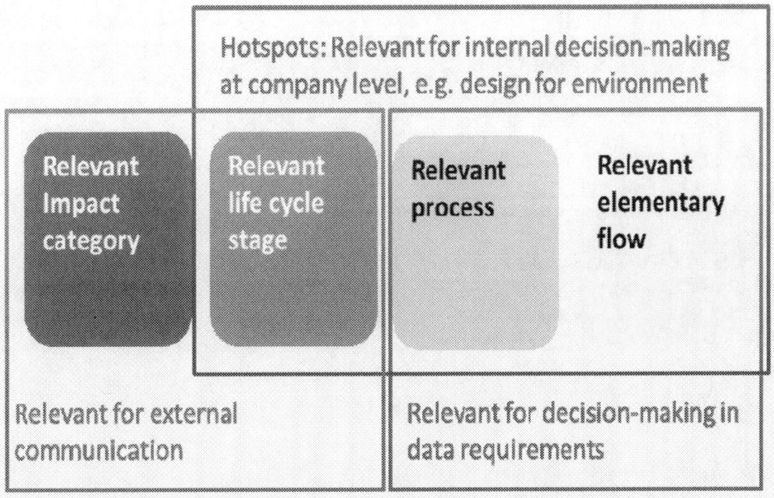

HOTSPOTS : Virtual beer

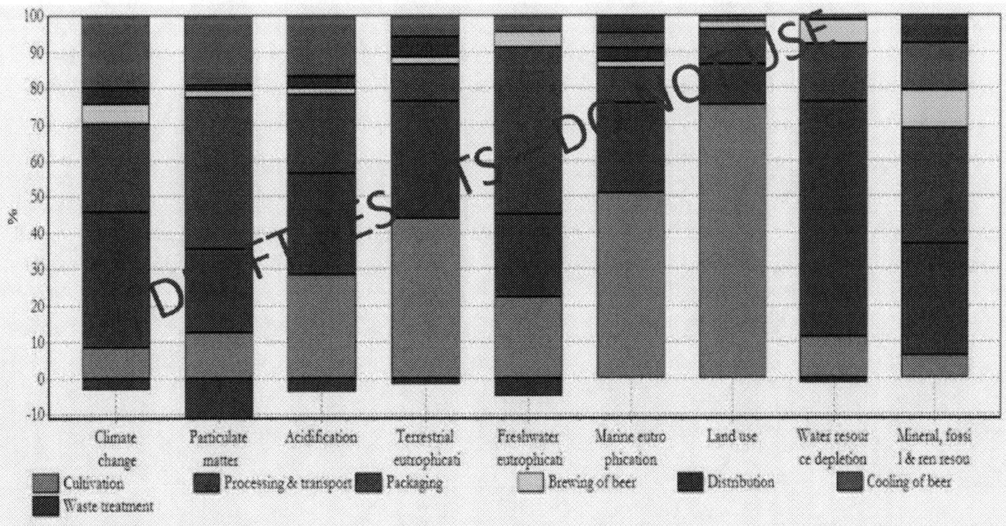

Analyzing 1 p 'Total life cycle virtual beer';
Method: ILCD 2011 Midpoint+ (for use in PEF/OEF pilots) V1.04 / EU27 2010, equal weighting / Characterization

The Brewers of Europe

HOTSPOTS : summary

Life cycle stages	Most relevant processes	Most relevant Impact categories			
		CC	A	ME	W
Cultivation	• Fertilization	✓	✓	✓	
	• Energy from diesel	✓			
	• Fertilizer production	✓			
	• Irrigation of crops				✓
Processing of raw materials	• Transport of ingredients from farm to processing	✓	✓	✓	
	• Energy consumption for the processing of raw and malted cereals	✓	✓	✓	
Brewery operations	• Thermal energy consumption	✓			
	• Inbound distribution	✓			
Packaging material production	• Packaging material production	✓	✓	✓	✓
Use stage	• Electricity consumption for cooling	✓	✓		

CC = Climate Change / A = Acidification / ME = Marine Eutrophication / W = Water resource depletion

The Brewers of Europe

BEER PEFCR: content

- Clear definition of scope
- Definition of representative product(s)
- Benchmark(s)
- Identification of the most relevant life cycle stages, impacts, and processes (for which primary data shall be produced)
- **Calculation rules**
- Additional environmental information

The Brewers of Europe

CALCULATION RULES

Life Cycle Stage	Data requirements
Cultivation	Secondary data if no primary data is available
Processing of raw materials	A minimum of 60% of beer ingredients must be primary data
Brewery operations	100% must be based on company specific
Packaging material production	A minimum of 80% of the primary packaging material to be based on primary data
Use stage	Secondary data if no primary data is available

Non-compliance = no PEF

Compliance = Input from suppliers

The Brewers of Europe

CALCULATION RULES: impact for maltster

- **Data on bill of materials:** total malt output; co-product output – per type; total raw material input – per grain
- **Thermal energy use:** per type of source; heat export
- **Electrical energy use:** renewable or not; from grid or own production; electricity export
- **Water use:** per type of source
- **Effluent:** direct or indirect
- **Other emissions:** e.g. to air, Nox, SO2

The Brewers of Europe

CALCULATION RULES: impact for glass manufacturer

- **Data on bill of materials:** total glass output; total raw material input – per product
- **Thermal energy use:** per type of source; heat export
- **Electrical energy use:** renewable or not; from grid or own production; electricity export
- **Water use:** per type of source
- **Effluent:** direct or indirect
- **Other emissions:** e.g. to air, Nox, SO2

The Brewers of Europe

BEER PEFCR: content

- Clear definition of scope
- Definition of representative product(s)
- Benchmark(s)
- Identification of the most relevant life cycle stages, impacts, and processes (for which primary data shall be produced)
- Calculation rules
- **Additional environmental information** (finally decided to leave it out)

The Brewers of Europe

COMMUNICATION: B2B

- Discussions are ongoing as this also impact on B2C communication: will be a policy, legislative decision
- EU Commission public consultation in the next months
- Template declaration:

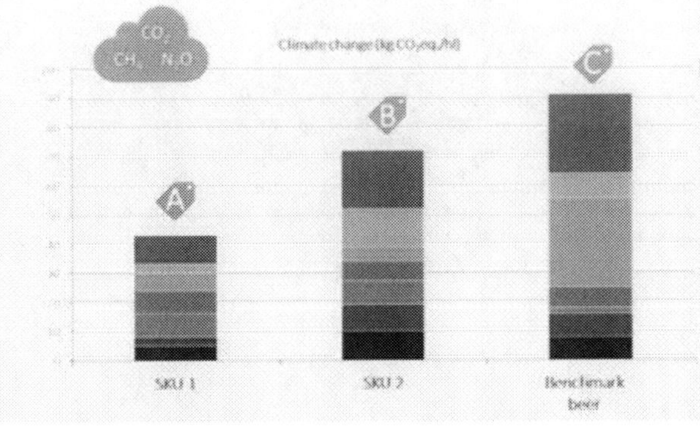

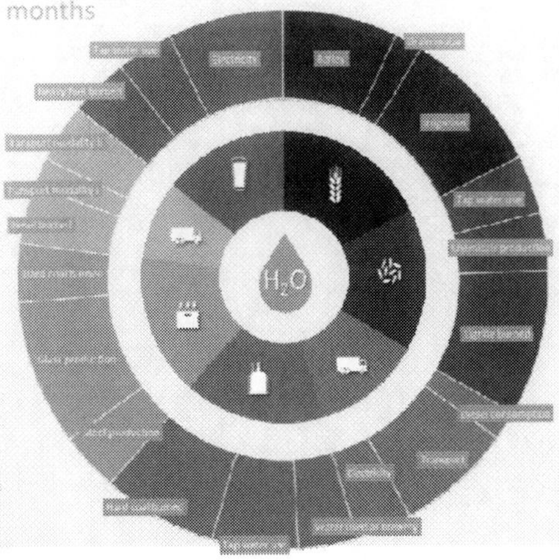

The Brewers of Europe

COMMUNICATION: B2C

- Discussions are ongoing: will be a policy, legislative decision
- Examples of communication vehicles tested

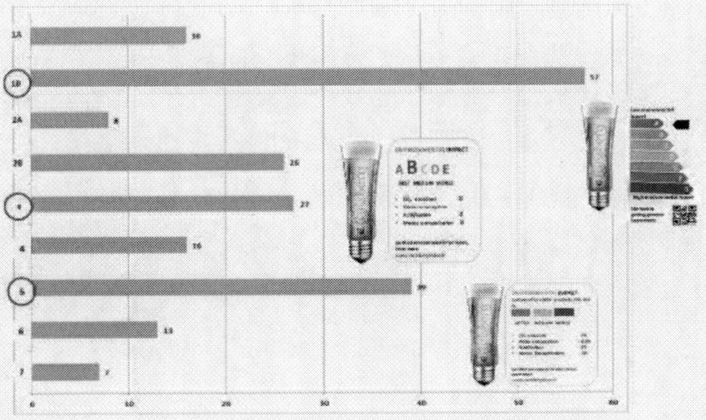

The Brewers of Europe

IT TOOL: practical application

- The draft PEFCR is text, not always easy to understand and not all brewers have the time to translate this in a tool.
- An IT tool is being elaborated.
- Testing will be required:
    - Functionality
    - Language
    - User friendliness
    - Output

The Brewers of Europe

GREEN BEER: result of a collaborative action of the entire value chain

- Shared approach of all brewers, whatever size
- Relies on data transparency
- Relies on supply chain partner
- Helps to identify where improvements are possible
- Contributes to the realisation of the green market

The Brewers of Europe

MORE INFORMATION

- European Commission
  http://ec.europa.eu/environment/eussd/smgp/policy_footprint.htm
- PEF Pilot Wikipage
  http://ec.europa.eu/environment/eussd/smgp/pdf/Wiki_reginstr.pdf
- The Brewers of Europe
  info@brewersofeurope.org

THANK YOU FOR YOUR ATTENTION

30

Multinationals & Craft: A Work in Progress

Larry Nelson

BREWERS' Guardian

What is craft beer?

The US Brewers Association definition: small, independent, traditional.

- **Small:** six million barrels or less in annual production. (Note: this used to be two million barrels.)
- **Independent:** less than 25% of the craft brewery is owned or controlled by an alcoholic beverage industry member who is not themselves a craft brewer.
- **Traditional:** either an all malt flagship (the beer which represents the greatest volume amongst that brewer's brands) or has at least 50% of its volume in either all malt beers or in beers which use adjuncts to enhance rather than lighten flavour.

BREWERS' Guardian

What is craft beer – alternative definition

GlobalData's consumer understanding focused definition:

- **Perception of being small scale:** either independent, or a small scale brewery within a larger organisation
- **Perception of being independent:** even if this is not actually the case
- **A portfolio of beers that encompass different styles:** in other words, not just mass lager
- **A relatively short history**: breweries that are perceived as 'craft' outside of America typically have been founded in the last 10 years.

BREWERS' Guardian

Craft beer: leading markets (vol. hl)

China	319,900
Sweden	326,250
Japan	362,000
New Zealand	690,500
France	784,050
Brazil	910,300
Canada	935,000
Poland	1,634,000
Australia	1,961,000
UK	2,495,060
USA	32,332,800

Source: GlobalData

BREWERS' Guardian

Craft beer: leading markets (market %)

Greece	2.3
Denmark	3.4
France	3.8
Poland	4.2
Canada	4.7
UK	5.5
Sweden	7.0
Australia	11.3
USA	13.5
New Zealand	22.9

Source: GlobalData

BREWERS' Guardian

US craft brewing growth

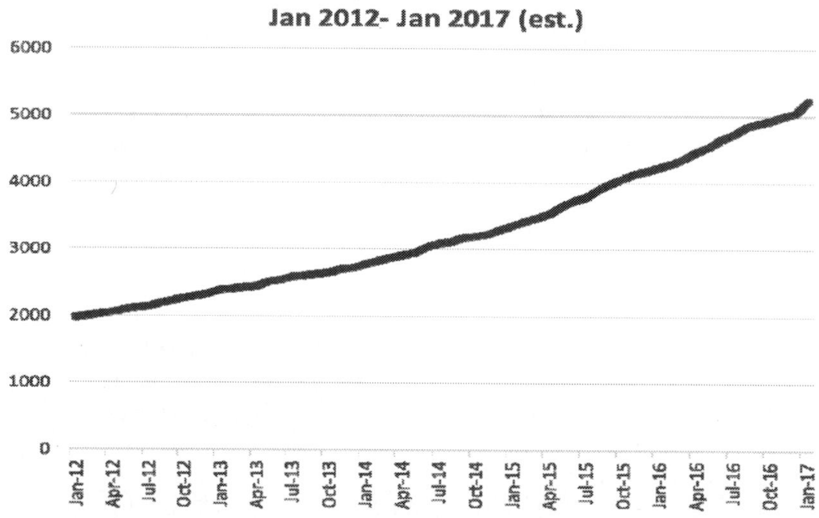

BREWERS' Guardian

Product attributes trump brewery perception

How important is each attribute when choosing a craft beer to purchase?
(% craft beer drinkers saying very / somewhat important)

- 60% Locally Made
- 63% Made by Ind. Brewer
- 63% Appearance ⬆
- 68% Bitterness ⬇
- 75% Ingredients
- 78% Aroma
- 95% Freshness

Source: Nielsen's Craft Beer Insights Poll (CIP) conducted June 2016 by Harris Poll (1,018 craft drinkers, 555 weekly craft drinkers)

BREWERS' Guardian

The craft brewer – large brewer symbiosis

Craft brewer obtains:

- Exit strategy – craft founders secure future for their brewery.
- Access to capital – rapid expansion now possible; new routes to market nationally and internationally.
- Access to brewing capabilities – possibilities of enhanced training for brewers, operating systems allowing more consistent results.

Large brewer obtains:

- Brands and beer styles - typical of craft brewing creativity 'street cred'.
- Value – easier to acquire expertise than to start from scratch.
- Growth potential - mixing and matching brands in international markets.

BREWERS' Guardian

Carlsberg Group

- As of 2016, craft, speciality and non-alcoholic brands account for of group beer volumes and 10% of beer net revenue.

- Most prominent craft initiative is its partnership with Brooklyn Beer, with brewhouses in Stockholm and the E.C. Dahls brewery in Trondheim, Norway. Also imports Brooklyn portfolio in 15 countries.

- As part of SAIL'22, a promise to pursue 'strong positions' in craft, e.g., Jacobsen and Brooklyn
  e.g., Grimbergen, 1664 Blanc (speciality beers, brand extensions)

BREWERS' Guardian

- Heineken could have been described as an 'accidental' craft brewer: All of Caledonian, Monteith's, Devassa, Eisenbahn and Baden Baden were parts of larger transactions.

- As with Carlsberg, in recent years an increased emphasis on craft and variety beers, e.g., H41 and H71 (brewed with wild yeast), Birra Moretti (regional Italian variants making use of local ingredients)

- The Lagunitas transaction(s) – remaining 50% of shares acquired this month with Lagunitas founder Tony Magee to take advisory role within Heineken on global and local craft strategy.

BREWERS' Guardian

ABInBev

US brewery acquisitions 2011 -

DATE	BREWERY	LOCATION
2011		
March	Goose Island	Chicago, IL
2014		
February	Blue Point Brewing	New York, NY
November	10 Barrel Brewing	Bend, OR
2015		
January	Elysian Brewing	Seattle, WA
September	Golden Road Brewing	Los Angeles, CA
December	Breckenridge Brewing	Breckenridge, CO
December	Four Peaks Brewing	Tempe, AZ
2016		
April	Devil's Backbone	Roseland, VA
November	Karbach Brewing Co	Houston, TX
2017		
May	Wicked Weed Brewing	Asheville, NC

BREWERS' Guardian

The BA's point: the 'illusion of choice' before

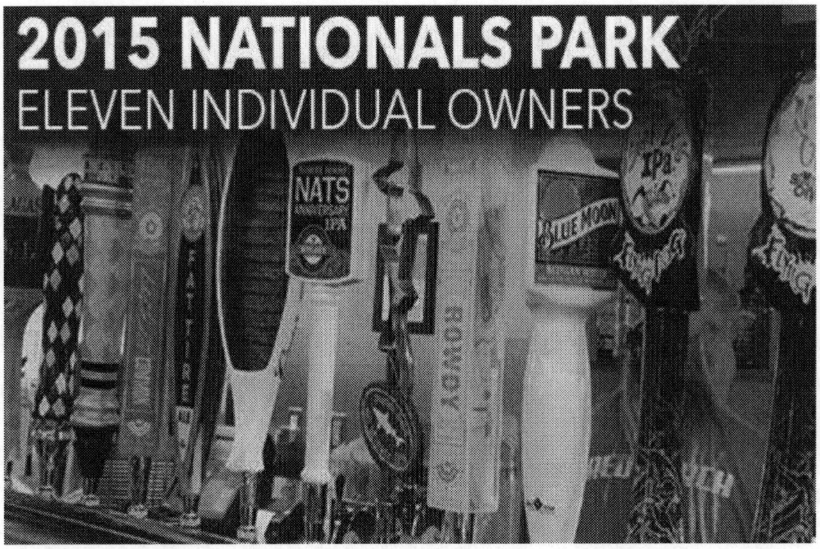

Source: Brewers Association

12

The BA's point: the 'illusion of choice' after

Source: Brewers Association

13

ABInBev International acquisitions 2015 -

DATE	BREWERY	LOCATION
2015		
December	Camden Town	London, UK
2016		
February	Beer Hawk *	Wetherby, UK
April	Birra del Borgo	Borgorose, Italy
April	Microbrasserie Archibald	Quebec, Canada
September	Bosteels Brewery	Buggenhout, Belgium
October	Ginette **	Walloon Brabant, Belgium
????	Kaiba	Shanghai, China
2017		
January	Cervezas La Virgen	Madrid, Spain
March	Boxing Cat Brewery	Shanghai, China

** On-line beer retailer ** Organic beer brand*

BREWERS' Guardian

UK brewery & brand owner starts

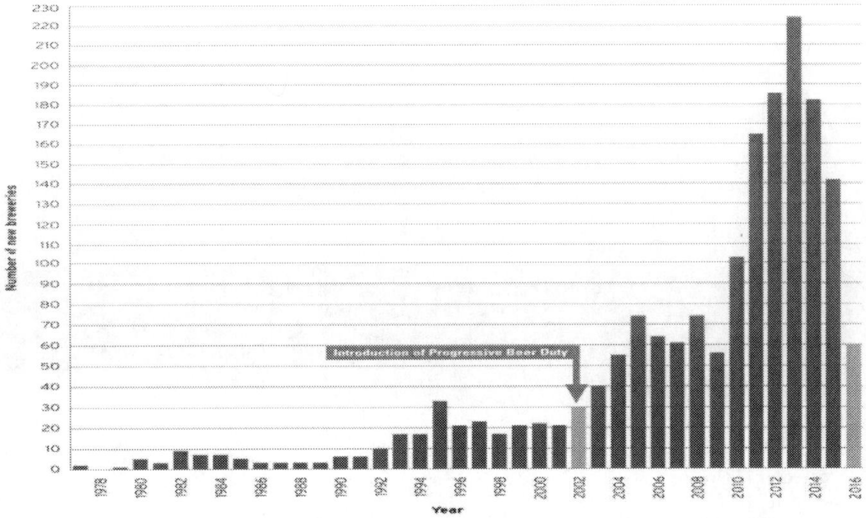

BREWERS' Guardian

Special thanks to:

- Bob Pease, Brewers Association
- Kevin Baker, GlobalData

BREWERS' Guardian

Identification of factors involved in the gushing of beer in order to develop a technique for early detection of "gushing risk"

EBC 2017 ... 14-18 MAY LJUBLJANA ... SLOVENIA

iFBM
THE REFERENCE IN THE BARLEY TO THE BEER

Julien BILLARD
Phd student

Thesis director IFBM:
- Marc Schmitt
- Patrick Boivin

Thesis director LRGP:
- Ivan Marc
- Romain Kapel

LRGP Laboratoire Réactions et Génie des Procédés
CNRS
Université de Lorraine

iFBM

Outline of this presentation

2

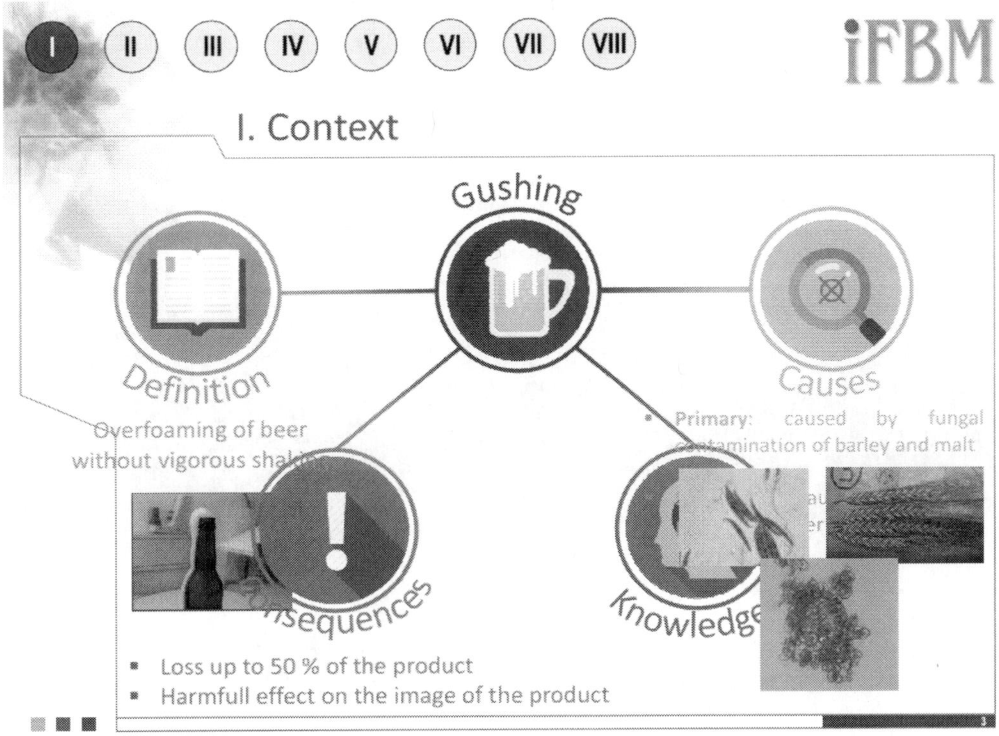
I
II
III
IV
V
VI
VII
VIII
iFBM
I. Context
Gushing
Definition
Overfoaming of beer
without vigorous shak
Causes
Primary: caused by fungal
amination of barley and malt
Consequences
Knowledge
Loss up to 50 % of the product
Harmfull effect on the image of the product
3

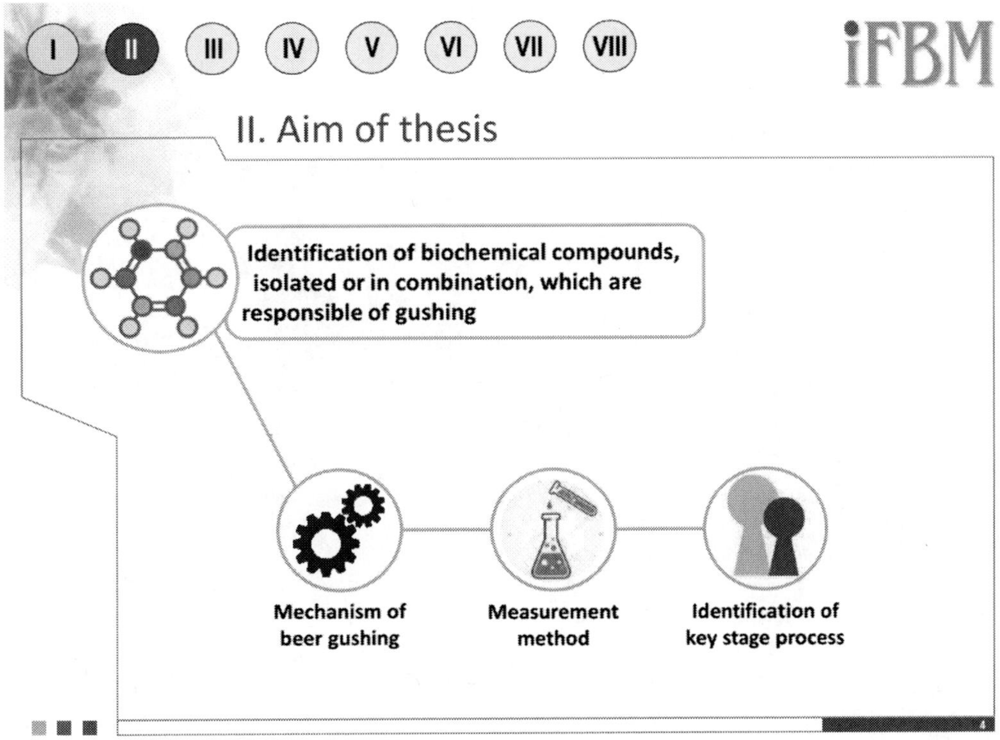
I
II
III
IV
V
VI
VII
VIII
iFBM
II. Aim of thesis
Identification of biochemical compounds, isolated or in combination, which are responsible of gushing
Mechanism of beer gushing
Measurement method
Identification of key stage process
4

I II III IV V VI VII VIII

III. State of art

a. Mechanism of gushing - nanobomb theory

Compounds with amphiphile properties and good liaison with carbon dioxide

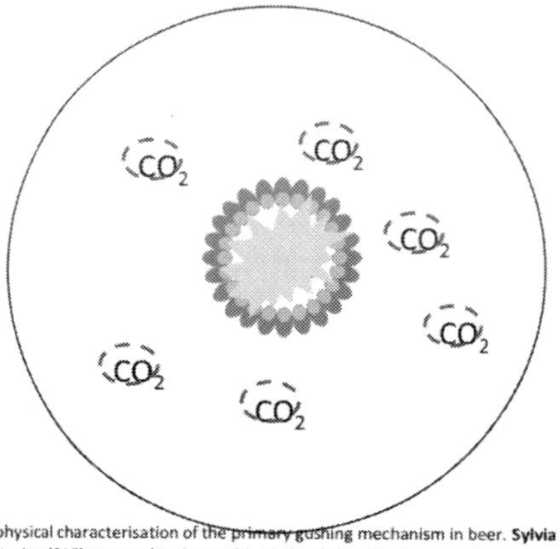

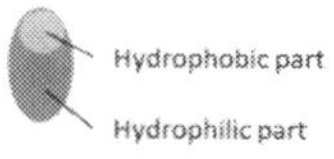

- Modeling and biophysical characterisation of the primary gushing mechanism in beer. **Sylvia Deckers. 2013**
- Dynamic light scattering (DLS) as a tool to detect CO_2-Hydrophobin structures and study the primary gushing potential of beer. **Deckers et al. 2011**
- CO_2-hydrophobin structures acting as nano bombs. **Deckers et al. 2010**

5

I II III IV V VI VII VIII

III. State of art

b. Hydrophobins hypothesis

Structure

Promote gushing

Function

- produced by fungus (10 – 15 kDa)
- needed for the aerial growth
- adhering to solid suface

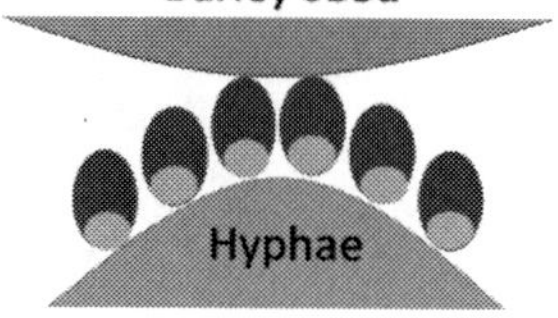

- Hydrophobins, beer foaming and gushing. **Shokribousjein et al. 2010**
- Hydrophobins: Proteins that self assemble at interfaces. **Linder, 2009**

6

I II III IV V VI VII VIII

iFBM

III. State of art

c. Production and purification of hydrophobins

Strain	Production	Extraction	Purification
T. reesei	Fermentor	Foam fractionation from media	15 RPC HPLC
Saccharomyces cerevisiae (modified)	Fed-batch fermentation	Overfoaming of media	HPLC C4 vydac
F. poae, T. reesei, Nigrospora	Shake flasks	From mycelium and foam fractionation of media	Preparative RP C4
Trichoderma reesei	Fermentor	From mycelium (tris-HCl+SDS)	Hydrophobic interaction and anion exchange column
T. reesei	Fed-batch fermentation	From mycelium (tris-HCl+SDS)	RPC HPLC C4

- A novel method for hydrophobin extraction using CO2 foam fractionation system. **Khalesi et al. 2013**
- Characterisation of HFBII biosurfactant production and foam fractionation with and without antifoaming agents. **Winterburn et al. 2011**
- Fungal hydrophobin as predictors of the gushing activity of malt. **Sarlin et al. 2005**
- Overproduction, purification and characterization of T. reesei hydrophobin. **Askolin et al. 2011**
- Modeling and biophysical characterization of the primary gushing mechanism in beer. **Sylvie Deckers 2013**

I II III IV V VI VII VIII

iFBM

IV. Strategy

G+: gushing positive
G-: gushing negative

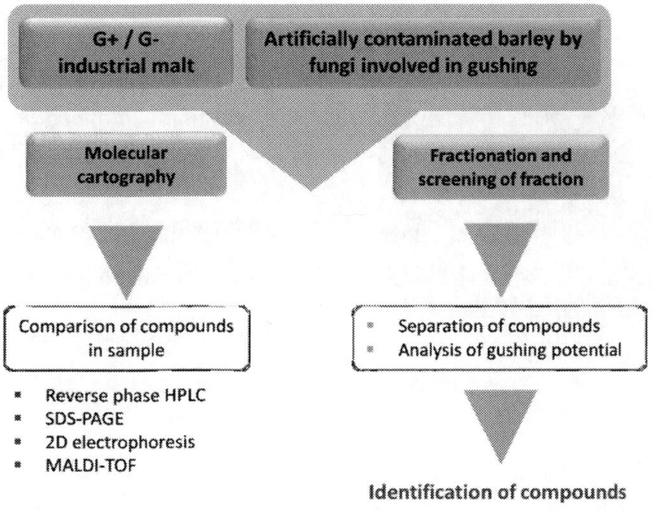

I II III IV V VI VII VIII

iFBM

V. Solid phase fermentation of barley seed for the extraction of compounds involved in beer gushing

Previous work at IFBM

	Gushing on barley	Gushing on grits	Gushing on beer
Species	Barley artificially infected by the fungi species	37.5 % of infected barley and 62.5 % of non gushing malt	Beer produced on 40 l brewery plant with 37.5 % of infected barley and 62.5 % of non gushing malt
Control (barley non artificially contaminated	non gushing	non gushing	non gushing
Fusarium graminearum	non gushing	117 g	95 g
Fusarium tricinctum	non gushing	61.3 g	100 g
Microdochium nivale	50.9 g	131 g	108 g

- Production of gushing positive beer with barley artificially infected
- 2 phenomena observed :
  - gushing on barley
  - gushing on a mix of barley and negative gushing malt

Is it possible to extract compounds directly from the surface of barley ?

9

I II III IV V VI VII VIII

iFBM

V. Solid phase fermentation of barley seed for the extraction of compounds involved in beer gushing

1. Solid phase fermentation of barley

F. tricinctum

- Isolating of Fungi strains from malting barley
  - *Microdochium nivale*
  - *Fusarium graminearum*
  - *Fusarium tricinctum*
  - *Alternaria*
- Production of spores on Agar
- Autoclaved malting barley
- Infection of sterile barley by fungi spores
- Incubation to 1 – 2 weeks at 25°C

10

I II III IV V VI VII VIII

iFBM

V. Solid phase fermentation of barley seed for the extraction of compounds involved in beer gushing

2. Water extraction of compounds on barley artificially infected by fungus

Problems:

Study the possibility to extract the compounds involved in gushing directly on the surface of barley seeds

Experimentation:

- Extraction of compounds in surface of barley with ultrapure water
- Analysis of gushing potential according to the modified Carlsberg test in beer

11

I II III IV V VI VII VIII

iFBM

V. Solid phase fermentation of barley seed for the extraction of compounds involved in beer gushing

2. Water extraction of compounds on barley artificially infected by fungus

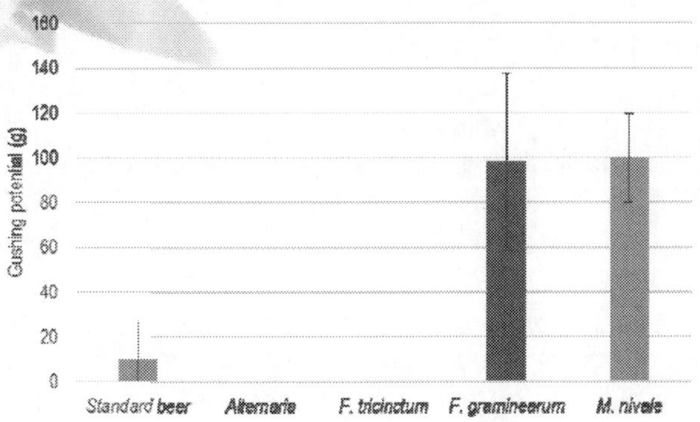

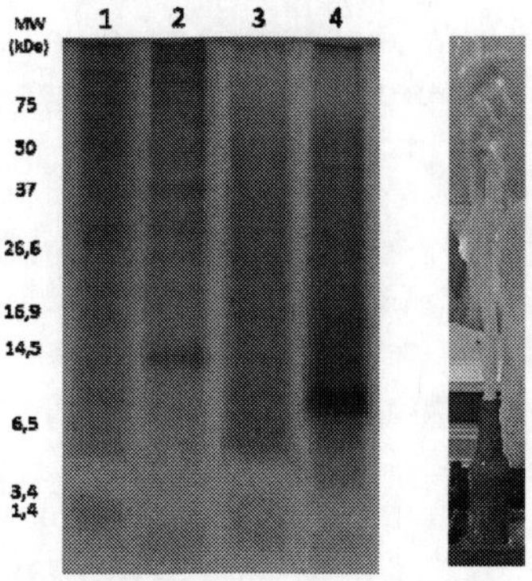

1: *Alternaria* 3: *F. graminearum*
2: *F. tricinctum* 4: *M. nivale*

Extraction of compounds involved in gushing is possible directly from surface of barley artificially infected by fungus

12

I II III IV V VI VII VIII

iFBM

V. Solid phase fermentation of barley seed for the extraction of compounds involved in beer gushing

3. Mix with gushing negative malt extract

- Water extraction on *F. tricinctum* barley
- Production of gushing negative malt extract according to the modified Carlsberg test
- Boiling of *F. tricinctum* extract with gushing negative malt extract
- Filtration
- Gushing analysis of extract on beer

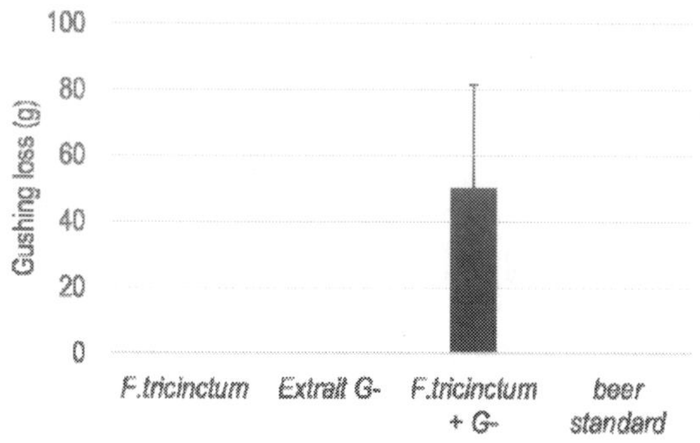

Phenomena close to IFBM previous experiments

13

I II III IV V VI VII VIII

iFBM

V. Solid phase fermentation of barley seed for the extraction of compounds involved in beer gushing

Conclusion

Two phenomena observed:

- production of gushing extract from surface of *M. nivale and F. graminearum* barley
- association /modification of compounds occured with *F. tricinctum* and malt negative extract

Extraction of compounds involved in gushing as been possible directly from surface of barley artificially infected by fungus especially by ***Microdochium nivale*** and ***Fusarium graminearum***

14

I II III IV V VI VII VIII

iFBM

VI. Purification of extracted compounds by RPC

1. Fractionation and analysis of gushing potential

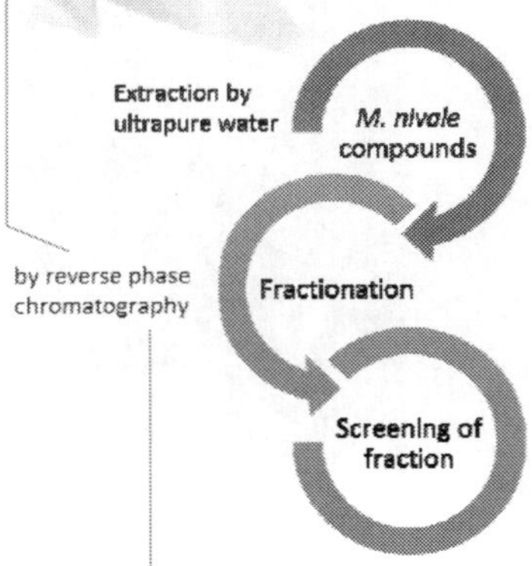

Objectives:

fractionation of compounds from *M. nivale* extract and identification of a gushing fraction

Experiment:

- extraction of compounds from barley artificially contaminated by *M. nivale*
- fractionation of compounds by reverse phase HPLC
- concentration of fractions
- analysis of fraction (electrophoresis, gushing test)

15

I II III IV V VI VII VIII

iFBM

VI. Purification of extracted compounds by RPC

1. Fractionation and analysis of gushing potential

- **Pool 1**: which contains carbohydrate and non-protein nitrogen
- **Pool 2**: which contains proteins (with high absorbance)
- **Pool 3**: which contains proteins (with low absorbance)

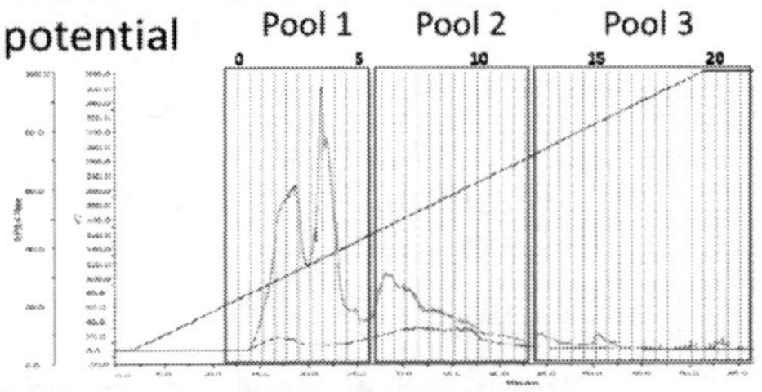

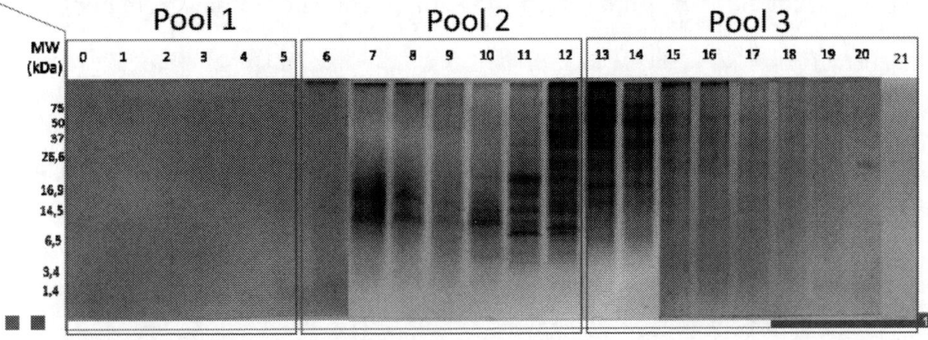

19

I II III IV V VI VII VIII

iFBM

VI. Purification of extracted compounds by RPC

1. Fractionation and analysis of gushing potential

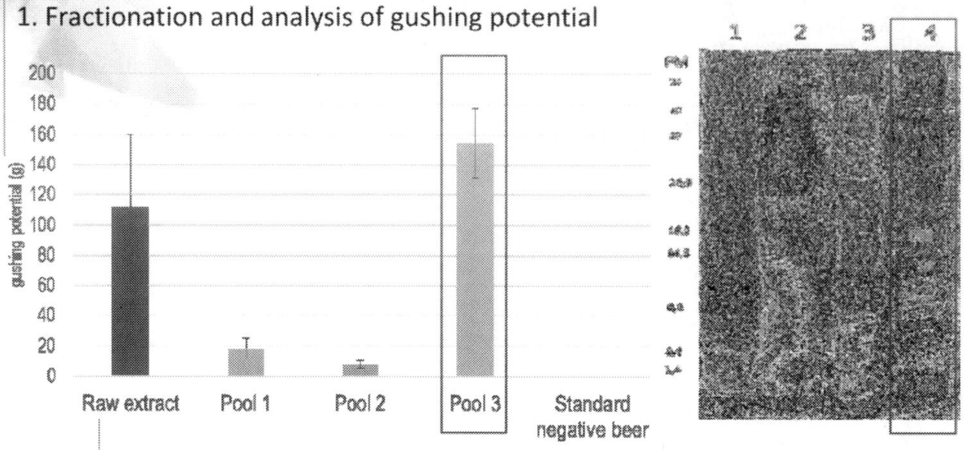

1: raw extract
2: pool 1
3: pool 2
4: pool 3

Identification of gushing fraction

20

I II III IV V VI VII VIII

iFBM

VI. Purification of extracted compounds by RPC

3. Characterization of gushing fraction

Problems:

No information about proteins from pool 3

Objectives:

Visualization, purification and identification of proteins from pool 3

Strategy:

- Production of *M. nivale* extract
- Fractionation of *M. nivale* extract and recovering of pool 3
- Analysis by 2D electrophoresis

19

I II III IV V VI VII VIII

VI. Purification of extracted compounds by RPC

3. Characterization of gushing fraction

- tris-tricine gel with 15 % of acrylamide
- coloration with colloidal blue G-250

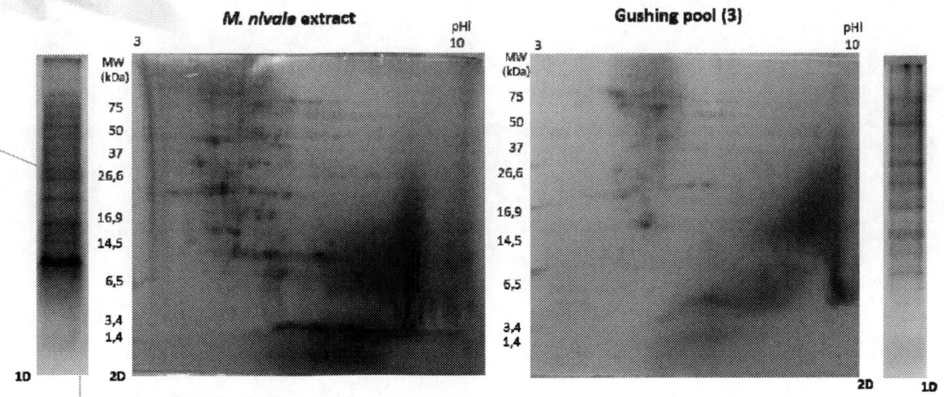

- Decrease of complexity of sample
- Spot can be analyzed for their identification

22

I II III IV V VI VII VIII

VI. Purification of extracted compounds by RPC

4. Identification by Peptide Mass Fingerprint

- Spot excision
- Trypsine digestion
- MALDI-TOF analysis
- Database : MASCOT

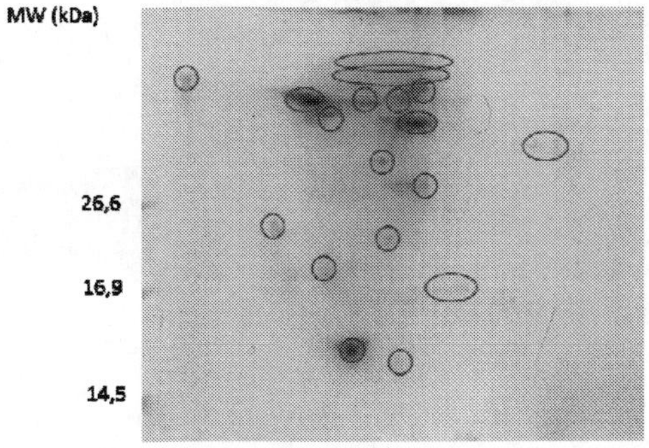

Peptide Mass Fingerprint provides a first idea of the proteins present in sample

Spots separate by 2DE has been identified as proteins from *Microdochium*

20

I II III IV V VI VII VIII

iFBM

CONCLUSION

Solid phase fermentation of barley seed

Extraction of compounds involved in gushing as been possible directly from grains surface artificially infected by fungus especially by ***Microdochium nivale***

Fractionation of compounds by reverse phase chromatography

- Preparative fractionation provides a good separation of compounds with preservation of gushing power
- Identification of a gushing pool (3)

Characterization of gushing fraction

Identification of proteins from *Microdochium* which have gushing capacity

21

I II III IV V VI VII VIII

iFBM

Prospect

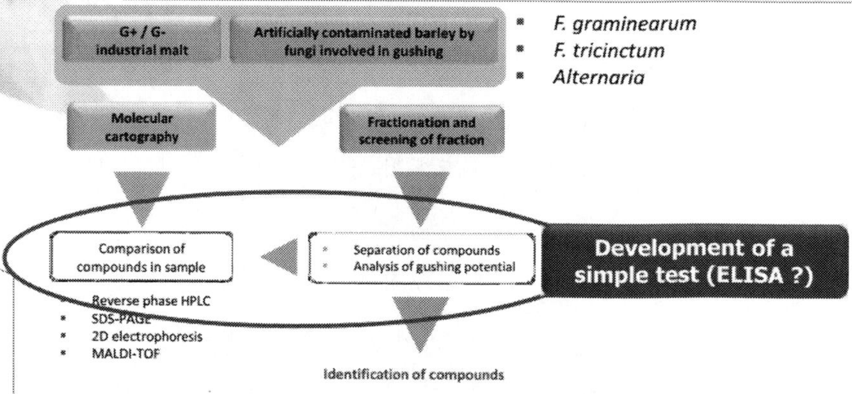

Apply the purification strategy by reverse phase HPLC to:
- barley artificially contaminated by other fungus involved in gushing
- Industrial malt gushing positive and negative

Compare protein profil of each fungus and malt sample
Identify common compounds

22

iFBM

Acknowledgment

Qualmalt UMT:

AGRO SUP DIJON

ensaia

Coopérative Agricole

anses

Thesis commitee:

Guy Derdelinckx
KU Leuven
Associate professor
Center for food and microbial technology

Didier Marion
INRA Nantes
Associate professor
Biopolymère, Interactions, Assemblages

Jean-Hughes Renault
Université de Reims
Associate professor
Isolement et structure (IS)

IFBM team :

LRGP team :

23

iFBM

Thank you for your attention

Contact:

IFBM

Julien Billard: julien.billard@qualtech-groupe.com

Thanks to the support of:

Malteurs de France & Brasseurs de France

MALTEURS DE FRANCE

BRASSEURS DE FRANCE

24

Development of novel hop-derived bitter acid oxides with body fat-reducing effect

15 May 2017
36th European Brewery Convention

KIRIN

○Takahiro Yamazaki[1],
Yumie Morimoto-Kobayashi[1],
Yoshimasa Taniguchi[2] (P009),
and Mikio Katayama[1]

[1]Research Laboratories for Health Science & Food Technologies
[2]Central Laboratories for Key Technologies

Hops are attractive materials with multiple functions

Hops

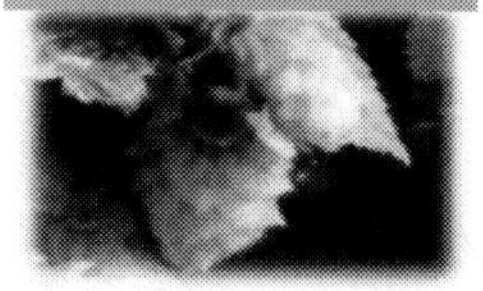

Fresh bitterness and flavor

Multiple health benefits

Relax effect
as herbal medicines

Gastrointestinal health
as crude medicines

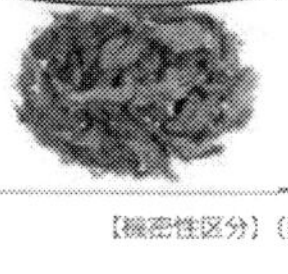

Lots of studies have been performed about hops.

Body fat-reducing effect of iso-α-acids

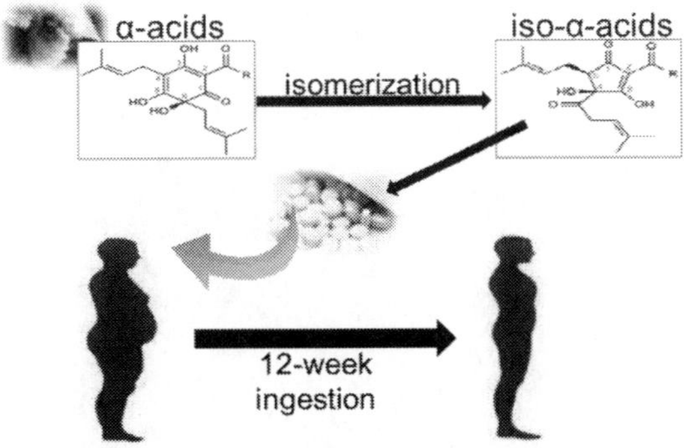

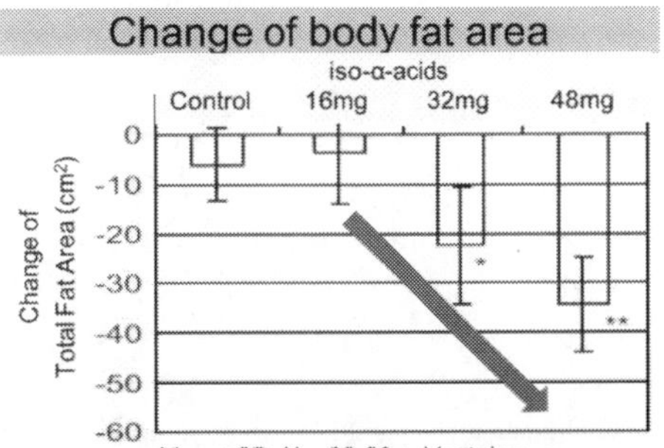

Mean±SD, N = 20-22 subjects/group
Dunnett's post hoc test, *P<0.05, **P<0.01
Clinical Nutrition 2009, 28, 278

Application is limited due to its strong bitterness…

Bitter acid oxides are generated during maturation process

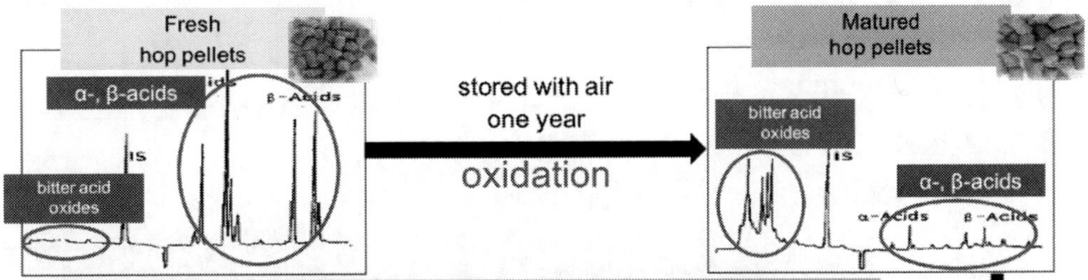

Beer brewed with matured hop pellets
➡ little iso-α-acids and milder bitterness!

Journal of the American Society of Brewing Chemists 1987, 45, 61

Do bitter acid oxides have health benefit like iso-α-acids?
➡ Hop material with mild bitterness and health benefit can be developed!

Contents

1. Chemical properties of matured hop

Details are presented by Dr. Taniguchi Poster No. P009

2. Health benefits on animals

3. Mass production of matured hop

4. Health benefits on humans

5. Flavor

Reduction of maturation time

Fresh hop pellets 

20°C → 10 months

40°C → 1 month

60°C → 2 days!

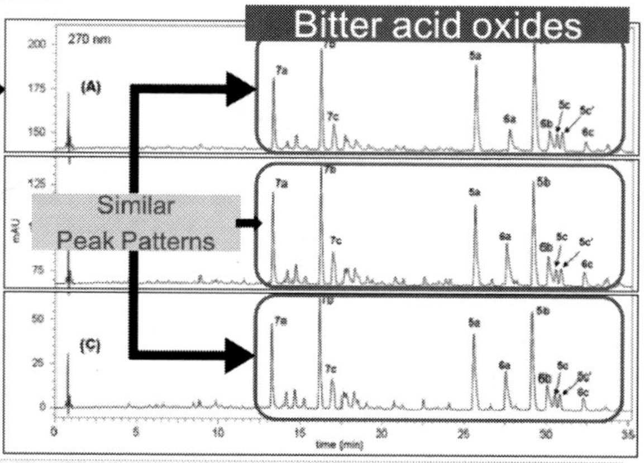

Hop breed: Hallertau Perle
J. Agric. Food. Chem. 2013, 61, 3121-3130.

Matured hop pellets can be prepared by heating at 60°C!

Generation pathways of bitter acid oxides

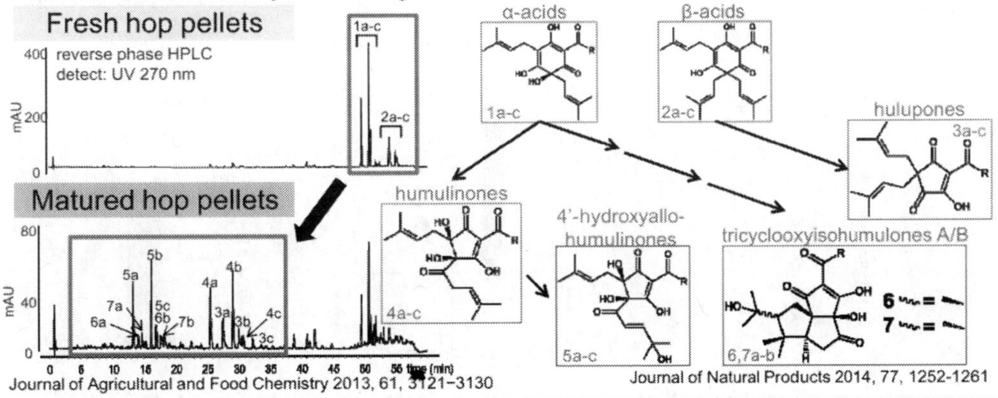

Bitter acid oxides were generated from α-, β-acids

Structural characteristics of bitter acid oxides

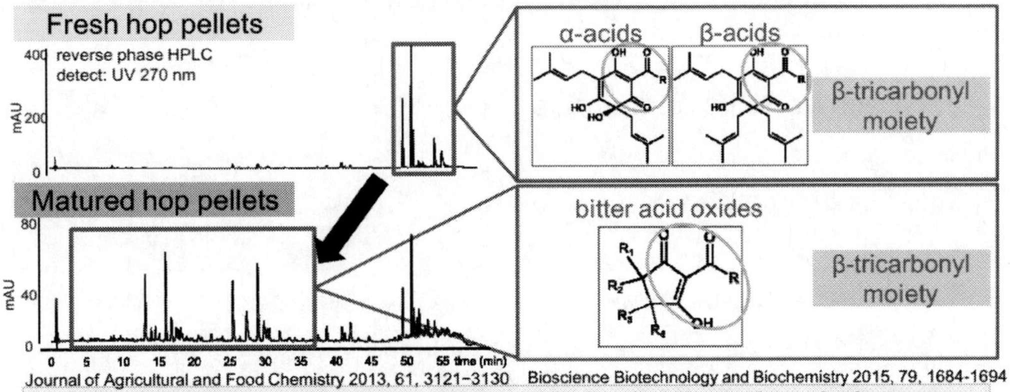

We defined α-, β-acid oxides with β-tricarbonyl moieties as matured hop bitter acids, "MHBA"

Removal of α-, β-acids

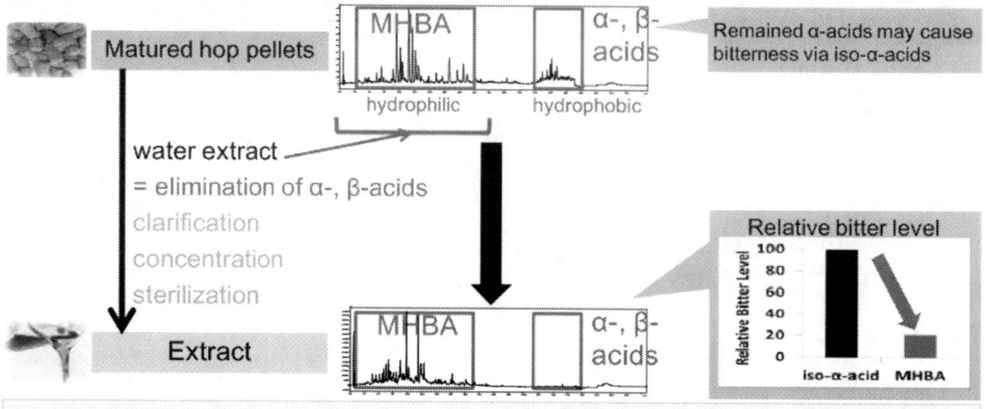

MHBA were selectively extracted from matured hop pellets!

Contents

1. Chemical properties of matured hop

2. Health benefits on animals

3. Mass production of matured hop

4. Health benefits on humans

5. Flavor

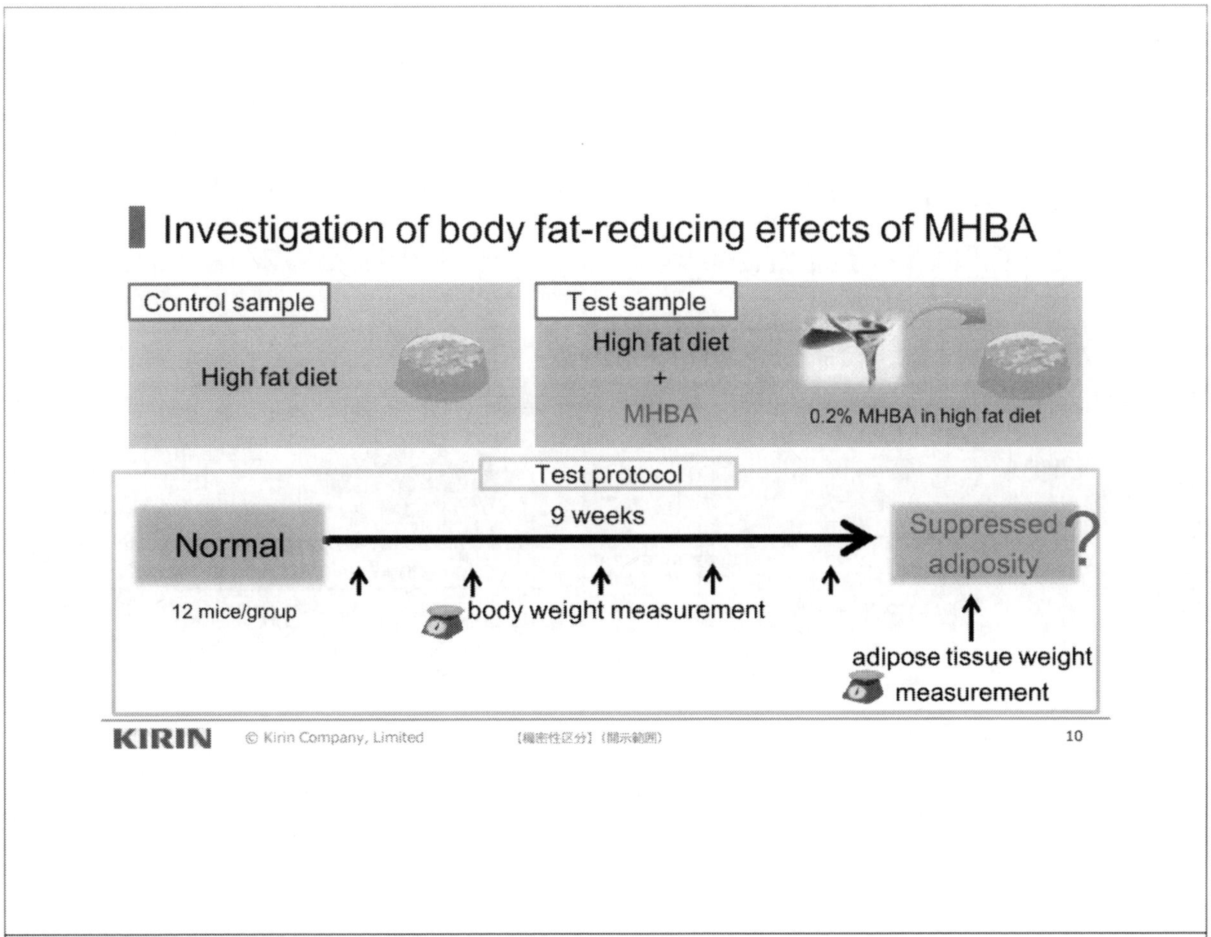
Investigation of body fat-reducing effects of MHBA
Control sample
High fat diet
Test sample
High fat diet
+
MHBA
0.2% MHBA in high fat diet
Test protocol
Normal
9 weeks
Suppressed adiposity ?
12 mice/group
body weight measurement
adipose tissue weight measurement
KIRIN
© Kirin Company, Limited
【機密性区分】(開示範囲)
10

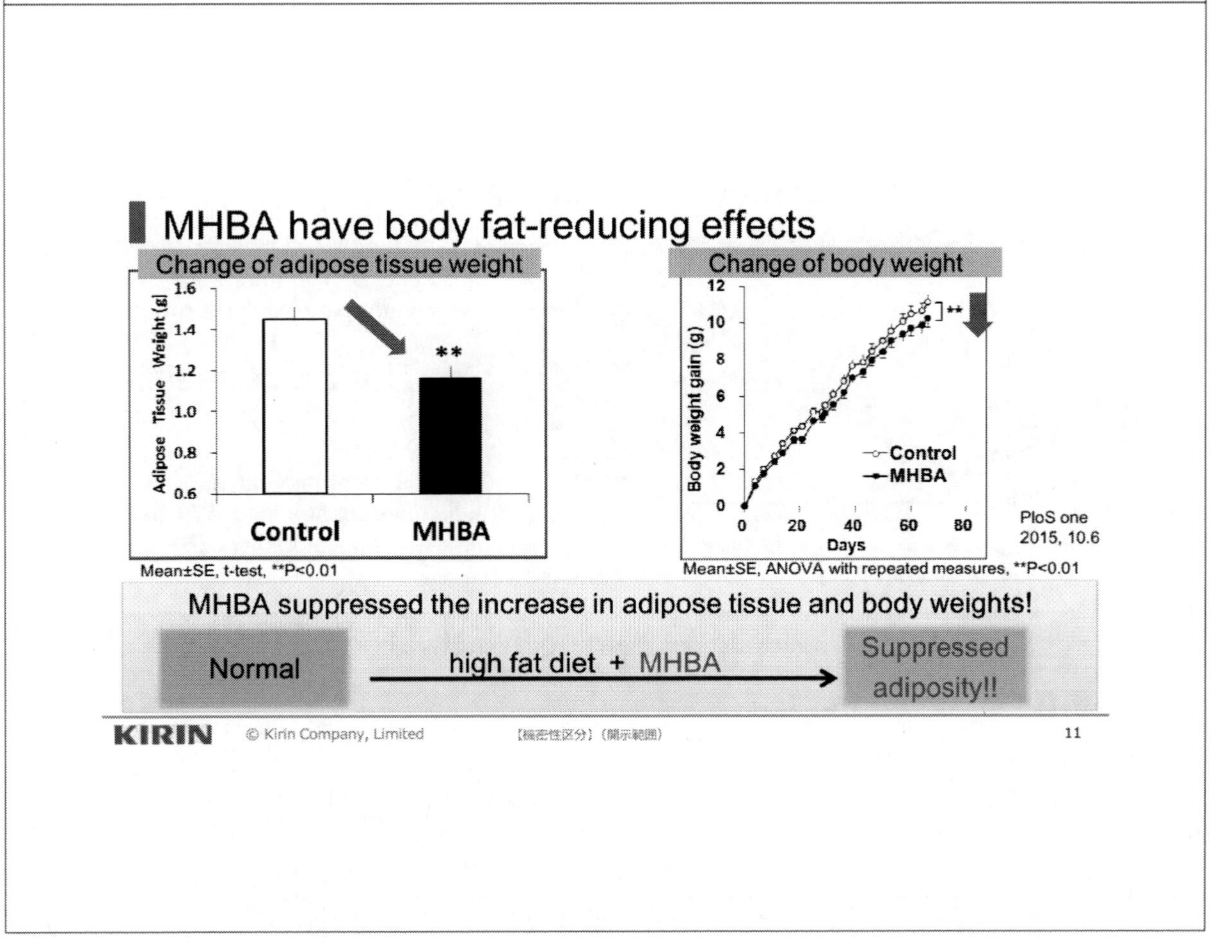
MHBA have body fat-reducing effects
Change of adipose tissue weight
Adipose Tissue Weight (g)
1.6
1.4
1.2
1.0
0.8
0.6
**
Control
MHBA
Mean±SE, t-test, **P<0.01
Change of body weight
Body weight gain (g)
12
10
8
6
4
2
0
0
20
40
60
80
Days
**
Control
MHBA
PloS one 2015, 10.6
Mean±SE, ANOVA with repeated measures, **P<0.01
MHBA suppressed the increase in adipose tissue and body weights!
Normal
high fat diet + MHBA
Suppressed adiposity!!
KIRIN
© Kirin Company, Limited
【機密性区分】(開示範囲)
11

How do MHBA show body fat-reducing effects?

Expected mechanism

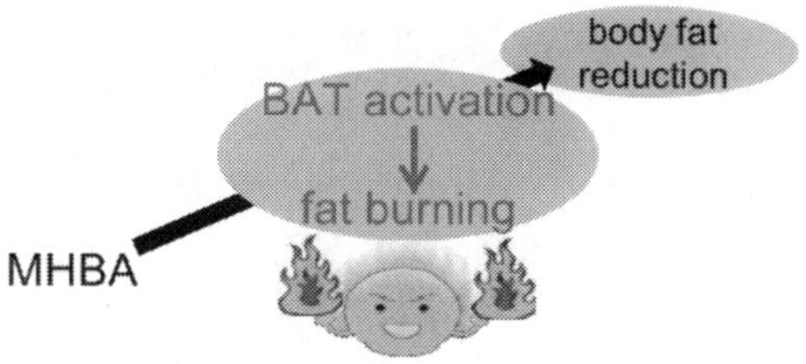

SCIENCE, 2010, 328, 1113
Journal of Clinical Investigation, 2013, 123, 8, 3404

Thermogenic adipose tissue

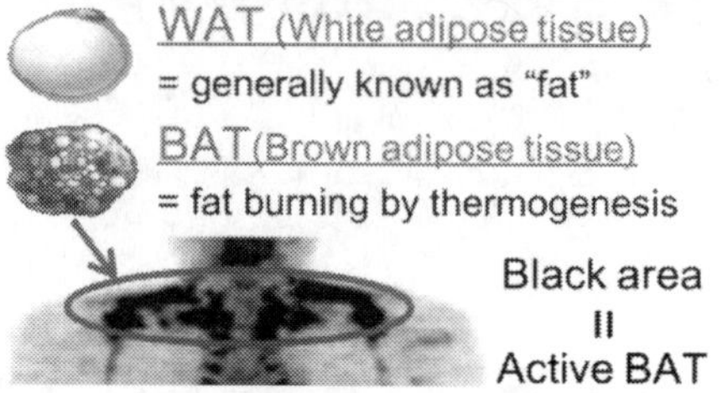

Researches about anti-obesity focusing on BAT are getting attention around the world

MHBA may induce fat burning by activating BAT, leading to body fat reduction

MHBA induce thermogenesis in BAT

Change of BAT temperature

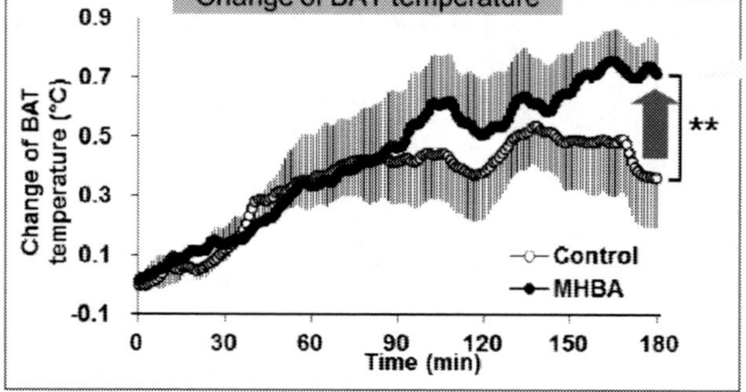

Mean±SD, N = 4 rats/group
ANOVA with repeated measures, **P<0.01
PloS one 2015, 10.6

BAT temperature increased in MHBA group
II
Thermogenesis!!

+ other evidences demonstrating BAT was activated by MHBA

MHBA induced fat burning by activating BAT!

Contents

Mass production of MHBA-rich material

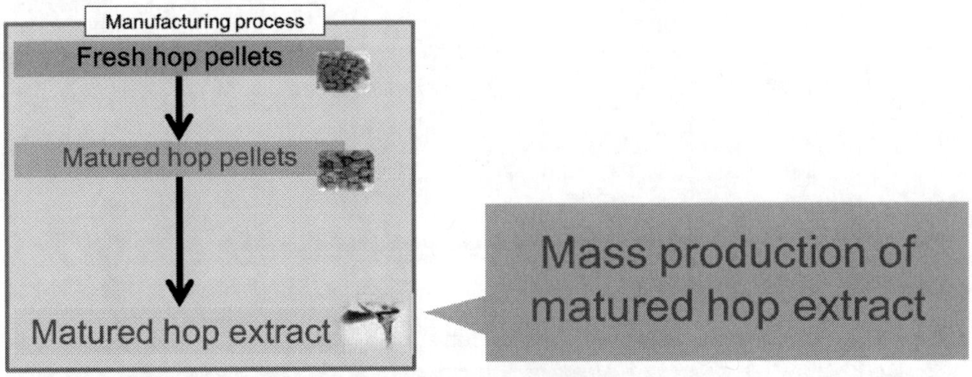

Patents
WO 2012081675 : HOPS OXIDATION PRODUCT, AND PRODUCTION METHOD AND USE THEREFOR
WO 2013191264 : CARBONATED BEVERAGE CONTAINING HOPS OXIDATION REACTION PRODUCT EXTRACT
WO 2013191258 : BEVERAGE CONTAINING AQUEOUS MEDIUM EXTRACT OF HOPS USED IN OXIDATION

Contents

Evaluation of body fat-reducing effects on humans

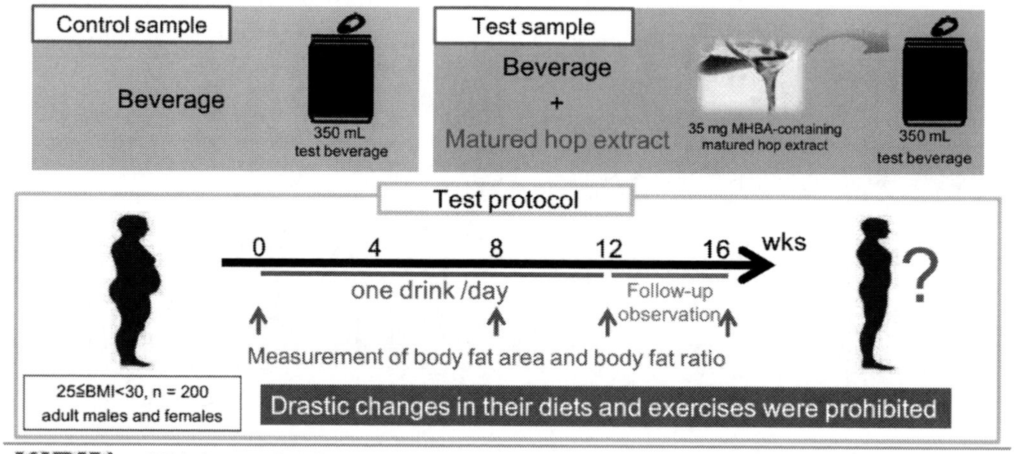

Matured hop extract have body fat-reducing effects on humans

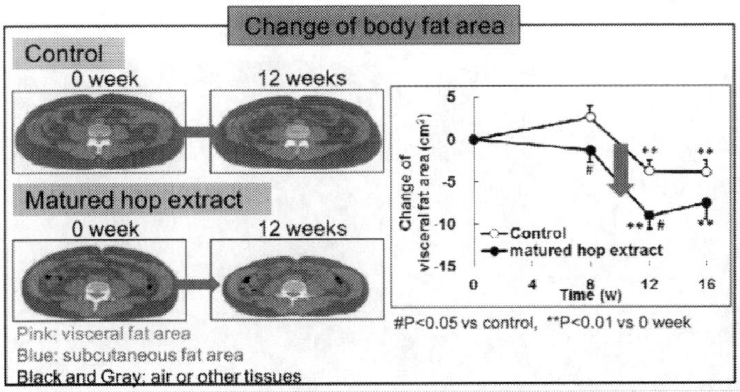

#P<0.05 vs control, **P<0.01 vs 0 week

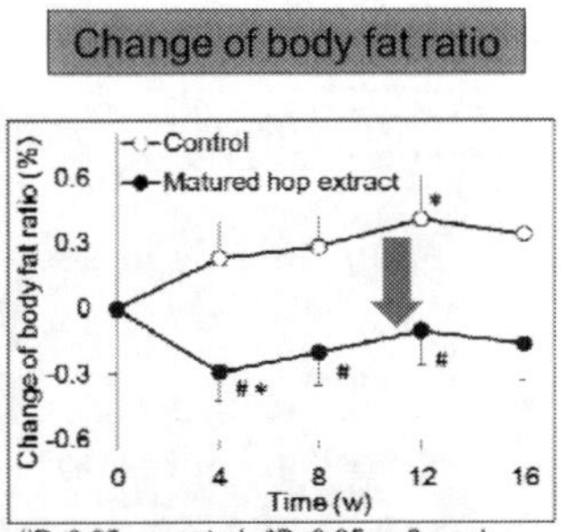

#P<0.05 vs control, *P<0.05 vs 0 week

*No adverse events were detected

Nutrition journal 2016, 15, 25

Matured hop extract safely reduced body fat on humans without changing their lifestyles!!

Contents

Flavor of beverage with matured hop extract

Preliminary sensory evaluation of beverage with/without matured hop extract was performed by three panelists

Samples

Control sample

Beer-taste beverage

Japanese brand "PERFECT FREE"

Test sample

Beer-taste beverage
+
matured hop extract

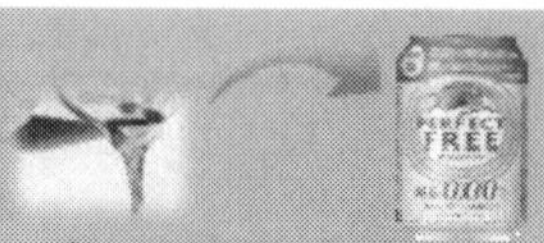

Matured hop extract was added (60 ppm MHBA)

Results

- No strong, but thick and complex bitterness
- No negative flavors

Summary

1st Stage	→ Stage Up!	2nd Stage	→ Stage Up!	3rd Stage

Chemical properties of MHBA

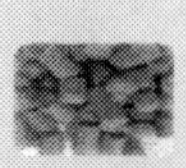

Health benefits of MHBA on animals

Mass production of matured hop extract

Health benefits on humans

Flavor

Solution of social problems with matured hop extract

Beer and food are the best combination!

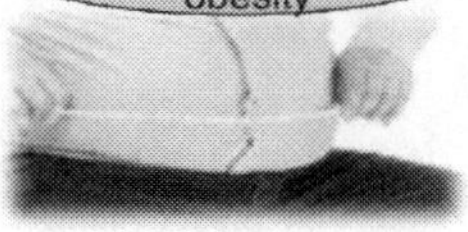

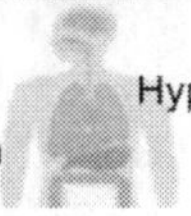

We developed novel hop material with body fat-reducing effects

"Matured hop extract"

Future prospects of matured hop extract

Matured hop extract

Body fat-reducing effect | Novel flavor

Functional foods

These products will conform to functional food system in Japan!

Applicable to Beers and beverages

Matured hop extract can make a healthy life with beers and beverages!!

Acknowledgements

We'd like to express our profound gratitude to the following people.

- Kazutoshi Shindo (Japan Women's University)
- Katsuya Nagai (ANBAS Co., Ltd. and Osaka University)
- Yoshihiro Ogawa (Kyushu University)
- Frank-Jürgen Methner (Technical University of Berlin)
- TTC Co., Ltd.
- Members of Shiga Factory (Kirin Brewery Co., Ltd.)
- Members of Research Laboratories for Alcoholic Beverage Technologies (Kirin Co., Ltd.)

 24

KIRIN

Quantitative analyses of MHBA in commercial beers

Table 4. Concentrations of MHBA and iso-α-acids in commercial beers.

Brand	Type of beer	MHBA concentration (mg/L)	Iso-α-acid concentration (mg/L)
A	Lager	19.1	18.3
B	Lager	31.8	23.6
C	Lager	30.4	26.6
D	Lager	24.1	15.7
E	Lager	38.0	26.7
F	Lager	20.1	16.5
G	India Pale Ale	210	62.6
H	India Pale Ale	181	64.0
I	India Pale Ale	152	41.5
J	Lambic	100	3.34
K	Lambic	151	6.39

Bioscience Biotechnology and Biochemistry 2015, 79, 1684-1694

Flavor of beer with matured hop extract

Sensory evaluation of beer with/without matured hop extract was performed by three experienced panelists

Samples

Control

Beer

Japanese brand "KIRIN Ichiban Shibori"

Matured hop extract

Beer
+
matured hop extract

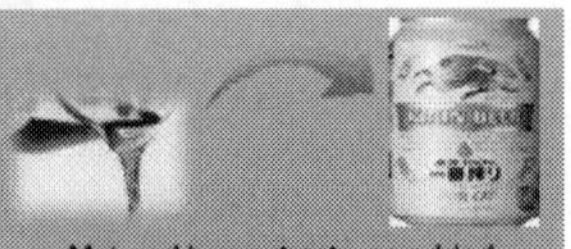

Matured hop extract was added (60 ppm MHBA)

"KIRIN Ichiban Shibori" { Type: pilsner type; Ingredient: malt, hop

Results

- No strong bitterness was detected, but thick and complex bitterness appeared
- No negative flavor related to oxidization was detected (oxidized flavor, cheese-like flavor)

A BETTER FUTURE FOR
THE BEER MARKET
EBC - MAY 2017 - Ljubljana
ABInBev

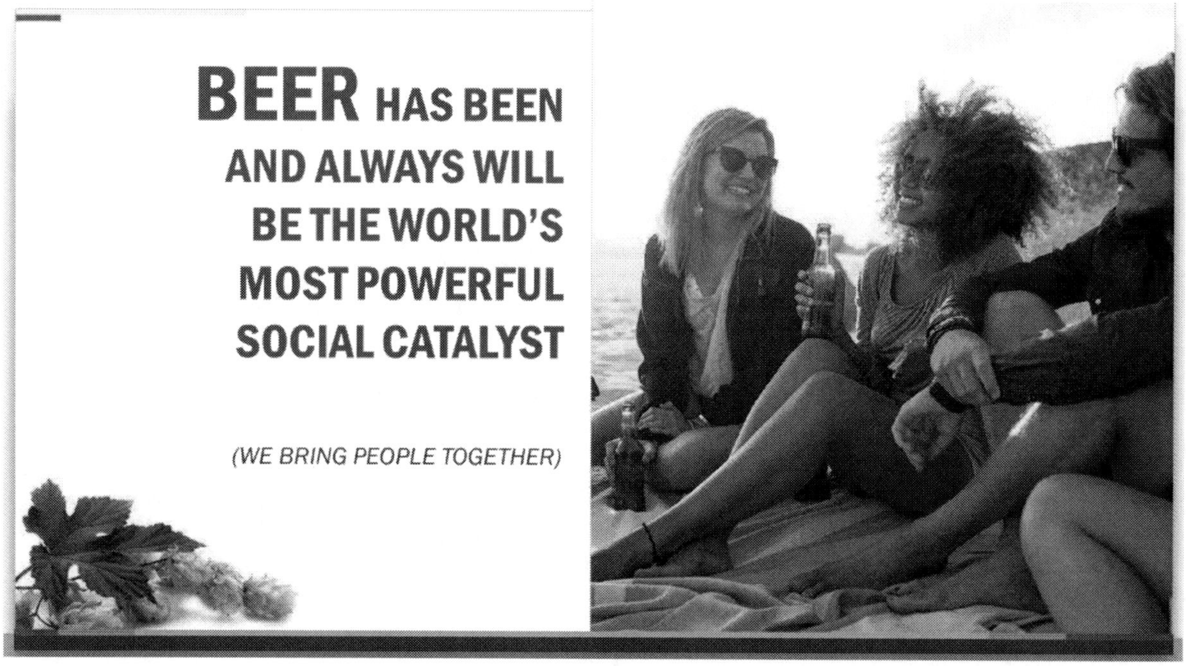
BEER HAS BEEN
AND ALWAYS WILL
BE THE WORLD'S
MOST POWERFUL
SOCIAL CATALYST
(WE BRING PEOPLE TOGETHER)

THIS CONTINUES TO BE AN
INCREDIBLY IMPORTANT
VALUE FOR THE CATEGORY

BUT CONSUMERS AND CONSUMPTION HAVE CHANGED

HOW CONSUMERS CHOOSE THEIR FOOD AND DRINKS

Eater's Digest: The Future of Food – Prosumer Report, Havas Worldwide - 2016

ABInBev

People are paying more attention to what is in what they eat/drink

I am much more aware of the nutritional/health value of the food I eat than I used to be

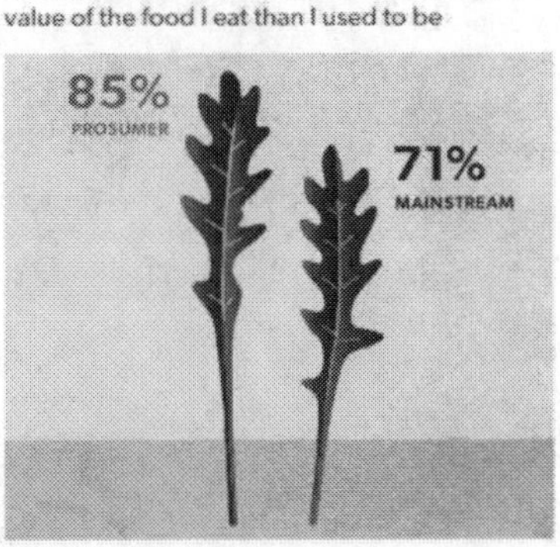

And that is impacting their purchase habits: the growing importance of nutritional value and ingredients over price

Please rank each of these criteria in terms of how important it is in your food-purchasing choices.

PERCENT RANKING AS #1:

30% 26% 23%

8% 7% 6%

PERCENT RANKING AS #1 OR #2:

52% 52% 45%

20% 17% 13%

HOW CONSUMERS CHOOSE THEIR FOOD AND DRINKS

Eater's Digest: The Future of Food – Prosumer Report, Havas Worldwide - 2016

ABInBev

And many of the trends that rise from this, directly impact our business.

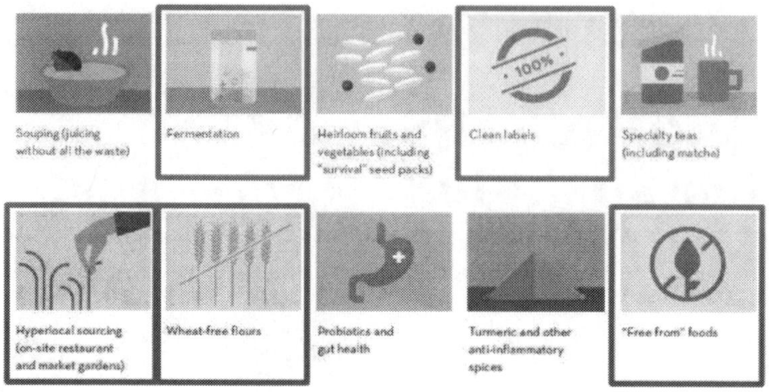

HOWEVER, BEER IN CONSUMERS' MINDS IS QUITE DISTANT FROM THIS REALITY

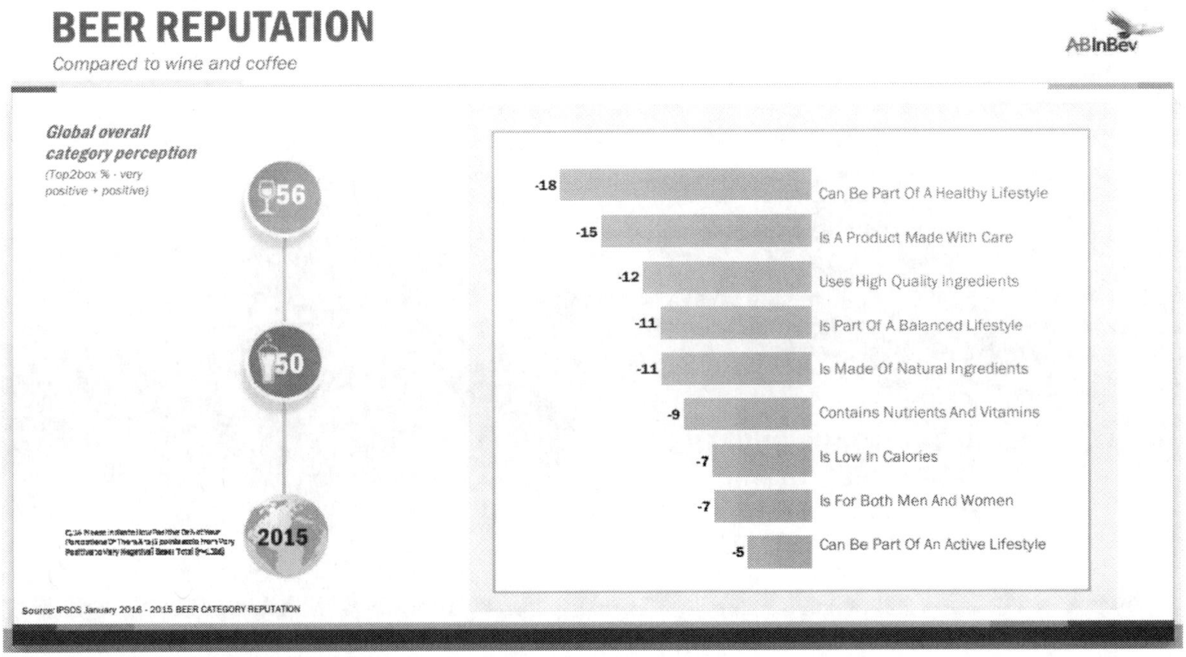
BEER REPUTATION
Compared to wine and coffee
AB InBev
Global overall category perception
(Top2box % - very positive + positive)
56
50
2015
-18 Can Be Part Of A Healthy Lifestyle
-15 Is A Product Made With Care
-12 Uses High Quality Ingredients
-11 Is Part Of A Balanced Lifestyle
-11 Is Made Of Natural Ingredients
-9 Contains Nutrients And Vitamins
-7 Is Low In Calories
-7 Is For Both Men And Women
-5 Can Be Part Of An Active Lifestyle
Source: IPSOS January 2016 - 2015 BEER CATEGORY REPUTATION

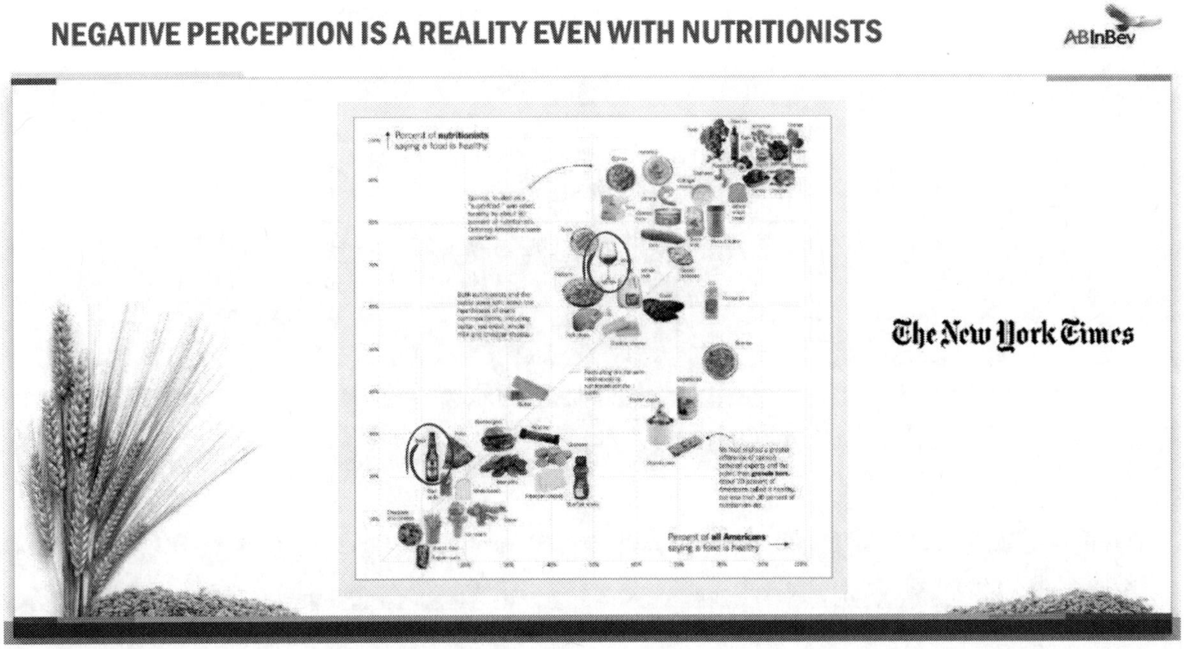
NEGATIVE PERCEPTION IS A REALITY EVEN WITH NUTRITIONISTS
AB InBev
Percent of nutritionists saying a food is healthy
Percent of all Americans saying a food is healthy
The New York Times

OUR CHALLENGE
CHANGE THE STATUS QUO, AND
POSITION BEER AS PART
OF WELL BEING AND
A HEALTHY LIFESTYLE

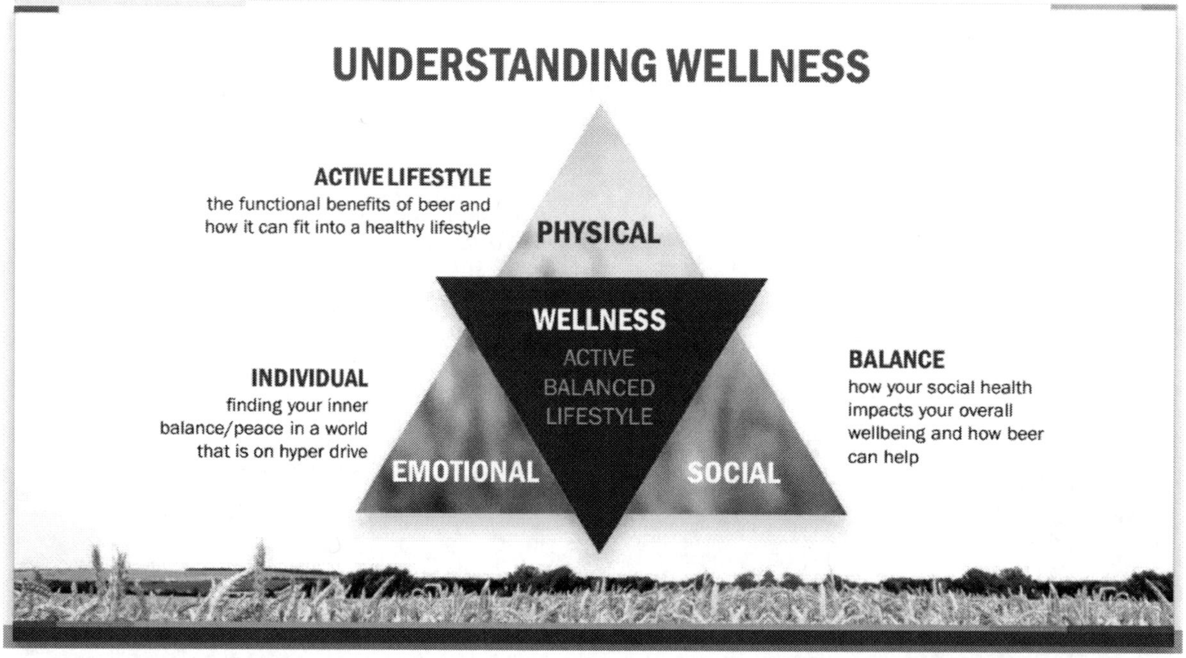
UNDERSTANDING WELLNESS
ACTIVE LIFESTYLE
the functional benefits of beer and
how it can fit into a healthy lifestyle
PHYSICAL
WELLNESS
ACTIVE
BALANCED
LIFESTYLE
INDIVIDUAL
finding your inner
balance/peace in a world
that is on hyper drive
EMOTIONAL
SOCIAL
BALANCE
how your social health
impacts your overall
wellbeing and how beer
can help

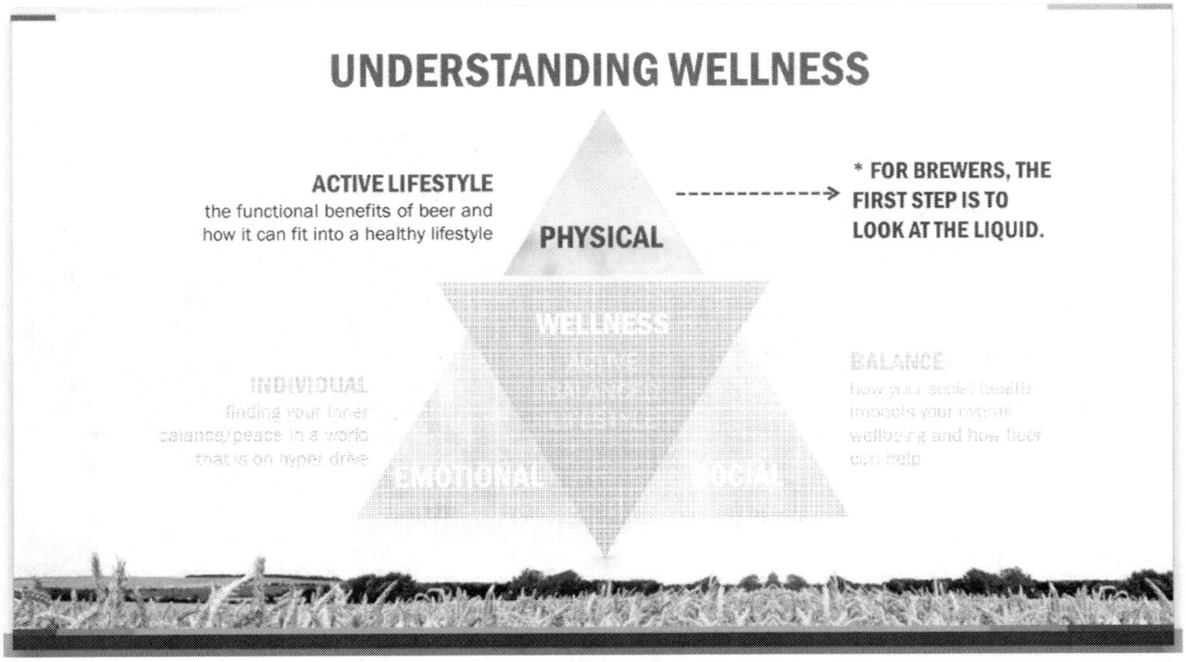
UNDERSTANDING WELLNESS
ACTIVE LIFESTYLE
the functional benefits of beer and how it can fit into a healthy lifestyle
PHYSICAL
* FOR BREWERS, THE FIRST STEP IS TO LOOK AT THE LIQUID.
WELLNESS
INDIVIDUAL
finding your inner balance/peace in a world that is on hyper drive
BALANCE
how your social health impacts your overall wellbeing and how beer can help
EMOTIONAL

BACK TO BASICS:
56% OF THE WORLD'S LDA POPULATION DOES NOT KNOW BEER IS MADE OF NATURAL INGREDIENTS

BACK TO BASICS:
AND EVEN LESS KNOW
THE PROCESS IT TAKES TO
TURN THOSE
INGREDIENTS INTO BEER

THE KEY TO THE FUTURE:
BUILD BEER
KNOWLEDGE TO
ALL, RESIGNIFYING
THE CATEGORY

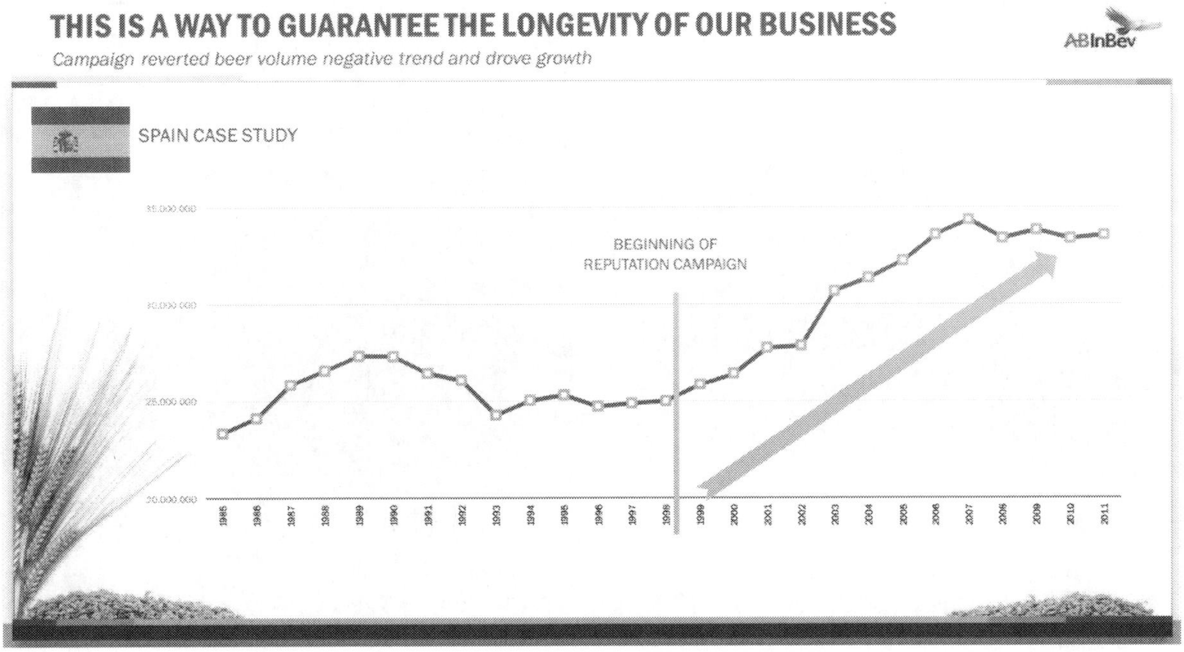
THIS IS A WAY TO GUARANTEE THE LONGEVITY OF OUR BUSINESS
Campaign reverted beer volume negative trend and drove growth
ABInBev
SPAIN CASE STUDY
BEGINNING OF
REPUTATION CAMPAIGN

With a different balance of bitterness and creaminess
in a surprising variety of flavors.

HERE IS A BIT OF WHAT WE HAVE BEEN DOING TO CHANGE THIS.

SHOWING BEER INGREDIENTS IN A NEW LIGHT

ABInBev

SHOWCASING BEER

WITH A DIFFERENT KIND OF BEER EXPERIENCE

In the meantime, we continue to listen to the Consumer:
People Want to Live Long, Happy and Healthy Lives
Beer contributes to this by bringing people together
But harmful use of alcohol can have a very negative impact

What is harmful use?
Harm to yourself or your community
Binge
Driving drunk
Addiction
Excess
Anti-social behavior
Underage
Disease
Alcohol-related violence

What do we want?

A Healthier World

Where every experience with beer is a positive one, for lives well lived

We believe the harmful use of alcohol is a serious global health issue

We have an opportunity -and a responsibility - to lead the world in reducing the harmful use of alcohol

Global Smart Drinking Goals

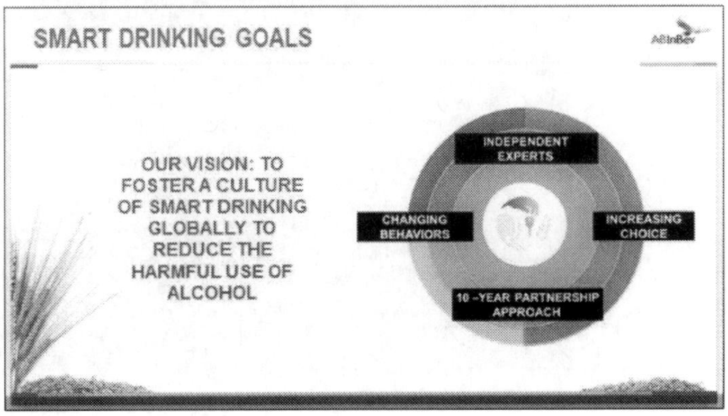

AB InBev

Global Smart Drinking Goals

Our Vision

To foster a culture of smart drinking globally to reduce the harmful use of alcohol

Consumer | Customer
Communities | Employees

HOW ARE WE GOING TO DO IT?

CHANGING BEHAVIORS THROUGH SOCIAL NORMS

1. MULTI-YEAR PILOTS

-10%

Reduce the harmful use of alcohol by at least 10% in nine cities by the end of 2020.

Implement the best practices globally by the end of 2025.

2. SOCIAL NORMS

$1B

Influence social norms and individual behaviors to reduce harmful alcohol use by investing at least 1 billion USD across our markets in dedicated social marketing campaigns and related programs by the end of 2025.

EMPOWERING CONSUMERS THROUGH CHOICE

3. PRODUCT PORTFOLIO

Ensure No- or Lower-Alcohol beer products represent at least 20% of AB InBev's global beer volume by the end of 2025.

4. ALCOHOL HEALTH LITERACY

Place a Guidance Label on all of our beer products in all of our markets by the end of 2020.

Increase alcohol health literacy by the end of 2025.

Integrated into business strategy - Cross-functional work

Operates with Independence

EXPERTS & TECHNICAL ADVISORY GROUP (TAG) | AB InBev FOUNDATION | 10 -YEAR PUBLIC PRIVATE PARTNERSHIP APPROACH

24

AB InBev

Empowering Consumers Through Choice

New Global Brands launches in 2016

INTRODUCING
Corona Cero *

NO ALCOHOL. NO BOUNDARIES.

Initial focus – only MEX

INTRODUCING
Prohibition Brew

ALL BUDWEISER. NO ALCOHOL *.

* Not including the US

25

ABInBev

BRAND COMMUNICATION

TV SPOT – TARGETED TO BEER DRINKERS

26

ABInBev

BRAND COMMUNICATION
TV SPOT – TARGETED TO BEER DRINKERS

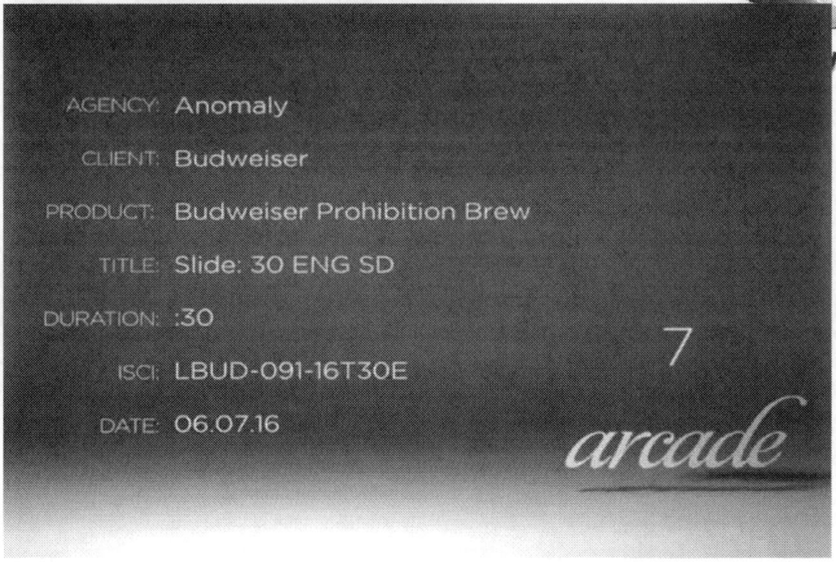

27

ABInBev

Changing behavior through social norm:

Investing at least 1 billion USD across our markets in dedicated social marketing campaigns and related programs by the end of 2025

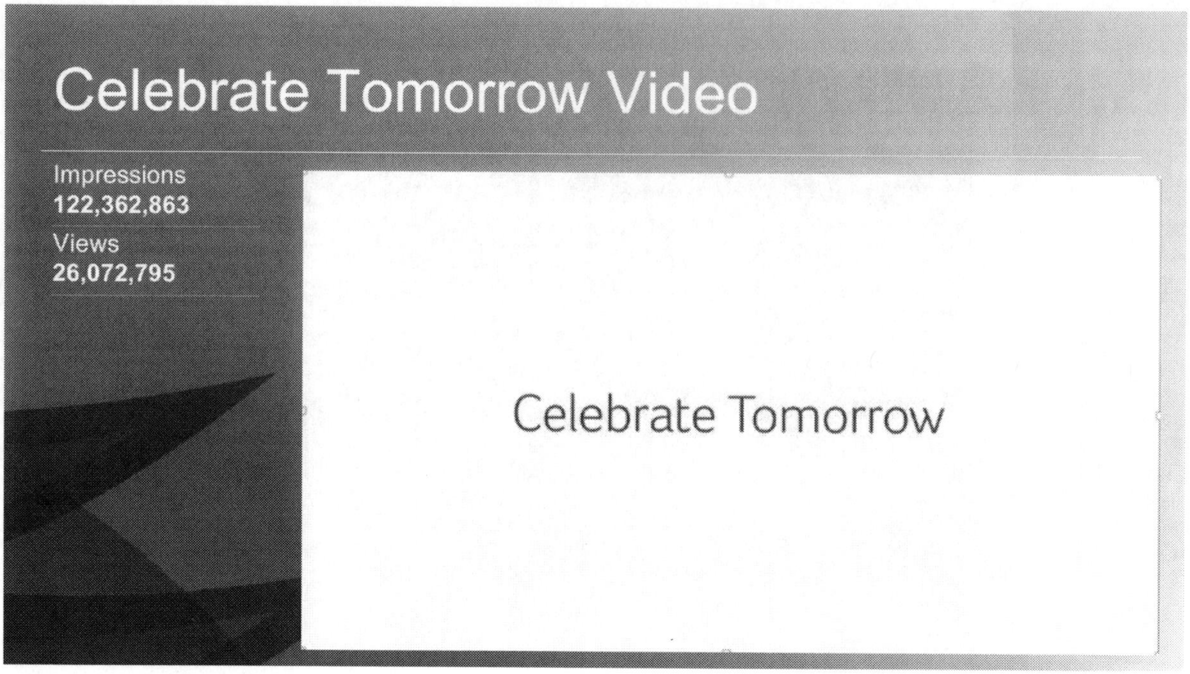
Celebrate Tomorrow Video
Impressions
122,362,863
Views
26,072,795
Celebrate Tomorrow

ANY QUESTIONS? THANKS AND CHEERS!

AB InBev

Construction of novel lager yeast strains through a fertile allotetraploid intermediate

36th EBC Congress
May 14-18, 2017 – Ljubljana, Slovenia
Kristoffer Krogerus

Background

- Pale lager is the most popular beer style worldwide
- Cold fermentation and clean flavour profile
- Made with lager yeast: *Saccharomyces pastorianus*

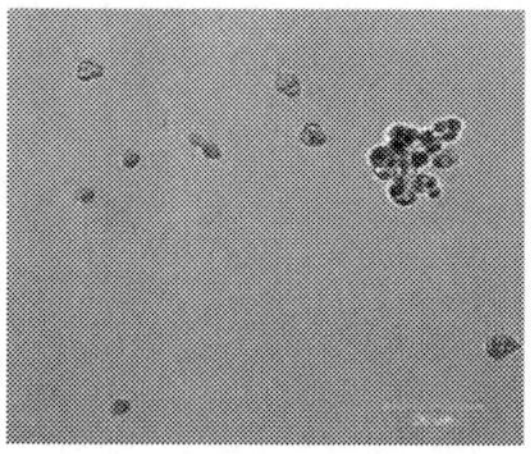

24/05/2017 2

Background

- Lager yeast, *S. pastorianus*, is a natural hybrid between *S. cerevisiae* and *S. eubayanus*:

Saccharomyces cerevisiae	*Saccharomyces pastorianus*	*Saccharomyces eubayanus*
Good wort fermentation	**Good wort fermentation**	**Poor wort fermentation**
Usually cold-sensitive	**Cold-tolerant**	**Cold-tolerant**

24/05/2017 3

Background

Liger

Zonkey

Peppermint

24/05/2017 4

Background

- Poor diversity among the traditional lager yeasts
- Belong to one of two distinct lineages:
  - Saaz
  - Frohberg

Group 1 (Saaz)	Group 2 (Frohberg)
• Weak fermentation • Very cold-tolerant • *S. eubayanus* genome dominates	• Strong fermentation • Cold-tolerant • *S. cerevisiae* genome dominates

Background

- Can we increase the diversity of lager yeast by creating new lager yeast hybrids?
  - Mating *S. eubayanus* with selected *S. cerevisiae* parents?

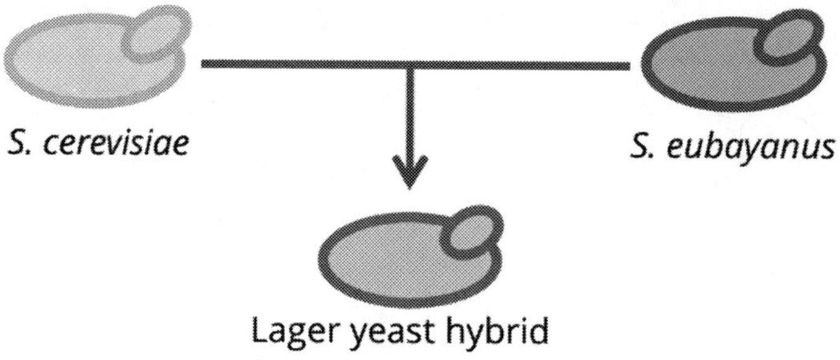

Our previous research

- *De novo* lager yeast hybrids can
  - outperform their parent strains during fermentation
  - produce beer with higher concentrations of aroma compounds

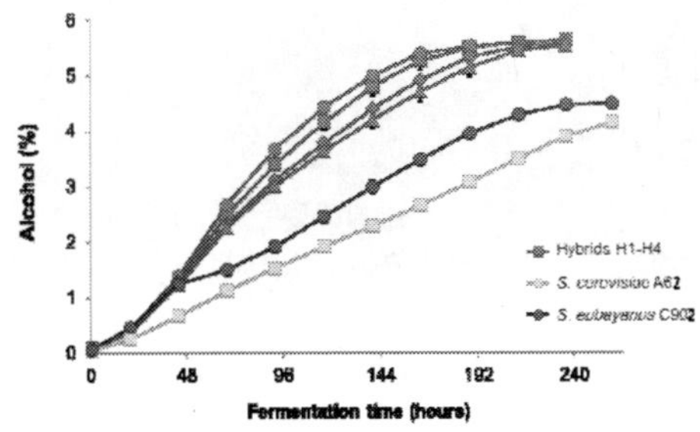

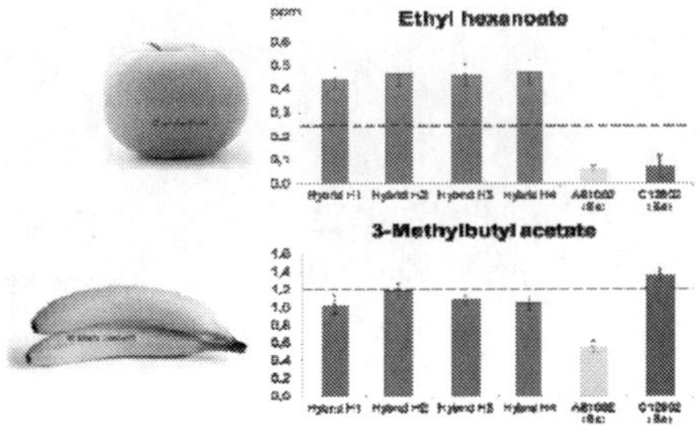

7

Controlling the properties of the hybrids

- How does the contribution of the parental subgenomes affect important phenotypic traits?
  - Fermentation performance
  - Aroma production
  - Stress tolerance

Group 1 (Saaz)

- Weak fermentation
- Very cold-tolerant
- *S. eubayanus* genome dominates
- **Triploid (3n)**

Group 2 (Frohberg)

- Strong fermentation
- Cold-tolerant
- *S. cerevisiae* genome dominates
- **Tetraploid (4n)**

24/05/2017

8

Controlling the properties of the hybrids

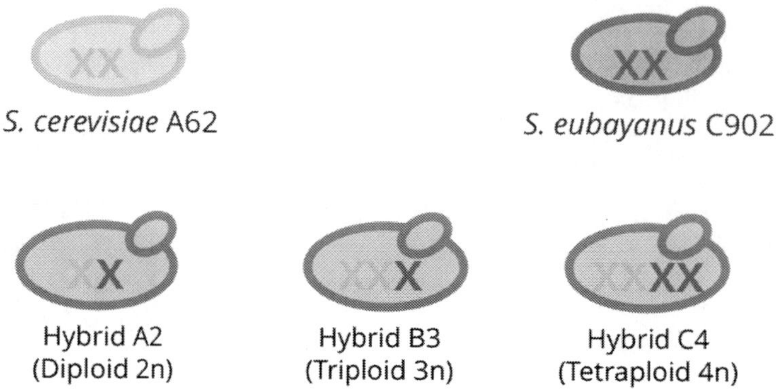

Fermentation performance (15 °C, 15 °P)

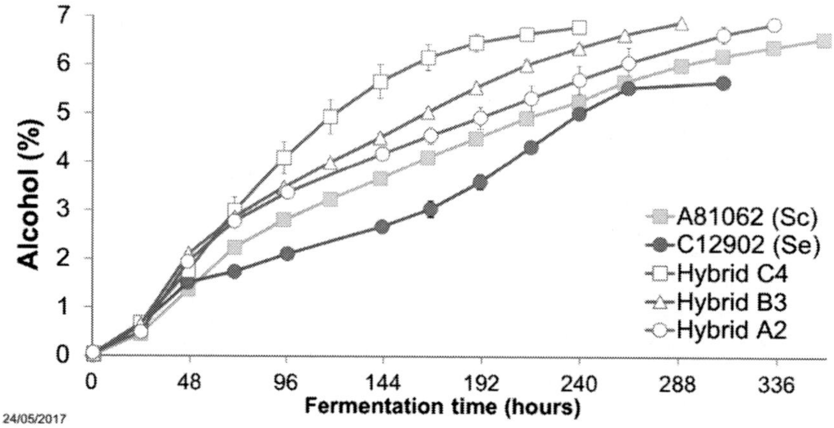

Aroma compounds (15 °C, 15 °P)

Ethyl hexanoate

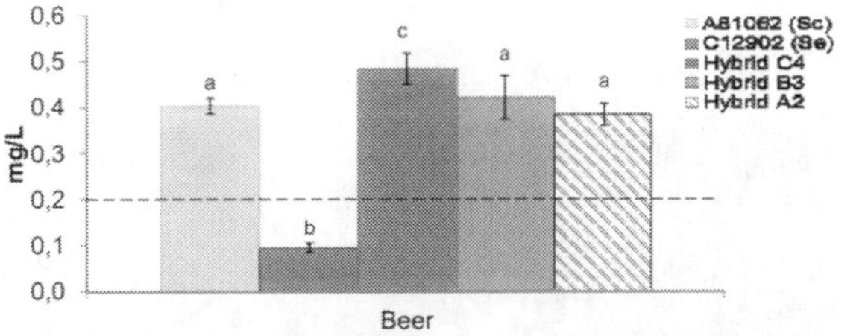

3-methylbutyl acetate

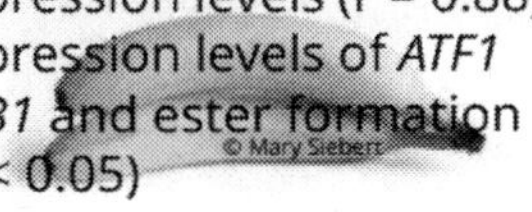

Positive correlation between

- gene copy numbers and expression levels (r = 0.88)
- expression levels of *ATF1* *EEB1* and ester formation ($p < 0.05$)

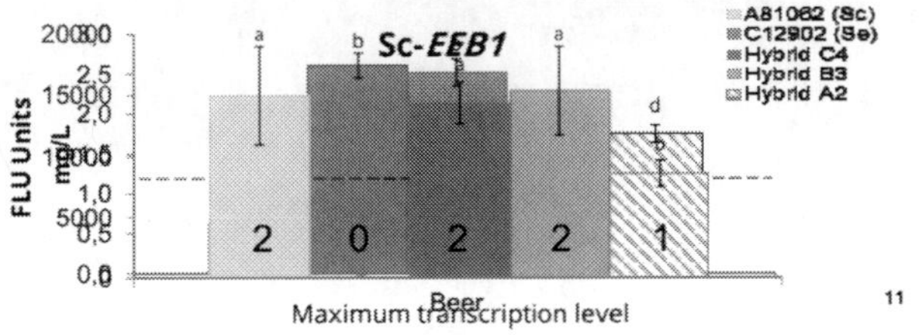

Fertility of the hybrids

- Interspecies hybrids tend to be sterile
  - Cannot produce viable spores
- However, research suggests that sterility is typically limited to diploid hybrids.

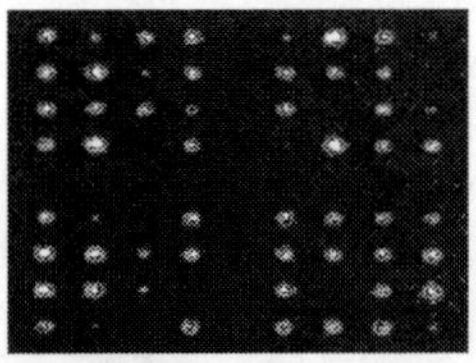

Hybrid A2 (Diploid 2n)	Hybrid B3 (Triploid 3n)	Hybrid C4 (Tetraploid 4n)
0%	**17%**	**55%**

Spore viability

How can we utilize hybrid fertility?

- Introduce variation and diversity via meiosis

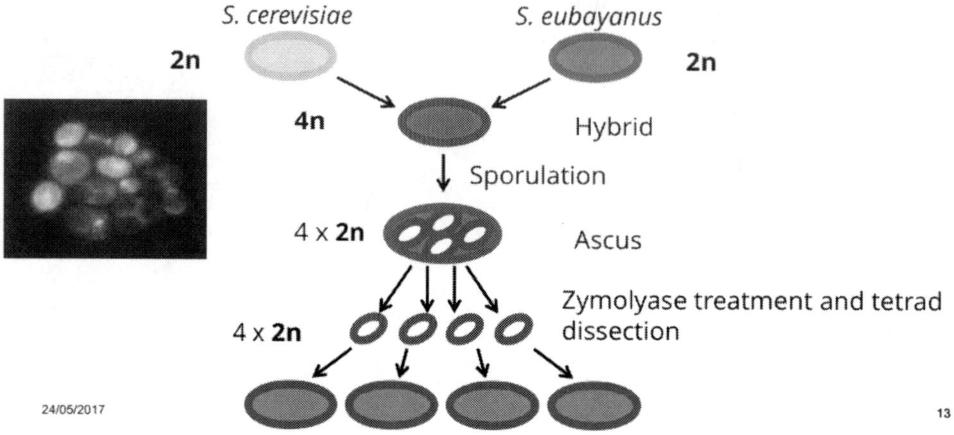

Comparison of spore clones

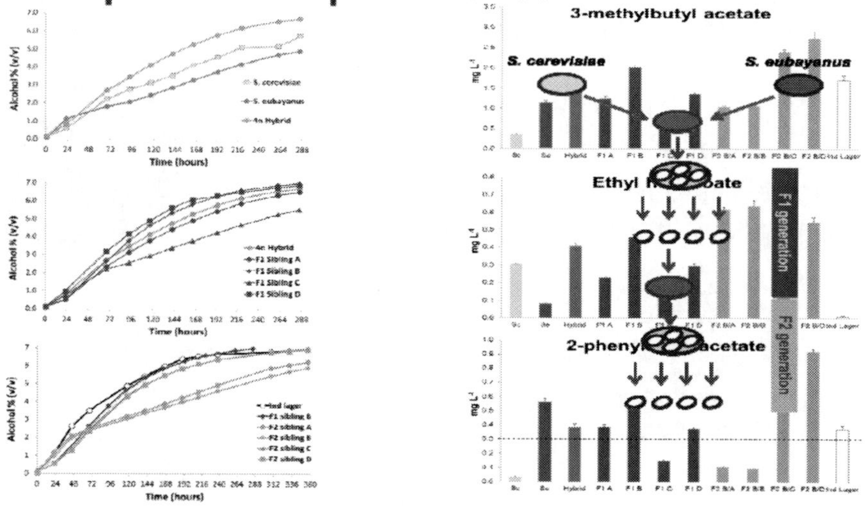

How can we utilize hybrid fertility?

- Construction of complex multi-parent hybrids:

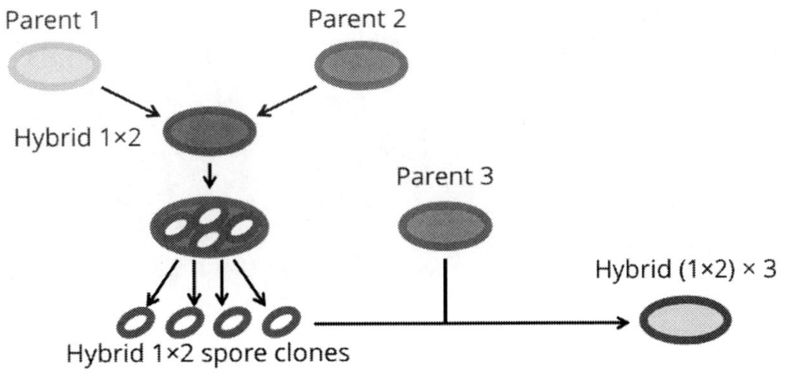

24/05/2017

15

Combining multiple strains

- Can we 'cherry-pick' traits from multiple parent strains?

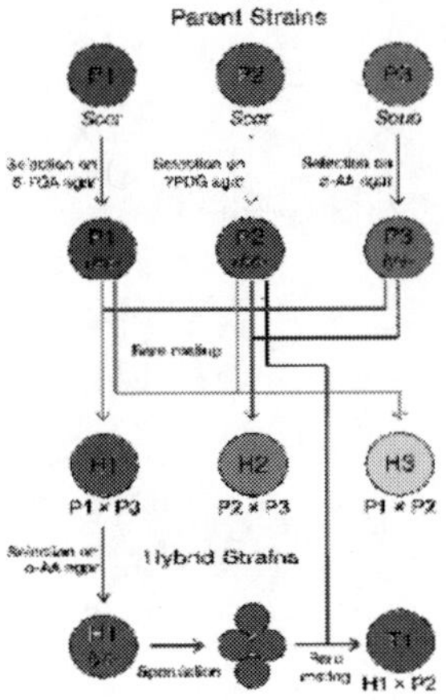

• Maltotriose use • POF+ • Cold sensitive	P1
• No maltotriose use • POF- • Cold sensitive	P2
• No maltotriose use • POF+ • Cold tolerant	P3
• Maltotriose use • POF+ • Cold tolerant	T1

24/05/2017

16

How can we utilize hybrid fertility?

- Removal of unwanted traits through meiosis
  - Phenolic off-flavour (4-vinyl guaiacol)
  - Sulfur compounds
  - (Flocculation)

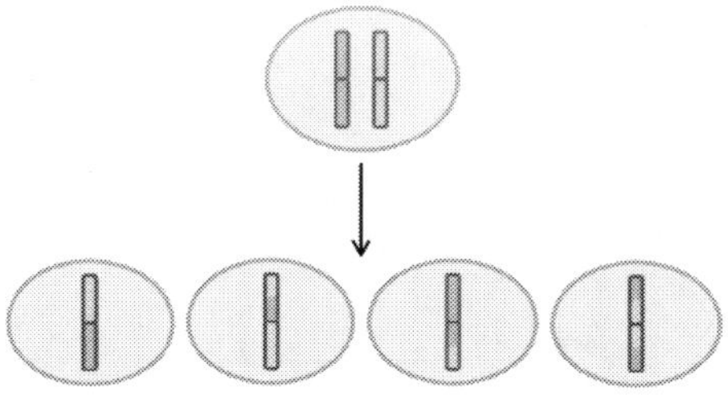

Removing 4-vinyl guaiacol formation

- 4-vinyl guaiacol formation is attributed to the adjacent *PAD1* and *FDC1* genes

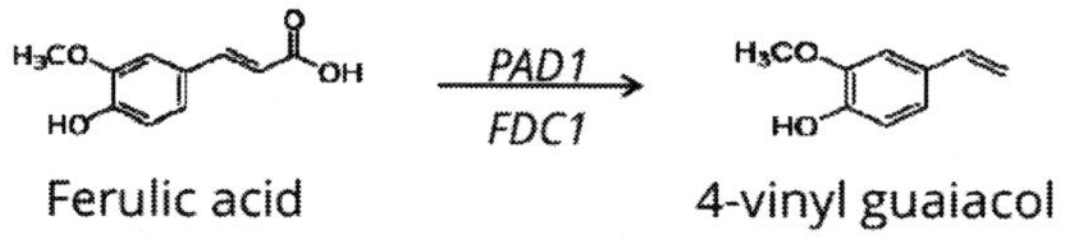

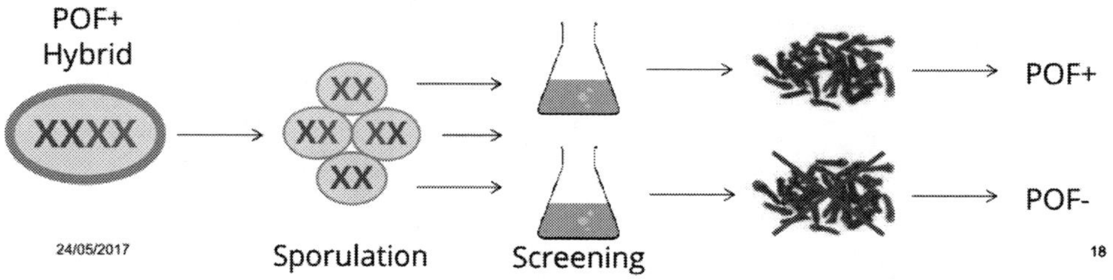

Obtaining a POF- lager hybrid

- T1, the 3-parent interspecifc hybrid, is also tetraploid
- T1 is fertile (38% spore viability) and was sporulated
- 12 segregants were screened for POF phenotype
- 3 segregants of these segregants were **POF-**

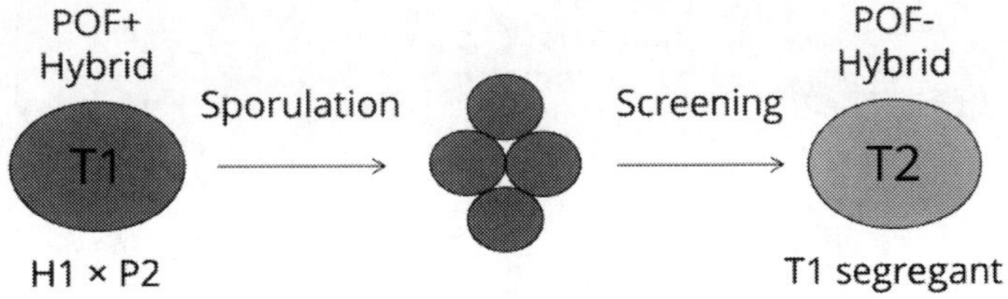

Fermentations at 15 °C in 15 °P wort

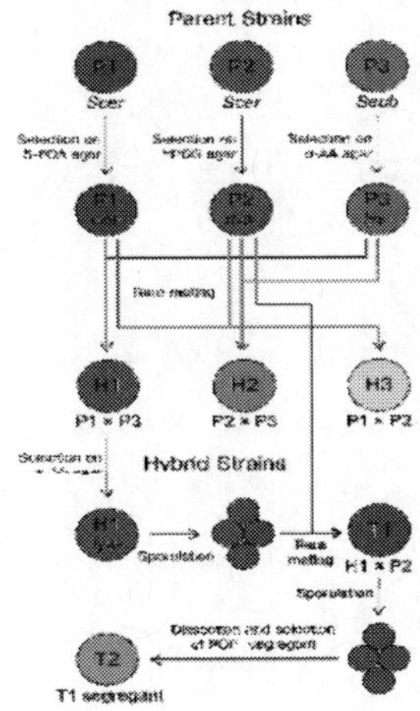

- Fastest fermentation by P2 and T1
  - Slows down as maltose is used up
- T2 reaches the highest ABV%

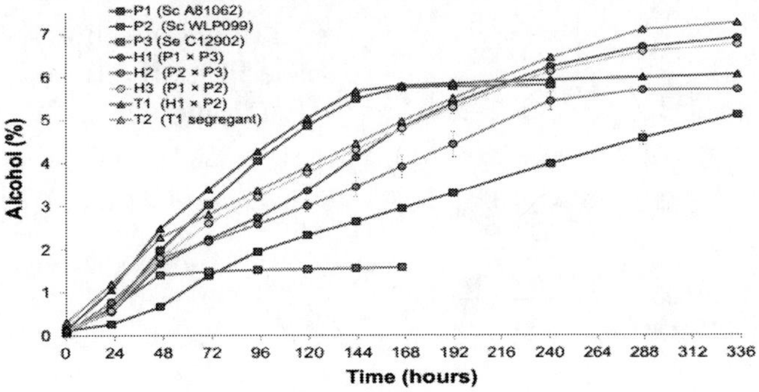

4-vinyl guaiacol and maltotriose

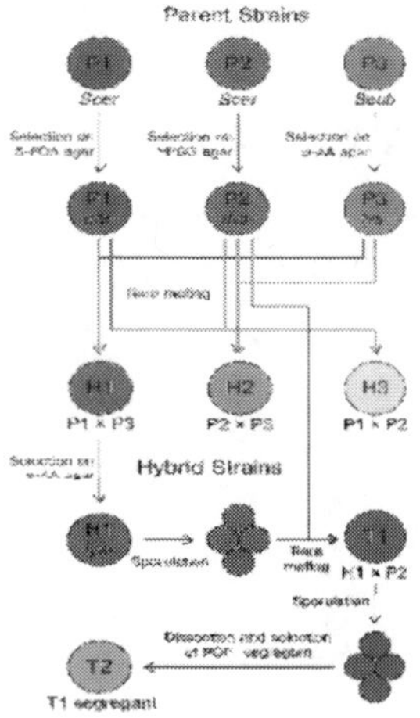

- Confirmation of POF- phenotype in T2

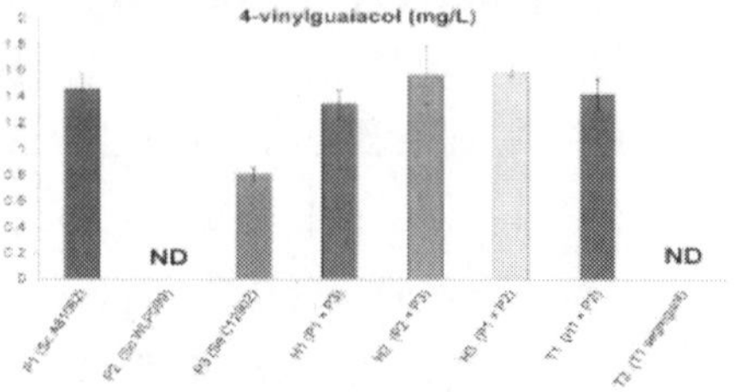

- Superior maltotriose use by T2 compared to T1.

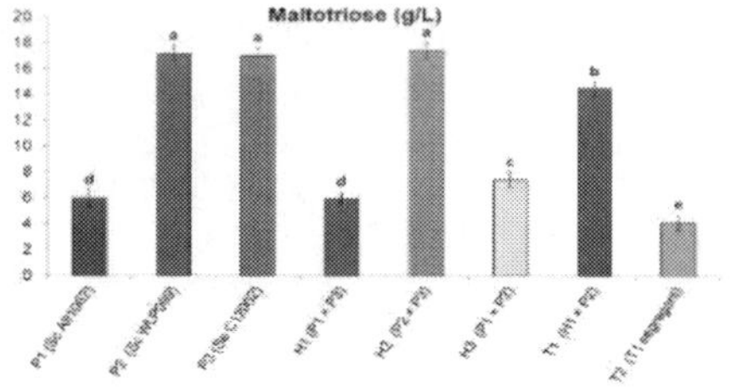

PAD1 alleles

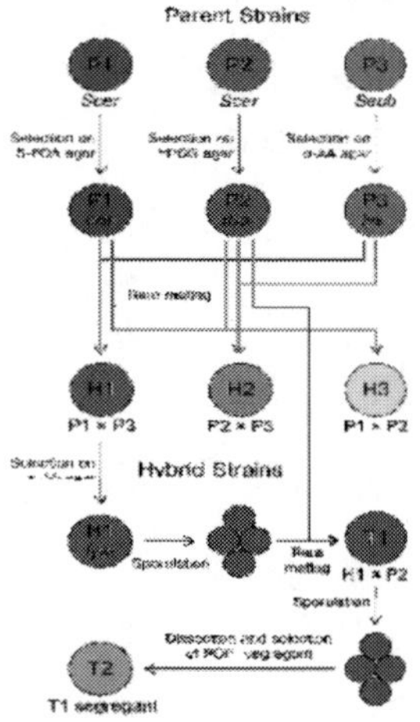

- Whole genome sequencing revealed:
  - T1 contained three different *PAD1* alleles
  - T2 only contained the P2-derived *PAD1* allele

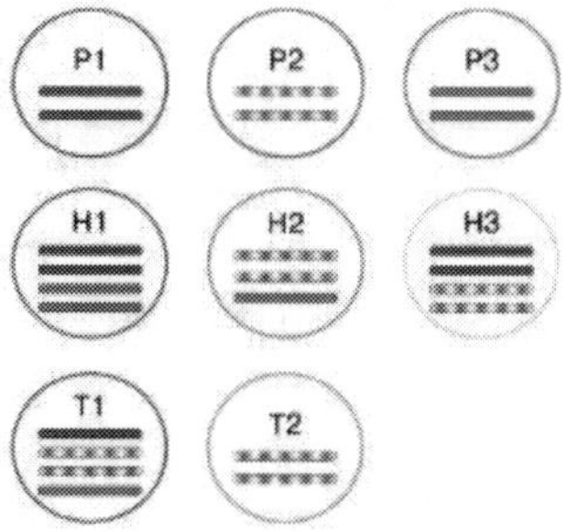

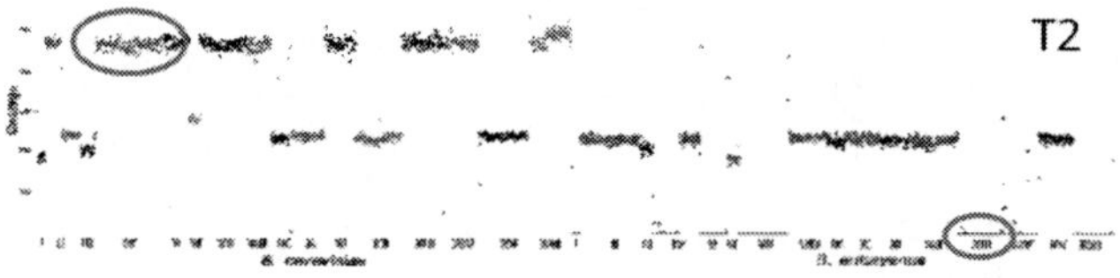

Conclusions

- Possible to create diverse lager yeast strains
- Properties of hybrid strains are to some extent affected by their ploidy and subgenome inheritance
- Tetraploid lager hybrids are fertile
    - Introduce variation through meiosis
    - Remove unwanted phenotypic traits
    - Create complex multi-parent hybrids
- No genetic engineering involved

24/05/2017 23

Acknowledgements

VTT:
- Brian Gibson
- Mikko Arvas
- Sirpa Jylhä
- Frederico Magalhães
- Merja Oja
- Virve Vidgren
- Annika Wilhelmson

- Aila Siltala
- Arvi Wilpola
- Eero Mattila

IRCAN:
- Gianni Liti
- Matteo De Chiara
- Jia-Xing Yue

Funding:
- Alfred Kordelin Foundation
- Svenska Kulturfonden
- PBL Brewing Laboratory
- SABMiller
- Academy of Finland
- FP7 Marie-Curie ITN YEASTCELL

Thank you for your attention!

24/05/2017 24

Video Presentation in Original e-media

24/05/2017

25

Stability of hybrid strains

- A62×C902 F2 hybrid B/D was serially repitched in ten 15 °P wort fermentations.
- The original hybrid, the G10 population and three G10 isolates were compared in a 15 °P fermentation:

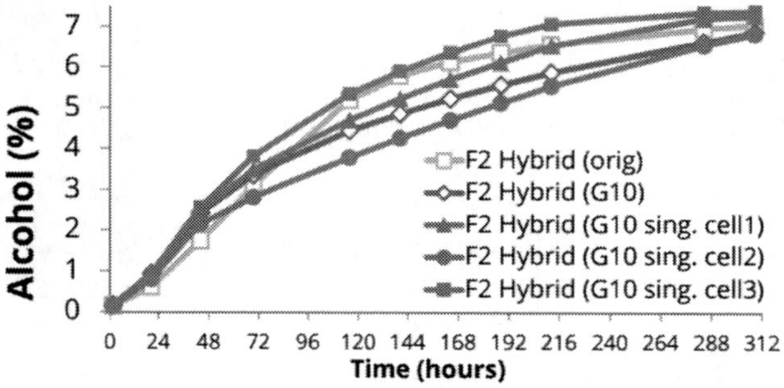

24/05/2017

26

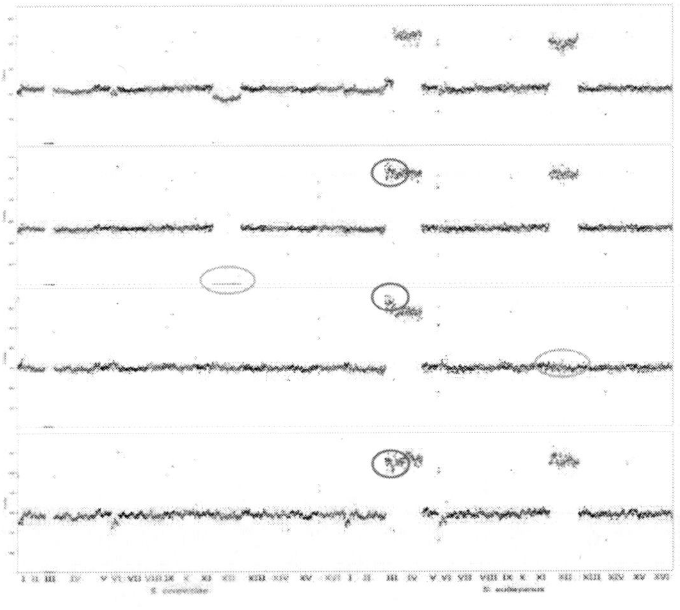

F2 Hybrid B/D

G10 isolate 1
Lost two copies of Sc-XII
Gained two copies of Se-III

G10 isolate 2
Lost two copies of Se-XII
Gained two copies of Se-III

G10 isolate 3
Gained two copies of Se-III

27

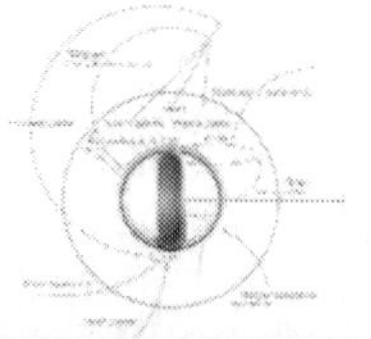

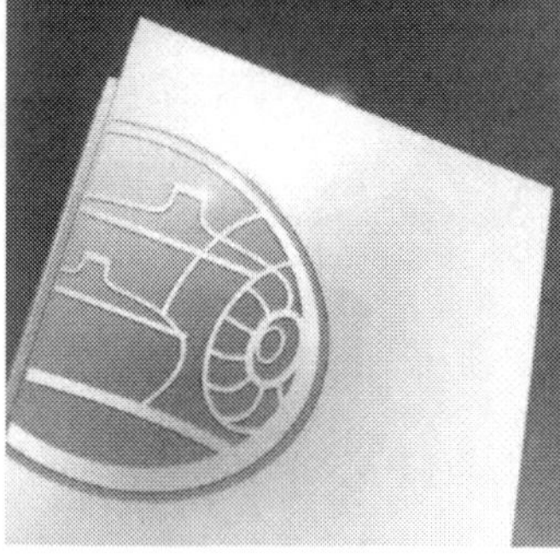

ZIEMANN HOLVRIEKA

21st Century Brewery by Ziemann®

ZIEMANN HOLVRIEKA GmbH
Technology, Research and Development
Ludwigsburg / Germany

21st Century Brewery

BY ZIEMANN ®

Basic aims of the 21st Century Brewery

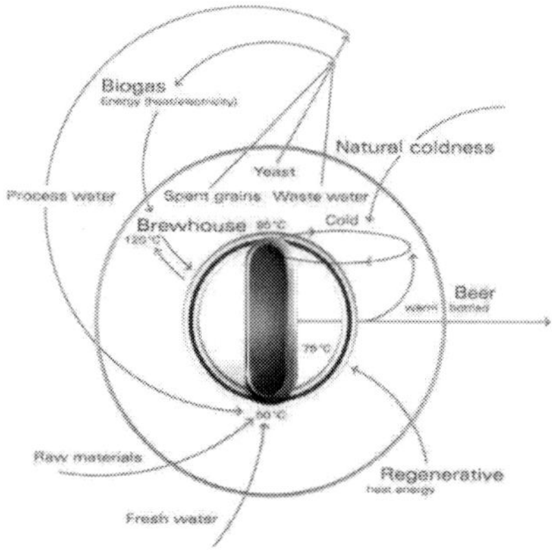

- If possible, without primary energy
- Natural sources of energy
- Energy from waste water
- Conversion of heat into cold
- Reduction of peak loads
- Use of natural cold

2

21st Century Brewery

BY ZIEMANN ®

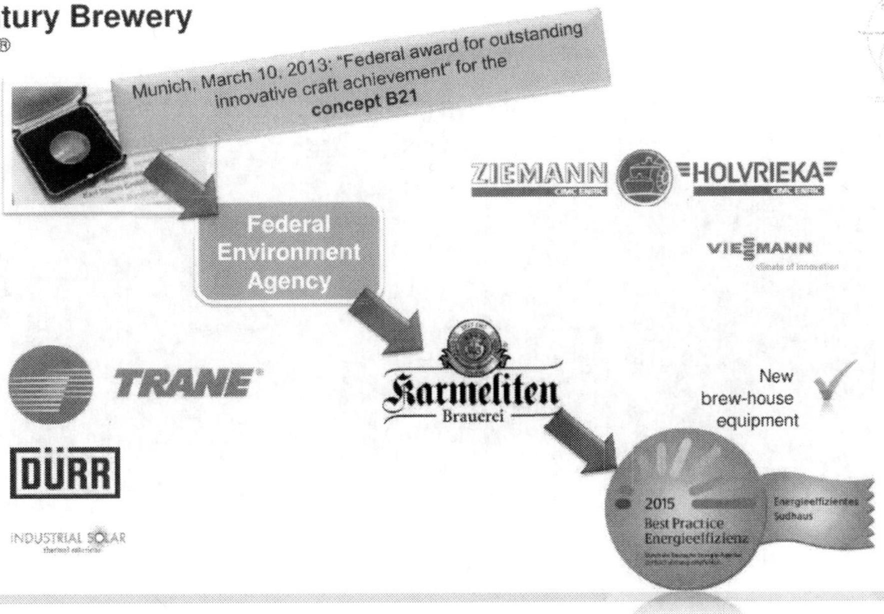

3

21st Century Brewery

BY ZIEMANN ®

ZIEMANN HOLVRIEKA

General three steps to success

1. New machine and plant technology in the brewhouse

- **Installation of cascade brewhouse**
  - Colibri by Ziemann®
  - Lotus by Ziemann®
  - Shark
- **Installation of vapour condenser**
  Energy recovery
- **Reduction of the evaporation rate**
- **Adjustment of heat transfer surfaces**

✓

2. Optimization of periphery and supply technology

- **New heating boiler**
  (old boiler 1 T-Level at 140…180 °C for all processes)
- **Recirculation of heat**
- **Minimization of waste heat due to insulation**
- **Replacement of old control technology of the boiler**

✓

3. Cross-linking of further machines and equipment

Already implemented:

- **Heating boiler**
- **Microgasturbine**
- **Exhaust heat exchanger**
- **Absorption cooling machine**
- **Ice storage**
- **Ice Age®**
- **Propane refrigeration machine**

✓

Still pending:

- **Fresnel lens**
- **Low-temperature storage**
- **Use of biogas**

4

21st Century Brewery

BY ZIEMANN ®

Former brewhouse

5

21st Century Brewery

BY ZIEMANN ®

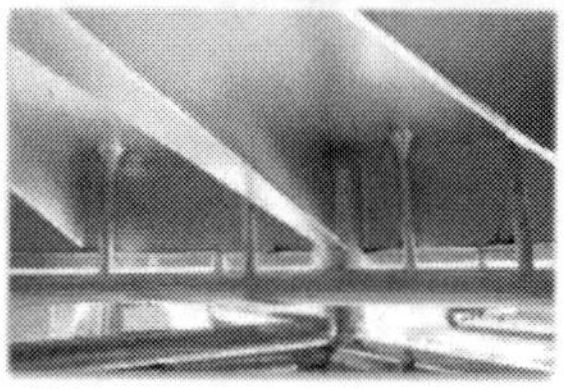

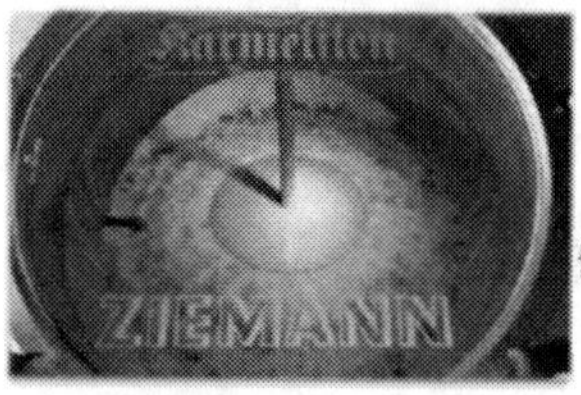

New brewhouse

7

21st Century Brewery
BY ZIEMANN ®

Adjustment of supply technology
Example: refrigeration system

European regulation (until 2025)

BEFORE
CFC*-containing refrigerant
3 Compression refrigeration systems

Advantages of CFCs
Non-combustible, nontoxic

*CFC – chlorofluorocarbons

AFTER
Natural refrigerating agent: propane gas

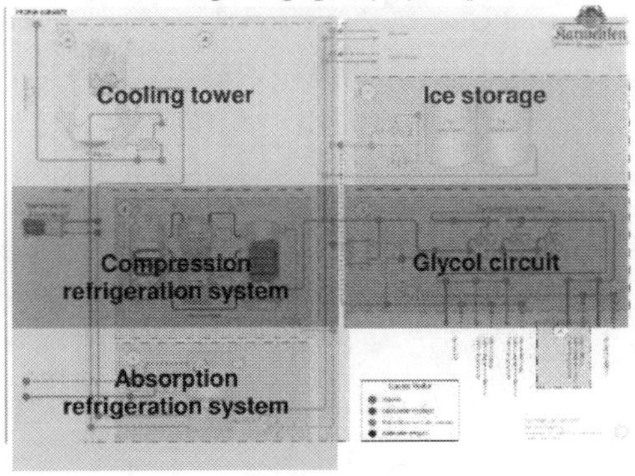

„Natural refrigerants" – naturally occurring in nature (carbon dioxide, ammonia, propane – combustible, toxic → additional monitoring required)

21st Century Brewery
BY ZIEMANN ®

1. New machine and plant technology in the brewhouse

- **Installation of cascade brewhouse**
  - Colibri by Ziemann®
  - Lotus by Ziemann®
  - Shark
- **Installation of vapour condenser**
  Energy recovery
- **Reduction of the evaporation rate**
- **Adjustment of heat transfer surfaces**

2. Optimization of periphery and supply technology

- **New heating boiler**
  (old boiler 1 T-Level at 140…180 °C for all processes)
- **Recirculation of heat**
- **Minimization of waste heat due to insulation**
- **Replacement of old control technology of the boiler**

3. Cross-linking of further machines and equipment

Already implemented:
- **Heating boiler**
- **Microgasturbine**
- **Exhaust heat exchanger**
- **Absorption cooling machine**
- **Ice storage**
- **Ice Age®**
- **Propane refriger machine**

Still pending:
- **Fresnel lens**
- **Low-temperature storage**
- **Use of biogas**

21st Century Brewery

BY ZIEMANN ®
Before – cross-linking

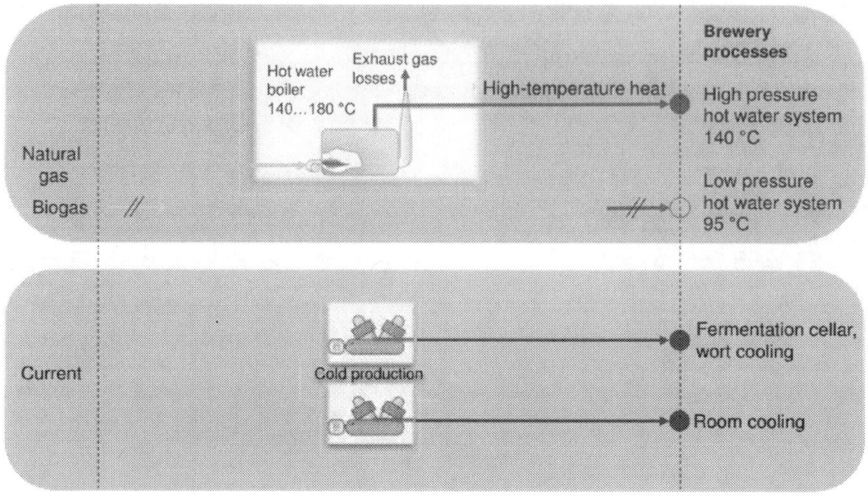

21st Century Brewery

BY ZIEMANN ®
After – cross-linking

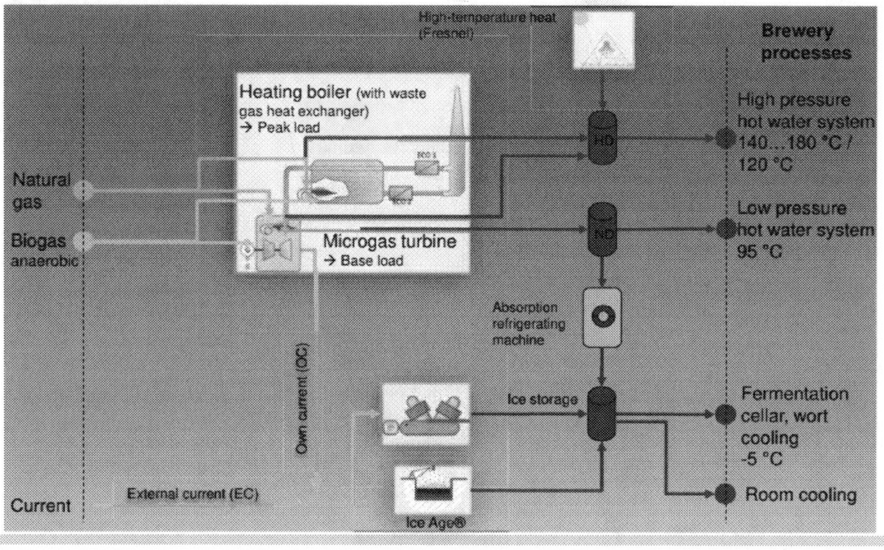

21ST CENTURY BREWERY
TAKING CARE OF BREWING

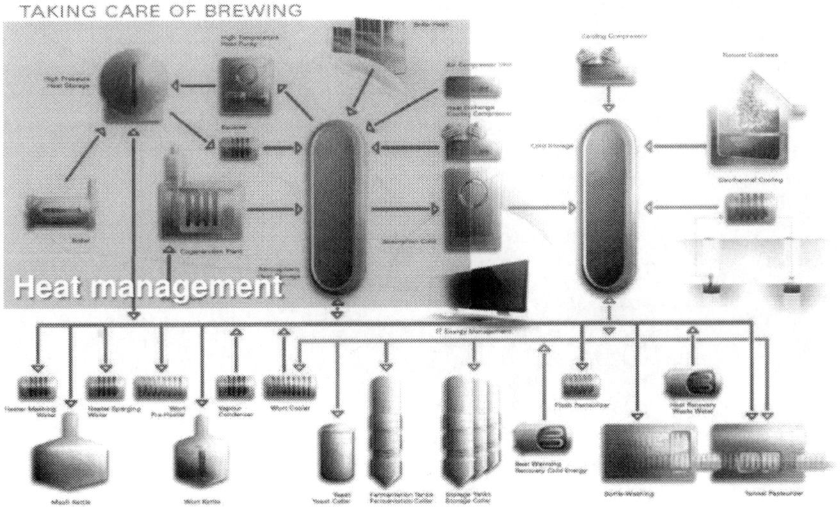

12

21st Century Brewery
BY ZIEMANN ®

Heating boiler

BEFORE

LOOS
Bosch Group

AFTER

VIESSMANN

12

21st Century Brewery
BY ZIEMANN ®

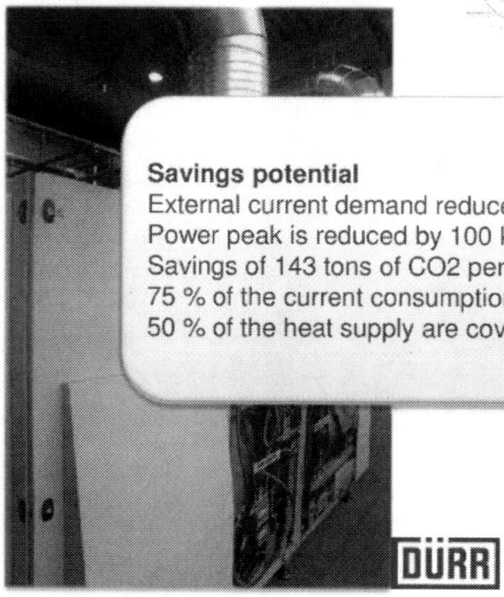

Savings potential

External current demand reduced by 500.000 $kWh_{electrical}$ per year
Power peak is reduced by 100 kW
Savings of 143 tons of CO_2 per year
75 % of the current consumption is covered by own current
50 % of the heat supply are covered

13

21st Century Brewery
BY ZIEMANN ®

- Flat plate collectors
- Vacuum tube collectors
- → Non-concentrating systems (< 100 °C)

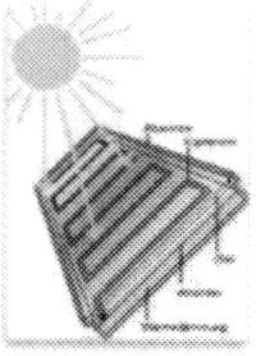

Fresnel collectors

→Concentrating system (well above 100 °C)

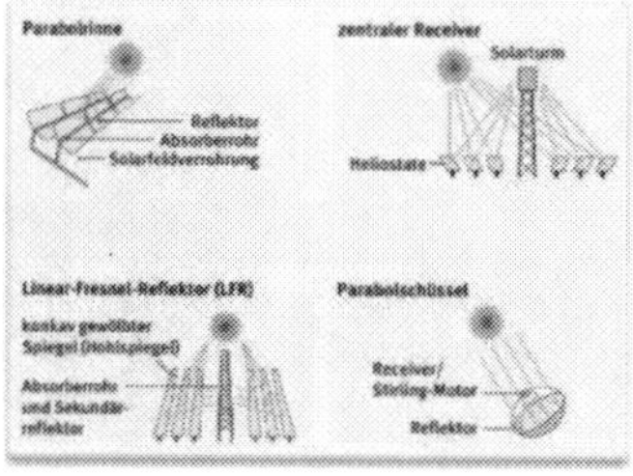

www.ikz.de
http://www.bothe-haustechnik.de

21st Century Brewery

BY ZIEMANN ®

ZIEMANN HOLVRIEKA

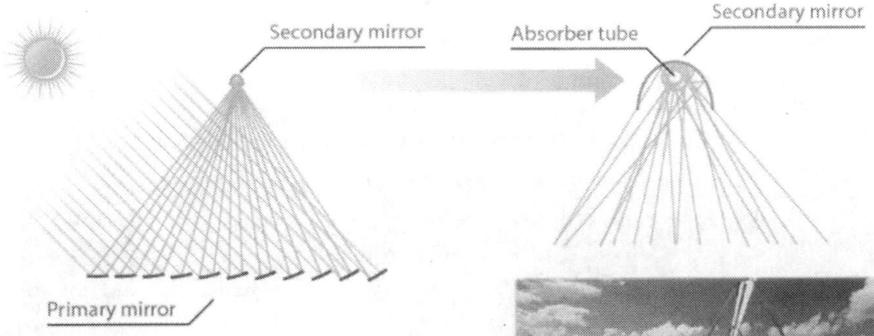

Fresnel reflector collectors
Solar thermal systems

INDUSTRIAL SOLAR

16

21st Century Brewery

BY ZIEMANN ®

ZIEMANN HOLVRIEKA

Objective: water temperature of 180 °C

What to do in the eve

…**insufficient** lighti
→ regulation of the

Savings potential
at a gross collector surface of 550 m²
Annual output ~128 $MWh_{thermal}$
Reduction of primary energy 175.000 kWh of natural gas per year
Savings of 35.000 kg of CO_2 per year

…**too intensive** solar radiation
→ turning away the collectors

17

21st Century Brewery

BY ZIEMANN ®

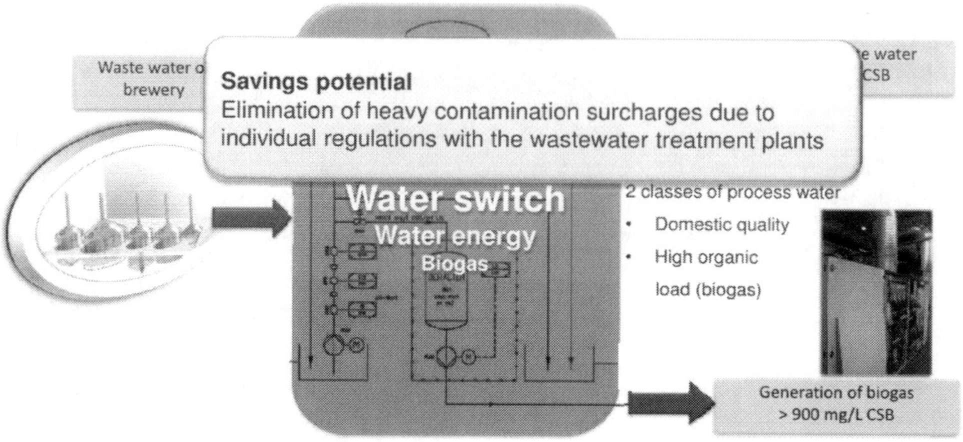

18

21ST CENTURY BREWERY

TAKING CARE OF BREWING

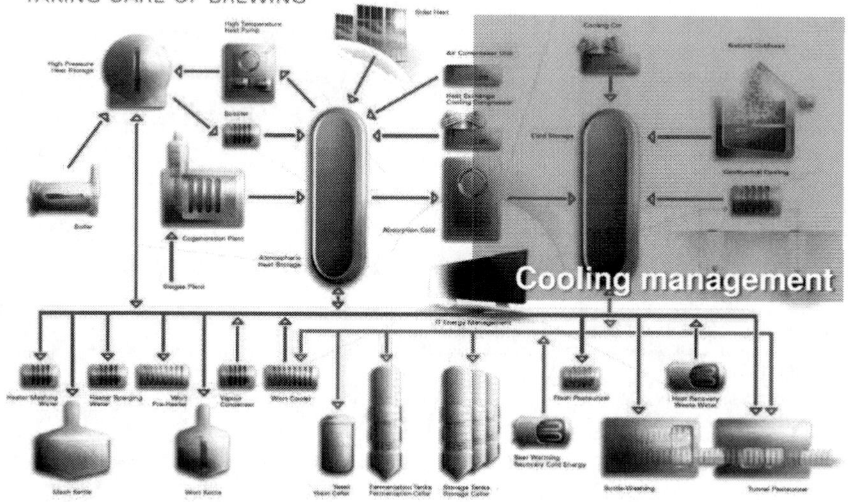

19

21st Century Brewery
BY ZIEMANN ®

Absorption refrigeration system

Heat sink
Cooling tower / room heating
25 °C up to 40 °C

Cold generation
process cooling
-5 °C up to 15 °C

Heat source
waste heat
microgas turbine
90 °C up to 95 °C

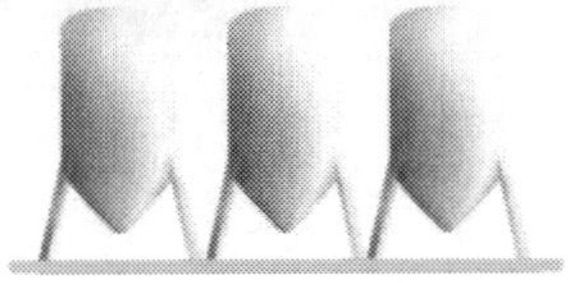

Fermentation and storage tanks / wort cooler

Refrigerating agent	Water
Solvent	Lithium bromide

20

21st Century Brewery
BY ZIEMANN ®

Absorption refrigeration system
Make: TRANE Thermax
Dimensions (l x w x h) = 2,85 x 1,57 x 2,31 m
Operating weight = 5.000 kg
EER* = 0,5

Savings potential
Extended exhaust gas heat recovery of the micro gas turbine of approx. 200.000 kWh per year

Thereof conversion of heat into cold ~ 70 %
Savings of 27.900 kg of CO2 per year

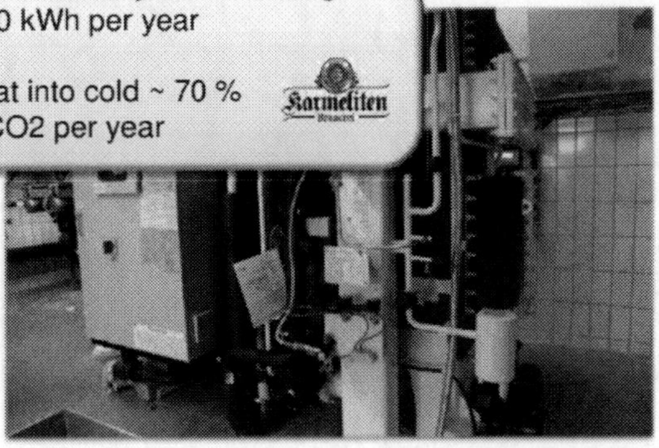

*EER = Energy Efficiency Ratio

Source: Karmeliten Brauerei

20

21st Century Brewery
BY ZIEMANN ®

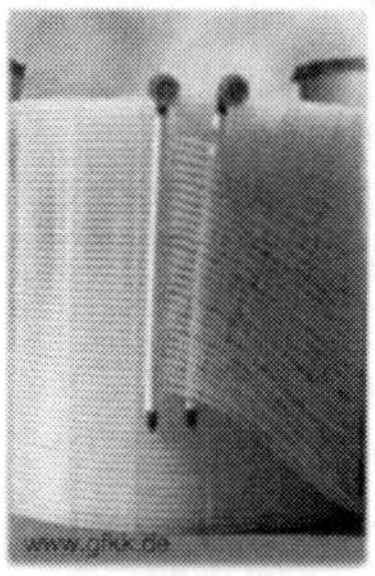

Ice storage tank
Calmac

Construction

Savings potential
Reduction of peak load by 25 – 30 kW

Charge temperature -5°C
Discharge temperature 2°C

Principle
- Latent heat accumulator
- Phase transition of water → ice = charge of the storage
- Phase transition of ice → water = discharge of the storage

→ Covering the cooling peak loads with the ice storage system

22

21st Century Brewery
BY ZIEMANN ®

Ice Age®

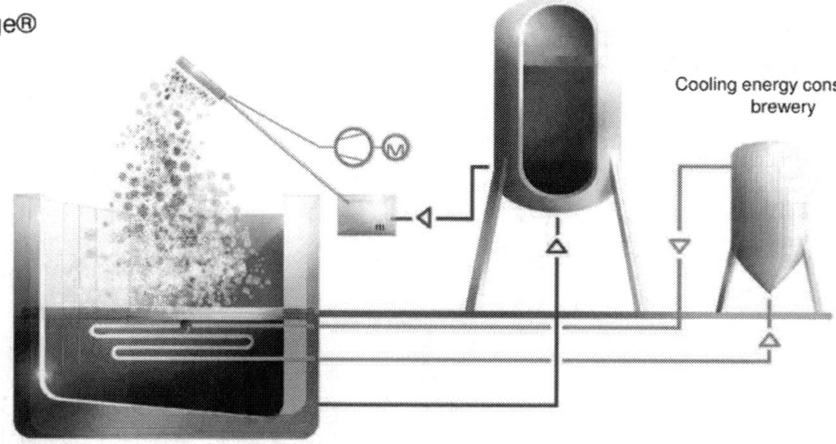

Source: Ziemann

22

21st Century Brewery
BY ZIEMANN ®

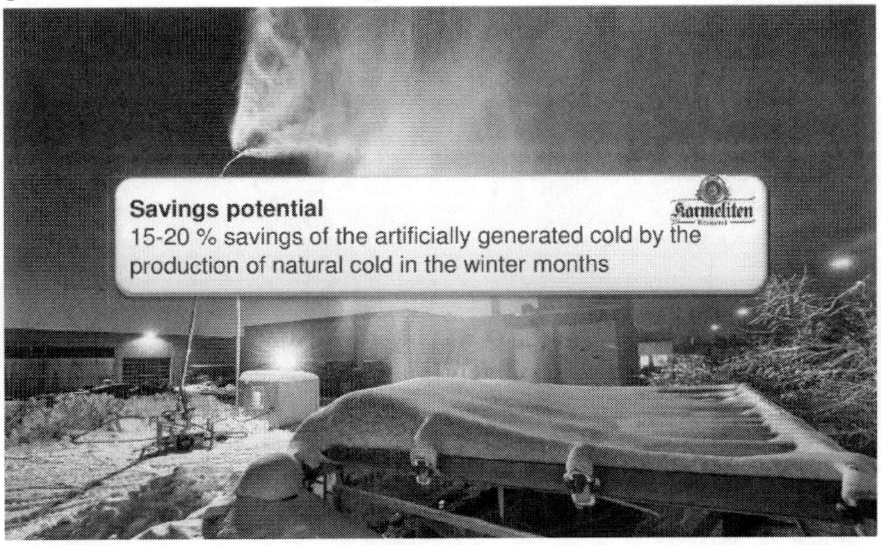

23

21st Century Brewery
BY ZIEMANN ®

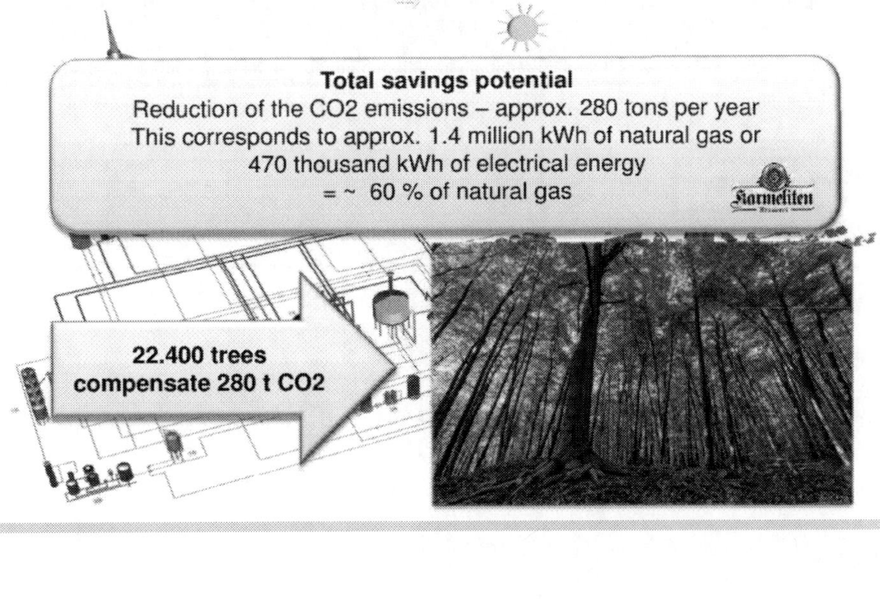

25

ZIEMANN
HOLVRIEKA
Dr. Verena Blomenhofer
ZIEMANN HOLVRIEKA GmbH, Ludwigsburg
Verena.Blomenhofer@ziemann-holvrieka.com
Phone: +49(0)7141408433
www.ziemann-holvrieka.com
© ZIEMANN HOLVRIEKA

High-throughput Mechanical Measurement of Yeast Cells Using Atomic Force Microscopy

Ryota Aritomo[1], Ryosuke Tanaka[2], Toru Shioi[1], Takaharu Okajima[2]
1. Sapporo Breweries LTD., Japan
2. Graduate School of Information Science and Technology, Hokkaido University, Japan

1/32

Agenda

I. Introduction
II. Materials & Methods
III. Results and Discussion
   1. Optimizing AFM measurements for yeast cells
   2. Quantifying the elastic modulus of top- and bottom-fermenting yeasts
   3. Investigation of the effect of ethanol on yeast elastic modulus
IV. Conclusions
   1. Conclusions
   2. Future work " Development of yeast selection system "

2/32

Agenda

SAPPORO

I. Introduction
II. Materials & Methods
III. Results and Discussion
   1. Optimizing AFM measurements for yeast cells
   2. Quantifying the elastic modulus of top- and bottom-fermenting yeasts
   3. Investigation of the effect of ethanol on yeast elastic modulus
IV. Conclusions
   1. Conclusions
   2. Future work " Development of yeast selection system "

3/32

ıtroduction

Our goal

SAPPORO

ep 1 To develop a method to evaluate elasticities of hundreds of yeasts using AFM

ep 2 To investigate brewing yeast properties

ep 3 To develop a yeast selection system

ep 4 Brewing test with selected yeast

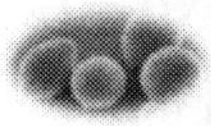

ur
oals **To improve product quality**
To develop new products using selected yeast

4/32

troduction

Significance of yeast quality management

SAPPORO

- Prevents fermentation delay
- Ensures beer foam stability
- Improves flavour and taste

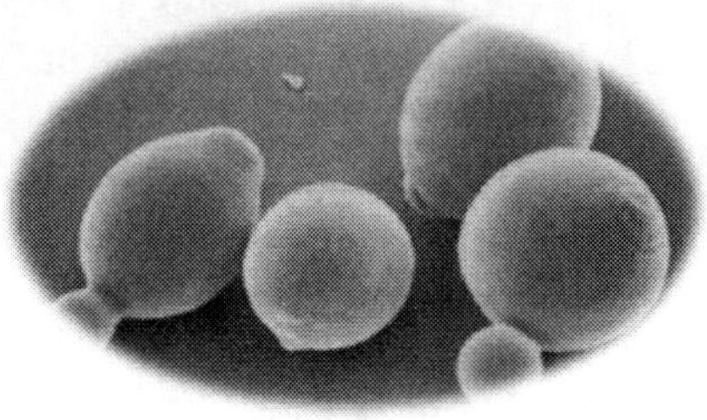

5/32

troduction

Mechanical property test for quality management of brewing yeast

SAPPORO

Biochemical methods

- ✓ Fermentation test
- ✓ Methylene blue stain
- ✓ Acidification Power Test etc.

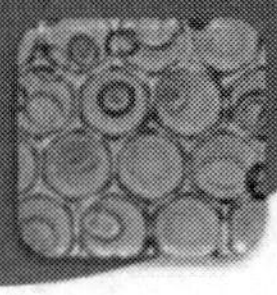

Mechanical methods

?

6/32

ıtroduction

Why do we focus on mechanical properties?

SAPPORO

ɜviously reported :

ʔelationship between yeast cell walls and high-gravity brewing

(Xin-er Li *et al.*, J. Soc. Brew. Chem., 2014)

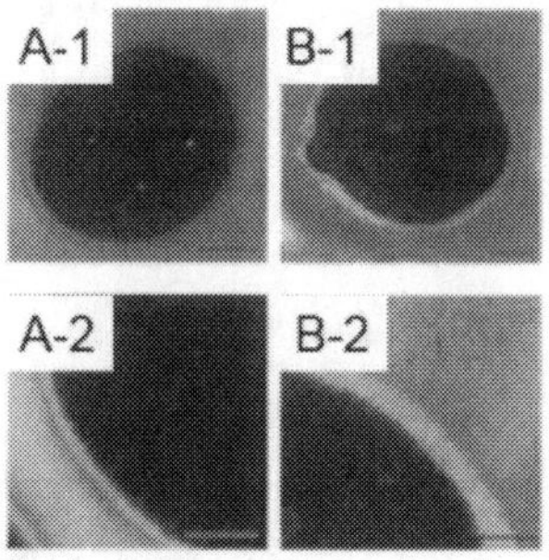

TEM images of yeast cell walls

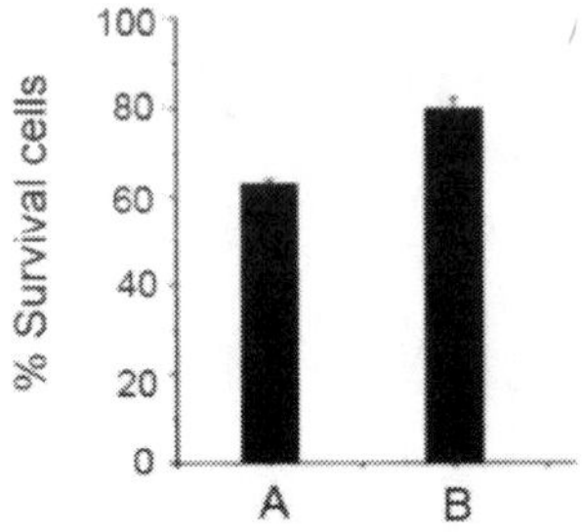

Cell survival rate at the end of fermentation (18° , 15°C)

Mechanical properties could be a new parameter for quality evaluation and selection of brewing yeast

7/32

ıtroduction

Atomic-force Microscopy （AFM）

SAPPORO

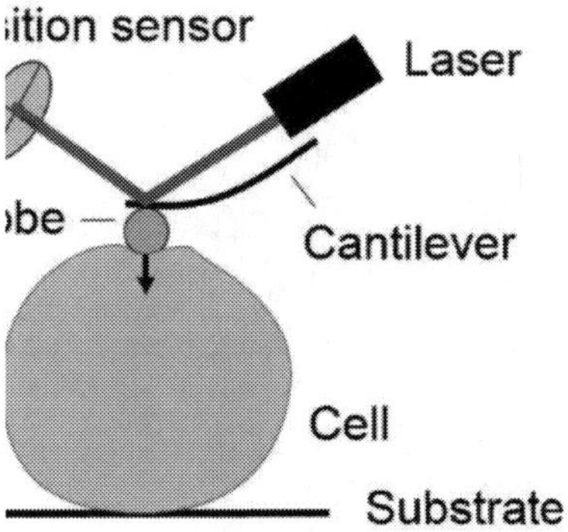

Height (μm)

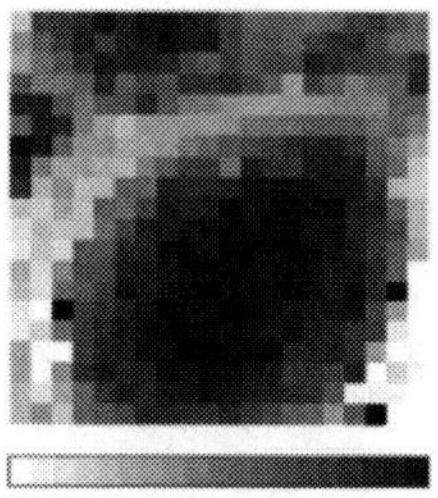

Elasticity (Pa)

ᖴM is applicable for…

- ✓ Living cells
- ✓ Single cells
- ✓ Cultivating cells after measurement

8/32

ıtroduction

Previous measurements of yeast mechanical properties using AFM

Ethanol tolerance

Untreated Ethanol treated

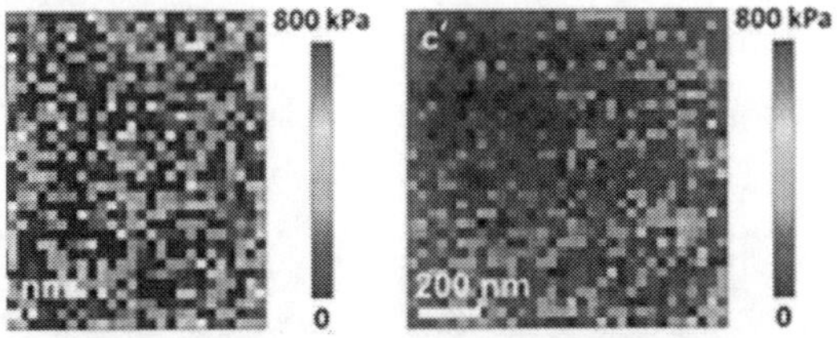

Schiavone M *et al.*, Appl Environ Microbiol, 2016

2) Heat tolerance

Untreated Heat treated

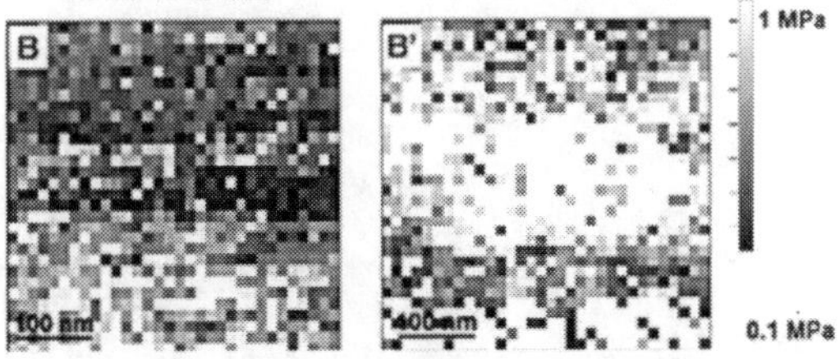

Pillet F *et al.*, BMC Biol, 2014

)ifferences between individual cell is unclear.

Necessity of "measuring a large number of cells" and
"measuring distribution of mechanical properties"

9/32

ıtroduction

Objectives of our study

SAPPORO

ep 1 To develop a method to evaluate elasticities of hundreds of yeasts using AFM

ep 2 To investigate brewing yeast properties

- ✓ Types of brewing yeast
- ✓ Ethanol stress

ep 3 To develop a yeast selection system

ep 4 Brewing test with selected yeast

)ur
oals **To improve product quality**
To develop new products using selected yeast

10/32

Agenda

SAPPORO

I. Introduction
II. Materials & Methods
III. Results and Discussion
   1. Optimizing AFM measurements for yeast cells
   2. Quantifying the elastic modulus of top- and bottom-fermenting yeasts
   3. Investigation of the effect of ethanol on yeast elastic modulus
IV. Conclusions
   1. Conclusions
   2. Future work “ Development of yeast selection system ”

11/32

Materials & Methods

AFM measurements

SAPPORO

Substrate selection & treatment → Treatment of yeast → Fixation of yeast → Measurement by AFM → Data analysis

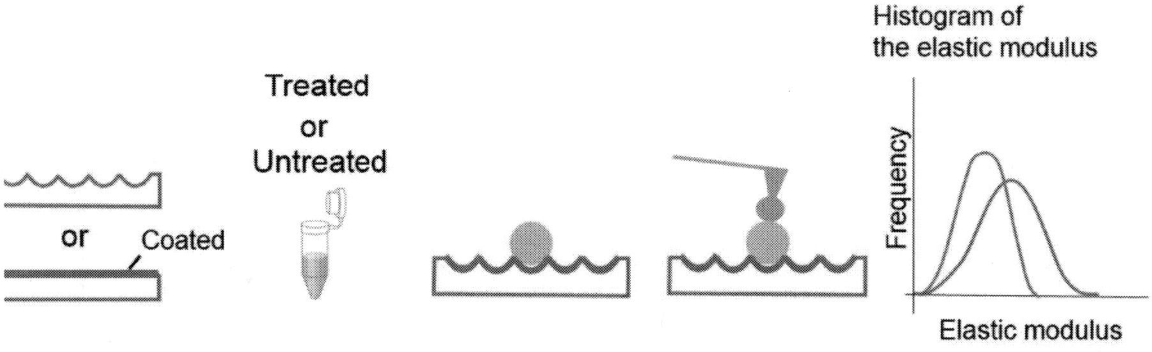

12/32

Materials & Methods

Determination of substrates and treatment of yeast

SAPPORO

Substrate selection & treatment > Treatment of yeast > Fixation of yeast > Measurement by AFM > Data analysis

< Substrate >

Poly-L-lysine (PLL) coated (40 μg/cm²)

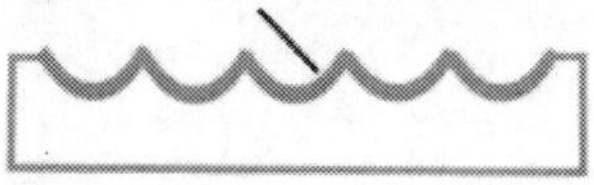

Glass micro-well plate
(φ 15 μm , ozone-washed)

< Yeast >

Wash in physiological saline
(3 times)

13/32

Materials & Methods

Independent fixation of yeast cells on well-plate

SAPPORO

Substrate selection & treatment > Treatment of yeast > Fixation of yeast > Measurement by AFM > Data analysis

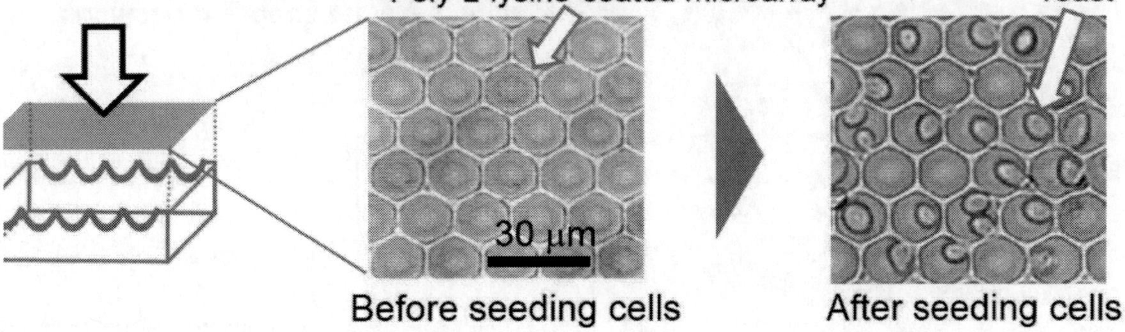

Accurate measurement without affecting other yeast
Applicable for isolation since it is a detachable fixed method

14/32

Materials & Methods

AFM system

SAPPORO

Substrate selection & treatment | Treatment of yeast | Fixation of yeast | **Measurement by AFM** | Data analysis

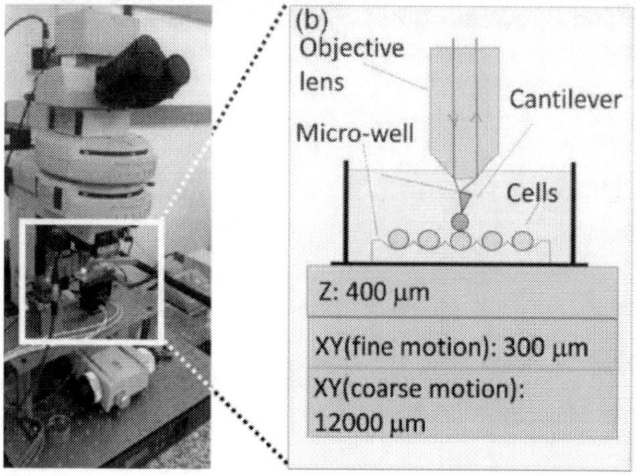

FM instrument (b) Schematics of AFM system

- ✓ An AFM equipped with an upright optical microscope
- ✓ Measure elastic modulus (Young's modulus, E) and height of single cells
- ✓ Measurement over a much wider range than normal AFM

Materials & Methods

Measuring elastic moduli of many single yeast cells using AFM

SAPPORO

Substrate selection & treatment | Treatment of yeast | Fixation of yeast | **Measurement by AFM** | Data analysis

1) Specifying positions from microscopic image

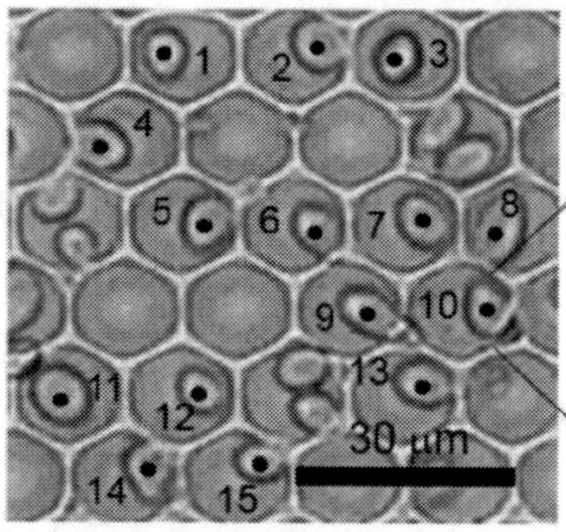

2) Mapping around specified positions

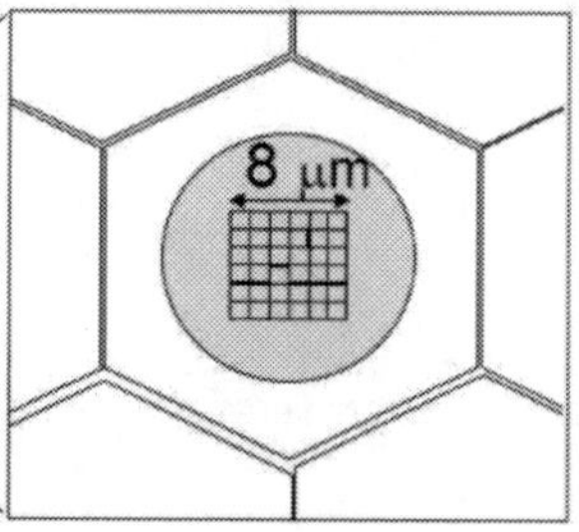

Ability to measure many targeted cells automatically at one time
Ability to select yeast from measured data

Agenda

SAPPORO

I. Introduction
II. Materials & Methods
III. Results and Discussion
  1. Optimizing AFM measurements for yeast cells
  2. Quantifying the elastic modulus of top- and bottom-fermenting yeasts
  3. Investigation of the effect of ethanol on yeast elastic modulus
IV. Conclusions
  1. Conclusions
  2. Future work " Development of yeast selection system "

17/32

Results

Optimizing AFM measurements for yeast cells

SAPPORO

Substrate selection & treatment → Treatment of yeast → Fixation of yeast → Measurement by AFM → Data analysis

to obtain the intrinsic mechanical property of single cells
by eliminating the effect of the cell shape on the AFM measurement

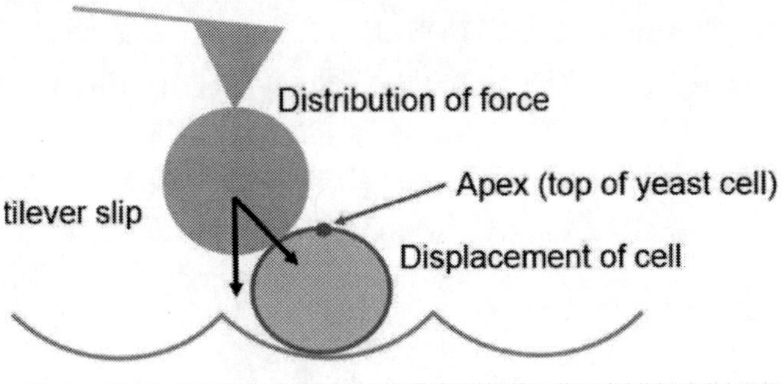

✓ ***Influence of inclination on measurement***

We measured yeast by AFM to investigate the effect of cell shape **18**/32

Results

Elastic modulus of cells as a function of the distance from the cell apex (n=100)

SAPPORO

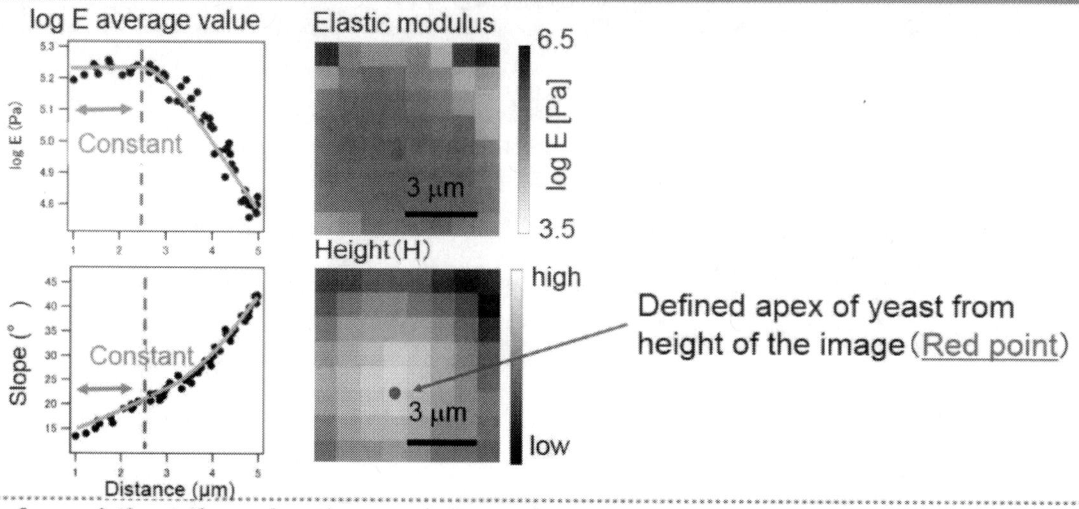

We found that the elastic modulus of each yeast cells can be recisely measured as an average value within the slope of 20°

19/32

Agenda

SAPPORO

I. Introduction
II. Materials & Methods
III. Results and Discussion
   1. Optimizing AFM measurements for yeast cells
   2. **Quantifying the elastic modulus of top- and bottom-fermenting yeasts**
   3. Investigation of the effect of ethanol on yeast elastic modulus
IV. Conclusions
   1. Conclusions
   2. Future work " Development of yeast selection system "

20/32

Results

Measurement of top- and bottom-fermenting yeasts

SAPPORO

ɔ investigate cell number distribution of different types of brewing yeast

)- or
tom-fermenting yeast

Lager Tank
(2 kl scale)

AFM measurement

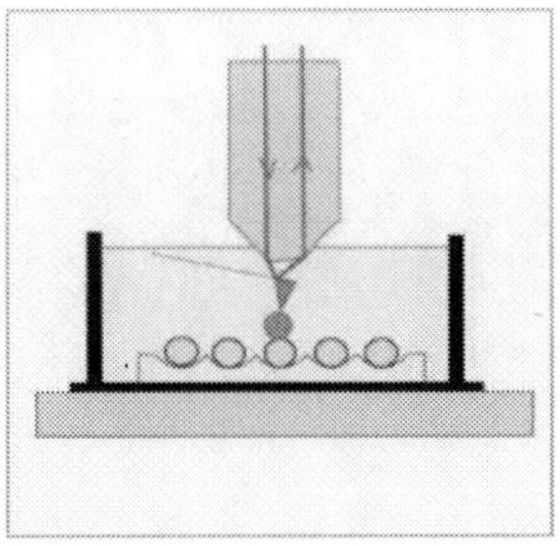

Histogram of
the elastic modulus

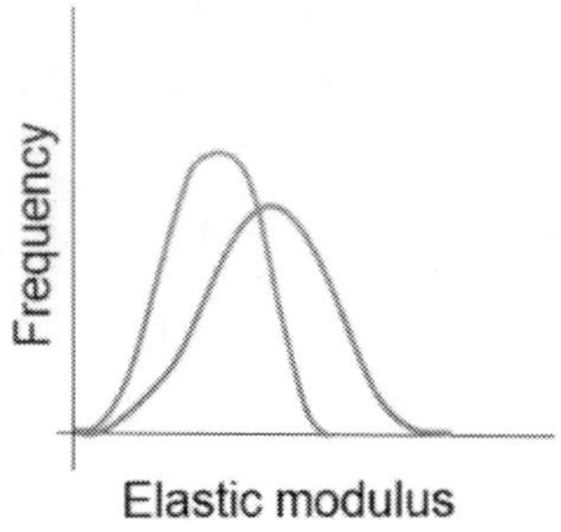

Results

Ѵleasurement of top- and bottom-fermenting yeasts (n=300)

SAPPORO

ermenting yeast (n= 300)

30 µm

ıt

c
lus

ım-fermenting yeast (n= 300)

ıt

:
lus

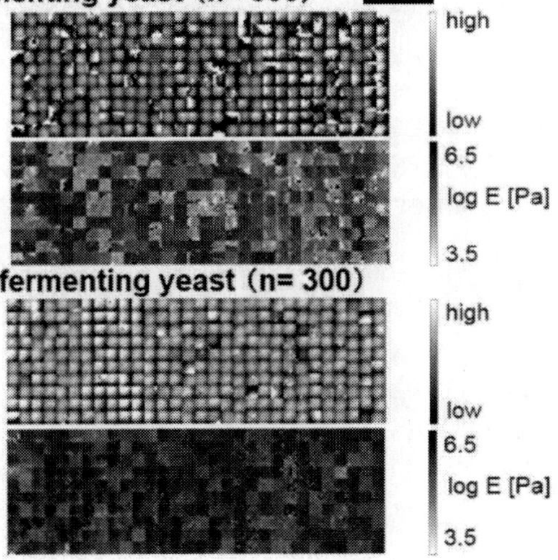

Histogram of the elastic modulus of yeast

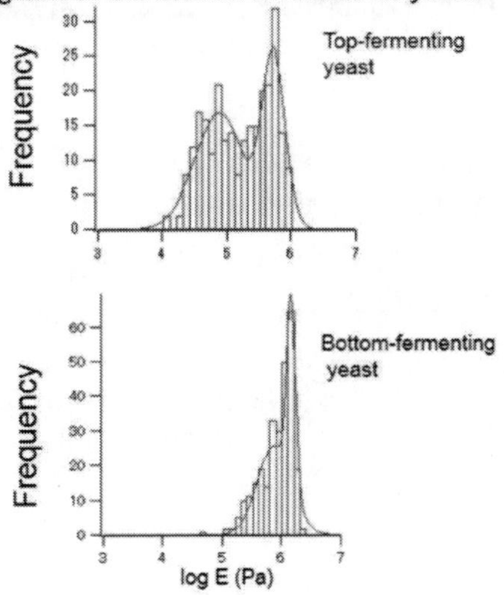

The elastic modulus of both yeast types exhibited two lognormal distributions

Agenda

SAPPORO

I. Introduction
II. Materials & Methods
III. Results and Discussion
   1. Optimizing AFM measurements for yeast cells
   2. Quantifying the elastic modulus of top- and bottom-fermenting yeasts
   3. Investigation of the effect of ethanol on yeast elastic modulus
IV. Conclusions
   1. Conclusions
   2. Future work " Development of yeast selection system "

23/32

Results

estigation of the effect of ethanol on yeast elastic modulus

SAPPORO

- To investigate the number distribution due to external stimulus (ethanol stimulation)

Dead cell identified by methylene blue staining

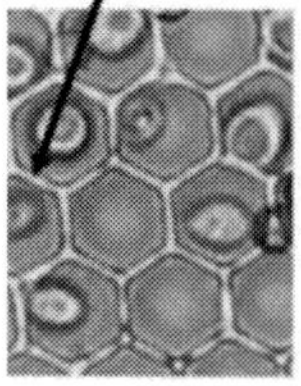

Ethanol treatment
15%, 2 h, 30 °C (86 °F)

AFM measurement (living cells) → **AFM measurement (same cells)** →

Dead cells after ethanol treatment

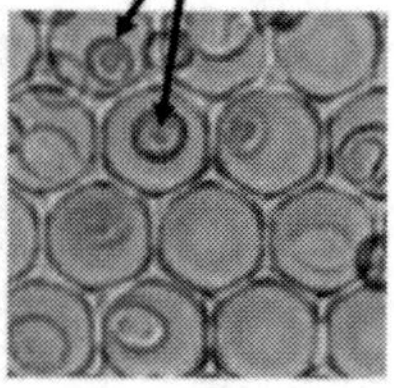

24/32

Results

estigation of the effect of ethanol on yeast elastic modulus

SAPPORO

Distribution of elastic modulus before and after ethanol treatment

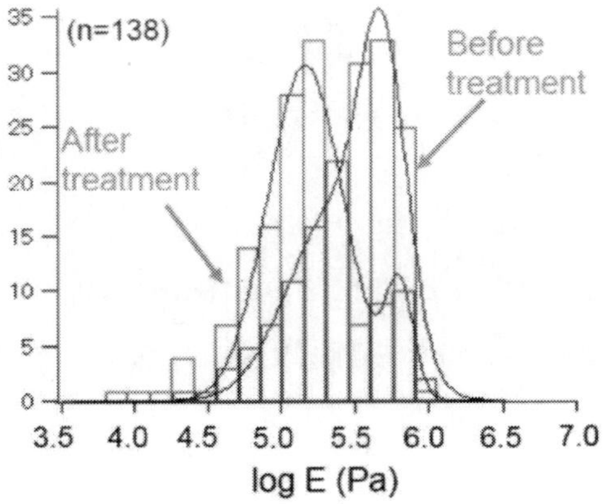

Change in elastic modulus with ethanol treatment

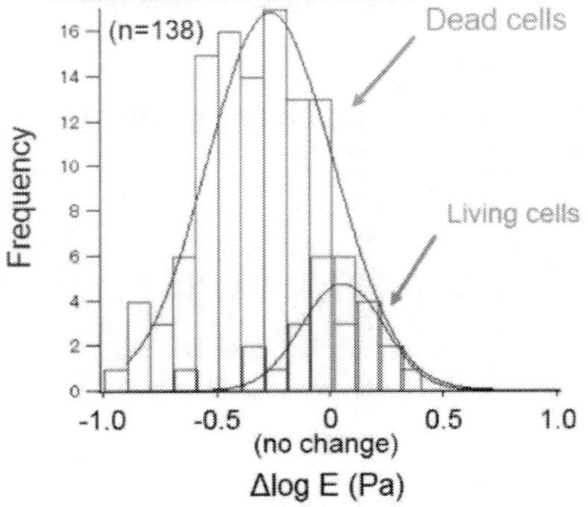

Yeast elastic modulus declined with death from ethanol treatment

25/32

Agenda

SAPPORO

I. Introduction
II. Materials & Methods
III. Results and Discussion
   1. Optimizing AFM measurements for yeast cells
   2. Quantifying the elastic modulus of top- and bottom-fermenting yeasts
   3. Investigation of the effect of ethanol on yeast elastic modulus
IV. Conclusions
   1. Conclusions
   2. Future work " Development of yeast selection system "

26/32

Conclusions

SAPPORO

- We developed a high-throughput AFM measurement for evaluating the number distribution of elastic modulus of yeast
- The number distribution of the elastic modulus of yeast cells exhibited two lognormal distribution functions
- Our observation indicated that single cells had a large variation of ethanol stress so that some had no change in ethanol stress

27/32

Agenda

SAPPORO

I. Introduction
II. Materials & Methods
III. Results and Discussion
   1. Optimizing AFM measurements for yeast cells
   2. Quantifying the elastic modulus of top- and bottom-fermenting yeasts
   3. Investigation of the effect of ethanol on yeast elastic modulus
IV. Conclusions
   1. Conclusions
   2. Future work " Development of yeast selection system "

28/32

Yeast selection model using AFM

. Measurement using AFM

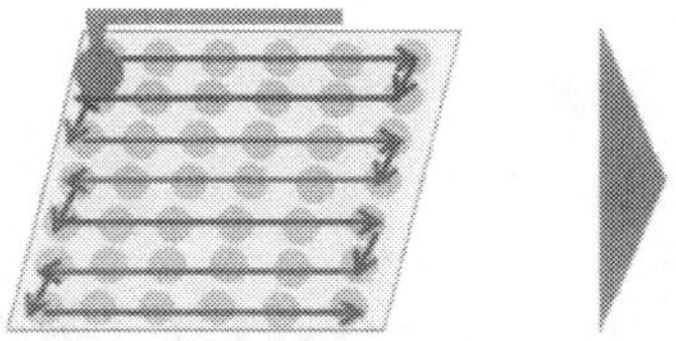

2. Selection of yeast cells by elasticity

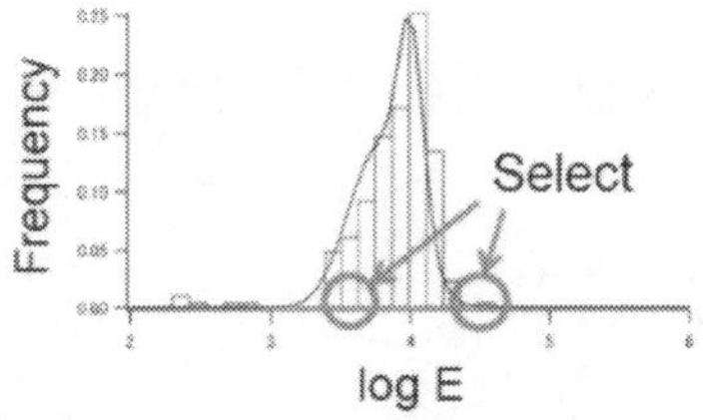

. Picking up cells

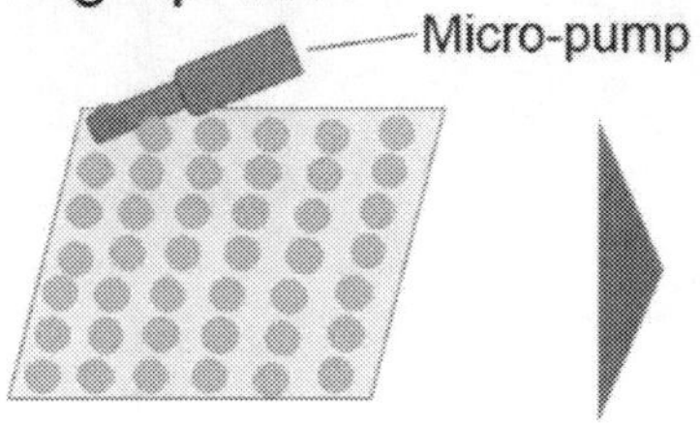

4. Cultivation

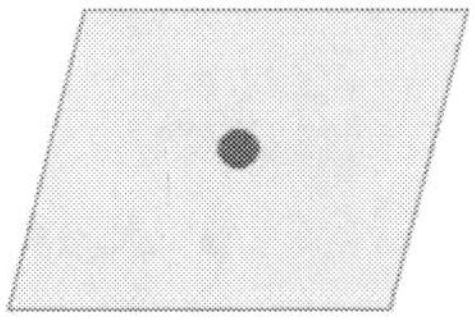

Measurement using AFM

Developing an isolation system combined with AFM system

solation of a yeast cell
ɹsing a micro-pump

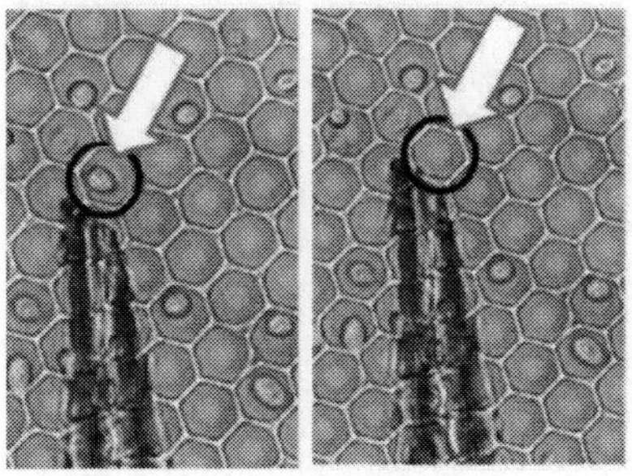

Cultivation

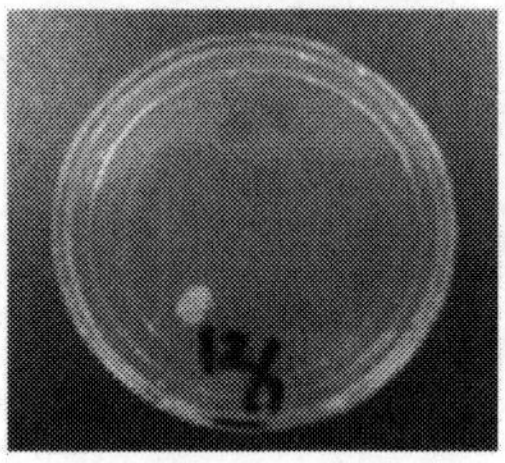

Measurement of elasticity using AFM

and

Evaluation by brewing

- We isolated and cultivated one targeted cell

onclusions & Future Works

Our goal

SAPPORO

ep 1 To develop a method to evaluate elasticities of hundreds of yeast using AFM

ep 2 To investigate brewing yeast properties

ep 3 To develop a yeast selection system

ep 4 Brewing test with selected yeast

Future Works

Our goals **To improve product quality**
To develop new products using selected yeast

31/32

SAPPORO

HOKKAIDO UNIVERSITY FOUNDED 1876

Thank you for your kind attention.

32/32

Reuse of second steeping water improves eco-efficiency of malting

PBL Brewing Laboratory

EBC 15.5.2017
Anneli Ritala, VTT Ltd

Malting is a water intensive process

- Malthouse utilizes 2-7 m^3 water / ton of malt
- Worldwide malthouse water consumption is 70-250 million m^3 / year to produce ca. 35 million tonnes of malt
- However,
  - 70% of world fresh water reservoirs is used by agriculture, 20% by industry and only 10% by households

Every drop counts!

PBL Brewing Laboratory

2

Water reuse in malting

Malthouse wastewater **reuse** in steeping is restricted by issues related to malt quality:

- Flavour and aroma
- Microbiological stability
- Germination inhibition caused by salinity and/or potential organic inhibitors

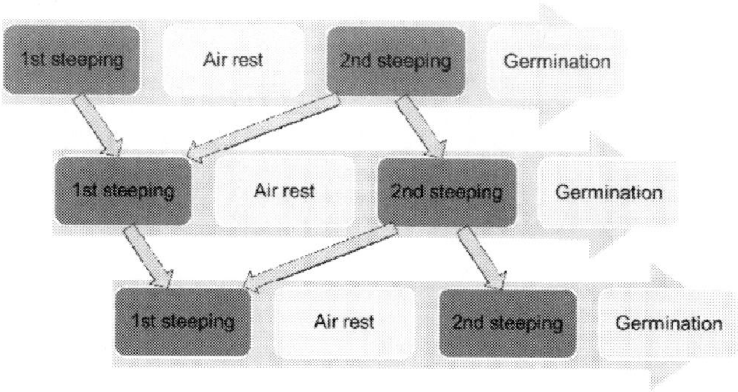

PBL Brewing Laboratory

3

Water reuse in malting

Hypothesis:

- **There are germination inhibitors that accumulate to steeping waters**

Can we see germination inhibition if 1st steeping water is reused as 1st steeping water?

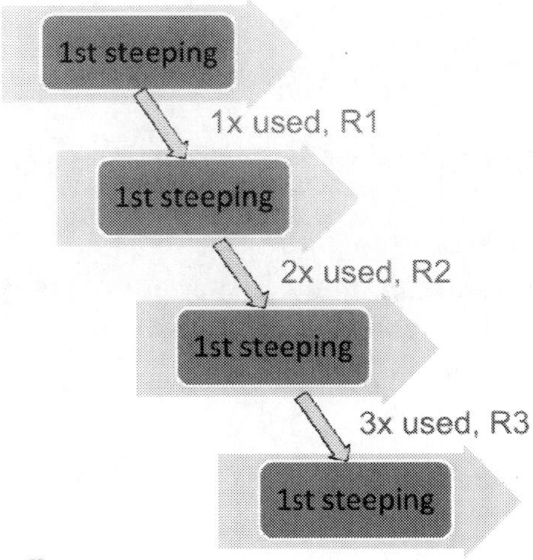

- Germination and water sensitivity assays performed with steeping waters reused 1- 3 times in industrial pilot settings
- Two varieties, Barke and Marthe were used in steeping trials
  - to produce the used steeping waters
  - to perform plate germination assays

PBL Brewing Laboratory

5

We did not observe germination inhibition if 1st steeping water was reused

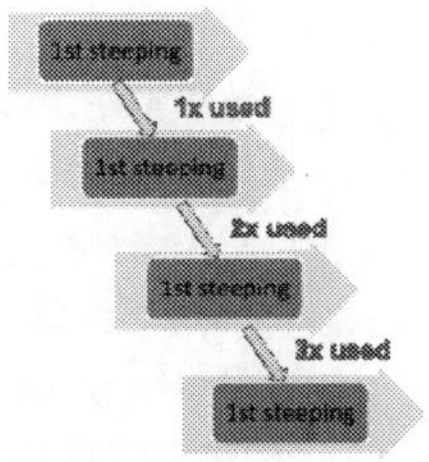

- In basic germination tests
- Only slight drop in germination if an assay indicating water sensitivity was used

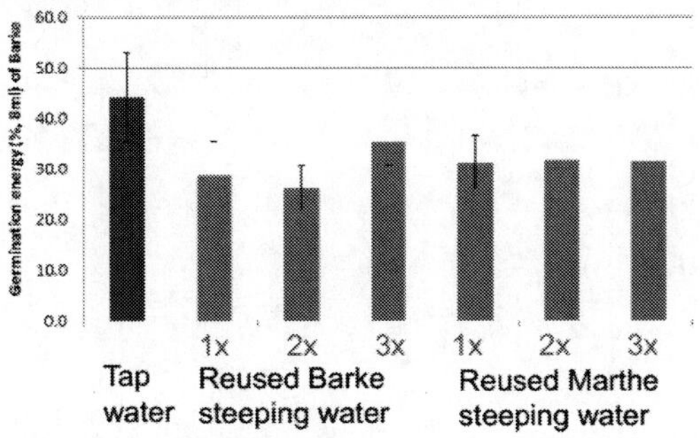

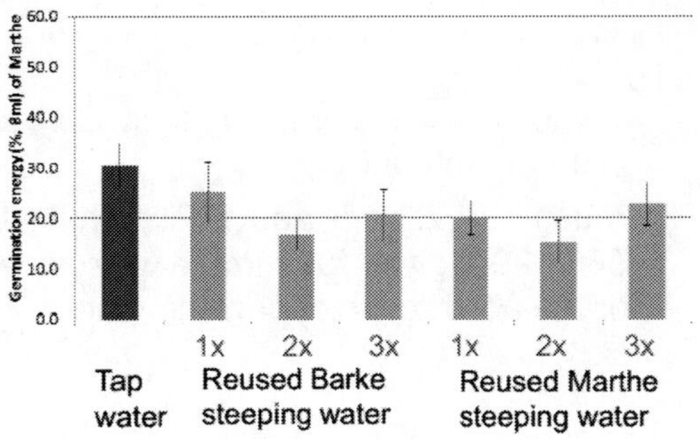

PBL Brewing Laboratory

6

Reusing 2nd steeping water as 2nd steeping water

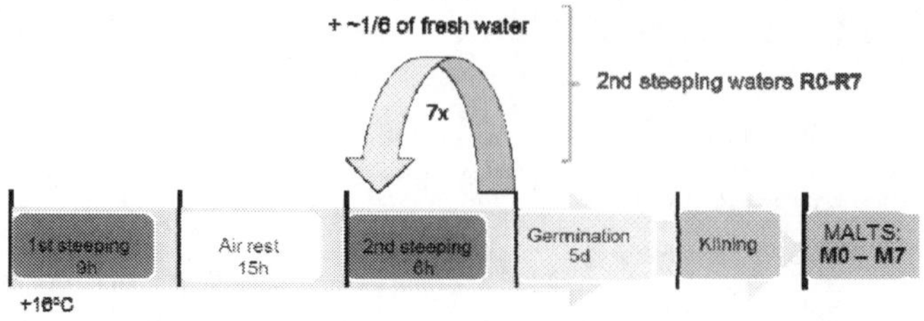

Laboratory scale experiment lacking microbial flora present in a malthouse

1x to 7x used 2nd steeping water used in plate germination assays:

- No inhibition of germination in plate germination assays was observed
- Only slight drop in germination was seen in assays indicating water sensitivity

PBL Brewing Laboratory

7

Quality of reused 2nd steeping water

- Conductivity stays below the recommended limit (250 mS/m) given for steeping water
- Chloride, sulphate and sodium below limit (98/83/EC) given for potable water
- Potassium: No given limit, recommended daily adult dose 3300 mg

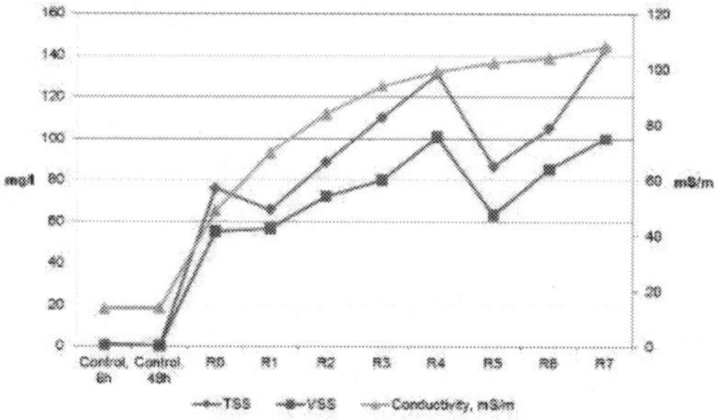

- Ammonium: Clearly above limit (98/83/EC) given for potable water
- Phosphate: Clearly above Finnish limit (366/1994) given for surface water intended for the abstraction of drinking water

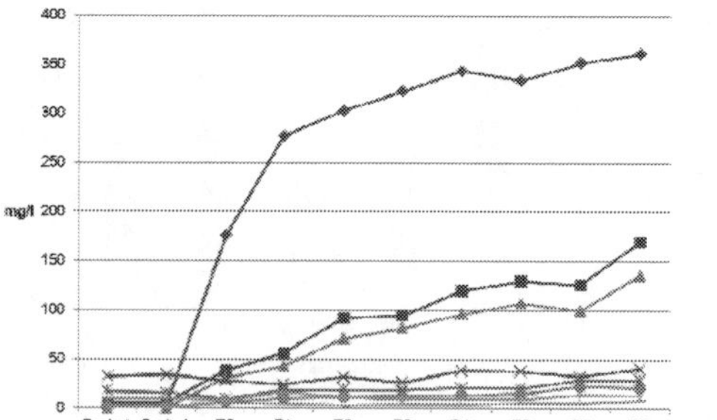

PBL Brewing Laboratory

Quality of reused 2nd steeping water

- The counts of bacteria and yeasts increased
- Accumulation of slimy polysaccharides partly due to microbial metabolism
  - Bacterial exopolysaccharides may have a strong impact on wort separation performance

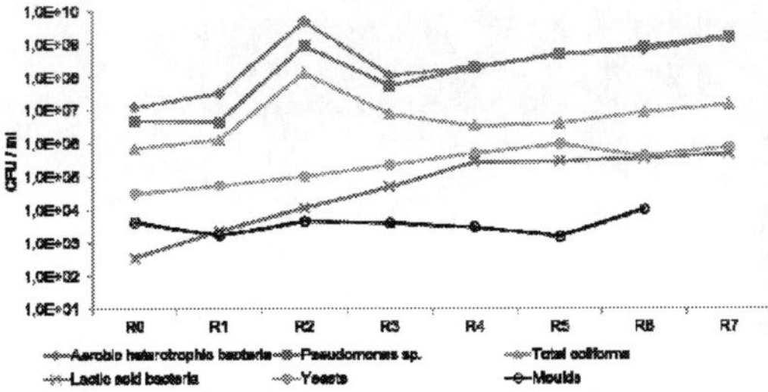

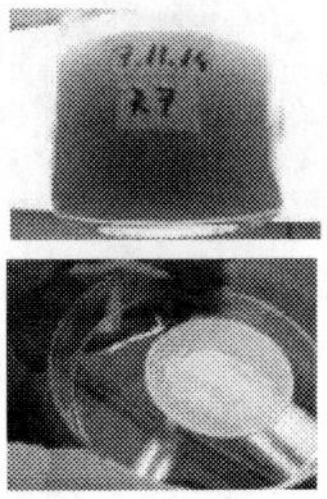

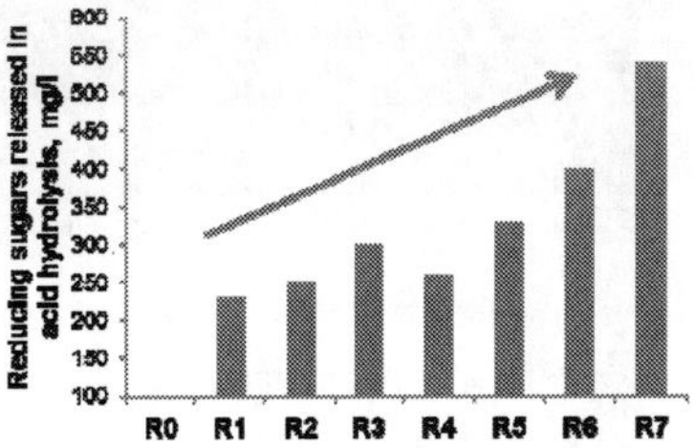

PBL Brewing Laboratory

9

Quality of malt produced with reused 2nd steeping water

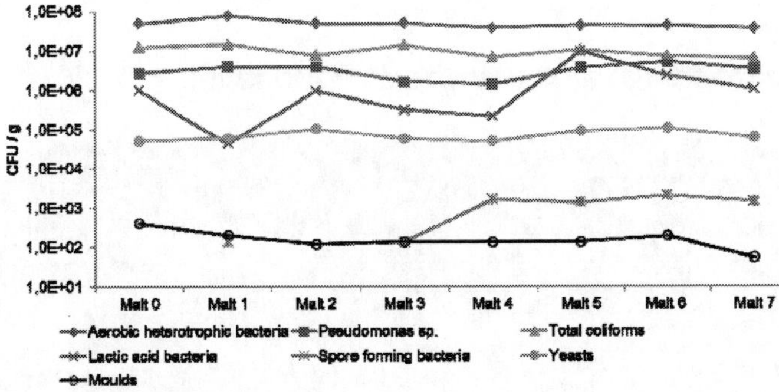

- No significant effect on the microbial quality - Most of the microbial counts remained at constant level
- Spore forming bacteria increased 10 x if water was reused more than 4 times
  - May have an effect on the end product quality if malt is used in the production of non-beer beverages or in baking
- Lactic acid bacteria counts fluctuated
- The amount of *Fusarium* DNA remained at low level

PBL Brewing Laboratory

10

Quality of malt produced with reused 2nd steeping water

- Filterability or other Congress wort parameters stayed at good levels

Line	Filtration 15 min (g)	Filtration Final (g)	Extract % (m/m)	Colour 25 mm (°EBC)	Viscosity (cP)	pH	Haze (EBC)	Wort β-glucan (mg/l)
M0	255.8	341.3	84.3	2.2	1.53	6.03	1.4	185
M7	269.2	343.6	84.6	2.2	1.52	5.98	1.3	163
r_{95}	-	-	0.6	0.3	0.03	0.02	0.01	11
						*	*	*

Line	Malt protein (% d.m.)	FAN (mg/l)	Soluble N (mg/100g)	Kolbach (%)	Friability (%)	PUG (%)	WK (%)
M0	10.1	168	817	45.5	89.1	1.2	0.5
M7	9.9	157	798	45.5	92.5	1.1	0.2
r_{95}	0.3	11	90	1.4	2.3	0.3	-
					*		

r_{95} = repeatability of the method

* = difference not explained by the repeatability of the method

PBL Brewing Laboratory

11

Quality of malt produced with reused 2nd steeping water

- High gravity mashing values stayed at good levels

Line	Saccharification (min)	Wort gravity (% m/m)	Theoretical extract yield (% d.m.)	Colour 25 mm (°EBC)	pH	Haze (EBC)	Büchner area (g/h)	Filtration 15 min	Filtration 30 min	Filtration 60 min
M0	10	16.92	87.1	4.5	5.66	1.4	87	66.7	98.0	121.2
M7	10	16.87	86.9	4.5	5.67	1.8	92	73.2	106.0	123.8
r_{95}	-	0.03	0.6	0.3	0.02	0.02	-	-	-	-
						*				

r_{95} = repeatability of the method

* = difference not explained by the repeatability of the method

Büchner filtration curves

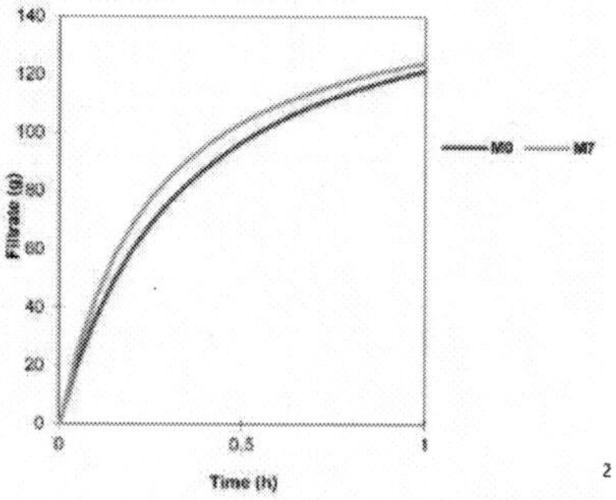

PBL Brewing Laboratory

2

Quality of malt produced with reused 2nd steeping water

Sensory evaluation of malts M0 and M7

- Test method: A triangle test, where panellists were asked to smell and taste the samples and mark the sample that differed from the other two, similar samples
- Malts were microwaved before the evaluation to ensure their microbiological quality
- Evaluation as:
  - Whole grains
  - Milled and mixed with water
- Samples covered with lids
- n = 18

PBL Brewing Laboratory

Quality of malt produced with reused 2nd steeping water

Whole grains

- Sample preparation: Boiling water poured on grains ~1h before evaluation (mod. Analytica-EBC, 2006)

8 panellists out of 18 found the differing sample -> **No statistically significant differences between the samples**

Milled grains mixed with water

- Sample preparation: Grains milled with EBC-mill, flour mixed with cold water ~1h before evaluation (mod. Murray *et al.*, 1999)

Only 4 panellists out of 18 found the differing sample -> **No statistically significant differences between the samples**

The reuse of 2nd steeping water did not have a statistically significant effect on the sensory quality of the malts

PBL Brewing Laboratory

14

Reuse of second steeping water
Concluding remarks

- **Germination inhibition is not a problem in steeping water recycling**
- **High ammonium and phosphate levels and formation of slimy polysaccharides need to be considered when water purification technologies are selected**
- **Good quality malt can be obtained with reused steeping water**

Acknowledgements

Steering:
Timo Huttunen, Viking Malt
Pekka Reinikainen, Viking Malt
Raimo Koljonen, Viking Malt
Mika Mäkelä, Viking Malt
Jari Olli, Viking Malt
Tapio Kangas-Heiska, Sinebrychoff
Rauno Sillanpää, Sinebrychoff
Finn Lok, Carlsberg
Riitta Saleva-Sjöblom, Hartwall
Jorma Rasi, Hartwall
Silja Hylkinen, Olvi
Petteri Selin, Olvi
Ville Vilen, Laitilan Wirvoitusjuoma
Reino Aikasalo, Boreal
Merja Veteläinen, Boreal
Hanna Haikka, Boreal
Raija Lantto, VTT

VTT team:
Ulla Holopainen-Mantila
Tuija Sarlin
Kaisu Honkapää
Johannes Jermakka
Elina Merta
Arvi Wilpola
Tarja Wikström
Eero Mattila
Tuulikki Seppänen-Laakso
Airi Hyrkäs
Arja Laitila
Annika Wilhelmson

Financing:
PBL Brewing Laboratory
Tekes
Boreal

VTT

PBL Brewing Laboratory

Thank You!

16

TECHNOLOGY FOR BUSINESS
PBL Brewing
Laboratory

Shelf-life and consistency of active dry yeast for breweries

Yves Gosselin, Laurence Van Nedervelde, Annick Boeykens, Philippe Janssens

THE OBVIOUS CHOICE FOR BEVERAGE FERMENTATION

Meurice R&D

Fermentis
LESAFFRE FOR BEVERAGES

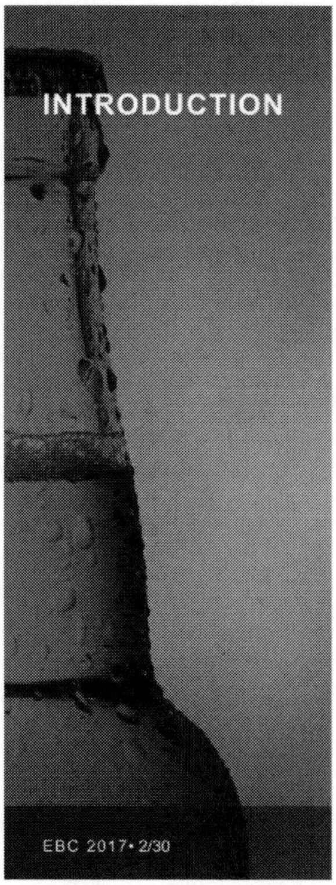

THE DRYING PROCESS

- MANUFACTURING PROCESS
- YEAST CELL CYCLE PHYSIOLOGY
- MEMBRANE DURING DRYING

ACTIVE DRY YEAST IN THE BREWERY INDUSTRY

Fermentis
LESAFFRE FOR BEVERAGES

INTRODUCTION : THE DRYING PROCESS

MANUFACTURING PROCESS : A DAY TO DAY INCREMENTAL IMPROVEMENT

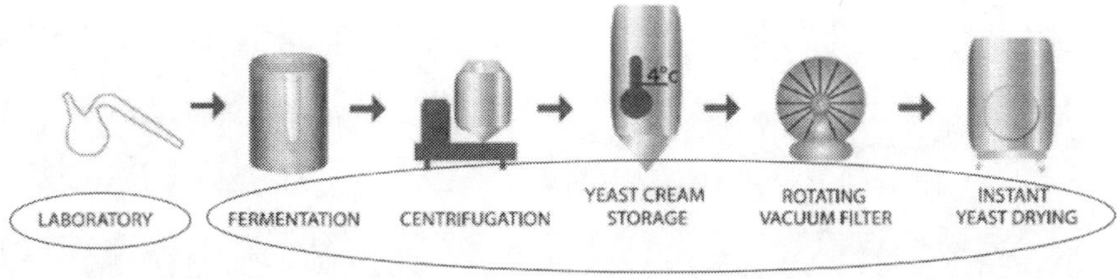

EBC 2017• 3/30

Fermentis
LESAFFRE FOR BEVERAGES

INTRODUCTION : THE DRYING PROCESS

YEAST CELL CYCLE PHYSIOLOGY

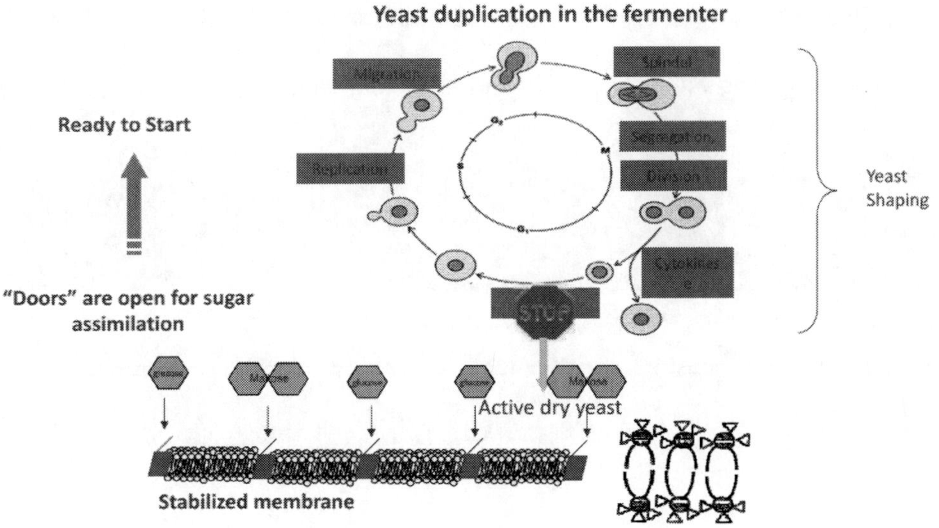

EBC 2017• 4/30

Fermentis
LESAFFRE FOR BEVERAGES

INTRODUCTION : THE DRYING PROCESS

MEMBRANE DURING DRYING

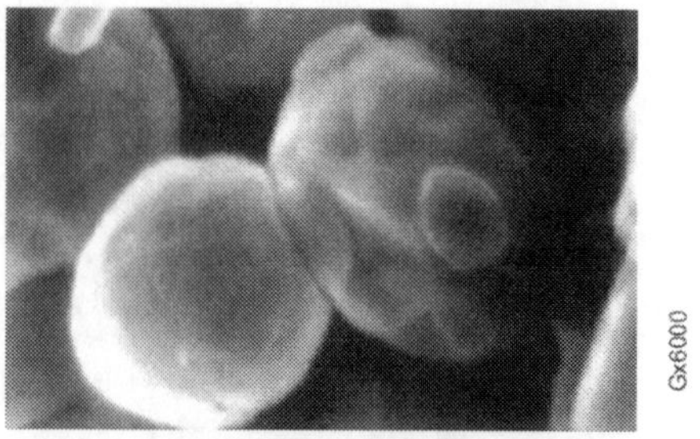

Before drying

- 25-30% Dry Matter
- Smooth Cell Surface

After drying

- 94-96,5% Dry Matter
- Uneven Cell Surface

INTRODUCTION : THE DRYING PROCESS

MEMBRANE DURING DRYING

<u>Membrane</u> = Phospholipids + Proteins + Carbohydrates + Lipids + ...

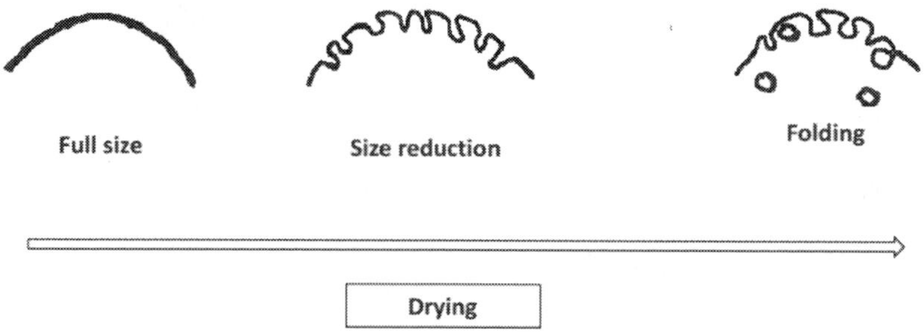

INTRODUCTION : THE DRYING PROCESS

MEMBRANE DURING DRYING

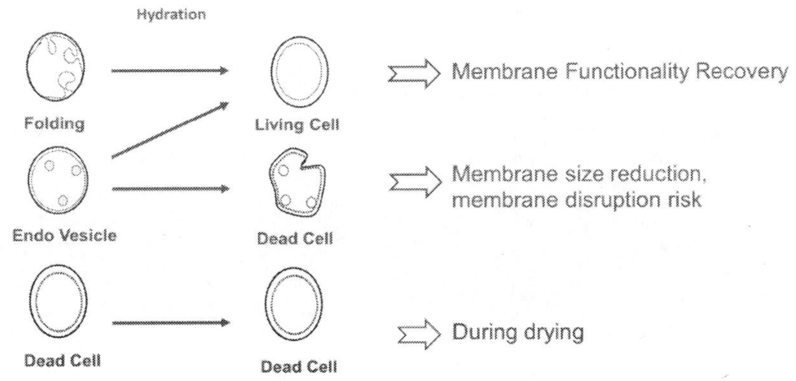

The yeast manufacturer's know-how makes the difference

Fermentis
LESAFFRE FOR BEVERAGES

INTRODUCTION : THE DRYING PROCESS

KEY POINTS

Fermentation

→ shape the yeast to the best physiological stage for:
- drying
- fermentation readiness

→ specific adaptation of recipe to the yeast strain

Drying

→ maintaining the membrane integrity

→ specific process adaptation to the yeast strain

Fermentis
LESAFFRE FOR BEVERAGES

INTRODUCTION : ACTIVE DRY YEAST IN THE BREWERY INDUSTRY

USE OF ACTIVE DRY YEAST (ADY) IN THE BREWING INDUSTRY IS BECOMING INCREASINGLY ATTRACTIVE DUE TO SEVERAL KEY FEATURES

- Capacity to be stored for extended periods of time
- Speed at which it can be prepared for use
- Use for primary and secondary (conditioning in bottle) fermentations
- Facilitate practices for seasonal beer production and franchise brewing

ABUNDANT SCIENTIFIC LITERATURE (SINCE 1996)

Fermentation with dried yeast – S. Fels, Y. Gosselin, B. Taidi – BRFI Quaterly, 1996, October, 17-19

The use of dry yeast in the brewery industry – A. Debourg, L. Van Nedervelde – Proceeding of the 27th EBC Congress, 1999, 751-760,

Pratical use of dried yeast in the brewery industry – S. van den Berg, A. Van Landschoot – Cerevisia, 2003, 28(3), 25-30

Bottle refermentation of high alcohol-beers – Y. Gosselin – XIV Chair J. De Clerck Special Beers The Belgian paradox, 2012

Fermentis
LESAFFRE FOR BEVERAGES

DETERMINE THE SHELF-LIFE AND CONSISTENCY OF ACTIVE DRY YEAST FOR BREWERIES

1. FORCED AGEING TEST
   - ✓ 10 ADY (3 lagers and 7 ales)
   - ✓ Different production batches for each yeast
   - ✓ Forced ageing corresponding to **2**, **3** and **4** years of natural ageing

Meurice R&D

2. NATUREL AGEING
   - ✓ 2 ADY (1 lager and 1 ale)
   - ✓ Natural ageing during **3 years and 8 months** at -20, 5, and 25°C

Fermentis
LESAFFRE FOR BEVERAGES

FORCED AGEING TESTS

1. Fermentative power and viability determined before and after forcing test
2. Flask fermentation performances before and after forcing test

   λ = lag time (h)
   R_f = rate of fermentation (% ethanol (v/v)/h)
   T_f = time of end of fermentation (h)

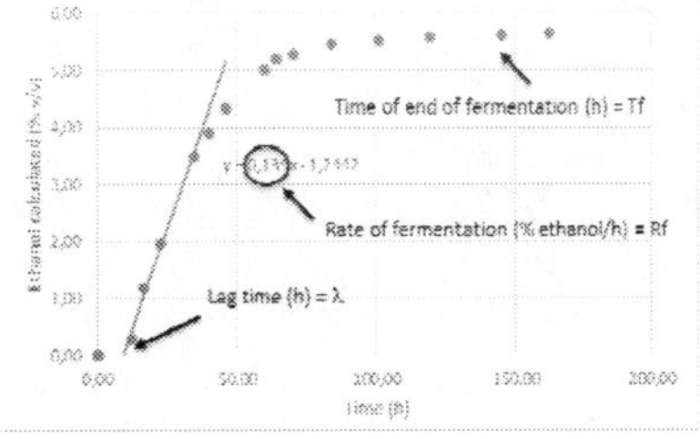

Analyses at the end of fermentation: ethanol, volatile compounds (acetaldehyde, esters, higher alcohols and vicinal diketones), residual sugars

Meurice R&D

Fermentis LESAFFRE FOR BEVERAGES

RESULTS : FORCED AGEING TEST

1. RELATIONSHIP BETWEEN FERMENTATIVE POWER DETERMINED JUST AFTER PRODUCTION (FPI) AND AFTER SOME TIME OF STORAGE (FPT)

FPi = 8,9 + 0,997*FPt   r = 0,966

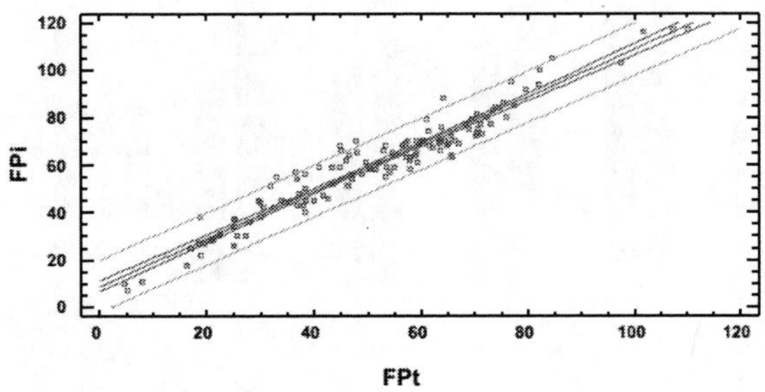

Fermentative power is expressed in mL CO_2/h.g yeast dry weight

EBC 2017• 12/30

Meurice R&D

Fermentis LESAFFRE FOR BEVERAGES

RESULTS : FORCED AGEING TEST

2. Influence of forced ageing on yeast fermentative power

Ale brewing yeast strains

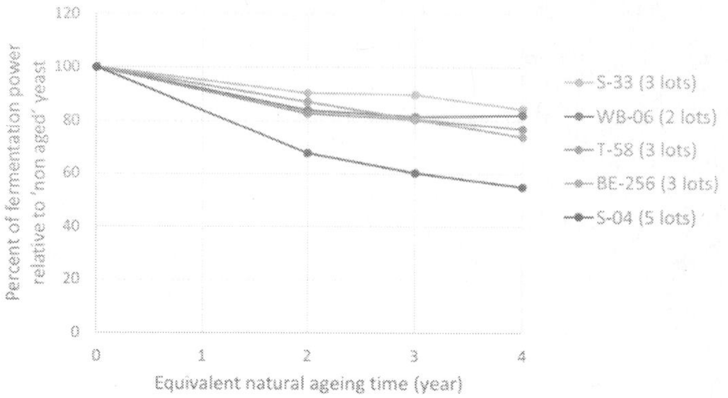

 Meurice R&D

Fermentis
LESAFFRE FOR BEVERAGES

RESULTS : FORCED AGEING TEST

3. Fermentative power and viability determined before and after forcing test

Ale brewing yeast strain SafAle S-04 (5 production batches)

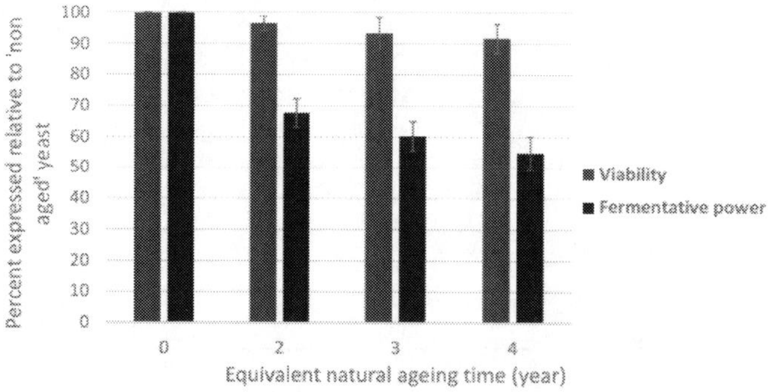

⇨ **Fermentation performances on wort**

 Meurice R&D

Fermentis
LESAFFRE FOR BEVERAGES

RESULTS : FORCED AGEING TEST

4. Fermentation performances

SafAle S-04 – Kinetics comparison of fresh yeast (5 production batches)

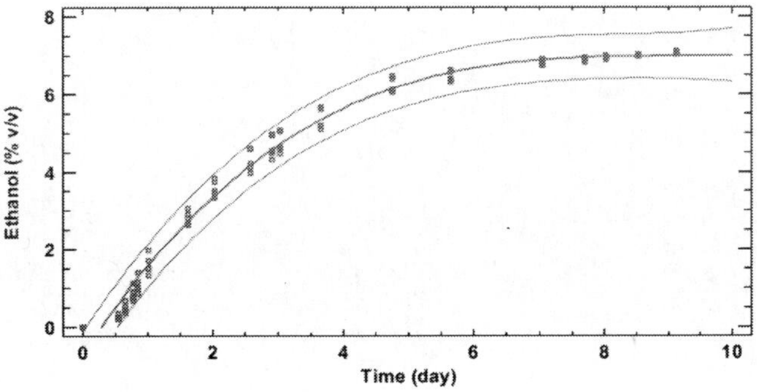

Fermentation conditions for Ale brewing yeast strains :
Pitching rate : 50 g/hL, standard wort (Weyermann syrup) at 15°P, temperature 20°C

EBC 2017• 15/30 Meurice R&D

Fermentis LESAFFRE FOR BEVERAGES

RESULTS : FORCED AGEING TEST

4. Fermentation performances

SafAle S-04 – Kinetics comparison of yeast aged 4 years (5 production batches)

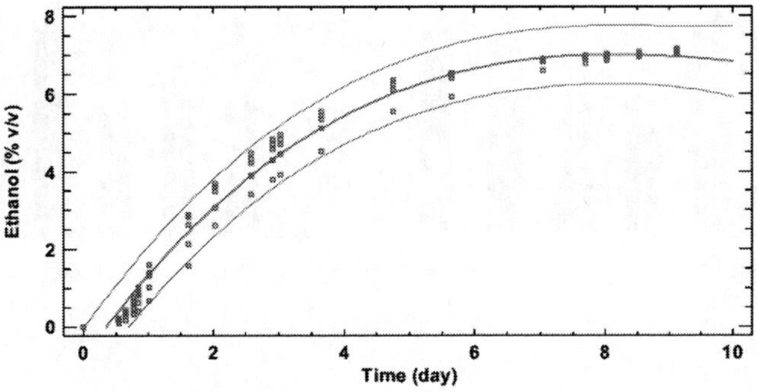

Fermentation conditions for Ale brewing yeast strains :
Pitching rate : 50 g/hL, standard wort (Weyermann syrup) at 15°P, temperature 20°C

EBC 2017• 16/30 Meurice R&D

Fermentis LESAFFRE FOR BEVERAGES

RESULTS : FORCED AGEING TEST

4. FERMENTATION PERFORMANCES

SafAle S-04 – Kinetics comparison of fresh yeast vs yeast aged 4 years (5 batches)

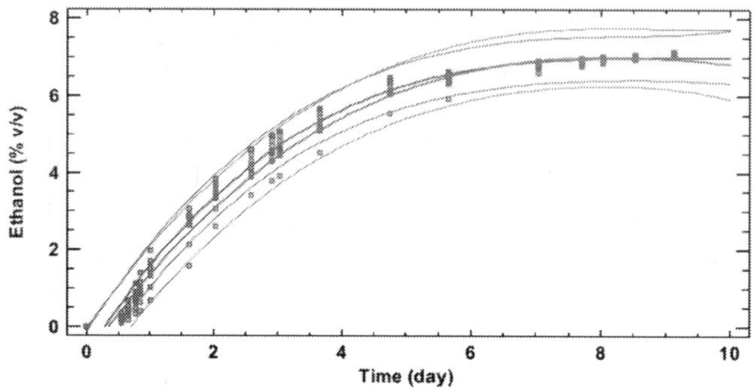

→ No significant difference between fresh and forced aged yeast

EBC 2017• 17/30 Meurice R&D

Fermentis
LESAFFRE FOR BEVERAGES

RESULTS : FORCED AGEING TEST

4. FERMENTATION PERFORMANCES

Ale strains – Impact of accelerated ageing on time of end of fermentation and fermentation rate

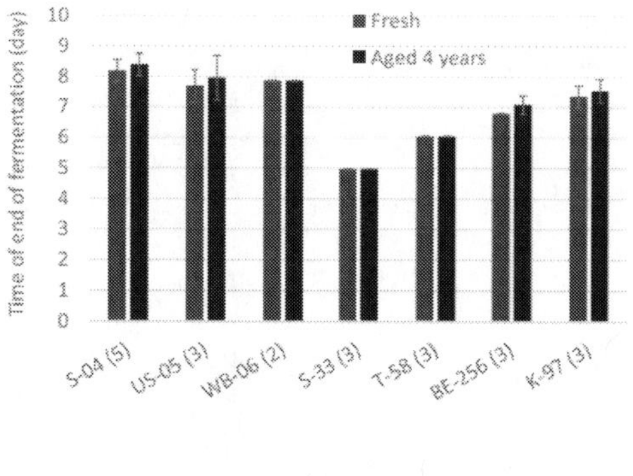

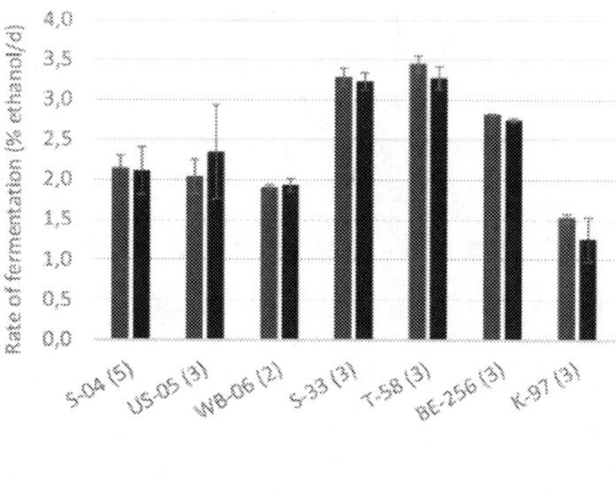

EBC 2017• 18/30 Meurice R&D

Fermentis
LESAFFRE FOR BEVERAGES

RESULTS : FORCED AGEING TEST

4. FERMENTATION PERFORMANCES

Ale strains – Impact of accelerated ageing on volatile compound production

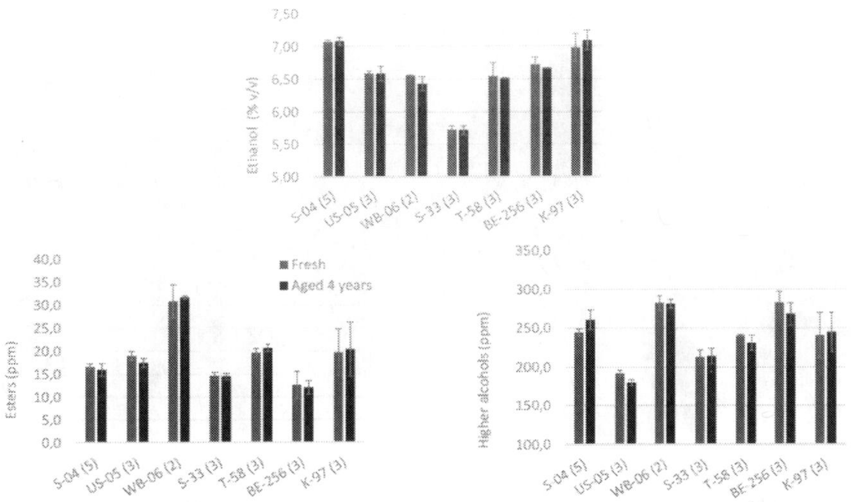

EBC 2017• 19/30 Meurice R&D

Fermentis LESAFFRE FOR BEVERAGES

RESULTS : FORCED AGEING TEST

4. FERMENTATION PERFORMANCES

SafLager S-34/70 – Kinetics comparison of fresh yeast vs yeast aged 4 years (5 batches)

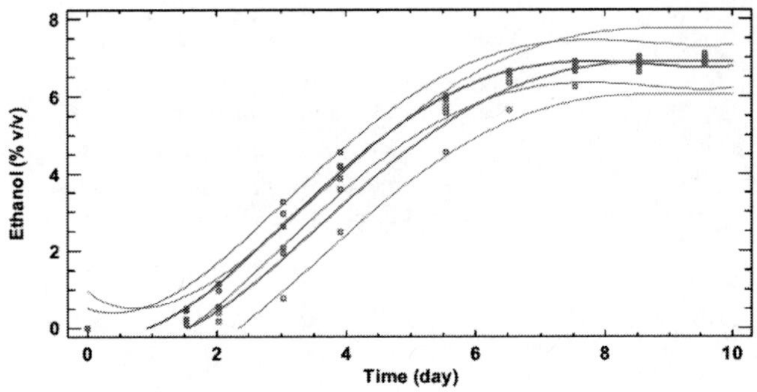

Fermentation conditions for Lager brewing yeast strains :

Pitching rate : 100 g/hL, standard wort (Weyermann syrup) at 15°P, temperature 14°C

EBC 2017• 20/30 Meurice R&D

Fermentis LESAFFRE FOR BEVERAGES

RESULTS : FORCED AGEING TEST

4. FERMENTATION PERFORMANCES

Lager strains – Impact of accelerated ageing test on final ethanol concentration, fermentation rate and volatile compound production

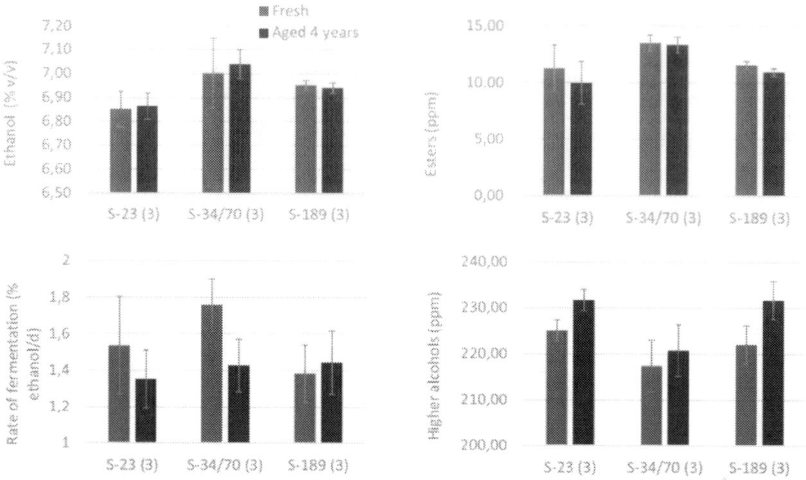

Meurice R&D

Fermentis
LESAFFRE FOR BEVERAGES

NATUREL AGEING

1. Fermentation performances after storage during 3 years and 8 months at different temperatures (-20, 5 and 25°C)
   - ✓ 2 ADY (lager yeast W-34/70 and ale yeast S-04)
   - ✓ Comparison with a new fresh batch of yeast
   - ✓ Pitching by dry yeast weight and viable yeast cell count
   - ✓ Same wort for all assays from pilot brewery of KU Leuven (15°P pure malt), fermenter 50L
   - ✓ Analyses at the end of fermentation: ethanol, acetaldehyde, esters, higher alcohols, vicinal diketones, sulfites, residual sugars

2. Beer tasting by an expert panel

Odisee HOGESCHOOL

Fermentis
LESAFFRE FOR BEVERAGES

RESULTS : NATUREL AGEING

1. Fermentation performances

SafAle S-04 – Kinetics comparaison of fresh yeast vs yeast stored 3 years and 8 months at -20°C, 5°C, 25°C

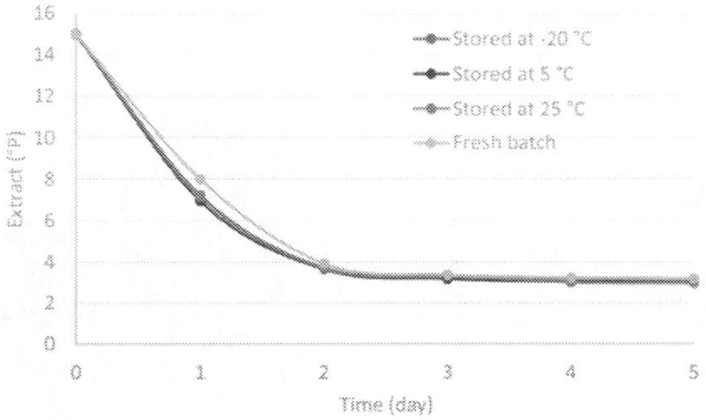

Yeast sample	Viability (expressed in % relative to fresh yeast)
Stored at -20°C	98
Stored 5°C	94
Stored at 25°C	94

Pitching rate 40 g/hL
T° 23°C

Fermentis
LESAFFRE FOR BEVERAGES

EBC 2017• 23/30

RESULTS : NATUREL AGEING

1. Fermentation performances

SafAle S-04 – FAN reduction of fresh yeast vs yeast stored 3 years and 8 months at -20°C, 5°C, 25°C

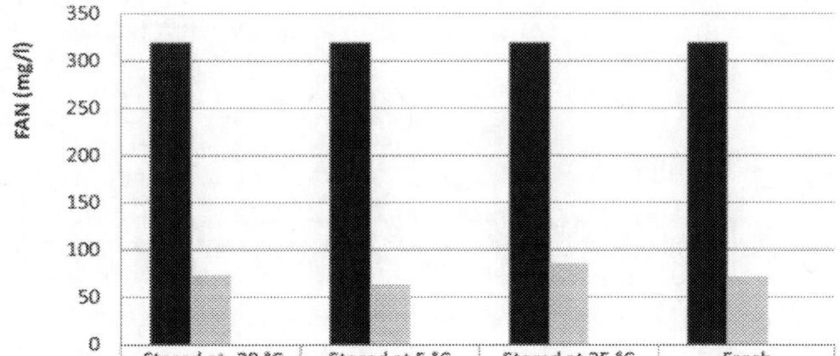

	Stored at -20 °C	Stored at 5 °C	Stored at 25 °C	Fresh
FAN wort	319	319	319	319
FAN beer	73	63	85	72
% decrease	77	80	73	77

Fermentis
LESAFFRE FOR BEVERAGES

EBC 2017• 24/30

RESULTS : NATUREL AGEING

1. Fermentation performances

SafAle S-04 – Sulfite production of fresh yeast vs yeast stored 3 years and 8 months at -20°C, 5°C, 25°C

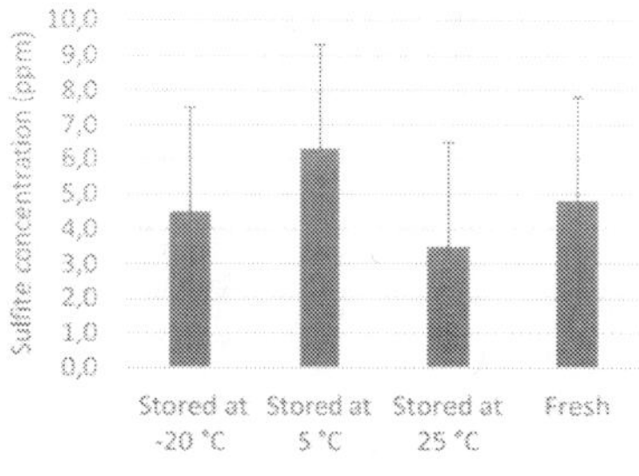

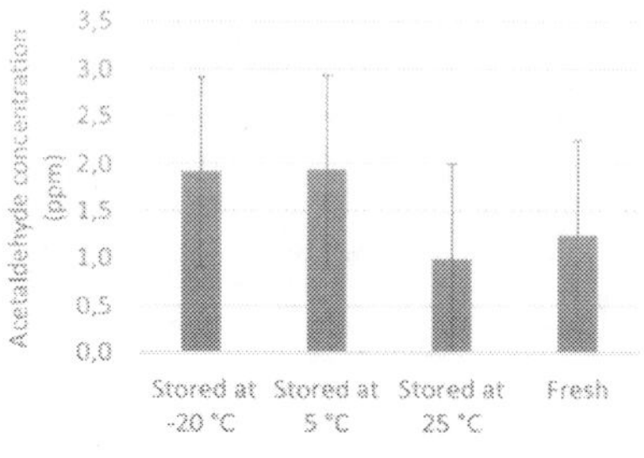

Fermentis
LESAFFRE FOR BEVERAGES

RESULTS : NATUREL AGEING

1. Fermentation performances

SafLager W-34/70 – Kinetics comparaison of fresh yeast vs yeast stored 3 years and 8 months at -20°C, 5°C, 25°C

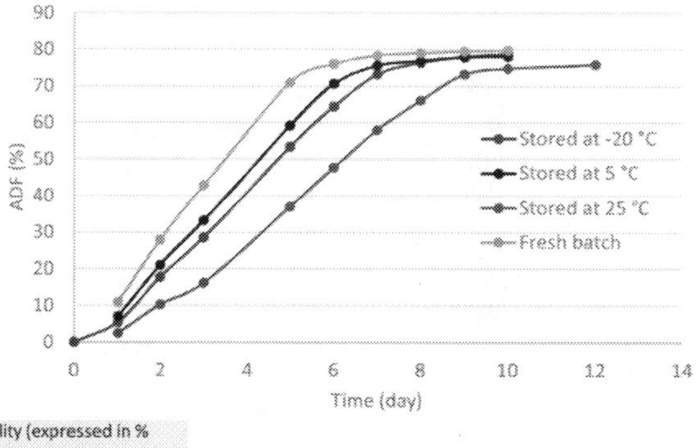

Yeast sample	Viability (expressed in % relative to fresh yeast)
Stored at -20°C	81
Stored 5°C	96
Stored at 25°C	59

Pitching rate 120 g/hL
T° 14°C

Fermentis
LESAFFRE FOR BEVERAGES

RESULTS : NATUREL AGEING

2. BEER TASTING BY AN EXPERT PANEL

SafAle S-04 – Testing of beer obtained with fresh yeast vs yeast stored 3 years and 8 months at 5°C and 25°C

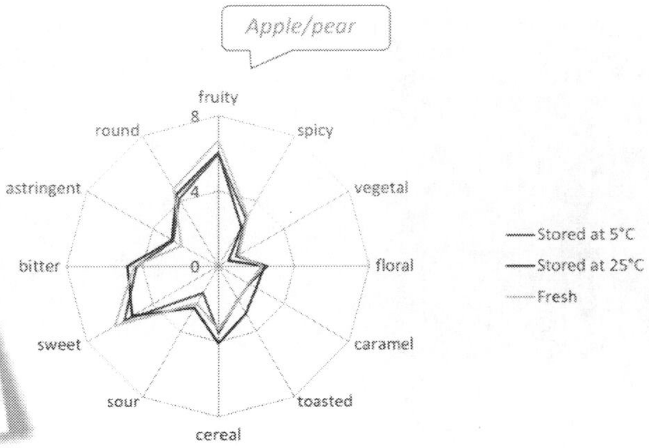

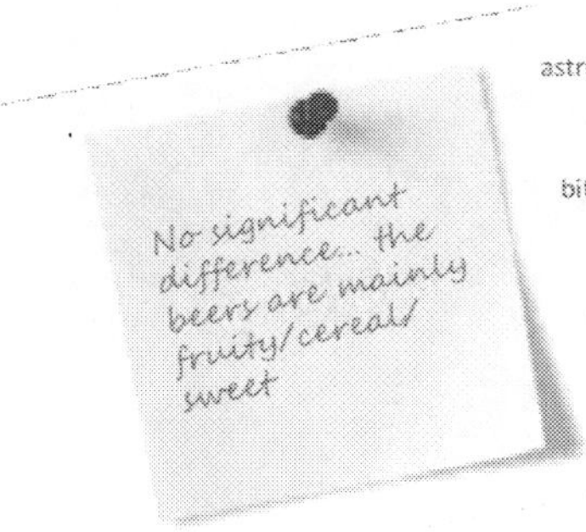

EBC 2017• 27/30

Fermentis
LESAFFRE FOR BEVERAGES

RESULTS : NATUREL AGEING

2. BEER TASTING BY AN EXPERT PANEL

SafLager W-34/70 – Testing of beer obtained with fresh yeast vs yeast stored 3 years and 8 months at 5°C and 25°C

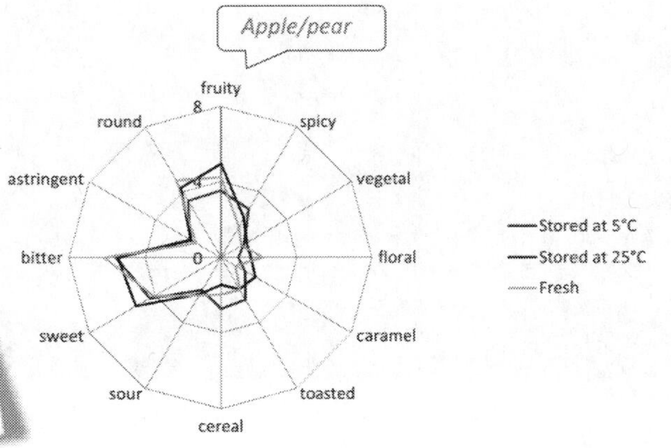

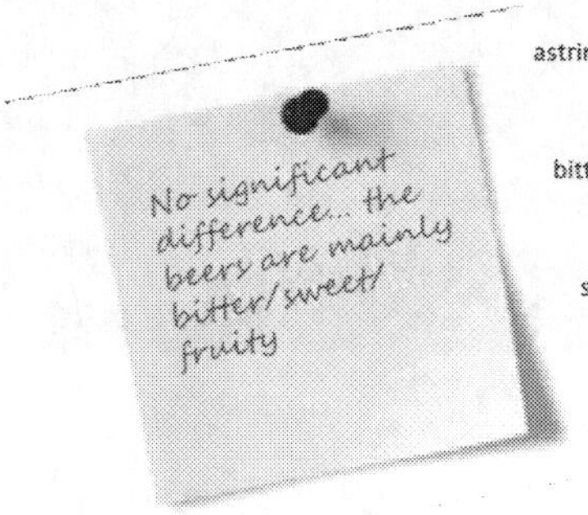

EBC 2017• 28/30

Fermentis
LESAFFRE FOR BEVERAGES

- ✓ NO SIGNIFICANT DIFFERENCES BETWEEN FRESH AND AGED YEAST
- ✓ REAL SHELF LIFE OF 3 TO 4 YEARS FOR ALE AND LAGER BREWING YEASTS DOES NOT AFFECT THE QUALITY
- ✓ STORAGE AT COOL AND DRY CONDITIONS IS HIGHLY RECOMMENDED

Fermentis
LESAFFRE FOR BEVERAGES

Fermentis
LESAFFRE FOR BEVERAGES

BDI

explore

Spent Grain to Energy – Biogas Plant Göss

36th CONGRESS EBC SLOVENIA 2017 14-18 MAY LJUBLJANA EUROPEAN BREWERY CONVENTION

BDI BioEnergy International AG

---

explore
the land of new technology

BDI-BIOENERGY INTERNATIONAL:

- Highly professional plant engineering and construction company
- Patented technologies implemented in tailor-made, turn-key solutions
- Industrial BioDiesel & BioGas plants for waste-to-value applications
- 20 years of experience with more than 40 reference plants on 4 continents

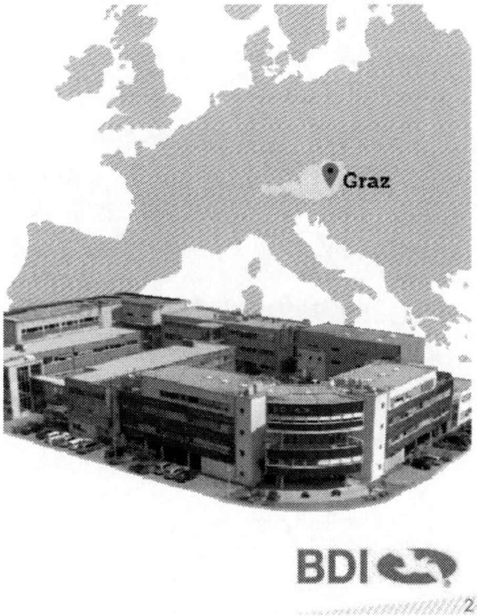

from waste to value

BDI

2

explore
the land of new technology

ANAEROBIC DIGESTION:

Anaerobic digestion (AD) is a biological process of organic matter degradation under the absence of air forming biogas

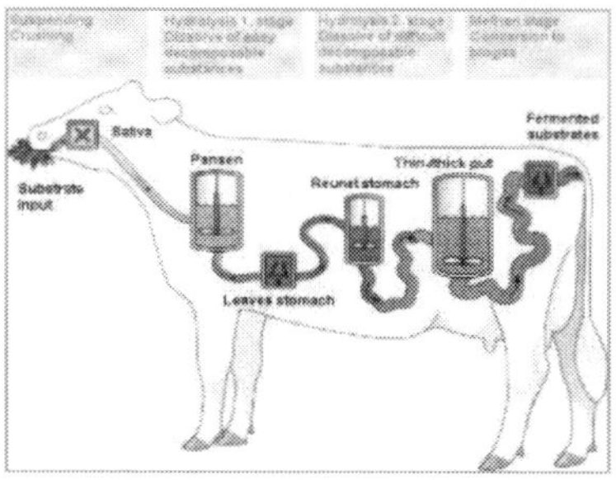

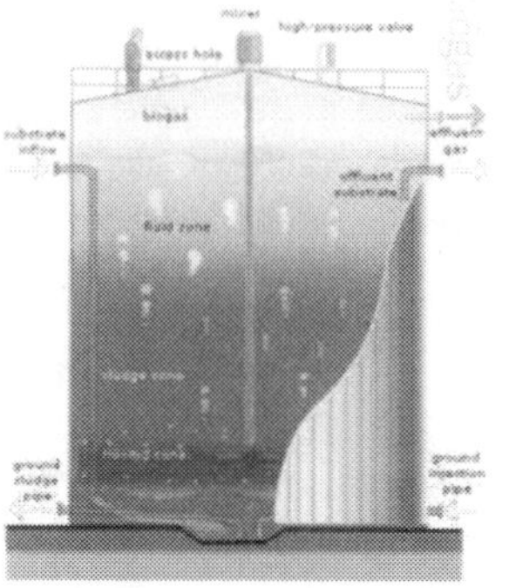

Most suitable for various complex organic residues

from waste to value

BDI

3

explore
the land of new technology

INPUT:

- **Spent Grain** (15 kg/hl brewed beer)
  - ➢ TS ~ 21%, VS ~ 20%, TKN ~ 8 kg/t
- **Surplus Yeast** (1,5 kg/hl brewed beer)
  - ➢ TS ~ 15%, VS ~ 14%, TKN ~ 9 kg/t

OUTPUT:

- **BioGas** (~ 125 Nm³ per ton spent grain)
  - ➢ CH4 ~ 58%, LHV ~ 21 MJ/Nm³
- **Fertilizer** (~ same amount of feedstock)
  - ➢ TS ~ 6%, TKN ~ 0,8%, → for agro use

from waste to value

BDI

4

explore
the land of new technology

MULTISTAGE PROCESS:

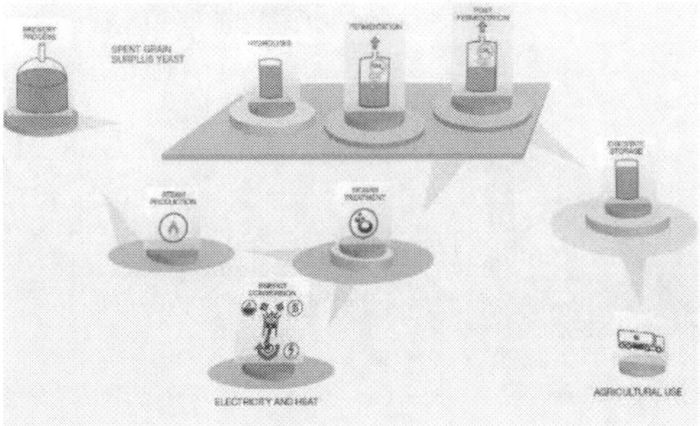

- **Hydrolysis:**
  Break-down of long chain components
- **Fermenter:**
  Main biodegradation and production of BG
- **Post-fermenter:**
  Maximization of yield and digestate quality

from waste to value

BDI

5

explore
the land of new technology

REFERENCE PLANT Brauunion Göss (1 Mio hl)

Location:	Leoben, Austria
Substrate:	Brewery Residues
Input:	~ 18.000 t/a
Output:	Biogas for Steam Boiler ~ 0,7 MW_{th} CHP ~ 0,45 MW_{el} & ~ 0,47 MW_{th}
Digestate:	Onsite storage; used as fertilizer
Energy:	Biogas to steam boiler and CHP
Start-up:	December 2015
BDI Scope:	Engineering, construction, operation

from waste to value

BDI

6

GÖSS – Energy Balance

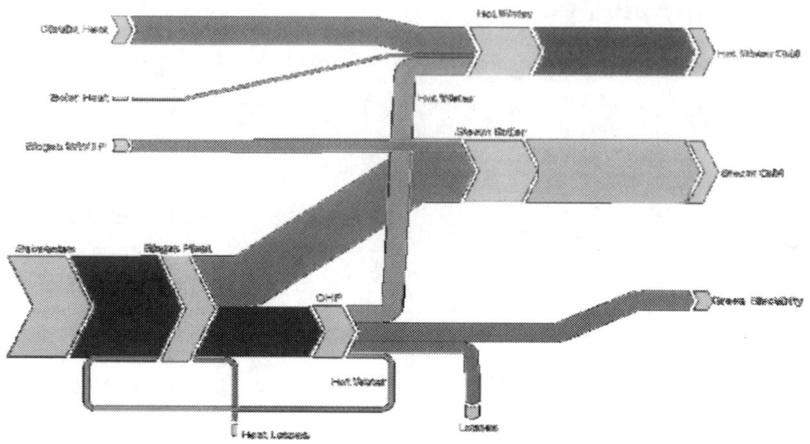

GÖSS – Main Dimensions

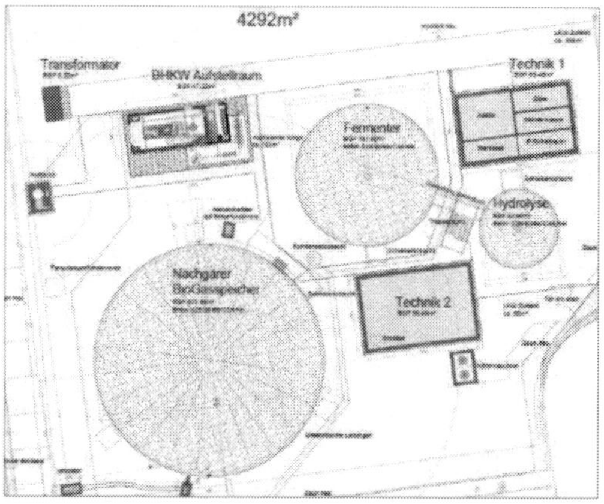

Hydrolysis		
Diameter	8	m
Height	9	m
Volumen	452	m³
Fermenter		
Diameter	15	m
Height	14,5	m
Volumen	2562	m³
Post Fermenter		
Diameter	25	m
Height	7,5	m
Volumen	3682	m³
Digestate Storage		
Diameter	37,5	m
Height	7,5	m
Volumen	8283	m³

explore
the land of new technology

GÖSS – Operation experience

- High level of nitrogen: Tackled by proper feeding and mixing, combined with the addition of nutrients and removal of H2S.
- Digestion efficiency: Even the husks of the spent grain can be reasonably digested thanks to adequate sizing and premixing.
- Brewery load changes: Balanced quite effectively by means of robust plant design including sufficient buffer capacity.

Energy Globe Austria Award 2016

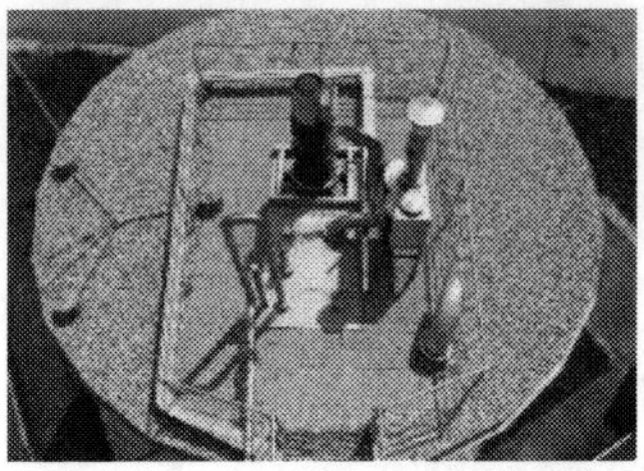

from waste to value

BDI

9

explore
the land of new technology

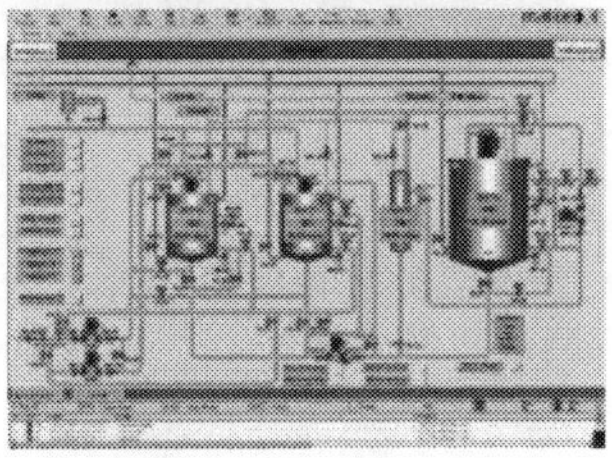

BIOGAS PILOT PLANT:

- Downscaled container plant, consisting of four equipped tanks (hydrolysis, fermenter, buffer)
- Ongoing test run with spent grain and yeast to simulate and optimize the Göss plant process
- Test runs with spent grain have already been executed in 2008 for brewery Weihenstephan
- Capability to execute test runs with various substrates for a first evaluation of gas yield

from waste to value

BDI

10

explore
the land of
new technology

- **Brewery Market:**
  More than 400 breweries > 1 mio hl output
- **Valuable Digestate:**
  Rich in nutrients to replace chemical fertilizer
- **Green Brewery:**
  Reduction of carbon footprint by renewables
- **Reliable Energy :**
  Own power production secures energy supply
- **Design Matrix:**
  Standard sizes for 20 to 100 kt/a spent grain

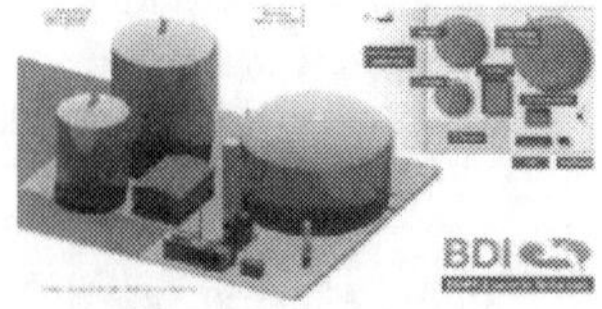

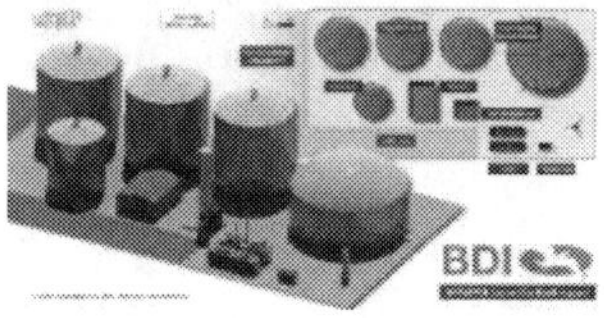

from waste to value

BDI

11

BDI

Thank you for your attention

explore
the land of
new technology

Elmar Offenbacher
Director Sales BioGas
8074 Grambach, Austria

36th CONGRESS EBC SLOVENIA 2017 14-18 MAY LJUBLJANA EUROPEAN BREWERY CONVENTION

Yeast & fermentation 1
L13

Elucidation of the formation mechanism of 2-mercapto-3-methyl-1-butanol (2M3MB) in beer

○Shigekuni Noba, Nana Yako, Taku Irie,
Minoru Kobayashi, Tetsuya Watanabe

Asahi Breweries, ltd
Research Laboratories for Alcohol Beverages

ASAHI “SUPER DRY”

- No.1 beer in the Japanese market
- “Karakuchi” taste: refreshingly crisp, clear taste
- Continually working to improve the flavor

Our technologies

- **Yeast**
  L16 : Oomuro et al.

- **Ingredients**
  P39 : Kishimoto et al.

- **Flavor**
  my presentation
  P90 : Irie et al.

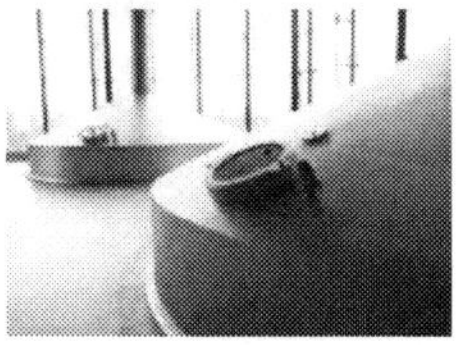

2M3MB

Odor

Onion, Sweat

Threshold

0.13 ppb

Origin

Hops

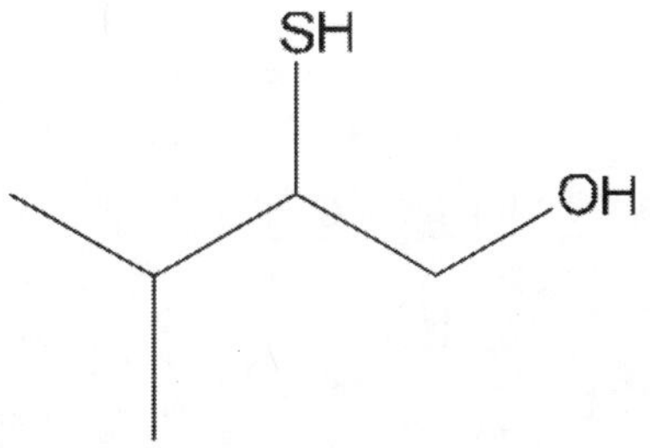

2-mercapto-3-methyl-1-butanol

2M3MB formation during fermentation

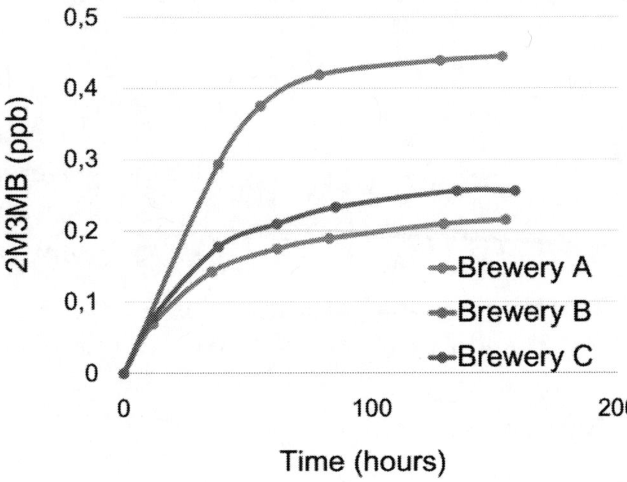

Kobayashi, ASBC Meeting, 2006

2M3MB precursor

Origin

Hops

Generation factor

Hot aeration of wort

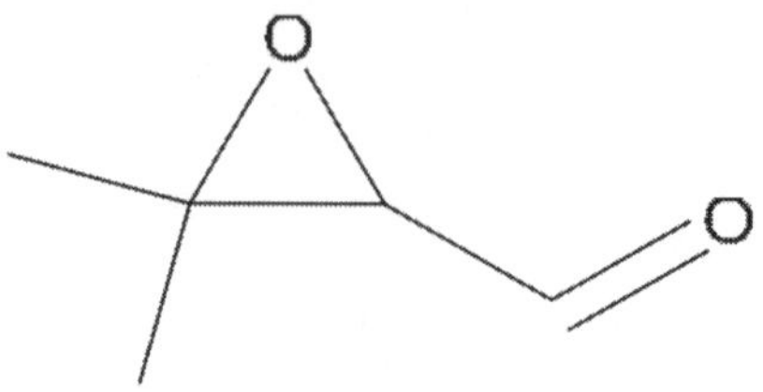

2,3-epoxy-3-methyl-butanal (EMB)

✓The discovery of the precursor in food was first reported in 2016.

Noba, WBC, 2016

Behavior of the precursor during fermentation

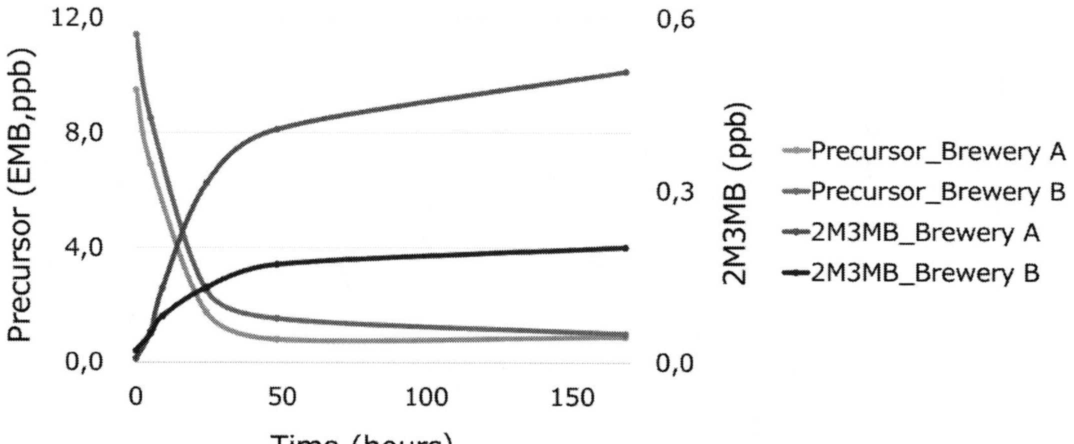

Research objective

To identify the factors necessary for
2M3MB formation

Hypothetical pathway of 2M3MB formation

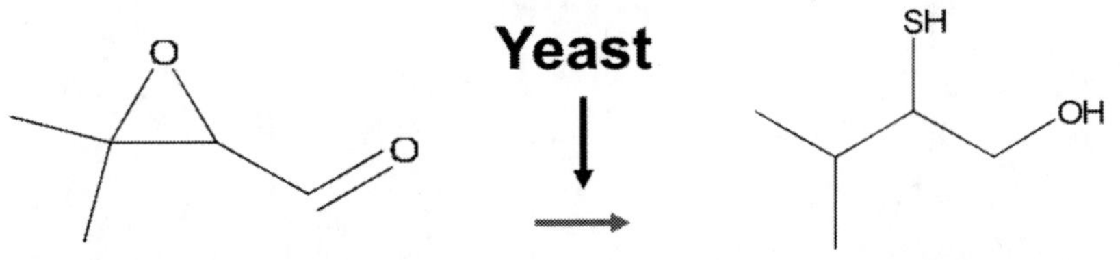

Precursor (EMB) **2M3MB**

The role of yeast

✓ Reduction of aldehyde

✓ Substitution of an SH group

Research design

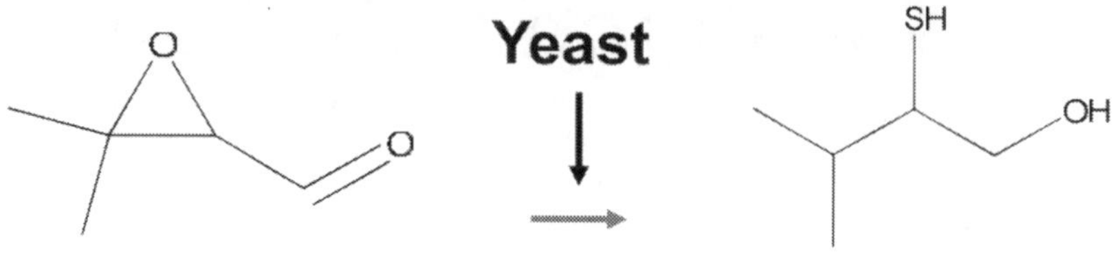

Precursor (EMB) **2M3MB**

Yeast
(1) Enzymes
(2) Metabolites

Research design

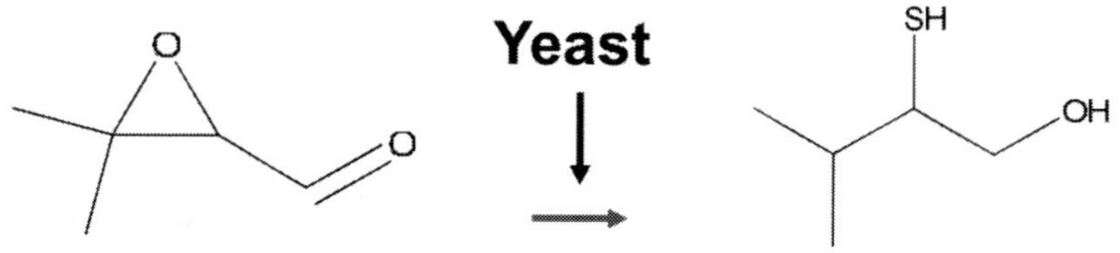

Precursor (EMB) **2M3MB**

Yeast
(1) Enzymes
(2) Metabolites

(1) Yeast enzyme

Yeast addition test

↓

Yeast extract addition test

Yeast addition test

20 ppm precursor + 0.2 g yeast in acetate buffer (pH5)

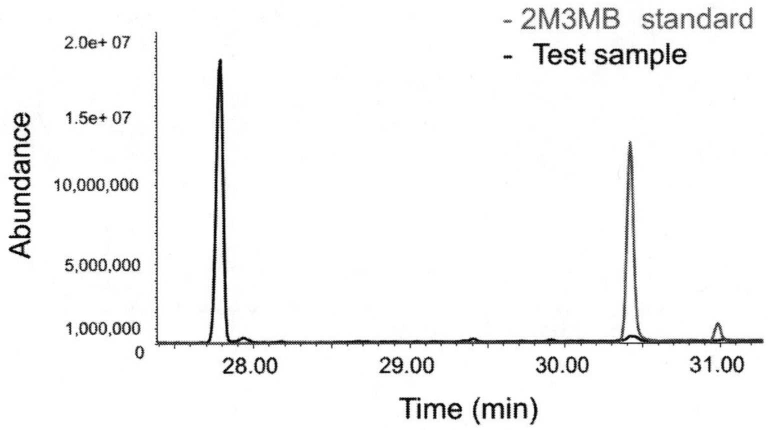

Yeast addition test

20 ppm precursor + 0.2 g yeast in acetate buffer (pH5)

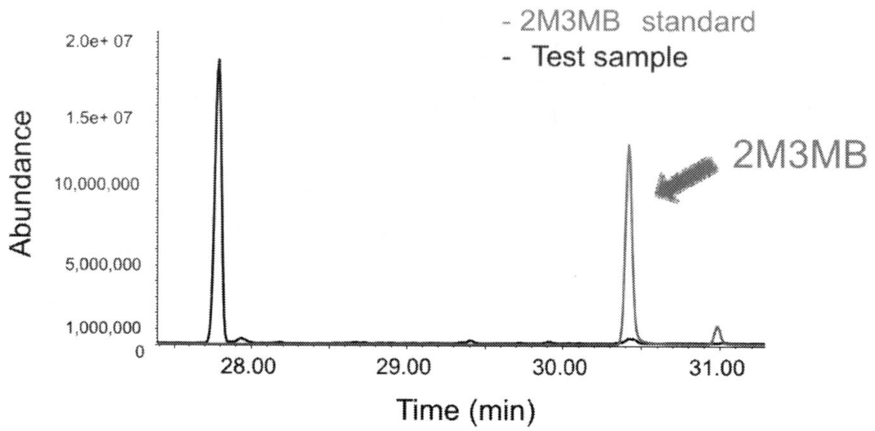

Yeast addition test

20 ppm precursor + 0.2 g yeast in acetate buffer (pH5)

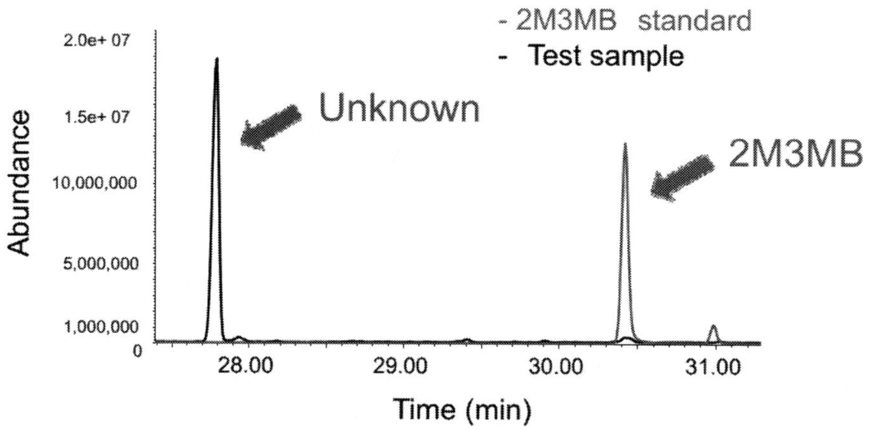

Yeast addition test

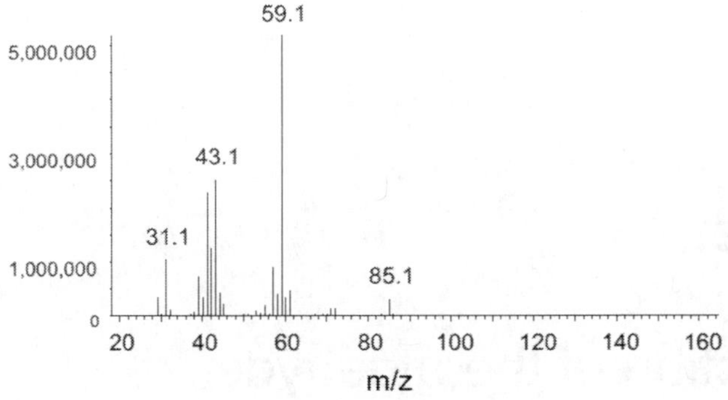

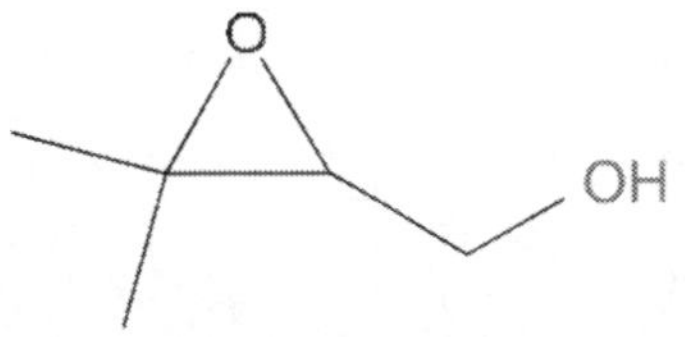

2,3-epoxy-3-methyl-butanol

- ✓ This compound was identified by comparison with a synthesized reagent.

Yeast extract addition test

20 ppb precursor + Crude protein extracted using a DEAE column

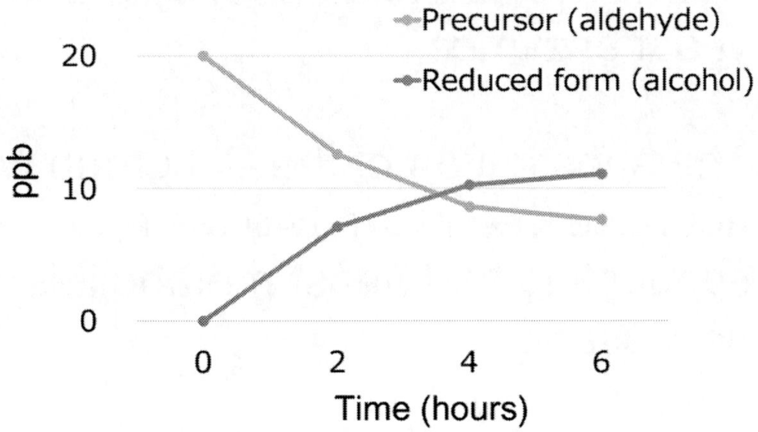

Hypothetical pathway of 2M3MB formation

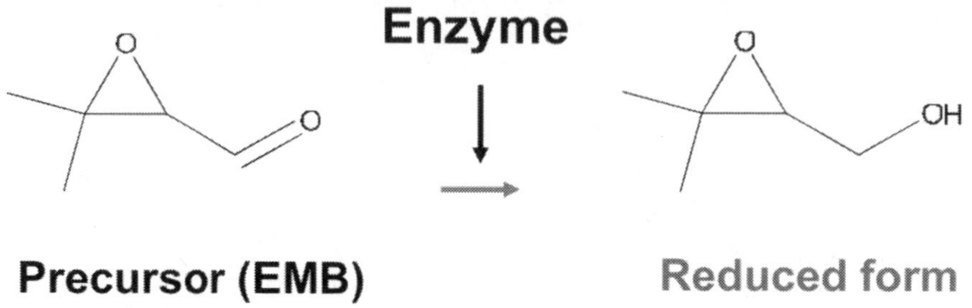

✓ Reduction of the aldehyde (by alcohol dehydrogenase?)

Short summary 1

- ✓ The terminal aldehyde of the precursor was converted to alcohol by crude yeast enzymes.
- ✓ The substitution of the SH group was not observed in the two tests; suggesting that yeast metabolites are necessary.

Research design

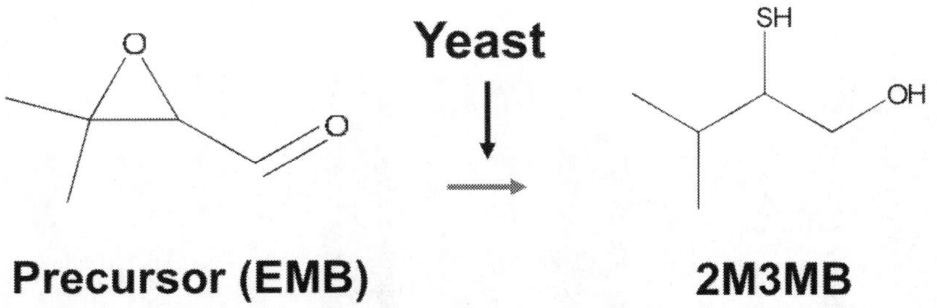

Yeast
(1) Enzymes
(2) Metabolites

(2) Yeast metabolites

Comparative test of culture media

↓

Screening of medium components

- Addition test
- Omission test

↓

Metabolome analysis

↓

Fermentation test using mutant strain

Comparative test of culture media

40 ppb of precursor was added to the media.

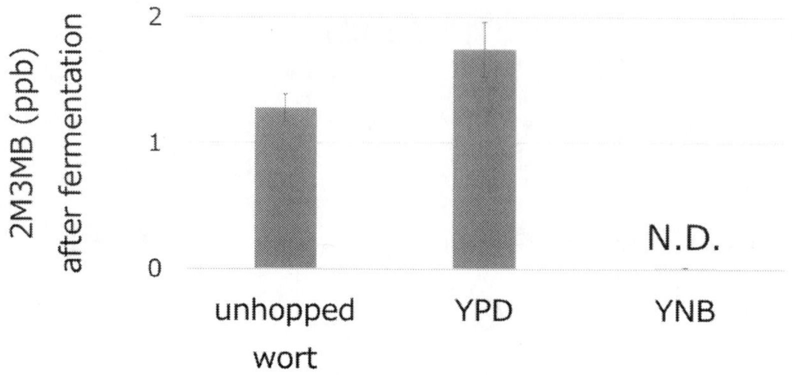

✓ 2M3MB was produced only from wort and YPD medium.

Comparative test of culture media

YPD medium		YNB medium
yeast extract peptone glucose	⇔	ammonium sulfate salts vitamins minerals glucose

✓ amino acids?

Screening of medium components : Addition test

Amino acids concentration in wort: J. Inst. Brew. 111(1), 51–60, 2005

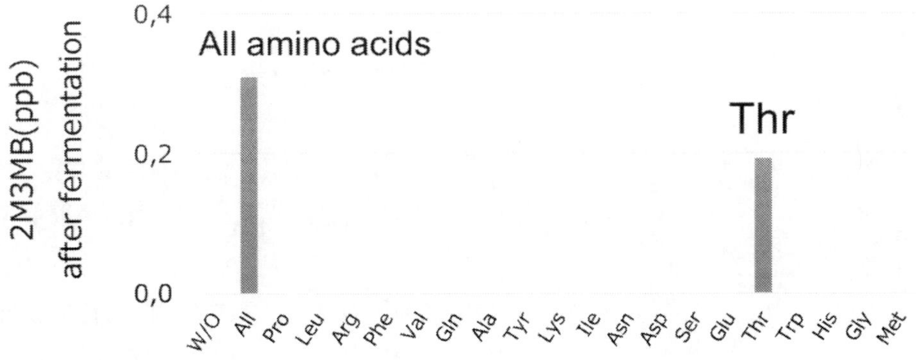

✓YNB medium + Thr was needed for the formation of 2M3MB

Screening of medium components : Omission test

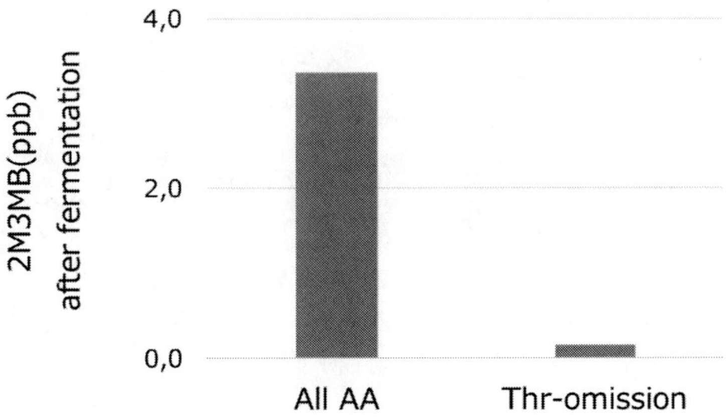

Metabolome analysis

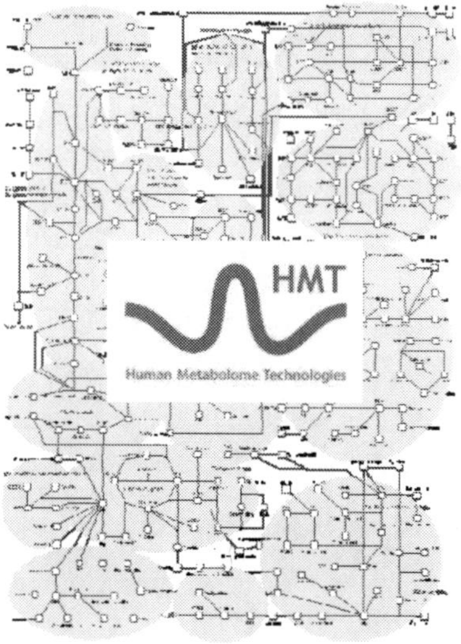

- ✓ Yeast at 6 hours of fermentation was used.
  Control: YNB + all AA
  Test: Thr omission
- ✓ 228 intracellular metabolites were detected by CE-tof/MS
- ✓ Only sulfur metabolism of yeast was affected during growth in Thr-omission medium

Yeast sulfur metabolism

C : all amino acids
T : Thr omission

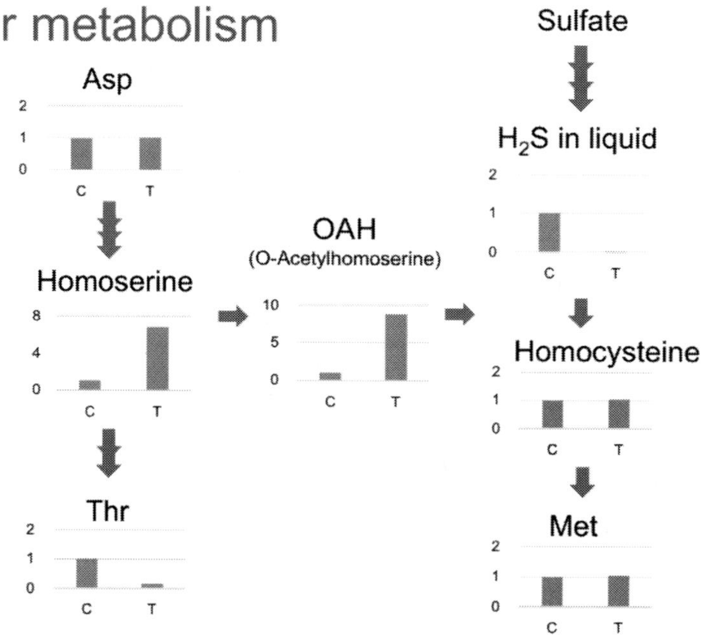

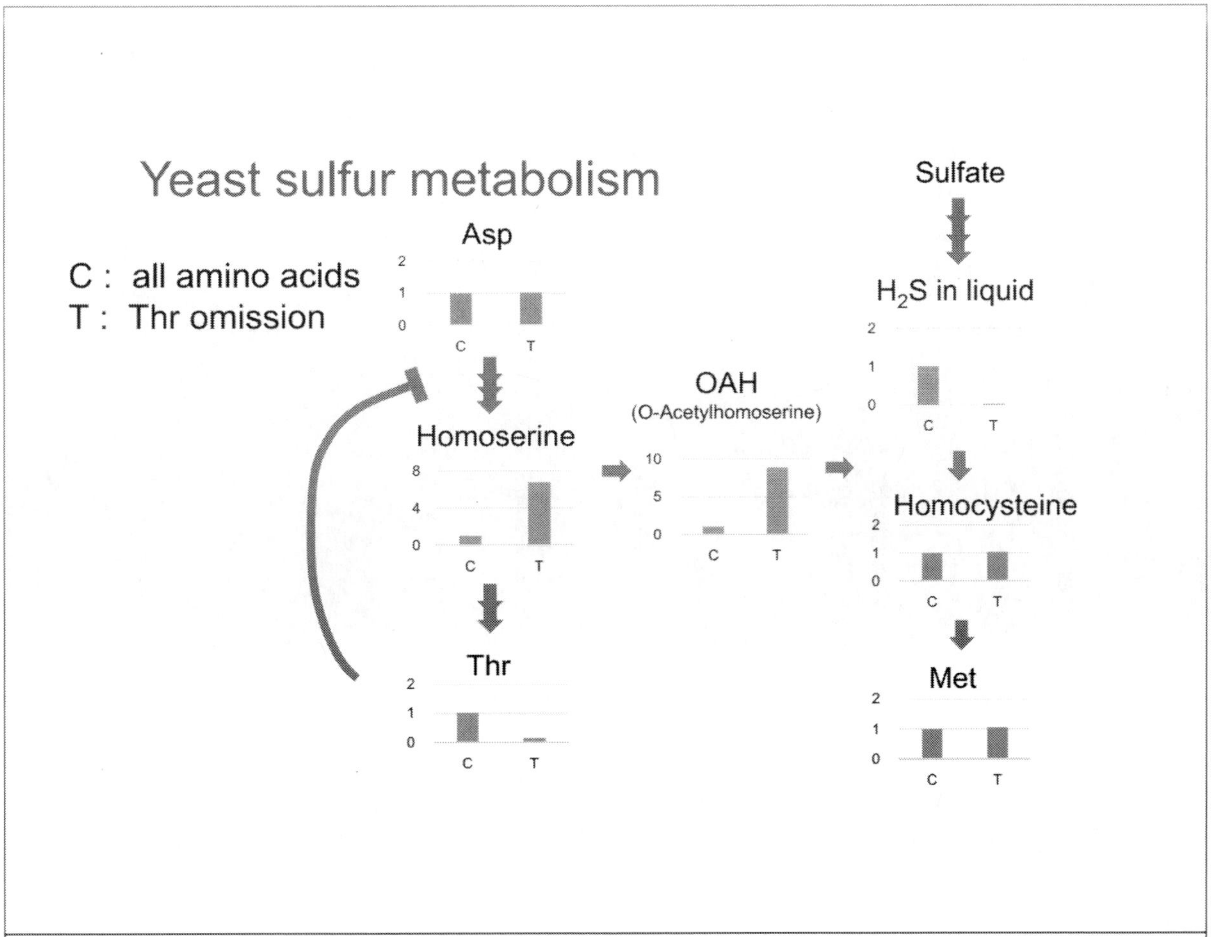
Yeast sulfur metabolism
C : all amino acids
T : Thr omission
Asp
Homoserine
Thr
OAH
(O-Acetylhomoserine)
Sulfate
H_2S in liquid
Homocysteine
Met

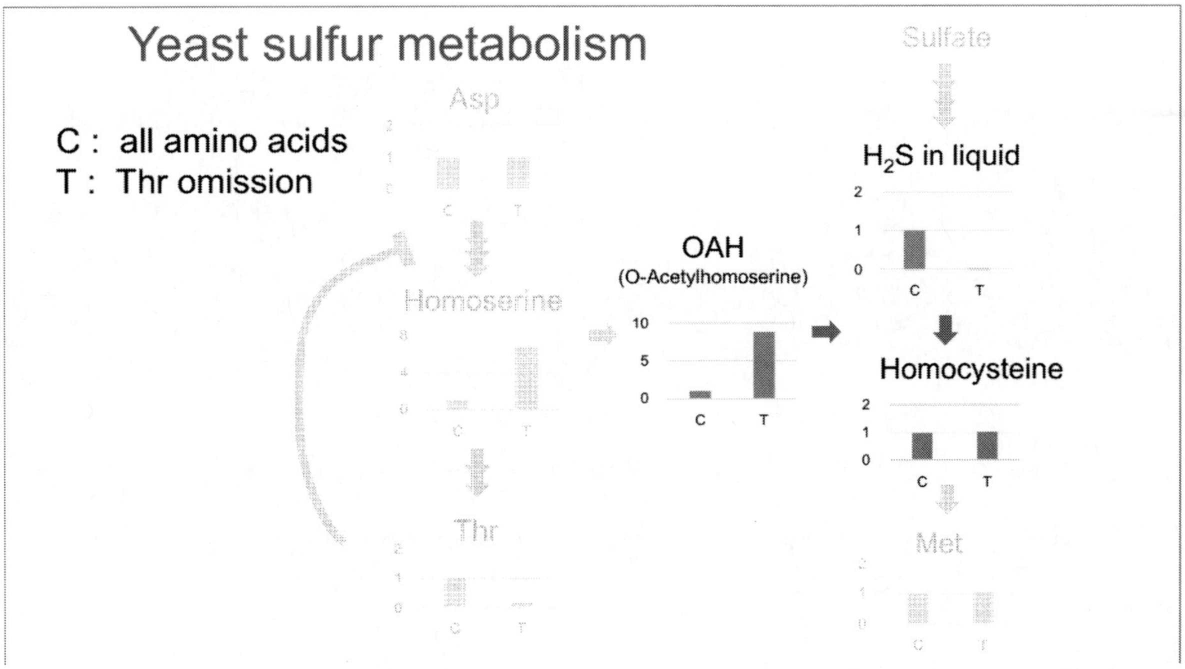
Yeast sulfur metabolism
C : all amino acids
T : Thr omission
OAH
(O-Acetylhomoserine)
H_2S in liquid
Homocysteine

Yeast sulfur metabolism

C : all amino acids
T : Thr omission

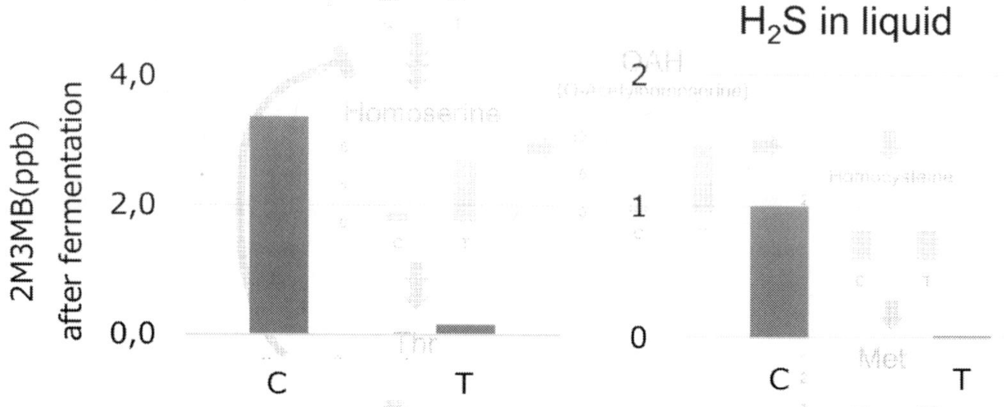

H_2S behavior during fermentation in Thr-omission test

All amino acids / Thr-omission in YNB (5 L)

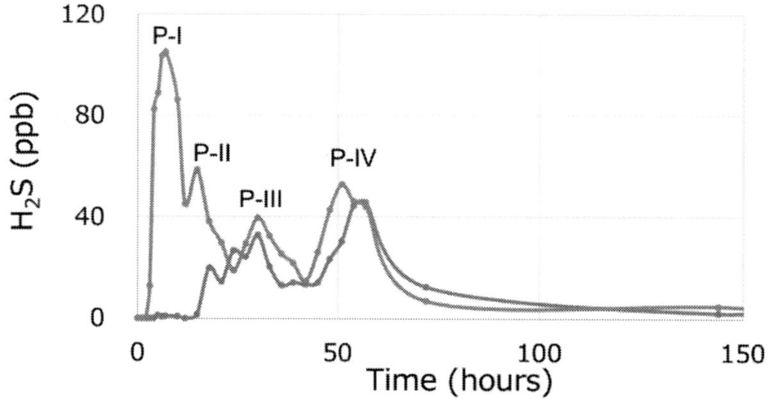

✓ There was a large difference in the initial concentration of H_2S (P-I) during fermentation.

H_2S non-producing mutant

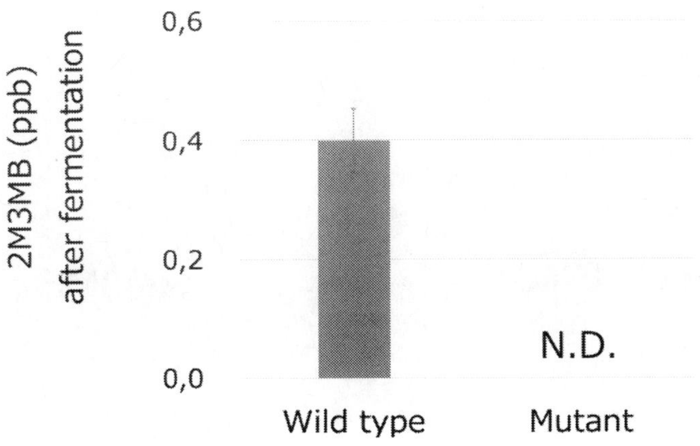

Short summary 2

- ✓ 2M3MB was formed only in the YNB medium supplemented with threonine.
- ✓ Metabolome analysis revealed that only sulfur metabolism, particularly H_2S concentration, was affected by threonine omission.
- ✓ 2M3MB was not formed by the H_2S non-producing yeast.

Conclusions : Two factors

Reduction by yeast enzyme

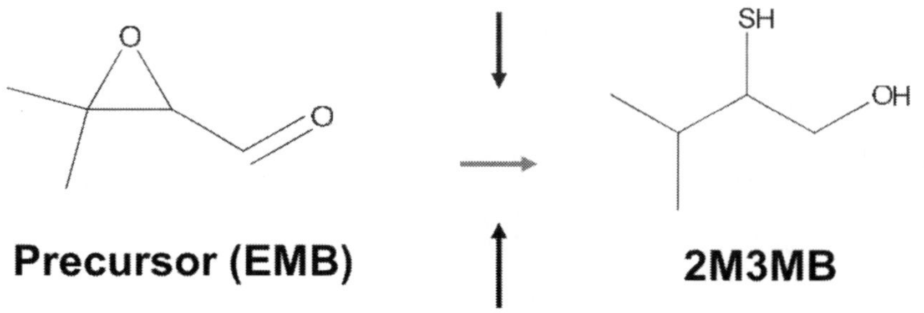

Substitution of an SH group by H_2S

Future work

- ✓ Detailed investigation of the 2M3MB formation factors, identified in this study, is being conducted at our brewery.
- ✓ Some of the results are shown on the poster P90 by Irie et al.

PENTAIR

THE ROLE OF A BEER MEMBRANE FILTER IN AN ENVIRONMENTAL FRIENDLY BREWERY

15TH MAY 2017

DR. ROLAND FOLZ – HYGIENIC PROCESS SOLUTIONS- PENTAIR

FOOD AND BEVERAGE PROCESS SOLUTIONS

A GROWING & DYNAMIC NEW WORLD....

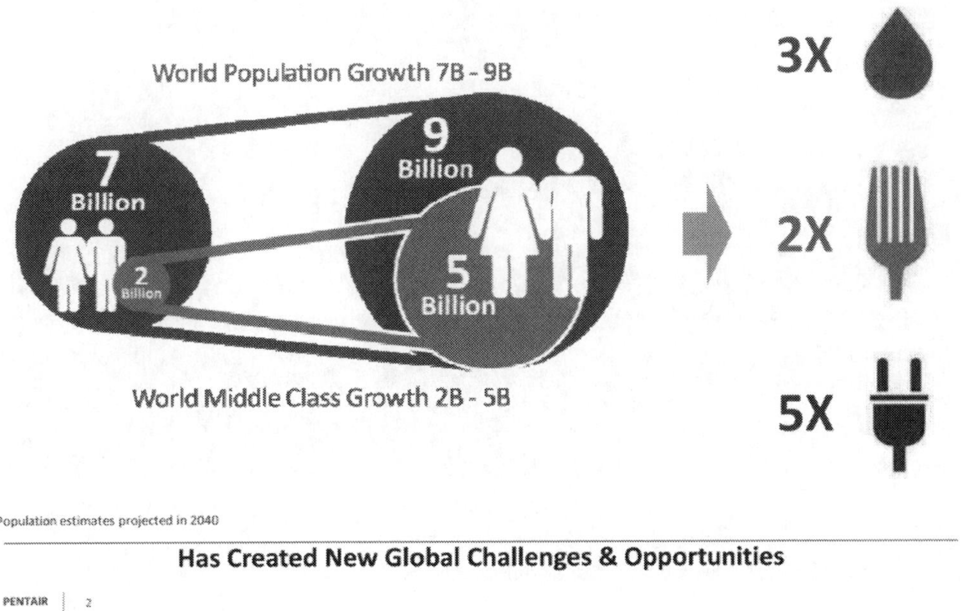

Population estimates projected in 2040

Has Created New Global Challenges & Opportunities

PENTAIR | 2

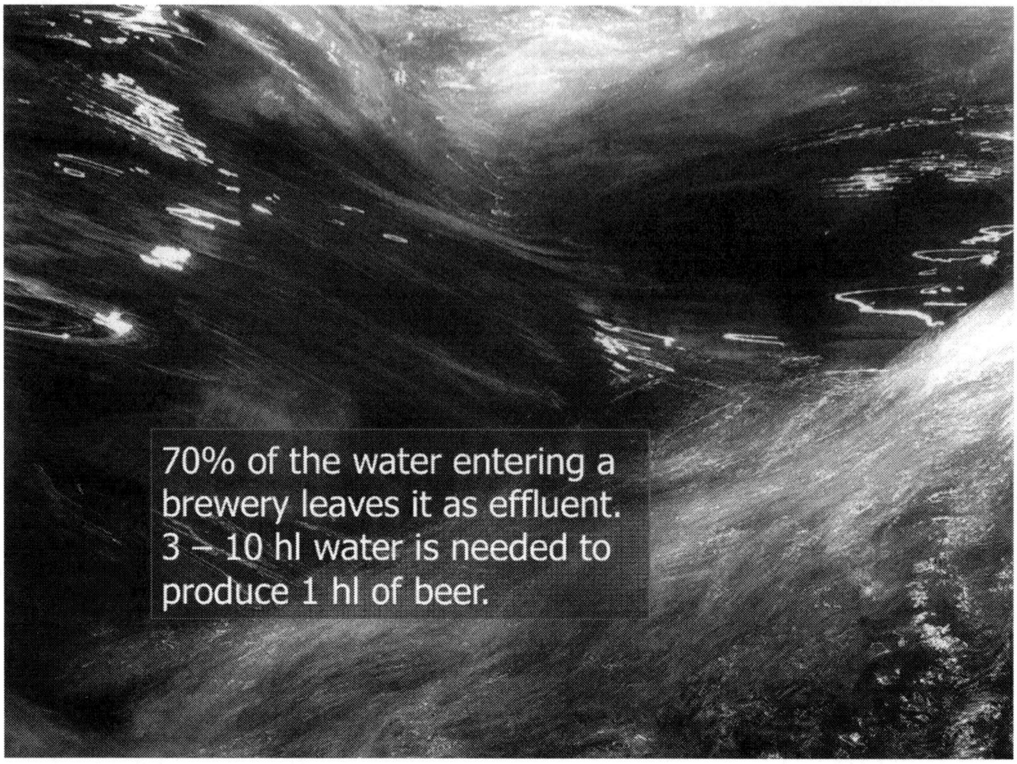
70% of the water entering a brewery leaves it as effluent. 3 – 10 hl water is needed to produce 1 hl of beer.

Brewery processes are intensive users of both electrical and thermal energy. 3 to 8 % of production costs of beer is energy & water.

BREWING INDUSTRY & SUSTAINABILITY

More than a decade sustainability is high on the agenda of the brewer of today

How does Beer Membrane Filtration support?

CORPORATE SOCIAL RESPONSIBILITY LEADERSHIP

IN PHILANTHROPIC CONTRIBUTIONS
TOTAL MORE THAN
$60 MILLION

PROVIDING ACCESS
TO SAFE, CLEAN, AFFORDABLE
SOURCES OF WATER
TO NEARLY
3,000,000
PEOPLE

EXPANDED SUSTAINABLE
MANUFACTURING PRACTICES FOR
RENEWABLE ENERGY- SOURCES, WATER REUSE, AND ZERO WASTE
AT OUR MANUFACTURING FACILITIES

TEAM PENTAIR
LAUNCHED WITH EMPLOYEE VOLUNTEERS, MENTORS, AND EXPERTS TO HELP CAUSES AND CHARITIES AROUND THE WORLD

TECHNICAL EXPERTS – THE HEART OF WHAT WE DO

330 Professionals

- Chemical, Mechanical & Electrical Engineers
- PhD Chemists
- Lab Technicians

Global Presence

- USA, Mexico
- Netherlands, Italy, UK, Germany
- China, India
- Australia, New Zealand

Capabilities

- Separations Science
- Membrane Technology
- Gas Treatment
- Hydraulic Design
- Material Science

- Polymer Synthesis
- Material Characterization
- Analytical Chemistry
- Modeling & Prototyping
- Fabrication
- Mechanical Engineering

INNOVATION STRATEGY

Focus on Differentiated Solutions
- Products, Systems, Services

R&D Teams Aligned with Market Needs

Innovation Targets
- Vitality Index > 20%

Process Discipline
- "3D" Stage Gate Process (Discover, Develop, Deploy) & Rapid Growth Process

Flexible Innovation Model
- In-house R&D: Proprietary Solutions (230 active patent families)
- External R&D: Funded University Research, Consortia & Contract Research
- Partnerships: Joint Development with Key Customers or Value Chain Partners

SOCIETAL NEEDS & TECHNOLOGY DRIVING CHANGE

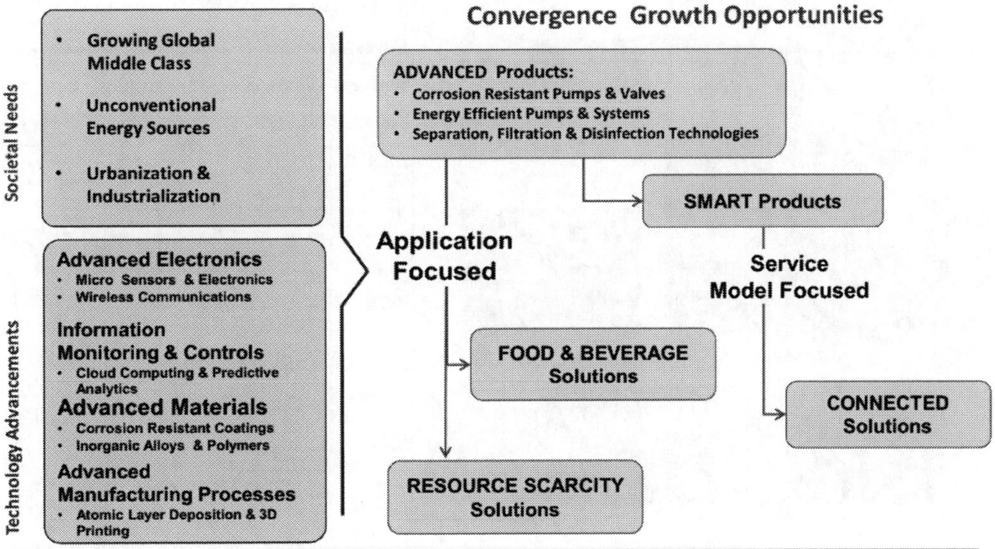

Market Back & Technology Forward Driven Innovation Opportunities

A LEADING PORTFOLIO OF WATER TECHNOLOGIES

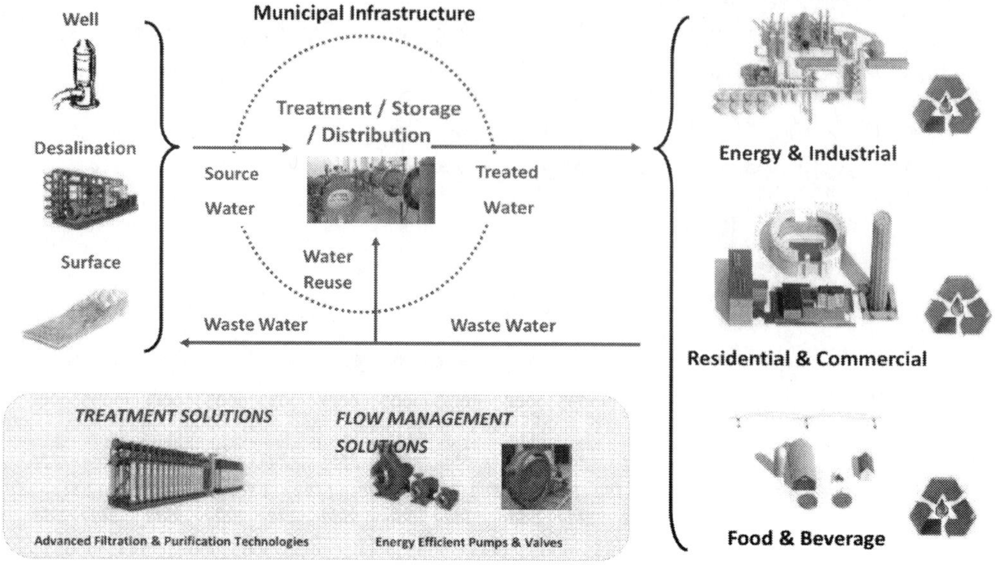

Innovations To Relieve Scarcity, Improve Quality & Improve Energy Efficiency

MEMBRAINS & MEMBRANES

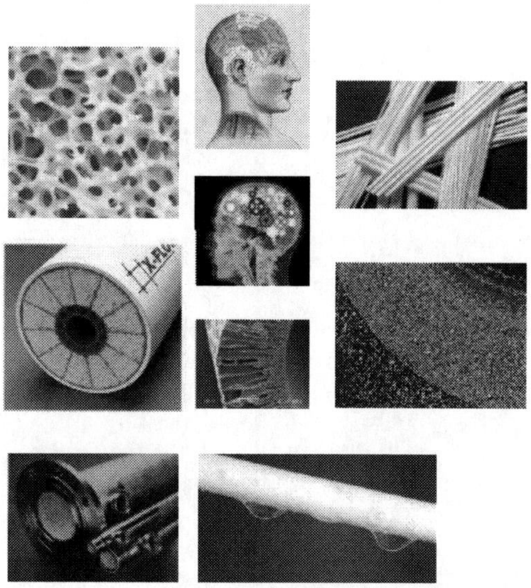

- **Pentair R&D membrane filtration**
- **Membranes**
- **Modules**
- **Water & Waste Water Treatment**
- **Beverage Filtration – Beer, Wine & Cider**

Membrane separation	*Membrane type*	*Driving force*	*Application*
Microfiltration	Symmetric and asymmetric microporous	Hydrostatic pressure	Clarification, sterile filtration
Ultrafiltration	Asymmetric microporous	Hydrostatic pressure	Separation of macro-molecular solutions
Nanofiltration	Asymmetric	Hydrostatic pressure	Separation of small organic compounds and selected salts from solutions
Reverse osmosis or hyperfiltration	Asymmetric, composite with homogeneous skin	Hydrostatic pressure	Separation of micro-solutes and salts from solutions
Gas permeation	Asymmetric or composite homogeneous or porous polymer	Hydrostatic pressure, concentration gradient	Separation of gas mixtures
Dialysis	Symmetric microporous	Concentration gradient	Separation of micro solutes and salts from macromolecular solutions
Pervaporation	Asymmetric, composite non-porous	Concentration gradient, vapour pressure	Separation of mixtures of volatile liquids
Vapour permeation	Composite non-porous	Concentration gradient	Separation of volatile vapours from gases and vapours
Membrane distillation	Microporous	Temperature	Separation of water from non volatile solutes
Electrodialysis	Ion exchange, homogeneous or microporous polymer	Electrical potential	Separation of ions from water and non-ionic solutes
Electrofiltration	Microporous charged membrane	Electrical potential	De-watering of solutions of suspended solids
Liquid membranes	Microporous, liquid carrier	Concentration, reaction	Separation of ions and solutes from aqueous solutions

ADVANCED MEMBRANE TECHNOLOGIES

Material Science Is The Key To Continued Innovation

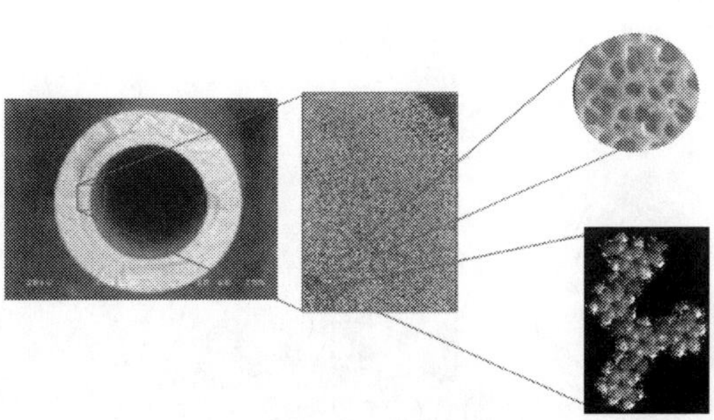

Nano to Micro Polymer Structures
Proprietary Polymer Formulations
&
Advanced Casting Methods

Molecular Chemical Structures
Advanced Coating
&
Synthesis Technologies

Critical Material Properties
- Strength & Flexibility
- Porosity
- Pore Size & Distribution
- Surface Chemistry

Creating Value to Breweries
- Lower Energy Requirements
- Low Fouling & Improved Separation Efficiency
- Longer Membrane Life
- New Applications

Filtration Innovations Enable Many Hygienic & Resource Scarcity Solutions

MODULE DESIGN INNOVATIONS ... Advanced Hydrodynamics

Tubular Membrane Modules
- Challenging Industrial WW & Process Applications
- Self Supporting Tubes w/ 5 – 8 mm ID

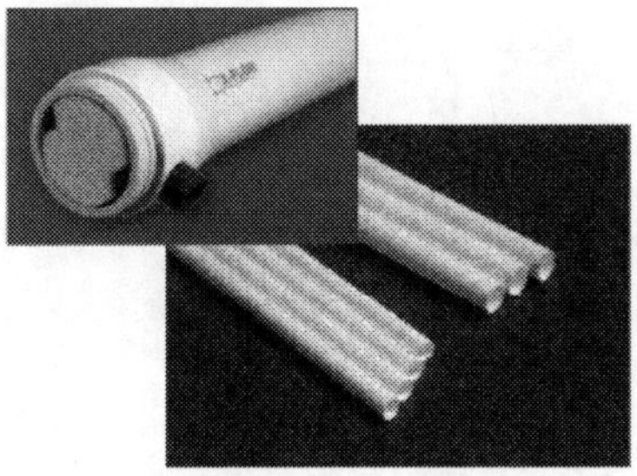

Excellent Process Performance ... but CAPEX & OPEX Challenged due to Energy Requirements

Helical Tubular Membranes
- The Process Benefits of Tubular Membranes
- Breakthrough Efficiency Due to Spiral Flow Pattern

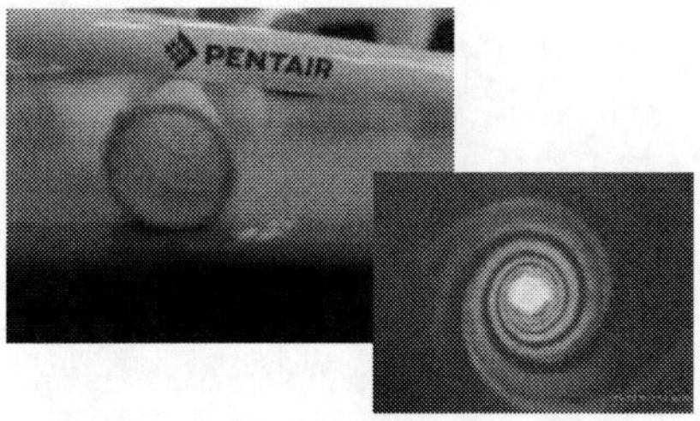

Advanced Hydraulic Design Enables:
- *Up to 2X Increase in Capacity per Membrane Area*
- *Up to 50% Reduction in OPEX & CAPEX*

Helical Tubular Module ... A Significant Breakthrough for Industrial Reuse

TECHNOLOGY ROADMAP - BMF

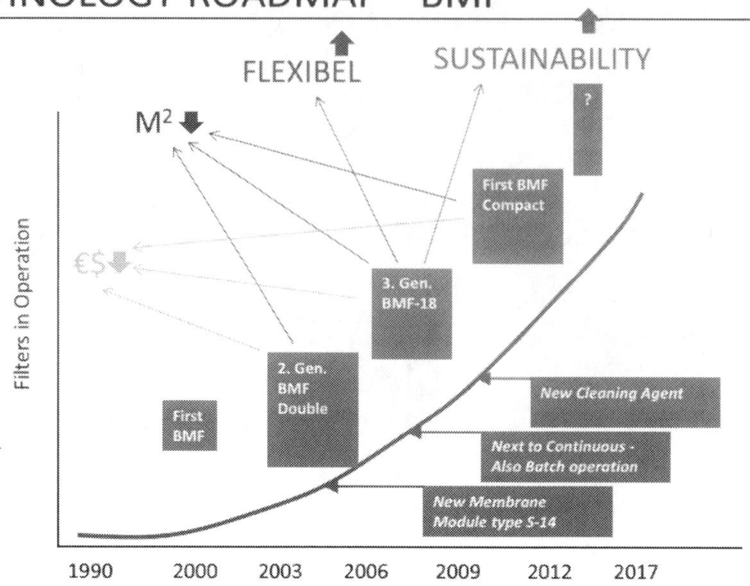

SUSTAINABILITY KPI'S

- CO2 Footprint
- Water Consumption
- Solid Waste Stream
- Beer Loss
- Shelf Life

CO2 FOOTPRINT (1)

- An investigation was executed to come to grounded number for CO2 footprint for membrane filtration and DE filtration of beer
- For the calculation of CO2 footprint the following directive of the British Standards Organization BSI was used – 'Guide to PAS2050: How to asses the carbon foot print of goods and services
- The CO2 footprint is calculated over the whole supply chain, so also energy consumption in the production and shipping of DE and membranes is taken into account
- The energy consumption in making membranes has been taken from the Pentair factory in Enschede

CO2 FOOTPRINT (2)

DE Filtration

- Production of DE
- Shipping DE to a brewery 1000 km
- Use of energy for pumps and hot cleaning
- Caustic for cleaning
- Shipping for disposal 250 km

Membrane Filtration

- The production of PES
- The energy consumption of the membrane manufacturing plant
- Shipping to a brewery 1000 km
- Use of energy for pumps and hot cleaning
- Caustic for cleaning
- Shipping back to factory 1000 km

CO2 FOOTPRINT (3)

Diatomaceous Earth Beer Filtration

	kg CO2eq per hl beer
Production of diatomaceous earth	0.11
Transport of diatomaceous earth to brewery	0.0060
Filtration of beer, steam & power	0.073
Filtration of beer, caustic	0.000017
Transport of used diatomaceous earth to place of disposal	0.0060
Total	**0.19**

Beer Membrane Filtration

	kg CO2eq per hl beer
Production of PES	0.00094
Manufacturing of membrane modules	0.0035
Transport of modules to brewery	0.000017
Membrane filtration of beer, steam & power	0.10
Membrane filtration of beer, caustic	0.0045
Transport of modules back to Enschede	0.000017
Total	**0.11**

CO2 FOOTPRINT (4)

CO2 Footprint of Beer

Source	kg CO2eq per hl beer
Fosters Brewery Group, Australia	10
Heineken, the Netherlands	10
SABMiller, South-Africa	14
Sapporo, Japan	6
New Belgium, USA	8.5
Average	**9.7**

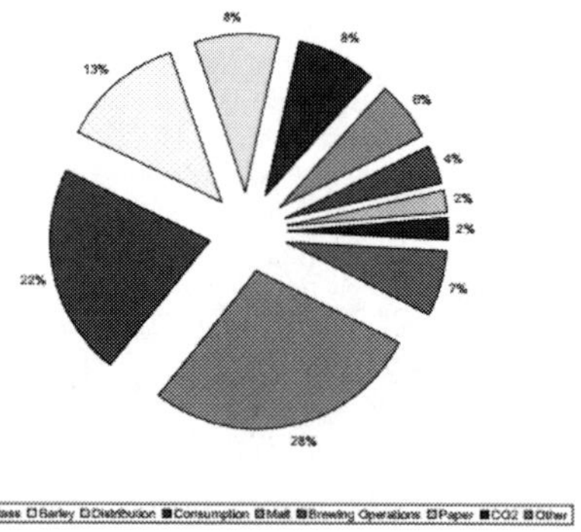

WATER CONSUMPTION

DE Filtration

- Water in a DE filter is used for flushing out the DE, cleaning and sterilization
- Consumption depends on run volumes
- Typical consumption is 0.1 – 0.2 hl/hl

Membrane Filtration

- Water in a BMF is used for periodical back flushing with cold caustic and regeneration of the membranes
- Consumption depends on achieved run volumes
- Typical consumption is 0.2 hl/hl, lower for beers produced with adjuncts and use of extra enzymes

SOLID WASTE STREAM

- Typical solid waste stream of a brewery is 0.8 kg/HL
- DE is responsible for 50% being 0.4 kg/HL
- For a brewery with an annual output of 1,000,000 HL it means the following:
  - DE Filter: 400,000 kg/yr
  - BMF: 180 kg/yr (replacement of membranes)

BEER LOSS (1)

- With a Membrane Filter the beer is washed out and directly dosed back in the beer stream – No Head & Tail Tank needed with complicated logistics because of dosing back procedures and limitations in mixing

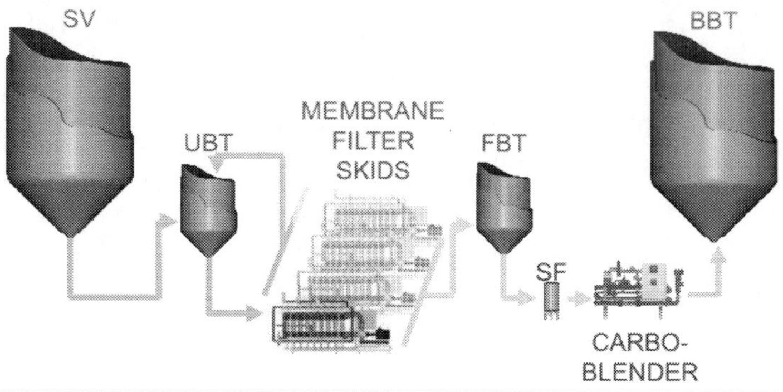

PENTAIR | 'The role of a beer membrane filter in an environmental friendly brewery'

23

BEER LOSS (2)

- System Volume is lower vs. DE filter - 30 HL vs. 70 HL
- Especially in high gravity processes and high run volumes, the membrane filter can reach down to 0.1%
- A DE filter can reach this number seldom, only in a brewery with very low SKUs and brand changes
- As an average the beer loss for DE filtration is double the beer loss of membrane filtration

PENTAIR | 'The role of a beer membrane filter in an environmental friendly brewery'

24

ADDED VALUE BMF IN SUSTAINABILITY

- Lower CO2 footprint – 40% reduction vs. DE
- 50% Lower Solid Waste Stream – Strong reduction
- Water Footprint – Comparable level vs. DE
- Average 0.2% Beer Loss – DE has double
- Shelf Life - Better taste stability – No iron pick-up

PENTAIR

MEMBRANE & SEPERATION TECH FOR A CIRCULAR ECONOMY

RE-IMAGINING THE WATER TREATMENT CYCLE

Anaerobic Membrane Bioreactor ... *Advanced Treatment Technology*

Waste from Beverage Processing Plant

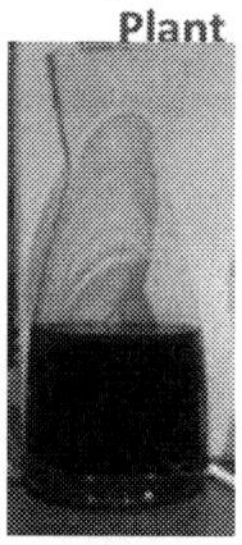

X-Flow UF Membrane Systems

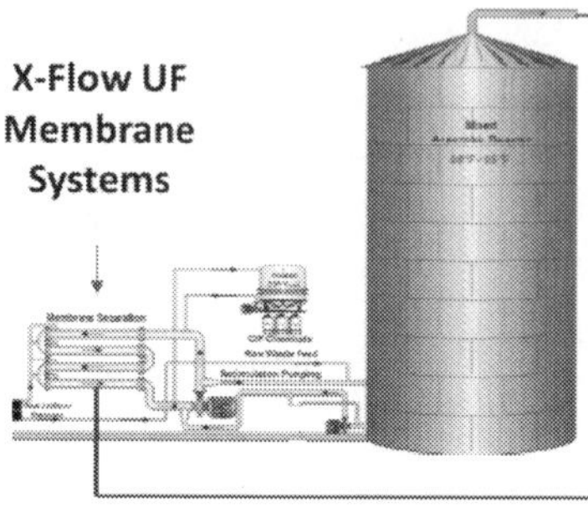

Methane Used To Offset The Natural Gas supply

UF Treated Water Suitable For Reuse after Nano-Filtration or RO

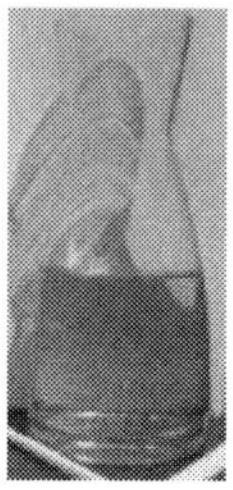

Innovative Technologies ...

20 to 30% Higher Methane Production Than Conventional Treatment While Producing Water Suitable For Reuse

From Waste to a Source of Energy & Water ... New Opportunities Possible

BIOGAS UPGRADING

WWTP

Biomass Digester

Biogas

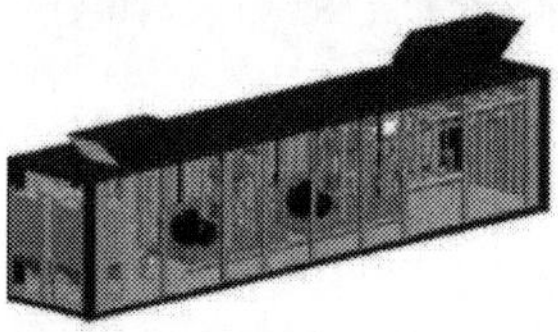

Biomethane to Grid, CNG

Optionally $BioCO_2$

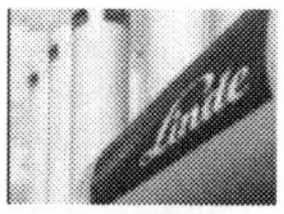

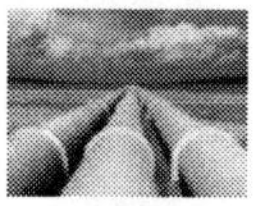

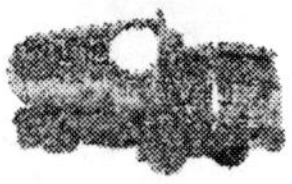

- **Biomethane Production with optional CO_2 recovery**
- **up to 100 % CH_4 recovery –> CNG (and eventual LNG)**
- **Byproduct CO_2 for Industrial, Greenhouse or Food-grade Use (Liquid/Gas form)**

SMART PRODUCTS & CONNECTED SOLUTIONS ROADMAP

SMART Products

- Valves & Actuators
- Pumps & VFD's
- Heating Cables & Controllers
- Cooling Technologies & Controllers
- Filtration Products
- Sensors & Instrumentation (e.g., Flow Meters, CO2 Sensors, Thermistors, Leak Detection, etc.)

CONNECTED Solutions

SYSTEM Solutions | **ENTERPRISE Solutions**

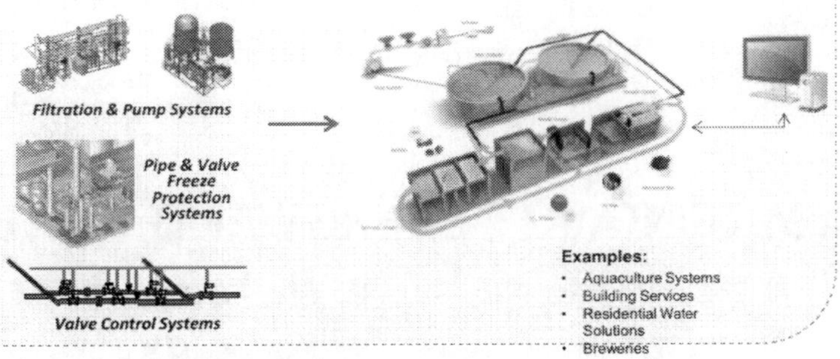

Examples:

- Aquaculture Systems
- Building Services
- Residential Water Solutions
- Breweries

SMART Products:
Devices with Advanced Features to Enable Component Level Diagnostics, Preventative Maintenance & Asset Management Services

CONNECTED Solutions:
Solutions Integrating **SMART Products** to Enable Component & System Level Performance Monitoring, Diagnostics, Preventative Maintenance & Asset Management Services, as well as Simplify Integration into an Enterprise Solution

2015 | 2016 | 2017 | 2018

SMART PRODUCTS ... A KEY STEP TO EXECUTING OUR ROADMAP

TAKE HOME MESSAGE

- With the BMF the brewing industry has an alternative for beer filtration with a clearly better contribution to a sustainable beer brewing & production process

- Especially the 50% reduction of solid waste stream, lower beer loss and better shelf life of a membrane filter is supporting the brewery on sustainability KPI's

BALANCING CO_2, ENERGY AND WATER

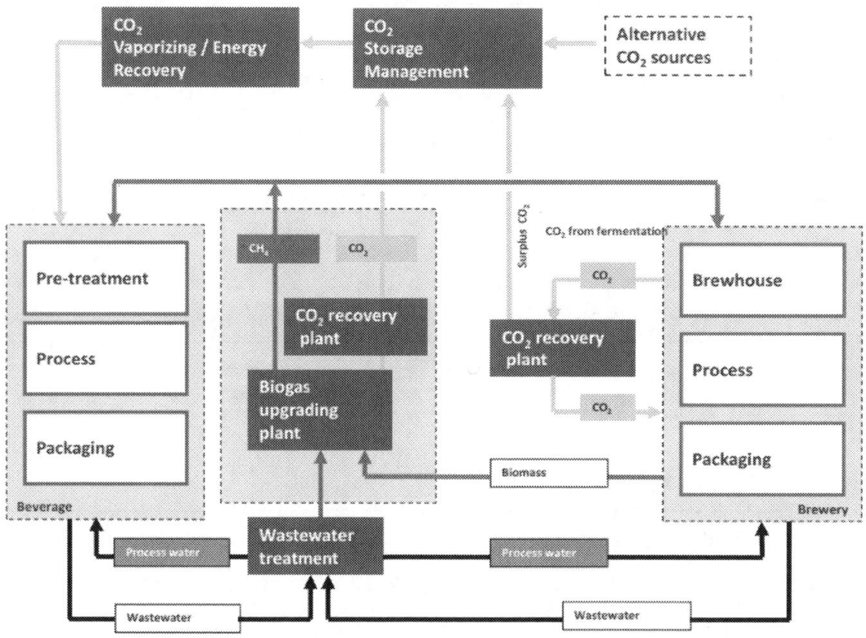

PENTAIR | A HOLISTIC R&D APPROACH OF A GLOBAL COMPANY

16/May/2017 L16

Genetic analysis of bottom-fermenting brewer's yeast imparting good *KIRE* (crispness) to beer taste

○Mayu Oomuro, Yasuo Motoyama, Tetsuya Watanabe
Research Laboratories for Alcohol Beverages, Asahi Breweries Ltd.

Outline

- Introduction
- Research objective
- Results
- Conclusion

Outline

- ➢Introduction
- ➢Research objective
- ➢Results
- ➢Conclusion

"SUPER DRY"

- ➢Our Leading Product
- ➢Good "*KIRE*"
- ➢Uses Original Yeast

"SUPER DRY"

- **<u>Our Leading Product</u>**
- **Good "*KIRE*"**
- **Uses Original Yeast**

ASAHI "SUPER DRY"

- **<u>No.1</u> beer in the Japanese market**
- **Debut in 1987 as Japan's first "*<u>Karakuchi</u>*"* beer**
- **Continually working to improve the flavor**

* "*<u>Karakuchi</u>*" taste:
good "*KIRE*"
refreshingly crisp, clear taste

"SUPER DRY"

- Our Leading Product
- **Good "*KIRE*"**
- Uses Original Yeast

What is "*KIRE*" ?

- **Japanese word for crispness or smoothness of taste**
- **Major characteristic of "SUPER DRY"**
- **Important characteristic for Japanese beer market**

Voices of the Japanese customers

"I want to drink a lighter, more refreshing beer,"
"I like a beer light enough that I can drink several glasses,"
"I want a beer that goes well with any kind of dish."

What is "*KIRE*" ?

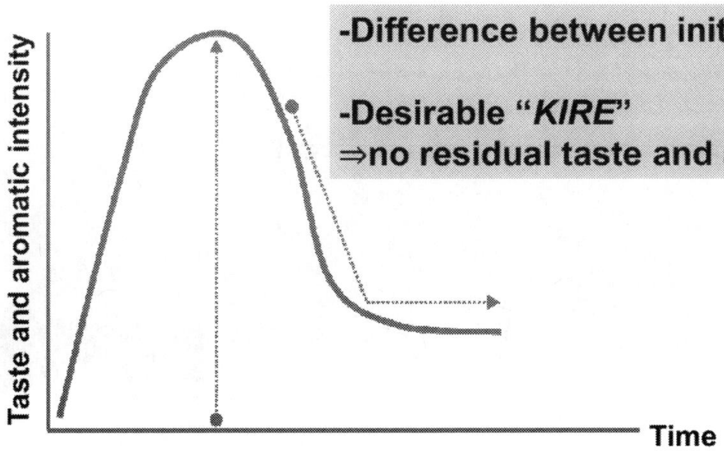

-Difference between initial and final mouthfeel

-Desirable "*KIRE*"
⇒no residual taste and aromas after drinking

Seiko Miyashita, et. al., WBC2016
Hirotaka Kaneda, et. al., *J. Am. Soc. Brew. Chem.* 60(2):71-76, 2002

Asahi's Challenge to Achieve "*KIRE*"

Mission

Make consistently high quality "SUPER DRY" by maintaining and/or improving "*KIRE*"

Efforts

◆Production : Quality control of Ingredients, wort, all processes of brewing

◆R&D : Clarification of factors impacting "*KIRE*"

- **Compositional profile of beer flavor**
- **Bitterness depends on iso-alpha acid**
- **Residual sugar content of beer**

Our recent research into "*KIRE*"

- ➢Compositional profile of beer flavor
- ➢Bitterness depends on iso-alpha acid
- ➢Residual sugar content of beer

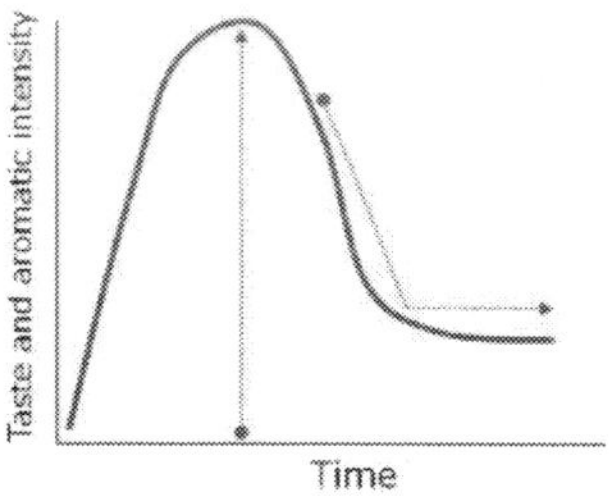

Our recent research into "*KIRE*"

- ➢Compositional profile of beer flavor

Miyashita et. al., WBC2016

WORLD BREWING CONGRESS
August 13–17, 2016 • Denver, Colorado, U.S.A. | #ElevateBeer

Sensory and instrumental analyses of compounds affecting the *KIRE* (crispness) of beer

Research Laboratories for Alcohol Beverages
ASAHI BREWERIES, Ltd.

Seiko Miyashita, Minoru Kobayashi, Kaori Kikuchi
Naoki Kato, Susumu Masuda, Masayuki Aizawa

Asahi

A) Effect of all candidate compounds in combination on *KIRE*

Compound	VIP	Relation to "*KIRE*"
C8 Fatty acid	2.44	Negative
C10 Fatty acid	1.53	Negative
C8 Fatty acid ethyl ester	4.59	Negative
C10 Fatty acid ethyl ester	3.54	Negative
β-Myrcene	2.70	Negative
Phenylethyl acetate	1.84	Negative
Isoamyl alcohol	1.23	Positive

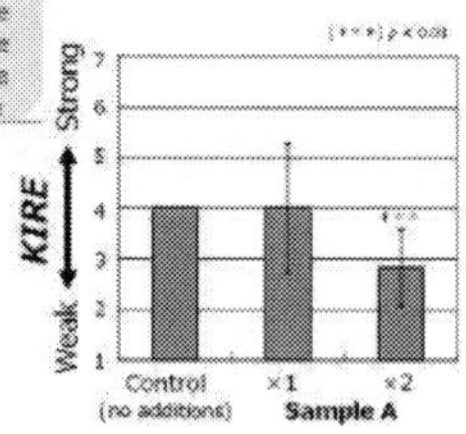

Seiko Miyashita, et. al., WBC2016

Our recent research into "*KIRE*"

➢**Bitterness depends on iso-alpha acid**

Kishimoto et. al., EBC2017 poster P39

Bitterness with higher ratio of
iso-co-humulone contributes to the "*KIRE*" of beer

Our recent research into "*KIRE*"

➢**Residual sugar content of beer**

This presentation

Fermentation ability (sugar uptake, fermentation rate)
of yeast is very important

"SUPER DRY"

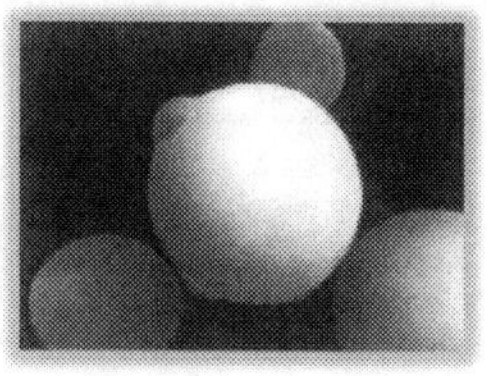

- Our Leading Product
- Good "*KIRE*"
- **<u>Uses Original Yeast</u>**

Fermentation process affects beer profile

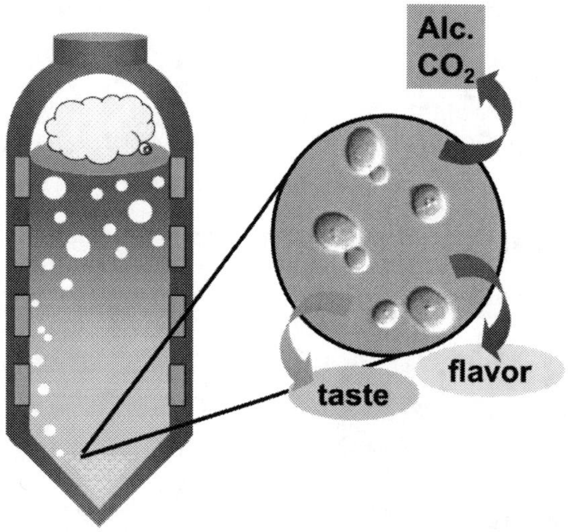

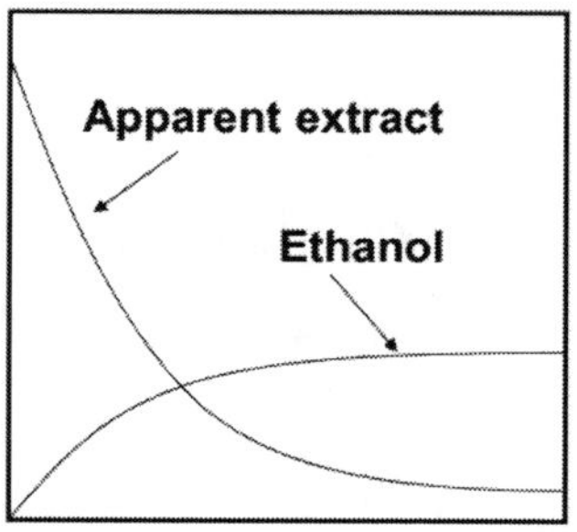

Fermentation process affects beer profile

- Metabolizes fermentable sugar into ethanol and CO_2
- To provide good "*KIRE*" taste, the sugar content in beer must be rather low

⇒Fermentation ability
(sugar uptake, fermentation rate)
of yeast is very important

Best yeast for "SUPER DRY"

- **We previously selected "SUPER DRY" yeast exhibiting high fermentation performance (sugar uptake, fermentation rate)**

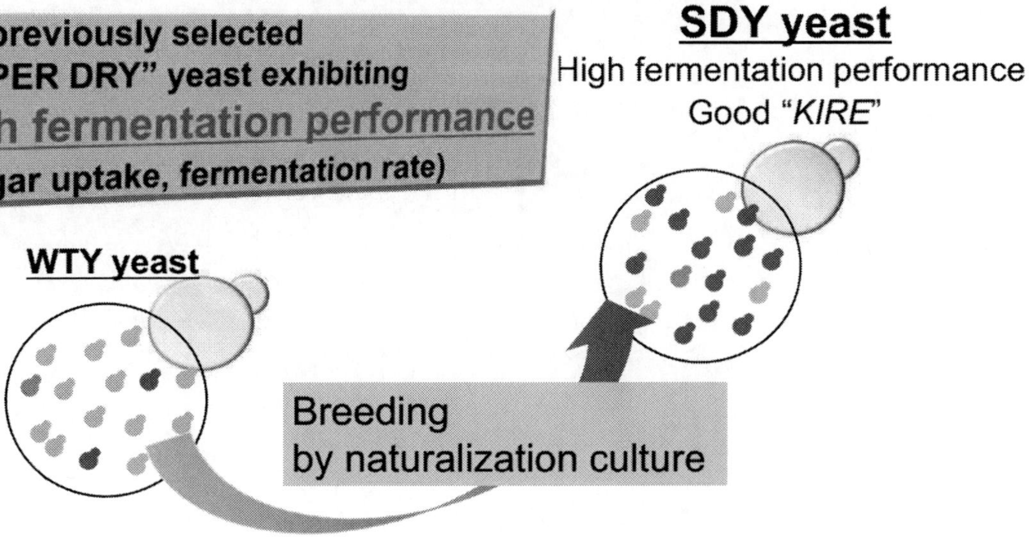

Why gives SDY its fermentation ability ?

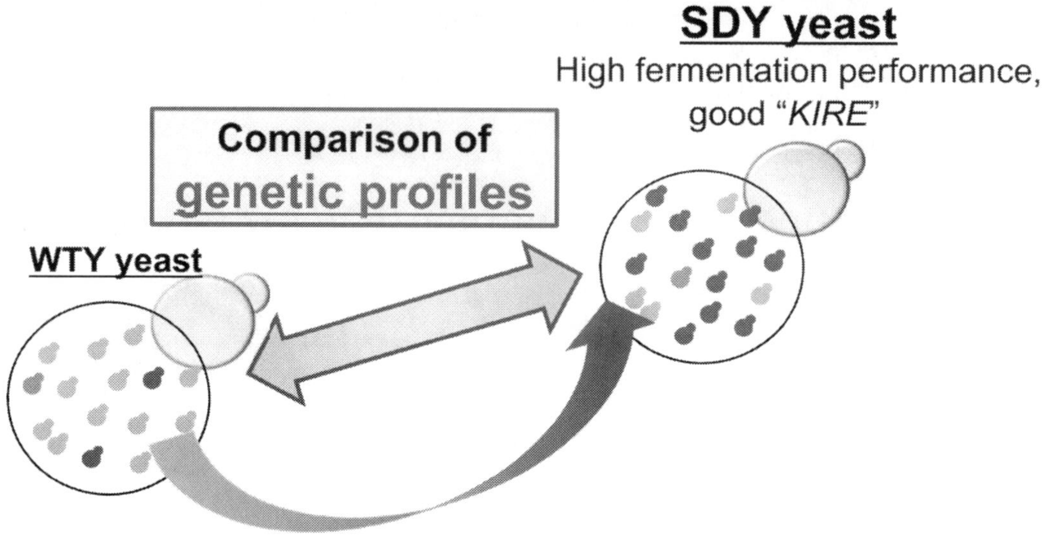

Outline

- ➢Introduction
- **➢Research objective**
- ➢Results
- ➢Conclusion

Research objective

Investigation of gene profiles related to high fermentation performance in bottom-fermenting yeast imparting good "*KIRE*"

Outline

- Introduction
- Research objective
- **Results**
- Conclusion

Results

1 : Characteristics of SDY
2 : Genetic Analysis
3 : Pilot-scale Fermentation Test

Results 1

Characteristics of SDY

SDY exhibited higher fermentation rate

Apparent extract during fermentation

5-L scale

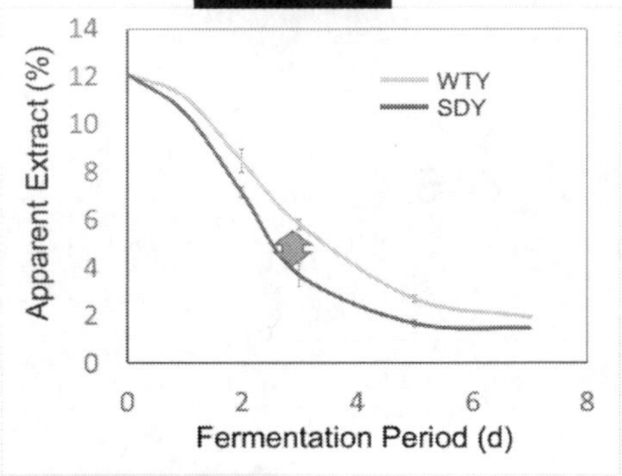

N=3

Condition:
Temperature : 13℃
DO. : 10 ppm
Period : 7 days

20-L scale

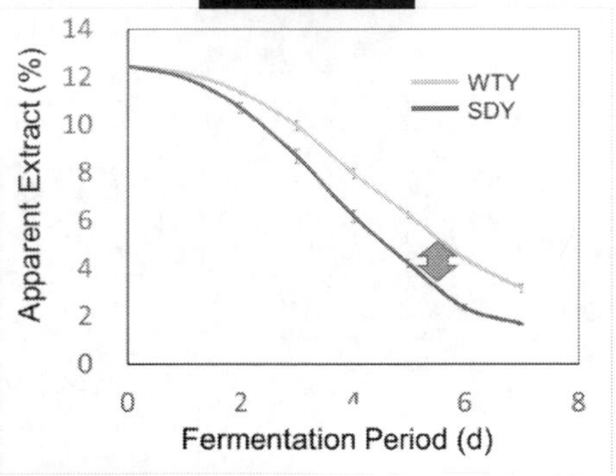

N=3

Condition:
Temperature : 10.5℃
DO. : 8 ppm
Period : 7 days

Beer made with SDY imparted good "*KIRE*"

"*KIRE*" analysis by taste sensor (AISSY Inc.)

5-L scale

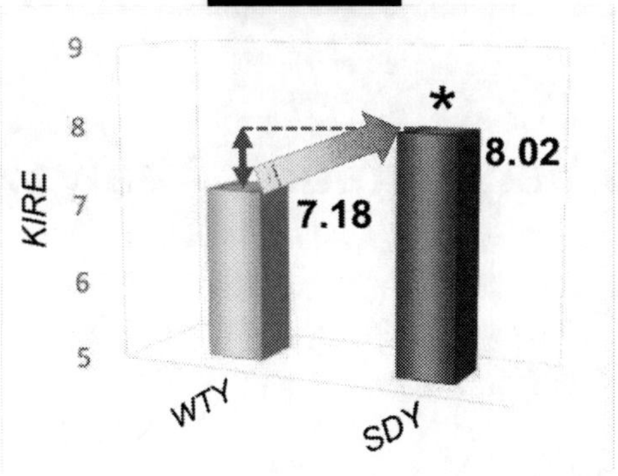

20-L scale

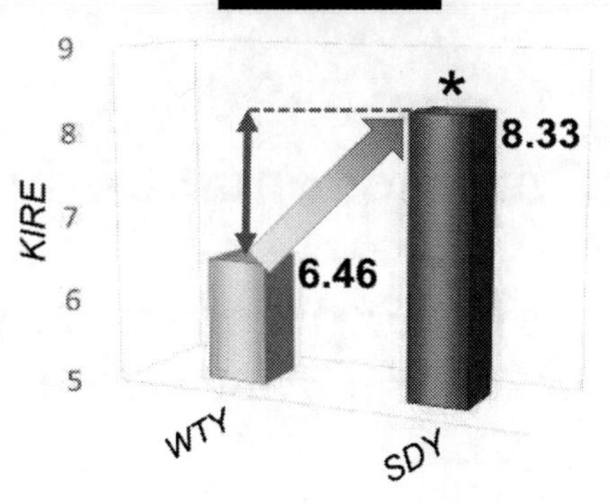

*** Δ >0.2 ,**
Significant difference

Results 2

Genetic Analysis

Genetic Analysis Methods

Next-generation sequencing

- **Chromosomal Copy Number Variation Analysis**
- **Gene Expression Analysis**
  0, 3, 7 days of 5-L scale fermentation

Chromosomal copy number had changed

WTY < SDY

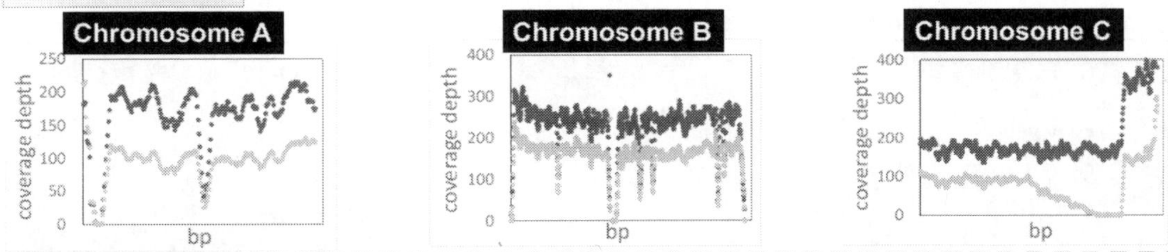

WTY > SDY

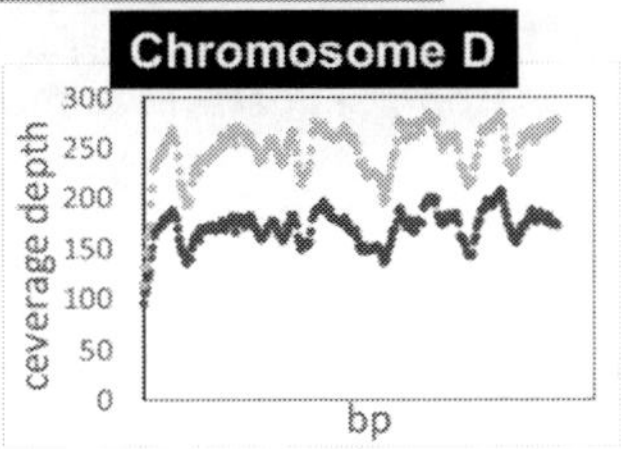

Gene expression had also changed

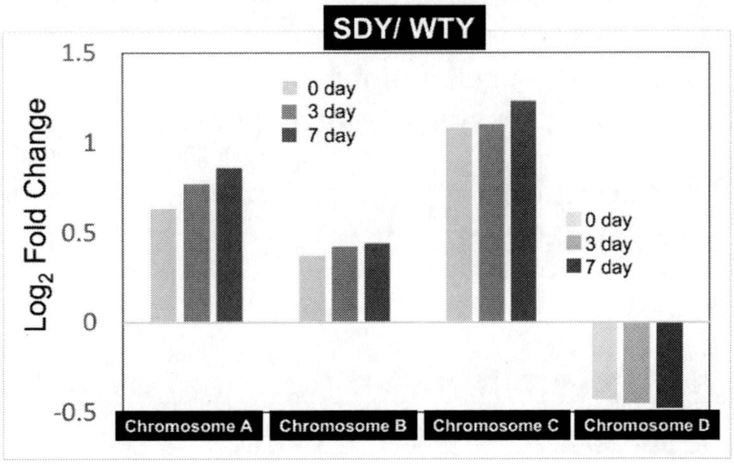

We focused on genes related to glycolysis

☆*YCK1* gene
⇒Glc. sensor
inducing expression of
Glc. transporter genes

☆*HXK1* gene
⇒rate-determining step
of glycolysis

※Other
rDNA gene
⇒protein synthesis proliferation

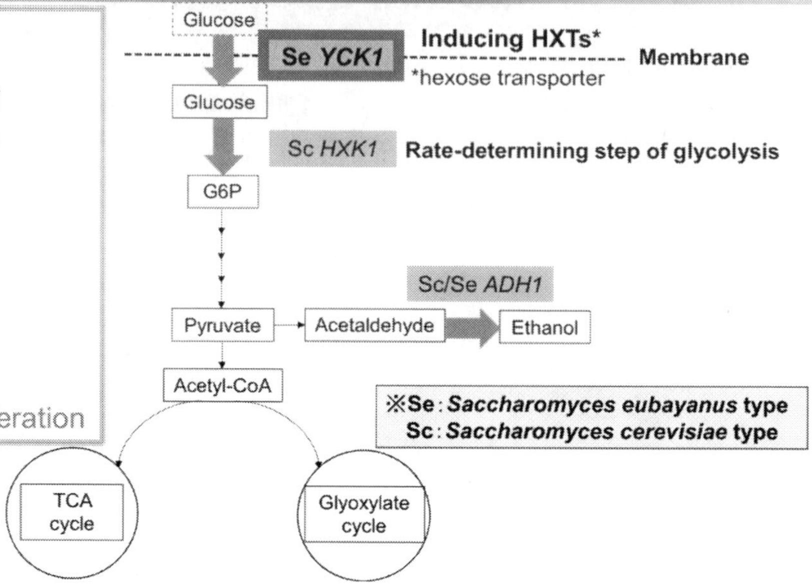

YCK1 gene is important for glycolysis pathway signaling

☆*YCK1 gene*
- Expressed on cell membrane
- Induces expression of glucose transporter genes

Moriya H et. Al., *Proc. Natl. Acad. Sci. USA*, **101**, 1572-1577 (2004)

⇒Glucose uptake ↑↑
⇒Fermentation rate ↑↑

Constructed bottom-fermenting yeast over-expressing Se *YCK1* gene and evaluated fermentation rate

Construction of Se *YCK1* gene-overexpressing strain

strain:
***Saccharomyces pastorianus* Weihenstephan 34/70 uracil auxotrophs**

Murakami N et. al., *Yeast*, **29**, 155-165 (2012)

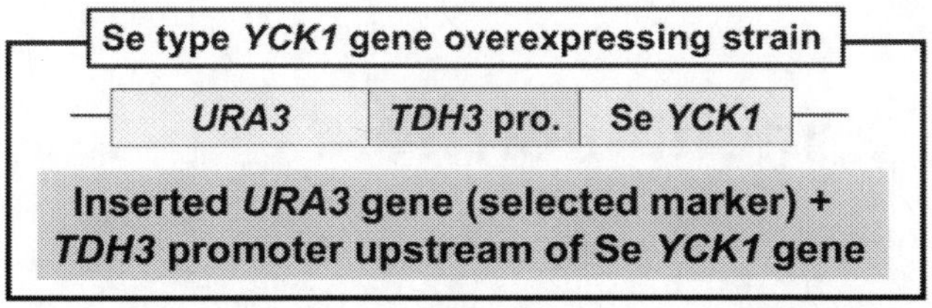

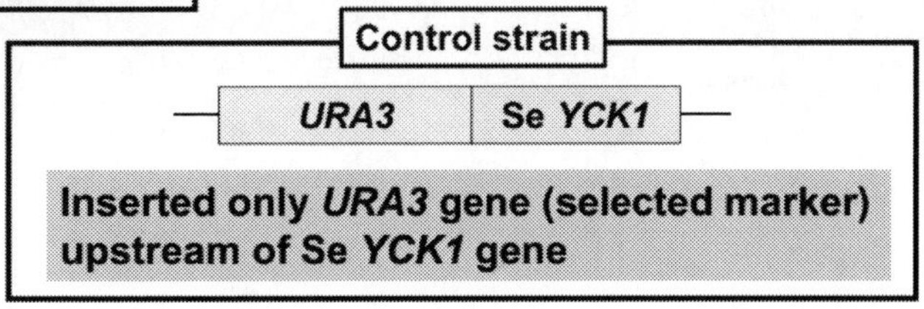

Constructed strain exhibited higher fermentation rate

Apparent extract during fermentation

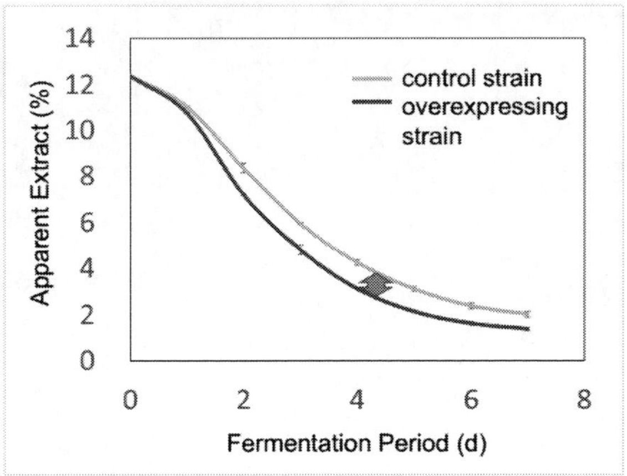

Condition:
5-L scale
Temperature : 13℃
DO. : 10 ppm
Period : 7 days

N=3

Results 3

Pilot-scale Fermentation Test

SDY exhibited higher fermentation rate

Apparent extract during fermentation

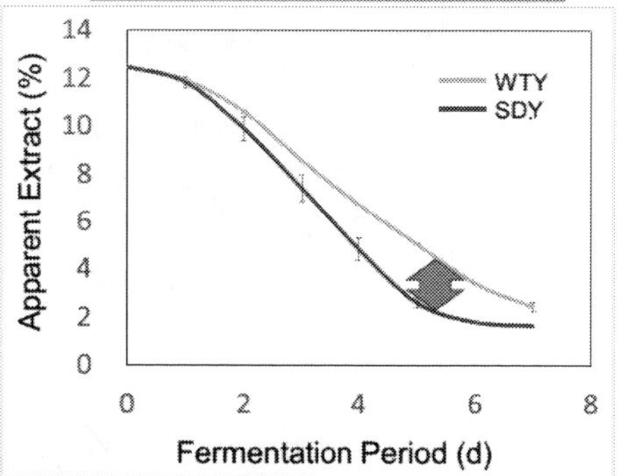

N=3

Condition:
3-kL scale
Temperature : 10.5℃
Period : 7 days

Residual sugar content at fermentation end

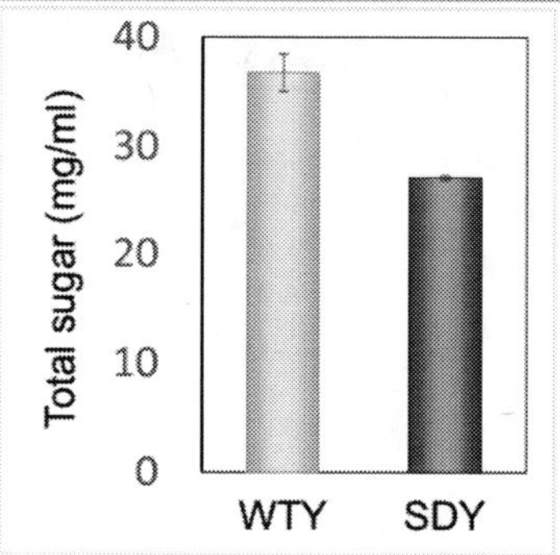

N=3

Increasing copy number and expression of Se *YCK1* gene in SDY

Copy number analysis by qPCR (day 0)

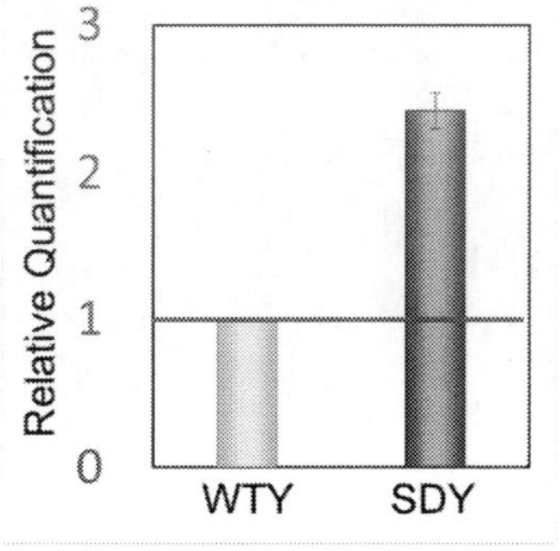

N=3

Gene expression analysis by qRT-PCR (day 0, 1, 3 and 7)

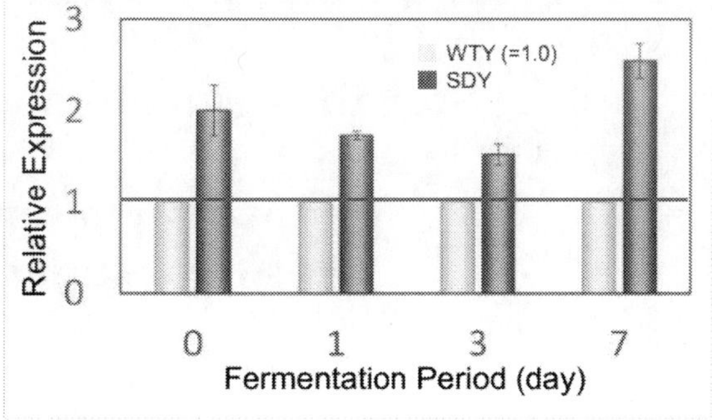

N=3

Reference gene
Sc type *ALG9* gene

Outline

- Introduction
- Research objective
- Results
- **Conclusion**

YCK1 gene ⇔ "*KIRE*" (hypothesis)

YCK1 gene expression ↑

↓

Glucose uptake ↑

↓

Growth→fermentation shift ↑

↓

Fermentation rate ↑
Residual sugar ↓

↓

"*KIRE*" ↑

Conclusion

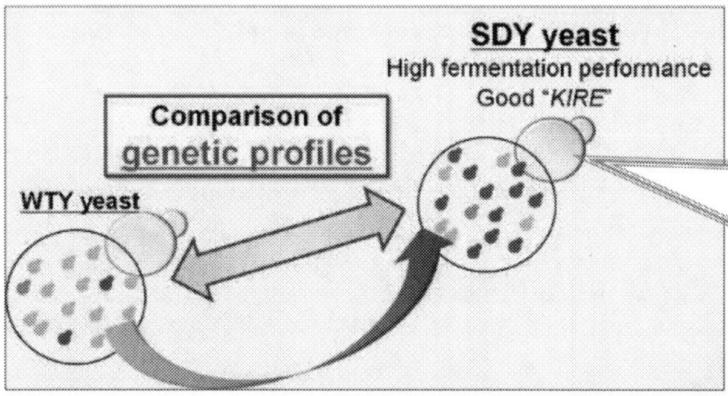

Evidence in support of high fermentation performance of yeast for "SUPER DRY"

Genes related to sugar uptake↑
(copy number & expression)
⇓
Fermentation ability ↑
Residual sugar ↓

Next approach

Examine detailed relationship between differences in fermentation performance of yeast and "*KIRE*"

Asahi's Challenge Continues

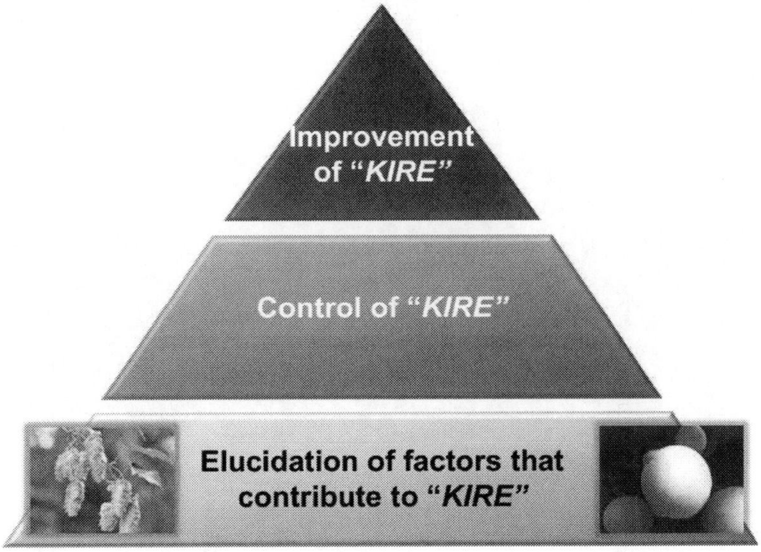

Acknowledgement

Life technologies Japan Ltd.

Thank you for your kind attention !

36th CONGRESS
EBC
2017 EUROPEAN BREWERY CONVENTION
14-18 MAY LJUBLJANA
SLOVENIA

Scaling and Logistics Rethought: Multi-function Mash-Tun, Kettle, & Whirlpool

developed by:

SIMATEC impiantibirra

www.breweryplant.com

SIMATEC STAFF

- Salvatore Sparacio
  - Sales & Marketing Manager

Business background with strong technical expertise in brewing Salvatore set up a **microbrewery** in Turin in 2007 and In 2012 he sold this business. His background is based in manufacturing and planning environment, including IT issues, that makes him able to feel comfortable in every aspect of business. At the moment Salvatore has accepted the challenge to drive SIMATEC impiantibirra from having an Italian based market to a **worldwide operating company**.

ABSTRACT

How big should my brewhouse be?

An oversized brewhouse will output stagnant product, and in turn an inability to experiment with smaller batches. Too small of a brewhouse transforms your success into a burden- equipment wear & labor skyrocket-while this relatively young brewery must embark on the process upgrading the brewery. SIMATEC Impiantibirra has patented two designs that allow for a pragmatic solution to this question. First, a whirlpool paddle that converts a kettle into a brewhouse. This paddle is unique in itself for its simple yet elegant design, and for reducing hot side aeration of the wort by eliminating an auxillary pump. But it also opens up possibilities for scaling. Our second patent is held over the modular brewhouse: because the kettle is multi-functional as a whirlpool and a mash-tun, we can exponentially expand the brewery with the existing tanks. Simatec can add 1, 2, or 3 multi-function tanks to the existing tank and lauter tun, stagger brews very closely together.

OVERVIEW

1) Dimension problems – how to choose?
2) Investments issues
3) Modular Brewhouse
4) Productivity chart

"How big should the brewhouse be?"

The single most important question a brewery owner can ask.

And it carries serious implications:

- An oversized brewhouse yields stagnant product, in turn an inability to pivot, customize, and experiment with smaller batches.
- Too small of a brewhouse makes success a burden: equipment wear & labor skyrocket-while this relatively young brewery must embark on upgrading the brewery: permits, construction, fundraising, scaling, starting from square one (while still operating a brewery).

Investments Benefits of Modular approach

- Brewery management can invest on production equipment only when needed
- No need for oversize production capacity
- Each capacity upgrade expenditure < new bigger equipment expenditure

MULTI-TANK CONCEPT

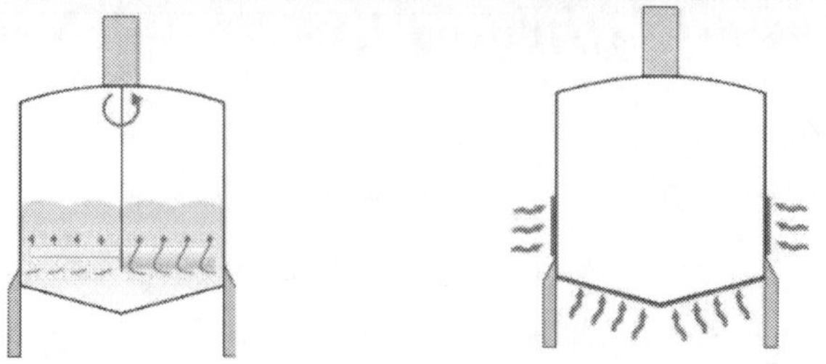

MULTI-TANK = Mash Tun + Kettle + Whirlpool tank

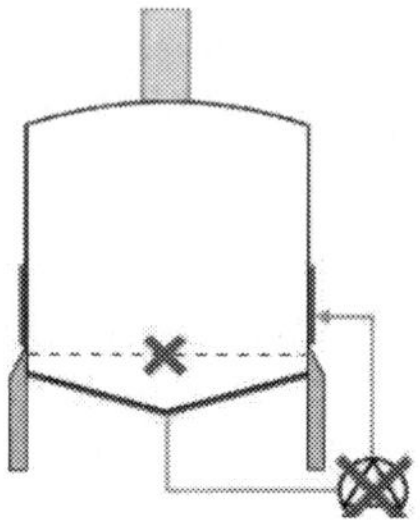

MULTI-TANK CONCEPT

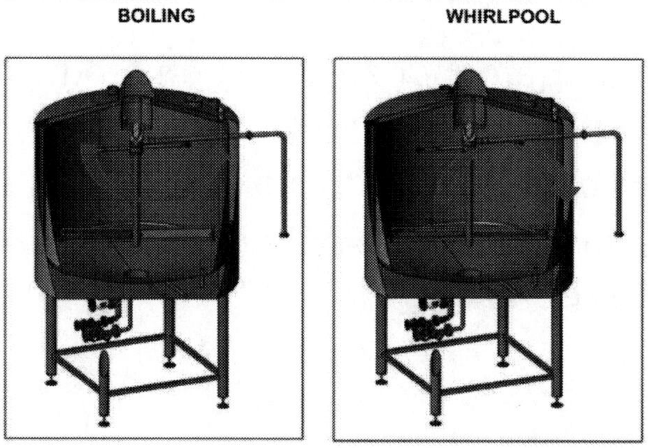

MULTI-TANK CONCEPT

• Paddle morphology – **SIMATEC PATENT**

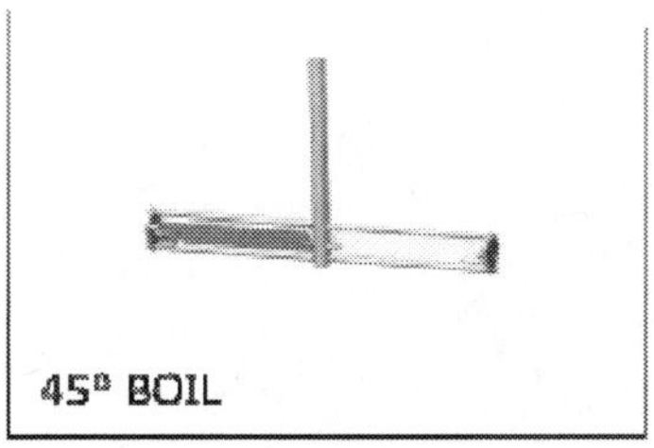

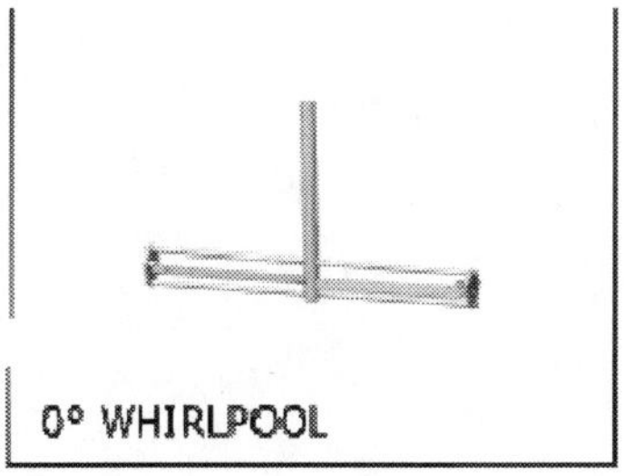

MODULAR SYSTEM

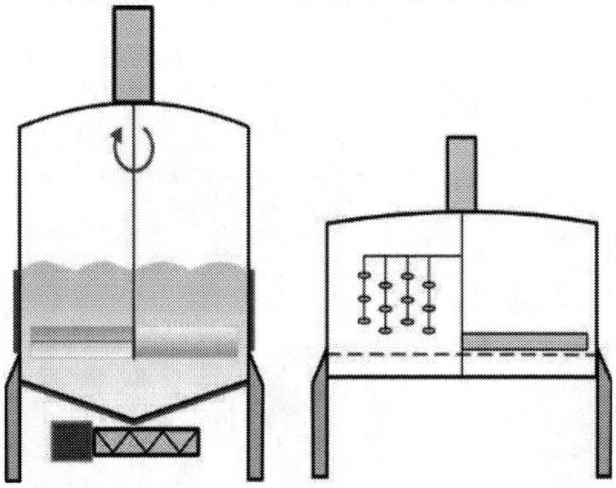

Base modular Brewhouse has 2 components:

Multifunctional Tank + Lauter Tun

PROCESS

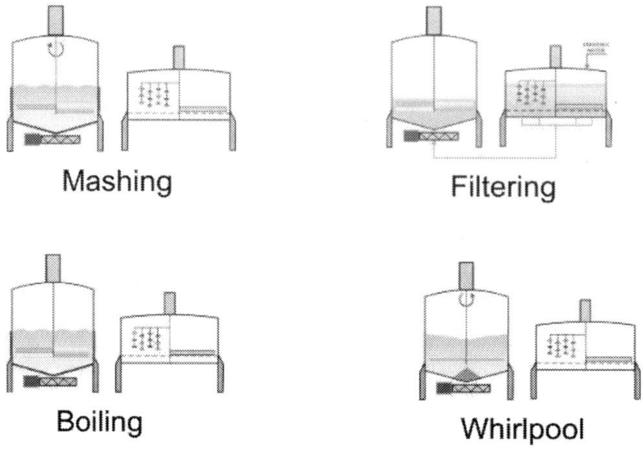

FULL SCALING

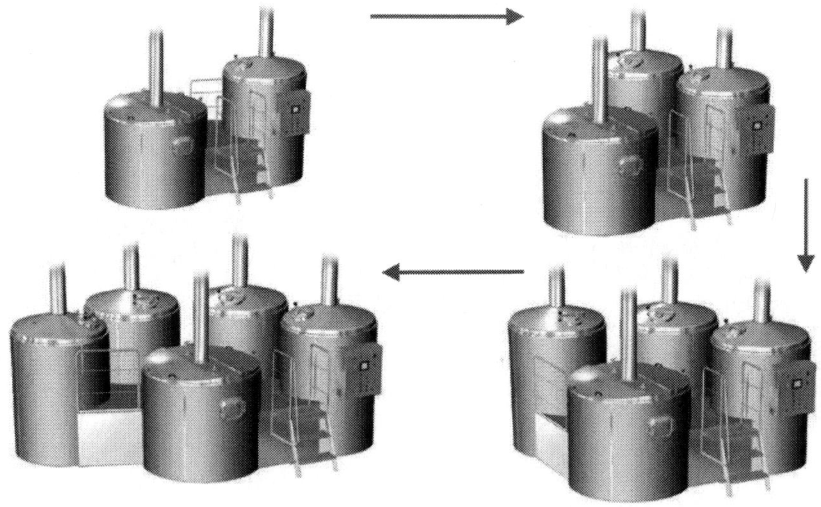

Single scaled Brewhouse

Fully scaled Brewhouse

BABY 5 BBL + 5 BBL

48HL x 4 KETTLES

3 Vessels – 1st Upgrade

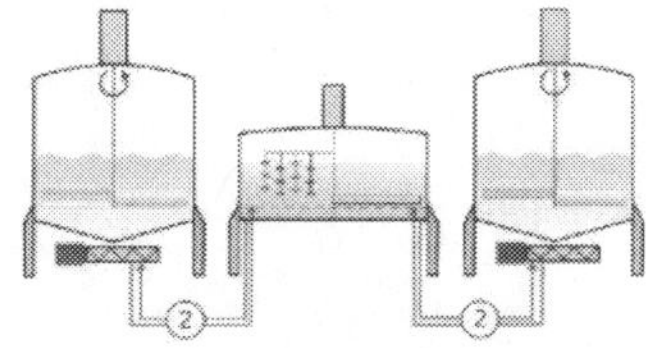

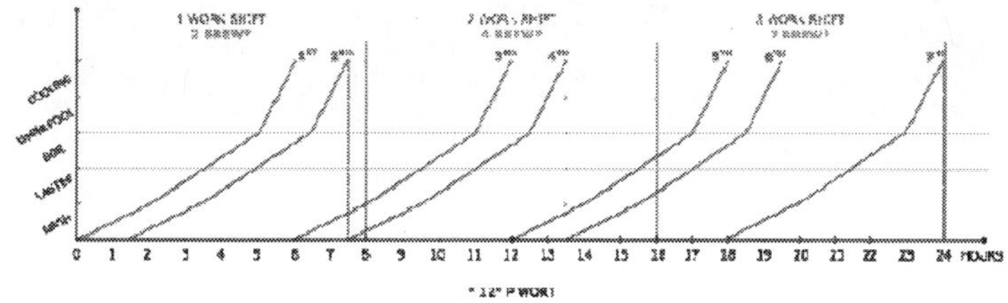

4 Vessels – 2nd Upgrade

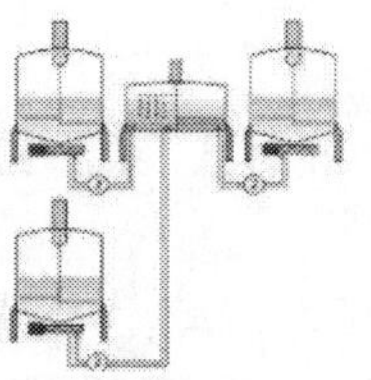

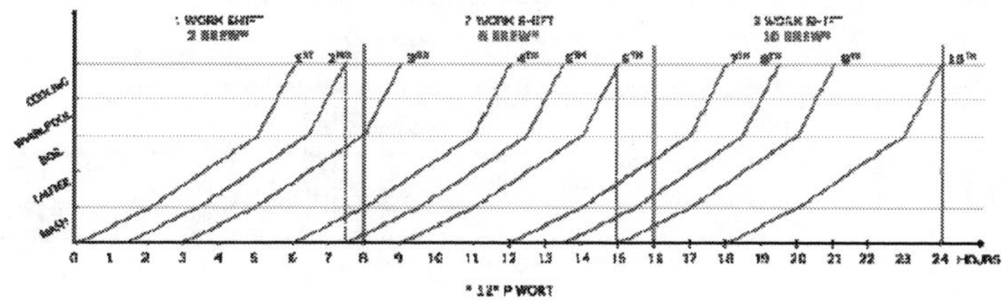

5 Vessels – 3rd Upgrade

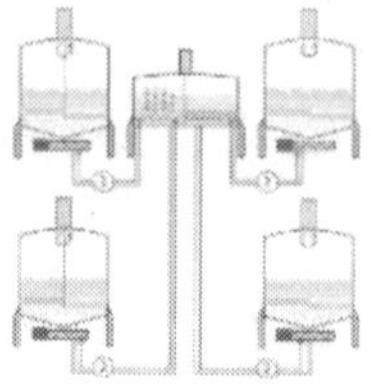

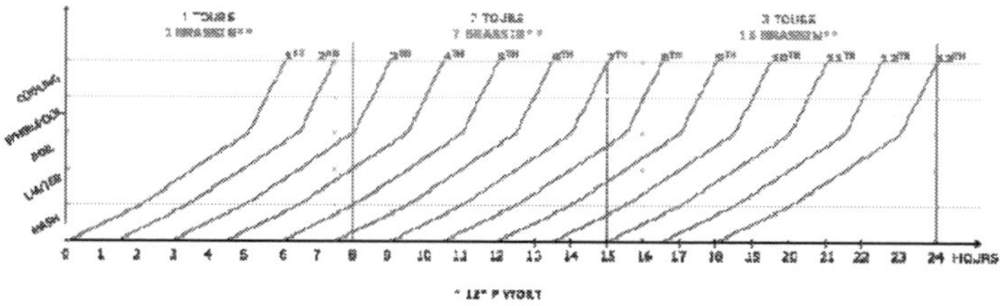

PROCESS TIMELINE

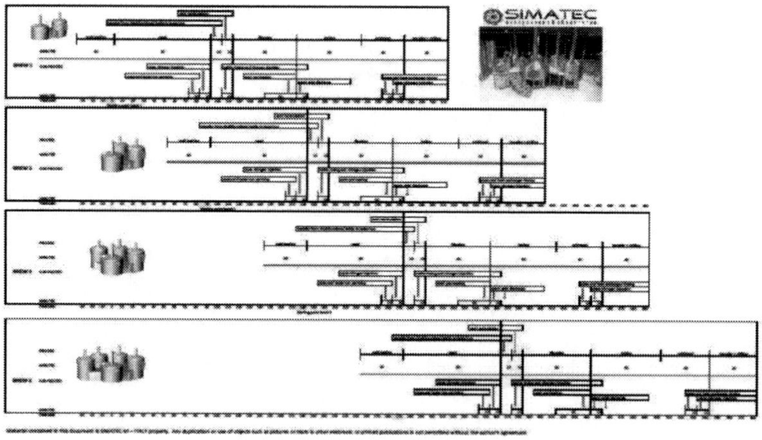

CHEERS & BEERS!

SIMATEC
impiantibirra

Understanding the impact hopping rate has on the aroma quality and intensity of beer dry hopped with Cascade

Scott Lafontaine

Ph. D. Advisor: **Tom Shellhammer**

Oregon State University

36th European Brewery Convention

May 16th, 2017

Oregon State University

Rationale for this study

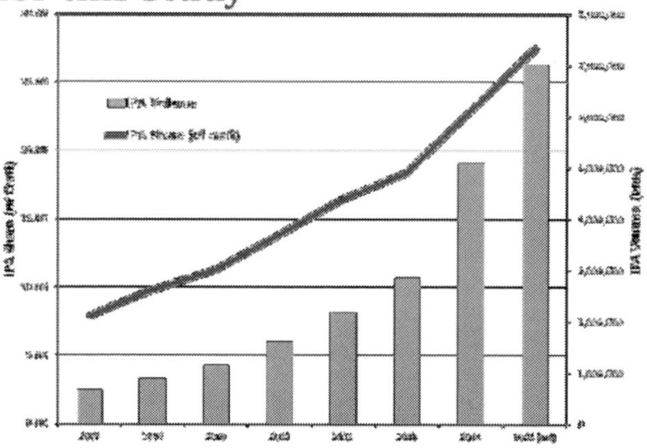

*Source: Brewers Association and IRI

Oregon State University

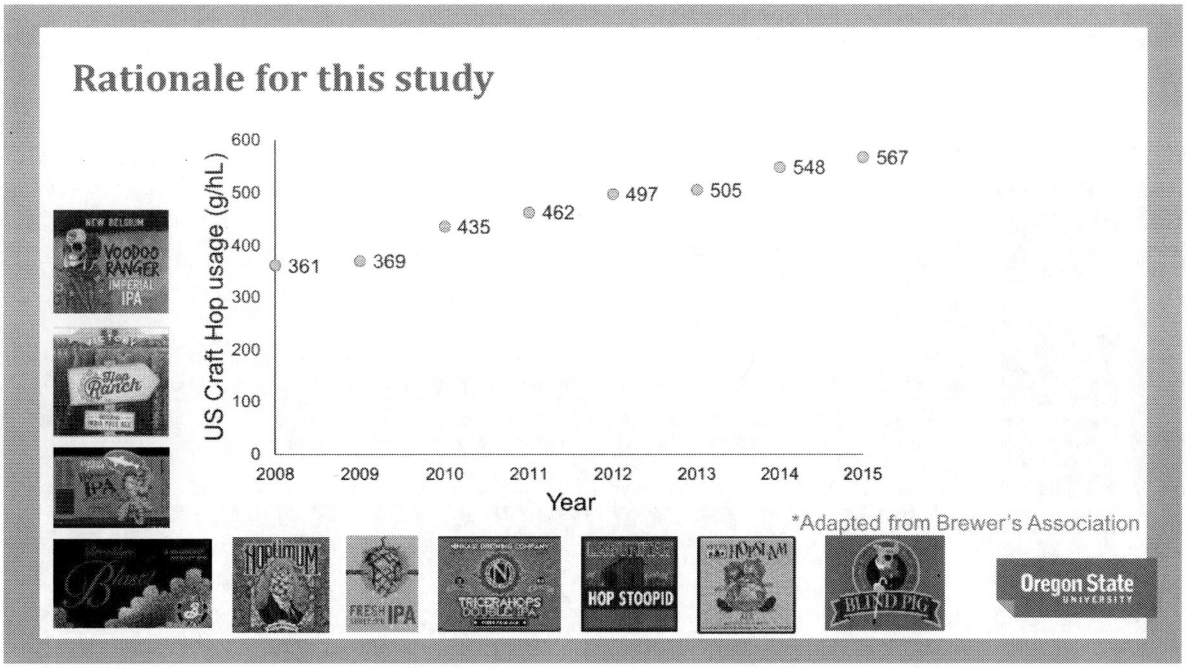
Rationale for this study
US Craft Hop usage (g/hL)
600
500
400
300
200
100
0
361
369
435
462
497
505
548
567
2008
2009
2010
2011
2012
2013
2014
2015
Year
*Adapted from Brewer's Association
NEW BELGIUM
VOODOO RANGER
IMPERIAL IPA
HOP STOOPID
FRESH IPA
TRICERAHOPS DOUBLE IPA
BLIND PIG
Oregon State
UNIVERSITY

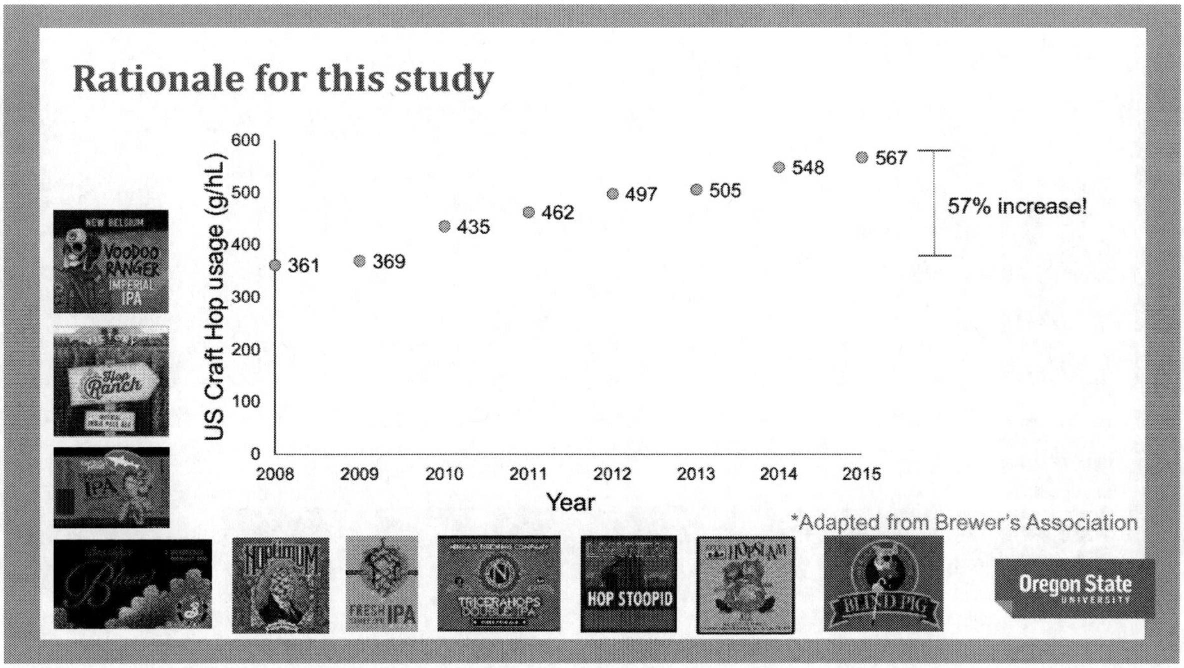
Rationale for this study
US Craft Hop usage (g/hL)
600
500
400
300
200
100
0
361
369
435
462
497
505
548
567
57% increase!
2008
2009
2010
2011
2012
2013
2014
2015
Year
*Adapted from Brewer's Association
NEW BELGIUM
VOODOO RANGER
IMPERIAL IPA
HOP STOOPID
FRESH IPA
TRICERAHOPS DOUBLE IPA
BLIND PIG
Oregon State
UNIVERSITY

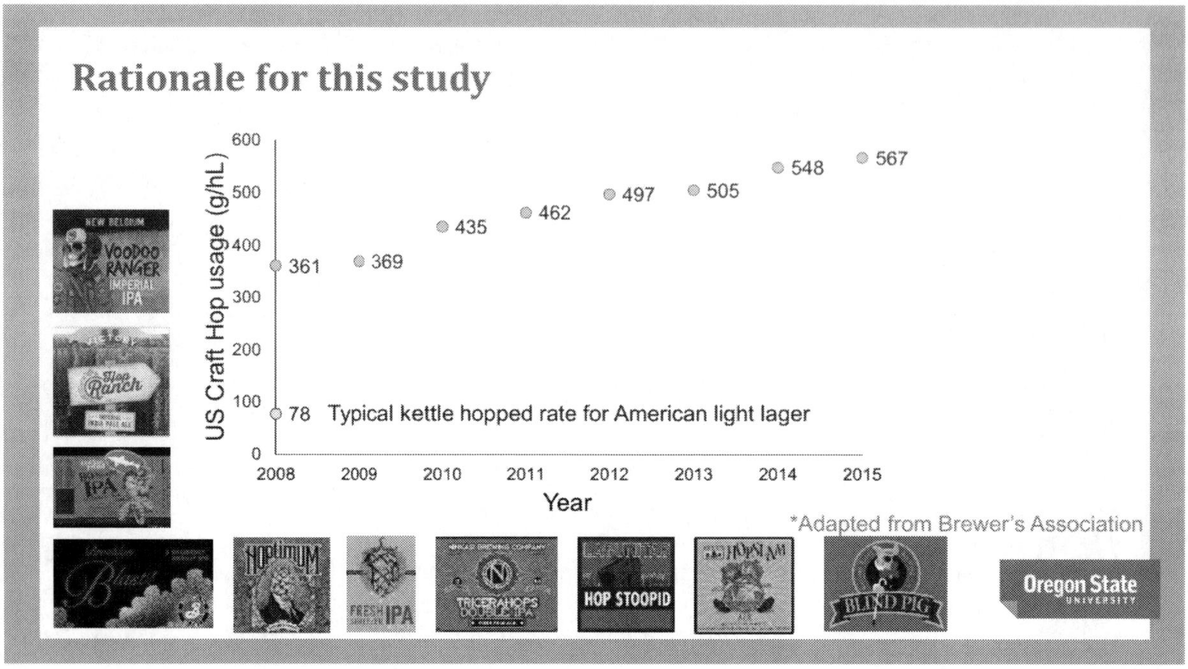
Rationale for this study
US Craft Hop usage (g/hL)
600
500
400
300
200
100
0
361
369
435
462
497
505
548
567
78 Typical kettle hopped rate for American light lager
2008
2009
2010
2011
2012
2013
2014
2015
Year
*Adapted from Brewer's Association
HOP STOOPID
Oregon State University

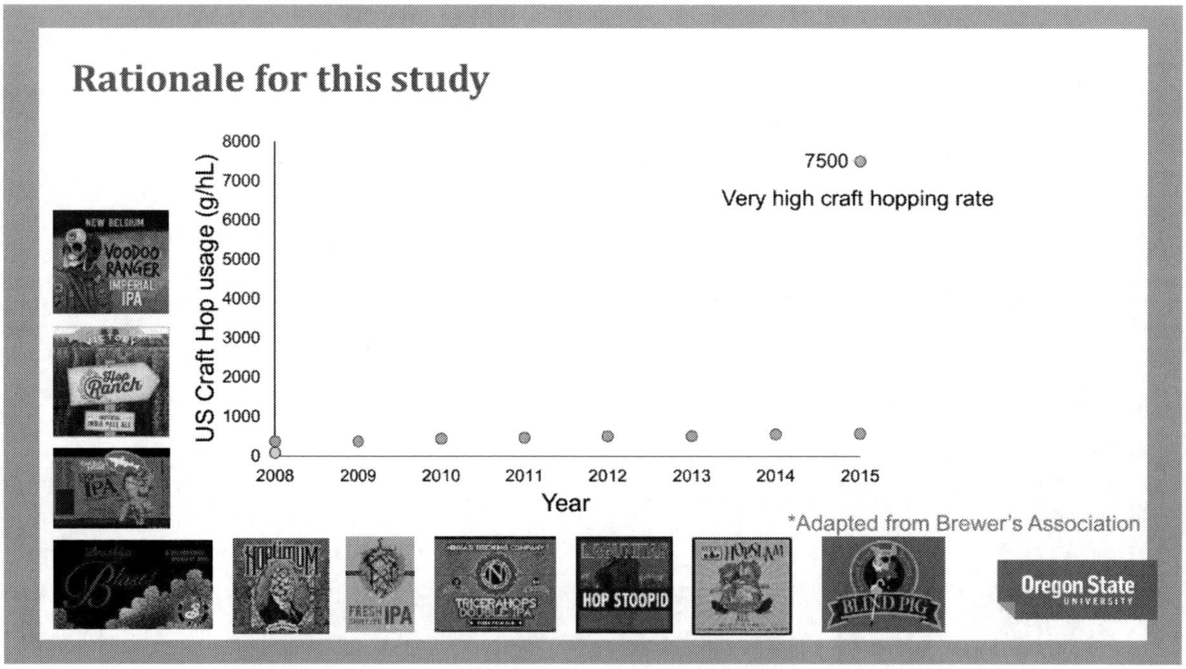
Rationale for this study
US Craft Hop usage (g/hL)
8000
7000
6000
5000
4000
3000
2000
1000
0
7500
Very high craft hopping rate
2008
2009
2010
2011
2012
2013
2014
2015
Year
*Adapted from Brewer's Association
HOP STOOPID
Oregon State University

Main objective

The main goal of this project was to determine a dose response curve for Cascade hop aroma.

Oregon State University

Main objective

The main goal of this project was to determine a dose response curve for Cascade hop aroma.

Does more hop material = more aroma?

Oregon State University

Background

2014 Harvest

	Region	OSU Hop Oil (ml/100g)
CAS_01_15	WA	0.6
CAS_18_15	WA	0.7
CAS_02_15	OR	0.7
CAS_11_15	WA	0.9
CAS_20_15	WA	1.0
CAS_06_15	WA	1.0
CAS_12_15	OR	1.0
CAS_05_15	OR	1.1
CAS_21_15	WA	1.2
CAS_14_15	WA	1.2
CAS_09_15	OR	1.3
CAS_03_15	OR	1.4
CAS_10_15	WA	1.5
CAS_04_15	OR	1.7
CAS_07_15	WA	1.7
CAS_13_15	WA	1.7
CAS_15_15	WA	1.7
CAS_16_15	WA	1.7
CAS_24_15	WA	1.8
CAS_08_15	OR	1.8
CAS_17_15	WA	1.9
CAS_22_15	WA	2.0

2015 Harvest

	Region	OSU Hop Oil (ml/100g)
CAS_12_16	WA	0.5
CAS_27_16	WA	0.6
CAS_21_16	ID	0.6
CAS_22_16	ID	0.6
CAS_24_16	WA	0.6
CAS_07_16	ID	0.7
CAS_09_16	ID	0.8
CAS_19_16	WA	0.8
CAS_04_16	WA	0.8
CAS_25_16	OR	0.8
CAS_26_16	WA	0.9
CAS_06_16	ID	0.9
CAS_05_16	WA	1.0
CAS_11_16	WA	1.0
CAS_16_16	WA	1.1
CAS_17_16	OR	1.1
CAS_15_16	WA	1.2
CAS_03_16	OR	1.2
CAS_23_16	WA	1.2
CAS_20_16	WA	1.3
CAS_28_16	WA	1.4
CAS_29_16	WA	1.4
CAS_02_16	OR	1.4
CAS_08_16	OR	1.5
CAS_13_16	WA	1.5
CAS_10_16	WA	1.5
CAS_01_16	OR	1.7
CAS_18_16	WA	1.7
CAS_14_16	WA	2.6

Thesis Project:

Examining the raw material factors that describe the dry hop aroma performance of hops in beer

Multi-year study:

Dry hopped different lots of the same cultivar

22 Sample lots in 2014

- *Work performed by Dan Vollmer*

29 Sample lots in 2015

Oregon State UNIVERSITY

Background

Cascade from 2015 Harvest

29 Sample lots

19 Farms

13 Unique oil values

Dry hopped at 3.8g/L

	Region	Farm (coded)	OSU Hop Oil (ml/100g)
CAS_12_16	WA	2	0.5
CAS_27_16	WA	1	0.6
CAS_21_16	ID	10	0.6
CAS_22_16	ID	10	0.6
CAS_24_16	WA	9	0.6
CAS_07_16	ID	7	0.7
CAS_09_16	ID	14	0.8
CAS_19_16	WA	20	0.8
CAS_04_16	WA	5	0.8
CAS_25_16	OR	13	0.8
CAS_26_16	WA	12	0.9
CAS_06_16	ID	7	0.9
CAS_05_16	WA	5	1.0
CAS_11_16	WA	2	1.0
CAS_16_16	WA	15	1.1
CAS_17_16	OR	17	1.1
CAS_15_16	WA	16	1.2
CAS_03_16	OR	4	1.2
CAS_23_16	WA	21	1.2
CAS_20_16	WA	19	1.3
CAS_28_16	WA	1	1.4
CAS_29_16	WA	11	1.4
CAS_02_16	OR	4	1.4
CAS_08_16	OR	8	1.5
CAS_13_16	WA	2	1.5
CAS_10_16	WA	2	1.5
CAS_01_16	OR	4	1.7
CAS_18_16	WA	18	1.7
CAS_14_16	WA	2	2.6

Oregon State UNIVERSITY

Background

Cascade from 2015 Harvest

29 *Sample lots*
19 Farms
13 Unique oil values
Dry hopped at 3.8g/L

	Region	Farm (coded)	OSU Hop Oil (ml/100g)
CAS_12_16	WA	2	0.5
CAS_27_16	WA	1	0.6
CAS_21_16	ID	10	0.6
CAS_22_16	ID	10	0.6
CAS_24_16	WA	9	0.6
CAS_07_16	ID	7	0.7
CAS_09_16	ID	14	0.8
CAS_19_16	WA	20	0.8
CAS_04_16	WA	5	0.8
CAS_25_16	OR	13	0.8
CAS_26_16	WA	12	0.9
CAS_06_16	ID	7	0.9
CAS_05_16	WA	5	1.0
CAS_11_16	WA	2	1.0
CAS_16_16	WA	15	1.1
CAS_17_16	OR	17	1.1
CAS_15_16	WA	16	1.2
CAS_03_16	OR	4	1.2
CAS_23_16	WA	21	1.2
CAS_20_16	WA	19	1.3
CAS_28_16	WA	1	1.4
CAS_29_16	WA	11	1.4
CAS_02_16	OR	4	1.4
CAS_08_16	OR	8	1.5
CAS_13_16	WA	2	1.5
CAS_10_16	WA	2	1.5
CAS_01_16	OR	4	1.7
CAS_18_16	WA	18	1.7
CAS_14_16	WA	2	2.6

Selected one Cascade lot and dry-hopped at 4 distinct levels

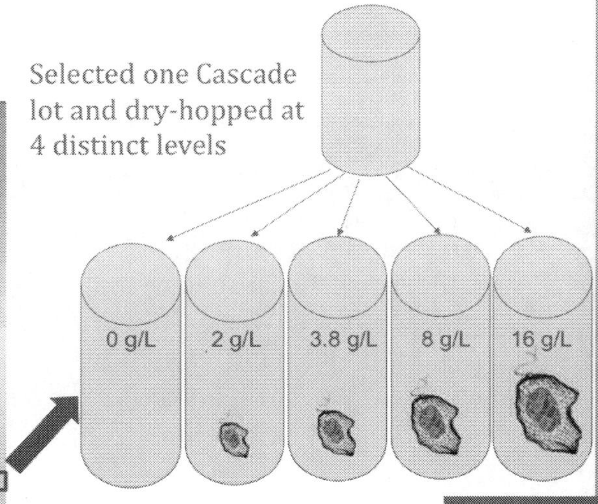

Oregon State University

Brewing "unhopped" beer

Beer Specifications:

- Grist:
  - 85% Pale 2-row
  - 13.5% Carmel 10L
  - 0.5% Carmel 120L
- **Original Gravity:** 10.6 P
- **Real Extract:** 3.16 P
- **BU** = 20 mg/L (iso-extract)
- **ABV** = 4.8 % ABV

BRIDGEPORT

Oregon State University

Hop Preparation and Dry-Hopping Parameters

Hop Preparation For Dry Hopping

- Blended whole cone hops of Cascade, Mixed and homogenized by grinding
  - Weighted into different ratios

  **sample also taken for oil analysis*

Dry Hopping Parameters

- Added to **40 L** finished beer at **0, 2, 3.8, 8 & 16 g/L** for **24 hrs**
  - CO_2 blanket during addition
- 18°C during dry-hopping

Oregon State University

Blending & filtration

- All dry-hop events occur in duplicate (40 L beer each)
- During filtration 2 kegs are blended during filtration into 1 keg
  - Oxygen monitoring

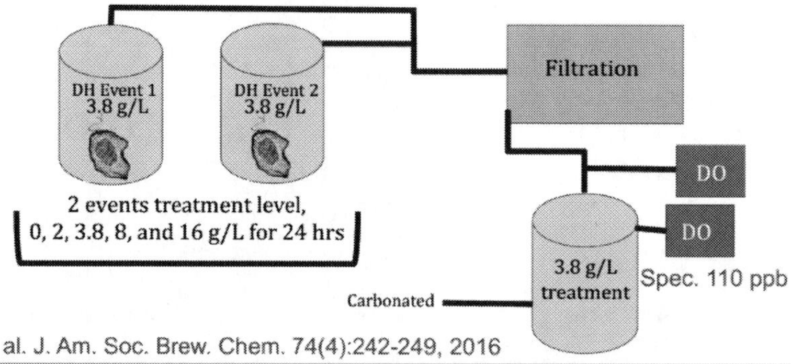

Vollmer, D. et. al. J. Am. Soc. Brew. Chem. 74(4):242-249, 2016

Oregon State University

Evaluations using draft beer

- Minimized total package oxygen
- Great for sensory testing implementation

Oregon State University

Sensory evaluation - descriptive analysis

Sensory protocol

- 12 panelists
  - (10 males, 2 female; 27-54 yrs. old)
- 4 training sessions in advance
- 33 products/treatments
- 4 replications
- Blocked by replication

Sensory Attributes Assessed

Overall Hop Aroma Intensity
Citrus
Herbal/Tea

Oregon State University

Sensory evaluation - descriptive analysis

Sensory protocol

- 12 panelists
  - (10 males, 2 female; 27-54 yrs. old)
- 4 training sessions in advance
- 33 products/treatments
- 4 replications
- Blocked by replication

Sensory Attributes Assessed

Overall Hop Aroma Intensity

Citrus

Herbal/Tea

Oregon State University

Sensory evaluation – descriptive analysis external controls

	Attributes	Base (No dry hop)	3.8 g/L	16 g/L	Ballast Point Grapefruit Sculpin	Citrus Mistress
Assess Descriptors Based on Aroma Only	**Overall Hop Aroma Intensity**	0	8-9	14-15	14-15	7-8
	Citrus	0	7-8	5-6	13-14	6-7
	Herbal/Tea	0	5-6	12-13	1-2	6-7

- **Panelists came to consensus for attributes on commercial and internally made samples**
- **References were served to panelists at each DA session**

Oregon State University

Sensory Results

Oregon State University

Panelist Performance Metrics with OHAI (on samples)

Source	DF	Sum of squares	Mean squares	F	Pr > F
Products	32	2717	84.9	10.3	< 0.0001
Assessors	9	1751	194.6	21.2	< 0.0001
Sessions	3	5	1.9	0.3	0.810
Products*Assessors	288	2371	8.2	1.6	< 0.0001
Products*Sessions	96	464	4.8	1.0	0.511
Assessors*Sessions	27	157	5.8	1.2	0.230
Error	864	4222	4.9		

- **10 Panelists could discriminate using OHAI attribute**
- **Panelist * Sample Interaction (p <0.0001):**
  - For the response OHAI, the rating given to a product depended on the assessor providing the rating
  - Issues with consensus among the panelists (or groups of panelists)

Oregon State University

Hop dose response – overall hop aroma intensity

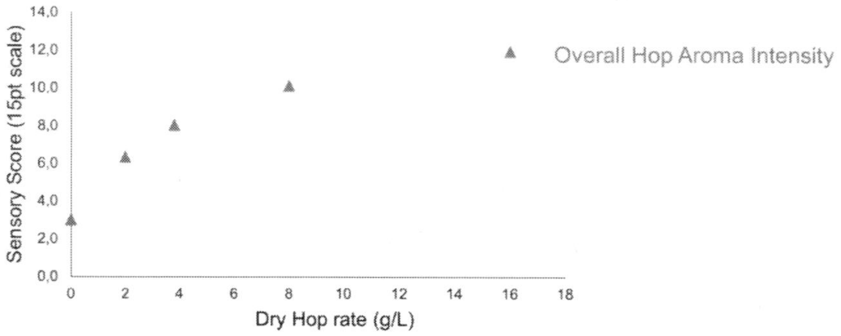

Oregon State University

Hop dose response – overall hop aroma intensity

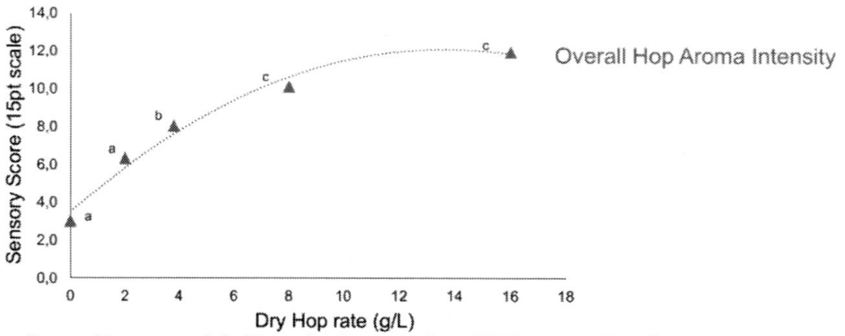

- **Panelists could discriminate the different dry hop rate samples (i.e. 0, 2, 3.8, 8, & 16 g/L)**

*Letters represent Tukey's HSD groupings

Oregon State University

Hop dose response – overall hop aroma intensity

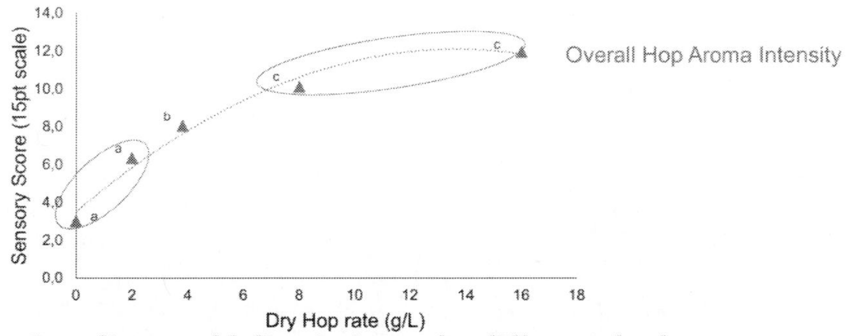

- **Panelists could discriminate the different dry hop rate samples (ie 0, 2, 3.8, 8, & 16)**

*Letters represent Tukey's HSD groupings

Oregon State University

Hop dose response – hoppy quality (citrus and herbal/tea)

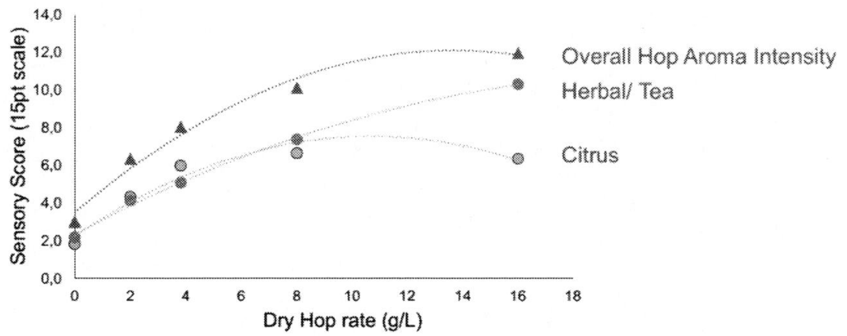

Oregon State University

Hop dose response – hoppy quality (citrus and herbal/tea)

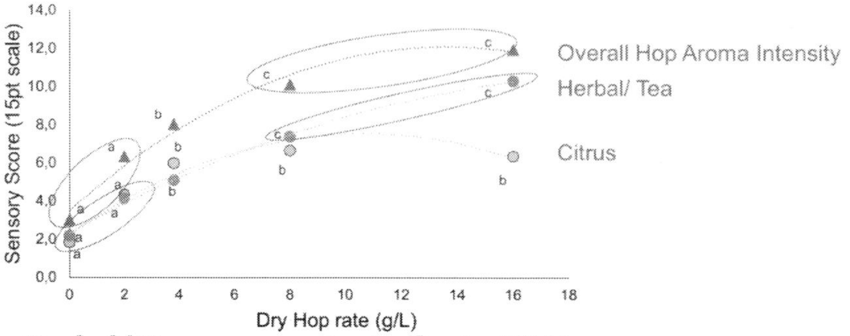

- **Herbal/ Tea response is similar to OHAI**

Oregon State UNIVERSITY

*Letters represent Tukey's HSD groupings

Hop dose response – hoppy quality (citrus and herbal/tea)

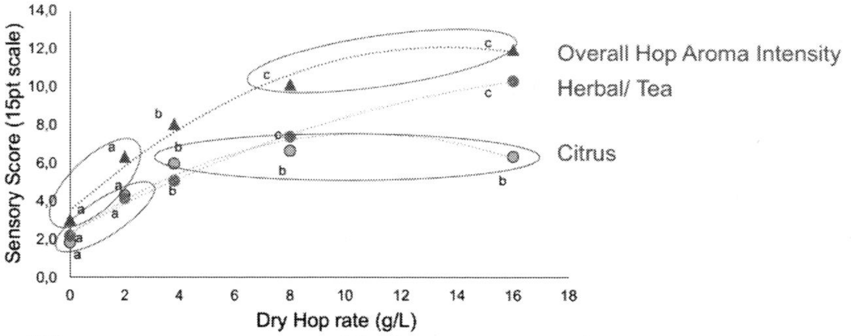

- **Citrus response seemed to be suppressed compared to OHAI and Herbal/Tea**

Oregon State UNIVERSITY

*Letters represent Tukey's HSD groupings

Aroma intensity/ quality as power functions (log-log plots)

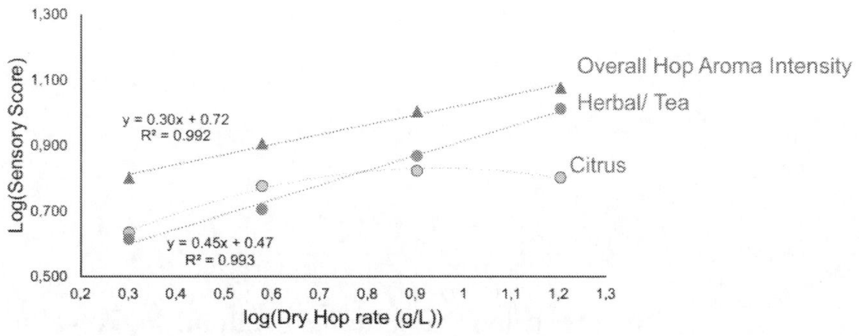

- **Overall hop aroma intensity and herbal/tea appear to be power functions**

Kamadia, V.V., et. al. Relationships between Odorant Concentration and Aroma Intensity. Journal of Food Science, 71: S193–S197

Oregon State University

Aroma intensity/ quality as power functions (log-log plots)

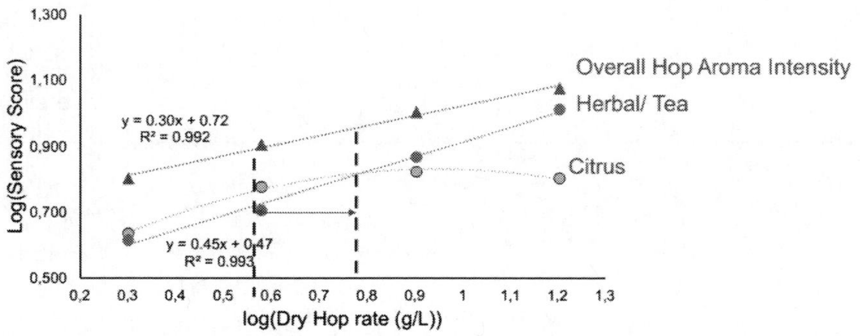

- **If citrus is important there is an optimal rate ~ 4-6 g/L**

Oregon State University

Hop dose response – overall hop aroma intensity

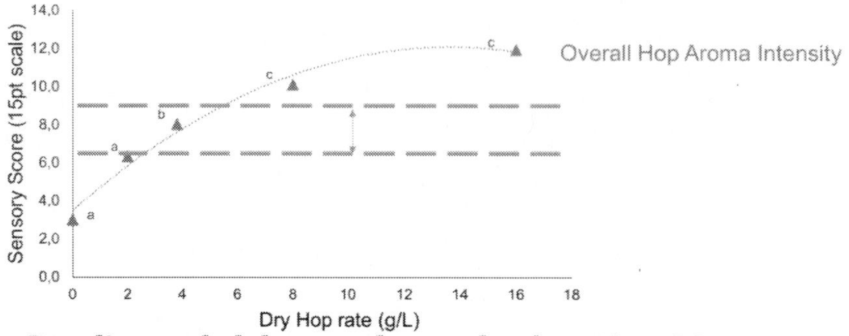

- **Panelists scaled the samples randomly against 28 other samples in the DA panel**
- **Range of DA OHAI ratings (i.e. ~6- 9.5) for 2015 and 2016 Cascade DA Panels**

*Letters represent Tukey's HSD groupings

Oregon State UNIVERSITY

Principal Component Analysis- 2015 Cascade Harvest Samples

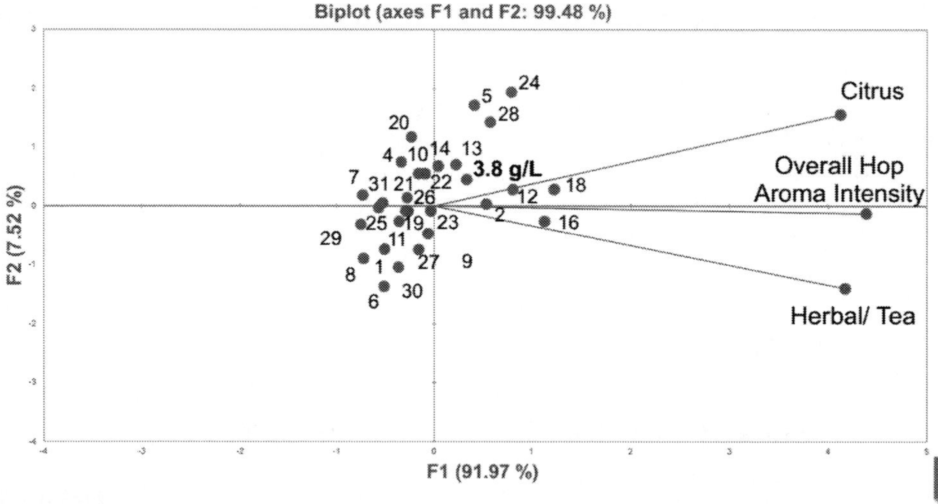

PCA Type: Pearson (n-1)

Oregon State UNIVERSITY

Principal Component Analysis- Hop dosage treatments

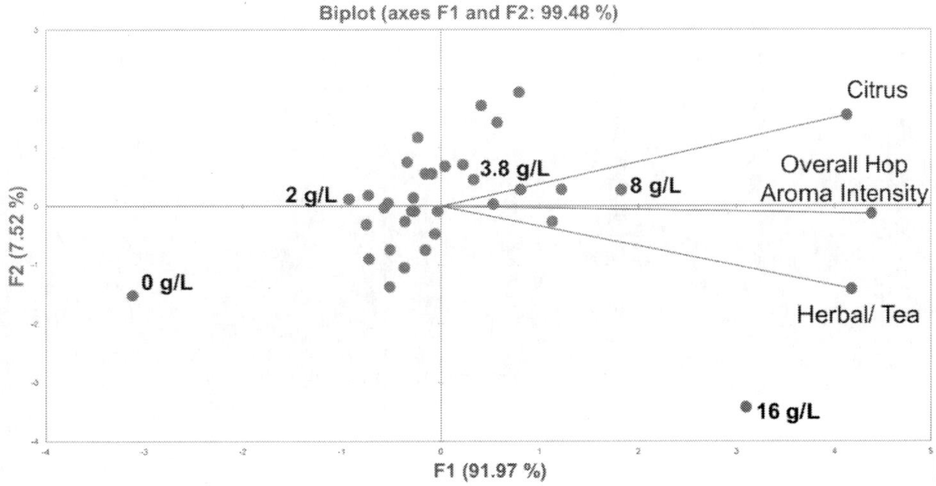

PCA Type: Pearson (n-1)

Chemistry Results

Solid phase micro extraction GC/MS – Hop volatiles

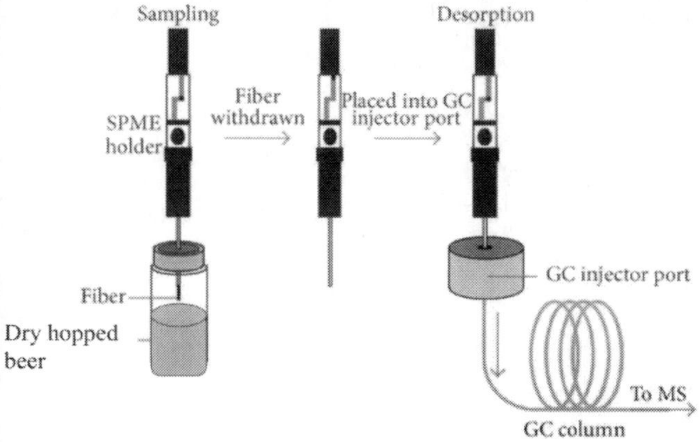

Internal Standard
4-octanol
Target Hop Analytes
Linalool
Terpinen-4-ol
α-terpineol
Nerol
Phenyl Acetate
Geraniol
Geranial-citral
Methyl Geranate
Geraniol Acetate
β-Caryophellyene
α-Humulene
β-Farnesene
Gernyal Isobutyrate

Oregon State University

Hop Compound Concentrations – Terpene Alcohols

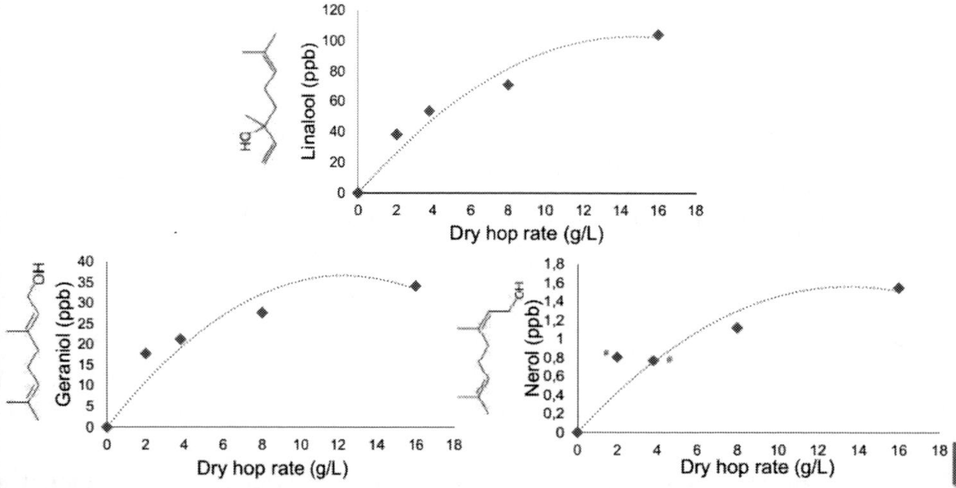

*Average of 4 SPME-GC/MS instrumental runs

#Estimated values lower than LOQ

Oregon State University

Hop compound concentrations – terpene alcohols

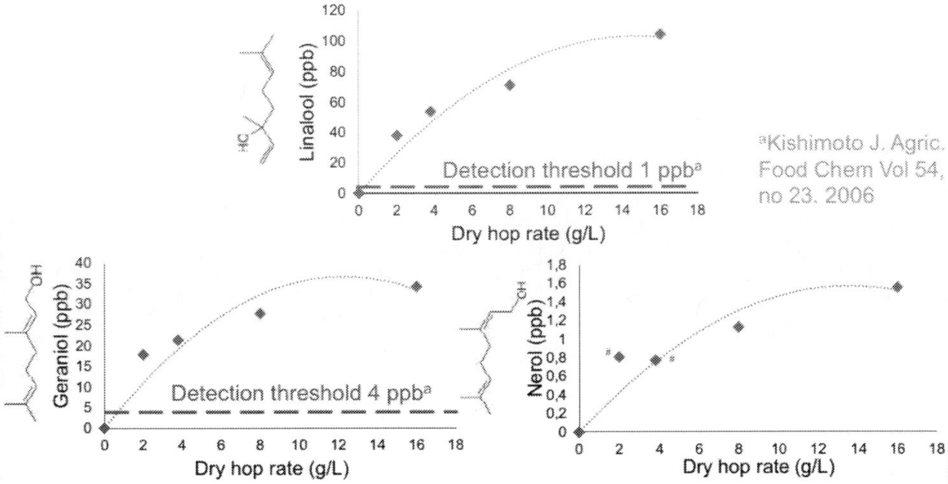

[a]Kishimoto J. Agric. Food Chem Vol 54, no 23. 2006

*Average of 4 SPME-GC/MS instrumental runs

#Estimated values lower than LOQ

Oregon State UNIVERSITY

Hop oil content and compositional analysis

- **ASBC Hops 13 – Steam Distillation**
  - Total oil content
- **ASBC Hops 17 – Hop Oil Compositional Analysis**
  - (GC-FID) for 2015 samples
  - (GC-MS) for 2016 samples
  - 42 hop analytes

Oregon State UNIVERSITY

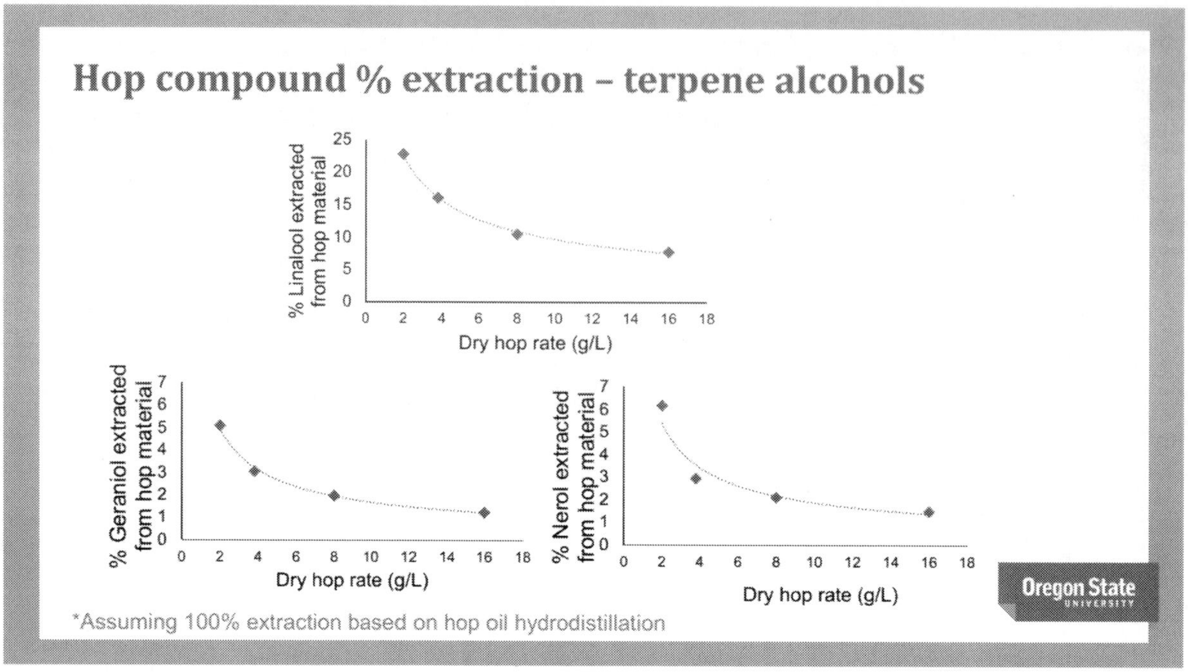
Hop compound % extraction – terpene alcohols
% Linalool extracted from hop material
Dry hop rate (g/L)
% Geraniol extracted from hop material
Dry hop rate (g/L)
% Nerol extracted from hop material
Dry hop rate (g/L)
*Assuming 100% extraction based on hop oil hydrodistillation
Oregon State University

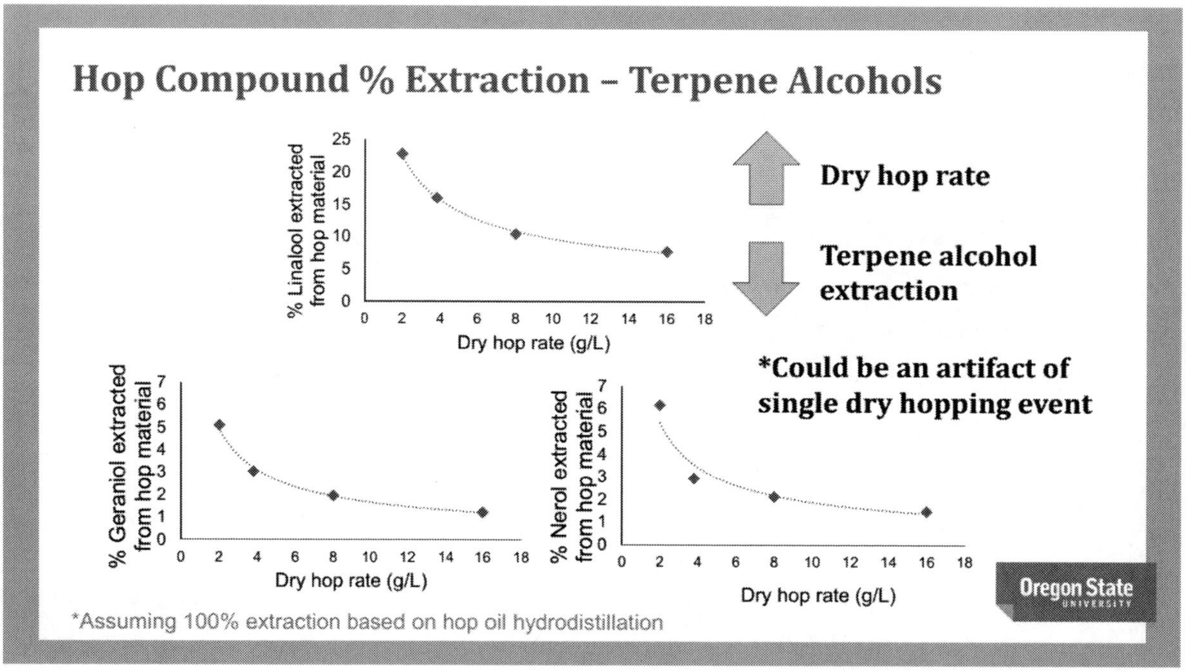
Hop Compound % Extraction – Terpene Alcohols
% Linalool extracted from hop material
Dry hop rate (g/L)
Dry hop rate
Terpene alcohol extraction
*Could be an artifact of single dry hopping event
% Geraniol extracted from hop material
Dry hop rate (g/L)
% Nerol extracted from hop material
Dry hop rate (g/L)
*Assuming 100% extraction based on hop oil hydrodistillation
Oregon State University

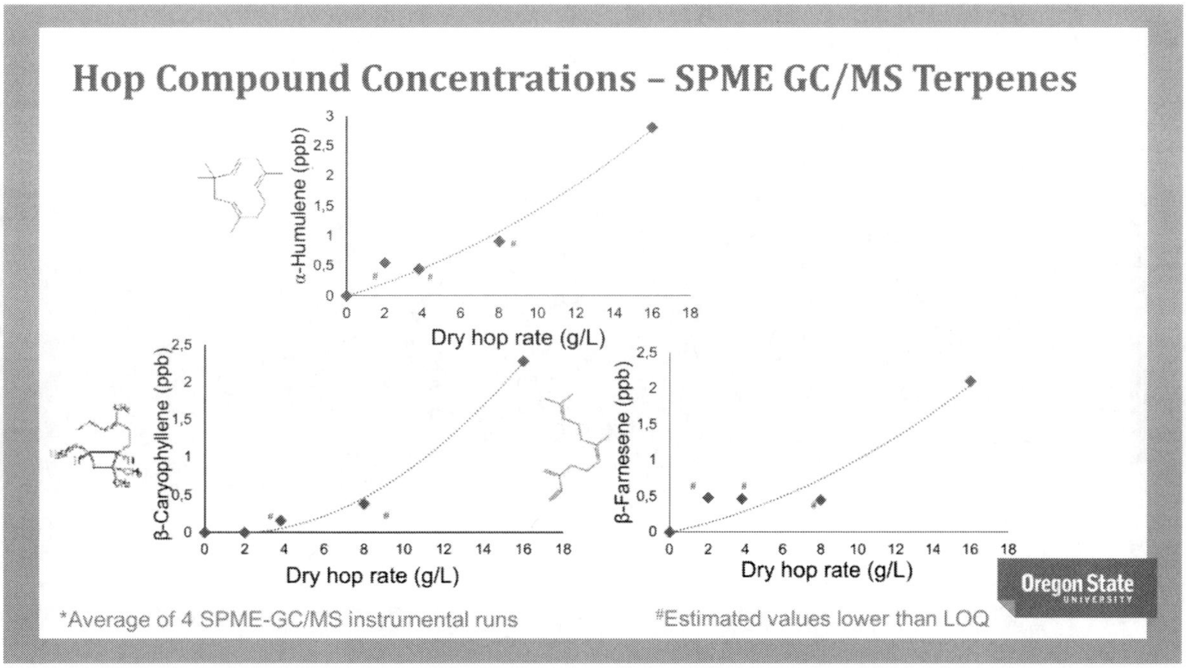
Hop Compound Concentrations – SPME GC/MS Terpenes
α-Humulene (ppb)
Dry hop rate (g/L)
β-Caryophyllene (ppb)
Dry hop rate (g/L)
β-Farnesene (ppb)
Dry hop rate (g/L)
Oregon State University
*Average of 4 SPME-GC/MS instrumental runs
#Estimated values lower than LOQ

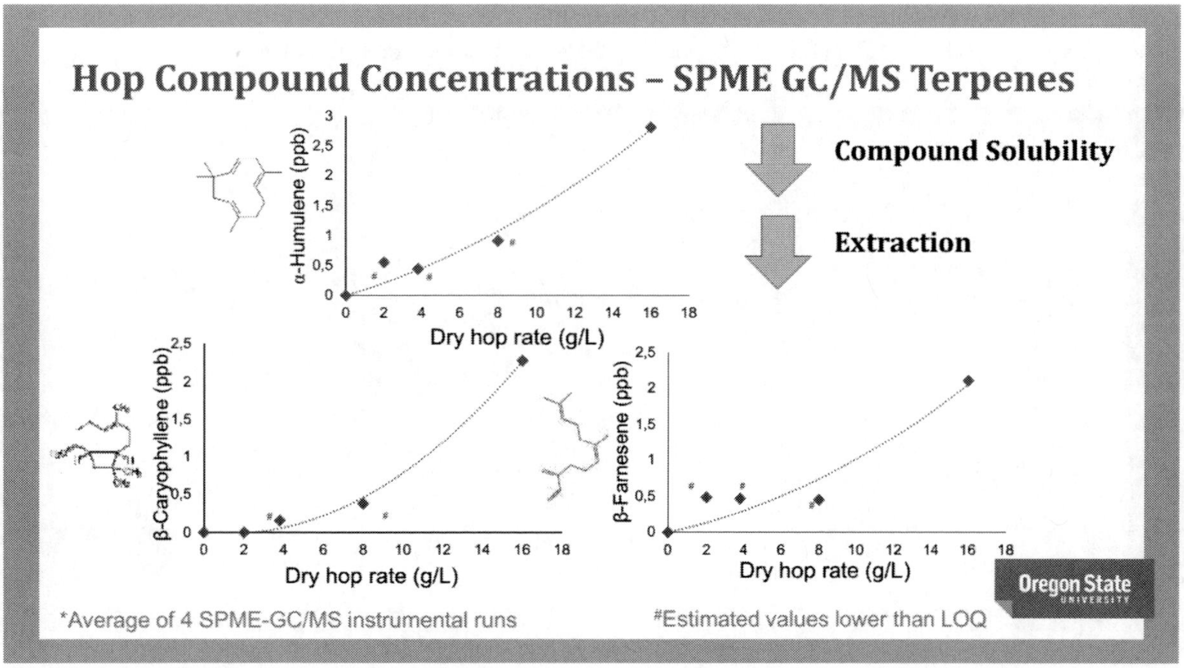
Hop Compound Concentrations – SPME GC/MS Terpenes
Compound Solubility
Extraction
α-Humulene (ppb)
Dry hop rate (g/L)
β-Caryophyllene (ppb)
Dry hop rate (g/L)
β-Farnesene (ppb)
Dry hop rate (g/L)
Oregon State University
*Average of 4 SPME-GC/MS instrumental runs
#Estimated values lower than LOQ

Principal component analysis – hop chemistry

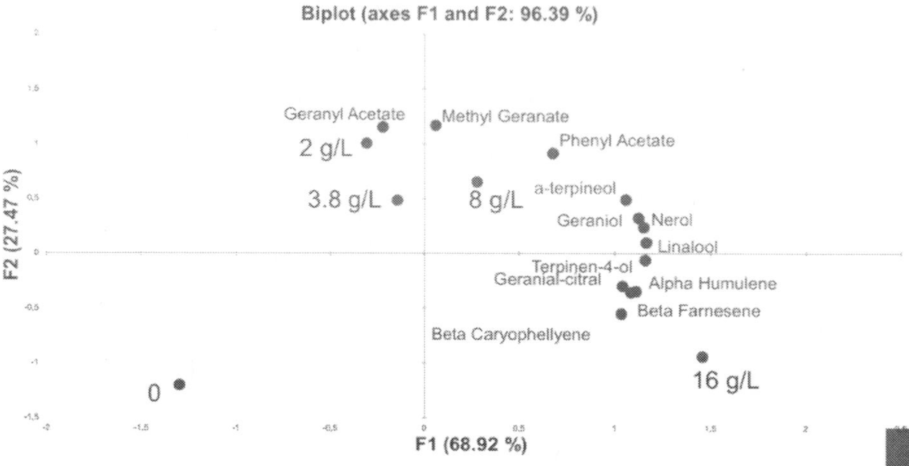

PCA Type: Pearson Correlation (n-1)

Oregon State University

Principal Component Analysis – Just Chemistry

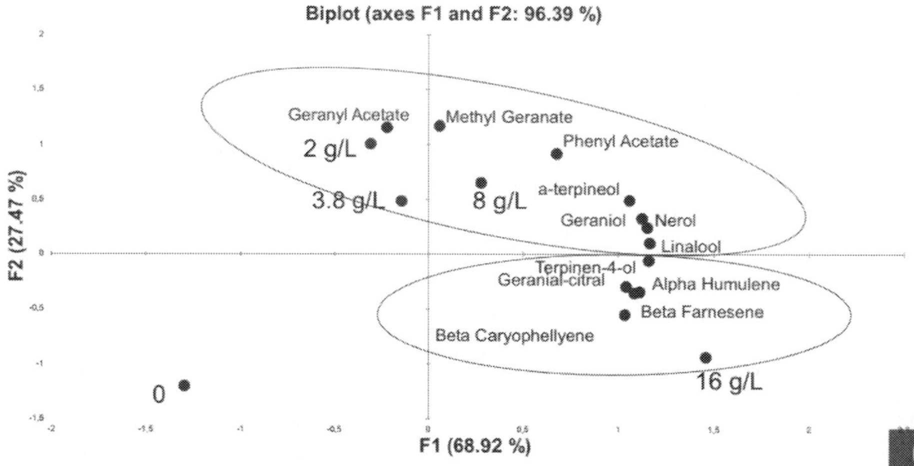

PCA Type: Pearson Correlation (n-1)

Oregon State University

Sensory + Chemistry Results

Oregon State UNIVERSITY

Principal Component Analysis – Chemistry + Sensory

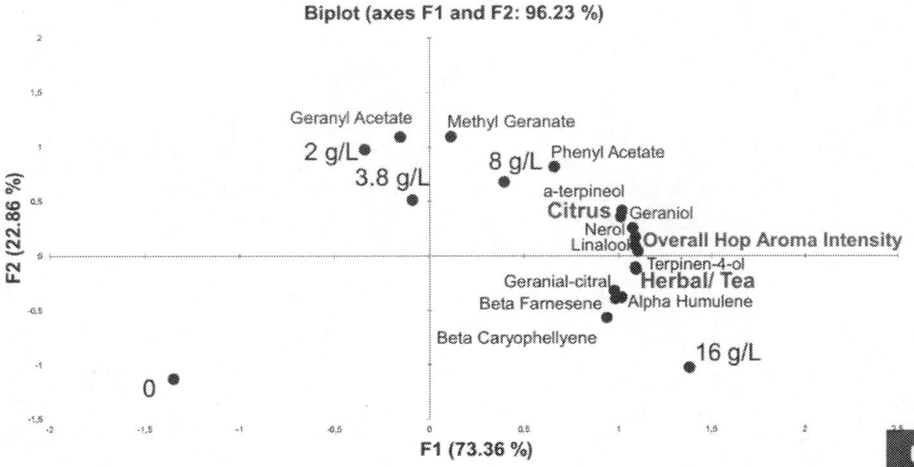

PCA Type: Pearson Correlation (n-1)

Oregon State UNIVERSITY

Principal Component Analysis – Chemistry + Sensory

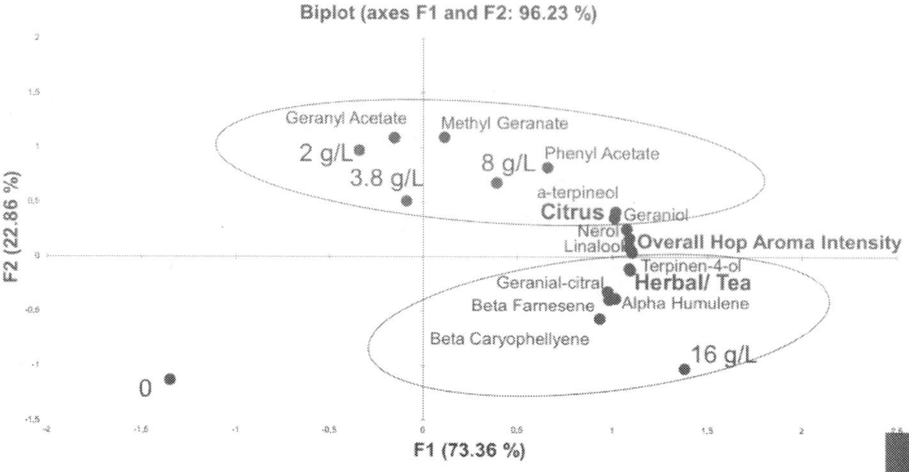

PCA Type: Pearson Correlation (n-1)

Oregon State University

Dry hopping having other impacts on beer

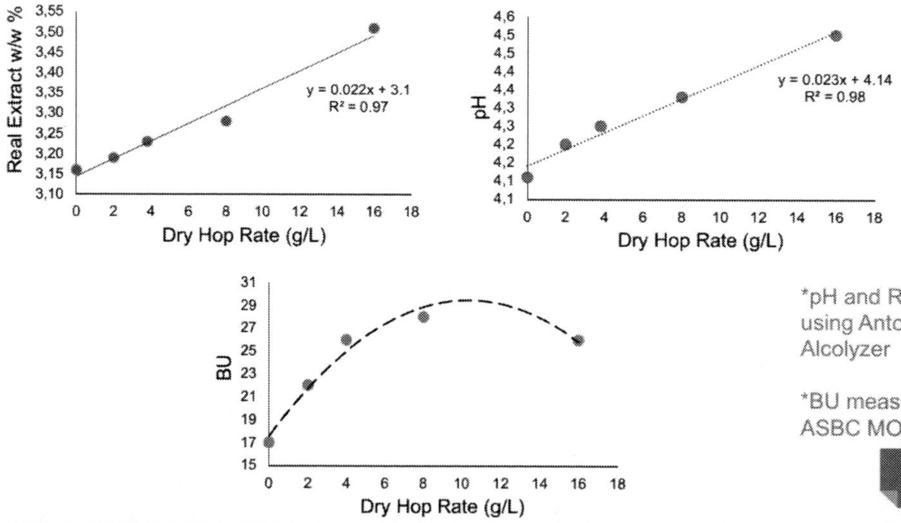

*pH and RE measured using Anton Paar Alcolyzer

*BU measured using ASBC MOA Beer-23

Oregon State University

Dry hopping having other impacts on beer

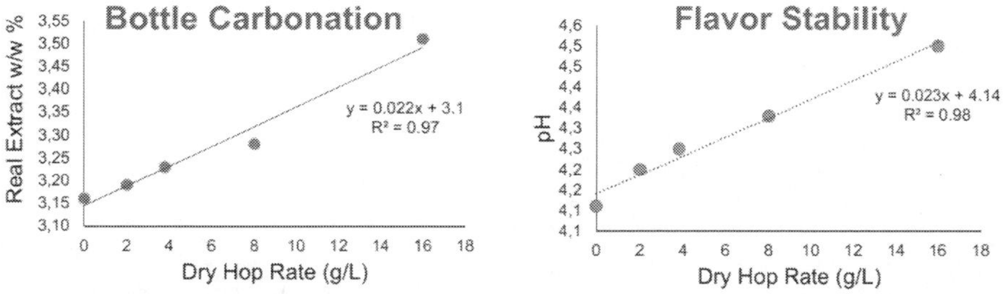

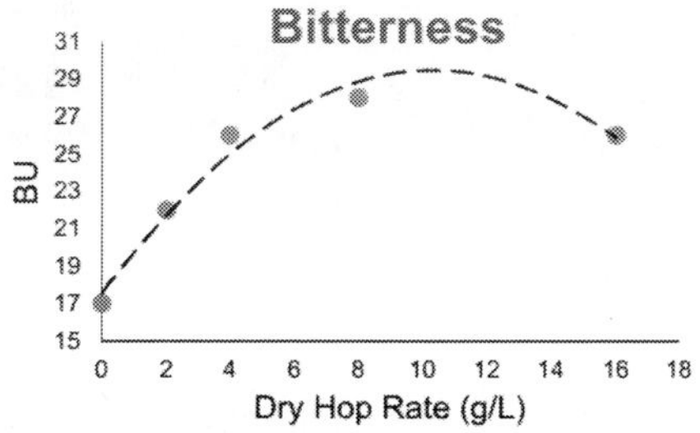

*pH and RE measured using Anton Paar Alcolyzer

*BU measured using ASBC MOA Beer-23

Oregon State UNIVERSITY

Conclusions

- More hops might not be better!
  - For Cascade there seems to be a optimal dry hopping rate ~4-6 g/L to maintain citrusy quality
- *Are there other ways to increase extraction efficiencies of desirable compounds?*
  - *Double/ Triple dry hopping*
- *Dry hopping is having more of an impact on beer than just flavor*

Oregon State UNIVERSITY

A special thanks to...

- Barth Haas Research Grant
- Ph. D. Advisor
  - **Tom Shellhammer**
- OSU Pilot Plant Manager
  - **Jeff Clawson**
- OSU Graduate Students:
  - **Daniel Vollmer, Christina Hahn, Kaylyn Kirkpatrick**
- OSU Undergraduates
  - **Brian Wagman**
- Crosby Hop Farm
  - **Staci Wallace**
- BridgePort Brewing Company
  - **Jeff Edgerton and Christian Engstrom**

BARTH-HAAS GROUP
FOR YOUR SUCCESS

BRIDGEPORT

CROSBY®
HOP FARM, LLC.

Oregon State University

Keynote presentations – Japan, Germany, Belgium, USA
Roundtable discussions – Dry-hopping, breeding/local, adv. products
Oregon hops farm tour – OSU breeding + Coleman Farms

hopsflavor2017.com

Oregon State University

Panelist Performance Metrics with OHAI (on controls)

Source	Type	DF	Sum of squares	Mean squares	F	Pr > F
Products	Fixed	4	2077.498	519.374	41.725	< 0.0001
Assessors	Random	9	113.968	12.663	1.173	0.358
Sessions	Fixed	3	5.740	1.913	0.448	0.721
Products*Assessors	Random	36	448.113	12.448	2.099	0.002
Products*Sessions	Fixed	12	54.599	4.550	0.767	0.683
Assessors*Sessions	Random	27	115.418	4.275	0.721	0.835
Error		108	640.558	5.931		

- **Panelist * Sample Interaction (p <0.0001):**
  - For the response OHAI, the rating given to a product depended on the assessor providing the rating
  - Issues with consensus among the panelists (or groups of panelists)
- **9 Panelists could discriminate using OHAI attribute**

Oregon State UNIVERSITY

The Centec DeAlcoTec System

"Centec offer fully automated, skid-mounted process units and high precision measurement technology from a single source."

Innovative Dealcoholization of Beer and Fermented Beverages

"Our systems and sensors are engineered to perfectly meet your requirements."

Dr. Robert Koukol, Owner & CEO

 www.centec.de Centec . Maintal . Germany

The Centec Group: German Headquarters

Centec GmbH
Wilhelm-Röntgen-Strasse 10
63477 Maintal. Germany
info@centec.de

- General Management
- Purchasing, Sales & Service
- Development, Design & Engineering
- Sensor Manufacturing

© Copyright 2016. All right reserved. www.centec.de Centec . Maintal . Germany

The Centec Group: Czech Subsidiaries

Prague Plant

- Detail Engineering
- Automation Engineering
- Switch Cabinet Manufacturing

Horatev Plant

- Materials Warehouse
- Tank & Skid Manufacturing
- Factory Acceptance Tests

 www.centec.de Centec . Maintal . Germany

The Centec Group: Horatev Industrial Park

 www.centec.de Centec . Maintal . Germany

The Centec Group: Global Presence

Centec subsidiaries and a network of sales partners

- Centec GmbH, Germany
- Centec automatica sro, Czech Republic
- Centec UK, Great Britain
- Centec LLC, USA
- Centec Technology Co Ltd, China
- Centec Predstavništvo u Beogradu, Serbia
- Centec RRR Systems & Sensors Pvt Ltd, India
- Centec América Latina Ltda, Brazil

DeAlcoTec: Basic Facts

- Applicable for beer, wine, cider and other alcoholic beverages and products
- Based on the principle of counter current vacuum distillation
- Gentle processing due to low pressure technology
- Selective for alcohol removal and minimized loss of aroma
- Minimized loss of beer
- Fully automated operation and process control

DeAlcoTec: Operating Principle

- After gentle pre-heating, the product is degassed (including CO_2 removal) and send to the distillation column
- Beer vapor flowing in counter current to the product removes the alcohol
- On top of the column the first stage of aroma recovery is included
- The resulting alcohol-free product is chilled, blended and carbonated
- The alcohol-rich exhaust vapor can be concentrated delivering sellable alcohol
- Aroma from vapor/CO_2 is recycled to the product by a recovery system

 www.centec .de Centec . Maintal . Germany

DeAlcoTec: Technical Data

- Capacity
  200 to 20.000 l/h;
  other capacities on request
- Operation
  24 h/day; CIP every 1 -2 weeks
- Dealcoholisation
  down to 0,01 vol. %
- Alcohol Concentration
  30 to 35 vol. %;
  75 to 90 vol. % (with add. rectification)
- Investment Cost
  between 250.000 and 3.000.000 EUR

 www.centec .de Centec . Maintal . Germany

DeAlcoTec: System Options

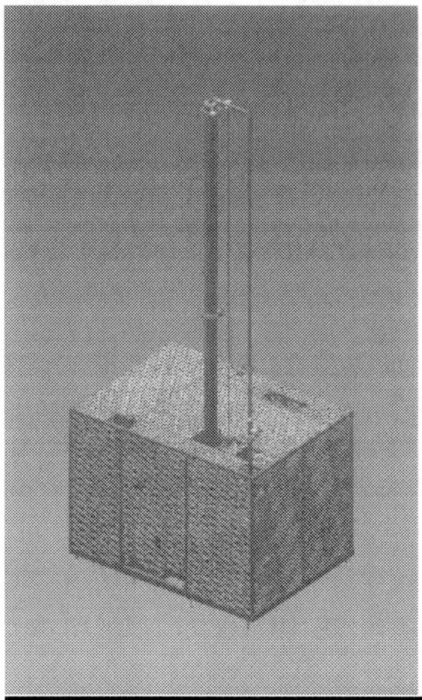

- Rectification column for alcohol concentration; alcohol storage tank
- Aroma re-blending
- Mother beer blending
- Carbonation/High Gravity Blending of final alcohol-free beer
- Online quality control
- Flash pasteurization or sterile filtration of alcohol-free beer
- Air chilled glycolset for chilling of the dealcoholization system
- Deaeration of water

DeAlcoTec: P&ID

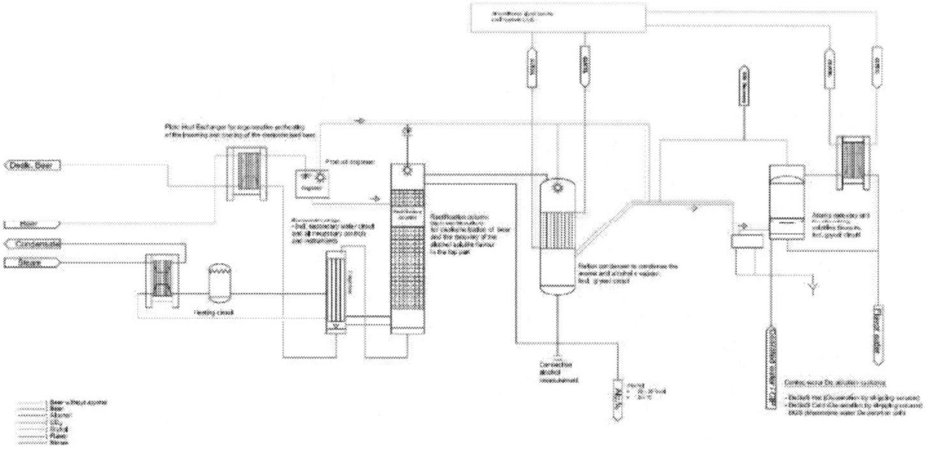

DeAlcoTec: Scheme of the Distillation Column

centec

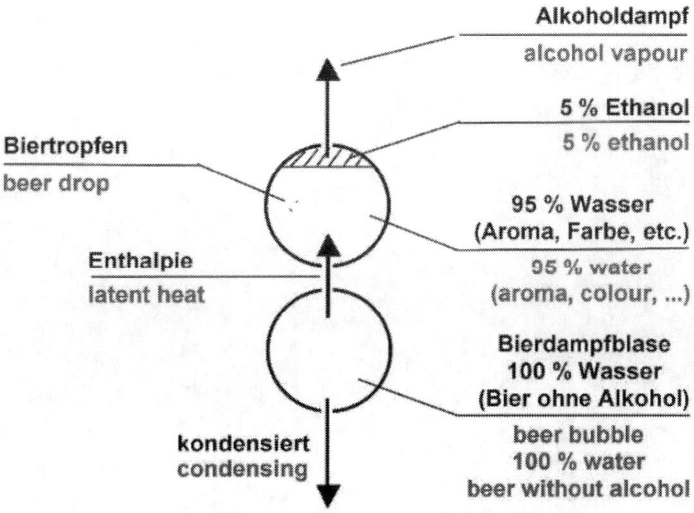

 www.centec.de Centec . Maintal . Germany

DeAlcoTec: System Datasheet

centec

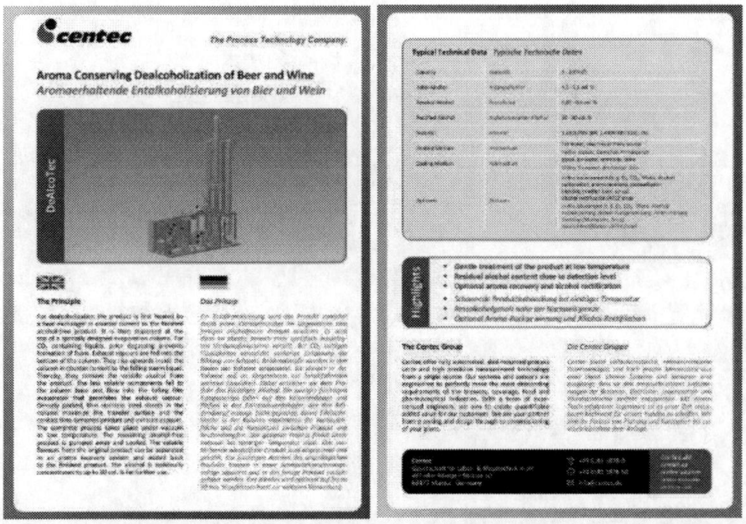
centec
The Process Technology Company.
Aroma Conserving Dealcoholization of Beer and Wine
Aromaerhaltende Entalkoholisierung von Bier und Wein
DeAlcoTec
The Principle
Das Prinzip
Typical Technical Data *Typische Technische Daten*
Highlights
The Centec Group
Die Centec Gruppe

 www.centec.de Centec . Maintal . Germany

DeAlcoTec for Puls Brewery, Uzbekistan

centec

 www.centec .de Centec . Maintal . Germany

DeAlcoTec for Gaffel Brewery, Germany

centec

 www.centec .de Centec . Maintal . Germany

DeAlcoTec for Gaffel Brewery, Germany

centec

 www.centec.de Centec . Maintal . Germany

DeAlcoTec for Gaffel Brewery, Germany

centec

© Copyright 2016. All right reserved. www.centec.de Centec . Maintal . Germany

DeAlcoTec for Gaffel Brewery, Germany

centec

 www.centec .de Centec . Maintal . Germany

DeAlcoTec for Gaffel Brewery, Germany

centec

© Copyright 2016. All right reserved. www.centec .de Centec . Maintal . Germany

DeAlcoTec for Gaffel Brewery, Germany

centec

 www.centec .de Centec . Maintal . Germany

DeAlcoTec for Gaffel Brewery, Germany

centec

 www.centec .de Centec . Maintal . Germany

DeAlcoTec for Gaffel Brewery, Germany

 www.centec.de Centec . Maintal . Germany

DeAlcoTec for Gaffel Brewery, Germany

 www.centec.de Centec . Maintal . Germany

The Use of Hops in Brewhouse and Cold Block

36th Congress EBC 2017 – Lecture L21

LJUBLJANA, 16 MAY 2017
DR. RUDOLF MICHEL, DR. LUDWIG SCHELLER & KORNEL VETTERLEIN
GEA BREWERY SYSTEMS GMBH; KITZINGEN

engineering for
a better world

Agenda

Brewhouse

- Classical hop dosing units
- Hop dosing units for craft brewing
- HOPSTAR™ Iso
- High hop applications in brewhouse

Cold Block – Dry Hopping

- Dry hopping for craft brewing
- HOPSTAR™ Dry – a perfect extraction method
- Automated dry hopping procedure for large tanks

engineering for
a better world

2

The classical system in a brewhouse

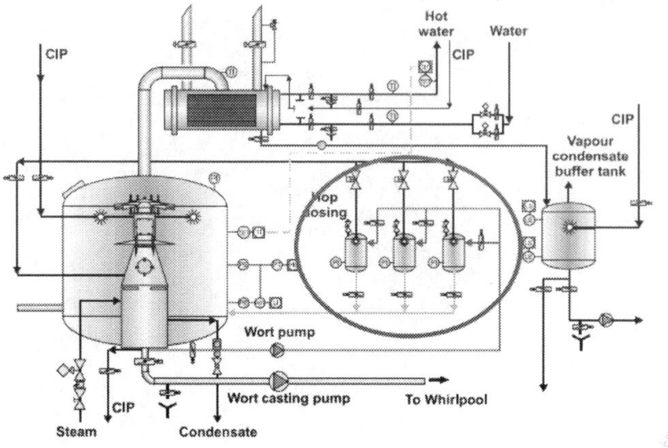

- Standard hop dosing vessels for pellets, extract in cans and whole hops
- Size: DN 600
- Volume: 235 ltr

Craft brewing

A 4 unit skid in a 110 hl COMPACT-STAR™

- Standard hop dosing vessels for pellets, extract in cans and whole hops
- Size: DN 400
- Volume: 100 ltr

A 2 unit skid in a 100 hl COMPACT-STAR™

HOPSTAR™ Iso

The innovative technology to improve hop utilization in the brewhouse

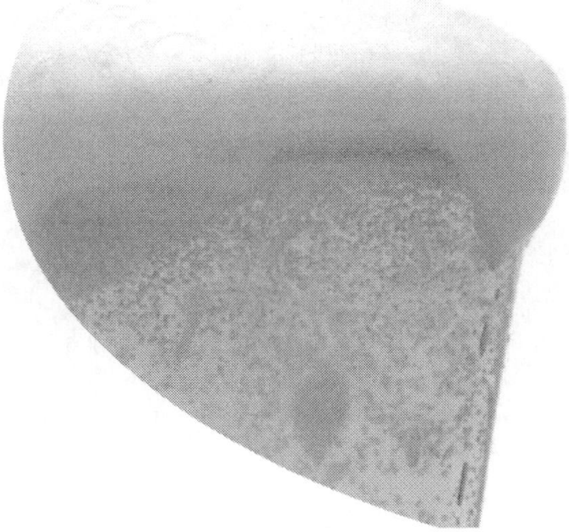

GEA engineering for a better world

Background

Driver: Losses of hop alpha acids during the brewing process

Present situation:

- Dosing of e.g. 8 g/hl alpha acids = 100 %
- Realized yield in wort = 66 %
- Realized yield in finished beer = 33 %

  ➔ **about 2/3 of the added hops are not utilized**

Reasons:

- Suboptimal isomerization procedures
- Losses caused by insolubility of alpha acids in watery solutions – alpha acids have a hydrophobic structure
- Losses with trub and yeast

GEA engineering for a better world

6

Hop isomerization during wort boiling

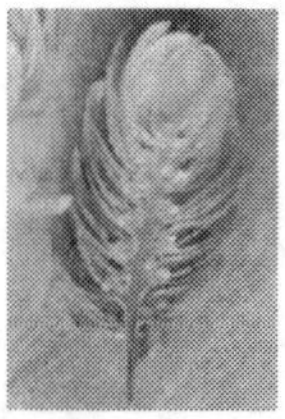

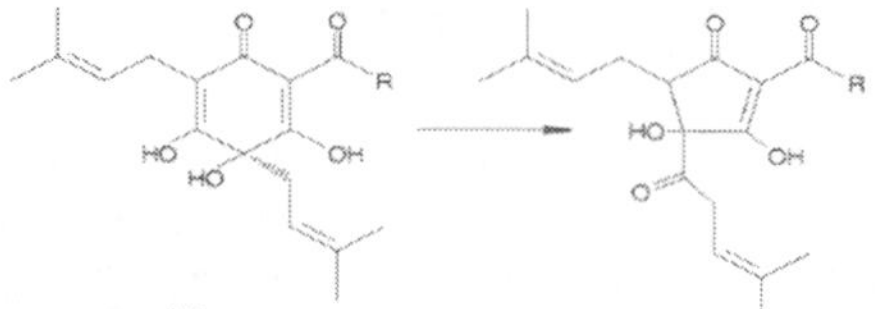

α-acid (hydrophobic) ⇨ iso-α-acid (hydrophilic)

- Alpha acids — valuable constituents in the lupulin of hops
- Wort boiling — ring changes from hexagonal to pentagonal structure
  isomerized bitter acids can be solved in the wort
  most important process step to create the bitterness profiles in the beer
- Isomerization — depending on
  temperature / reaction time / reactive surface / pH value

7

Influence of temperature

Isomerization rate doubles per every 10 K temperature increase

Comparable isomerization result can be achieved with:

100 °C	**about 140 min**
110 °C	**about 70 min**
120 °C	**about 35 min**
130 °C	**about 17 min**
140 °C	**about 8 min**

8

Influence of reactive surface

Particle size reduction

- **Dispersing** the hop mix with tip speed of about 21 m/s by means of high speed mixer
- **Homogenizing** the hop extract droplets at 250 bar by means of high pressure piston pump and homogenizing valve

Dispersing unit

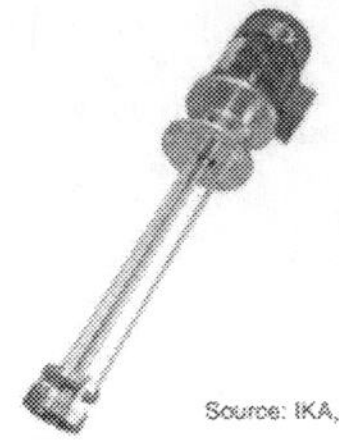

Source: IKA, Ultra Turrax

Homogenizing unit

Homogenizing valve

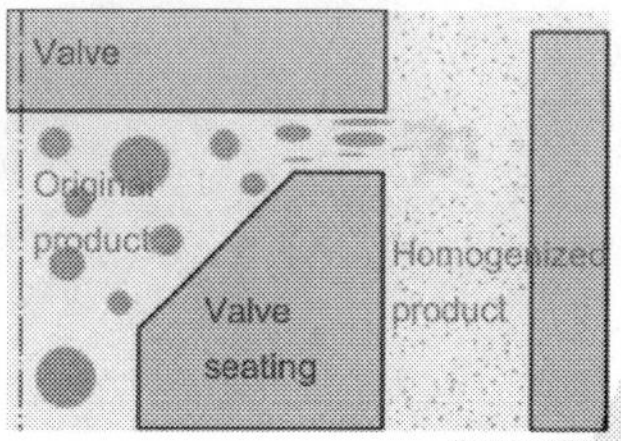

Source: GEA

9

The skid: HOPSTAR™ Iso

Equipment with two isomerization tanks for a controlled isomerization process

Possible process conditions

- 120 – 140 °C
- 15 – 20 min
- pH > 6
- Alternative: proportionate dosing of pellets
- Isomerization rate >> 92 %

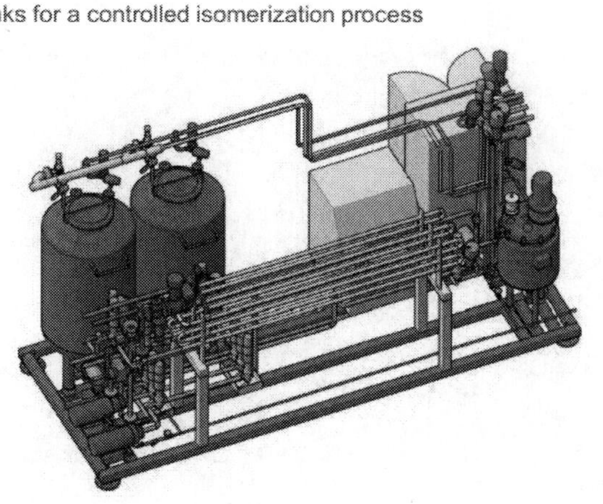

10

Hop extract suspension

Dispersed hop extract suspension (extract droplets visible)	**Homogenized** hop extract suspension	**Homogenized and isomerized** hop extract suspension (stable for months)
Suspension made from 35 % ethanol extract and 65 % CO_2 extract		

11

Pellets and extract − Large-scale trials

- Hop products, mixed (Extract and pellets)
  - 75 % CO_2 extract: disperser and homogenizer at 250 bar; isomerized at 140 °C
  - 25 % pellets: only dispersed with high speed mixer; isomerized at 140 °C
- Dosing of isomerized suspension mix
  - in wort kettle (end of boiling)
  - in Whirlpool (start of wort transfer)

Dosing rate blank: **8.1 g** alpha acid per hl

Dosing rate trial: **5.7 g** alpha acid per hl (= - 30 %)

- Bitterness in bottled beer
  - Blank 22 BE
  - Wort kettle 23 BE
  - Whirlpool 19 BE

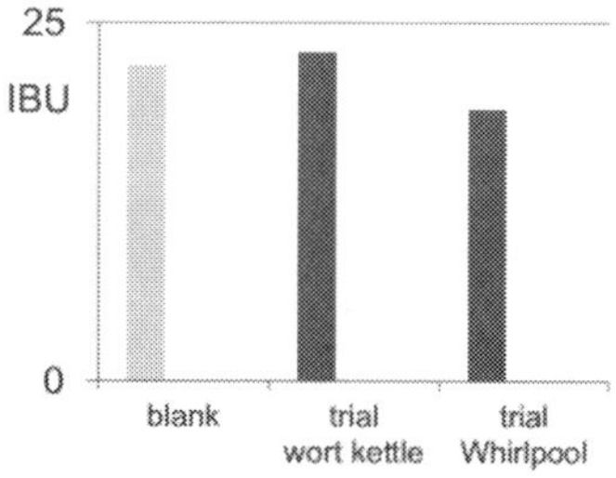

Result: comparable bitterness values with 30 % less hop dosing

12

Calculation of profitability (ROI)

Potential from hop yield and energy savings

Produced wort quantity	**2 million hl/year**
1. **Savings of hop extract** (dosing reduced by 30 %)	187,000 €/year
2. **Additional cost for HOPSTAR™ Iso**	
Electrical energy	- 8,000 €/year
Additional heating (only 1 % of the yearly wort production volume to be heated for isomerization in HOPSTAR™ Iso)	- 5.000 €/year
3. **Total Savings**	**174,000 €/year**
Equipment costs approx. 250 – 350 T€	

➔ Possible ROI is in the range of 1.5 – 2 years

GEA engineering for a better world

13

HOPSTAR™ Iso: Efficient process for breweries

- Yield increase of alpha acid quantity between 15 % up to 30 % depending on hop products and applied technology
  - ROI ~ 1,5 to 2 years
- Energy savings due to lower total evaporation rate (DMS-P to DMS transformation)
- Technological aspects
  - Improved flavour stability due to reduced thermal stress
  - Pleasant hop flavour in smell and taste
    No scratchy, harsh and astringent bitterness in the aftertaste (even with higher bitterness units)
- Flexibility when using a stable hop suspension
- Central hop production for affiliated breweries

GEA engineering for a better world

14

High Hopping in Brewhouse

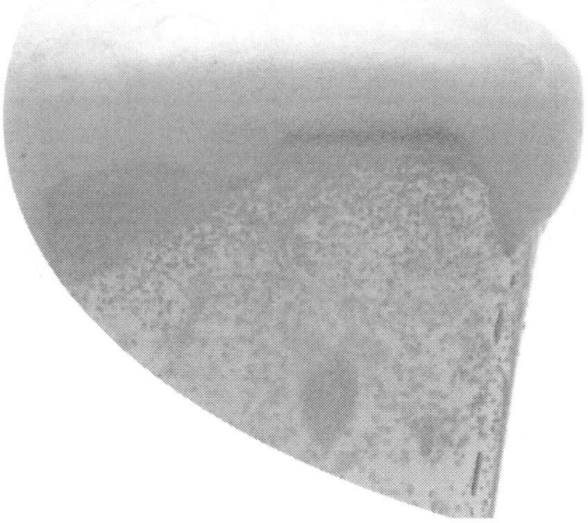

engineering for a better world

High hopping in brewhouse

- Request from Craft Brewers (i.e. IPA)
- Challenge – high hop load, acceptable wort losses
  The more hop pellets, the higher wort losses

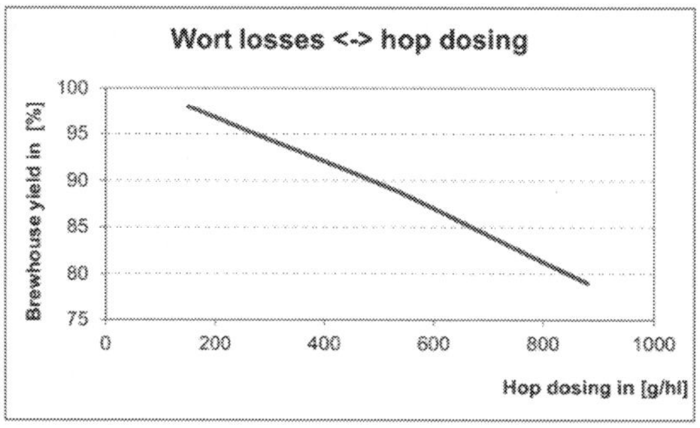

engineering for a better world

16

High hopping in brewhouse

Trails

- "Hop basket" to remain solids in hop dosing vessel
  - ➢ wort losses ok, high hop losses
- Curved Sieve between wort kettle and whirlpool
  - ➢ Mesh size between 100 – 250 µm
  - ➢ Reduction of solids about 20-30 %
  - ➢ Still high wort losses

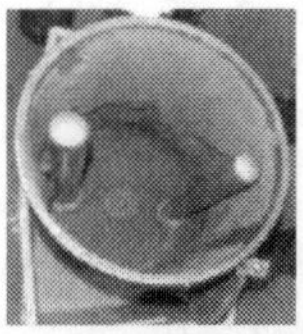

Imhoff (60')
before:130 ml behind: 97 ml

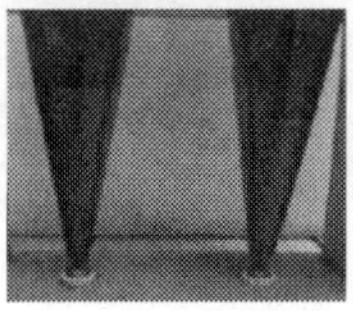

GEA engineering for a better world

17

High hopping in brewhouse

Solution – Centrifuge

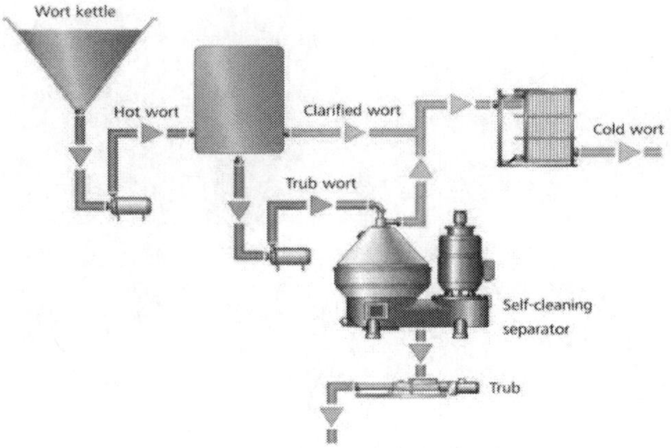

GEA engineering for a better world

18

High hopping in brewhouse

- Trub Wort Separation
  - Clarification by Whirlpool
  - Concentration of solids by centrifuge
  - Partial stream
- Hot Wort Clarification
  - Clarification and concentration by centrifuge
  - Wort kettle with mixing device or Whirlpool
  - Sizing or limiting factors: solids load and meanly flow
  - Full stream

➔ No change over between brews!
Clarified wort goes to same brew.

GEA engineering for a better world

19

Manual Dry Hopping

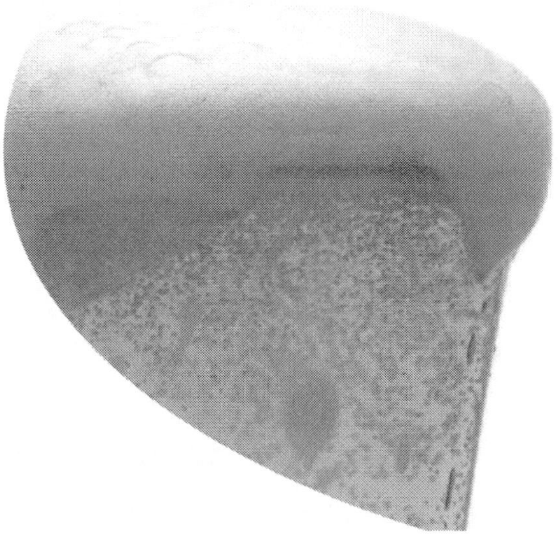

GEA engineering for a better world

Dry hopping vessel

- Easy, but reliable cold hopping technology
  - Moveable skid on rollers with manual valves
- Application both for hop pellets and hop cones
  - Additional basket to hold back hop cones
- Connection to CCT with (existing) beer hoses and mobile beer pump
- Purging with CO_2 prior use
- CIP connection

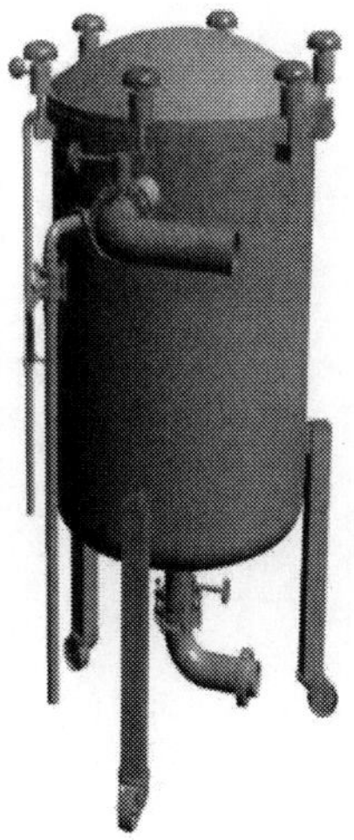

GEA engineering for a better world

21

Dry hopping vessel

- Easy, but reliable cold hopping technology
  - Moveable skid on rollers with manual valves
- Application both for hop pellets and hop cones
  - Additional basket to hold back hop cones
- Connection to CCT with (existing) beer hoses and mobile beer pump
- Purging with CO_2 prior use
- CIP connection

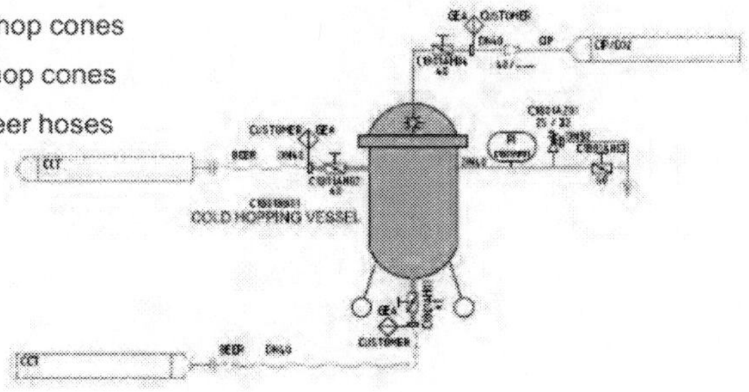

GEA engineering for a better world

22

HOPSTAR™ Dry

Advanced dry hopping system based on extraction

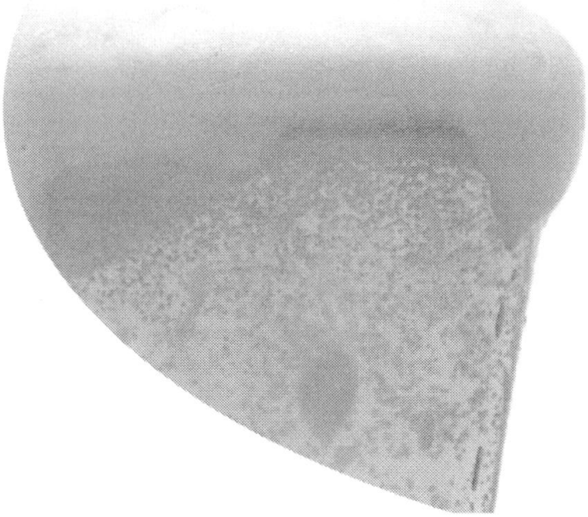

GEA engineering for a better world

Parameters for Dry Hopping

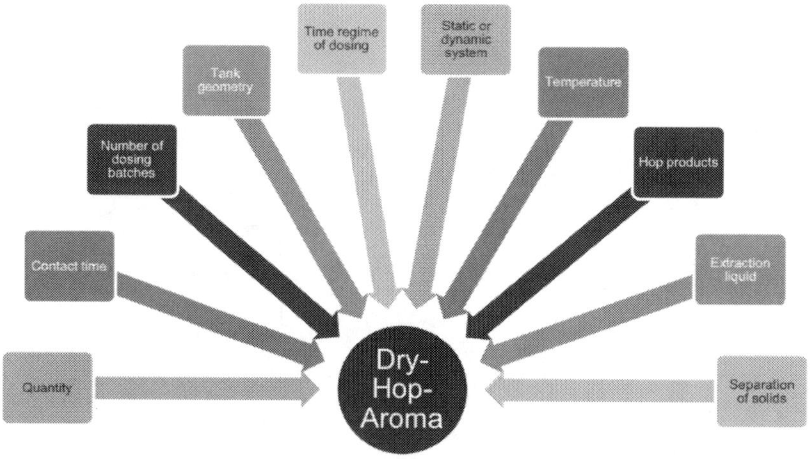

GEA engineering for a better world

24

What makes HOPSTAR™ Dry so special?

- Unique, fully automated GEA tool
- Very efficient pellet extraction reduces raw material demand and keeps cost lower for dry hopping
- Beer enriched with hop aroma is homogenized in the storage tank for uniform beer quality
- Automation allows flexible adjustment of process parameters to reflect pellet quality or hop variety
- Extraction efficiency can be optimized by repeating process steps
- Minimum solids are transferred into the storage tank
- No beer losses by pellet solid sludge in the storage tank
- Sustainable process with corresponding improvement of work safety

GEA engineering for a better world

25

Features of HOPSTAR™ Dry

- Skid-mounted unit:
  The HOPSTAR™ Dry will come as pre-piped, pre-wired and tested unit on a frame
- Design of extractor tank considers CCT volume and maximal pellet load
- Extractors will be installed separately beside the frame with all peripheral installation
- Pre-piped skid has a standard design, only extractor is adapted
- One HOPSTAR™ Dry system to be used for several storage tanks
- Process conditions (e.g. soaking or suspending of pellets) can be adjusted individually
- Process under CO_2 atmosphere to avoid oxidation

- Automation as stand-alone unit or integrated in existing installation
- Automation of all process steps and of CIP meets the highest standards of quality and hygienic conditions

GEA engineering for a better world

26

HOPSTAR™ Dry – Conditions of extractor

- Flow speed and sedimentation velocity

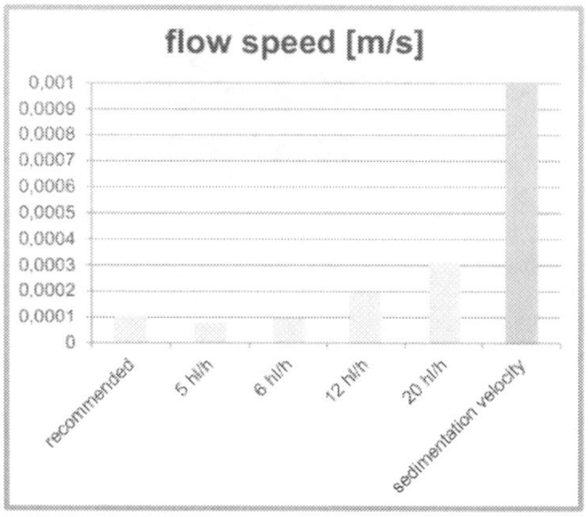

- Extraction vessel diameter and flow volume [hl/h] define flow speed (upward) contrary to sedimentation velocity of hop particles

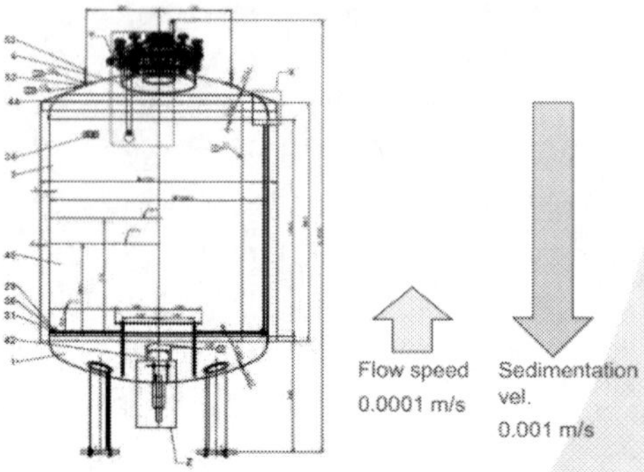

GEA engineering for a better world

27

HOPSTAR™ Dry – Process scheme

1. Extraction of hop ingredients

Circulation for soaking and dissolving pellets in extraction vessel

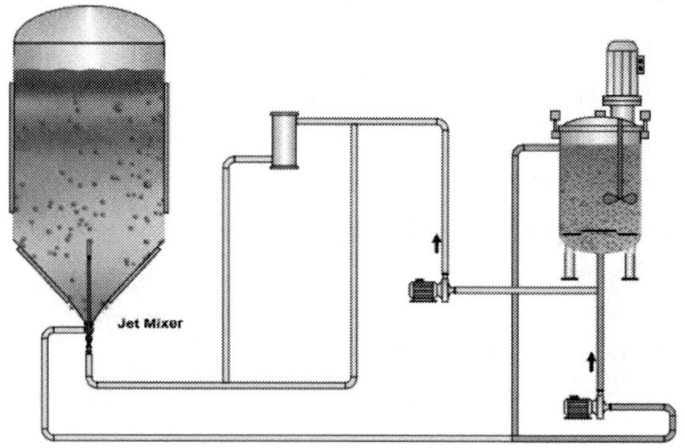

GEA engineering for a better world

28

HOPSTAR™ Dry – Process scheme

2. Transfer of hop aroma enriched beer to CCT

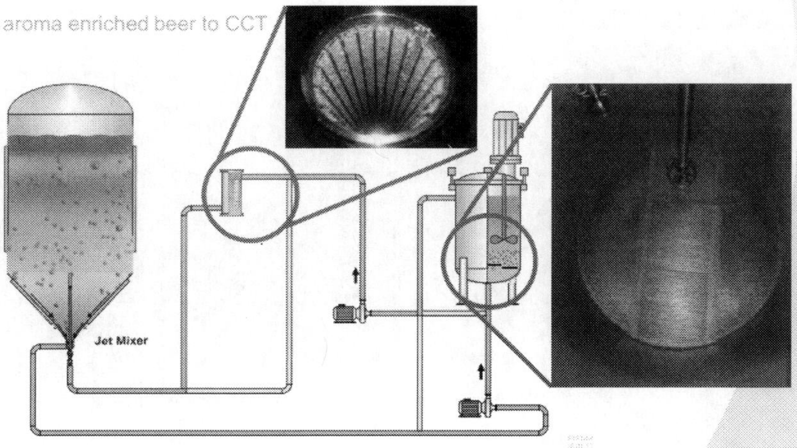

HOPSTAR™ Dry – Process scheme

3. Emptying of extraction vessel: hop aroma enriched beer to CCT

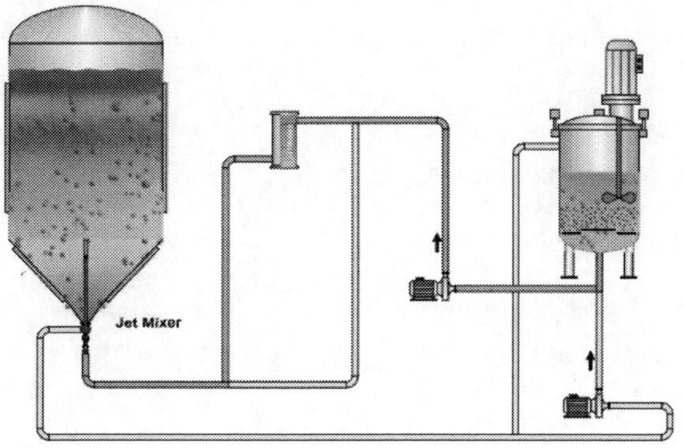

HOPSTAR™ Dry – Process scheme

4. Homogenizing of beer in CCT using a GEA jet mixer

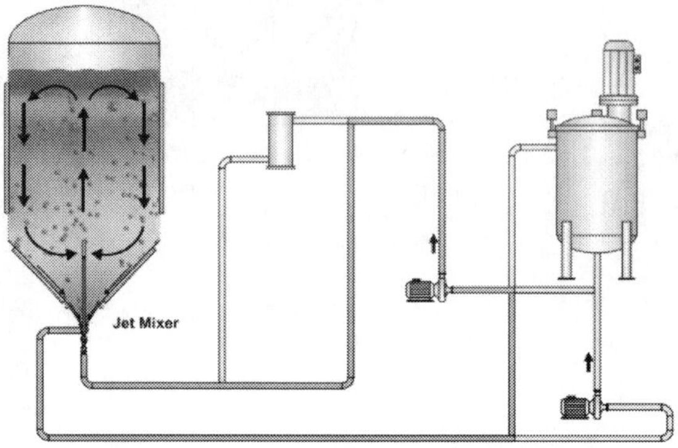

GEA engineering for a better world

31

HOPSTAR™ Dry – Conditions for basic design

Production parameters

• Beer transfer	5 hl/h
• Transfer time	6 h for 30 hl
• Crushing and mixing of pellets	1 h
• Sedimentation rest	1 h
• Total time per extraction- and transfer-step	7 h
• Number of transfer steps (assumption)	5
• Emptying of extraction vessel	6 h
• CIP	4 h
• Total production time per batch of 500 hl	46 h or 2 days
• Total production volume per week	2 x 500 hl = 1,000 hl

GEA engineering for a better world

32

Trial # 2 with pellets Type 90

Hop oil yield [%]

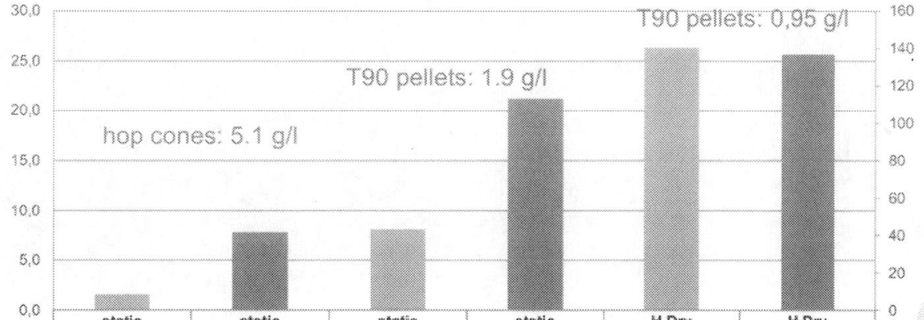

	static cones	static cones	static pellets	static pellets	H Dry pellets	H Dry pellets
Myrcene	1,6		8,1		26,3	
Linalool		41,6		113,2		137,0

33

HOPSTAR™ Dry – Process

Process parameters of realized installation

• Raw material:	Hop pellets type 90
• Dosing quantity:	200 – 600 g/hl
• Beer volume in CCT:	500 hl
• Number of CCTs for dry hopping:	1
• Extractor capacity :	33 hl
• Max. pellet quantity for one dry hopping batch:	approx. 300 kg
• Realized pellet load:	from 120 kg/250 hl up to 240 kg/500 hl
• Extraction and transfer cycles:	recommended up to 5 times for small batch, more for big batch

34

Dry Hopping – Continuous pellet dosing into large tanks

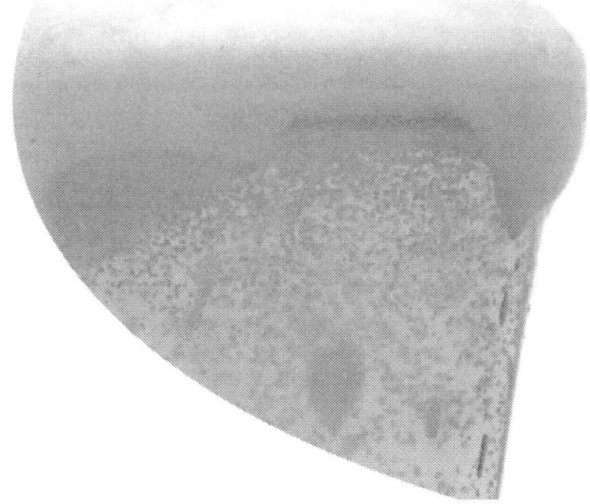

GEA engineering for a better world

Preparation of system

Process steps:

1. Filling of hops into hopper
2. Evacuation and inert gas flashing of system
3. Start circulation of beer

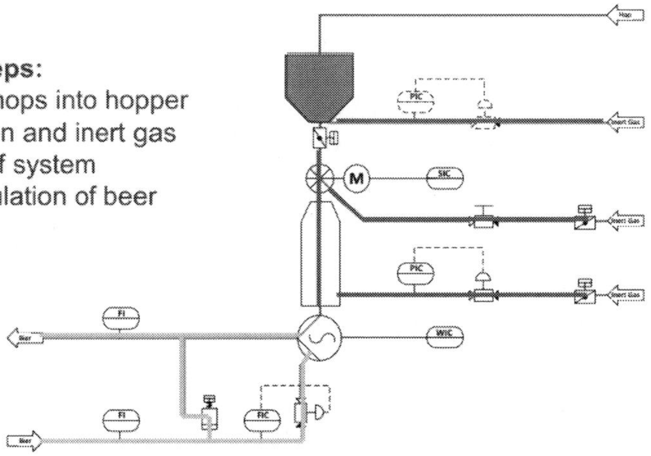

GEA engineering for a better world

36

Production mode

Process steps:
1. Filling of hops into hopper
2. Evacuation and inert gas flashing of system
3. Start circulation of beer
4. **Start hop dosing via pump**

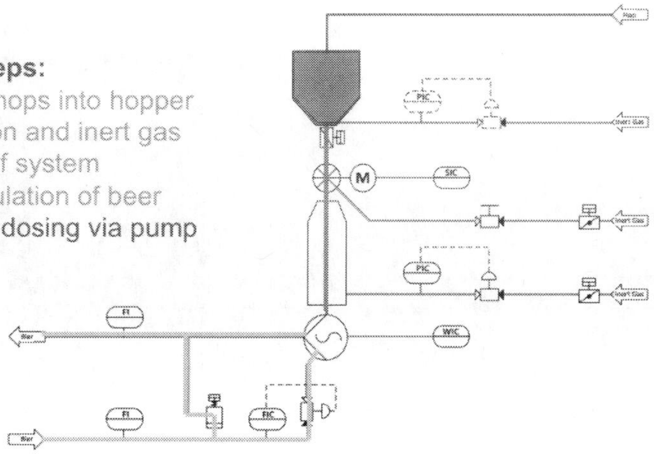

GEA

37

GEA engineering for a better world

56TH CONGRESS
EBC
2017
14-18 MAY
LJUBLJANA
EUROPEAN BREWERY CONVENTION
SLOVENIA

gea.com

Anton Paar

Alcolyzer Beer ME

Results of alcohol analysis on beer samples with an alcohol content lower than 0.4% v/v

www.anton-paar.com

Introduction

Anton Paar

Markus Peterherr
Product Management Beverage
Anton Paar GmbH

- :: Alcolyzer Beer M Alcohol Meter
- :: Alcolyzer Beer Analyzing System Generation M
- :: PBA-B Packaged Beverage Analyzer for Beer
- :: HazeQC ME Turbidity Meter
- :: pH ME ph measuring module

www.anton-paar.com

Overview

Anton Paar

- Alcolyzer Beer ME
  - measuring principle
  - field of application
  - current technology
  - latest improvements
    - ✓ hardware
    - ✓ software

➡ **Alcohol determination in alcohol free beers**

www.anton-paar.com

Anton Paar

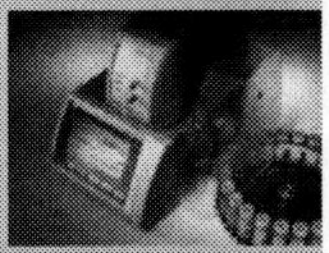

Measuring principle

Alcohol determination by NIR spectroscopy

www.anton-paar.com

Measuring principle – NIR Spectroscopy

Anton Paar

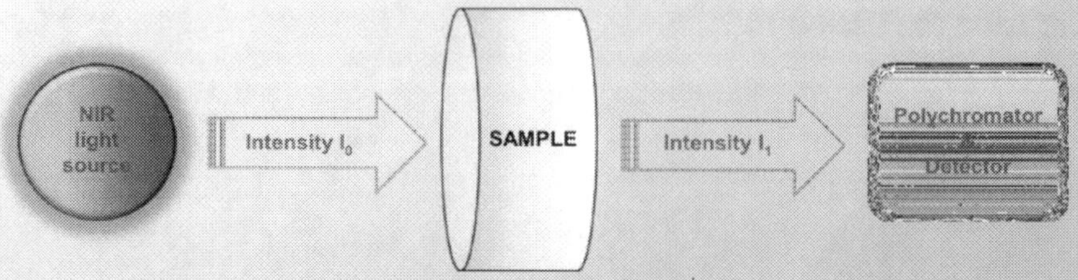

Loss in intensity (I_0:I_1) corresponds to the concentration of ethanol

www.anton-paar.com

Anton Paar

Field of application

Instrument overview & measured parameters

www.anton-paar.com

Field of application – Alcolyzer Beer Analyzing System

Anton Paar

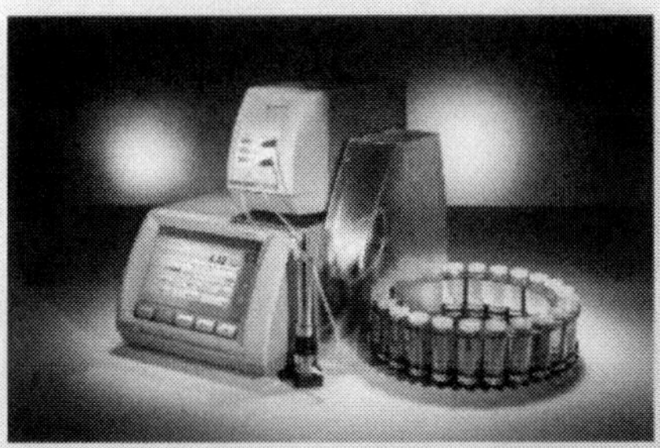

+ direct alcohol determination

- Measured Parameters:
  - Alcohol (% v/v)
  - Density
  - Optional: color, pH, turbidity
- Calculated Parameters:
  - Real, original and apparent extract
  - Real and apparent degree of fermentation
  - Alcohol (% w/w)
  - Calories
  - etc...

www.anton-paar.com

Field of application – PBA-B Generation M

Anton Paar

+ direct alcohol determination
+ no sample preparation

- Measured Parameters:
  - Alcohol (% v/v)
  - Density (g/cm^3)
  - CO_2 (g/l)
  - Optional: color, pH, turbidity, O_2 (ppb)
- Calculated Parameters:
  - Real, original and apparent extract
  - Real and apparent degree of fermentation
  - Alcohol (% w/w)
  - Calories
  - etc...

www.anton-paar.com

Anton Paar

Current technology

Alcolyzer Beer ME

www.anton-paar.com

Current technology - Alcolyzer Beer ME

Anton Paar

- Comparison study - Alcolyzer vs. distillation:
  - Period: 2000 – 2017
  - Samples: Standard-Beers / Beer mix drinks / Alcopops

Average deviation to distillation of 0.05%v/v

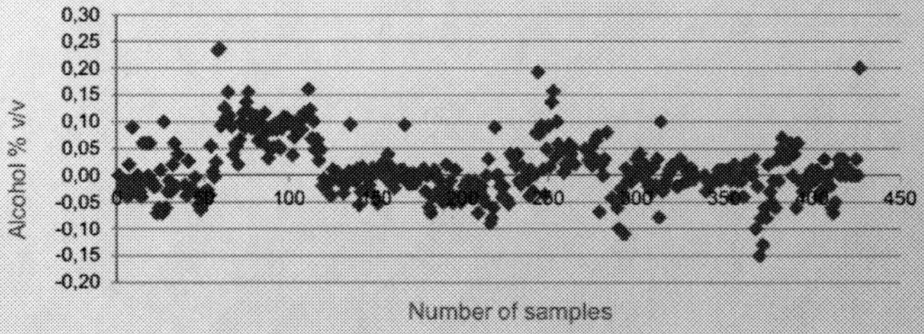

Valid for Standard Beers, Beer Mix Drinks and Alcopops!

www.anton-paar.com

Current technology - Alcolyzer Beer ME

Anton Paar

- Impact of fructose + glucose on the alcohol result
  - Common ingredients in „alcohol free beers“
  - Increasing fructose / glucose concentration leads to negative alcohol readings
  - Evaluation of impact: ***headspace gas chromatography***

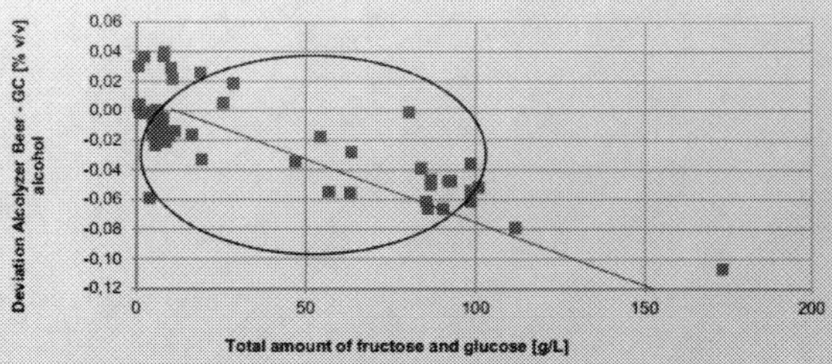

www.anton-paar.com

Anton Paar

Latest improvements

Alcolyzer Beer ME

www.anton-paar.com

Alcolyzer Beer ME: alcohol-free / non-alcoholic beers

Anton Paar

- Categorization of beers

	Standard beers + beer mix drinks	Low/non-alcoholic beers
Alcohol content	0.5 - 10% v/v	up to approx. 0.5% v/v
Extract content	25 - 100 g/L	up to 160 g/L
Maltose	< 5 g/L (up to 50 g/L on alcohol free beer)	up to approx. 50 g/L
Oligosaccharides	✓	✓
Fructose	✗ (exception: Radlers)	✓
Glucose	✗ (exception: Radlers)	✓
Saccharose	✗ (exception: Radlers)	✓
Citric acid	approx. 0.2 g/L (up to 3 g/L on Radlers)	approx. 1 -3 g/L
Other residues	< 5 g/L	< 5 g/L

www.anton-paar.com

Alcolyzer Beer ME: alcohol-free / non-alcoholic beers

Anton Paar

- **Breaking the limits:**
  - **Alcohol- free beers:** **up to 0.4% v/v alcohol**
  - **Non- alcoholic beers:** **0.0% v/v alcohol**
  - **Malt drinks**

➢ Correction

- **Improved technology on Alcolyzer Beer M/ME**

Alcohol content > 0.4% v/v ➡ **Correction for standard beers** ***current technology***

Alcohol content < 0.4% v/v ➡ **Correction for low/non-alcoholic beers** ***improved technology***

www.anton-paar.com

Alcolyzer Beer ME: alcohol-free / non-alcoholic beers

Anton Paar

- **Hardware improvements**
  - **Comparison of spectra of ethanol without / with improvements**

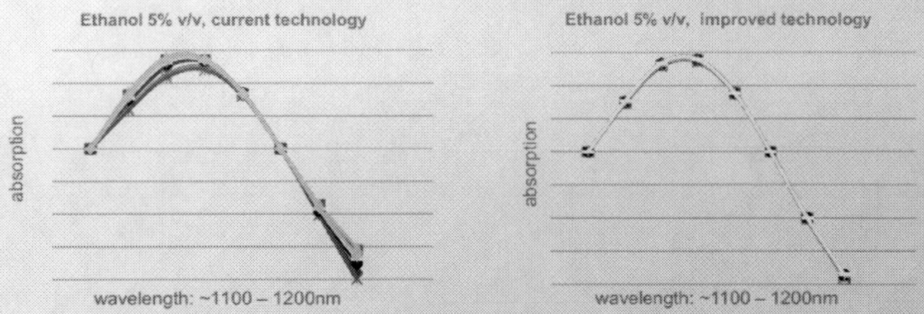

www.anton-paar.com

Alcolyzer Beer ME: analysis of low/non-alcoholic beer

Anton Paar

- **Reproducibility Study**
  up to ~0.5% v/v of alcohol and 100 g/L of extract

PEM7

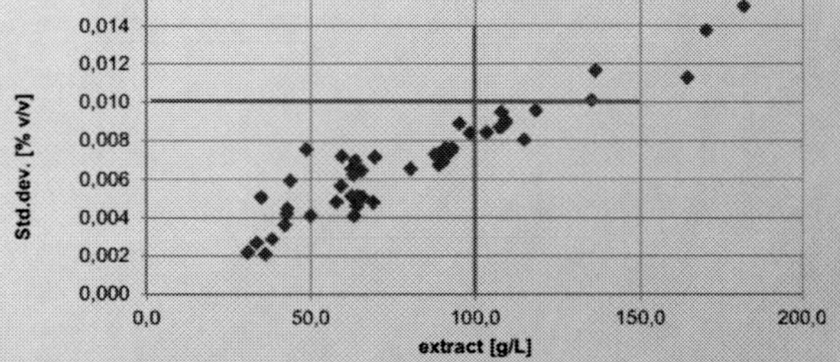

Standard deviation < 0.02% v/v

Reference: headspace gas chromatography

Total of 51 low/non-alcoholic beers

www.anton-paar.com

Folie 16

PEM7 Terminologie - ROMAN

ALKOHOL ODER ETHANOL

Durchziehen.
Markus Peterherr; 10.05.2017

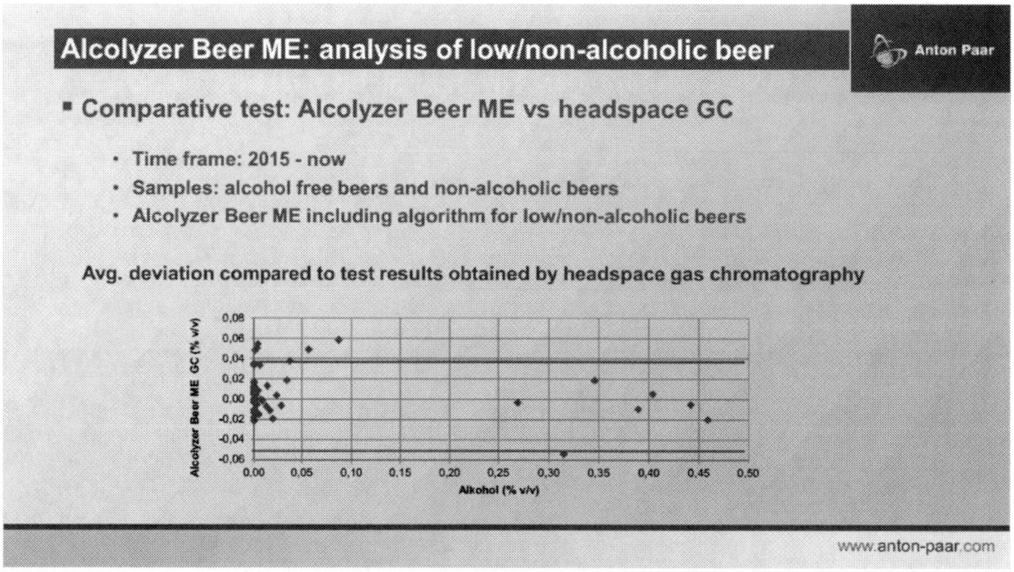

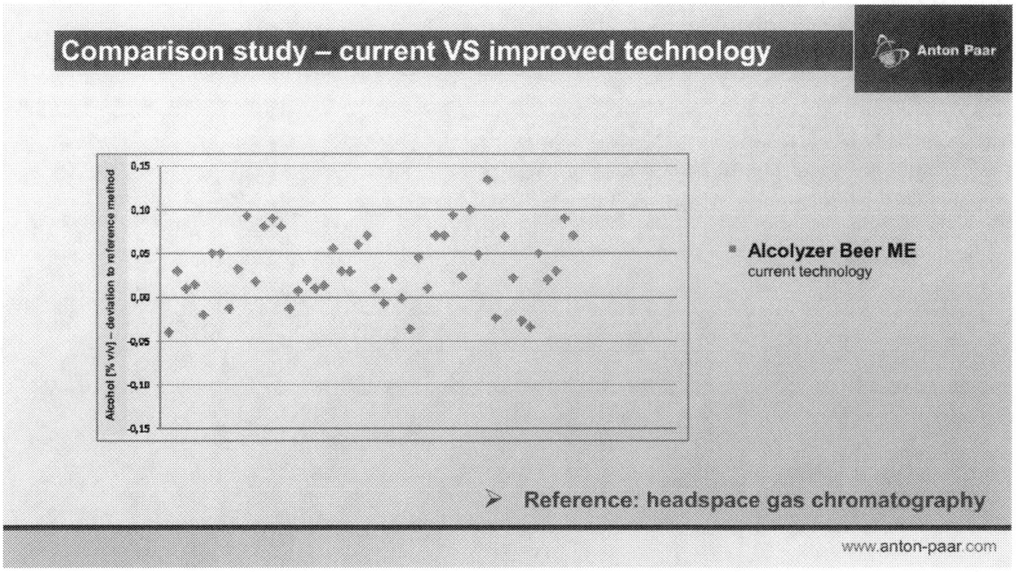
Comparison study – current VS improved technology
Anton Paar
Alcohol [% v/v] – deviation to reference method
0,15
0,10
0,05
0,00
-0,05
-0,10
-0,15
Alcolyzer Beer ME
current technology
Reference: headspace gas chromatography
www.anton-paar.com

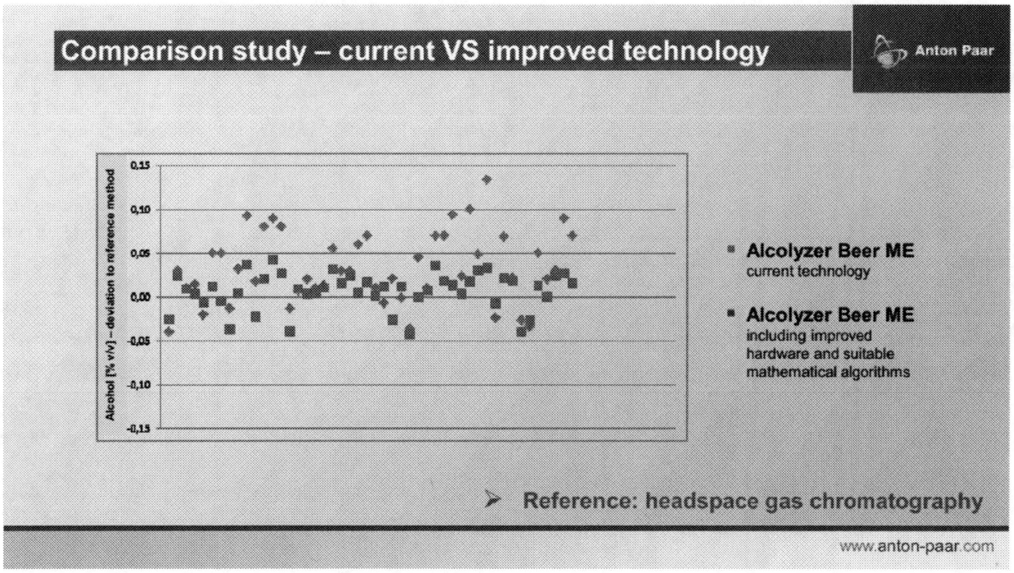
Comparison study – current VS improved technology
Anton Paar
Alcohol [% v/v] – deviation to reference method
0,15
0,10
0,05
0,00
-0,05
-0,10
-0,15
Alcolyzer Beer ME
current technology
Alcolyzer Beer ME
including improved hardware and suitable mathematical algorithms
Reference: headspace gas chromatography
www.anton-paar.com

Summary

Anton Paar

Alcolyzer Beer ME

Alcohol determination by NIR spectroscopy

- **Goal: Alcohol analysis on beer samples with an alcohol content lower than 0.4% v/v**
  - ✓ Hardware improvements
  - ✓ Software improvements
  - ✓ Implementaion of suitable mathematical algorithms
  - ✓ Verfication test: comparison of test results to headspace gas chromatography

www.anton-paar.com

Anton Paar

Thank you for your attention

Alcohol determination by NIR spectroscopy

www.anton-paar.com

56th CONGRESS EBC SLOVENIA 2017 EUROPEAN BREWERY CONVENTION 14-18 MAY LJUBLJANA

16 May 2017

[L-25]

The development of a non-alcoholic beer using a novel idea undertaken by casting off preconceptions about beer

Development Laboratories for Alcohol Beverages
ASAHI BREWERIES, Ltd.

Shinsuke Ito, Jun Kubota

Today's presentation

Asahi Dry Zero

No.1 in sales

in the Japanese market
for non-alcoholic beer taste beverages

Source: SRI survey by INTAGE, Non-alcoholic beer taste beverages market, January 2016-December 2016 (cumulative amount sold), in seven sales channels

Presentation outline

- ✓Introduction
- ✓Product design
- ✓Development of product content
- ✓Manufacturing techniques for mass production
- ✓Market changes after product release

Presentation outline

- ✓Introduction
- ✓Product design
- ✓Development of product content
- ✓Manufacturing techniques for mass production
- ✓Market changes after product release

World beer market (2005-2015)

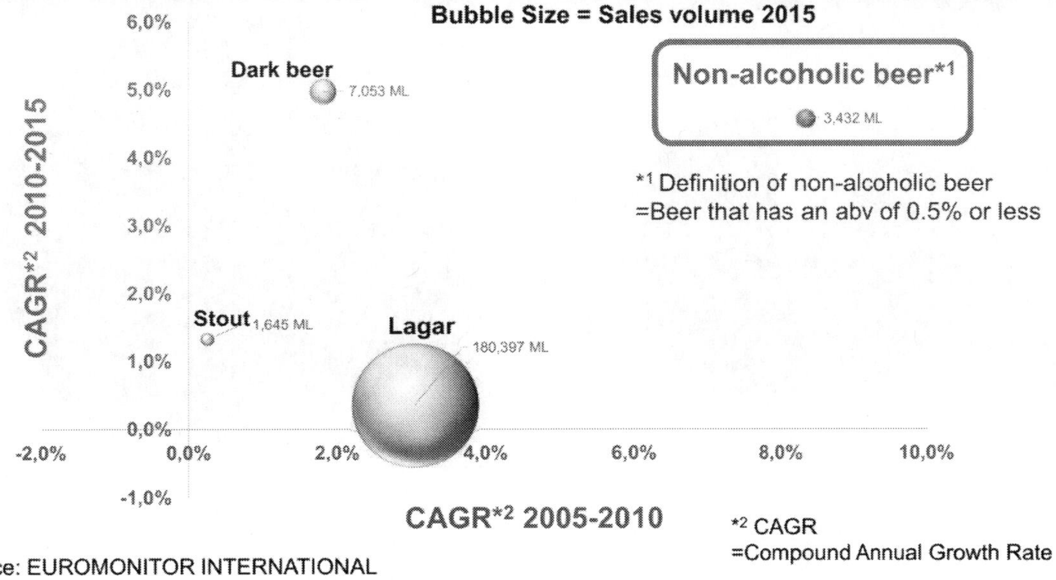

Source: EUROMONITOR INTERNATIONAL

Sales of non-alcoholic beer in Japan (2002-2011)

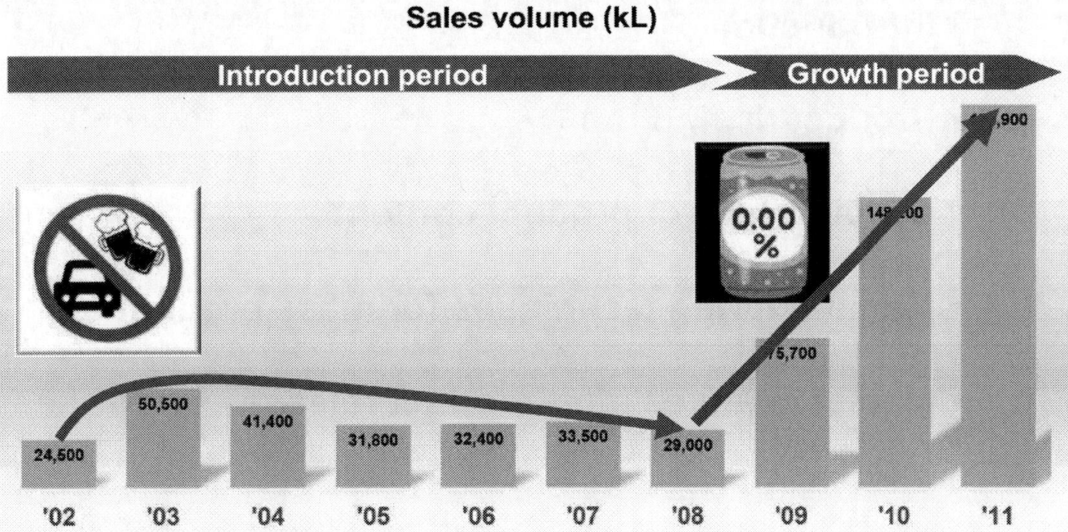

Source: Fuji Economy "2012 Food Marketing Handbook"

Customer requests for non-alcoholic beer

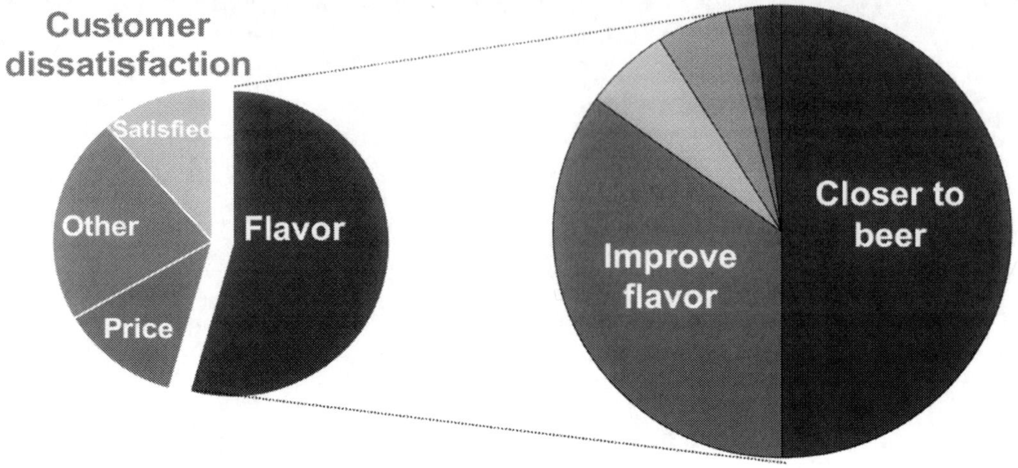

Source: Our survey (Sep 2011)
Target = 420 non-alcoholic beer consumers

Presentation outline

- ✓**Introduction**
- ✓**Product design**
- ✓**Development of product content**
- ✓**Manufacturing techniques for mass production**
- ✓**Market changes after product release**

Keyword

Customer value

What does "beer-like" mean?

Voice of the Customer (VOC)

■ Focus Group Interview

■ In-depth Interview

Customer Core Values

Start drinking without any unpleasant sensation

Stimulating feeling in the throat

Good aftertaste (Retronasal Aroma)

Quality characteristics

- **Thorough analysis of beer product**

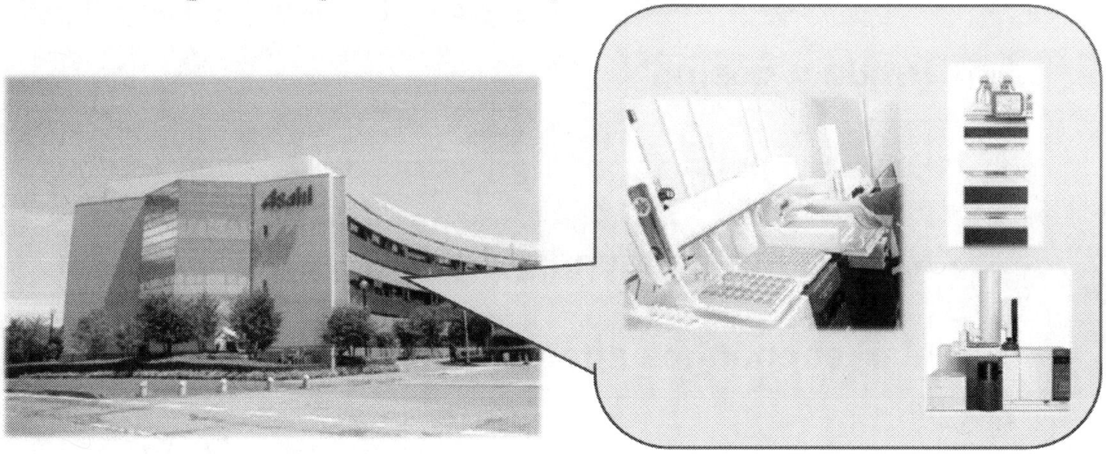

Asahi Group Research & Development Center

Quality Function Deployment (QFD)

How
(Quality characteristics)

What (Core values)	
	Start drinking without any unpleasant sensation
	Stimulating feeling in the throat
	Good aftertaste

Presentation outline

- ✓Introduction
- ✓Product design
- ✓Development of product content
- ✓Manufacturing techniques for mass production
- ✓Market changes after product release

Keyword

Break the Bias

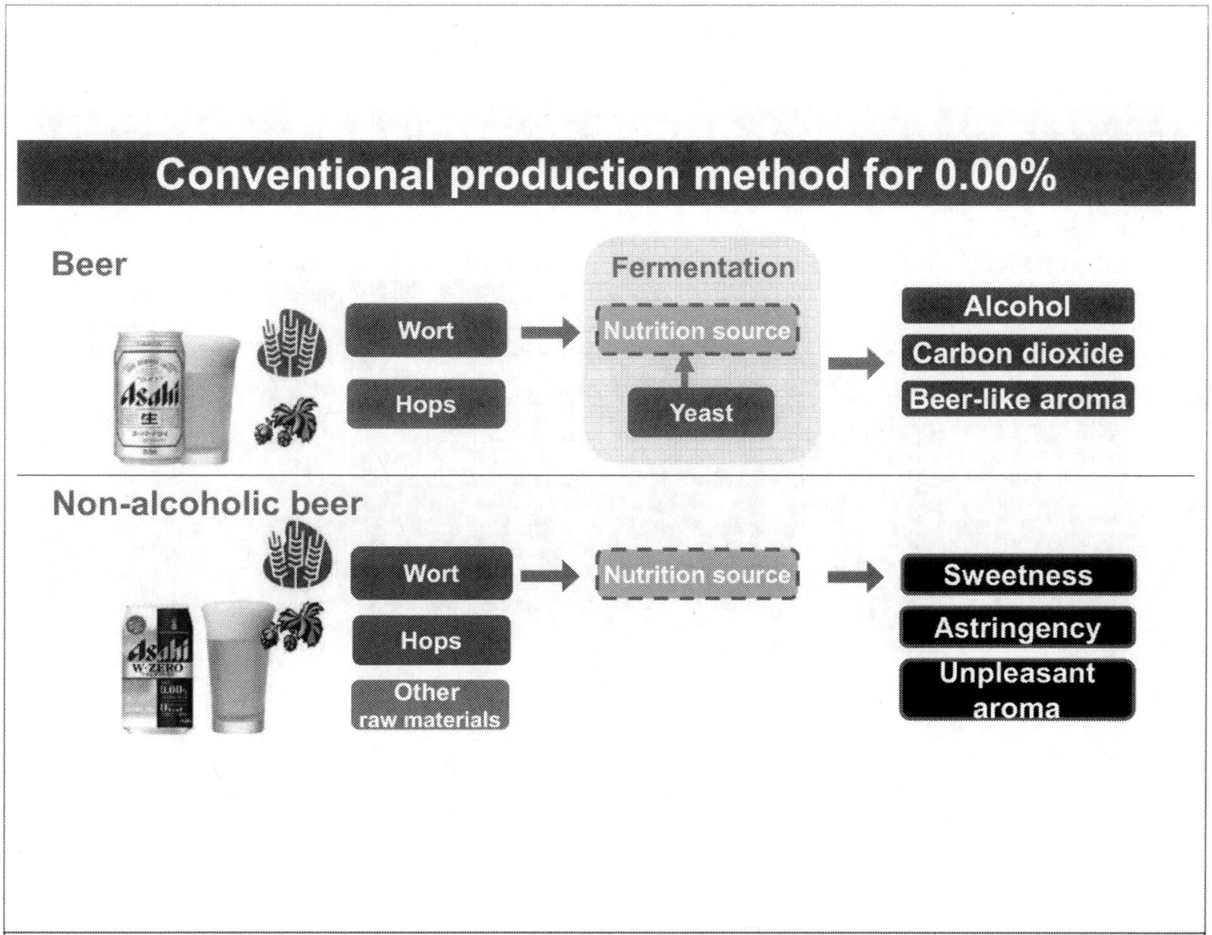
Conventional production method for 0.00%
Beer
Fermentation
Wort
Nutrition source
Hops
Yeast
Alcohol
Carbon dioxide
Beer-like aroma
Non-alcoholic beer
Wort
Nutrition source
Hops
Other raw materials
Sweetness
Astringency
Unpleasant aroma

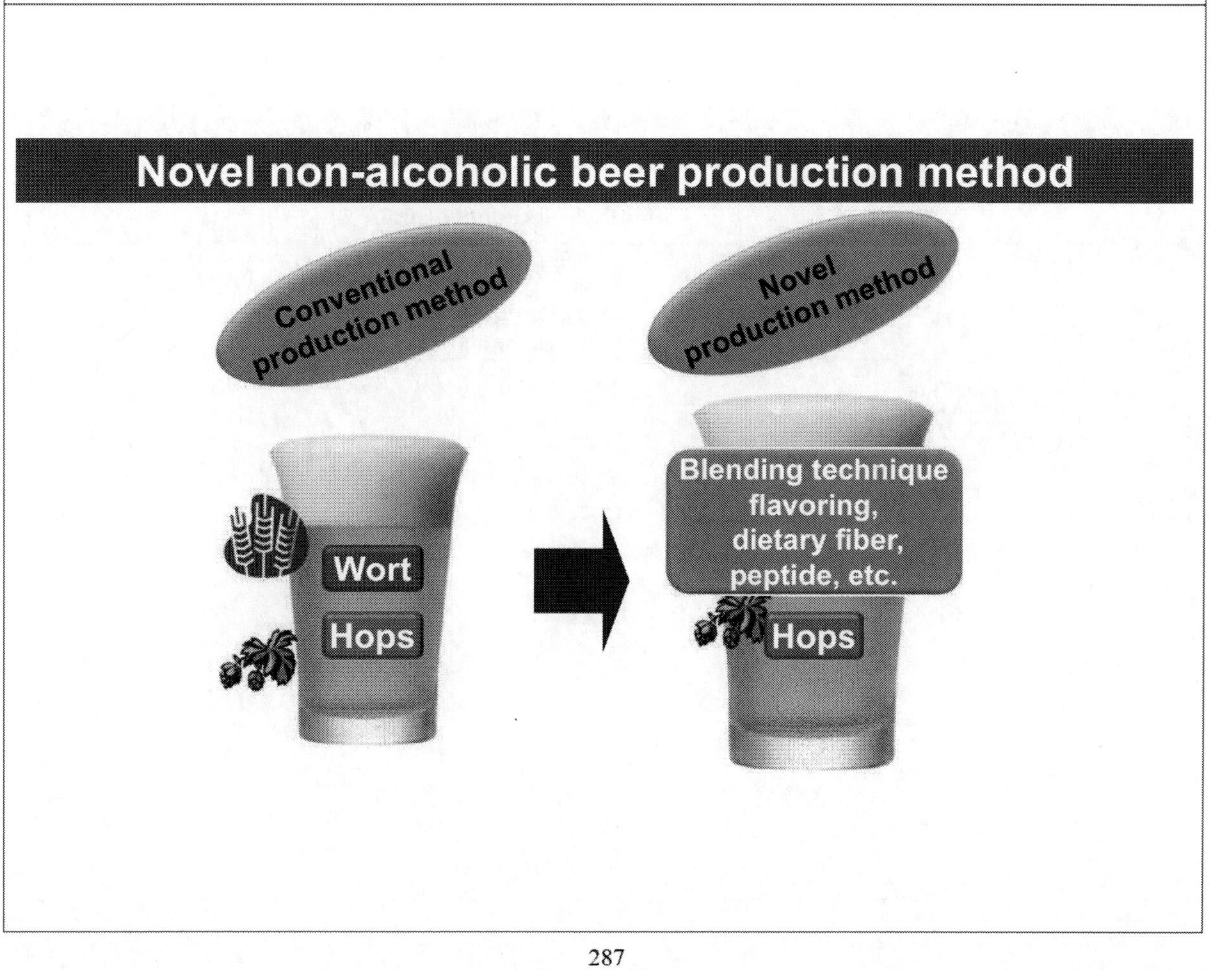
Novel non-alcoholic beer production method
Conventional production method
Novel production method
Wort
Hops
Blending technique flavoring, dietary fiber, peptide, etc.
Hops

Is it negative or positive ?

Is it negative or positive ?

<MBT>

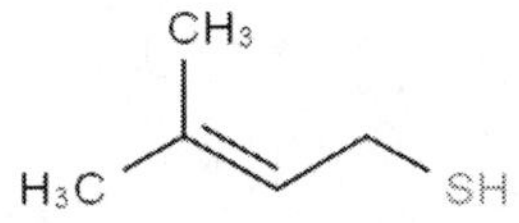

3-methyl-2-butene-1-thiol

Perception
skunky , sun-stuck flavor

Threshold
About 4ppt in beer

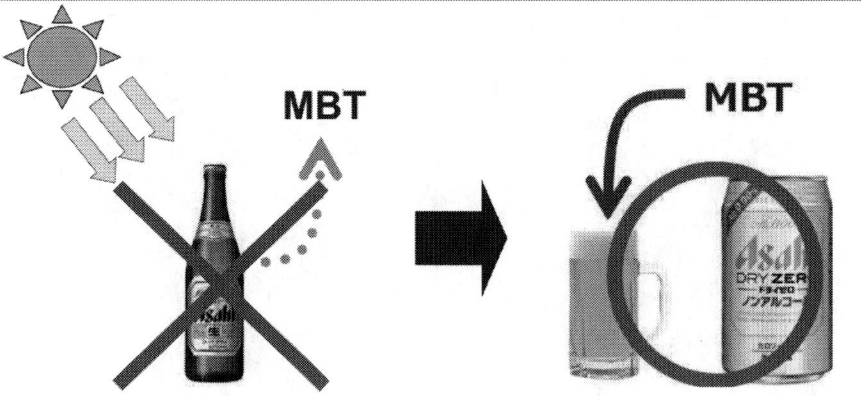

Customer survey results

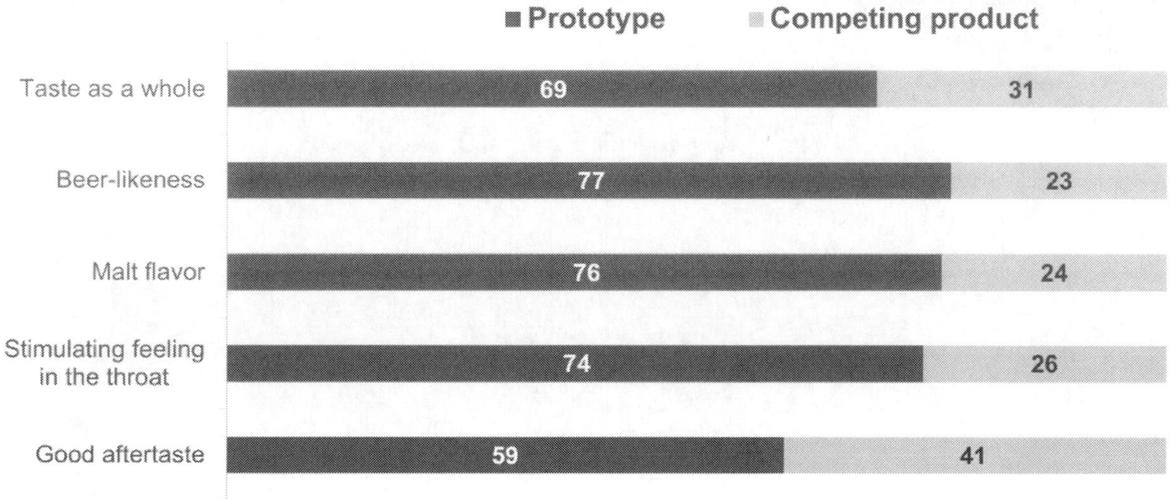

Source: Our survey (paired comparison test)
Target = 100 competing product consumers

Presentation outline

- ✓**Introduction**
- ✓**Product design**
- ✓**Development of product content**
- ✓**Manufacturing techniques for mass production**
- ✓**Market changes after product release**

Keyword

Control Point

Problems in mass production

<MBT>

3-methyl-2-butene-1-thiol

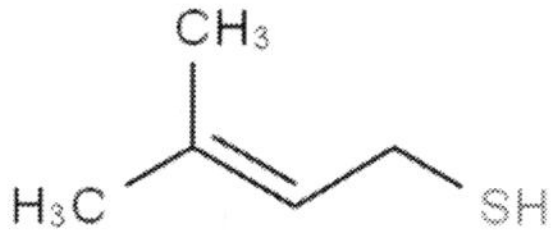

Perception
skunky, sun-stuck flavor
Threshold
About 4 ppt in beer

➢ **It does not smell**

<DMTS>

Dimethyl trisulfide

Perception
sulphury, rotten vegetable flavor
Threshold
About 10-100 ppt in beer

➢ **It smells**

Relationship between MBT and copper

MBT

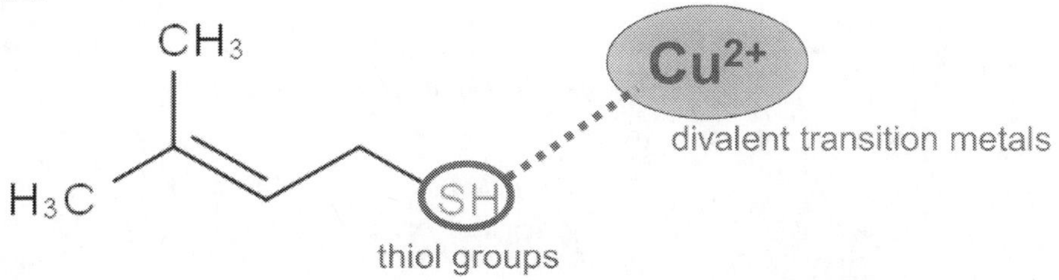

···· In general volatile thiols combine with metal (copper) ions, and the aromas are lost.

Kobayashi, H; Tominaga, T. ***J.ASEV Jpn.*** **2005**, *16*, 150-151)

Hypothesis testing (Model test)

[Method]

1. Added increasing amount of $CuSO_4$ to laboratory prototype.
2. Analyzed MBT amount.

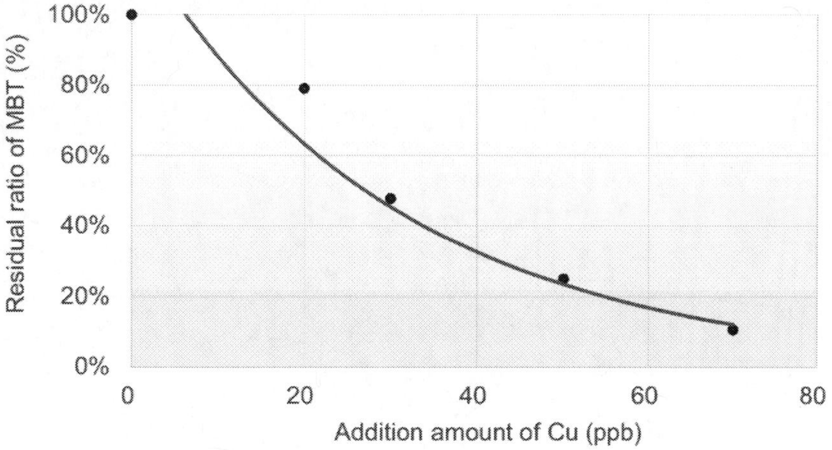

Relationship between DMTS and copper

- DMTS was generated when wort was at high temperatures over 90 °C.
- Cu promoted DMTS formation during whirlpool process.

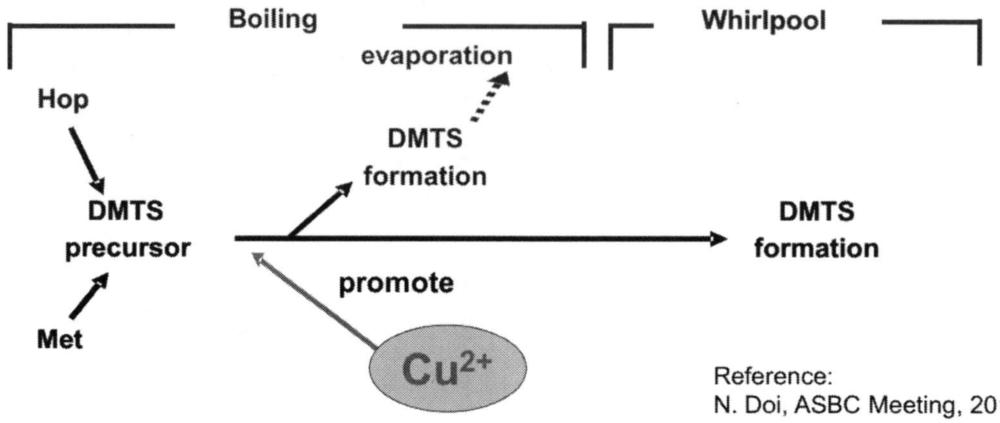

Reference:
N. Doi, ASBC Meeting, 2013

Hypothesis testing (Model test)

[Method]
1. Added increasing amount of $CuSO_4$ to the model solution.
2. Solution maintained at 95 °C for one hour and the amount of DMTS was analyzed.

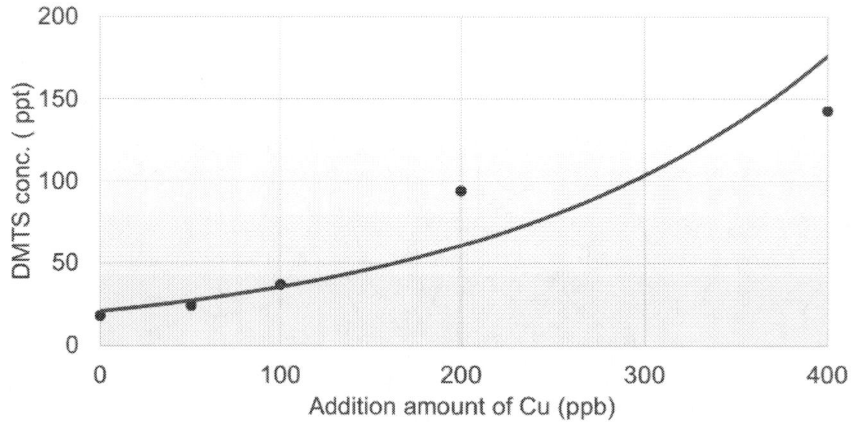

Conclusion of this section

- By controlling the elution of Cu during the process, we were able to achieve the target flavor quality

<MBT>

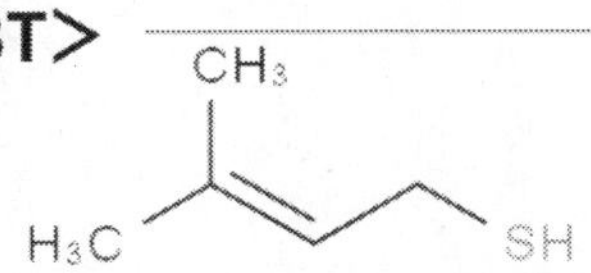

3-methyl-2-butene-1-thiol

- It does not smell ~~Cu^{2+}~~ ⇨ stably produce targeted flavor

<DMTS>

Dimethyl trisulfide

- It smells ~~Cu^{2+}~~ ⇨ It does not smell

Presentation outline

✓Introduction

✓Product design

✓Development of product content

✓Manufacturing techniques for mass production

✓Market changes after product release

Launch "Asahi Dry Zero" in 2012

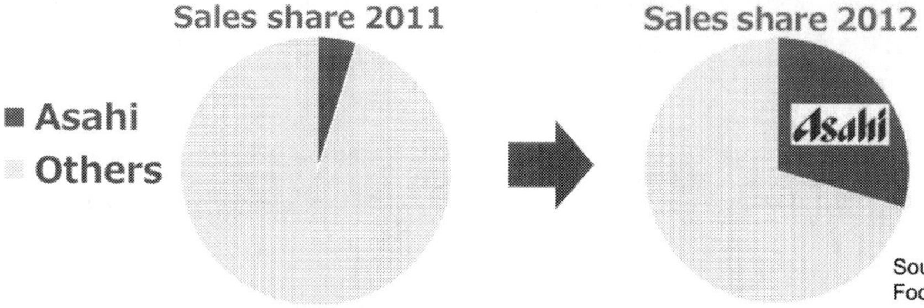

Source: Fuji Economy "2013,14 Food Marketing Handbook"

Sales of non-alcoholic beer in Japan (2011-2012)

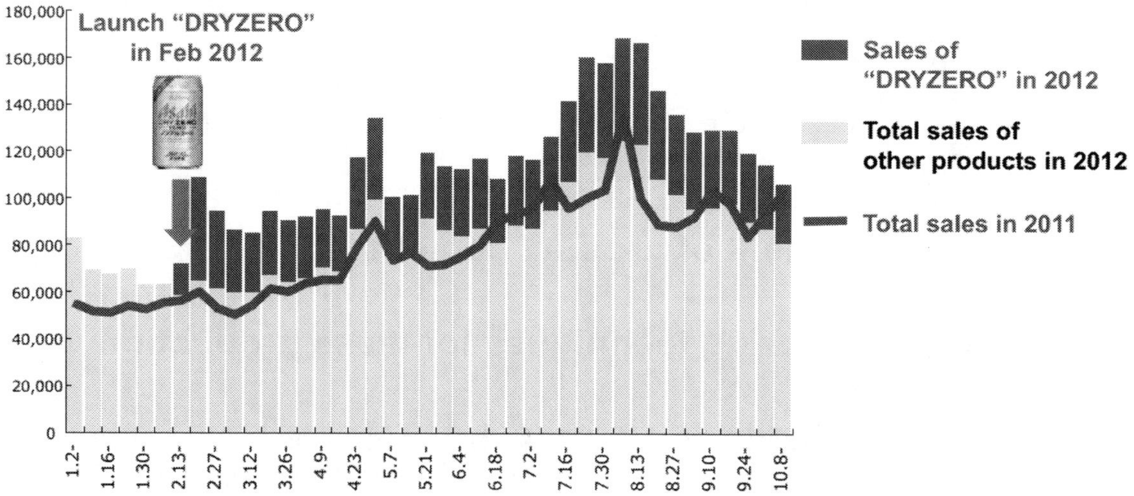

Source: MAI survey by INTAGE (NAB weekly performance trend in Japan) Capacity per store in 2011-2012

Sales of non-alcoholic beer in Japan (2011-2016)

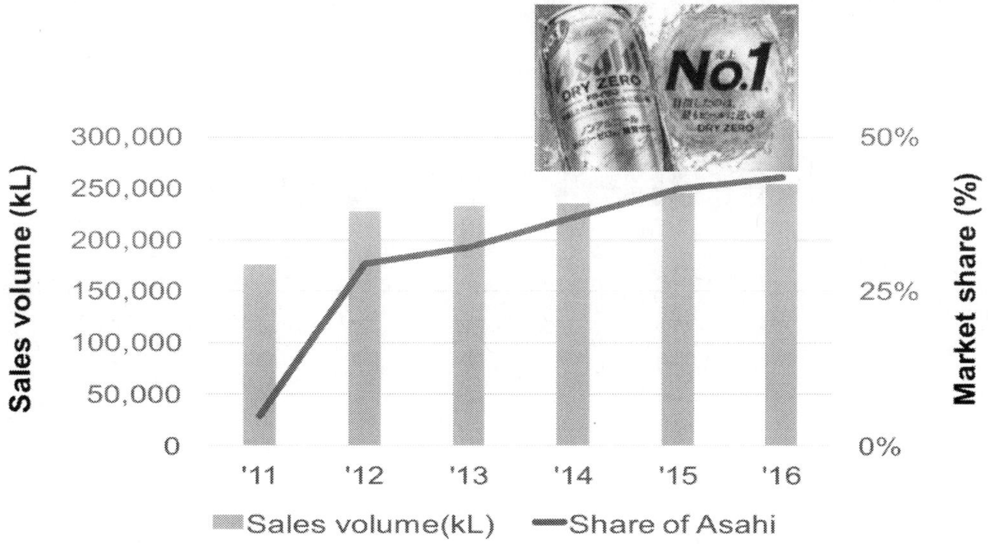

Source: Fuji Economy "2013-2017 Food Marketing Handbook"

Reference material

Customer Core Values

Start drinking without any unpleasant sensation

For example,

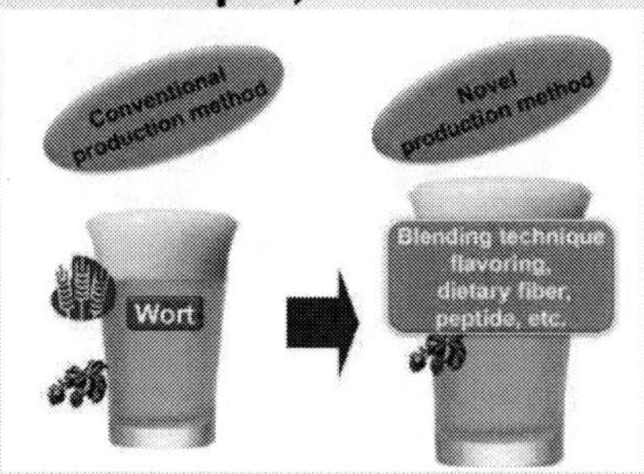

Customer Core Values

Stimulating feeling in the throat

For example,
we found that the stimulating feeling in the throat is increased by adding high-molecular-weight polysaccharides and dietary fiber.

Customer Core Values

Good aftertaste (Retronasal Aroma)

For example,
we built a flavor placing importance on retronasal aromas balance.

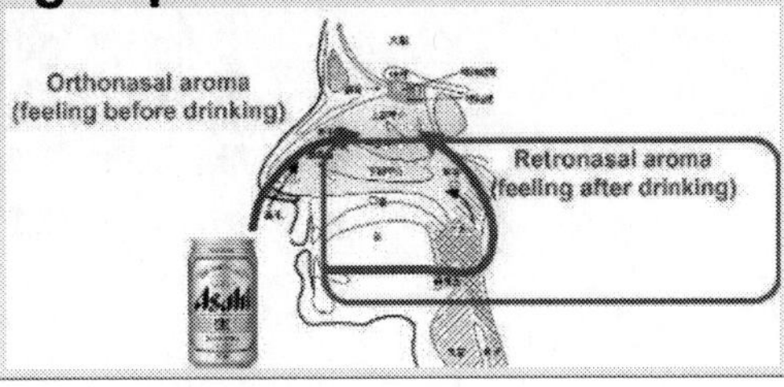

Product information

Asahi Dry Zero Non-Alcoholic Beer has been relaunched. It features a taste like a real beer and is free from alcohol, calories and sugar. This product retails in a 350ml recyclable can. Launched in late December 2016, open-priced. Also relaunched in a 500ml can and a 334ml bottle.

<Raw ingredients>
dietary fibre, soy peptide, hops, flavouring, acidifier, caramel colour, antioxidant (vitamin C), sweeteners (acesulfame potassium)

< Nutrition Information per100ml>
Calories : less than 0.5 calories
Total Carbs. : 1.1g
Dietary Fiber : 1.0g
Protein : 0.2g

Improvement of foam quality by soybean polysaccharide

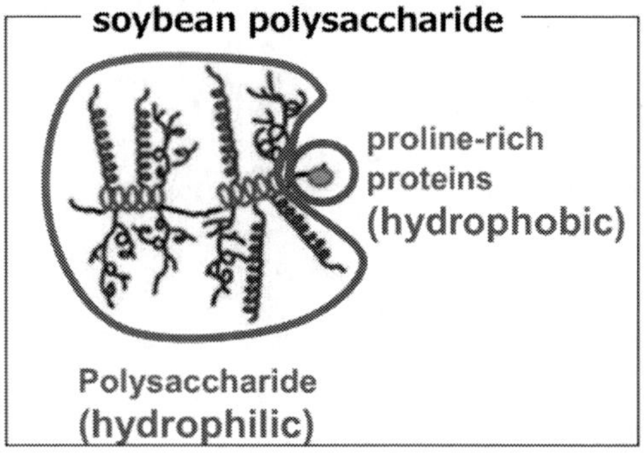

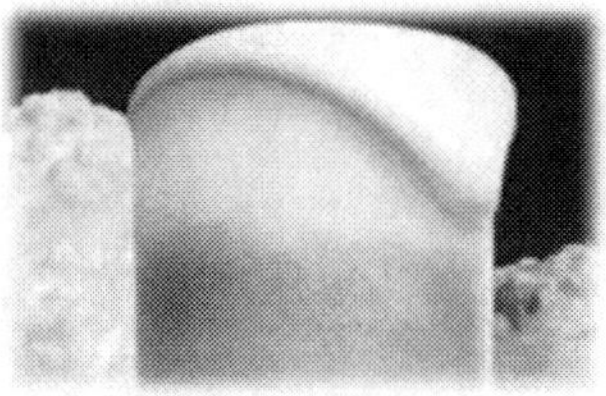

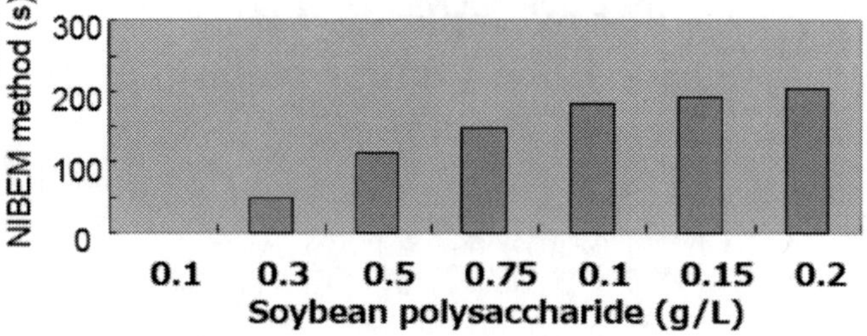

Journey into history – Analysis of century-old beer

Jana Olšovská, Dagmar Matoulková et al.

Brewery Záhlinice, Czech Republic, Raven Trading, Ltd.

Brewery Záhlinice, Czech Republic, Raven Trading, Ltd. – **April 8, 2015**

3

00 years

erature ~10 °C

36th EBC Congress, Ljubljana, May 2017 — Research Institute of Brewing and Malting, PLC

Brewery and Malt-house Záhlinice, 1895 - 1925

4

Mikoláš Aleš

36th EBC Congress, Ljubljana, May 2017 — Research Institute of Brewing and Malting, PLC

Brewery and Malt-house Záhlinice, 1895 - 1925

5

František Skopalík
(1822 – 1891)
native and mayor of Záhlinice,
deputy of the Moravian Land Assembly,
deputy of the Imperial Council (1871)

1875 – **Emanuel Proskowetz Jr.**
Proskowetz Hanácký Pedigree
(from Kvasický Hanácký barley)

- 1861 - 1895 – construction approval of the brewery and malt-house, gross construction
- 1896 – **First brew** – brewery with modern equipment – having already steam boilers!!
- 1921 – bought by the Přerov Brewery
- 1925 – closure of the brewery
- Floor malt-house – **still in operation**

Brewery and Malt-house Záhlinice – Today

Raven Trading Ltd., reconstruction, the first brew: June 26, 2016

Záhlinice – The museum of František Skopalík

History – The origin of lager beer in the Czech Republic

Historical photo of brewery in Pilsen

J. Groll

F.O. Poupě

- 18th century – Brewmaster František Ondřej Poupě, the father of Czech modern brewing (1753 – 1805)
- **1842** – Town brewery in **Pilsen was founded**
- Martin Kopecký, Burgomaster of Pilsen, invited a brewmaster Josef Groll from Bavaria
- The first brew was brewed using a **bottom fermentation**

Budějovický Budvar 1895

Holba Hanušovice 1874

Analysis of century-old beer

9

- Sensory
- Chemical
- Microbiology/Genetic

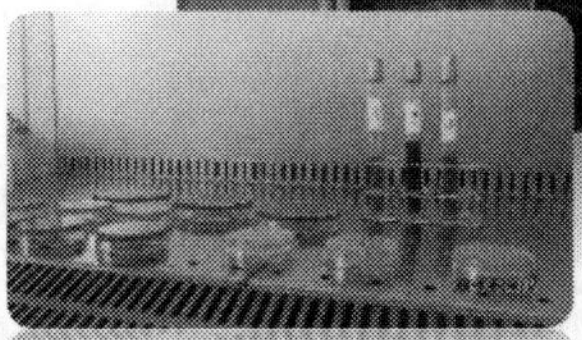

36th EBC Congress, Ljubljana, May 2017

Research Institute of Brewing and Malting, PLC

Performed analysis

10

- Original extract (EBC 9.4)
- Alcohol (EBC 9.2.6)
- Total acidity (ASBC Beer 8)
- pH (EBC 9.35)
- Color (EBC 9.6)
- Bitterness (EBC 9.8)
- Total nitrogens (EBC 9.9.1)
- FAN (EBC 9.10)
- Volatile compounds (EBC 9.39)
- Minerals and heavy metals (EBC 9.11-20)
- Carbonyl compounds (GC-MS, Čejka et al., 2013)
- Carbohydrates (HPLC-RI, Jurková et al., 2014)
- Fatty acids (GC-FID, Horák et al., 2013)
- Oxidation products of bitter acids (HPLC-HRMS, Dušek et al., 2014)
- Isolation of genomic DNA, PCR, sequencing (Olšovská et al., 2017)

36th EBC Congress, Ljubljana, May 2017

Research Institute of Brewing and Malting, PLC

Basic chemical parameters of century-old beer and current beers

11

Analyte	Units	Beer A	Beer B	Beer C	Beer D	Beer E	Beer F	Beer G	Beer H
		1900?*	1900?*	1900?*	1999*	2004*	2015*	2015*	2015*
Extract real	%	1.87	1.58	2.09	5.04	4.09	3.66	5.1	4.61
Alcohol (v/v)	%	5.45	3.85	5.38	4.41	4.79	5.2	4.3	5.16
Alcohol (w/v)	%	4.31	3.04	4.25	3.24	3.75	4.08	3.35	4.04
Original extract	%	10.30	7.62	10.40	11.71	11.39	11.59	11.61	12.42
Bitterness	IBU	15	8	22	15	20	31	32	25
Colour	EBC u.	10.0	60.2	17.2	24.0	22.4	8.4	12.0	11.3
pH	-	3.89	3.7	3.86	4.67	4.43	4.53	4.6	4.6
Total acidity	mL 1M NaOH/100 mL	5.08	12	4.34	2.6	2.86	2.16	2.4	3.63
Total nitrogen	(g/100 mL)	0.36	0.13	0.48	0.47	0.47	0.37	0.45	0.48
FAN	mg/L	149	12	180		99	198	180	257
Total polyphenols	mg/L	105	5	21		116	100	150	125

*year of beer production

- Beer D, 1999, stored in the dark at 20 °C, typical Czech lager (stabilized, pasteurized)
- Beer E, 2004, stored in the dark at 20 °C, typical Czech lager (stabilized, pasteurized)
- Beers F and G, fresh 2015, traditional and well-known Czech commercial lagers, which are chemically and sensorially different (stabilized, pasteurized)
- Beer H, fresh 2015, lager produced at our institute (non-stabilized, non-pasteurized).
- Probably because of a secondary fermentation or even some kind of contamination of century-old beers, **higher content of alcohol and very low real extract were detected**.

Beer A

12

- Original extract 10.30 %
- Alcohol 5.45 % (v/v)
- Real extract 1.87 %
- Low pH 3.89
- Higher value of total acidity 5.08 mL NaOH/100 mL (ref 2.73)
- Higher concentration of DMS - 66 ug/L (ref 20)
- Increased concentration of volatile compounds
  - ethyl acetate 70.4 mg/L (ref 10.71)
  - ethyl lactate 0.64 mg/L (ref 0.12)
  - furfuryl alcohol 0.32 mg/l (ref 0.06)

- Light color (10 u. EBC)
- Hazy
- Sulphuric and fecal

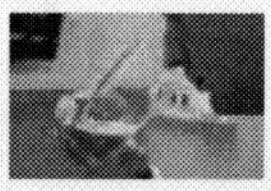

- *Coniochaeta*
- *Lecythophora sp.*
- *Dikarya*
- *Hypocreales sp.*
- *Staphylococcus hominis*
- *Streptomyces paucisporeus*

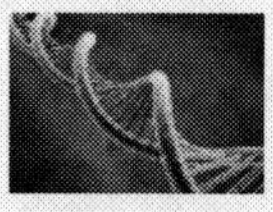

36th EBC Congress, Ljubljana, May 2017 Research Institute of Brewing and Malting, PLC

Beer A

13

- No DNA of *Saccharomyces* yeasts was detected, DNA was probably damaged during the storage.
- DNA of bacteria *Staphylococcus* and *Streptomyces* was identified.
- Microscopic analysis proved the presence of dead yeasts or yeast-like cells.

Amplification of DNA

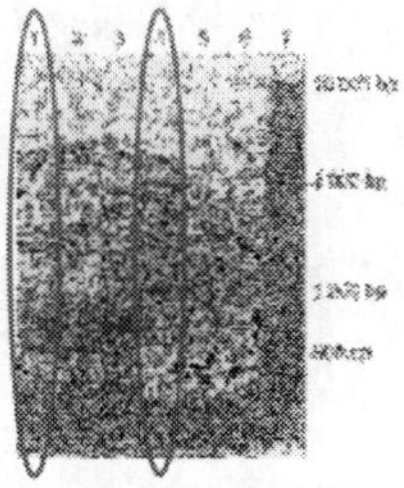

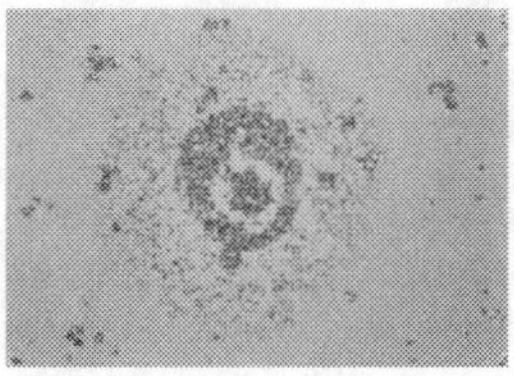

Microscopic sediment analysis

Beer B

14

- Original extract 7.62 %, alcohol 3.85 %, real extract 1.58 %
- Very low pH 3.7 (ref 4.5)
- Extremly high total acidity 12.00 mL NaOH/100 mL (ref 2.73)
- High acetic acid 9000 mg/L (ref 100)
- High ethyl acetate 0.92 mg/L (ref 0.12)
- Inverse ratio of alcohols/esters 0.15 (ref 5.72)
- Higher concentration of fatty acids
  - isobutyric, butyric, isovaleric (4 x ref)
- Extremly low FAN – 12 mg/L (ref 212)
- Extremly low (=no) total polyphenols 5.0 mg/L

- Dark color (60 u. EBC)
- Very sour (lambik-like)
- Madeira, fruity

- *Saccharomyces bayanus*
- *Saccharomyces cerevisiae x Saccharomyces eubayanus x Saccharomyces uvarum*
- *Saccharomyces pastorianus*
- *Saccharomyces uvarum*
- *Dekkera bruxellensis*

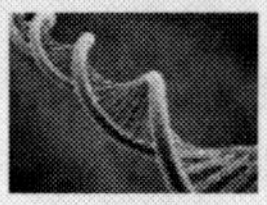

Beer B

15

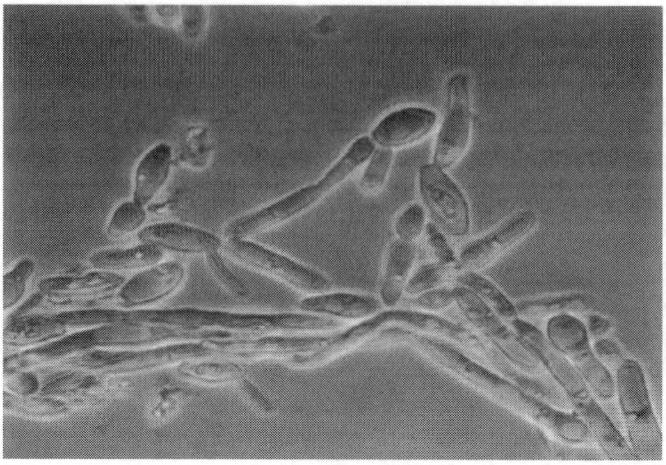

Beer B – DNA of *Saccharomyces* yeasts was identified

16

Saccharomyces bayanus/pastorianus

Identities: 185/186 (99 %)

```
Query  1    TAAATAAAATTGTTTGTGTTTGTTACCTCTGGGCTCGATTGCTCGAAACCCAAAGAAAAA  60
            ||||||||||||||||||||||||||||||||||||||||||||||||||||||||||||
Sbjct  308  TAAATAAAATTGTTTGTGTTTGTTACCTCTGGGCTCGATTGCTCGAAACCCAAAGAAAAA  249

Query  61   GTTGCAAAAGTATAAAAACTCCACAGTGTGTTGTATTGAAACGGTTTTAATTGTCCTATA  120
            ||||||||||||||||||||||||||||||||||||||||||||||||||||||||||||
Sbjct  248  GTTGCAAAAGTATAAAAACTCCACAGTGTGTTGTATTGAAACGGTTTTAATTGTCCTATA  189

Query  121  ACAAAAACAGAGAAGTCTCACTGTTTGGAATAGCAAGAAAGAAACTTGCAAGCCCTAGCA  180
            ||||||||||||||||||||||||||||||||||||||||||||||||||| ||||||||
Sbjct  188  ACAAAAACAGAGAAGTCTCACTGTTTGGAATAGCAAGAAAGAAACTTGCAA-CCCTAGCA  130

Query  181  AGACCG  186
            ||||||
Sbjct  129  AGACCG  124
```

An example of comparison of DNA from Beer B with GENBank NCBI database

Beer C

17

- Original extract 10.40 %
- Alcohol 5.38 %
- Real extract 2.09 %
- Low pH 3.86 (ref 4.5)
- High lactic acid 3500 mg/L (ref 200-500)
- High ethyl lactate 624 mg/L (ref 11)
- Inverse ratio of alcohols/esters 0.84 (ref 5.72)
- Extremly high amount of carbonyl compounds
    - 2-furfuraldehyde 6667 µg/L (ref 43)
- Interesting (detectable) profile of degradation products of bitter acids

- Light brown color (17 u. EBC)
- Beer-like, traces of CO_2
- Very slight bitterness
- Oxidized, papery, sweet, syrup

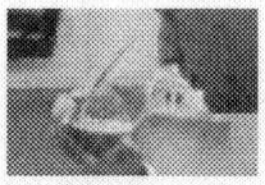

- *Debaryomyces hansenii*
- *Coniochaeta*
- *Dikarya*
- *Acremonium sp.*
- *Simplicillium sp.*
- *Bacillus simplex*
- *Streptomyces sp.*

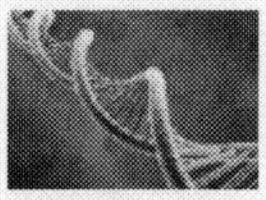

Beer C

18

- DNA of *Debaryomyces* and of *Bacillus* and *Streptomyces* bacteria was identified.
- No DNA of *Saccharomyces* yeast detected, DNA was probably damaged during the storage.
- Microscopic analysis proved the presence of dead yeasts or yeasts-like cells.

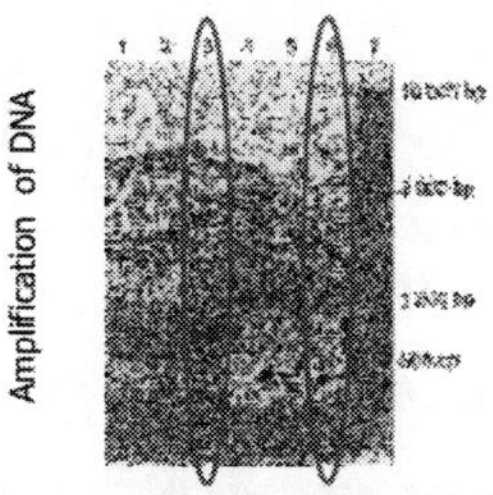

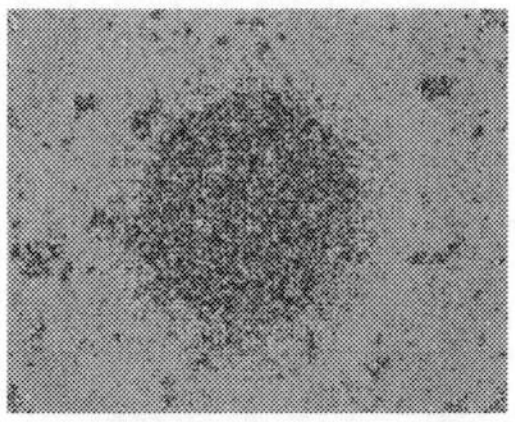

Microscopic sediment analysis

Carbonyl compounds – „Indicators of beer aging"

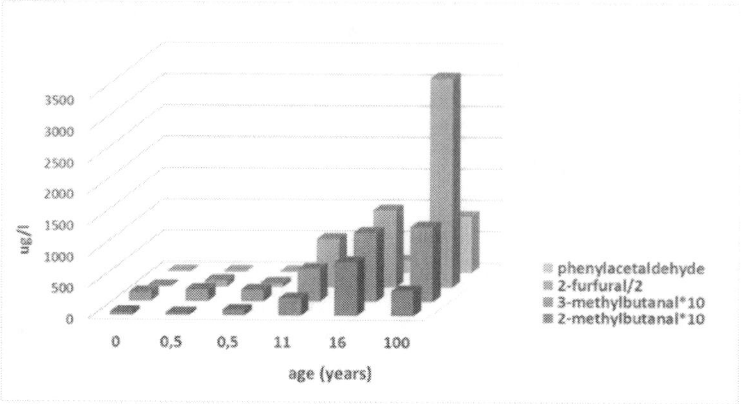

Transformation of bitter compounds (iso/iso – spiro?)

Analyte	Units	Beer A	Beer B	Beer C	Beer H
Bitter acids	IBU	15	8	22	25
Σ iso-alpha	mg/L	0.1	0.02	0.05	26.27
Bitterness	-	0	0	1	3

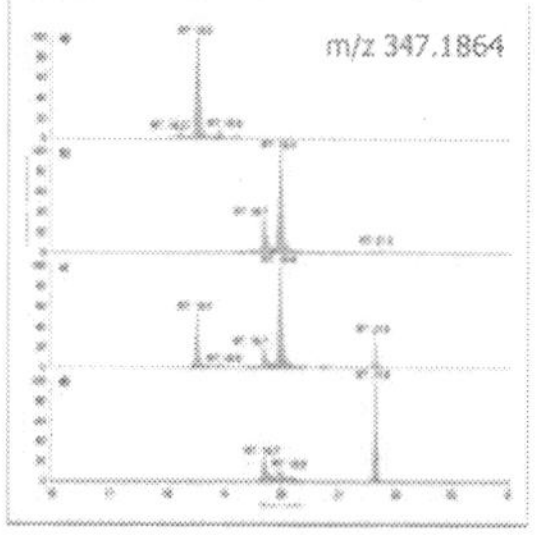

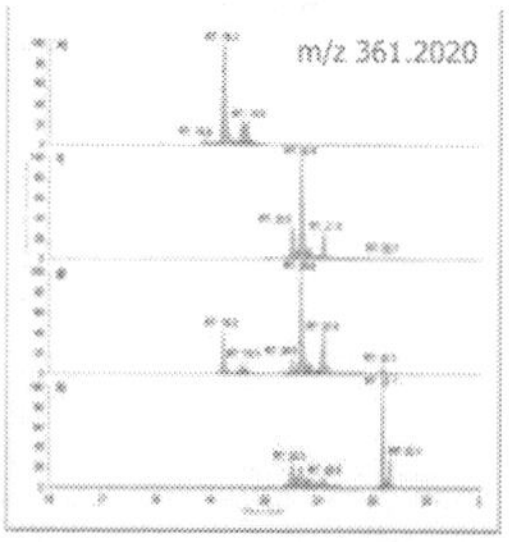

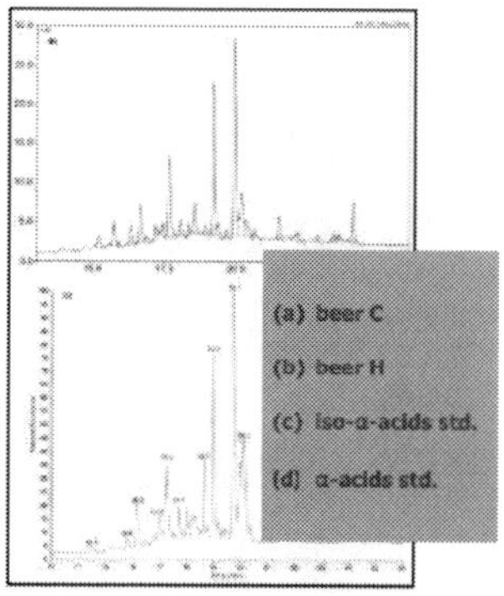

Isooctane extract
a) HPLC/UV 275 nm b) HPLC/MS TIC

21

Transformation α → iso-α → spiro-α

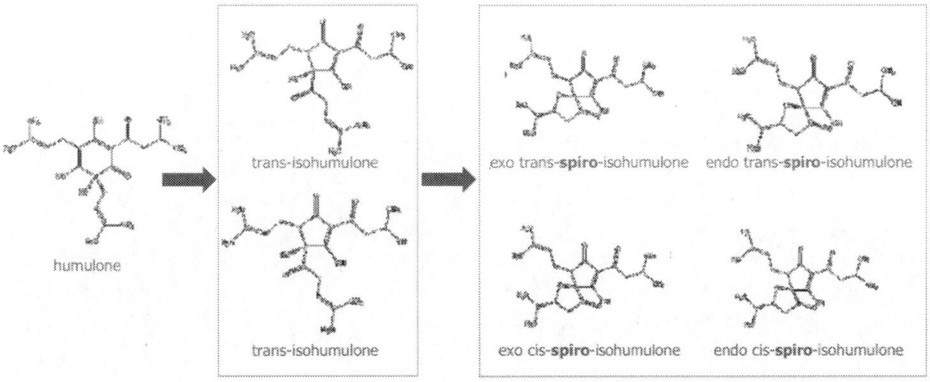

Verzele, M.; De Keukeleire, D. Volume 27. Chemistry and Analysis of Hop and Beer Bitter Acids, 1st Edition; Elsevier Sience Publishers B.V., Amsterdam, Netherlands, 1991.

22

Similar parameters of beers A, B, C

- **High concentration of metals**
  - from brewhouse equipment
  - from non-treated water
- **High concentration of myristic acid**
  - myristic acid is not a product of yeasts
  - the most probably origine from adjunct
  - varietal specifity
- **Low concentration of silicon**
  - probably consequence of adjunct (silicon originated from palea of barley)
  - varietal specificity of barley

Analyte (mg/l)	Beer A	Beer B	Beer C	Beer H
Fe	0.25	0.46	0.25	0.03
Mn	0.28	0.1	0.17	0.09
Zn	0.09	0.09	0.13	0.02
Cu	0.24	0.15	0.15	0.05
Si	9.7	4.7	6.4	35.4
Myriscic acid	0.94	1.12	1.41	0.41

Summary

23

- All beers were lager type (recipe, DNA results)
- Different contamination – various mechanisms of chemical transformation

- **Beer B**
  - *Dekkera bruxellensis* – corresponding sensorial and chemical profile
  - *Saccharomyces* yeasts
- **Beer C** – products of natural aging process
  - extremely high concentration of carbonyl compounds (2-furfuralaldehyde)
  - identification of naturally originated transformation products of iso-α-bitter-acids
- **Beers A, B, C**
  - high amount of Fe, Mn, Zn, Cu
    - similar raw material and technology
  - high ratio of adjuncts
    - low concentration of Si, high concentration of myristic acid

J. Agric. Food Chem., 2017, 65, 3341 - 3350

24

JOURNAL OF AGRICULTURAL AND FOOD CHEMISTRY

Analysis of 100-Year-Old Beer Originated from the Czech Republic

Jana Olšovská,[*,†] Dagmar Matoulková,[†] Martin Dušek,[†] Jürgen Felsberg,[§] Markéta Jelínková,[§] Pavel Čejka,[†] and Karel Štěrba[†]

[†]Research Institute for Brewing and Malting, PLC, Lípová 15, CZ-120 44 Prague, Czech Republic

[§]Institute of Microbiology, Academy of Sciences of the Czech Republic, v.v.i., Vídeňská 1038, CZ-142 20 Prague, Czech Republic

Supporting Information

ABSTRACT: Three bottles of different beers were found in 2015 during a reconstruction of the brewery of the Raven Trading s.r.o. company in Záhlinice, Czech Republic. Thanks to good storage conditions, it was possible to analyze their original characteristics. All three bottles contained most probably lager type beer. One beer had sulfuric and fecal off flavors; it was bright with the original extract of 10.3° Plato. The second beer, with an original extract of 7.6° Plato, was dark and very acidic, resembling Lambic. DNA analysis proved the presence of *Dekkera bruxellensis*, which corresponded to its chemical profile (total acidity, FAN, ethyl acetate, total esters). The third beer contained traces of carbon dioxide bubbles, was light brown and slightly bitter, with an original extract 10.4° Plato. Because it obviously underwent a natural aging process, sweetness, honey, and fruity off-flavors were detected and transformation products of iso-α-acids were found.

KEYWORDS: *beer aging, carbonyl compounds, transformation of isohumulones, Dekkera bruxellensis, Saccharomyces*

Acknowledgement

- Mr. Aleš Přinosil, the brewmaster of Záhlinice, Raven Trading Ltd.
- Co-autors and colleagues
  - Pavel Čejka
  - Martin Dušek
  - Martin Slabý
  - Marie Jurková
  - Karel Štěrba
  - Markéta Jelínková (IMB, AS CR)
  - Jürgen Felsberg (IMB, AS CR)
- Associate editor of JAFC, Prof. Dr. Peter Schieberle

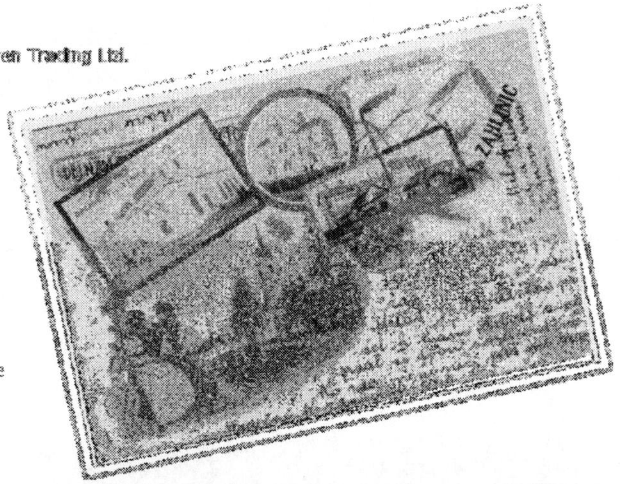

Thank you for your attention

BARTH-HAAS GROUP
FOR YOUR SUCCESS

The influence of yeast fermentation on dry hop aroma and flavour

Basic idea

BARTH-HAAS GROUP
FOR YOUR SUCCESS

- Over 400 aroma compounds in hop oil detected so far
- Hoppy aroma is far from beeing completely understood
- Growing importance of hop aroma compounds on the overall aroma of beer
- Influence of yeast on the overall beer aroma
- Proven intereaction of yeast and hop aroma compunds

Experimental setup

BARTH-HAASGROUP
FOR YOUR SUCCESS

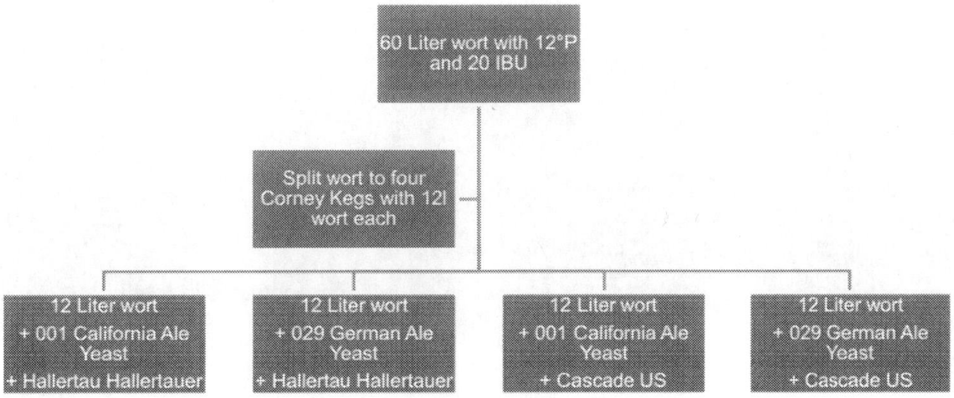

 3

Thiols and potential Thiols in hops

BARTH-HAASGROUP
FOR YOUR SUCCESS

➢ No decrease by oxidation in this case

	3MH	4MMP	A3MH	C3MH	C4MMP	G3MH	G4MMP
	(µg/Kg)	(µg/Kg)	(µg/Kg)	(µg/Kg)	(µg/Kg)	(µg/Kg)	(µg/Kg)
Hallertau Hallertauer (no oxygen)	3,2	0,3	0,7	582	1,5	13199	nd
US Cascade (no oxygen)	10,8	2,6	4,7	461	2,2	16620	nd
Hallertau Hallertauer (opened)	3,4	0,3	0,8	641	1,7	12217	nd
US Cascade (opened)	11,7	2,6	4,3	477	2,3	17206	nd

 4

Analysed aroma components

BARTH-HAAS GROUP
FOR YOUR SUCCESS

Thiols	3-mercaptohexan-1-ol (3MH)
Thiols	4-methyl-4-sulfanylpentan-2-one (4MMP)
Thiols	3-mercaptohexyl-acetate (3MHA)
Terpenols	α-terpineol
Terpenols	cis-rose oxide
Terpenols	citronellol
Terpenols	geraniol
Terpenols	linalool
Terpenols	nerol
Terpenols	trans-rose oxide
sotolon	sotolon
Pyrazines	2-isopropyl-3-methoxypyrazine
Pyrazines	2-methoxy-3-ethylpyrazine
Pyrazines	2-secbutyl-3-methoxypyrazine
Pyrazines	2-isobutyl-3-methoxypyrazine
Furaneols	furaneol
Furaneol	homofuraneol
Esters/Acetates	2-phenylethanol
Esters/Acetates	2-phenylethyl acetate
Esters/Acetates	butyl acetate
Esters/Acetates	ethyl 2-methyl butyrate
Esters/Acetates	ethyl 3-hydroxybutyrate
Esters/Acetates	ethyl butyrate
Esters/Acetates	ethyl cinnamate
Esters/Acetates	ethyl decanoate
Esters/Acetates	ethyl hexanoate
Esters/Acetates	ethyl isobutyrate
Esters/Acetates	ethyl isovalerate
Esters/Acetates	ethyl lactate
Esters/Acetates	ethyl laurate
Esters/Acetates	ethyl octanoate
Esters/Acetates	hexyl acetate
Esters/Acetates	isoamyl acetate
Esters/Acetates	isobutyl acetate
Esters/Acetates	methyl 2-methylbutyrate
Esters/Acetates	methyl isovalerate
Esters/Acetates	octyl acetate
Esters/Acetates	propyl acetate

DMS	DMS
C13	α-ionone
C13	β-ionone
C13	β-damascenone
C13	TDN
woody	4-methyl guaiacol
woody	5-methyl furfural
woody	cis-isoeugenol
woody	cis-whiskey lactone
woody	ethyl vanillin
woody	eugenol
woody	furfural
woody	guaiacol
woody	maltol
woody	o-cresol
woody	phenylacetaldehyde
woody	syringaldehyde
woody	syringol
woody	trans-2-nonenal
woody	trans-isoeugenol
woody	trans-whiskey lactone
woody	vanillin

60 components in total!

Analysed aroma components

BARTH-HAAS GROUP
FOR YOUR SUCCESS

- Most important components for hoppy aroma in beer

Thiols	3-mercaptohexan-1-ol (3MH)
Thiols	4-methyl-4-sulfanylpentan-2-one (4MMP)
Thiols	3-mercaptohexyl-acetate (3MHA)
Terpenols	α-terpineol
Terpenols	cis-rose oxide
Terpenols	citronellol
Terpenols	geraniol
Terpenols	linalool
Terpenols	nerol
Terpenols	trans-rose oxide

Example for reproducibility

BARTH-HAAS GROUP
FOR YOUR SUCCESS

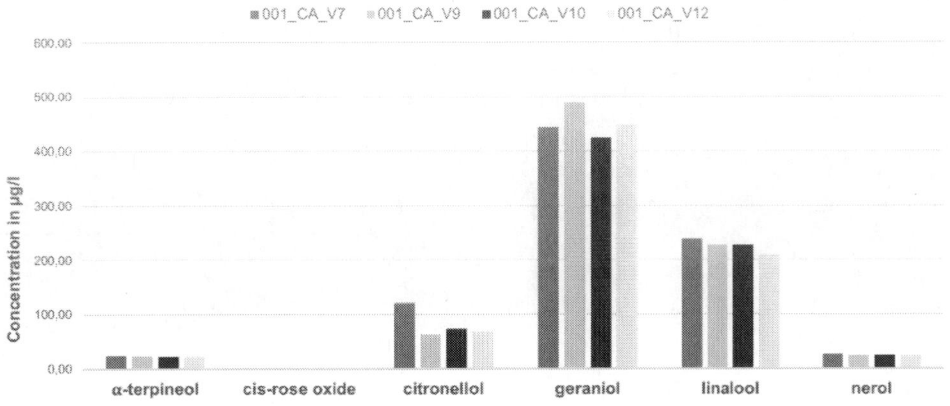

Concentration of 3MH & 4MMP

BARTH-HAAS GROUP
FOR YOUR SUCCESS

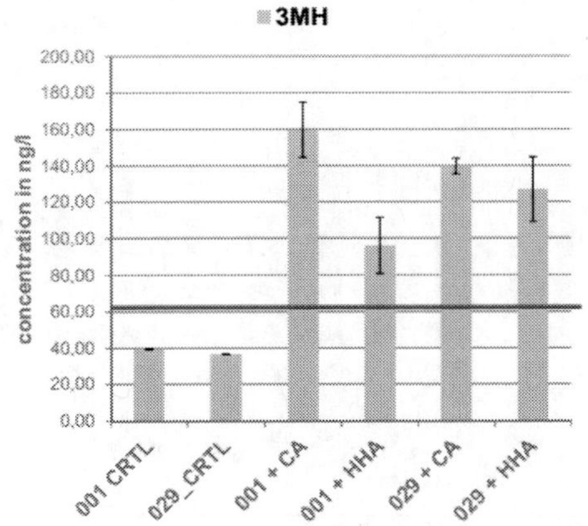

- Sensory threshold at 60 ng/l
- Trend: released 3MH depends more on hop variety than yeast
- Significant difference in released 3MH for yeast 001 and Cascade US hops

Concentration of 3MH & 4MMP

BARTH-HAAS GROUP
FOR YOUR SUCCESS

- Sensory threshold at 2 ng/l
- Trend: released 4MMP depends more on hop variety than on yeast
- 4MMP only found in 2 out of 4 batches Hallertau Hallertauer

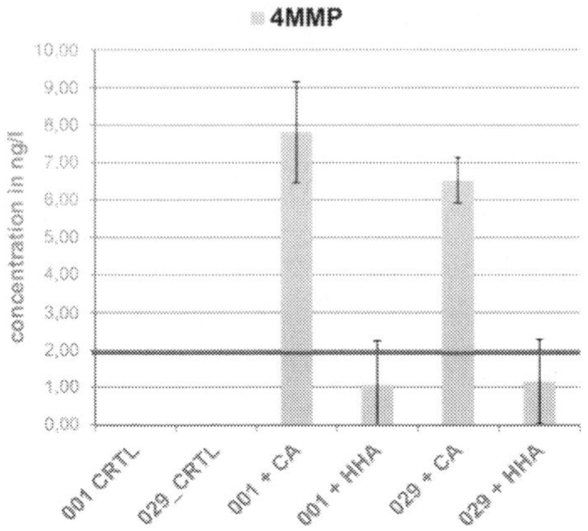

 9

Analysed geraniol and linalool content

BARTH-HAAS GROUP
FOR YOUR SUCCESS

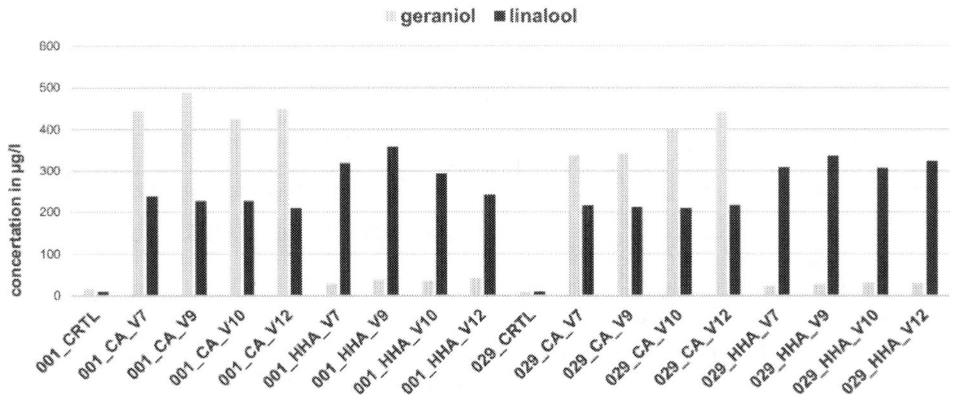

 10

Terpenols in total

BARTH-HAASGROUP
FOR YOUR SUCCESS

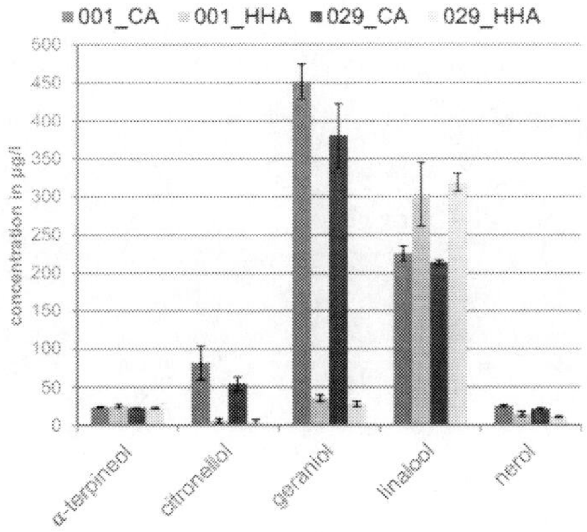

- Geraniol higher in Cascade beers
- Linalool higher in Hallertau Hallertauer beers
- Yeast 001 releases more geraniol and citronellol

 11

Terpenols in total

BARTH-HAASGROUP
FOR YOUR SUCCESS

- Cascade beers with higher terpenol concentration in total
- Beers with yeast 001 higher in concentration of terpenols

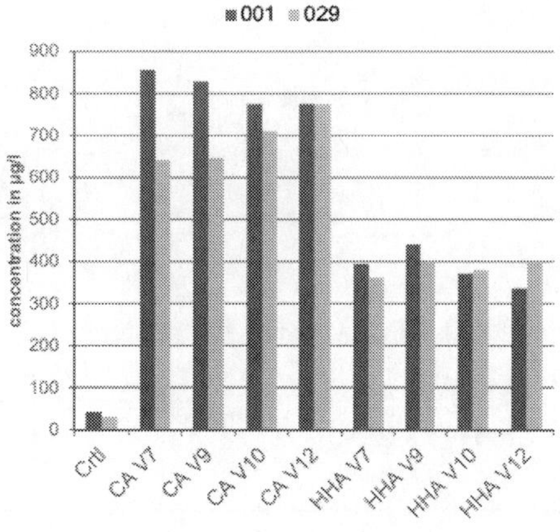

 12

Sensory profiling of the beers

BARTH-HAAS GROUP
FOR YOUR SUCCESS

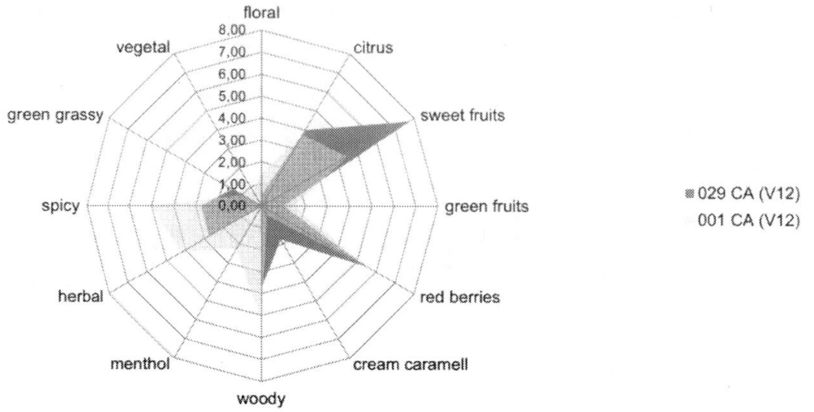

n=18

Sensory profiling of the beers

BARTH-HAAS GROUP
FOR YOUR SUCCESS

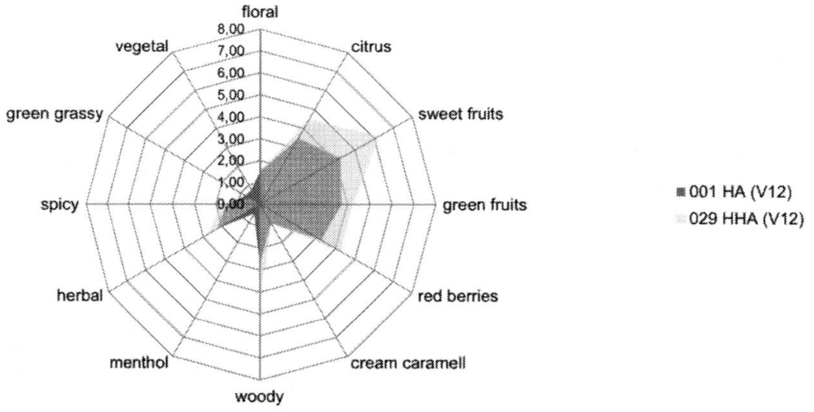

n=18

Sensory profiling of the beers

BARTH-HAAS GROUP
FOR YOUR SUCCESS

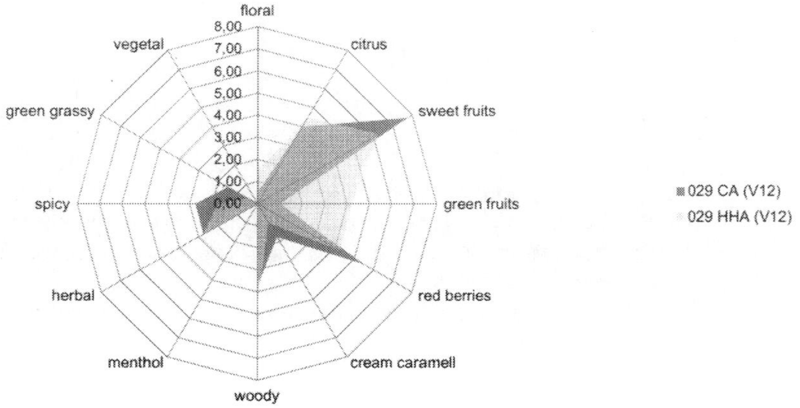

n=18

 15

Sensory profiling of the beers

BARTH-HAAS GROUP
FOR YOUR SUCCESS

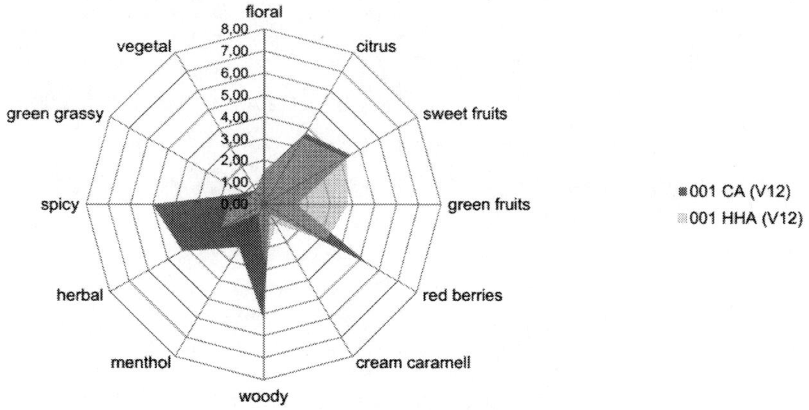

n=18

 16

Correlation between sensory and lab

BARTH-HAAS GROUP
FOR YOUR SUCCESS

	Sum of terpenols (µg/l)	Sum of esters (µg/l)	Floral (1-10)	Citrus (1-10)	Sweet fruits (1-10)	Green fruits (1-10)
001_CA_V12	775,0	2813,7	1,5	3,75	4,5	1,5
001_HHA_V12	336,6	3080,6	1,5	3,5	4,25	3,75
029_CA_V12	774,0	4330,4	0,5	4	7,75	1
029_HHA_V12	361,2	7043,1	1,5	4,5	6,25	4,25

- Sum of terpenols independent from sensory intensity of floral/citrus!
- High differences in geraniol have no influence on floral or citrus intensity; no clear correlation between analytical single components and sensory intensities
- Sum of yeast esters correlates with intensitiy of sweet fruit aroma (maybe also green fruits)

Correlation between sensory and lab

BARTH-HAAS GROUP
FOR YOUR SUCCESS

	4MMP (ng/l)	Sum of terpenols (µg/l)	Citrus (1-10)	Sweet fruits (1-10)	Green fruits (1-10)	Red berries (1-10)
001_CA_V12	7,6	775,0	3,75	4,5	1,5	5,5
001_HHA_V12	nd	336,6	3,5	4,25	3,75	3,25
029_CA_V12	7,3	774,0	4	7,75	1	5,5
029_HHA_V12	2,1	361,2	4,5	6,25	4,25	4,25

- Higher amounts of 4 MMP also show higher intensities in red berry aroma

Correlation between sensory and lab

BARTH-HAASGROUP
FOR YOUR SUCCESS

	Sum of woody components	Woody (1-10)	Menthol (1-10)	Herbal (1-10)	Spicy (1-10)	Green grassy (1-10)	Vegetal (1-10)
001_CA_V12	757,7	5,25	2,25	4,25	5	1	0,75
001_HHA_V12	730,5	2,75	0,5	2,25	1,5	0,75	0,75
029_CA_V12	753,9	3,75	0	2,75	2,75	1,5	0
029_HHA_V12	888,7	4,5	0	2,75	1,75	0,5	0

- The sum of woody components matches the intensities for woody characters
- „woody" components found: guaiacol, maltol, vanillin

 19

BARTH-HAASGROUP
FOR YOUR SUCCESS

Questions & Answers

Benedikt Matsche

Benedikt.matsche@simplyhops.co.uk

www.BarthHaasGroup.com

 20

FlavorActiV™

"OFF-FLAVOURS DUE TO MICROBIOLOGICAL PROBLEMS IN SPECIALITY BEERS IN EUROPE"

Dr Boris Gadzov

36th EBC Congress Ljubljana, Slovenia 2017

Detected off-flavours in Beer?

Contents

- Sensory Evaluation Currently
- Study
- Results
- Conclusions
- Beer Judging – Sensory Best Practices

Sensory Evaluation Currently

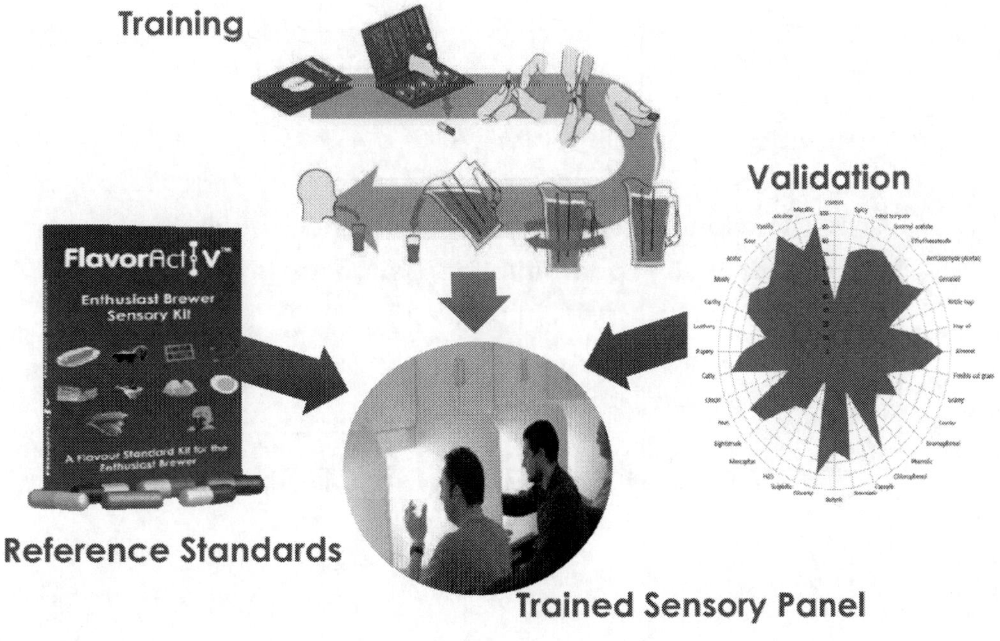

Use a Sensory Panel to Reduce Risks

- **Why humans?** Sensory properties cannot be completely and thoroughly measured by instruments, but require the interaction between the beverage and the individual
- **Why a panel?** Individuals have different sensitivities and may have an "off" evaluation; thus a panel of assessors is required

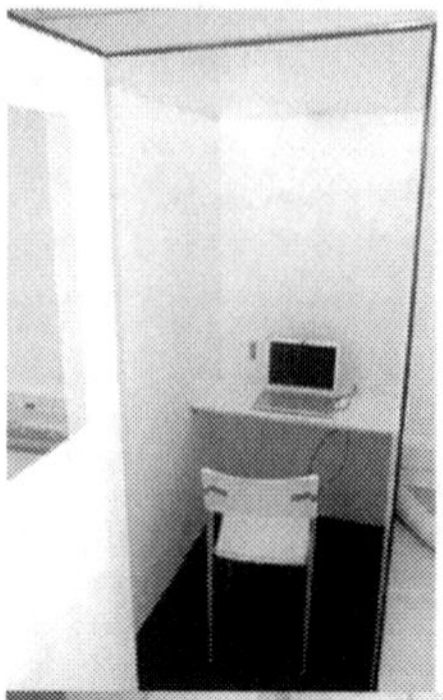

Mechanisms of perception:

- **Vision : appearance**
- **Touch : consistency & texture**
- **Olfaction : odour, aroma, fragrance, smell**
- **Trigeminal : pain, irritation, pungency [mouthfeel]**
- **Gustation : taste, [mouthfeel]**
- **Hearing : sound**
- **Temperature : cool - heat**

Chemical senses

Tasting Techniques

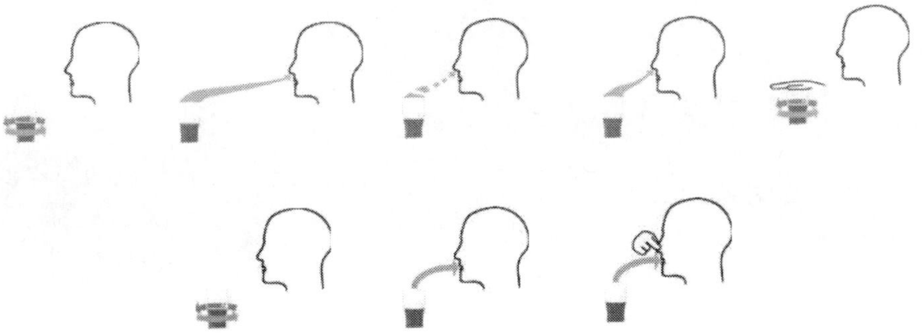

Good Sensory Practice Requires Attention to…

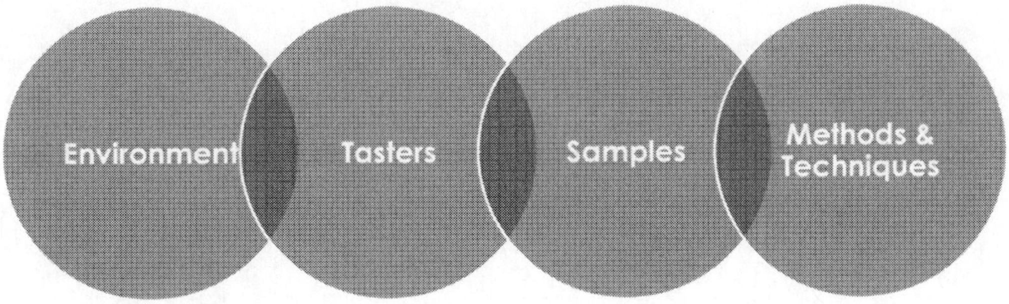

Study

278 brands

12 countries

165 judges

Beer Categories

- Wheat beers
- Porter
- Stout
- Ale
- Honey beers
- API
- Sour beers

Beer Judges

- Trained on Flavour Recognition & Intensity
- Validated several times per year
- Beer Category & Tasting form familiarization
- Individual and group sessions
- Evaluate 3 samples per sessions & repetitions
- Monitoring score on repetitive samples & Calculate Standard Deviation

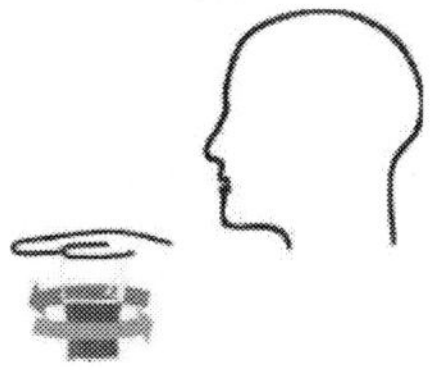

Results - Micro Contaminations in %

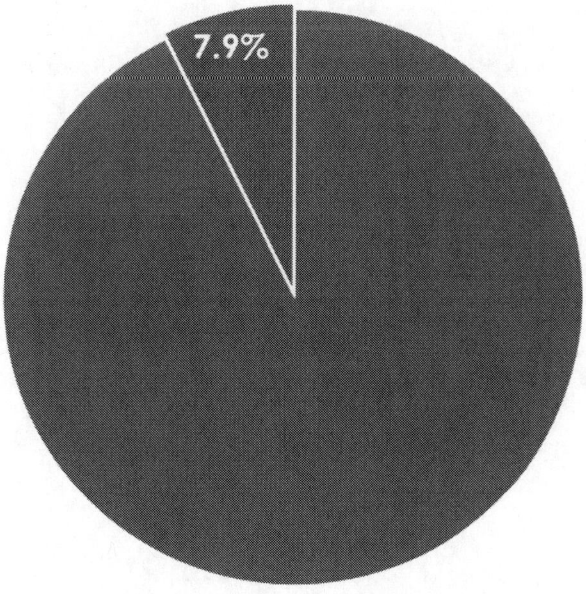

Micro Contaminations per beer style

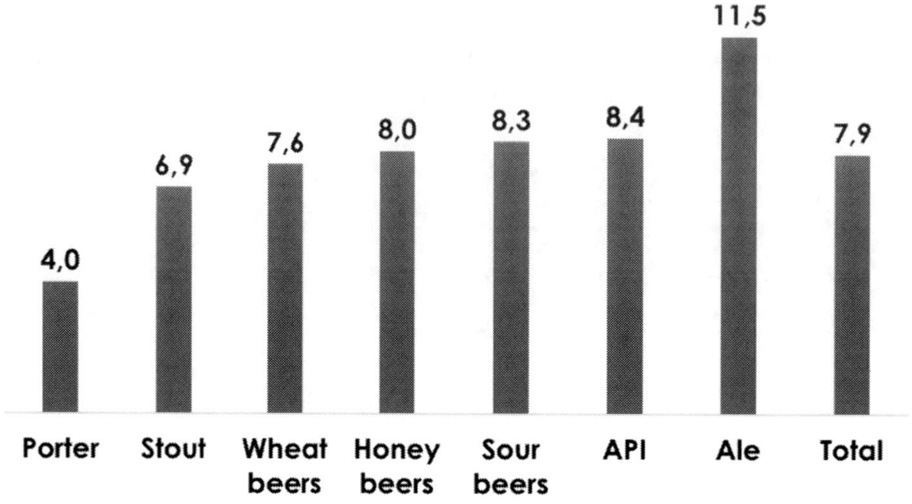

Beer Style	Most common off-flavours detected
Wheat beers	Yeasty, DMS, Indole
API	Acetic, Lactic, Phenolic, Plastics, Isovaleric, Metallic
Sour beers	Mercaptan, Phenolic
Porter	Acetic, Lactic, Phenolic
Stout	Acetic, Lactic
Honey Beers	Butyric, Rancid
Ale	Diacetyl, Acetic, Butyric, Acetaldehyde

Some Critical Control Points...

Production stage	Microorganisms	Off-flavours
Worth *Oxygenated, pH 5, high in carbohydrates, and contains some available protein*	Yeast and bacteria *(Lactobacillus spp, Pediococcus, Acetic acid bacteria; Enterobacteriaceae)*	**Acetic, Lactic, Butyric, Diacetyl, DMS, Other Sulphurs, Indole...**
Fermentation *Fresh air and Oxygen critical*	*Yeast (Saccharomyces cerevisiae; Saccharomyces carlsbergensis);Wild yeast (Brettanomyces bruxellensis) bacteria - open fermentation (Lactobacillus spp, Acetic acid bacteria; Enterobacteriaceae; Acetobacter; Zymomonas); mold*	**Ethanol, CO2, Esters - Fruity, Acetaldehyde, Phenolic, Plastics, Acetic, Lactic, Diacetyl, H2S, SO2, DMS, Mercaptan, Other Sulphurs, Indole, Earthy...**
Packaging, Maturation & Storage *Good hygiene critical*	Secondary infections - bacteria *(Megasphaera, Pectinatus spp, Lactobacillus spp, Pediococcus, Acetic acid bacteria; Enterobacteriaceae); mold*	**Caprylic, Yeasty, Metallic, Acetic, Lactic, Diacetyl, H2S, SO2, DMS, Other Sulphurs, Indole, Earthy, Musty...**

Sensory as diagnostic tool...

- **Phenolic & Plastics ⟶ SPOILAGE BY *WILD YEAST***
- **DMS & Indole ⟶ *BACTERIAL CONTAMINATION IN FERMENTATION***
- **Acetic & Butyric ⟶ *BACTERIAL GROWTH IN MASH***
- **Ethyl Hexanoate & H2S & Mercaptan ⟶ *POOR YEAST HEALTH***
- **Isovaleric ⟶ *OLD DEGRADED HOPS***

Conclusions

- 22 out of 278 of Brands showed Micro problems within 2-6 weeks from production
- Ale beer style showed largest number of Micro problems, followed by API, Sour, Honey and Wheat Beers
- Most of the off-flavours detected originates from the Fermentation
- Tasting In-process samples will prevent and reduce the risks for Micro problems in early stage of the production

Good Sensory Practice for Professional Beer Judges

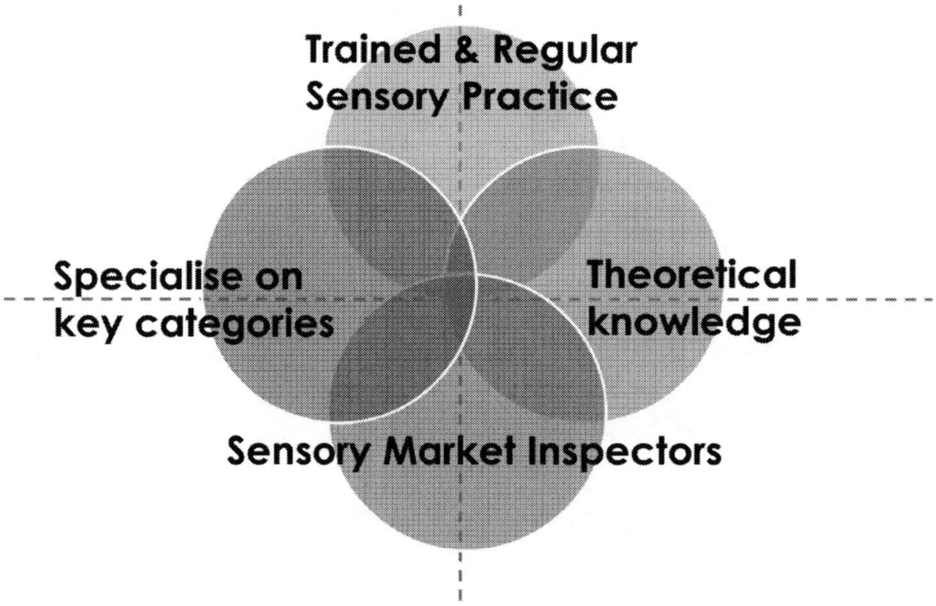

Would like to thank:

- Dr Heidi Grimmer – FlavorActiV, Johannesburg, South Africa
- Mag. Alexander Guzhiev – FlavorActiV, Oxfordshire, UK
- Mag. Tina Tian – FlavorActiV, Oxfordshire, UK
- Marie Pohler – FlavorActiV, Oxfordshire, UK
- Evelyne Canterranne – FlavorActiV, Oxfordshire, UK
- M.S. Marcia Waters – FlavorActiV, Oxfordshire, UK
- drs Ronald Nixdorf – FlavorActiV, The Hague, Netherlands
- Aruna Singh – FlavorActiV, Johannesburg, South Africa
- Dr Binod K. Maitin – FlavorActiV, Bangalore, India
- Amanda Wang – FlavorActiV, Shanghai, China
- Dr Katia Jorge – FlavorActiV, Rio de Janeiro, Brazil
- M.S. Sue Langstaff – FlavorActiV, Sacramento, California, USA

and

FlavorActiV's Global Customers

FlavorActiV™

36th Congress of the European Brewery Convention Ljubljana, Slovenia, May 14th – 18th 2017

„Genetic and phenotypic characterization of different top-fermenting *Saccharomyces cerevisiae* ale yeast isolates“

Tim Meier-Dörnberg

Maximilian Michel

Ronald Steven Wagner

Fritz Jacob

Mathias Hutzler

36TH CONGRESS EBC 2017 EUROPEAN BREWERY CONVENTION 14-18 MAY LJUBLJANA SLOVENIA

Forschungszentrum Weihenstephan für Brau- und Lebensmittelqualität

Technische Universität München

Background

Yeast -flavor engine of the brewing industry-

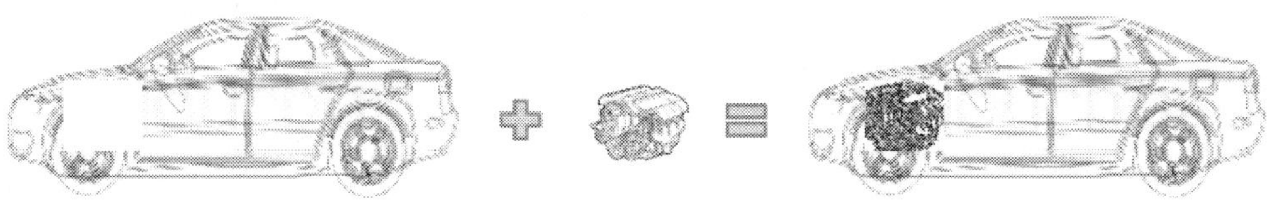

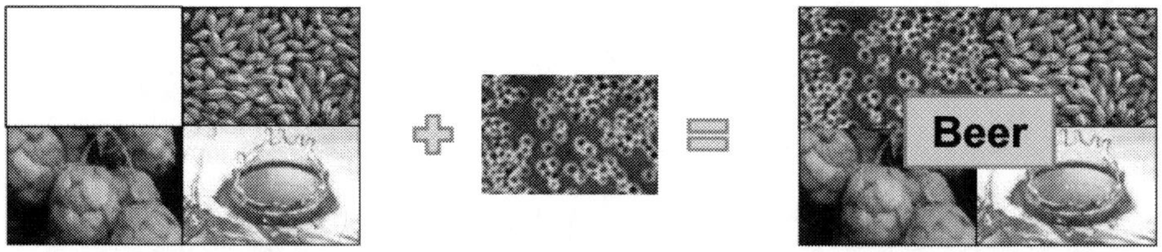

Source: Blog.uvinum.de; colourbox.de

Background

Selction criteria for an optimized application of brewing yeast

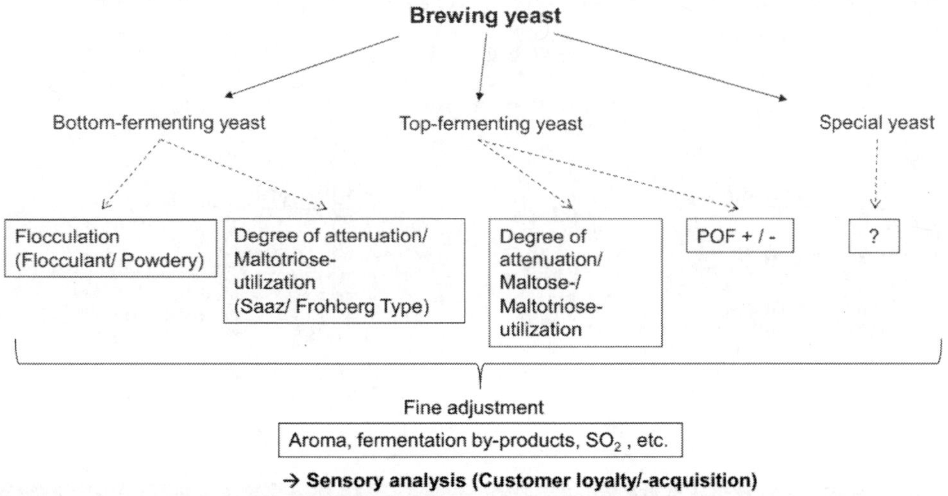

Method -Yeast isolates-

Practical yeast characterization for industrial application

Yeast isolates (TUM identifier)	Industrial application
TUM 210	**ale production**
TUM 211	**ale production**
TUM 213	**ale production**
TUM 506	**ale production**
TUM 511	**ale production**
LeoBavaricus - TUM 68®	*wheat beer production*
Frisinga - TUM 34/70®	*lager beer production*

5 ale yeast isolates with unknown origin

2 reference strains to compare with

Method

? Yeast Strain or Brewing Yeast ?

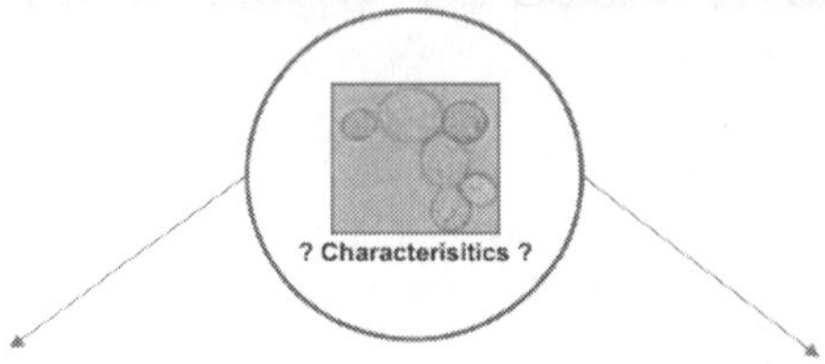

Genetic-

- Identification
  (Real time PCR)
- Sequencing
  (ITS1-5.8S-ITS2 and D1/D2 26S ribosomal rRNA gene)
- Strain determination
  (IGS2-314 PCR-capillary electrophoresis)

Phenotypic-

- Brewing properties
  (Fermentation performance, sugar utilization, amino acid utilization, cell growth, flocculation behaviour, change in pH value, sulfur dioxide, fermentation by-products)
- Sensorial testings
  (Expected beer type, DLG-scheme for beer, descriptive sensory analysis, triangle test, phenolic off-flavor)

Method -Brewing trials-

Yeast characterization at standard conditions

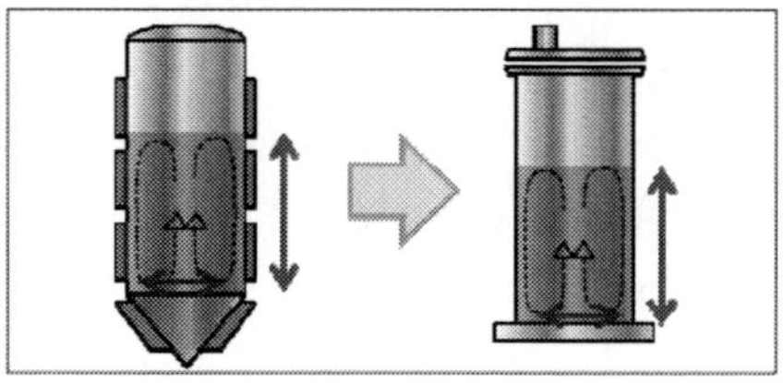

Stainless steel vessels
Ratio diameter to high: 1:3
20 % headspace
Nominal-/Filling quantity: 2,5 / 2 Liter
Pressure proofed

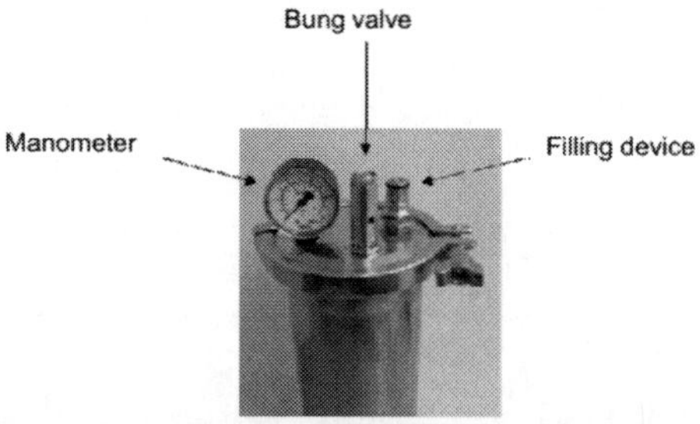

- TÜV SV 15-2055 safety bung valve, spring loaded, 0.5 bar, to ensure practical fermentation conditions
- Oxygen and contamination-free filling
- Evaluation in triplicate

Method -Brewing trials-

Tempered cooling chamber for constant fermentation temperature

1 additional fermentation vessel for daily measurement of extract, pH-value und yeast cells in suspension

➔ The system will not be influenced by taking the samples

Method -Brewing trials-

Experimental beer production parameters

Parameter	Set value
pressure	0.5 bar
medium	stand. wort 12.4 °P (base malt extract)
propagation	standardised and in parallel (20 °C)
pitching cell concentration	15 x 10^6 cells/ml
temperature (pit./ ferm./mat.)	20 °C
volume	2 l (mass adjustment 2000 g)
system number	variabel
brief mixing	N_2
reference strain	*S. cerevisiae* TUM 68
Enforced fermentation and maturation	Fermentation until final attenuation (2 d constant) /maturation: 5 d lager conditions: 7 d; 0°C

36th EBC Congress Ljubljana, Slovenia, May 15th 2017

Method -Brewing trials-

Analysis

Parameter	Wort	Pitching wort	Daily analysis	Finished beer
Original gravity	×		×	×
pH	×		×	×
Alcohol				×
Degree of attenuation				×
Nitrogen (N_2) soluble/ coagulable	×			
Thiobarbituric acid (TBZ)	×			
Iodine test spectrophotometric	×			
Beta glucans	×			
Zinc	×			
Dimethyl sulfide (DMS) total and free/ DMS-P	×			×
Alpha-amino nitrogen (FAN)	×	×		×
Total amino acids	×	×		×
Sugar composition	×	×		×
EBC-bittering units	×			
Yeast cells in suspension		×	×	×
Total yeast cell count		×		×
Diacetyl			×	
Sulfur dioxide (SO_2)				×
Fermentation by-products				×
Vicinal diketones total / free				×
4-Vinylguajacol				×
Sensory analysis				×

Tim Meier-Dörnberg | Slide 9 of 24 | Research Center Weihenstephan for Brewing and Food Quality

36th EBC Congress Ljubljana, Slovenia, May 15th 2017

Method -Phenolic off-flavor-

Phenolic off-flavor (POF)

- Yeast isolates were spread on a YM-agar plate containing one of the aroma active compounds (Precursor) : ferulic acid, coumaric acid and cinnamic acid
- Evaluation by sniffing after 3 days of incubation at 24° C

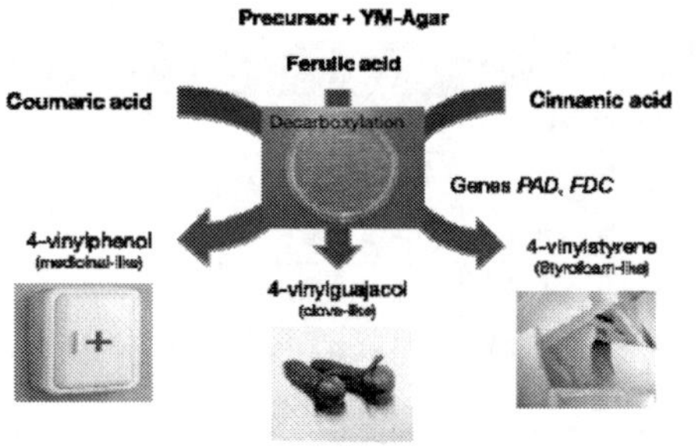

Precursor 100ppm coumaric acid, cinnamic acid, ferulic acid in YM-Agar

Tim Meier-Dörnberg | Slide 10 of 24 | Research Center Weihenstephan for Brewing and Food Quality

Method -Sensory analysis-

Sensory analysis

- Expected beer type

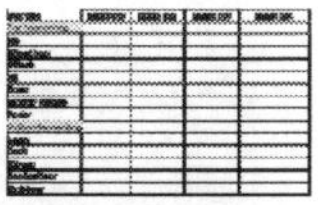

- DLG-scheme for beer

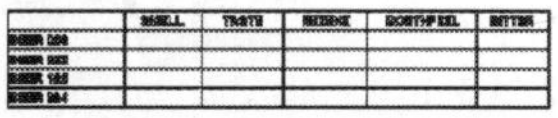

- Triangle test

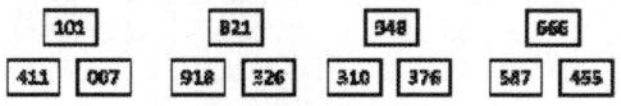

- Descriptive sensory analysis

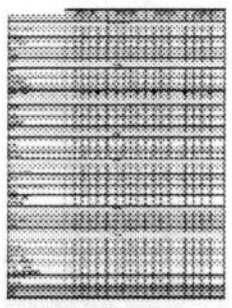

36th EBC Congress Ljubljana, Slovenia, May 15th 2017

Method -Sensory analysis-

Sensory analysis

- Expected beer type
- DLG-scheme for beer
- Triangle test
- Descriptive sensory analysis

3.) Descriptive sensory analysis (aroma)

Please mark the appropriate aroma impression according to their intensity
(1 ≙ slightly noticeable, 5 ≙ extremely noticeable)
Pleaser write perceived non-listed flavor impressions in the category called „Other".

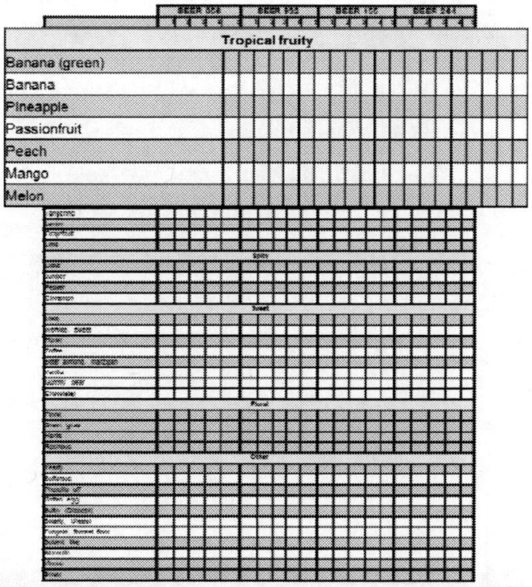

Results -Genetic profiling-

Genetic profiling -part 1-

DNA-Extraction → Real-time-PCR → Analysis

e. g. Real Time PCR Screening for *Saccharomyces* spp.

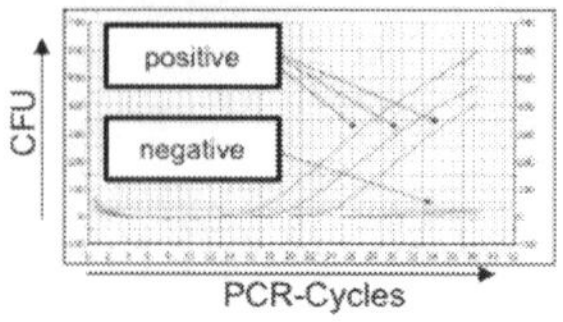

Qualitative results of the real-time PCR systems used for the investigated ale yeast isolates and the reference strains to differentiate *Saccharomyces* sensu stricto species; positive (+), negative (–)

		RT-PCR-System						
Species	Yeast isolates / reference strains	Sc-GRC3	Sce	TF-COXII	Sbp	BF-LRE1	BF-300	Sdia
S. cerevisiae	TUM 210	+	+	+	–	–	–	–
	TUM 211	+	+	+	–	–	–	–
	TUM 213	+	+	+	–	–	–	–
	TUM 503	+	+	+	–	–	–	–
	TUM 506	+	+	+	–	–	–	–
	TUM 511	+	+	+	–	–	–	–
	LeoBavaricus - TUM 68*	+	+	+	–	–	–	–
S. pastorianus	Frisinga – TUM 34/70*	+	+	–	+	+	+	–

Results -Genetic profiling-

Genetic profiling -part 2-

DNA-Extraction → PCR-DNA sequencing

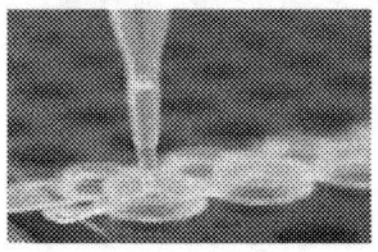

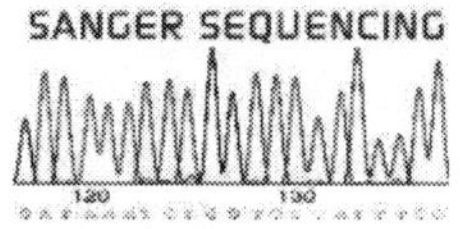

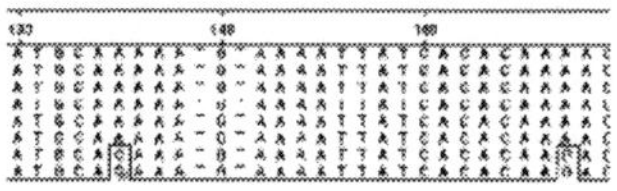

D1/D2 26S
ITS1-5.8S-ITS2 rDNA

Analysis

DNASequences | Translated Protein Sequences

Species/Abbrv	Gr	******* ***************************************
1. TUM 210		CGGCAAAAGCTCAAATTTGAAATCTGGTACCTTCGGTGCCCGAGTTGT
2. TUM 211		CGGCAAAAGCTCAAATTTGAAATCTGGTACCTTCGGTGCCCGAGTTGT
3. TUM 213		CGGCAAAAGCTCAAATTTGAAATCTGGTACCTTCGGTGCCCGAGTTGT
4. TUM 506		CGGCAAAAGCTCAAATTTGAAATCTGGTACCTTCGGTGCCCGAGTTGT
5. TUM 511		CGGCAAA-GCTCAAATTTGAAATCTGGTACCTTCGGTGCCCGAGTTGT
6. LeoBavaricus - TUM 68		CGGCAAAAGCTCAAATTTGAAATCTGGTACCTTCGGTGCCCGAGTTGT
7. Frisinga - TUM 34/70		CGGCAAAAGCTCAAATTTGAAATCTGGTACCTTCGGTGCCCGAGTTGT

S. cerevisiae 99 %

☑

36th EBC Congress Ljubljana, Slovenia, May 15th 2017

Results -Genetic profiling-

Genetic profiling -part 3-

DNA-Extraction

Capillary electrophoresis

IGS2-314 rDNA curve-based genetic relationship in percentage

Analysis

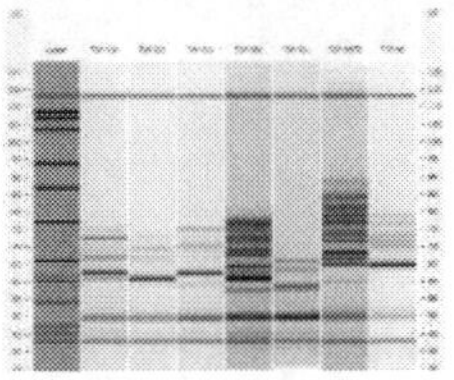

Capillary electrophoresis IGS2-314 rDNA patterns using lab on a chip, Agilent Bioanalyzer 2100, Agilent Technologies, Santa Clara, CA, USA

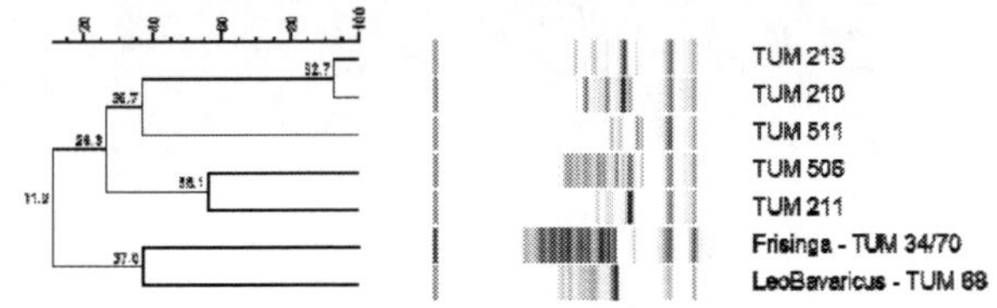

IGS2-314 rDNA curve-based genetic relationship in percentage using Bionumerics 7.6, Applied Maths, Belgium

36th EBC Congress Ljubljana, Slovenia, May 15th 2017

Results -Brewing trials-

Fermentation performance

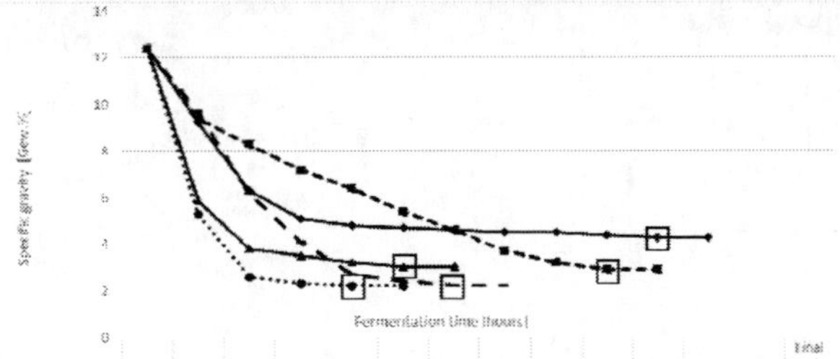

	0	24	48	72	96	120	144	168	192	216	240	264	Final gravity Ø ± 95 %
TUM 211	12.4	9.2	6.3	5.1	4.8	4.7	4.6	4.5	4.5	4.4	4.3	4.3	0.06
TUM 210	12.4	5.8	2.6	2.3	2.2	2.2							0.49
TUM 213	12.4	5.9	3.8	3.5	3.2	3	3						0.2
TUM 506	12.4	9.9	8.3	7.2	6.3	5.4	4.6	3.7	3.2	2.9	2.9		0.18
TUM 511	12.4	9.7	6.2	4.1	2.7	2.4	2.2	2.2					0.06

Drop in specific gravity measured in a single reference vessel compared with the average in final gravity (marked with box) measured in triplicate; confidence level 95 %

Apparent attenuation (AA %) of the final beer compared with specific time for primary fermentation for the investigated yeast strains TUM 210, TUM 211, TUM 213, TUM 506 and TUM 511; confidence level 95 %

	Apparent attenuation (AA %) of the final beer				
	TUM 210	TUM 211	TUM 213	TUM 506	TUM 511
AA (%)	82.73 ± 3.52	66.13 ± 0.51	74.77 ± 0.38	77.37 ± 1.34	82.7 ± 0.42
Fermentation time (hours)	96	240	120	216	144

Results -Brewing trials-

Total sugar utilization

Mean percentage of total wort sugar utilization in beer, measured in triplicate after lagering; confidence level 95%

	TUM 210	TUM 211	TUM 213	TUM 506	TUM 511
Glucose	99.16 ± 0.05	98.55 ± 0.47	99.13 ± 0.00	98.23 ± 0.12	99.42 ± 0.12
Fructose	99.35 ± 0.21	98.57 ± 0.21	98.83 ± 0.00	98.05 ± 0.36	98.05 ± 0.00
Sucrose	98.51 ± 0.48	98.51 ± 0.48	98.21 ± 0.00	99.11 ± 0.00	95.54 ± 0.00
Maltose	98.89 ± 1.23	87.23 ± 0.82	99.39 ± 0.07	98.48 ± 0.93	99.28 ± 0.09
Maltotriose	75.01 ± 8.39	26.66 ± 0.26	38.96 ± 0.46	59.28 ± 0.81	83.91 ± 0.71

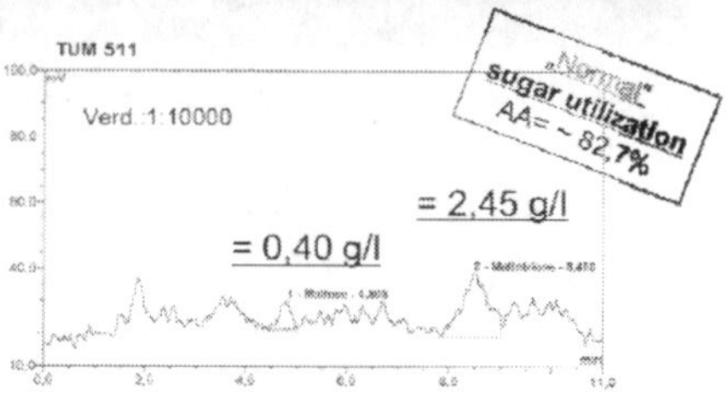

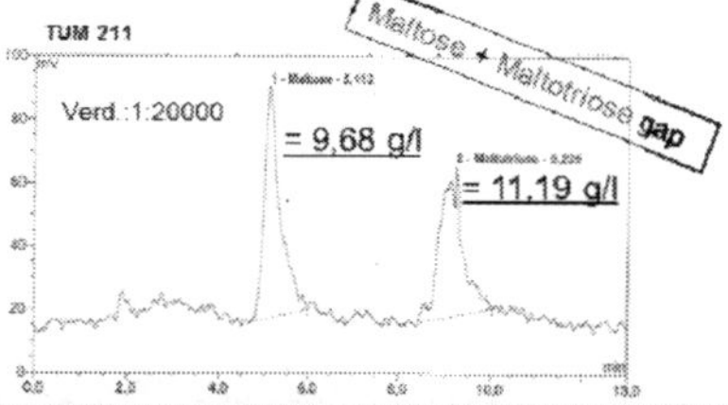

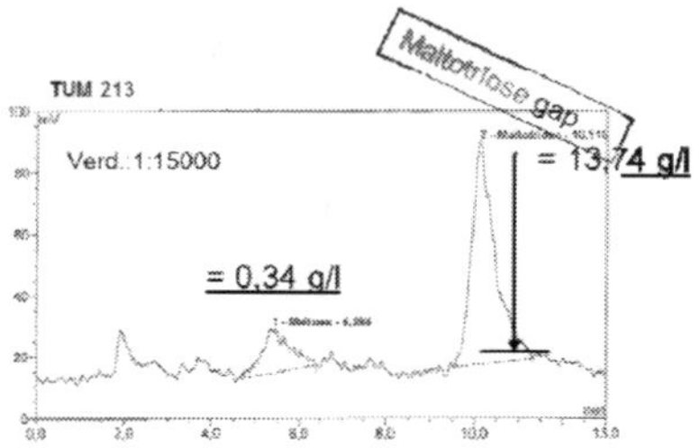

Results -Brewing trials-

Phenolic off-flavor (POF) test

POF results of the investigated yeast strains

Product/ Precursor	POF-test/Sniffing perception of TUM 210	TUM 211	TUM 213	TUM 506	TUM 511
4-vinylguajacol/ ferulic acid	–	–	+	–	+
4-vinylphenol/ coumaric acid	–	–	+	–	+
4-vinylstyrene/ cinnamic acid	–	–	+	–	+

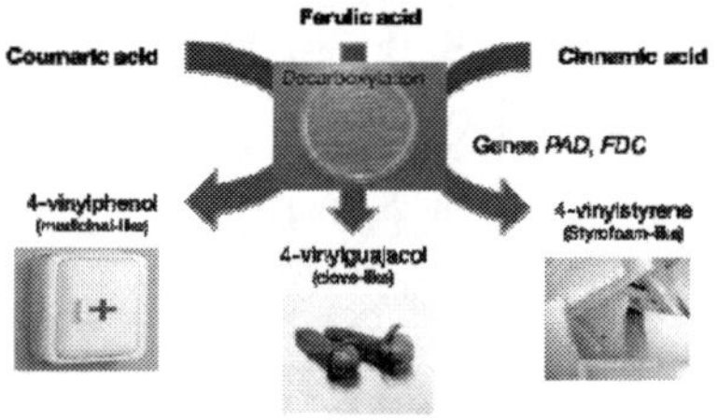

4-vinylguajacol measured in beer after lagering

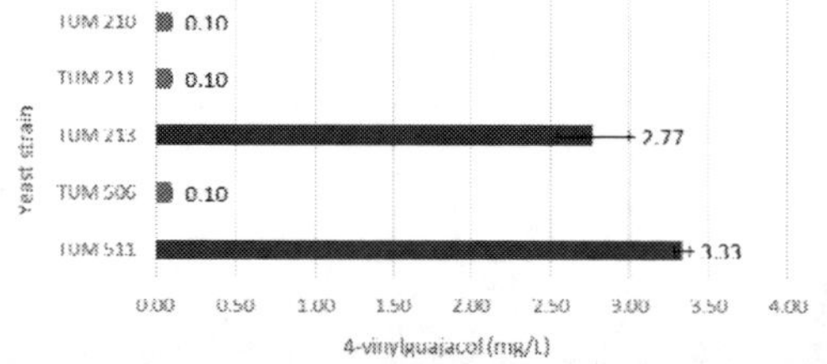

4-VG (clove)
= Bavarian wheat beer

Results -Sensory analysis-

Average of main flavor intensity

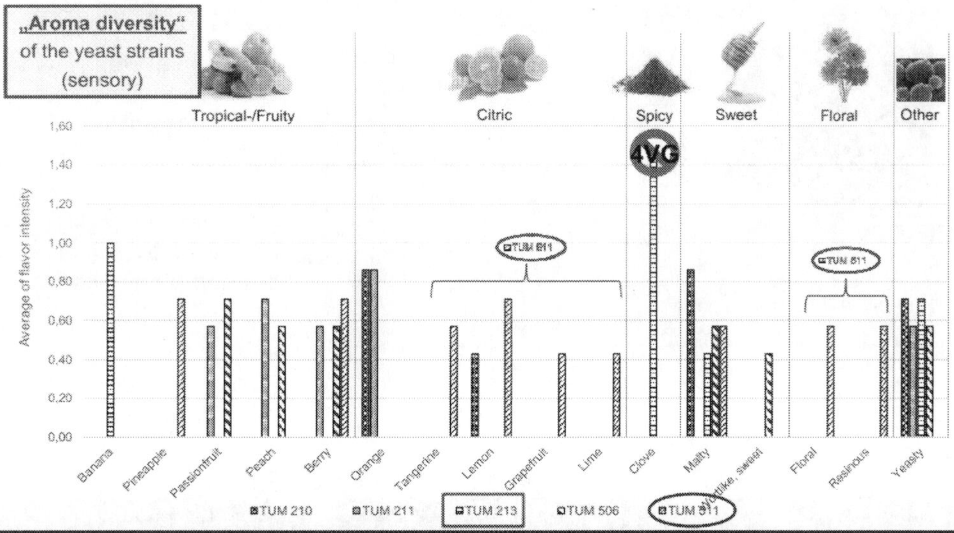

Results -Sensory analysis-

Sensory analysis

- All produced beers are good in taste
- 2 of 5 investigated yeast strains (TUM 213, TUM 511) are capable to produce phenolic off-flavors and referred to wheat beer
- Except TUM 211 and TUM 506 all yeast strains could be clearly distinguished by the tasters

Sensory analysis

Yeast strain	TUM 211	TUM 506	TUM 210	TUM 213	TUM 511
Beer type	Ale beer	Ale beer	Ale beer	Wheat beer	"Bavarian ale"
DLG (taste)	4.7	4.5	4.7	4.8	4.5
Triangle test	57%		100%	100%	100%

Detailled strain profiling -TUM Yeast Poster-

Visualized strain profiling for optimized brewing application

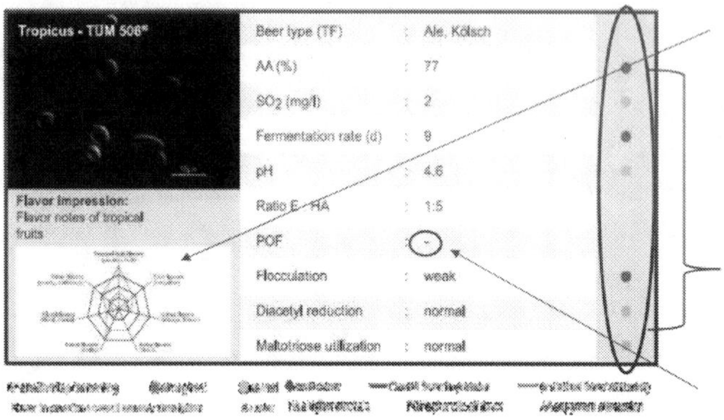

Sensory profil
Provides information about the overall sensory impression of the beer and the characteristics of special aromas

Traffic light system
For fast detection of brewing properties

POF
Phenolic off-flavor ability
Important for bavarian wheat beer production

Detailled strain profiling -TUM Yeast Poster-

TUM Yeast Poster

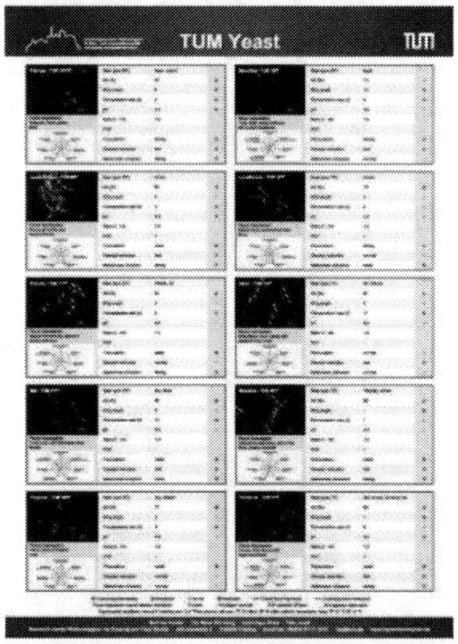

The ten most important TUM-Yeast strains of the Research Center Weihenstephan BLQ have been published by the Fachverlag Hans Carl.
See **www.carllibri.com**

Results –Additional brewing data-

Additional brewing data

FAN – amino acid uptake

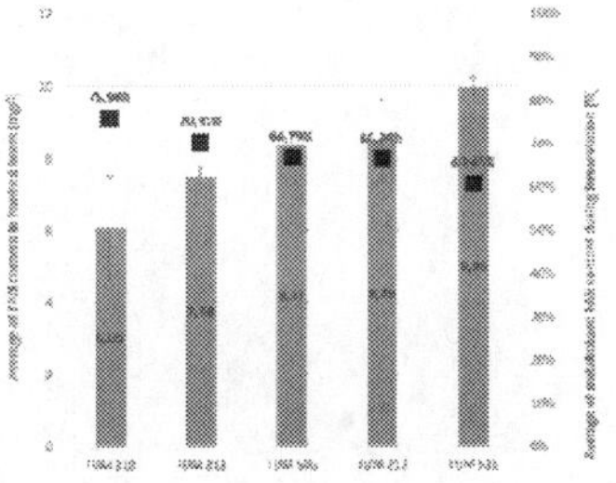

pH drop

pH value decrease during primary fermentation	TUM 210	TUM 211	TUM 213	TUM 506	TUM 511
0 h	5.2	5.2	5.2	5.2	5.2
24 h	4.6	4.5	4.5	4.6	4.6
48 h	4.4	4.4	4.5	4.5	4.4
72 h	4.4	4.4	4.5	4.5	4.4
96 h	4.4	4.4	4.5	4.5	4.4
After primary fermentation	4.4	4.4	4.5	4.5	4.4
After maturation	4.5	4.5	4.5	4.6	4.4
Final Beer (after lagering)	4.6 ± 0.01	4.7 ± 0.01	4.6 ± 0.01	4.7 ± 0.01	4.4 ± 0.01
ΔpH	0.6	0.5	0.6	0.5	0.8

fermentation-by-products

Fermentation by-products (mg/l)	TUM 210	TUM 211	TUM 213	TUM 506	TUM 511	Threshold (Lager beers)*
Isoamyl acetate	2.40 ± 0.32	2.00 ± 0.05	4.77 ± 0.25	1.57 ± 0.11	2.89 ± 0.05	1.2
Ethyl acetate	31.57 ± 2.79	37.83 ± 0.31	91.57 ± 3.35	22.87 ± 1.31	54.48 ± 0.01	30
Σ Ester (E)	34.37 ± 4.38	38.87 ± 0.14	98.32 ± 3.68	24.43 ± 1.30	57.32 ± 0.65	n.v.
n-Propanol	27.50 ± 0.83	15.38 ± 0.98	38.08 ± 1.48	28.87 ± 1.01	28.77 ± 0.65	800
i-Butanol	21.57 ± 4.28	18.28 ± 0.37	13.88 ± 0.88	28.80 ± 1.40	13.13 ± 0.25	200
Amyl alcohols	103.67 ± 5.65	59.27 ± 1.32	79.87 ± 7.38	86.50 ± 5.50	74.87 ± 1.34	70
Σ Higher alcohols (HE)	156.73 ± 12.21	92.77 ± 1.80	131.79 ± 8.85	139.07 ± 6.12	109.67 ± 1.53	n.v.
4-Vinylguajacol	0.15 ± 0.08	0.48 ± 0.08	2.77 ± 0.23	0.18 ± 0.00	3.30 ± 0.05	0.3
Diacetyl	0.08 ± 0.01	0.18 ± 0.01	0.07 ± 0.30	0.12 ± 0.01	0.08 ± 0.00	0.15
2,3-Pentandione	0.02 ± 0.01	0.02 ± 0.03	0.02 ± 0.01	0.02 ± 0.00	0.01 ± 0.00	0.9
Σ Vicinal diketones	0.11 ± 0.01	0.12 ± 0.01	0.09 ± 0.01	0.14 ± 0.01	0.07 ± 0.00	n.v.
Acetaldehyde	2.83 ± 0.58	4.88 ± 0.48	8.60 ± 1.81	8.83 ± 0.53	4.83 ± 0.48	25
Ratio (ΣE : ΣHE)	1 : 4.57	1 : 2.35	1 : 2.88	1 : 5.40	1 : 1.88	

*(Meilgaard 1975)

SO_2

SO_2 concentration of the finished beers (mg/l)	TUM 210	TUM 211	TUM 213	TUM 506	TUM 511
SO_2	2.63 ± 1.64	2.60 ± 0.96	2.63 ± 1.521	2.23 ± 1.02	0.50 ± 0.00

Source :
Meier-Dörnberg, T., Michel, M., Wagner, R. S., Jacob, F. and Hutzler, M.: "Genetic and Phenotypic Characterization of Different Top-fermenting Saccharomyces cerevisiae Ale Yeast Isolates" BrewingSience, January/February 2017 (Vol.70), 2017

Thanks for your attention…

and thanks to:

36TH CONGRESS EBC 2017 14-18 MAY LJUBLJANA EUROPEAN BREWERY CONVENTION SLOVENIA

Team Research Center Weihenstephan
Prof. Dr. Ronald Steven Wagner
(Central Washington University)
Konrad Müller-Auffermann
(Krones AG)
Dr. Erika Hinzmann
(Wissenschaftsförderung der Deutschen Brauwirtschaft e.V.)

Dipl.-Ing. Tim Meier-Dörnberg
Alte Akademie 3
85354 Freising-Weihenstephan
Phone: +49 (0) 8161 / 71-3417
Fax: +49 (0) 8161 / 71-4181
Email: tmd@wzw.tum.de
www.blq-weihenstephan.de

Parts of this work were funded by the German Ministry of Economics and Technology and the Wifö (Wissenschaftsförderung der Deutschen Brauwirtschaft e.V., Berlin, Germany) in project **AiF 17698 N.**

iFBM

New insights on fermentation : looking for key amino acids in lager fermentation

36TH CONGRESS
EBC
2017 EUROPEAN BREWERY CONVENTION
14–18 MAY LJUBLJANA
SLOVENIA

EBC 36th Congress – Ljubljana – May 2017

Marc SCHMITT – L33

1

iFBM

Presentation overview

1. Introduction
2. Reminder previous works
3. Experimental strategy
4. First results
5. Conclusions & perspectives

Introduction

iFBM

- Free amino nitrogen (FAN) is used as general index of healthy yeast growth, viability, vitality & fermentation efficiency
- FAN is composed of :
  - Amino acids (60 to 80%)
  - Ammonium ions (2 to 5%)
  - Small peptides
- Is FAN enough to predict appropriate fermentation performance ?

IFBM / EBC Ljubljana / May 2017

Previous works

iFBM

- Looking for AA composition of pilot brews as a possible key for fermentati

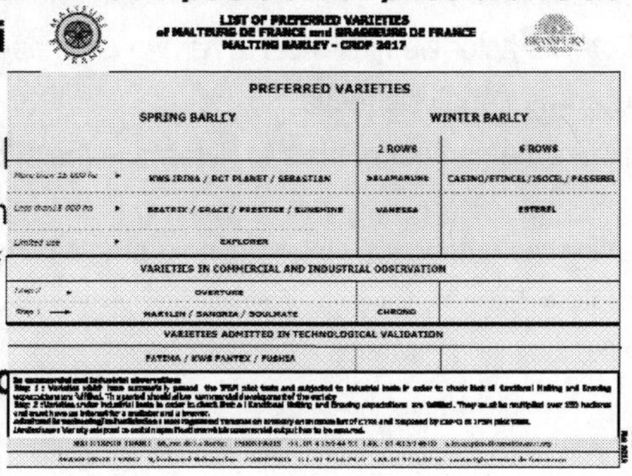

LIST OF PREFERRED VARIETIES
of MALTEURS DE FRANCE and BRASSEURS DE FRANCE
MALTING BARLEY - CROP 2017

PREFERRED VARIETIES			
	SPRING BARLEY	WINTER BARLEY 2 ROWS	WINTER BARLEY 6 ROWS
	KWS IRINA / RGT PLANET / SEBASTIAN	SALAMANDRE	CASINO/ETINCEL/ISOCEL/PASSEREL
	BEATRIX / GRACE / PRESTIGE / SUNSHINE	VANESSA	ESTEREL
Limited use	EXPLORER		
VARIETIES IN COMMERCIAL AND INDUSTRIAL OBSERVATION			
	OVERTURE		
	MARYLIN / SANGRIA / SOULMATE	CHRONO	
VARIETIES ADMITTED IN TECHNOLOGICAL VALIDATION			
	FATIMA / KWS FANTEX / FUSHIA		

- Impact of l
  - Variety h
  - Effect of
- Looking fo
  - "Time to ... ɲt adapted for lager yeasts

IFBM / EBC Ljubljana / May 2017

iFBM

Looking for key AA in lager fermentation

- **Objectives :**
  - **Studying the influence of single amino acid supplementation to wort (and interactions)**
    - Wort produced with several malts
    - Fermentations with several yea
  - **Statistical identificati ect fermentation kinetics**
  - **Looking for t impact on beer flavour**

DESIGN OF EXPERIMENT

iFBM

Technical strategy

- **DOE : Design of experiment**
  - Two major components:
    - **Predictor** : Factor we hypothesize having an impact on an…
    - … **Outcome** : Analysis result
  - Optimisation of the number of trials to determine the effect of one or several factors
    - Predicting the outcome by introducing a precise change of the predictor parameters.
  - Time and costs saving strategy

iFBM

Technical strategy

⇨ **DOE to determine the impact of AA composition**

11 predictors = 11 selected amino acids

2 levels = 2 amino acid concentration (min / max)

12 experiments = 12 fermentations

OUTCOMES
- Fermentation kinetics
- Aromatic profile

iFBM

Technical strategy

Experimental variability
- Determination of the experimental variability

Bibliography
- State of the art

Amino acid supplementation
- Preparation of low FAN wort
- Supplementation matrix

Experiment
- Pitching of wort and monitoring of the 12 supplemented wort fermentations
- Analyses

iFBM

Technical strategy

Experimental variability

- Determination of the experimental variability

- State of the art

- Preparation of 12 low FAN worts
- Adding of amino acid in worts according to the matrix

- Pitching of wort and monitoring of the 12 supplemented wort fermentations
- Analyses

iFBM

Experimental variability

Experimental variability

- Determination of the experimental variability

⇨ **Monitoring replicates from a control wort fermentation**

- ⇨ 15 experiments = 15 fermentations = 15 replicates
- ⇨ Same wort, same yeast, same conditions of fermentation
- ⇨ Monitoring of gravity, pH, OD, fermentability

Experimental variability

iFBM

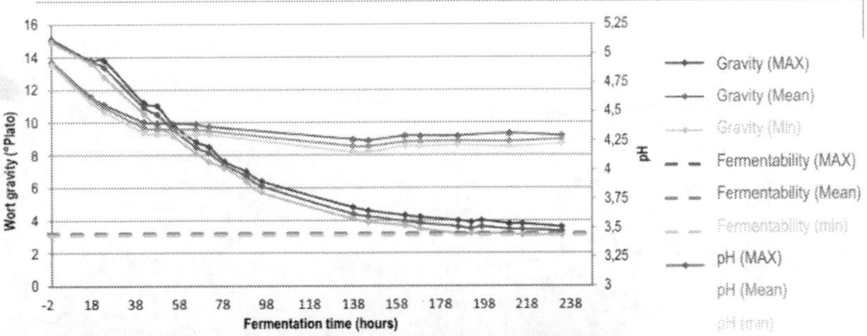

Experimental variability (%)			
Optical density (OD)	Gravity (Plato°)	pH	Fermentability
11,5	4,9	0,74	0,3

Technical strategy

iFBM

- Determination of the experimental variability
- State of the art
- Preparation of 12 low FAN worts
- Adding of amino acid in worts according to the matrix
- Pitching of wort and monitoring of the 12 supplemented wort fermentations
- Analyses

iFBM

Bibliography

- State of the art

Group A Fast absorption	Group B Intermediate absorption	Group C Slow absorption	Group D Little or not absorption
Glutamate	Valine	Glycine	Proline
Aspartate	Methionine	Phenylalanine	
Asparagine	Leucine	Tyrosine	
Glutamine	Isoleucine	Tryptophan	
Serine	Histidine	Alanine	
Threonine		Ammonia	
Lysine			
Arginine			

➢ **AA classification by Jones & Pierce (1964)**

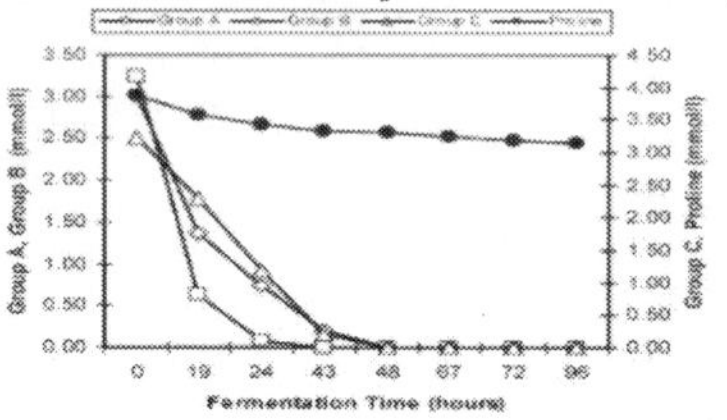

Class 1 *No matter*	Class 2 *Critical*	Class 3 *Exogenous - essential*
MET	VAL	LYS
ASP	ILE	LEU
ASN	PHE	ARG
GLU	GLY	HIS
SER	TYR	
THR		
PRO		

➢ **AA classification by Pierce (1987)**

iFBM

Bibliography

- State of the art

⇨ **Lekkas and al. 2003, 2005, 2007**

- ⇨ Lysine : Shortened the fermentation time
- ⇨ Methionine : Extended fermentation time

⇨ **Krogerus & Gibson (2013)**

- ⇨ Influence of VAL and other AA on VDK levels during fermentation

iFBM

Bibliography

- ⇨ **Methodology for AA selection**
  - ⇨ **AA according yeast ability to produce their ceto-acid precursor**
    - ⇨ Class 1 / 2 / 3 (Importance : 3 > 2 > 1)
  - ⇨ **AA biosynthesis pathway**
    - ⇨ 6 families :
      - ⇨ Glutamate family
      - ⇨ Aspartate family
      - ⇨ Aromatic AA family
      - ⇨ Serine family
      - ⇨ Pyruvate family
      - ⇨ Histidine family

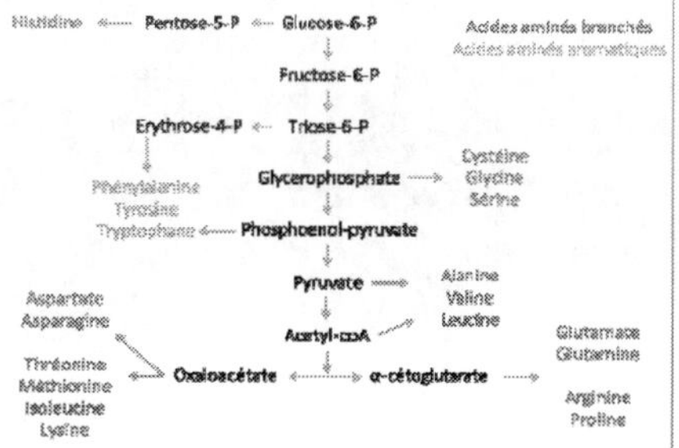

iFBM

Bibliography

- ⇨ **AA first selection :**
  - ⇨ **Histidine** : class 3 / family biosynthesis
  - ⇨ **Lysine** : class 3 / « first key AA »
  - ⇨ **Méthionine** : « inhibitor AA »
  - ⇨ **Glutamate** : family biosynthesis
  - ⇨ **Aspartate** : family biosynthesis
  - ⇨ **Phenylalanine** : class 2 / aromatic amino acid / family biosynthesis
  - ⇨ **Serine** : family biosynthesis
  - ⇨ **Glycine** : class 2
  - ⇨ **Alanine** : family biosynthesis / increase concentration inhibits the yeast
  - ⇨ **Isoleucine** : class 2
  - ⇨ **Ammonia** : biosynthesis pathway dependent / first nutrition of the yeast

Technical strategy

iFBM

- Determination of the experimental variability
- State of the art

Amino acid supplementation

- Preparation of 12 worts
- Adding of amino acid in worts according to the matrix

- Pitching of wort and monitoring of the 12 supplemented wort fermentations
- Analyses

Wort production and supplementation

iFBM

Amino acid supplementation

- Preparation of 12 low FAN worts
- Adding of amino acid in worts according to the matrix

⇨ **Microbrewing plant :**

- ⇨ Production of wort with low FAN malts
  - ⇨ Mash-in at 65°C
  - ⇨ Boiling for 60 min
- ⇨ Correction to 15°P with glucose syrup

⇨ **Initial wort at 14,4°P & +/- 100 mg/L FAN**

Wort production and supplementation

iFBM

⇨ **Supplementation matrix (12 fermentations)**

- ⇨ Level « **-1** » = [min] : Initial concentration of AA in wort
- ⇨ Level « **+1** » = [max] : Initial concentration of AA in wort **X3**
  - ⇨ by adding each individual amino acid in the corresponding wort

N° of experiments	1	2	3	4	5	6	7	8	9	10	11	12
HIS	1	-1	1	-1	-1	-1	1	1	1	-1	1	-1
LYS	1	1	-1	1	-1	-1	-1	1	1	1	-1	-1
MET	-1	1	1	-1	1	-1	-1	-1	1	1	1	-1
GLU	1	-1	1	1	-1	1	-1	-1	-1	1	1	-1
ASP	1	1	-1	1	1	-1	1	-1	-1	-1	1	-1
PHE	1	1	1	-1	1	1	-1	1	-1	-1	-1	-1
SER	-1	1	1	1	-1	1	1	-1	1	-1	-1	-1
GLY	-1	-1	1	1	1	-1	1	1	-1	1	-1	-1
THR	-1	-1	-1	1	1	1	-1	1	1	-1	1	-1
ILE	1	-1	-1	-1	1	1	1	-1	1	1	-1	-1
Ammonia	-1	1	-1	-1	-1	1	1	1	-1	1	1	-1

Technical strategy

iFBM

- Determination of the experimental variability monitoring replicates from a control wort fermentation
- State of the art
- Preparation of 12 low FAN worts
- Adding of amino acid in worts according to the matrix
- Experiment
  - **Pitching of wort and monitoring of the 12 wort fermentations**
  - **Analyses**

iFBM

Experiment

Experiment

- Pitching of wort and monitoring of the 12 fermentations
- Analyses

⇨ **Fermentations parameters**

⇨ Active dry yeast

⇨ Pitching at 20-25°C at 2,77 x 10^7 cells / mL

⇨ O_2 : 8 ppm

⇨ FAN content of worts : 95 to 150 mg/l

⇨ Fermentation at 13°C in « EBC » tubes

⇨ Monitoring of OD, gravity, pH, etc…

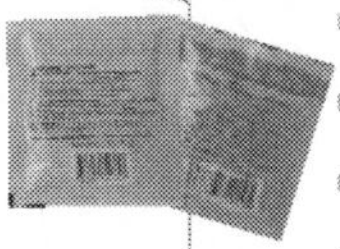

iFBM

Data analysis

⇨ **Data retrieval**

⇨ Matrix supplementation

⇨ Outcomes

⇨ **Analysis by the « NemrodW® » software**

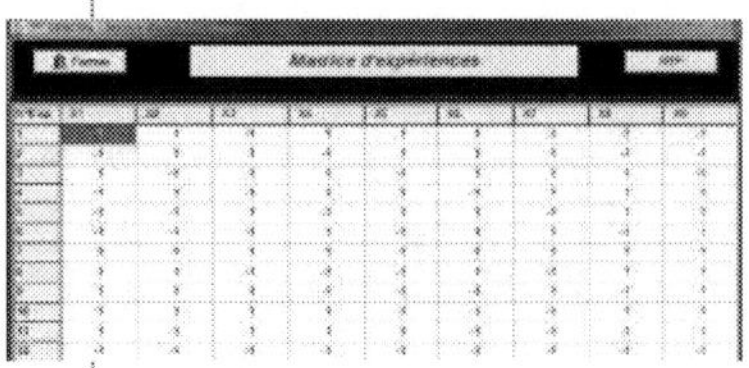

NemrodW®

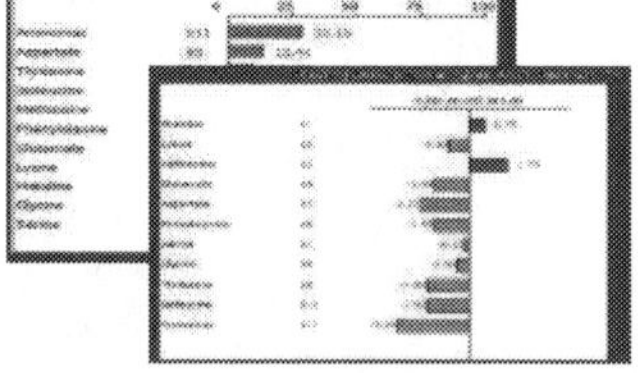

Analytical results

iFBM

- ⇨ **Not expected but interesting :**
  - ⇨ Contribution of each amino acid in the FAN determination with ninhydrin method (EBC 8.10)

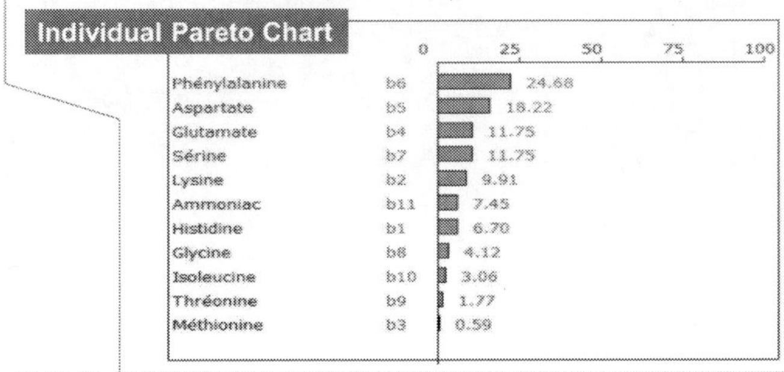

- ➢ **AA didn't have the same contribution to [FAN]**

SafLager™ S-23 (Fermentis)

iFBM

- ⇨ Lager fermentation yeast from Germany (Berlin)
- ⇨ Production of « fruity » and « estered » beers
- ⇨ Low Fermentability potential > round mouthfeel

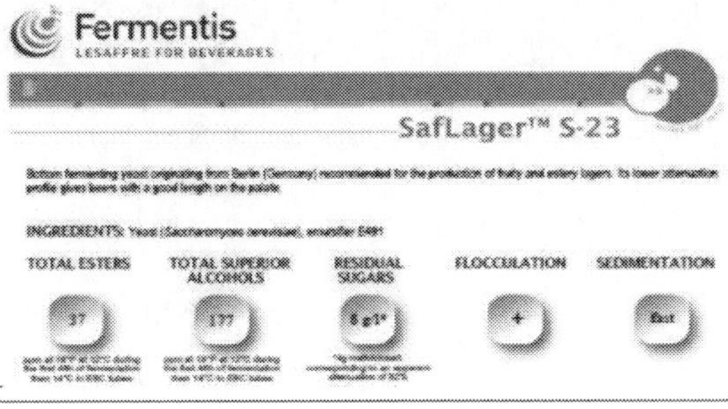

iFBM

Experimental trials on S-23

⇨ **DOE to determine the impact of AA composition**

11 predictors = 11 selected amino acids

2 levels = 2 amino acid concentration (-1/1)

12 experiments = **12 fermentations**

Low FAN wort

Lager yeast **Saflager S23**

Monitoring parameters *: Plato gravity, OD, pH, FAN, ammoniac, amino acid, fermentability, dicetones*

OUTCOMES

- Fermentation time
- Generation time of yeast
- FAN measurement

→ Fermentation kinetics

- Production of volatile compounds

→ Aromatic profile

iFBM

SafLager™ S-23 fermentation

⇨ **Effect on fermentation duration :**

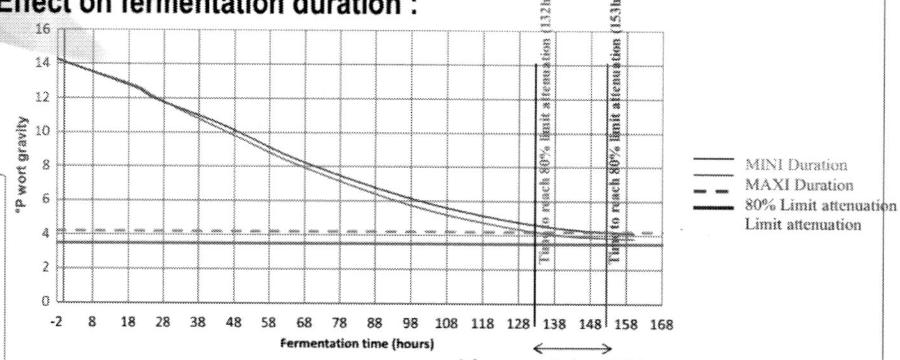

Δ(max-min) = 18 hours

SafLager™ S-23 fermentation

⇨ **AA effect on fermentation duration :**

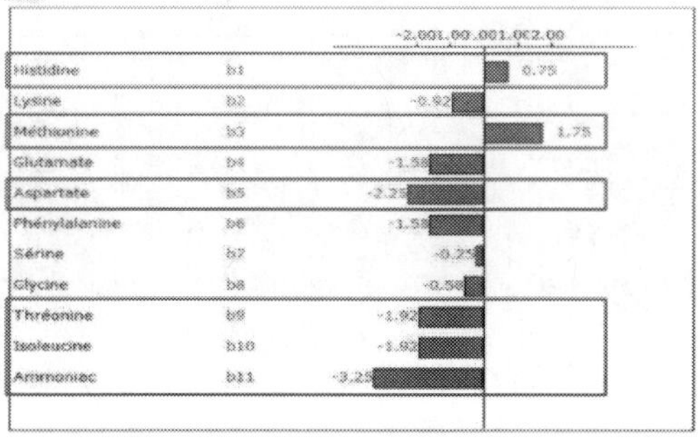

Lengthen the fermentation: **MET**, HIS

Decrease the fermentation duration: **NH3**, **ASP**, THR, ILE

IFBM / EBC Ljubljana / May 2017

SafLager™ S-23 fermentation

⇨ **AA effect on yeast generation time :**

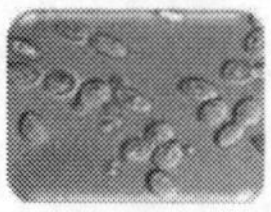

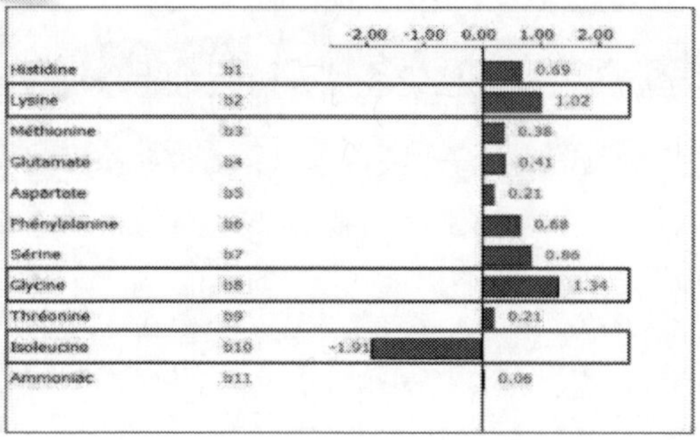

Lengthen the yeast generation: **GLY**, LYS

Decrease the Yeast generation: **ILE**

IFBM / EBC Ljubljana / May 2017

SafLager™ S-23 fermentation

iFBM

⇨ **AA effect on VDKs at 80% attenuation :**

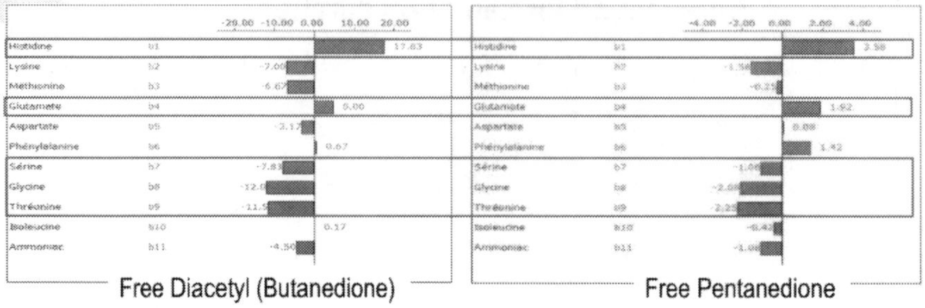

Promote the production of diketones : HIS, GLU
Reduce the production of diketones : THR, GLY, SER

SafLager™ S-23 fermentation

iFBM

⇨ **Selection of the most influent amino acids on SafLager™ S-23 according to our results**

	Time of fermentation	Yeast Generation time	Vicinal diketones
Methionine	↗	↗	↘
Histidine	↗	↗	↗
Glycine	↘	↗	↘
Lysine	↘	↗	↘
Isoleucine	↘	↘	-
NH3	↘	-	↘

SafLager™ W34/70 (Fermentis)

iFBM

- ⇨ Lager fermentation yeast from Germany (Weihenstephan)
- ⇨ Production of « fruity » and « floral » equilibrated beers

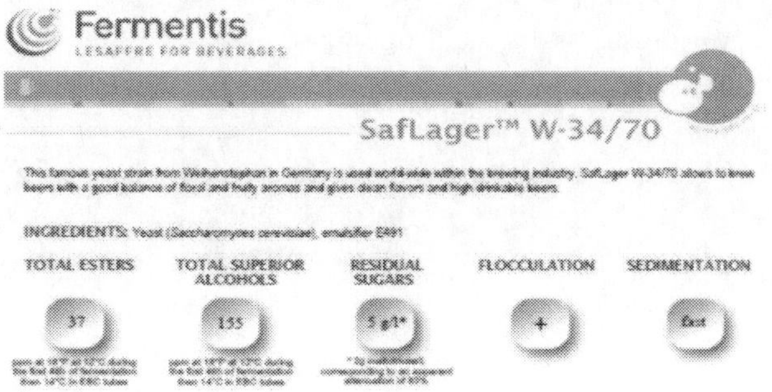

Experimental trials on W34/70

iFBM

- ⇨ **DOE to determine the impact of amino acids composition**

11 predictors = 11 selected amino acids

2 levels = 2 amino acid concentration (min / max)

12 experiments = **12 fermentations**

Low FAN wort

Lager yeast **Saflager W34/70**

Monitoring parameters *: Plato gravity, OD, pH, FAN, ammoniac, amino acid, fermentability, dicetones*

OUTCOMES

– Fermentation time – Generation time of yeast – FAN measurement	Fermentation kinetics
– Production of volatile compounds	Aromatic profile

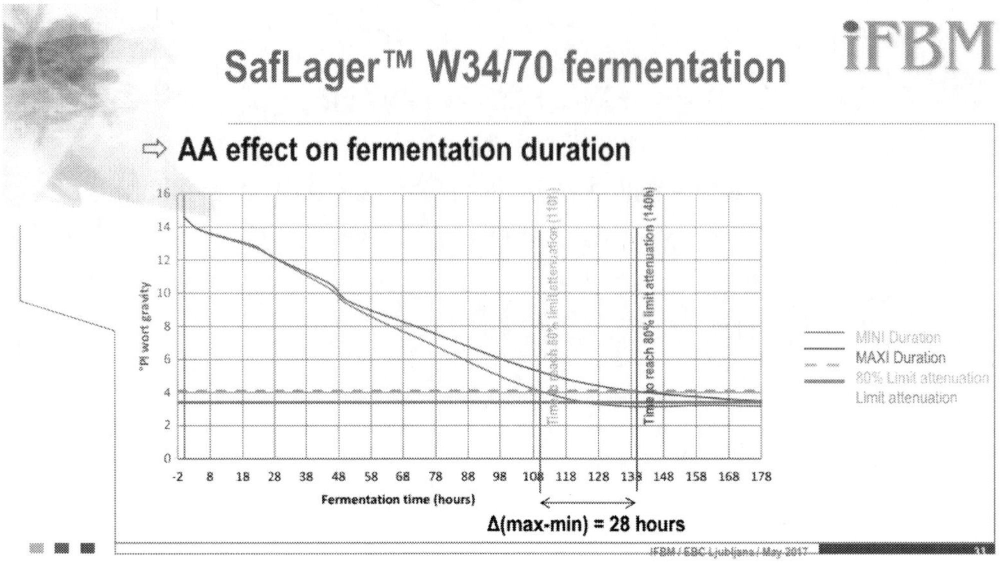
SafLager™ W34/70 fermentation
iFBM
⇨ AA effect on fermentation duration
°Pl wort gravity
16
14
12
10
8
6
4
2
0
-2 8 18 28 38 48 58 68 78 88 98 108 118 128 138 148 158 168 178
Fermentation time (hours)
Time to reach 80% limit attenuation (140h)
MINI Duration
MAXI Duration
80% Limit attenuation
Limit attenuation
Δ(max-min) = 28 hours
IFBM / EBC Ljubljana / May 2017

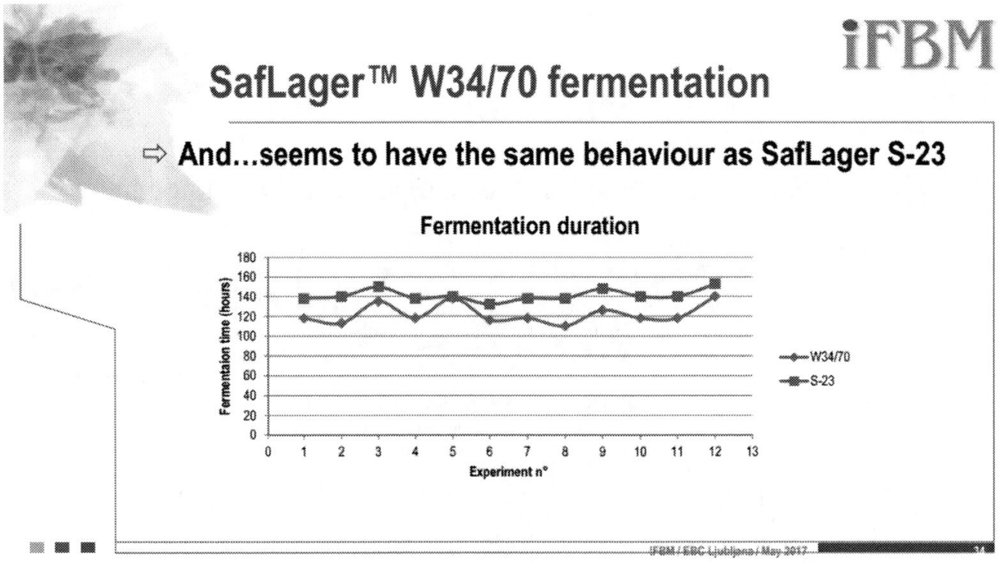
SafLager™ W34/70 fermentation
iFBM
⇨ And…seems to have the same behaviour as SafLager S-23
Fermentation duration
Fermentaion time (hours)
180
160
140
120
100
80
60
40
20
0
0 1 2 3 4 5 6 7 8 9 10 11 12 13
Experiment n°
W34/70
S-23
IFBM / EBC Ljubljana / May 2017

SafLager™ W34/70 fermentation

iFBM

⇨ **AA effect on fermentation duration :**

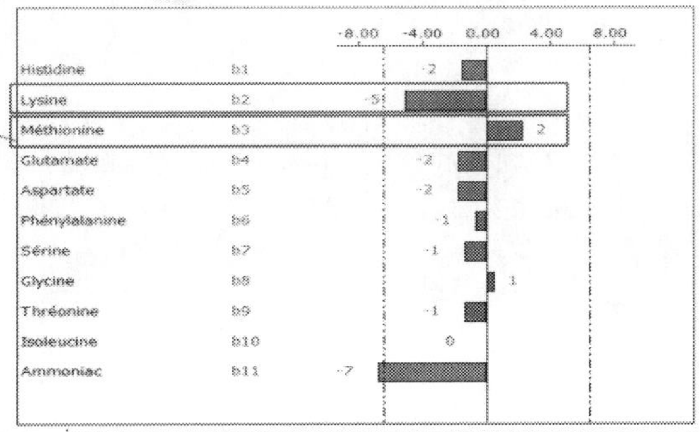

Lengthen the fermentation: **MET**

Decrease the fermentation duration: **NH3**, **LYS** (ASP / GLU THR SER)

SafLager™ W34/70 fermentation

iFBM

⇨ **AA effect on yeast generation time :**

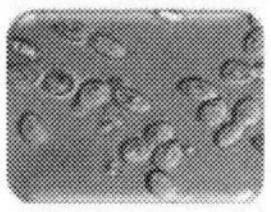

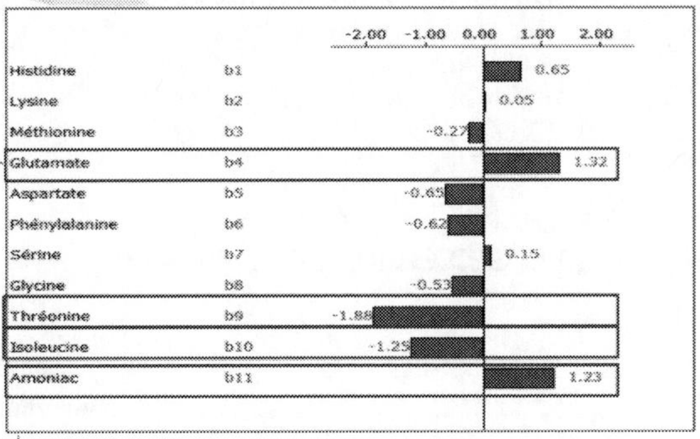

Lengthen the yeast generation: GLU, NH3

Decrease the Yeast generation: **THR**, **ILE**

SafLager™ W34/70 fermentation

iFBM

- Examples of AA effect on volatiles at 80% attenuation :

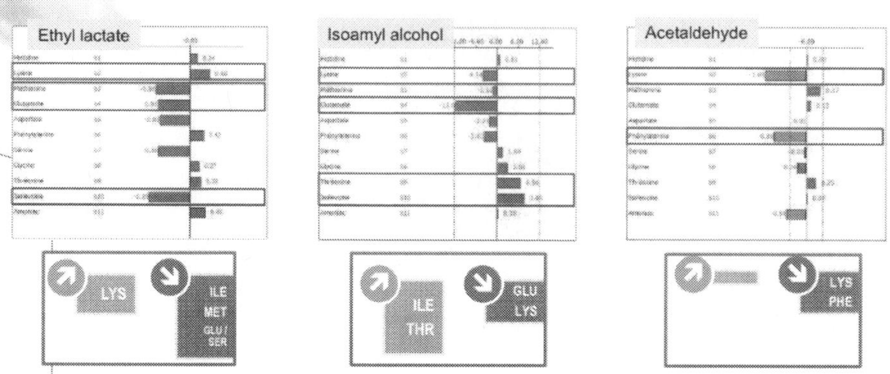

Experimental trials on other yeast

iFBM

- **Industrial yeast purchased by a brewer**
  - Same design of experiment (12 fermentations)
- **First results**
  - Fermentation duration impacted by AA
    - Delta higher than observed for S23 & W34/70
  - Analyses on course...

First conclusions

iFBM

- ⇨ **The objectives of this work were to study the influence of AA on lager yeast fermentation**
  - ⇨ It is the first step of a longer study
- ⇨ **DOE validate influence of wort AA composition on lager yeasts fermentation performances**
  - ⇨ Influence on fermentation duration
  - ⇨ Influence on volatiles compounds production
- ⇨ **Key AA in our experiments are : MET / HIS / LYS / (ILE)**

Perspectives

iFBM

- ⇨ **Finalize analyses on all tests**
  - ⇨ Data treatment & computation
- ⇨ **Additional design of experiments :**
  - ⇨ With the 20 AA in order to study all effects
- ⇨ **New DOE**
  - ⇨ Selection of key AA identified to enlarge the concentration range of the most influent AA

iFBM

Perspectives

- ⇨ **Regarding key AA composition, selection of very different profiles of worts :**
  - ⇨ Looking for their concentration in ≠ (barley varieties) malts
  - ⇨ Fermentation tests to « validate » the hypothesis
- ⇨ **Of course, fermentation of wort is linked to its AA composition & yeast used.**
- ⇨ **Of course, in most cases, AA concentration of wort is sufficient to avoid fermentation troubles**
- ⇨ **Can malting barley selection be improved by AA profile ?**

iFBM

Acknowledgements

- ⇨ **Pr Michel LINDER from Lorraine University**
  - ⇨ Support for DOE
- ⇨ **Yves GOSSELIN from FERMENTIS**
  - ⇨ Advices for yeasts management
- ⇨ **French Malsters & Brewers**
  - ⇨ Financial support
- ⇨ **Sophie SCHWEBEL & IFBM R&D team**

iFBM

EBC 2017 SLOVENIA

Thank you for your attention !

Contact : marc.schmitt@qualtech-groupe.com

ICBS International Centre for Brewing Science

The University of Nottingham UNITED KINGDOM · CHINA · MALAYSIA

New insights into bitterness perception of beer

56TH CONGRESS EBC 2017 14-18 MAY LJUBLJANA EUROPEAN BREWERY CONVENTION SLOVENIA

<u>David Cook[1*]</u>, Olayide Oladokun[1], Sue James[2], Katherine A. Smart[2] & Joanne Hort[1].

[1] University of Nottingham, International Centre for Brewing Science, Nottingham, UK

[2]SAB Miller Ltd, Woking, GU21 6HS

Context

The University of Nottingham UNITED KINGDOM · CHINA · MALAYSIA

- Bitterness is a key sensory property of beers. Largely, but not wholly, derived from the bitterness of iso-α-acids (hops).
- Bitterness is a diverse taste percept encompassing a range of qualitative sensations and mediated by a large family of taste receptors (25 x TAS2R receptors identified; *Intelmann et al., 2009*).
- Beer bitterness has both temporal (*Hughes & Bolshaw, 1995; Fritsch & Shellhammer, 2009*) and qualitative aspects (*McLaughlin et al., 2008*).
- Traditionally bitterness of beer has been indicated in terms of IBU's (1 IBU nominally = 1 mg/L isohumulones)
- **Prior studies and our sensory experience of different beers suggest that perceived bitterness is more complex than a number derived from absorbance of an extract of beer at 275 nm.**

Research Approach

- Develop a **sensory lexicon** to describe the full array of sensations associated with bitterness perception of lager beers
  - Diverse sample set of lager beers
  - Attribute generation
  - Rationalisation of terms
  - Defined lexicon
  - Train panel to rate terms and anchor scales

- ***Apply the new sensory lexicon to better understand beer bitterness perception and how it is influenced by beer composition and hop products usage***

Bitterness lexicon: consensus terms

Harsh: Tingly, painful, warning signal, irritating, raspy.

Acidic: Vinegary, fruit-associated acidity.

Sharp: Instant, bitterness taste at the tip of tongue.

Tart: Acidic with sour notes.

Vegetative: Cabbage, Sprout-like bitterness, Hop tea flavour.

Artificial: Chemically, unnatural beer flavour.

Metallic: Taste of tin/metal, silver coin taste.

Astringent: Dry, causing drying of the mouth.

Smooth: Velvety.

Rounded: Pleasant, not spiky, not harsh.

Progressive: Bitterness perception increases gradually.

Instant: Instantaneous bitterness perception.

Diminishing: Bitterness perception decreases quickly after ingestion.

ICBS International Centre for Brewing Science

The University of Nottingham
UNITED KINGDOM · CHINA · MALAYSIA

Contents lists available at ScienceDirect

Food Chemistry

journal homepage: www.elsevier.com/locate/foodchem

The impact of hop bitter acid and polyphenol profiles on the perceived bitterness of beer

Olayide Oladokun [a], Amparo Tarrega [a], Sue James [b], Katherine Smart [b], Joanne Hort [a], David Cook [a,*]

[a] International Centre for Brewing Science, Bioenergy and Brewing Science Building, University of Nottingham, School of Biosciences, Division of Food Science, Sutton Bonington Campus, Loughborough LE12 5RD, UK
[b] SABMiller Plc, SABMiller House, Church Street West, Woking, Surrey GU21 6HS, UK

ARTICLE INFO

Article history:
Received 27 October 2015
Received in revised form 7 March 2016
Accepted 8 March 2016
Available online 8 March 2016

Chemical compounds studied in this article:
Protocatechuic acid (PubChem CID:72)
Catechin (PubChem CID:73160)
Epicatechin (PubChem CID:72276)
Caffeic acid (PubChem CID:689043)

ABSTRACT

Thirty-four commercial lager beers were analysed for their hop bitter acid, phenolic acid and polyphenol contents. Based on analytical data, it was evident that the beers had been produced using a range of different raw materials and hopping practices. Principal Components Analysis was used to select a sub-set of 10 beers that contained diverse concentrations of the analysed bitter compounds. These beers were appraised sensorially to determine the impacts of varying hop acid and polyphenolic profiles on perceived bitterness character. Beers high in polyphenol and hop acid contents were perceived as having 'harsh' and 'progressive' bitterness, whilst beers that had evidently been conventionally hopped were 'sharp' and 'instant' in their bitterness. Beers containing light-stable hop products (tetrahydro-iso-α-acids) were perceived as 'diminishing', 'rounded' and 'acidic' in bitterness. The hopping strategy adopted by brewers impacts on the nature, temporal profile and intensity of bitterness perception in beer.

Global lager study

The University of Nottingham
UNITED KINGDOM · CHINA · MALAYSIA

- Survey of analytical bitterness composition versus perceived quality of bitterness for 33 fresh, globally/ regionally significant lager brands (*Right*).

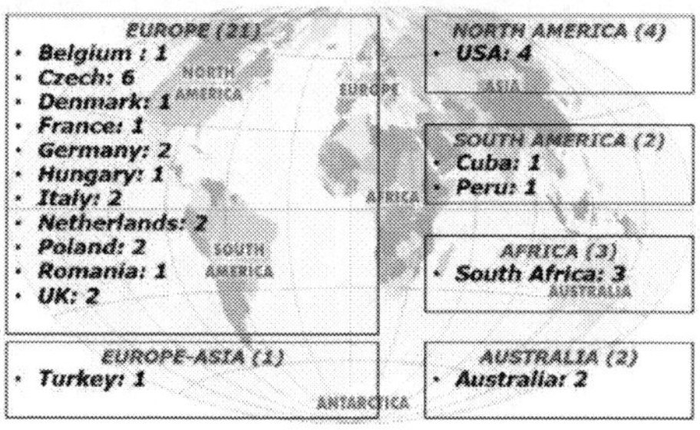

- Bitterness Units (ASBC-23) in range 8-36 BU.
- HPLC bitterness (sum iso-α-acids/ reduced side-chain iso-α-acids in range 8.7-43 mg/L. α-acids and humulinones also analysed (*Oladokun et al. 2016a*).
- Total polyphenols (ASBC-35), phenolic acids & flavanol monomers(HPLC) and amino acids (GC-MS) all analysed.
- **Sensory bitterness characteristics using the novel lexicon.**

PCA separation of the 33 beers according to their analytical profiles

The University of Nottingham
UNITED KINGDOM · CHINA · MALAYSIA

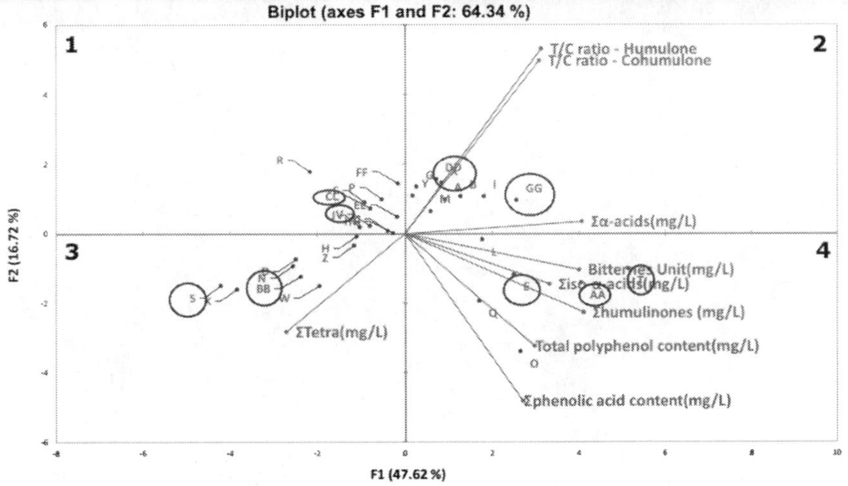

The ten circled beers were selected as representative of the diversity present in the analytical data set and were used in sensory studies of the quality of perceived bitterness.

Sensory evaluation of the 10 beers selected on the basis of their analytical diversity

The University of Nottingham
UNITED KINGDOM · CHINA · MALAYSIA

- Panel used 'Check All That Apply' (CATA) methodology to evaluate the beers.
- 9 of the 13 lexicon terms were scored significantly differently amongst the beers ($P<0.05$; Cochran's Q test).
  - **i.e. these terms were useful in distinguishing the bitterness characteristics present in the sample set.**

Attributes	*p-values**
Harsh	**0.000**
Acidic	0.491
Tart	0.219
Rounded	**0.000**
Metallic	**0.041**
Sharp	**0.008**
Smooth	**0.006**
Astringent	0.659
Artificial	0.517
Vegetative	**0.000**
Progressive	**0.009**
Instant	**0.020**
Diminishing	**0.002**

- **Correspondence analysis was then used to separate the beers according to their sensory bitterness qualities..........**

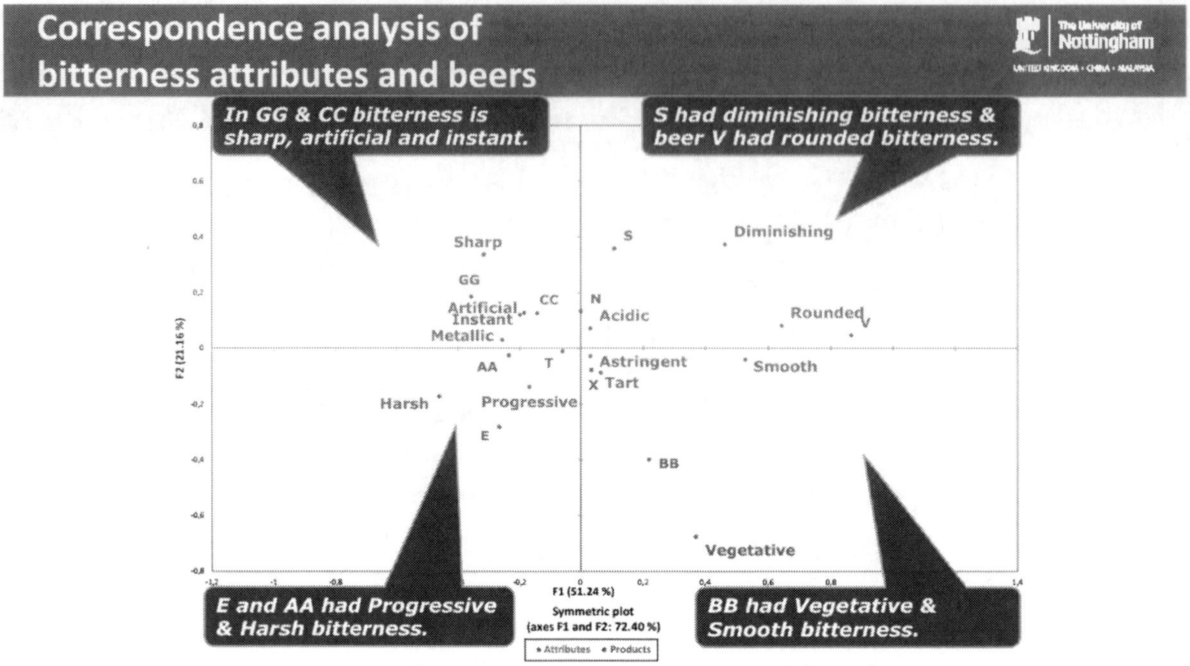
Correspondence analysis of bitterness attributes and beers
The University of Nottingham
UNITED KINGDOM · CHINA · MALAYSIA
In GG & CC bitterness is sharp, artificial and instant.
S had diminishing bitterness & beer V had rounded bitterness.
E and AA had Progressive & Harsh bitterness.
BB had Vegetative & Smooth bitterness.
Sharp
GG
Artificial
Instant
Metallic
CC
N
Acidic
S
Diminishing
Rounded
V
AA
T
Astringent
X
Tart
Smooth
Harsh
Progressive
E
BB
Vegetative
F2 (21.16 %)
F1 (51.24 %)
Symmetric plot
(axes F1 and F2: 72.40 %)
Attributes
Products

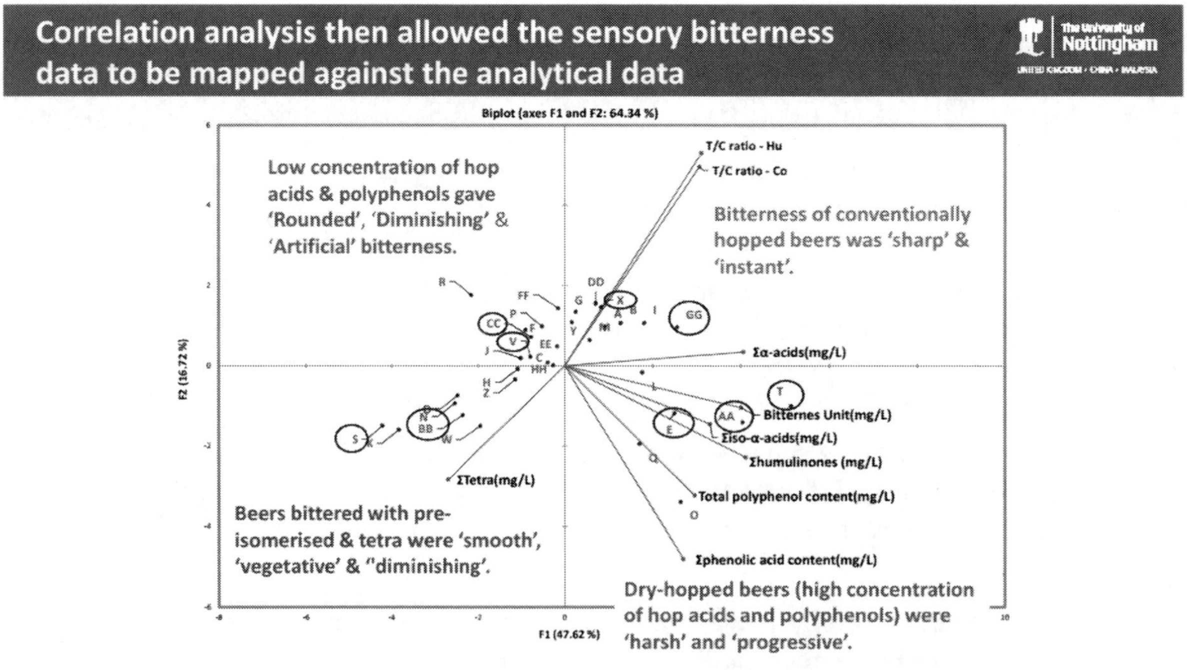
Correlation analysis then allowed the sensory bitterness data to be mapped against the analytical data
The University of Nottingham
UNITED KINGDOM · CHINA · MALAYSIA
Biplot (axes F1 and F2: 64.34 %)
Low concentration of hop acids & polyphenols gave 'Rounded', 'Diminishing' & 'Artificial' bitterness.
Bitterness of conventionally hopped beers was 'sharp' & 'instant'.
Beers bittered with pre-isomerised & tetra were 'smooth', 'vegetative' & "diminishing'.
Dry-hopped beers (high concentration of hop acids and polyphenols) were 'harsh' and 'progressive'.
T/C ratio - Hu
T/C ratio - Co
Σα-acids(mg/L)
Bitternes Unit(mg/L)
Σiso-α-acids(mg/L)
Σhumulinones (mg/L)
Total polyphenol content(mg/L)
Σphenolic acid content(mg/L)
ΣTetra(mg/L)
F2 (16.72 %)
F1 (47.62 %)

Predictive PLSR modeling of sensory bitterness qualities against analytical composition data.

The University of Nottingham
UNITED KINGDOM · CHINA · MALAYSIA

Table shows the coefficients of the terms for selected compounds/ group of compounds in the models for each bitterness quality

Bitterness Attribute	TPC	Iso-α-acids	Humulinones	Tryptophan	Tetrahydroiso-α-acids	α-acids	Model R^2
Intensity	+0.013	+0.048	+0.26	-0.020	-0.096	+0.21	0.90
Harsh	+0.017	+0.079	+0.073	-0.041	-0.078	+0.26	0.93
Smooth	-0.0064	+0.0060	-0.68	-0.017	+0.14	-0.12	0.95
Round	-0.0088	-0.053	+0.15	+0.014	+0.014	-0.13	0.95
Metallic	+0.0058	+0.10	-1.29	-0.01	+0.17	+0.061	0.96
Progressive	+0.010	+0.069	-0.24	-0.017	-0.0081	+0.16	0.97

Conclusions

The University of Nottingham
UNITED KINGDOM · CHINA · MALAYSIA

- **Hop products selection and hopping practice give distinct hop bitter acid & polyphenol profiles, which in turn impact on the perceived bitterness quality of beers.**
  - –Beers low in hop bitter acids and polyphenols rated as having artificial, rounded & diminishing bitterness.
  - –Conventionally hopped beers high in hop acids, α-acids were rated as having a sharp and instant bitterness.
  - –Beers bittered with a blend of tetra and pre-isomerised iso-α-acid products rated as having a smooth and diminishing bitterness.
  - –Bitterness of dry-hopped beers high in hop bitter acids and polyphenols was perceived as harsh and progressive.

ICBS International Centre for Brewing Science

The University of Nottingham
UNITED KINGDOM · CHINA · MALAYSIA

Contents lists available at ScienceDirect

Food Research International

journal homepage: www.elsevier.com/locate/foodres

Modification of perceived beer bitterness intensity, character and temporal profile by hop aroma extract

Olayide Oladokun [a], Amparo Tarrega [a], Sue James [b], Trevor Cowley [b], Frieda Dehrmann [b], Katherine Smart [b], David Cook [a], Joanne Hort [a,*]

[a] International Centre for Brewing Science, Bioenergy and Brewing Science Building, University of Nottingham, School of Biosciences, Division of Food Science, Sutton Bonington Campus, Loughborough LE12 5RD, UK
[b] SABMiller Plc, SABMiller House, Church Street West, Woking, Surrey GU21 6HS, UK

ARTICLE INFO

Article history:
Received 20 April 2016
Received in revised form 20 May 2016
Accepted 21 May 2016
Available online 31 May 2016

Keywords:
Beer
Bitterness intensity
Bitterness character
Taste-aroma interactions
Trigeminal sensation
Hop aroma extract

ABSTRACT

The effect of hop aroma on perceived bitterness intensity, character and temporal profile of beer was investigated. A hop aroma extract was added at 3 levels (0, 245, 490 mg/L) to beers at low, medium and high bitterness. Beers were evaluated for perceived bitterness intensity, harshness, roundedness and linger by a trained panel using a rank-rating technique at each bitterness level, with and without nose clips. The use of nose clips enabled the olfactory aspect to be decoupled from taste and mouthfeel aspects of bitterness perception. Results showed significant modification of perceived bitterness in beer by hop aroma depending on the inherent level of bitterness. These modifications were mainly driven by olfaction – in an example of taste-aroma interactions, as well as certain tactile sensations elicited by the hop aroma extract in the oral cavity. At low bitterness, beers with hop aroma added were perceived as more bitter, and of 'rounded' bitterness character relative to those without hop aroma. When judges used nose clips, this effect was completely eliminated but the sample was perceived to have a 'harsh' bitterness character. Conversely, at high bitterness, even when nose clips were used, judges still perceived beers containing hop aroma to be more bitter. These increases in bitterness perception with nose clips indicates the stimulating of other receptors, e.g. trigeminal receptors by hop aroma extract, which in tandem with the high bitterness, cause perceptual interactions enhancing bitterness intensity and also affecting bitterness character. Bitterness character attributes such as 'round' and 'harsh' were found to significantly depend on bitterness and aroma levels, with the second level of aroma addition (245 mg/L) giving a 'rounded' bitterness in low bitterness beers but 'harsh' bitterness in high bitterness beers. The impact of aroma on temporal bitterness was also confirmed with time-intensity measurements, and found to be mostly significant at the highest level of hop aroma addition (490 mg/L) in low bitterness beers. These findings represent a significant step forward in terms of understanding bitterness flavour perception and the wider impact of hop compounds on sensory perception.

Hop oil extracts add more to beer than just aroma!

The University of Nottingham
UNITED KINGDOM · CHINA · MALAYSIA

- This study examined the effects of varied addition rates of hop aroma oils to beers at different BU levels on perceived bitterness quality and intensity.
- An un-hopped base beer was prepared (70% pilsner malt/ 30% dextrose adjunct) which had an ABV of 5.0%, pH 4.23 and total polyphenol content (ASBC Beer-35) of 133 mg/L.
- Isohop and pure hop aroma (PHA) products were then added to the same base beer to produce the following sample matrix:

	Hersbrucker PHA (mg/L)			East Kent Goldings PHA (mg/L)		
Bitterness	*L0*	*L1*	*L2*	*L0*	*L1*	*L2*
Low (13 mg/L)	0	245	490	0	245	490
Medium (25 mg/L)	0	245	490	0	245	490
High (42 mg/L)	0	245	490	0	245	490

Experimental

The University of Nottingham
UNITED KINGDOM · CHINA · MALAYSIA

- After addition of isohop and pure hop aroma, beers recapped, mixed and stored for 2 days at 3°C.
- Beers evaluated by trained sensorial panel at 4±2°C.

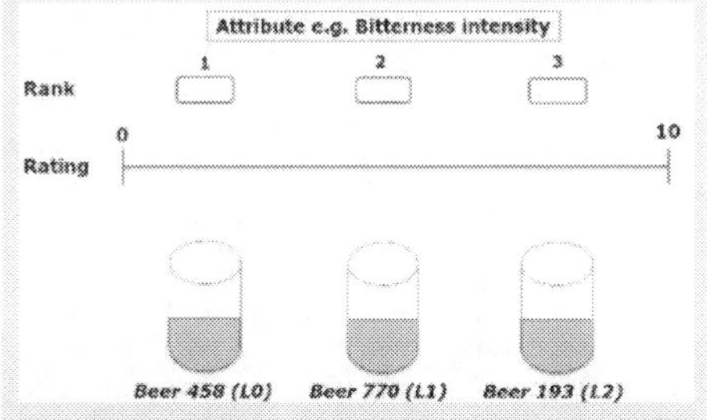

Rank-rating methodology

- Preliminary studies on the sensory bitterness of the beers was used to select 3 qualitative bitterness terms which best differentiated the samples:
  - *Harsh* (tingly, painful); *Round* (pleasant, smooth, non-spiky bitterness); *Lingering* (perceived bitterness intensity 10 seconds after beer consumption).

Impact of EKG aroma at low bitterness level (13 mg/L isohop)

The University of Nottingham
UNITED KINGDOM · CHINA · MALAYSIA

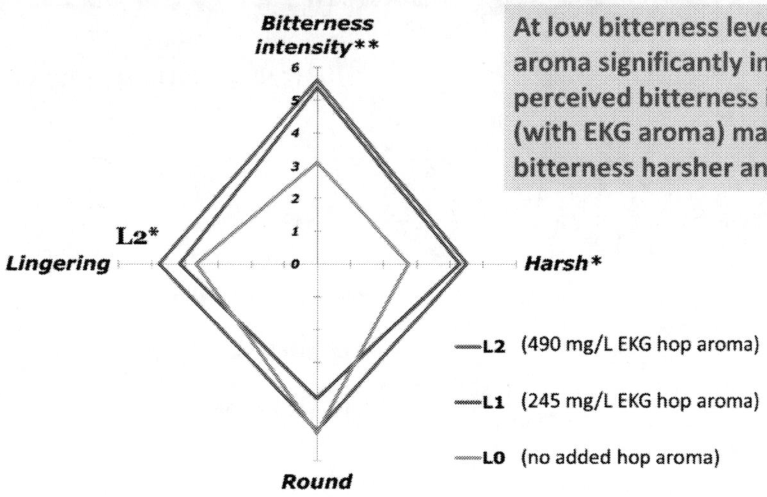

At low bitterness levels, added hop aroma significantly increased perceived bitterness intensity and (with EKG aroma) made the bitterness harsher and more lingering

Significance at 5%, 1%** based on ANOVA of intensity ratings and Rank analysis by Friedman's test ($p<0.05$); n= 7 trained panelists.*

Impact of EKG aroma at medium bitterness level (25 mg/L isohop)

The University of Nottingham
UNITED KINGDOM · CHINA · MALAYSIA

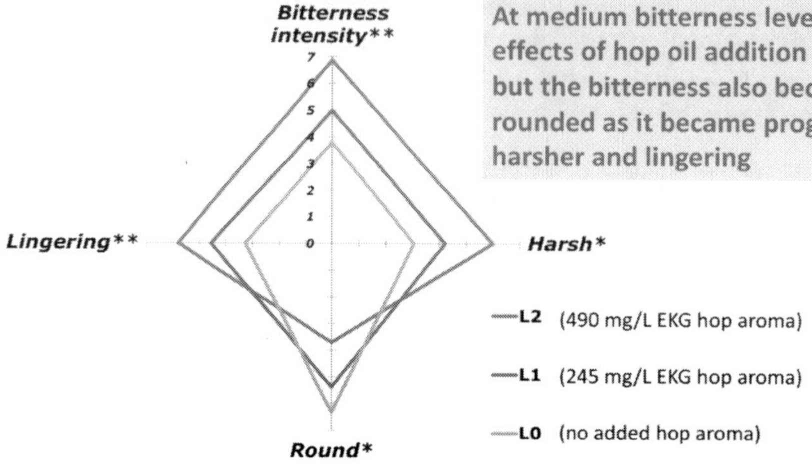

At medium bitterness levels, similar effects of hop oil addition were noted but the bitterness also became less rounded as it became progressively harsher and lingering

Significance at 5%, 1%** based on ANOVA analysis of intensity ratings and Rank analysis by Friedman's test (p<0.05); n=7 trained panelists.*

Impact of EKG aroma at high bitterness level (42 mg/L isohop)

The University of Nottingham
UNITED KINGDOM · CHINA · MALAYSIA

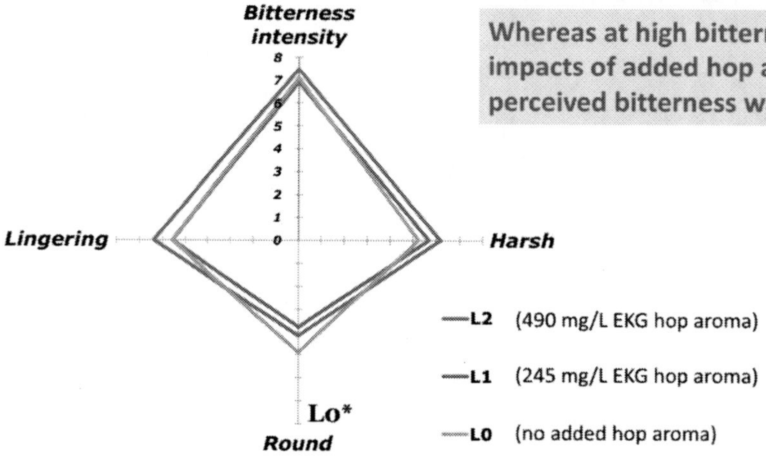

Whereas at high bitterness levels the impacts of added hop aroma on perceived bitterness was minimal.

Significance at 5%, 1%** based on ANOVA of intensity ratings and Rank analysis by Friedman's test (p<0.05); n= 7 trained panelists.*

Impact of Hersbrucker aroma at low bitterness level (13 mg/L isohop)

The University of Nottingham
UNITED KINGDOM · CHINA · MALAYSIA

Assessment without wearing nose-clips

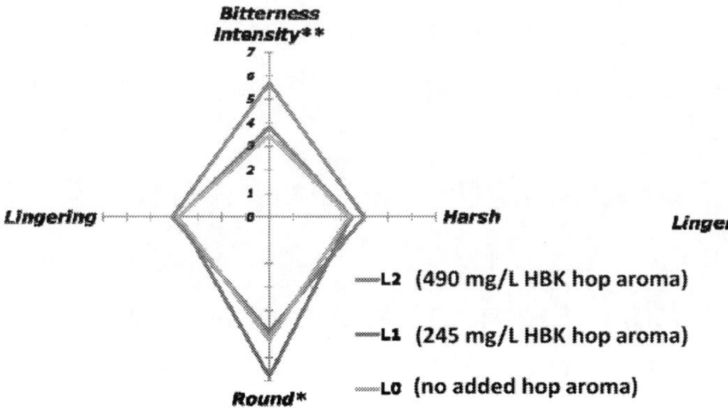

Same evaluation, panel wearing nose-clips to eliminate sense of smell

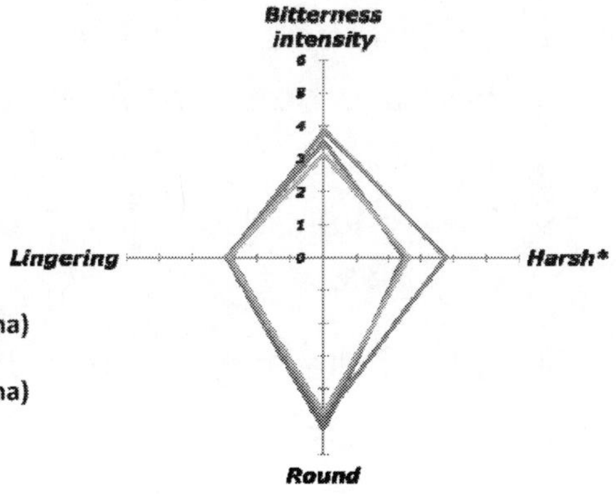

Significance at 5%, 1%** based on ANOVA of intensity ratings and Rank analysis by Friedman's test (p<0.05); n=7 trained panelists.*

Impact of Hersbrucker aroma at high bitterness level (42 mg/L isohop)

The University of Nottingham
UNITED KINGDOM · CHINA · MALAYSIA

Assessment without wearing nose-clips

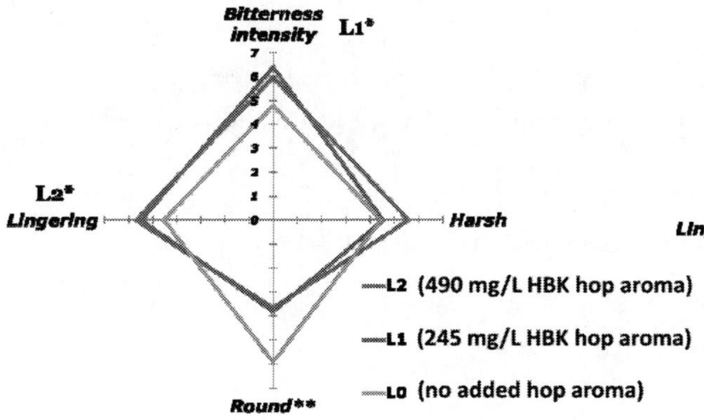

Same evaluation, panel wearing nose-clips to eliminate sense of smell

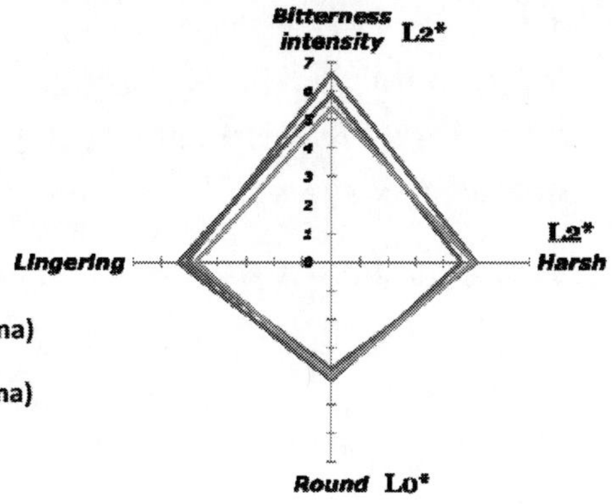

Significance at 5%, 1%** based on ANOVA of intensity ratings and Rank analysis by Friedman's test (p<0.05); n=7 trained panelists.*

Conclusions

ICBS International Centre for Brewing Science — The University of Nottingham, UNITED KINGDOM · CHINA · MALAYSIA

- Addition of hop aroma oils changed the perceived intensity and quality of bitterness.
- The magnitude of such effects was found to depend on the BU of the base beer and was greatest in beers of low to medium BU's.
  - Clear implications for beer design/NPD to achieve desirable bitterness
- At low bitterness the with/without nose-clip experiments indicated that perceived aroma was influencing bitterness via cross modal (taste-aroma) interactions.

Conclusions (2)

ICBS International Centre for Brewing Science — The University of Nottingham, UNITED KINGDOM · CHINA · MALAYSIA

- Even with nose clips, at high BU the panel could consistently differentiate the samples according to rate of aroma oil addition
  - Components of the Hersbrucker oil mediated trigeminal sensations in the mouth which can moderate perceived bitterness quality and intensity
  - **Hop oils do indeed add more to beer than purely aroma!**

ICBS International Centre for Brewing Science

The University of Nottingham
UNITED KINGDOM · CHINA · MALAYSIA

Food Chemistry 230 (2017) 215–224

Contents lists available at ScienceDirect

Food Chemistry

journal homepage: www.elsevier.com/locate/foodchem

Perceived bitterness character of beer in relation to hop variety and the impact of hop aroma

Olayide Oladokun, Sue James, Trevor Cowley, Frieda Dehrmann, Katherine Smart, Joanne Hort[1], David Cook[1,*]

International Centre for Brewing Science, Bioenergy and Brewing Science Building, University of Nottingham, School of Biosciences, Division of Food Science, Sutton Bonington Campus, Loughborough LE12 5RD, UK
SABMiller Plc, SABMiller House, Church Street West, Woking, Surrey GU21 6HS, UK

ARTICLE INFO

Article history:
Received 11 October 2016
Received in revised form 3 March 2017
Accepted 7 March 2017
Available online 8 March 2017

Chemical compounds studied in this article:
Protocatechuic acid (PubChem CID:72)
Catechin (PubChem CID:73160)
Epicatechin (PubChem CID:72276)
Caffeic acid (PubChem CID:689043)
Vanillic acid (PubChem CID:8468)

ABSTRACT

The impact of hop variety and hop aroma on perceived beer bitterness intensity and character was investigated using analytical and sensory methods. Beers made from malt extract were hopped with 3 distinctive hop varieties (Hersbrucker, East Kent Goldings, Zeus) to achieve equi-bitter levels. A trained sensory panel determined the bitterness character profile of each singly-hopped beer using a novel lexicon. Results showed different bitterness character profiles for each beer, with hop aroma also found to change the hop variety-derived bitterness character profiles of the beer. Rank-rating evaluations further showed the significant effect of hop aroma on selected key bitterness character attributes, by increasing perceived harsh and lingering bitterness, astringency, and bitterness intensity via cross-modal flavour interactions. This study advances understanding of the complexity of beer bitterness perception by demonstrating that hop variety selection and hop aroma both impact significantly on the perceived intensity and character of this key sensory attribute.

Perceived bitterness character in relation to hop variety and the impact of hop aroma

The University of Nottingham
UNITED KINGDOM · CHINA · MALAYSIA

- In this study 3 base beers were brewed similarly but bittered with 3 different hop varieties (*Hallertau Hersbrucker, East Kent Goldings, Zeus*) no aroma hop addition.
- Varietal hop aroma oils extracted from *Hallertau Hersbrucker, East Kent Goldings* were added to each base beer.
- Sensory bitterness characteristics were evaluated using the novel lexicon.
- <u>Learnings:</u>
  - Further evidence of the strong influence of hop aroma on perceived bitterness quality.
  - The sensory bitterness character of each base beer depended significantly on the particular varietal hop oil added to it.
  - Some limited support for the concept of *congruency* – i.e. adding each of the hop aroma oils to its' respective base beer gave a higher perceived bitterness intensity than when paired with bitterness of other varieties.

Bringing it all together.....

The University of Nottingham
UNITED KINGDOM · CHINA · MALAYSIA

- Beer bitterness is a multi-faceted sensory property with qualitative and temporal dimensions.
- Beer bitterness is the result of a combination of congeners:
  - Iso alpha acids, their oxidised or degradation products and light-stable counterparts
  - Polyphenols and simple phenolics
  - Bitter tasting amino acids
- The mixture of congeners, and thus the quality of beer bitterness is entirely dependent on brewing raw materials and their points of addition to the process.

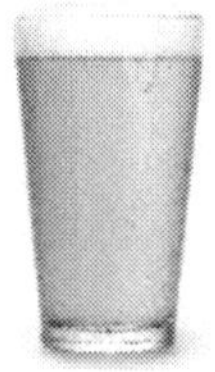

Bringing it all together.....

The University of Nottingham
UNITED KINGDOM · CHINA · MALAYSIA

- Human flavour perception is multisensory.
- Our work shows that interactions between the senses are dependent on bitterness level of the beer, and on the nature of the hop oil used to modulate the bitterness.
- Hop aroma oils such as Hersbrucker can contribute trigeminal sensations which influence bitterness perception.
- Use of different varietal hop aroma oils imparted different bitterness characteristics to the same base beer.
- **Perceived bitterness is much more complex than a number derived from absorbance of an extract of beer at 275 nm!!**

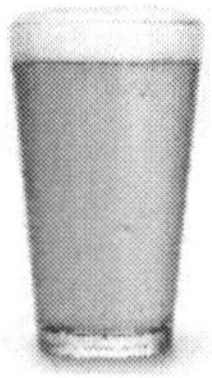

References

- Fritsch, A., & Shellhammer, T. H. (2009). The Bitter Qualities of Reduced and Nonreduced Iso-α-acids. *Journal of the American Society of Brewing Chemists, 67*(1), 8-13.
- Goiris, K., Ridder, M., de Rouck, G., Boeykens, A., Van Opstaele, F., Aerts, G., . . . Keukeleire, D. (2002). The oxygenated sesquiterpenoid fraction of hops in relation to the spicy hop character of beer. *Journal of the Institute of Brewing, 108*(1), 86-93.
- Hughes, P., & Bolshaw, L. (1995). *Time-dependent sensory responses to chemically-modified hop bitter acids.* Paper presented at the PROCEEDINGS OF CONGRESS-EUROPEAN BREWERY CONVENTION.
- Intelmann, D., Batram, C., Kuhn, C., Haseleu, G., Meyerhof, W., & Hofmann, T. (2009). Three TAS2R bitter taste receptors mediate the psychophysical responses to bitter compounds of hops (Humulus lupulus L.) and beer. *Chemosensory Perception, 2*(3), 118-132.
- McLaughlin, I. R., Lederer, C., & Shellhammer, T. H. (2008). Bitterness-modifying properties of hop polyphenols extracted from spent hop material. *Journal of the American Society of Brewing Chemists, 66*(3), 174-183.
- Oladokun, O., Smart, K., & Cook, D. (2016a). An improved HPLC method for single-run analysis of the spectrum of hop bittering compounds usually encountered in beers. *Journal of the Institute of Brewing, 122*(1), 11-20.
- Oladokun, O., Tarrega, A., James, S., Smart, K., Hort, J., & Cook, D. (2016b). The impact of hop bitter acid and polyphenol profiles on the perceived bitterness of beer. *Food Chemistry, 205*, 212-220.
- Oladokun, O., Tarrega, A., James, S., Cowley, T., Dehrmann, F., Smart, K. A., Cook, D. J., and Hort, J. (2016c) Modification of perceived beer bitterness intensity, character and temporal profile by hop aroma extract, *J. Food Res. Int., 86*, 104-111.
- Oladokun, O., James, S., Cowley, T., Dehrmann, F., Smart, K. A., Hort, J., and Cook, D. J. (2017) Perceived bitterness character of beer in relation to hop variety and the impact of hop aroma, *Food Chemistry, 230*, 215-224.
- Schönberger, C. (2006). Bitter is better. *Monatsschrift für Brauwissenschaft, 3*(4), 56-65.

ICBS International Centre for Brewing Science

The University of Nottingham — UNITED KINGDOM · CHINA · MALAYSIA

Thank you for your attention.......Questions?

With full acknowledgement to Ola Oladokun for his excellent and original research.

We gratefully acknowledge SABMiller Ltd and the University of Nottingham for funding this research.

e-mail: david.cook@nottingham.ac.uk

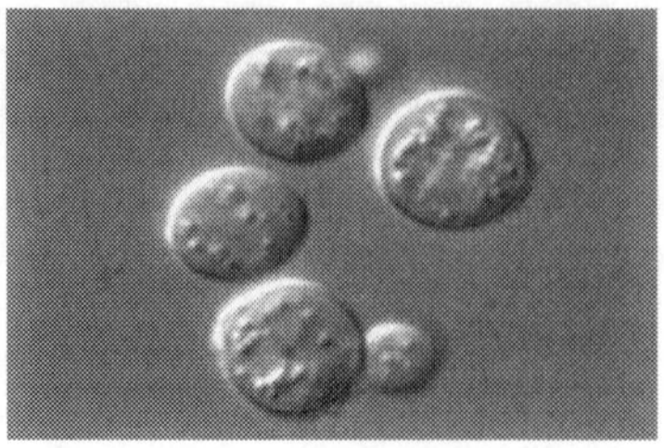

Characterization of *Kveik*, traditional *Saccharomyces* yeasts domesticated by Norwegian homebrewers

L35, EBC 2017 Ljubljana

Richard Preiss, Caroline Tyrawa, and George van der Merwe

What is *Kveik*?

- Historically, yeast was maintained & domesticated by repitching
- Commercial beer/yeast production abolished traditional yeasts in most of Europe
- Traditional yeast culture lives on in western Norway – *Kveik*
- Maintained through repitching and dry storage
- Some *Kveik* are reportedly hundreds of years old

www.garshol.priv.no

Rediscovery & Preservation of *Kveik*

- Lars Marius Garshol – Norwegian beer writer ("larsblog")
- Critical to characterize and preserve traditional yeasts like *Kveik*
- Some have been commercialized already
- Intriguing characteristics – reports of clean fermentation at 28-40ºC, highly flocculent, unique flavour
- No comprehensive study of pure culture *Kveik* yeasts exists to date

The Yeast Bay

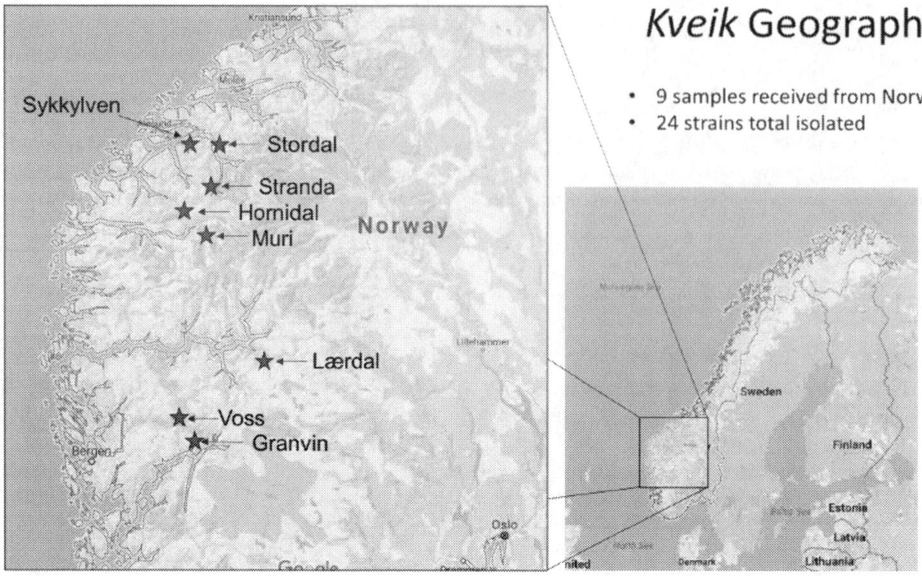

Kveik Geography

- 9 samples received from Norway
- 24 strains total isolated

Kveik are typically multi-strain *Saccharomyces cerevisiae* cultures

- Range from 1-9 putative strains in each *Kveik* culture based on WLN colony morphology
- Thus similar to traditional ale yeast cultures – multi-strain until purification techniques used

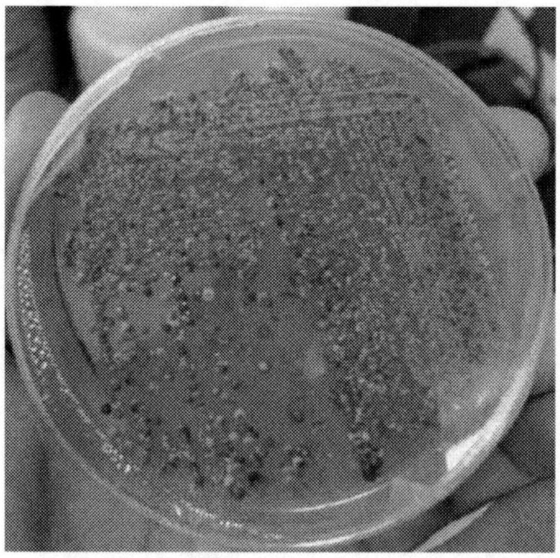

Traditional usage of *Kveik* yeasts

- Norwegian farmhouse ales – fairly broad range of techniques and resultant beers
- Very different from Franco-Belgian farmhouse ales
- Almost always juniper infused water (branches, not berries!)
- Raw vs boiled
- Gravity typically high (~1.080 SG)
- Non-phenolic beers (POF-), history and technology for yeast reuse

Skimming juniper infusion
http://www.garshol.priv.no/blog/291.html

Questions

- Are *Kveik* yeast genetically different from typical brewing yeasts?
- What kind of brewing characteristics do they have?
- How much do they differ amongst themselves?
- Are these yeasts useful for traditional beers only, or can they be applied to a range of beer styles?

Dried *Kveik*

Kveik post-rehydration
1000x DIC

Beer yeast family tree, Gallone *et al.* 2016

- Hallmarks of yeast domestication:
  - Efficient utilization of maltotriose
  - Loss of function in *PAD1* / *FDC1*: POF- phenotype
  - Copy number variations give rise to increased flocculation
- Where do *Kveik* fit in?

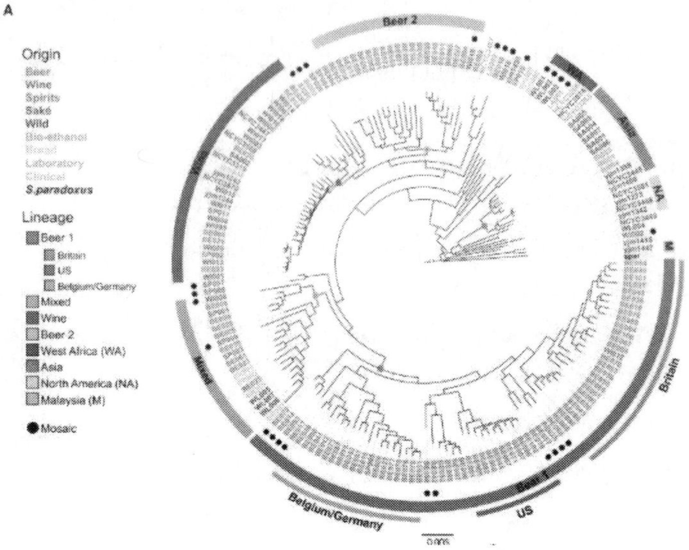

Gallone *et al.* Domestication and Divergence of *Saccharomyces cerevisiae* Beer Yeasts. **Cell**. 2016;166: 1397–1410

Phylogenetic analysis by composite interdelta PCR fingerprinting

1. Interdelta PCR
2. Agilent Bioanalyzer
3. GelJ (difference tables generated using Pearson correlation & UPGMA)
4. Composite analysis (hclust hierarchical clustering algorithm in R)
5. Visualization (FigTree v1.4.3)

	A	B	C	D	E
1	Strain Number	WLP570	WLP002	Sykkylven_1	Sykkylven_2
2	WLP570	0	0.45467189	0.58278362	0.68884985
3	WLP002	0.45467189	0	0.68280878	0.70817273
4	Sykkylven_1	0.58278362	0.68280878	0	0.18140723
5	Sykkylven_2	0.68884985	0.70817273	0.18140723	0

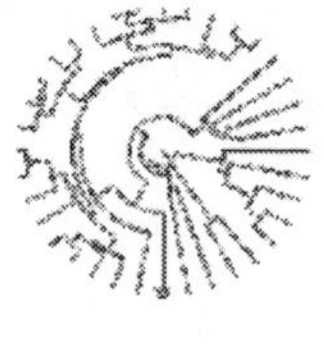

Genetic fingerprint analysis – *Kveik* yeast diversity

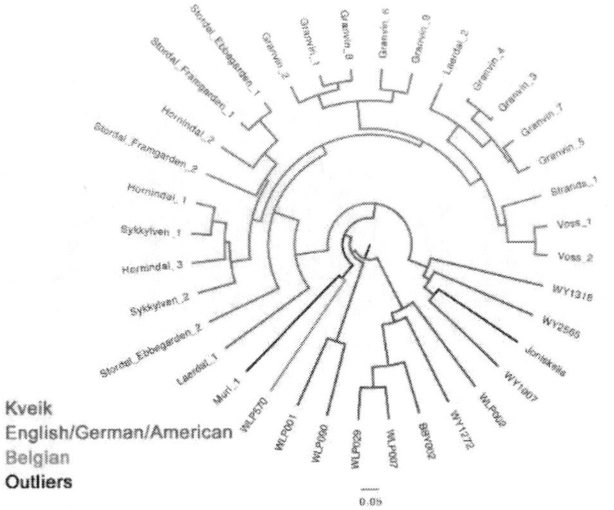

- *Kveik* yeasts more closely related to each other than to other ale yeasts
- *Kveik* may thus form a separate branch on the beer yeast family tree
- Most closely related to group consisting of two German ale and one English ale strain
- Also see what appears to be two major groups of *Kveik*

Mini Fermentation Trials

- 12.5ºP sterile, hopped wort (70mL)
- Fermentation at 30ºC
- Control ale strain included (US-05)
- Measuring CO_2 loss (fermentation rate), alcohol production (HPLC-RI), flavour compounds (HS-SPME-GC-MS)

Eg:

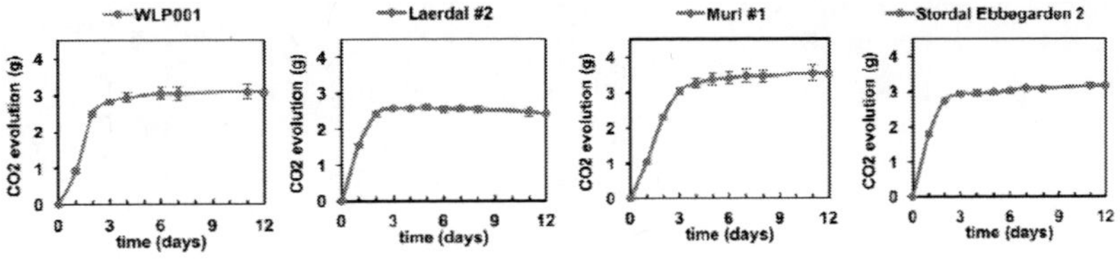

Kveik yeast fermentation kinetics

ESCARPMENT LABORATORIES

UNIVERSITY of GUELPH

CHANGING LIVES IMPROVING LIFE

- Most strains fermented faster in first 24h vs. control (shorter lag time)
- Several strains at complete attenuation by 72h
- Generally very good fermentation kinetics

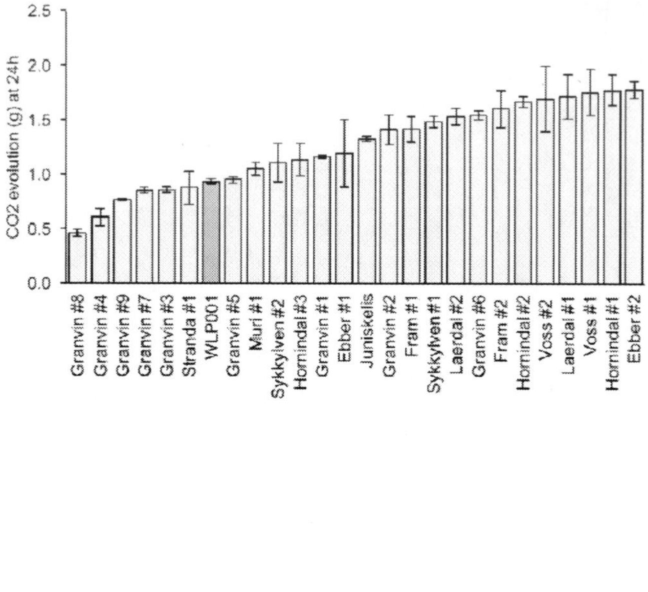

Kveik yeast ethanol production

ESCARPMENT LABORATORIES

UNIVERSITY of GUELPH

CHANGING LIVES IMPROVING LIFE

- Big range of apparent attenuation
  - 66%-95%
- Ethanol (%v/v) 4.4-6.4%

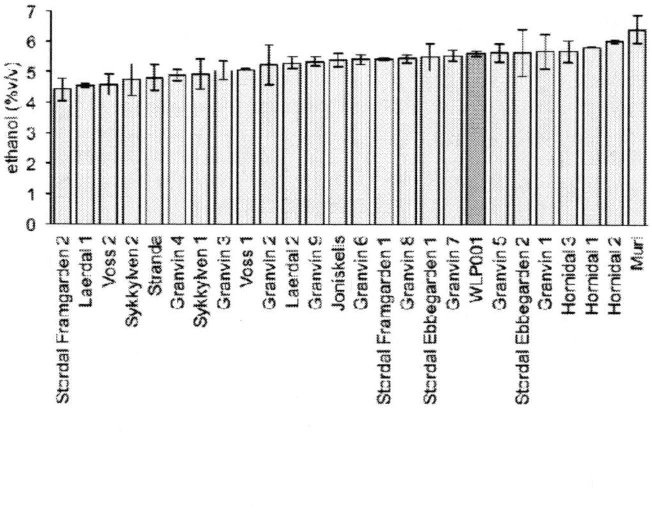

Kveik beer flavour analysis

- Flavour metabolites above threshold:
  - Ethyl caproate (pineapple)
  - Ethyl caprylate (pineapple, waxy, cognac)
  - Ethyl decanoate (red apple)
  - Phenethyl acetate (floral, honey)

- Only 2/24 strains phenolic (POF+)
  - These were "outliers" in the family tree
- Ethyl acetate, isoamyl acetate below threshold
- Isobutanol (fusel) ~50% of US-05
  - Suggestive of reported "cleaner" flavours at high temperatures

"Typical Kveik flavour profile"

Brewing Parameters

- Ethanol tolerance:
  - 22/24 strains tolerant to 13% ABV
  - 12/24 strains tolerant to 15% ABV
  - 8/24 strains tolerant to 16% ABV

- Temperature tolerance:
  - All Kveik tolerant to 40ºC
  - 16/24 tolerant to 42ºC

- Therefore *Kveik* yeast show very good ethanol tolerance and thermotolerance

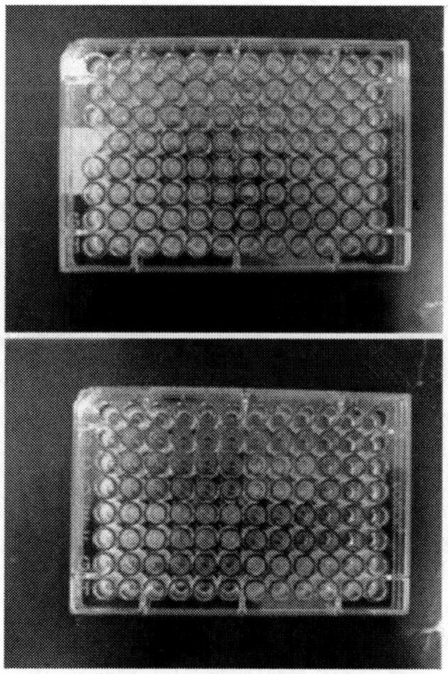

Brewing Parameters

- Flocculation (ASBCMOA-Yeast-11):
  - 16/24 strains classified flocculant
  - Many extremely flocculant

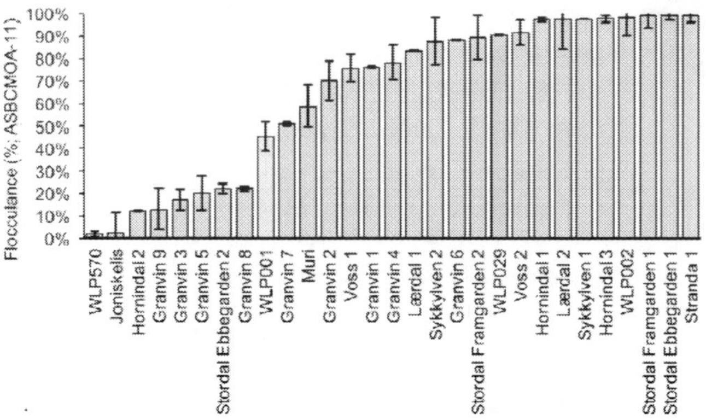

Key points

- Norwegian Kveik may represent a separate lineage of domesticated ale yeasts
- These yeasts display excellent fermentation kinetics, temperature tolerance, ethanol tolerance, and flocculation ability
- Rapid fermentation, POF-, and high flocculation are all suggestive of domestication events
- Kveik yeasts can be used to create new and distinctive beers
- Positive flavour at high temperature reduces need for fermentation tank cooling in hot/summer climates

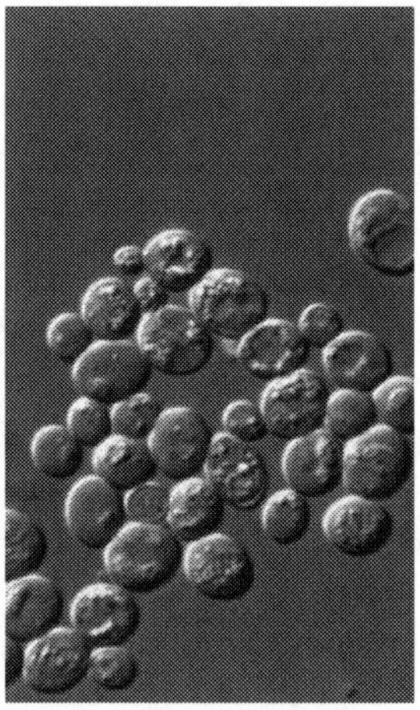

Commercialization example

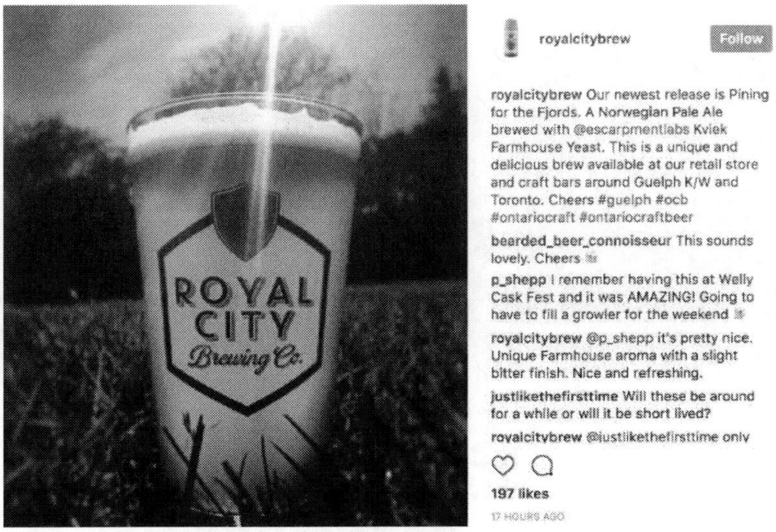

Royal City Brewing Co (Guelph, ON) – "Pining For the Fjords"

More on the way…

Acknowledgements

- Escarpment Labs:
  - Angus Ross, Nate Ferguson
- University of Guelph:
  - Caroline Tyrawa, George van der Merwe, Dyanne Brewer, Judith Vandeven
- Royal City Brewing
- Lars Marius Garshol & the tenacious Norwegian homebrewers

36TH CONGRESS
EBC
2017 EUROPEAN BREWERY CONVENTION
14–18 MAY LJUBLJANA SLOVENIA

MEURASTREAM

A global wort production concept with reduced energy and process water consumption

Frédérique Harmegnies
Meura R&D Manager

MEURA

Introduction

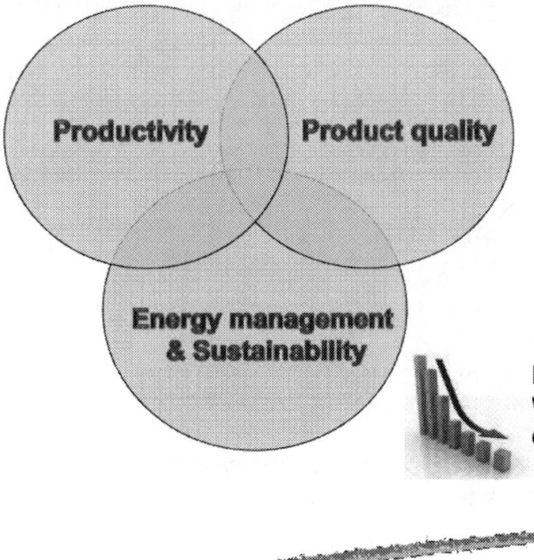

Energy
Water
Carbon footprint

MEURA

Meura's brewhouse concept

High gravity wort production based on fine grist technology coupled with thin bed membrane assisted mash filtration

CLASSICMILL

MECHAMASHER

MEURA 2001

Sieve	2001 mash filter Fine milling
1,25 mm	≤ 4%
1 mm	
0,5 mm	> 61%
0,25 mm	
0,125 mm	
bottom	≤ 35%

- High gravity brewing
- High extract recovery yield
- Reduced energy and water consumption

MEURA

Meura's brewhouse concept

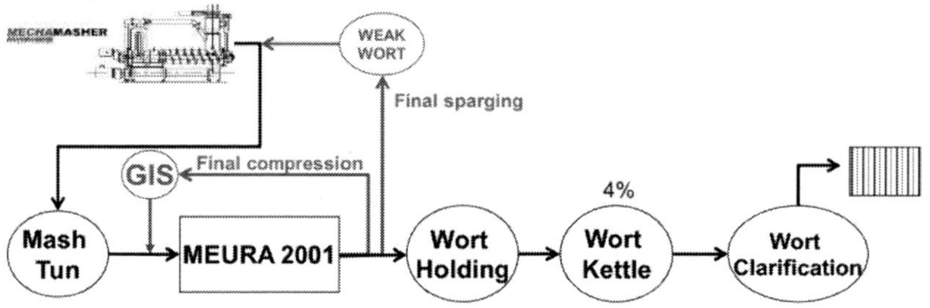

Mash Tun		Wort Holding	Wort Kettle
24 °P 400 hl	Conventional brewhouse with Meura2001	17 °P 565 hl	17,5 °P 548 hl
24,5 °P 391 hl	HGB Brewhouse with Meura 2001	20,0 °P 480 hl	20,8 °P 460 hl
26 °P 370 hl	VHGB Brewhouse with Meura 2001	24,0 °P 400 hl	25,0 °P 385 hl

HGB = High Gravity brewhouse

GIS = Gravity Increase System

VHGB = Very High Gravity brewhouse

With reduced volumes of vessels or increased capacity for an existing brewhouse, high gravity brewing is already a first step to decrease thermal energy / hl cold wort

Meura's brewhouse concept

Hot wort treatment based on a 3-stages de-intensified boiling process

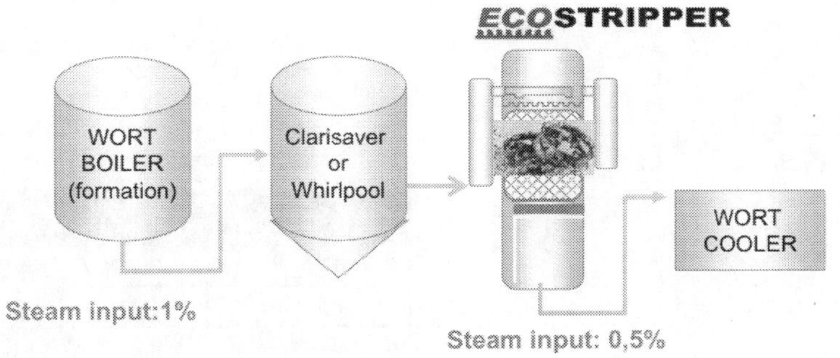

- Reduced energy consumption (max 1,5% steam input)
- Batch or continuous process
- Very efficient removal of volatiles
- Improved wort quality (low thermal load - no *de novo* DMS formation)

MEURA

Meura's brewhouse concept

How to design a brewhouse with the highest thermal efficiency with limited excess in hot water and brew water consumption ?

	Brewhouse without E recovery
Thermal E consumption	26,4 MJ/hl
Brewing water consumption	1,43 hl/hl
Excess hot water	0,37 hl/hl

For the annual capacity of 1 Mio hl 15°P wort, excess hot water amounts to **37 000 m³** per year !

MEURA

Energy recovery - Pfaduko

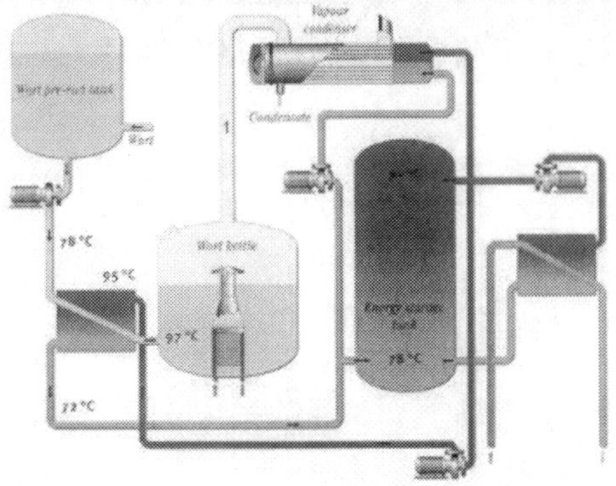

- The thermal energy balance requires around 4% evaporation rate at boiling for wort preheating
  → not applicable to de-intensified boilng
- Vapour condensation is not a solution for hot water excess

Brewhouse producing wort @15°P (evaporation 4%)

	Without E recovery	With Pfaduko
Thermal E	26,4 MJ/hl →	19,4 MJ/hl
Brewing water	1,43 hl/hl	1,43 hl/hl
Excess hot water	0,37 hl/hl	0,37 hl/hl

Pfaduko allows some thermal energy savings (26%) but it is not a solution to reduce process water consumption and hot water excess

Development of a new concept

Brewhouse concept applying de-intensified boiling

+

« Open » energy recovery system producing high temperature process water at counter flow of the brewing process

Energy recovery at wort cooling + Ecostripper

MEURASTREAM

MEURA

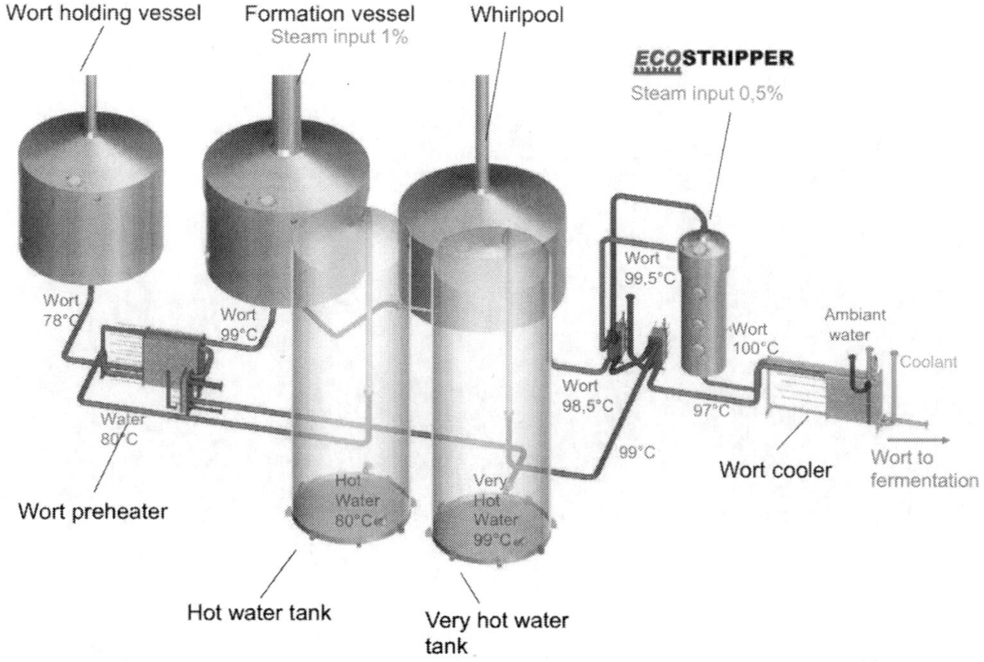

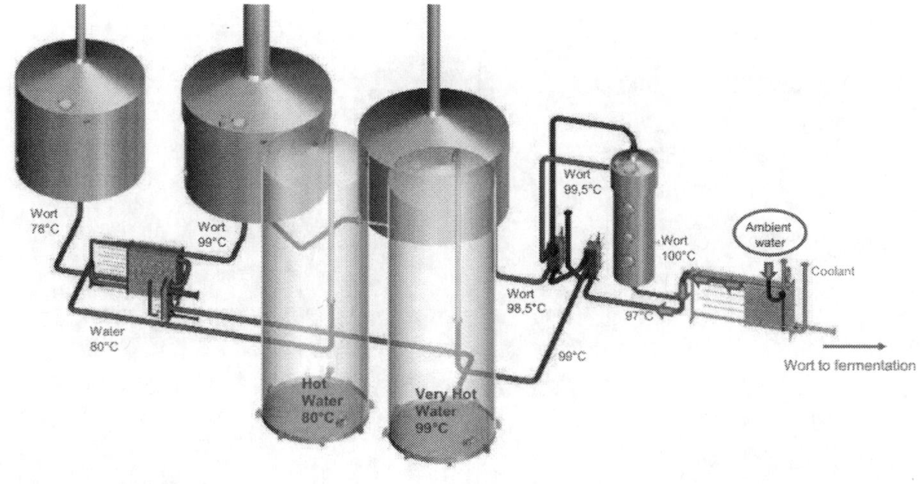

RECOVERY OF ENERGY
wort cooling → production of very hot water (97°C)

MEURA

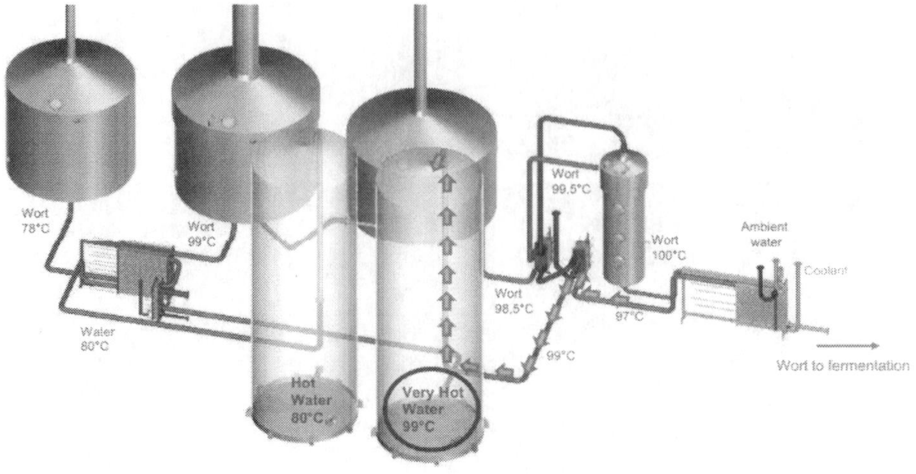

RECOVERY OF ENERGY
Ecostripper and water booster
→ production of Very Hot Water (99°C), stored in Very Hot WaterTank

MEURA

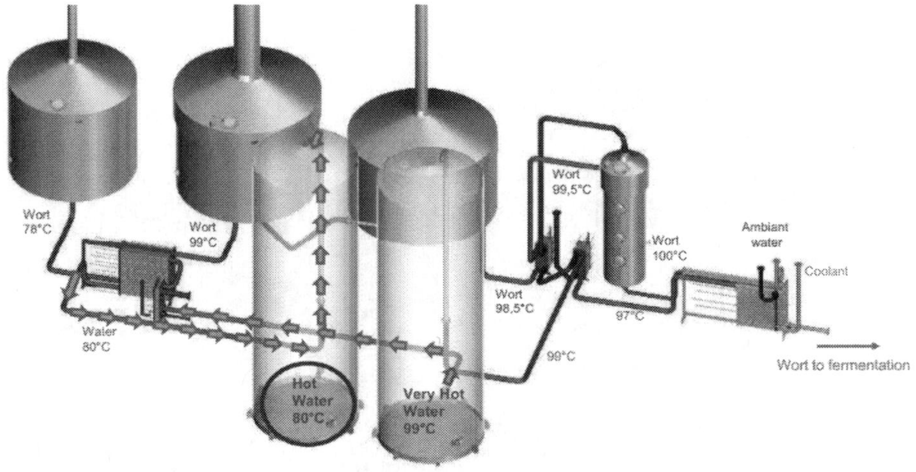

RE-USE OF ENERGY
Very Hot Water (99°C) used for wort pre-heating
→ production of Hot Water (80°C), stored in Hot Water Tank

MEURA

Case study - Opex savings

Thermal energy in the brewhouse

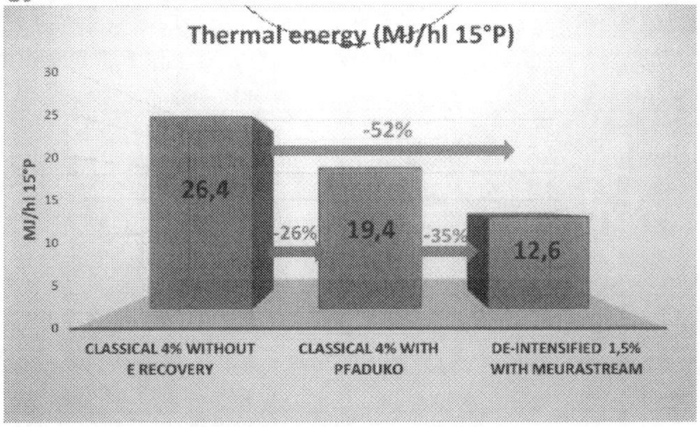

De-intensified boiling combined with MEURASTREAM allows reducing the thermal energy consumption in the brewhouse by 52%

→ 2 times more than E recovery with Pfaduko!

→ 35% savings compared to Pfaduko

Case study - Opex savings

Water

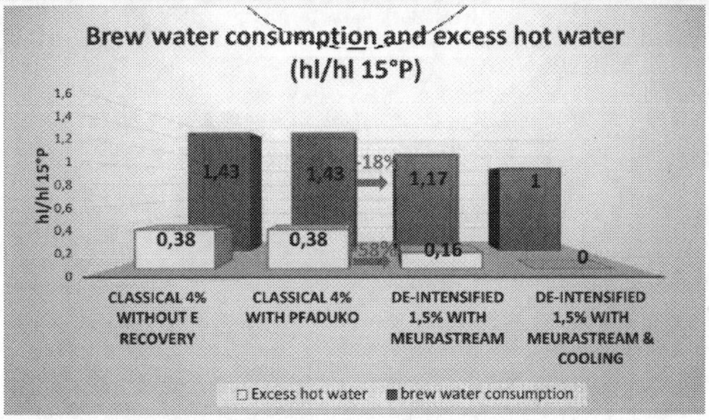

- Compared to classical boiling 4% (without E recovery or with Pfaduko), de-intensified boiling combined with MEURASTREAM allows reducing:
  - The brew water consumption by 18%
  - The excess hot water by 58%

- Further optimisation: cooling of excess hot water and use for mashing-in
  → Brew water consumption is reduced to 1 hl/hl wort 15°P
  → Excess hot water reduced to ZERO !

Case study - Opex savings

Cooling energy

	Conventional brewhouse	Meurastream concept
Wort boiling	4% in wort copper	1% in formation tank 0,5% stripping
Wort cooler	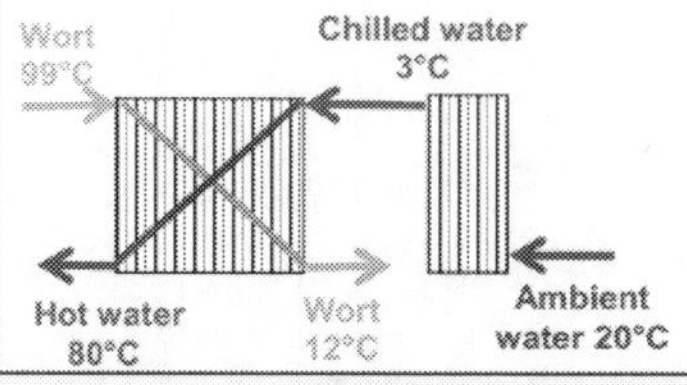	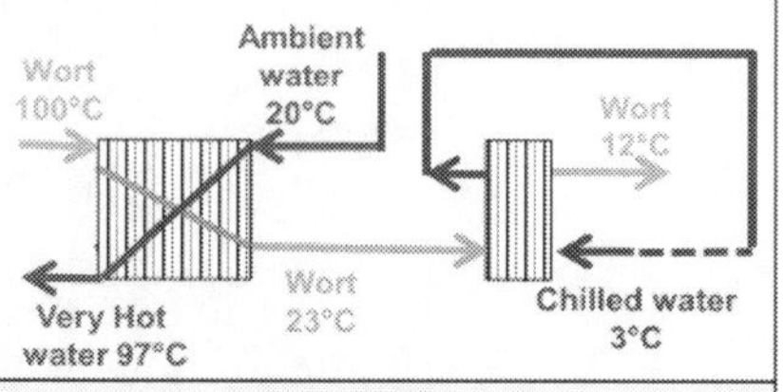
Chilled water for cooling (hl/hl)	1,13 hl/hl	1 hl/hl
Cooling energy (MJ/hl)	8 MJ/hl	- 42% → 4,6 MJ/hl

Case study - Opex savings

Brewhouse capacity of 1 Mio hl wort @ 15°P

	Classical brewhouse no E recovery	Meurastream concept	Annual savings
<u>Thermal energy in brewhouse</u>			
Total evaporation Stripping rate	4% -	1% 0,5%	
Thermal energy (MJ/hl 15°P)	26,4 MJ/hl	12,6 MJ/hl	
Annual thermal E cost (0,014 €/MJ)	369 600 €	176 400 €	**193 200 €**
<u>Process water</u>			
Brewing water (hl/hl 15°P)	1,43 hl/hl	1,17 hl/hl	
Annual cost (3 €/m³)	429.000 €	351.000 €	**78 000 €**
<u>Cooling energy</u>			
Cooling energy (MJ/hl 15°P)	8 MJ/hl	4,6 MJ/hl	
Electrical energy (KWh/hl 15°P)	0,89 KWh/hl	0,51 KWh/hl	
Annual cost (0,12 €/KWh)	106 800 €	61 200 €	**45 600 €**
Total savings			**316 800 €**

Project examples – industrial results

Water tanks 750 HL

Wort cooler

New brewhouses with Meurastream

Example with brewery in Ivory Coast:

- Capacity: 12 brews/day – 315 hl cold wort @ 17°P
- Main recipe : 75% malt 25% rice
- Mashing-in T°: 50°C in Mash tun
  52°C in Cereal Cooker
- Evaporation at boiling: 1,4% in Formation vessel
  0,6 % in Ecostripper

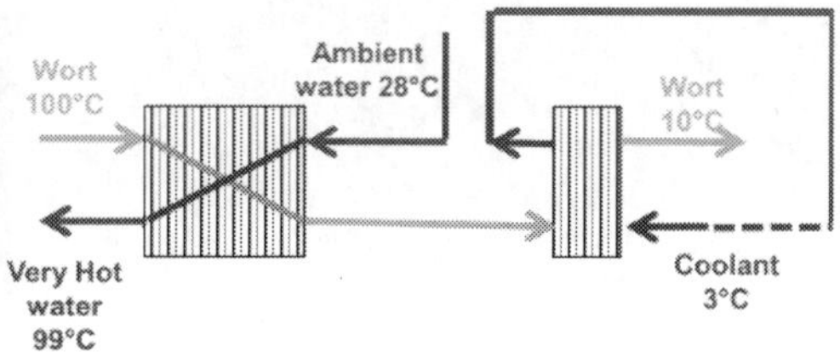

MEURA

Project examples – industrial results

New brewhouses with Meurastream

	Thermal E (MJ/hl 15°P)	Process water (hl/hl 15°P)
Brewery in Ivory Coast 315 hl cold wort 17°P (1 Mio hL/year 17°P wort)	**15,18 MJ/hl**	**1,24 Hl/hl**
Brewery in Cambodia 630 l cold wort 17,5°P (2 Mio hl / year 17,5°P wort)	**14,54 MJ/hl**	**1,14 Hl/hl**

MEURA

Industrial implementation

MEURASTREAM

✓ New brewhouse (batch or continuous)

✓ Existing brewhouse (revamping)

Phase 1: MEURASTREAM « light » → Phase 2: MEURASTREAM « full »

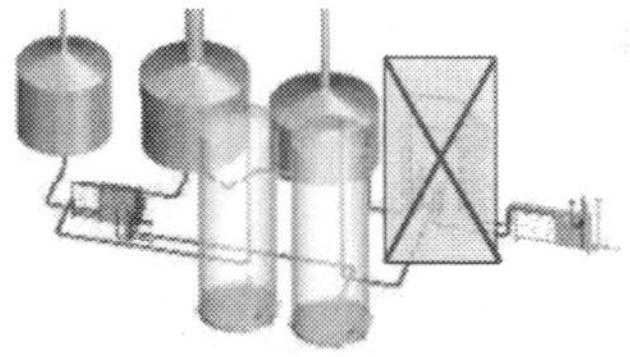

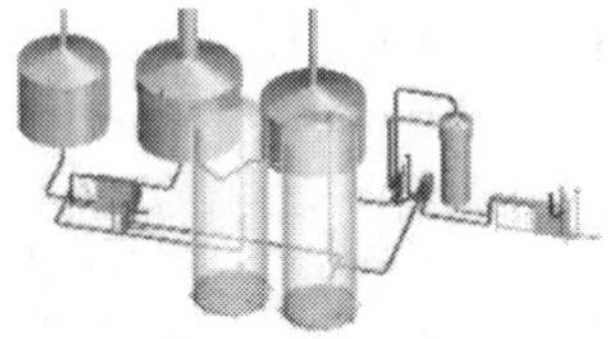

- 4% evaporation rate at boiling
- Thermal energy savings // Pfaduko
- Water consumption savings
- Reduced excess of hot water
- Cooling energy savings

- Further thermal energy savings by reducing evaporation rate at boiling (down to 1.5%) & implementing wort stripping
- All advantages of the Meurastream regarding energy and water savings

Case study - Opex savings

Brewhouse capacity of 1 Mio hl wort @ 15°P

	Classical brewhouse no E recovery	Meurastream « light » Phase 1	Meurastream « full » Phase 2
Thermal energy in brewhouse			
Total evaporation Stripping rate	4% -	4% -	1% 0,5%
Thermal energy (MJ/hl 15°P)	26,4	19,4	12,6
Annual thermal E cost (0,014 €/MJ)	369 600 €	271 600 € (-26%)	176 400 € (-52%)
Process water			
Brewing water (hl/hl 15°P)	1,43 hl/hl	1,17 hl/hl	1,17 hl/hl
Annual cost (3 €/m³)	429 000 €	351 000 € (-18%)	351 000 €
Cooling energy			
Cooling energy (MJ/hl 15°P)	8 MJ/hl	4,6 MJ/hl	4,6 MJ/hl
Electrical energy (KWh/hl 15°P)n	0,89 KWh/hl	0,51 KWh/hl	0,51 KWH/hl
Annual cost (0,12€/KWh)	106 800 €	61 200 € (-24%)	61 200 €
SAVINGS		221.600 €	316.800 €

MEURASTREAM Summary

- Combination of de-intensified boiling process with energy recovery at wort cooling producing very hot process water used for wort pre-heating
- MEURASTREAM leads to reduced
  - Thermal energy consumption (- 52% vs to classical brewhouse)
  - Brewing water consumption (- 18% vs classical & Pfaduko brewhouse)
  - Excess hot water losses (- 58% vs classical & Pfaduko brewhouse)
  - Cooling energy (- 42% vs classical brewhouse)
- Combination of proven technologies
- Short Pay-back period
- Integration of the concept possible in existing brewhouses (revamping in 1 or 2 phases)
- 9 industrial references since 2014

MEURA

References

9 references MEURASTREAM since 2014

New brewhouses

- Cambodia 630 hl cold wort 17,5°P
- Ivory Coast 315 hl cold wort 17°P
- Chili 465 hl cold wort 17,5°P
- Nigeria 700 hl cold wort 18°P
- Colombia 50 hl/h cold wort 18°P Continuous brewhouse

Revamping of existing brewhouses

- Vietnam 300 hl cold wort 16,5°P Meurastream full
- Haiti 200 hl cold wort 19°P Meurastream light
- Europe 530 hl cold wort 16,6°P Meurastream light
- Europe 575 hl cold wort 15,3°P Meurastream light

MEURA

Thanks to customers and Meura's colleagues for sites information & support
Thank you for your attention !
MEURA
MEURA
Always welcome in Belgium!

L39 / Innovative wort production –
Practical development and application

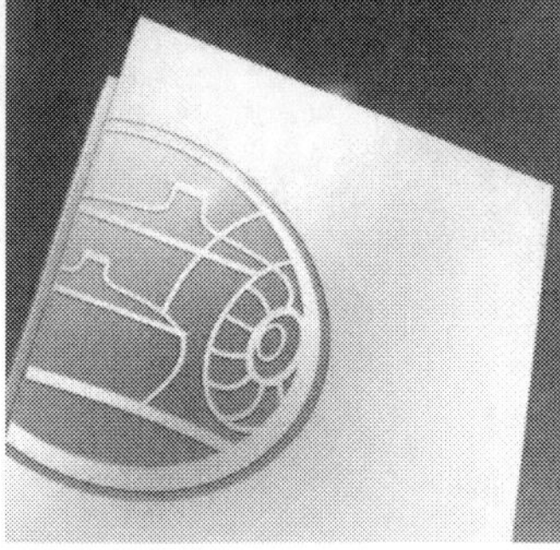

The transformation of old tradition – a novel brewhouse concept

Tobias Becher
ZIEMANN HOLVRIEKA GmbH,
Ludwigsburg / Germany

May 17th 2017

Just wanted to buy a new shirt

2

Lautering today

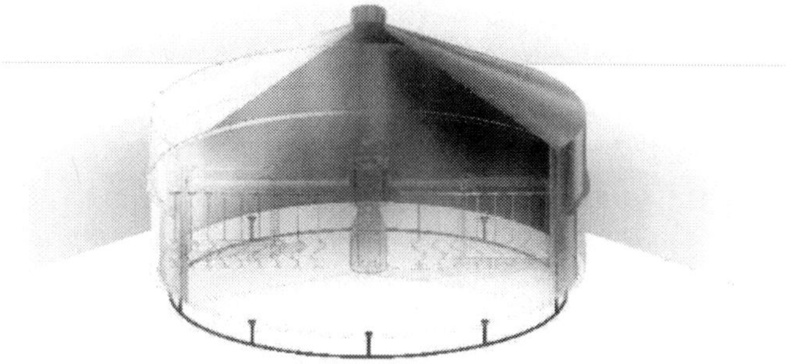

3

Desired

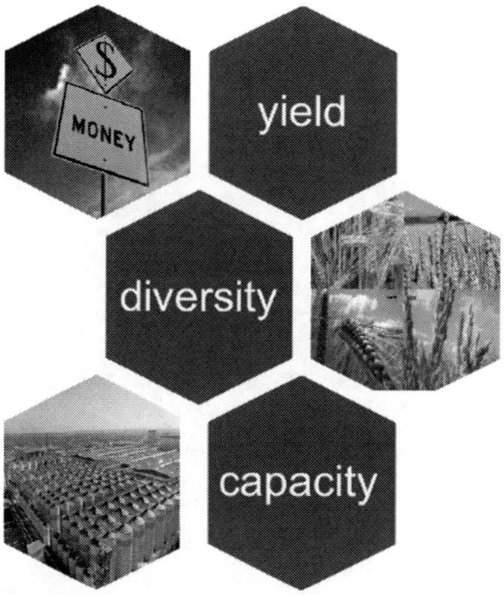

source: www.flickr.com & Ziemann

4

Limits

static process

limited flexibility

batch operation

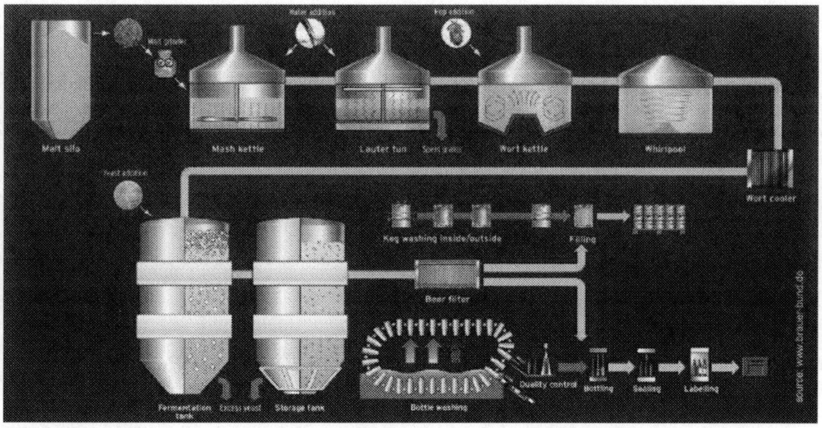

5

New way of mash filtration

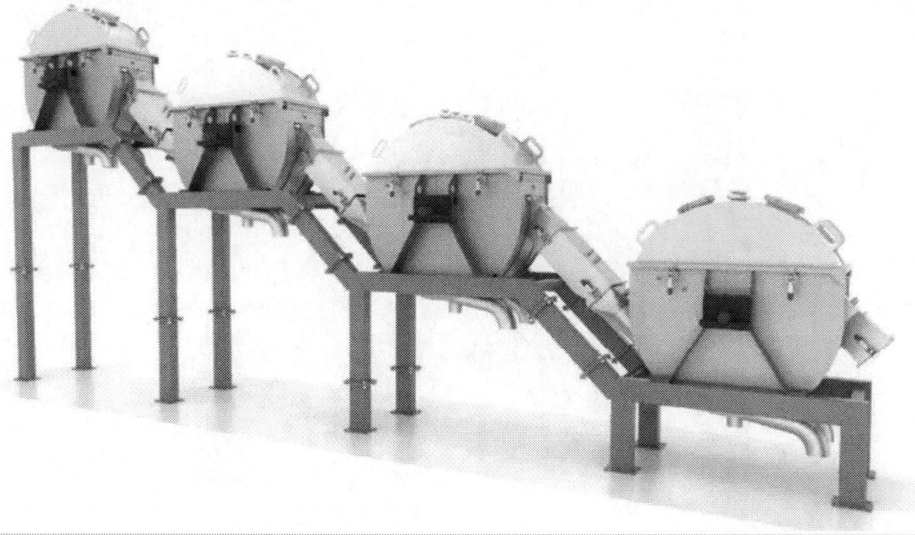

6

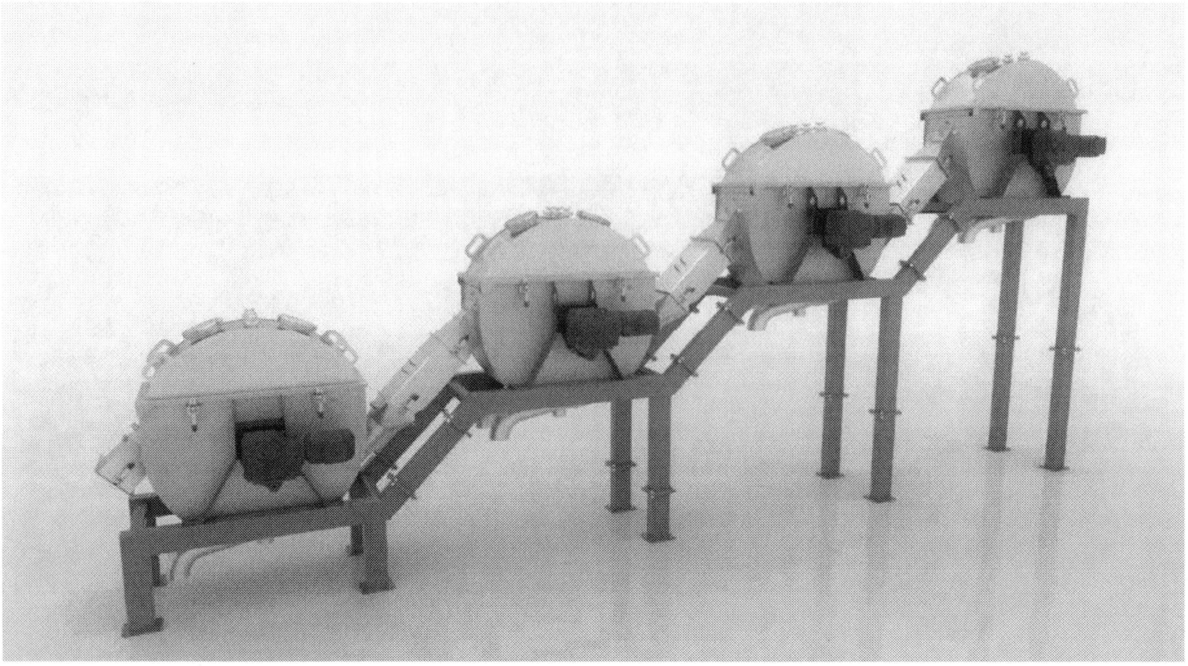

Sparging transfer

8

Malt extraction

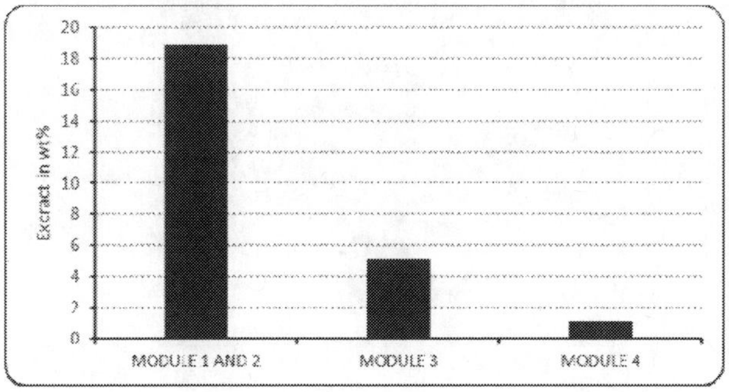

Parameter	NESSIE	Standard values
Water content	< 80 %	80 %
Total extract	0.8 %	1.6 %
Soluble extract (as is)	0.6 %	0.8 %
Convertible extract (as is)	0.2 %	0.8 %
Iodine value	1.8	< 3

18

Sluicing instead of leaching

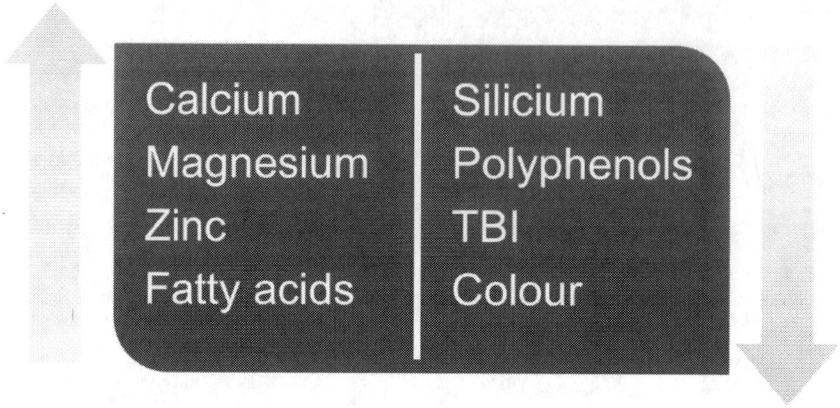

11

Challenge

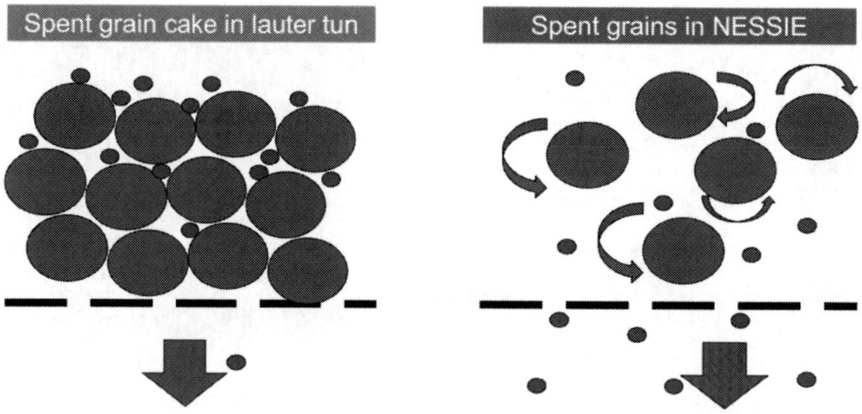

12

Solution

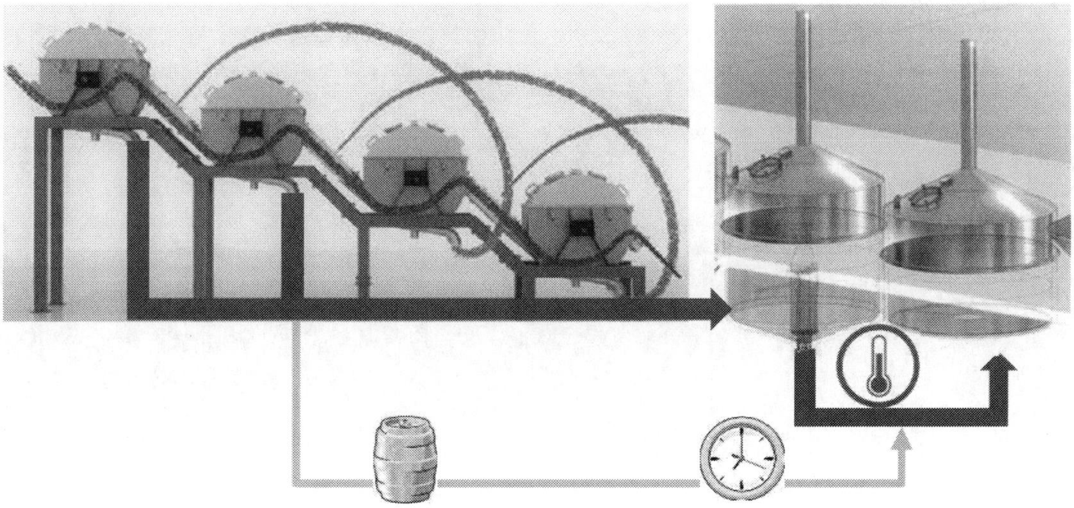

13

Literature

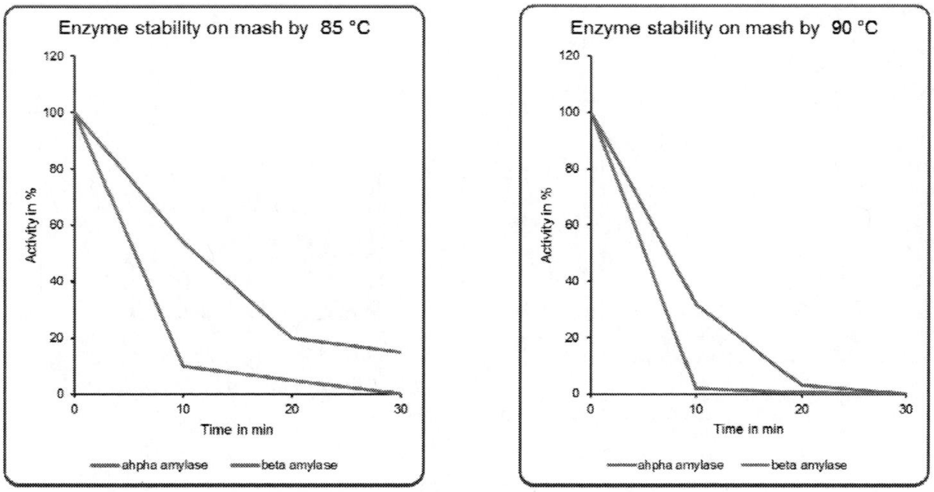

[Muller RE. 1991. The effects of mashing temperature and mash thickness on wort carbohydrate composition. J Inst Brew 97:85–92.]

14

Opportunity

15

Wort fractions

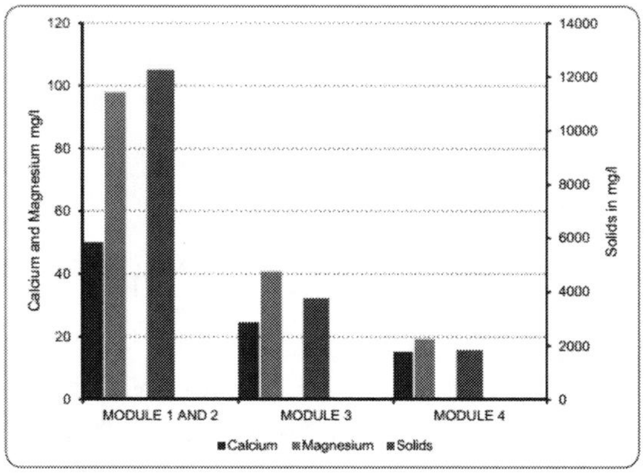

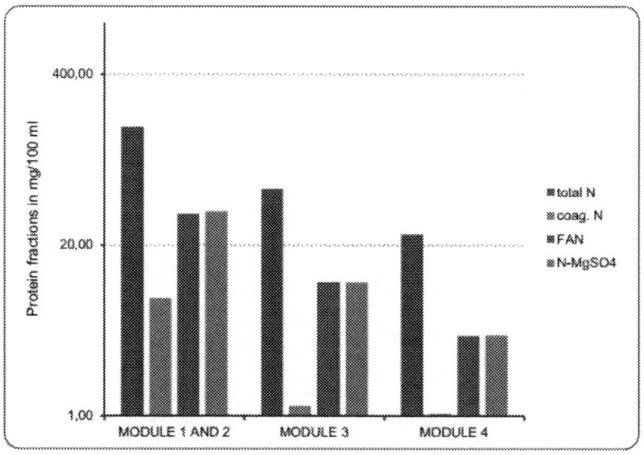

16

Fractional wort treatment

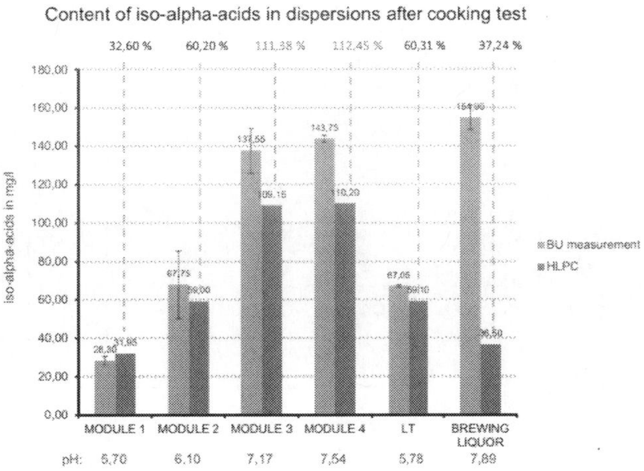

17

At the end, there is always beer

18

Fermentation

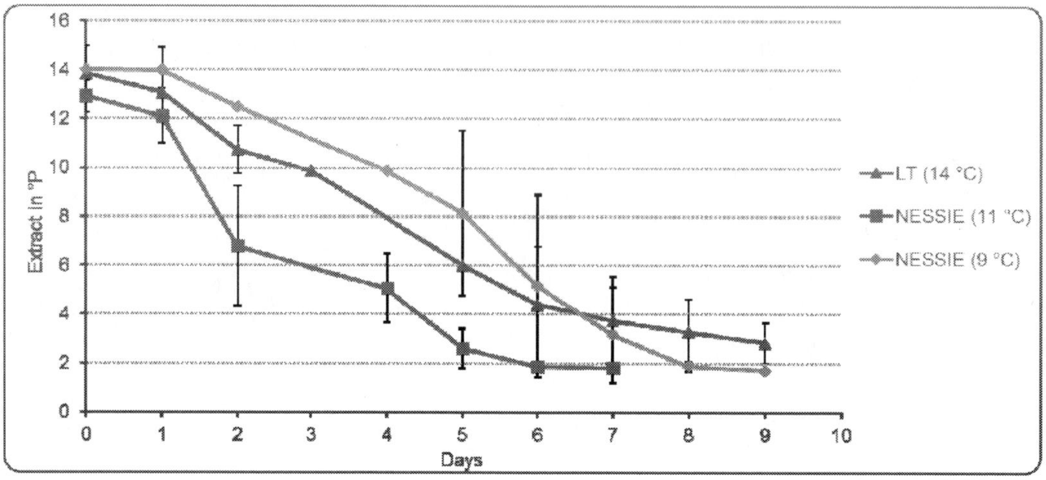

19

Yeast growth

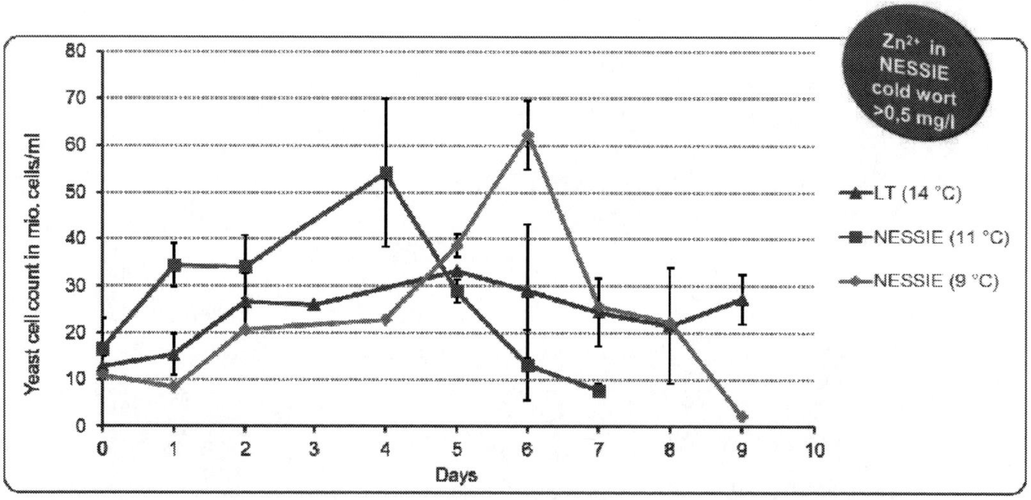

20

Good guys

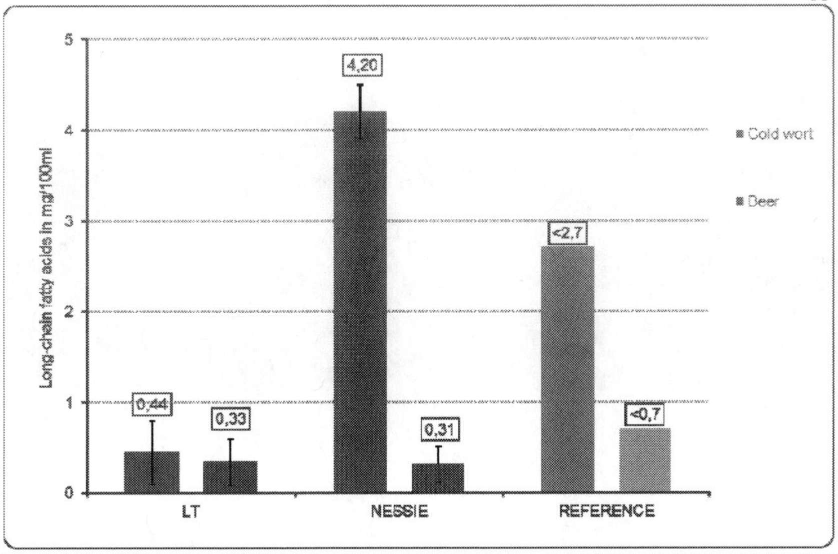

21

Bad guys

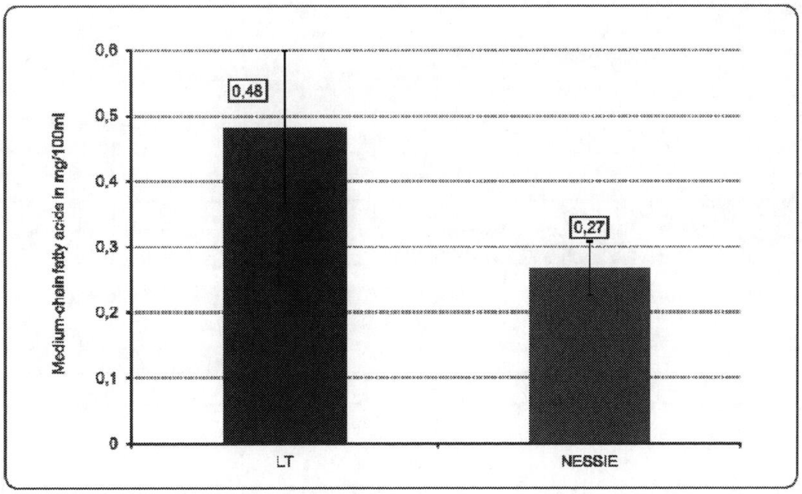

22

Just wanted to develop a new mash filter

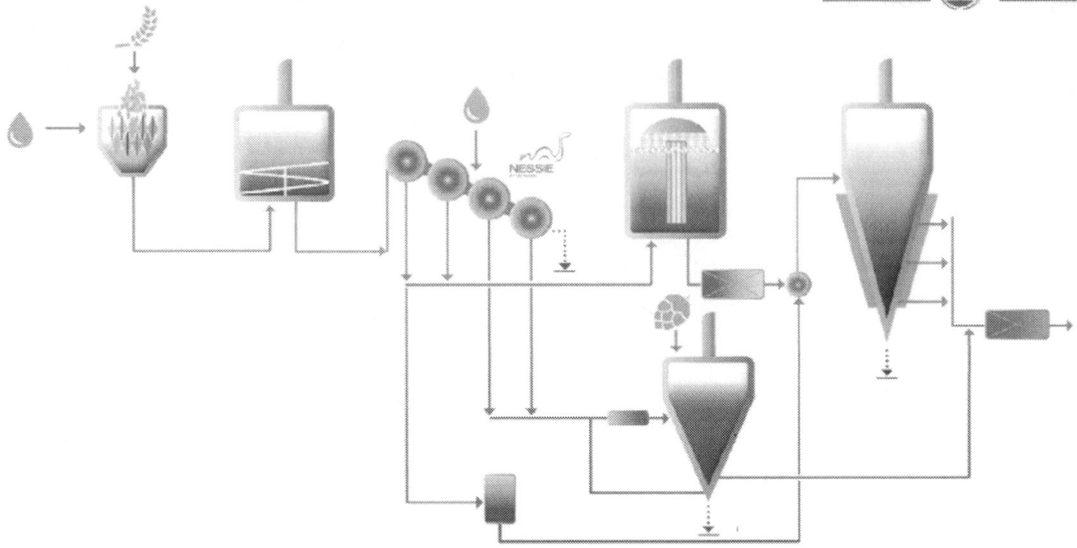

Process time

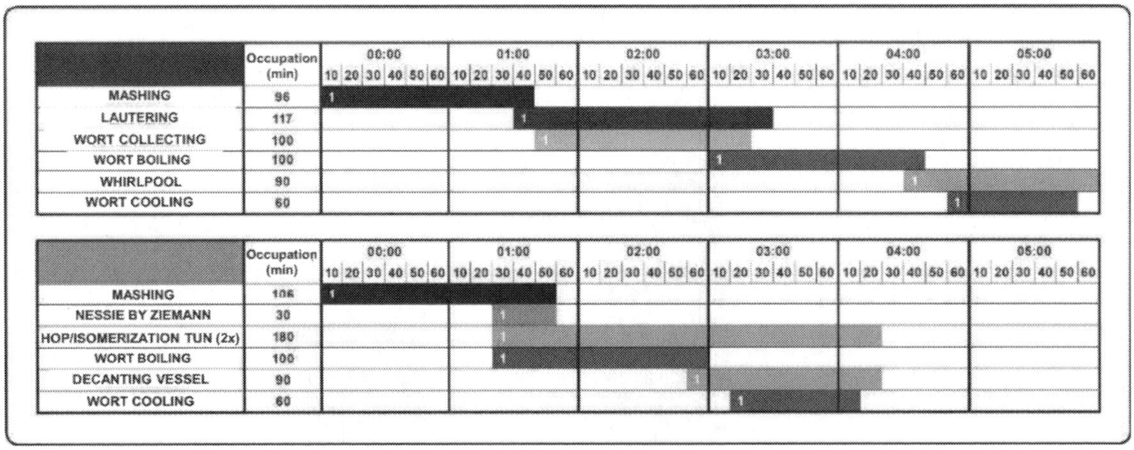

Benefits

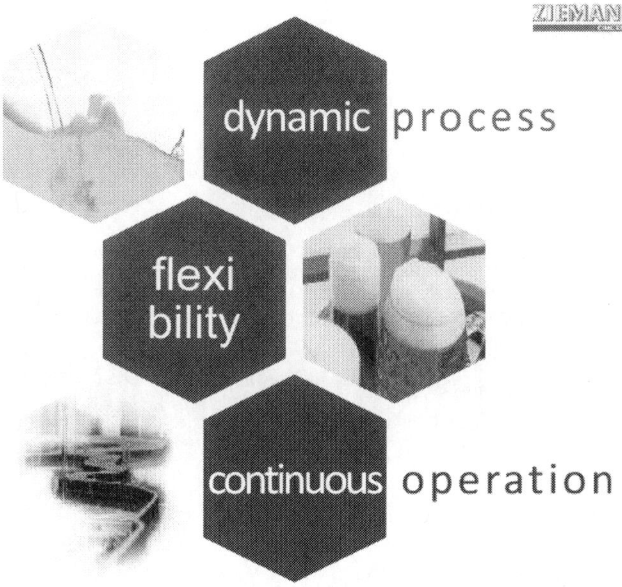

source: www.brauer-bund.de & Ziemann

25

Credits
Co-authors. Nele Bastgen, Klaus Wasmuht, Konstantin Ziller
R&D team. Verena Blomenhofer, Natalia Svedlund
Tom Benninghaus, Klaus Gehrig, Elmar Pongratz
Students. Hanna Konrad, Magdalena Öhlschläger, Beatriz Povedano
Isabel Wasmuht, Lorenz Hofbeck, Tobias Raiber
Scientists. Annette Schwill-Miedaner, Deniz Bilge, Josef Englmann, Fritz Jacob,
Martin Krottenthaler, Heinz Miedaner, Roland Pahl, Franz Preiss, Winfried Russ
ZIEMANN
HOLVRIEKA
EBC
2017
Tobias Becher
ZIEMANN HOLVRIEKA GmbH, Ludwigsburg
Tel.: 07141 408-244

36th EBC Congress 2017 Llubljana, Slovenia, May 14-18, 2017

36TH CONGRESS EBC 2017 14-18 MAY LJUBLJANA EUROPEAN BREWERY CONVENTION SLOVENIA

Influence of the grist fractions on the final beer quality

Martin Zarnkow, Stefan Biberger, Fritz Jacob

Forschungszentrum Weihenstephan für Brau- und Lebensmittelqualität

Technische Universität München

If you say today:

„Bier / beer /cerveja / cer-veza/ ビール (biru)/ 啤酒 (píjiǔ)/ bière/ пиво (pivo)/ बीयर (bīyara)/ 맥주(maegju)/ birra/ sör/ øl/ بيرة (bīra) …“

you mean a pale, carbonated and filtrated lager beer!

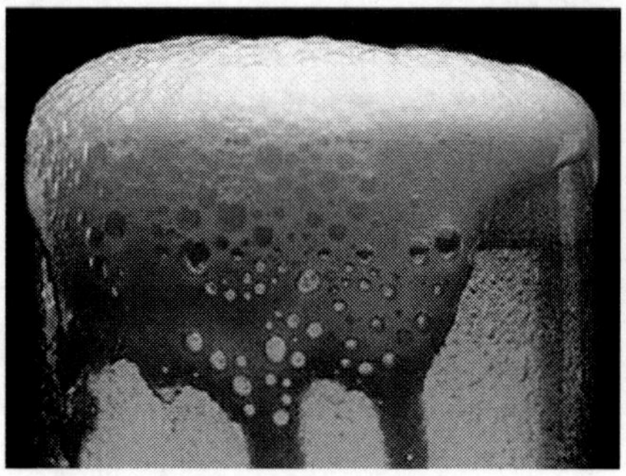

Photo: pixelio.de/tirot

Martin Zarnkow Folie 2 Research Center Weihenstephan for Brewing and Food Quality

Research Object

- Fractionating the grist (Pilsner malt quality) with 6-roller-mill
- Characterizing the grist fractions (see WBC 2016)
- Modify the brewing process due to the individual character of the fraction

- main aspect: pale Lager beer

Objective

- Pale, bottom-fermented lager beer
- Original gravity: 12 °P
- Alcohol: 5 ABV%
- Final fermentention degree: 81%
- Bitter units: 18 IBU
- pH: 4.5

Husks

Properties

- Low values for α- and β-amylase (according to EBC and MEBAK)
- Gelatinisation temperature not known
- Degree of protein modification: 38% → well solved
- Protein content: 11.6% → a bit too high
- Soluble protein: 0.70% → within the normal range
- Apparent final fermentation degree: 80%
- Extract: 55.75% water free

Husks

Procedures (planned)

- Modified congress mashing procedure according to MEBAK
- Mashing: 45 °C – 40 min
- Heating with 1 °C/min to 70 °C
- Short maltose rest at 62 °C – 15 min
- Keep temperature of 70 °C until saccharification is reached
- Mashing: 78 °C

Fraction 1 (1.6 mm)

Properties

- •Low value for α- and β-Amylase (according to EBC and MEBAK)
- •Gelatinisation temperature not known
- •Degree of protein modification 34.53% → moderate solved
- •Protein content: 7.65% → bit too low
- •Soluble protein: 0.42% → value too low
- •Apparent final fermentation degree: 71.25%
- •Extract: 29.95% water free

Martin Zarnkow | Folie 7 | Research Center Weihenstephan for Brewing and Food Quality

Fraction 1 (1.6 mm)

Mashing (planned)

- Mashing: 45 °C – 15 min
- β-glucan rest: 50 °C – 20 min
- 1. batch of the mash: 1: 2.5
  - Saccharification rest at 72 ° C for 20 min
  - Heating at 2 °C/min
  - Boiling for 20 min
  - Transferred to the vessel mash
  - Target temperature 62 °C
- 2. batch of the mash as thick mash (1: 2.5)
  - Saccharification at 72 °C for 20 min
  - Heating at 2 °C/min
  - Boiling for 15 min
  - Transfer to the vessel mash
  - Target temperature 78 °C
- Mash transfer

Martin Zarnkow | Folie 8 | Research Center Weihenstephan for Brewing and Food Quality

Fraction 2 (1.0 mm)

Properties

- Values of α- and β-amylase in normal range (according to EBC and MEBAK)
- Gelatinisation temperature not known
- Degree of protein modification: 36.75% → well solved
- Protein content: 12.05% → bit too high
- Soluble protein: 0.71% → value in normal range
- Apparent final fermentation degree: 72.75%
- Extract: 64.10% water free

Fraction 2 (1.0 mm)

Mashing (planned)

- Mashing: 45 °C – 30 min
- 1. batch of the mash: 1: 2.5
  - Saccharification rest at 72 °C for 20 min
  - Heating at 2 °C/min
  - Boiling for 20 min
  - Transfer to the vessel mash
  - Target temperature 62 °C
- 2. batch of the mash as thick mash (1: 2.5)
  - Saccharification at 72 °C for 20 min
  - Heating at 2 °C/min
  - Boiling for 15 min
  - Transfer to the vessel mash
  - Target temperature 78 °C
  - Mash transfer

Fraction 3 (0.63 mm)

Properties

- High values for α- and β-Amylase (according to EBC and MEBAK)
- Gelatinisation temperature at 61.4 °C → Maltose rest at 62 °C
- Degree of protein modification: 38% → well solved
- Protein content: 11.6% → bit too high
- Soluble protein: 0.70% → value in normal range
- Apparent final fermentation degree: 80%
- Extract: 84.65% water free

Martin Zarnkow | Folie 11 | Research Center Weihenstephan for Brewing and Food Quality

Fraction 3 (0.63 mm)

Mashing

- Bulk: 6.0298 kg
- Main pour: 27 L
- Mashing: 45 °C – 30 min
- Maltose rest: 62 °C – 30 min
- Saccharification rest: 72 °C – 40 min
- Transfer mashing: 78 °C

Lautering

- Addition of 0.6097 kg rice husks
- Digest: 15 L
- Distribution of the digest in two batches about 7.5 L

Boiling

- Boiling time: 45 min
- Hop dosage: 5 min after start boiling
- Hops: „Herkules“, bitter-hop

Martin Zarnkow | Folie 12 | Research Center Weihenstephan for Brewing and Food Quality

Fraction 4 (0.34 mm)

Properties

- Values for α- and β-amylase in normal range (according to EBC and MEBAK)
- Gelatinisation temperature at 62.7 °C → maltose rest at 63 °C
- Degree of protein modification: 37.3% → well solved
- Protein content: 11.7% → bit too high
- Soluble protein: 0.75% → value in normal range
- Apparent final fermentation degree: 81%
- Extract: 84.70% water free

Fraction 4 (0.34 mm)

Mashing

- Bulk: 6.0062 kg
- Main pour: 27 L
- Mashing: 45 °C – 30 min
- Maltose rest: 63 °C – 30 min
- Saccharification rest: 72 °C – 40 min
- Transfer mashing: 78 °C

Lautering

- Addition of 0.6147 kg rice husks
- Digest: 15 L
- Distribution of the digest in two batches about 7.5 L

Boiling

- Boiling time: 45 min
- Hop dosage: 5 min after start boiling
- Hops: „Herkules“, bitter-hop

Fraction 5 (0.18 mm)

Properties

- High values for α- and β-amylase (according to EBC and MEBAK)
- Gelatinisation temperature at 62.6 °C → maltose rest bei 63 °C
- Degree of protein modification: 42.3% → very well solved
- Protein content: 9.7% → bit too low
- Soluble protein: 0.65% → value in normal range
- Apparent final fermentation degree: 82%
- Extract: 89.95% water free

Fraction 5 (0.18 mm)

Mashing

- Bulk: 6.0384 kg
- Main pour: 28 L
- Mashing: 45 °C – 20 min
- β-Glucan rest: 50 °C – 20 min
- Maltose rest: 63 °C – 30 min
- Saccharification rest: 72 °C – 40 min
- Transfer mashing: 78 °C

Lautering

- Addition of 0.6016 kg rice husks
- Digest: 15 L
- Distribution of the digest in two batches about 7.5 L

Boiling

- Boiling time: 45 min
- Hop dosage: 5 min after start boiling
- Hops: „Herkules“, bitter-hop

Fraction 6 (0.125 mm)

Properties

- Values for α- and β-amylase in normal range (according to EBC and MEBAK)
- Gelatinisation temperature at 62.6 °C → maltose rest at 63 °C
- Degree of protein modification: 46.1% → very well solved
- Protein content: 6.75% → bit too low
- Soluble protein: 0.49% → value too low
- Apparent final fermentation degree: 81.5%
- Extract: 95.83% water free

Martin Zarnkow | Folie 17 | Research Center Weihenstephan for Brewing and Food Quality

Fraction 6 (0.125 mm)

Mashing

- Bulk: 5.0177 kg
- Main pour: 25 L
- Mashing: 45 °C – 15 min
- β-Glucan rest: 50 °C – 15 min
- Maltose rest: 62 °C – 30 min
- Saccharification rest: 72 °C – 40 min
- Transfer mashing: 78 °C

Lautering

- Addition of 0.5043 kg rice husks
- Digest: 10 L
- Distribution of the digest in two batches about 5 L

Boiling

- Boiling time: 45 min
- Hop dosage: 5 min after start boiling
- Hops: „Herkules", bitter-hop

Martin Zarnkow | Folie 18 | Research Center Weihenstephan for Brewing and Food Quality

Fraction 7 (< 0.125 mm)

Properties

- Low values for α- und β-amylase (according to EBC and MEBAK)
- Gelatinisation temperature at 61.8 °C → maltose rest at 62 °C
- Degree of protein modification: 51.2% → very well solved
- Protein content: 5.05% → too low
- Soluble protein: 0.41% → value too low
- Apparent final fermentation degree: 82%
- Extract: 91.13% water free

Fraction 7 (< 0.125 mm)

Mashing

- Bulk: 5.0162 kg
- Main pour: 26 L
- Mashing: 45 °C – 10 min
- β-Glucan rest: 50 °C – 15 min
- Maltose rest: 62 °C – 30 min
- Saccharification rest: 72 °C – 40 min
- Transfer mashing: 78 °C

Lautering

- Addition of 0.5023 kg rice husks
- Digest: 10 L
- Distribution of the digest in two batches about 5 L

Boiling

- Boiling time: 45 min
- Hop dosage: 5 min after start boiling
- Hops: „Herkules“, bitter-hop

Original gravity

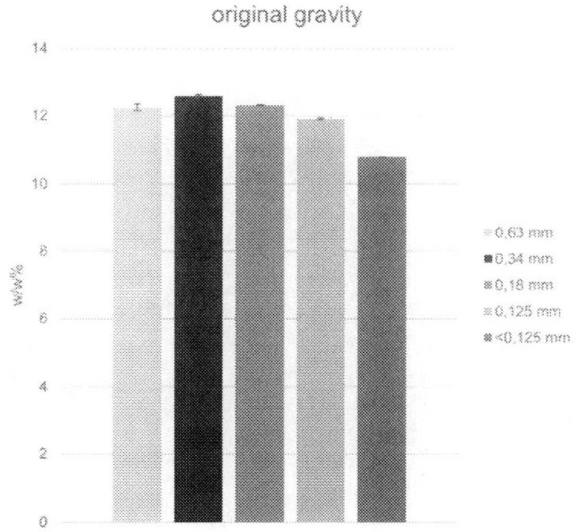

Sample	Original gravity (w/w%)
0.63 mm	12.26
0.34 mm	12.58
0.18 mm	12.31
0.125 mm	11.91
< 0.125 mm	10.78
Expected value	12.00

Martin Zarnkow Folie 21 Research Center Weihenstephan for Brewing and Food Quality

Alcohol

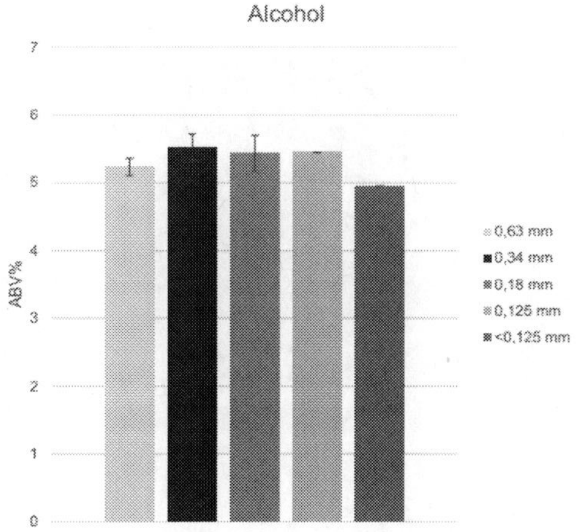

Sample	Alcohol (ABV%)
0.63 mm	5.23
0.34 mm	5.52
0.18 mm	5.43
0.125 mm	5.45
< 0.125 mm	4.95
Expected value	5.00

Martin Zarnkow Folie 22 Research Center Weihenstephan for Brewing and Food Quality

Extract

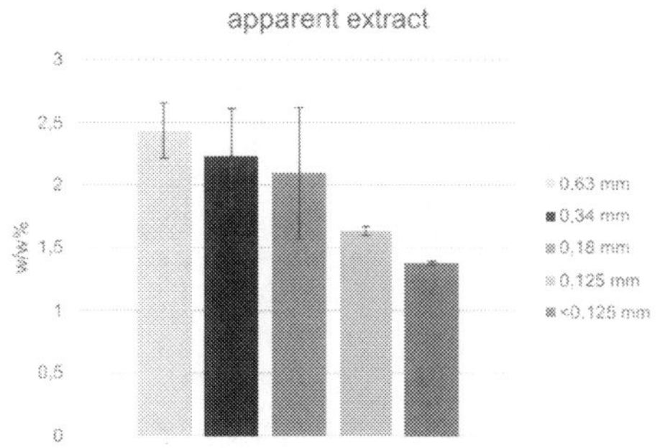

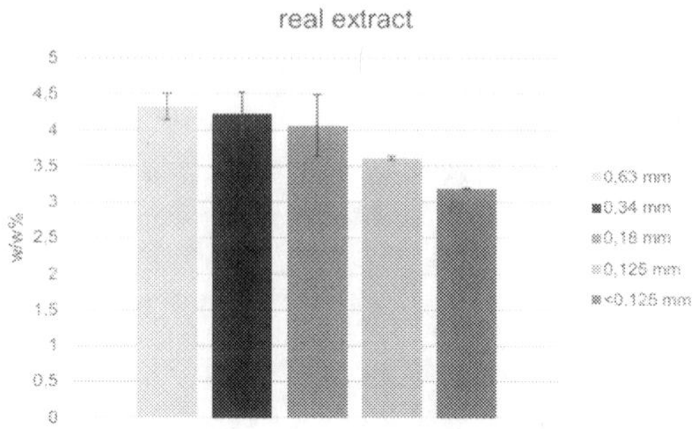

Sample	Apparent extract (w/w%)	Real extract (w/w%)
0.63 mm	2.43	4.33
0.34 mm	2.22	4.22
0.18 mm	2.10	4.10
0.125 mm	1.64	3.61
< 0.125 mm	1.38	3.19
Expected value	2.20	4.00

Martin Zarnkow Folie 23 Research Center Weihenstephan for Brewing and Food Quality

Attenuation

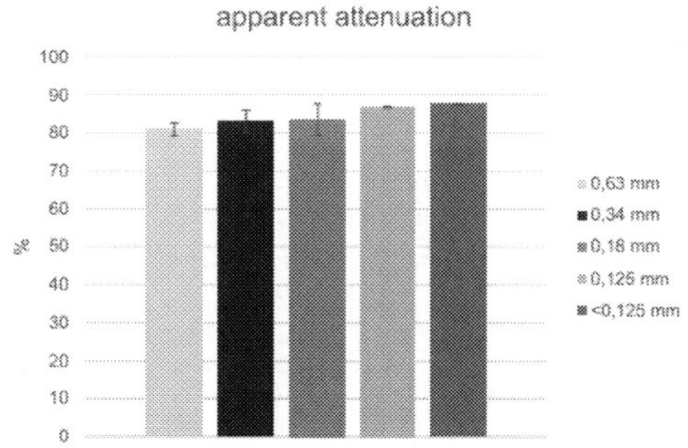

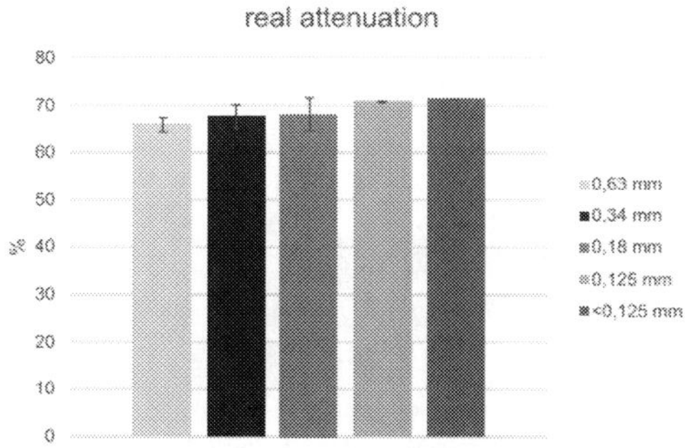

Sample	Apparent attenuation (%)	Real attenuation (%)
0.63 mm	80.90	65.83
0.34 mm	83.03	67.57
0.18 mm	83.63	68.08
0.125 mm	86.85	70.70
< 0.125 mm	87.65	71.30
Expected value	81.00	65.60

Martin Zarnkow Folie 24 Research Center Weihenstephan for Brewing and Food Quality

pH

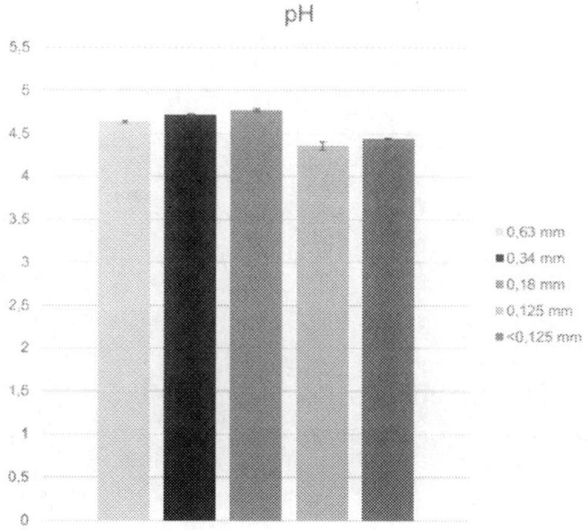

Sample	pH
0.63 mm	4.63
0.34 mm	4.71
0.18 mm	4.77
0.125 mm	4.36
< 0.125 mm	4.44
Expected value	< 4.50

Martin Zarnkow Folie 25 Research Center Weihenstephan for Brewing and Food Quality

Colour

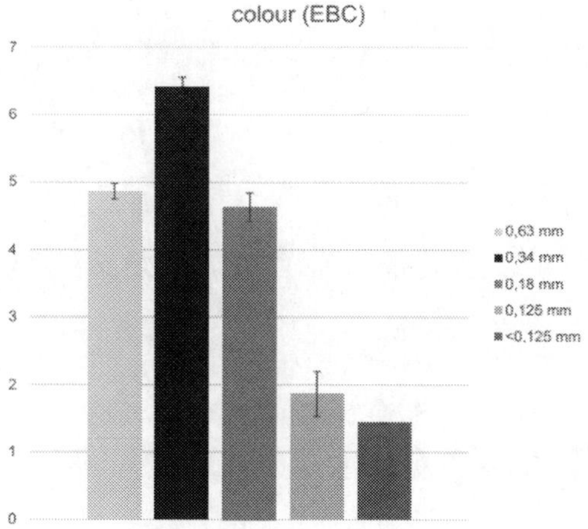

Sample	Colour (EBC)
0.63 mm	4.87
0.34 mm	6.42
0.18 mm	4.63
0.125 mm	1.87
< 0.125 mm	1.44
Expected value	7.80

Martin Zarnkow Folie 26 Research Center Weihenstephan for Brewing and Food Quality

Polyphenols

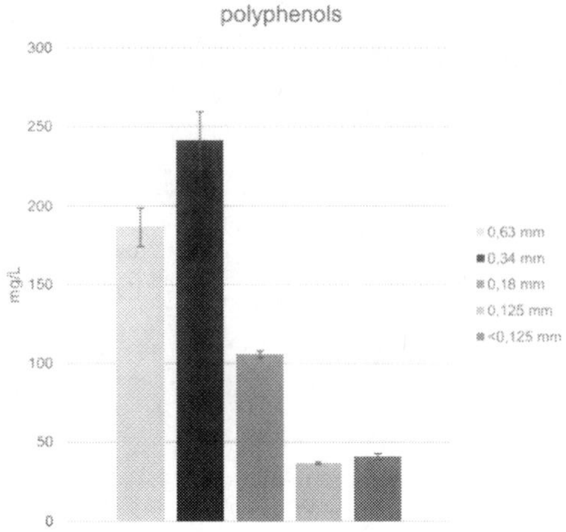

Sample	Polyphenols (mg/L)
0.63 mm	186
0.34 mm	241
0.18 mm	106
0.125 mm	36.5
< 0.125 mm	40.5
Expected value	150–200

Martin Zarnkow Folie 27 Research Center Weihenstephan for Brewing and Food Quality

Bitter Units

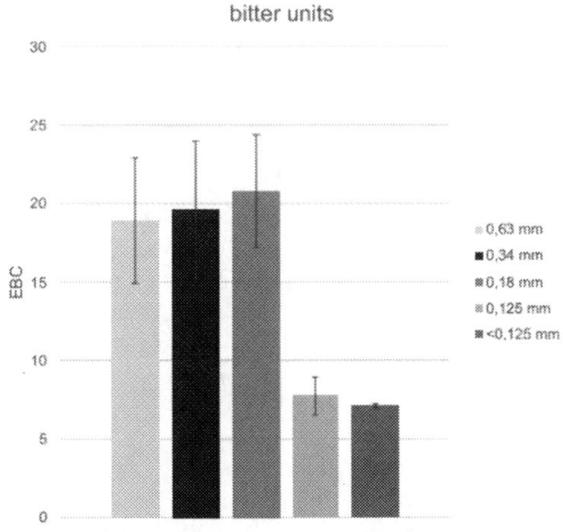

Sample	Bitter Units (EBC)
0.63 mm	19
0.34 mm	20
0.18 mm	21
0.125 mm	7.8
< 0.125 mm	7.1
Expected value	18

Martin Zarnkow Folie 28 Research Center Weihenstephan for Brewing and Food Quality

β-Glucan

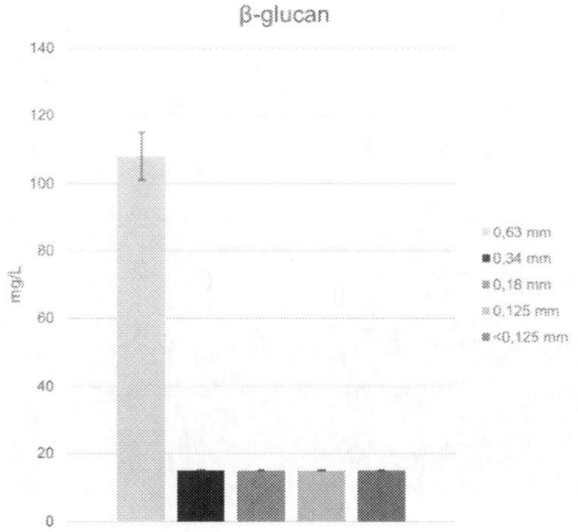

Sample	β-Glucan (mg/L)
0.63 mm	108
0.34 mm	< 15.0
0.18 mm	< 15.0
0.125 mm	< 15.0
< 0.125 mm	< 15.0
Expected value	< 200

Total Soluble Nitrogen

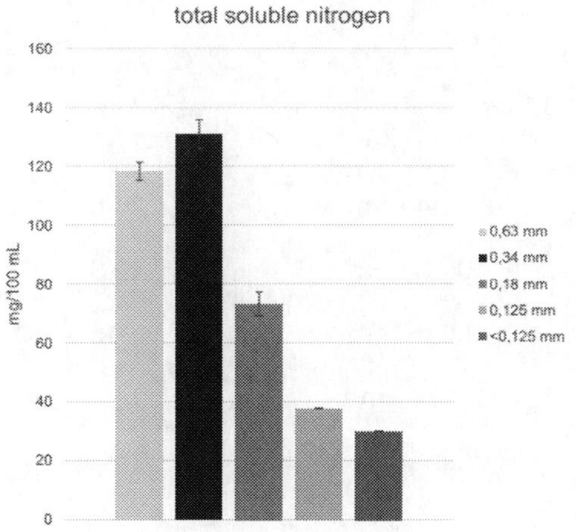

Sample	Total soluble nitrogen (mg/100 mL)
0.63 mm	118
0.34 mm	131
0.18 mm	73.3
0.125 mm	37.8
< 0.125 mm	30.1
Expected value	60–110

Coagulable Nitrogen

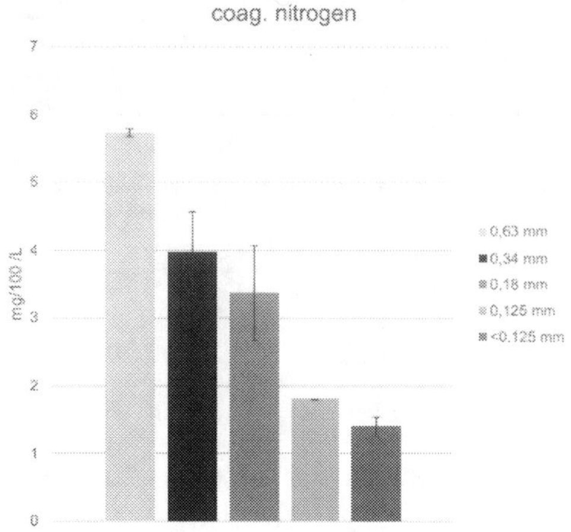

Sample	Coag. nitrogen (mg/100 mL)
0.63 mm	5.73
0.34 mm	3.97
0.18 mm	3.37
0.125 mm	1.80
< 0.125 mm	1.40
Expected value	2–3

Martin Zarnkow Folie 31 Research Center Weihenstephan for Brewing and Food Quality

Alpha-Amino-Nitrogen

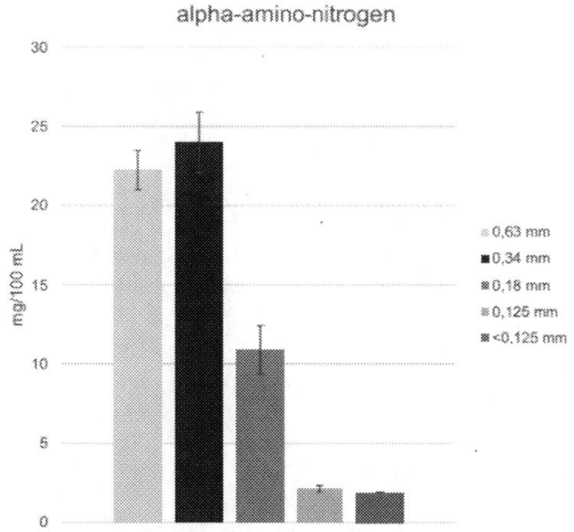

Sample	Alpha-amino-nitrogen (mg/100 mL)
0.63 mm	22.3
0.34 mm	24.0
0.18 mm	10.9
0.125 mm	2.15
< 0.125 mm	1.90
Expected value	8.0–15.0

Martin Zarnkow Folie 32 Research Center Weihenstephan for Brewing and Food Quality

Summary and Future Aspects

- the different fractions contain huge differences in quality malt attributes
- modification of technology with respect to lager beer with different fractions is possible
- with lauter tun additional filter aids are necessary
- but there are still more options!

- sensory evaluation is coming next

Martin Zarnkow Folie 33 Research Center Weihenstephan for Brewing and Food Quality

EBC 2017 – Ljubljana, May the 17th 2017

Many thanks for your kind attention!

many thanks BÜHLER

Dr. Martin Zarnkow
Alte Akademie 3
85354 Freising-Weihenstephan

Telephone: +49 (0) 8161 / 71- 3530
Telefax: +49 (0) 8161 / 71- 4181

E-Mail: Martin.Zarnkow@tum.de
www.blq-weihenstephan.de

Martin Zarnkow Folie 34 Research Center Weihenstephan for Brewing and Food Quality

Membrane processes in the brewing industry

Karolína Jastřembská

University of Pardubice,

Institute of Chemical and Environmental Engineering

Czech Republic

karolina.jastrembska@student.upce.cz

Content

- **Membrane separation processes**
- **Microfiltration**
- **Beer dealcoholization and aroma recovery methods**
  - o Reverse osmosis
  - o Pervaporation
  - o Membrane distillation

1

Questions

- How these processes work?
- Which type of materials and type of membrane module are used?
- Membrane processes in the brewing industry: Which advantages and disatvantages they have?

2

Membrane separation process

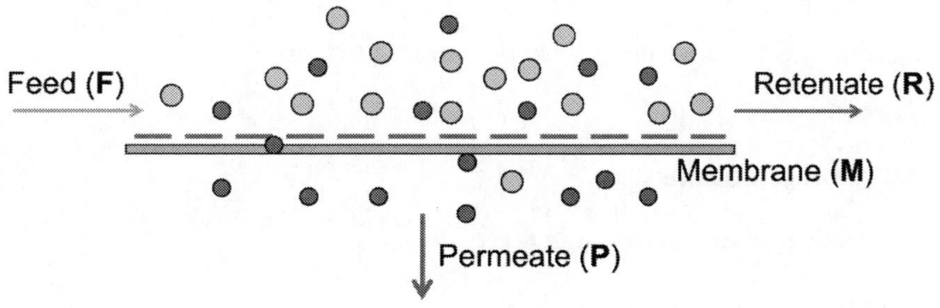

Parameters of the membrane efficiency

Permeate flux	Rejection factor
$J = \frac{1}{A}\frac{dV}{d\tau}$	$R = 1 - \frac{c_P}{c_F}$

3

Microfiltration
Theoretical backround

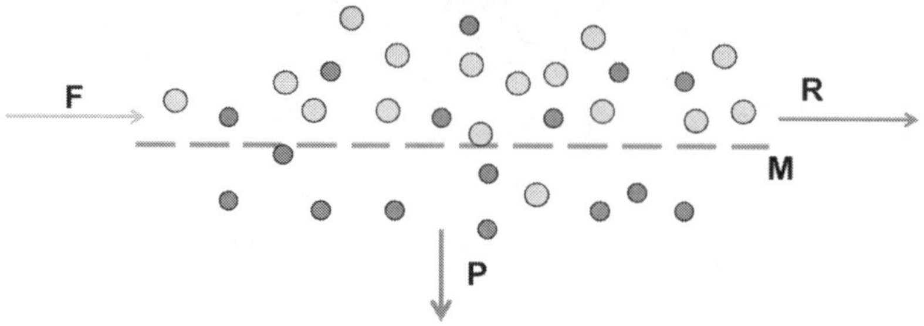

- **Driving force** of the process
  - Pressure difference (< 4 bar)
- **Separation principle**
  - **Sieving** mechanism (porous membrane)

4

Microfiltration
Theoretical backround

- **Membrane material and module**
  - Porous membrane, polymeric or ceramic material
  - Pore size: 0.2–2 μm
  - Membrane **module**:
    - Hollow fiber, tubular, flat sheet
- **Potential apllications of MF**
  - Green beer and rough beer recovery
  - Rough beer clarification
  - Cold sterilization

5

Microfiltration
Fouling mechanism

- Main foulants:
  - yest cells, protein-polyphenol aggregates, macromolecules

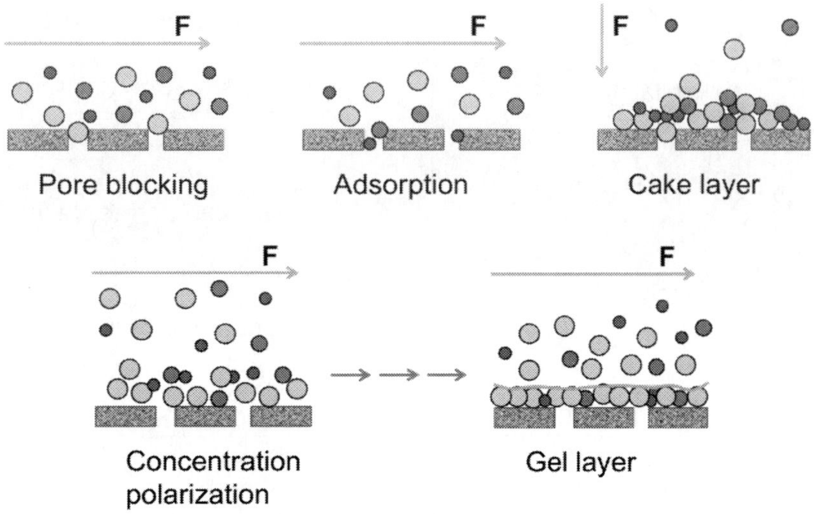

6

Microfiltration
Process intensification methods

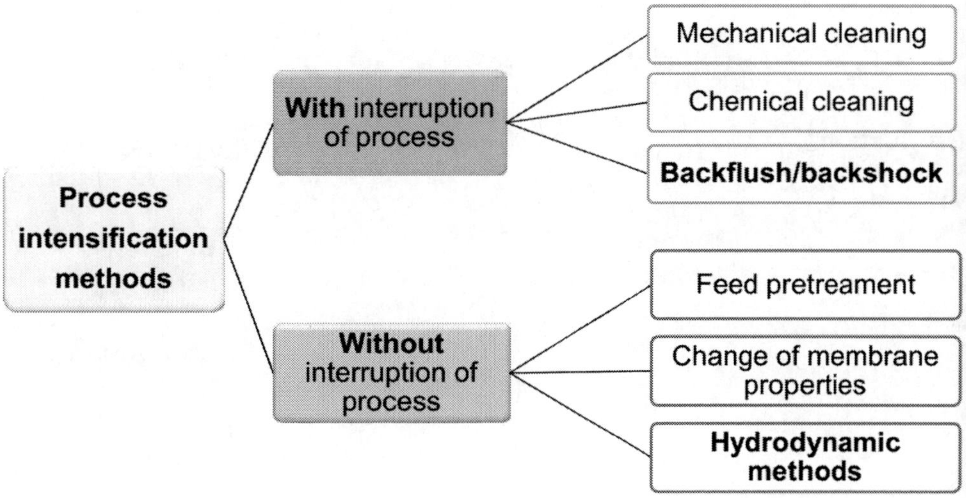

7

Microfiltration

Advantages

- Microbial stability
  Colloidal stability
- Reduction of beer losses
- Easy operation
  fully automatic units

Disadvantages

- Internal and external membrane fouling
- Variable permeate flux
- Changing beer quality (retention of aroma compounds)

8

Beer dealcoholization, aroma recovery

- **Main goals:** Ethanol removal, preservation of aroma compounds

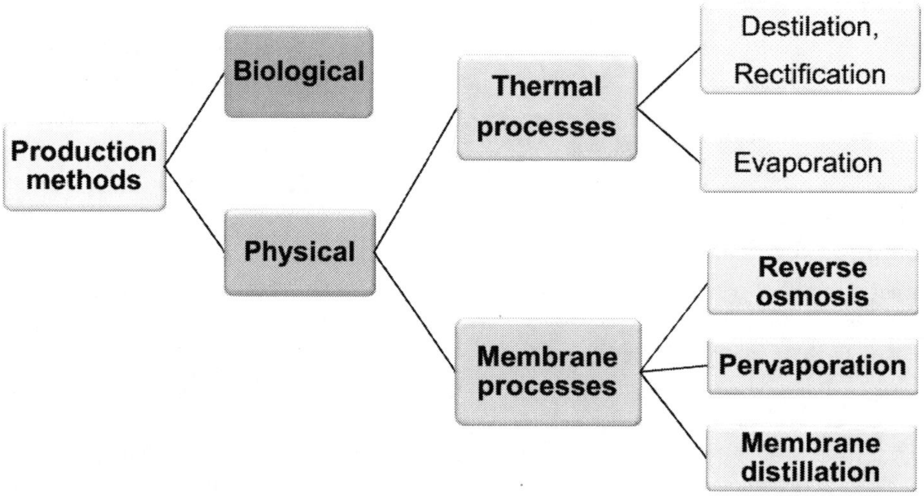

9

Reverse osmosis

Theoretical backround

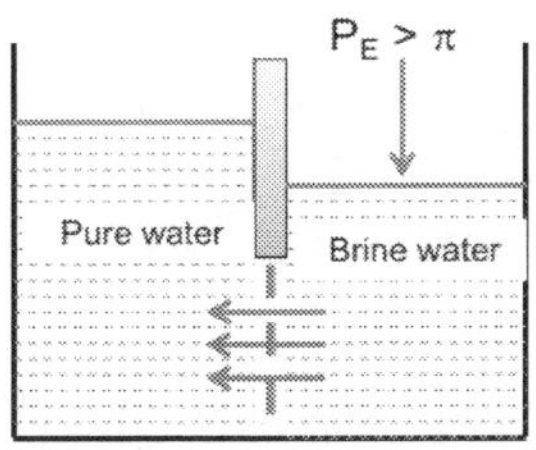

- **Driving force** of the proces
  - Pressure difference (20–100 bar)
  - Applied pressures are significantly higher then osmotic pressure!
- **Separation principle**
  - Solution-diffusion mechanism

10

Solution-diffusion mechanism

- **Transport mechanism** of the compounds through a **nonporous** membrane.

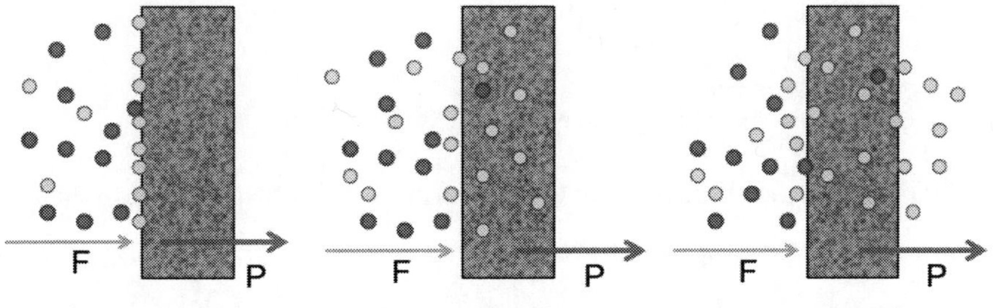

Selective absorption on the feed side

Selective diffusion through the membrane

Desorption on the permeate side

- **Important parameters:**
  - Permeability, solubility, diffusivity

11

Reverse osmosis

Process

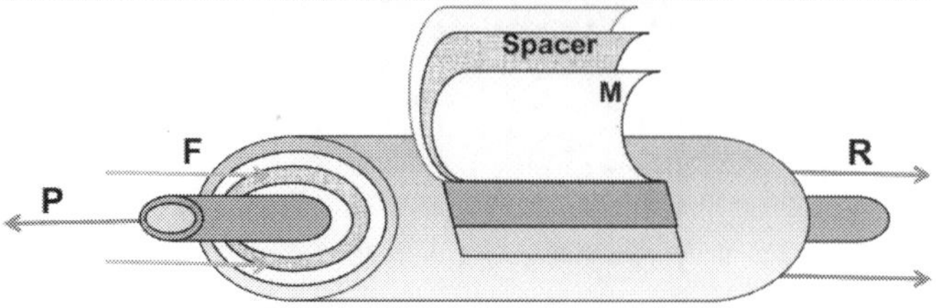

- **Membrane material and module**
  - Nonporous membrane: Cellulose acetate, PA
  - Spiral wound membrane
- **Separation parameters**
  - Feed temperature: 1–20°C, in practice: **1–5**°C
  - Applied pressure: 20–50 bar, in practice: **40–50** bar
  - R_{EtOH} =10–60%, Ethanol concentration in feed: **max. 0.5 vol.%**

12

Reverse osmosis in diafiltration mode

Process

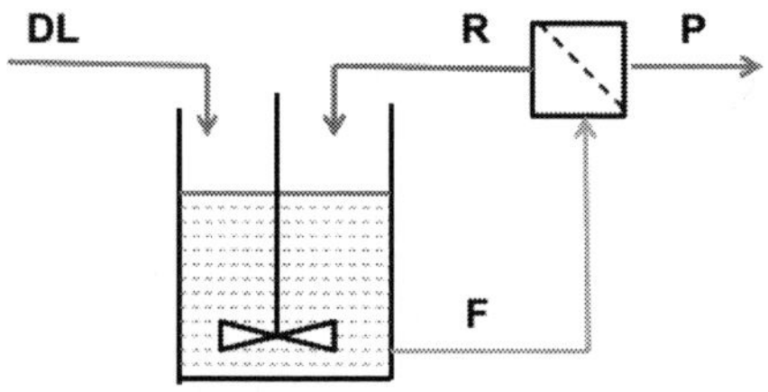

Diafiltration with constant volume in stirred tank

$$c_{etOH,M}=c_{etOH,0}e^{-\frac{V_{DL}(1-R_{etOH})}{V_T}}$$

- **Separation** technique for a separation between **high molecular** and **low molecular** substances.
- **Separation** principle
  - Feed is diluted with diafiltration liquid (water).
  - Low molecular substances are washed out.
- **High possibility to lowering ethanol content on feed side.**

13

Reverse osmosis

Advantages

- Sensitive to aroma compounds
- Non-alcoholic beer: Ethanol values below 0.5% vol.
- Easy operation, fully automatic units

Disadvantages

- Rejection of VOCs
- Low permeate flux
- High consumption of power and water

14

Pervaporation

Theoretical backround

VACUUM PV

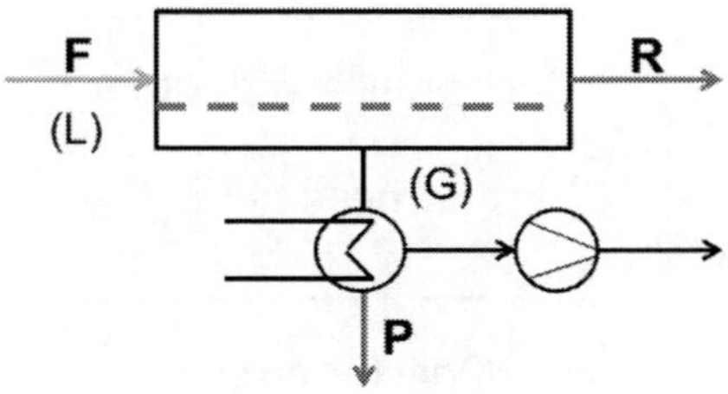

INERT CARRIER GAS PV

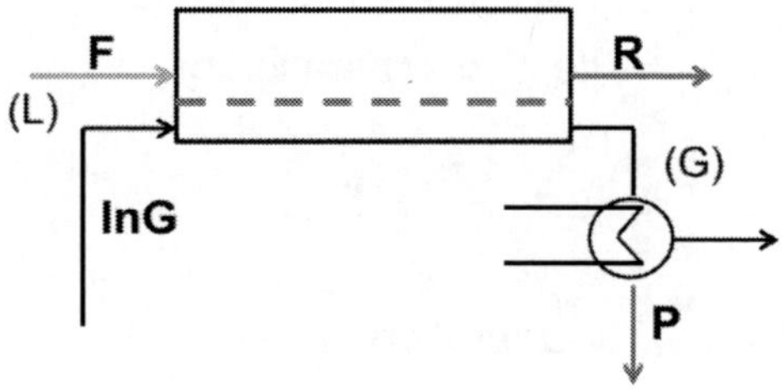

- **Driving force** of the process
  - Difference in partial vapour pressures of permeants
- **Separation principle**
  - Solution- diffusion mechanism

15

Pervaporation
Process

- **Membrane material and module**
  - *Hydro***philic** membrane: PVA, cellulose acetate
  - *Hydro***phobic** (Organophilic) **composite** membrane:
    - PDMS supported PET (flat-sheet)
    - POMS supported PEI (plate-and-frame)
- **Separation parameters**
  - Separation **with** hydro**philic** membrane
    - Sweep gas pervaporation (sweep gas included water vapor)
  - Separation **with** hydro**phobic** membrane
    - Vacuum pervaporation
    - Feed temperature: 25°C
    - Pressure on the **feed** side: atmospheric

16

Pervaporation

Advantages
- Effective separation of VOCs
- Operation at low temperatures
- Save energy

Disadvantages
- Cooperation with other processes
- Only for aroma recovery
- High capital costs

17

Membrane distillation

Theoretical backround

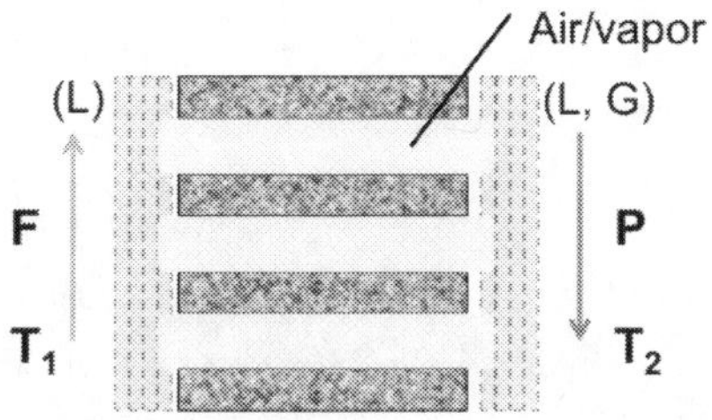

- **Driving force** of the process
  - Temperature difference, resulting vapor pressure difference
- **Separation principle**
  - Evaporation on the feed side.
  - Transport of vapor molecules through the pores of the membrane.
  - Condensation on the low-temperature side.

18

Concluding remarks

- **Microfiltration**
  - ✓ Clarification of rough beer, recovery of beer from the tank bottom sediment
  - ✓ Cost-saving process, environmental frendly
  - – Variable permeate flux, fouling of the membrane
- **Beer dealcoholization**
  - ✓ Ethanol removal, preserving of aroma compounds
  - ✓ Low energy consumption vs. conventional process, sensitive to aroma compounds
  - – Low permeate flux, rejection of aroma compounds, ethanol removal potentional
- **Membrane processes: Processes with potential of using in the brewing industry.**
- **The following research is necessary.**

19

Thank you for your attention!

Energy consumption of beverage bottling plants

TUM

Dipl.-Ing. Isabel Osterroth
Chair of Food Packaging Technology
Technical University of Munich

36th EBC congress
Ljubljana, 17 May 2017

TUM

Outline

- Motivation
- Scientific issue
- Analysis, modelling, simulation and validation
- Outlook: Industrial implementation and use of the model

Energy efficiency considerations

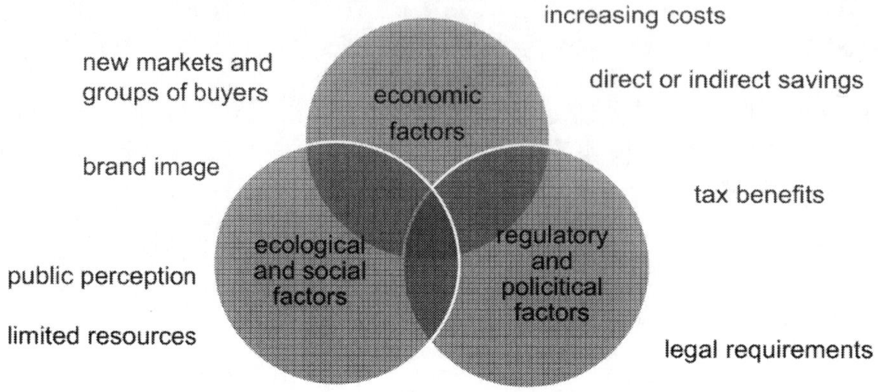

Application bottling plant

- 12-35% of the total electrical power for bottling
- Established **E**nergy **P**erformance **I**ndicator (EPI): e.g. kWh/ hl
  →influenced by machine efficiency
- energy demand during standby and idle times
- lack of detailed data for optimisation

Simulation on machine level
considering the energy demand

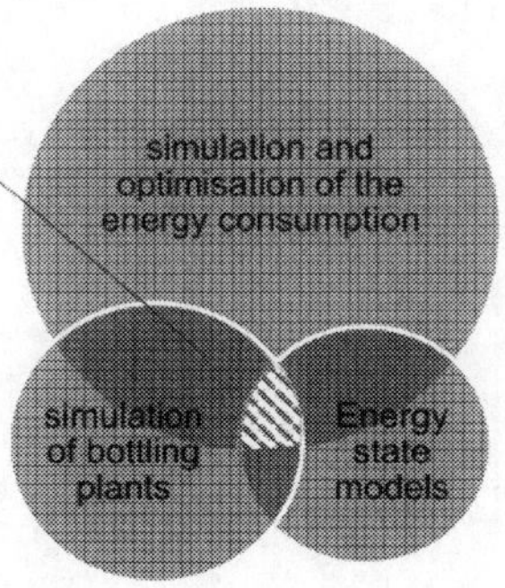

Beverages (heat, process, single machine optimisation)

E.g. „A utilities consumption model for real-time load identification in a brewery"
→bottling as a black box

Tooling machines

Target: efficiency enhancement

Scientific hypothesis

- Correlation between electrical energy consumption and operational state behaviour
- Hybrid simulation model as a instrument for a more energy efficient production

Scientific objective

- Generic model describing the energy behaviour of bottling machines
- Simulation tool for
  - Energy efficient planning
  - Resource orientated production control
  - Demand-driven supply

Scientific approach

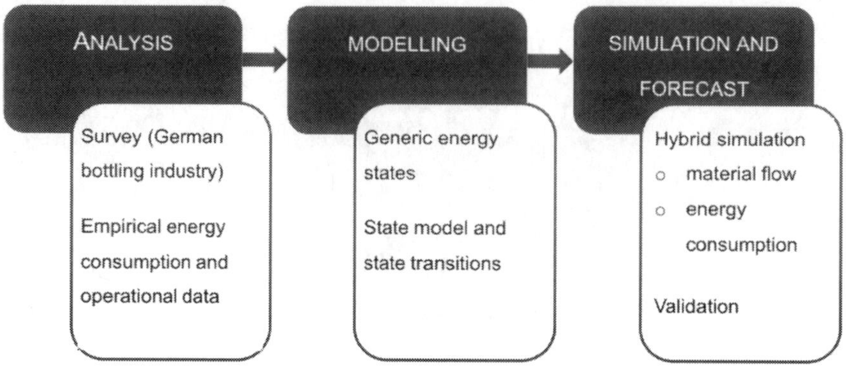

Scientific approach

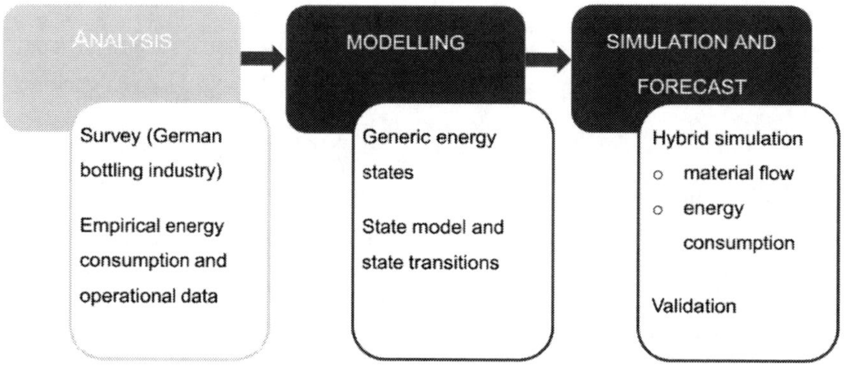

Installed load of bottling machines

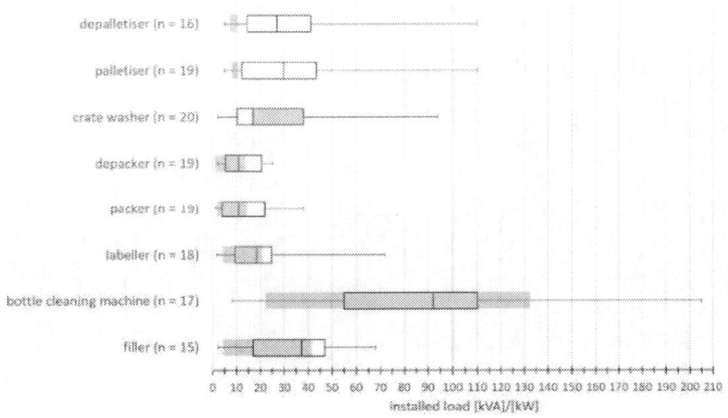

Result from our survey (2016) and literature data from the 1980s [Schreiner, 1982]

Demand during standby times

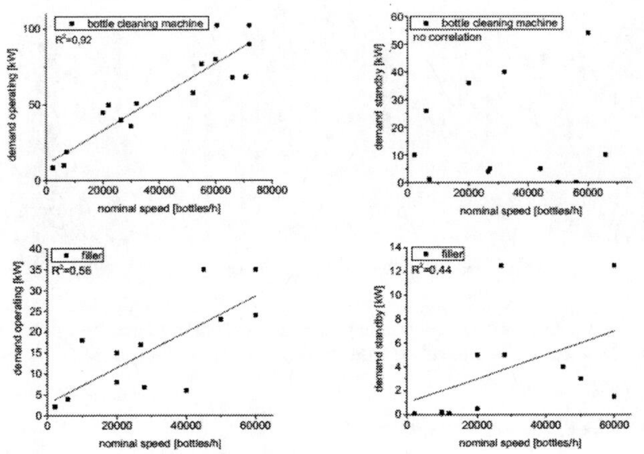

Data acquisition:

- primary data from industrial bottling plants on machine level, high level of detail (1 s/ 2 s)

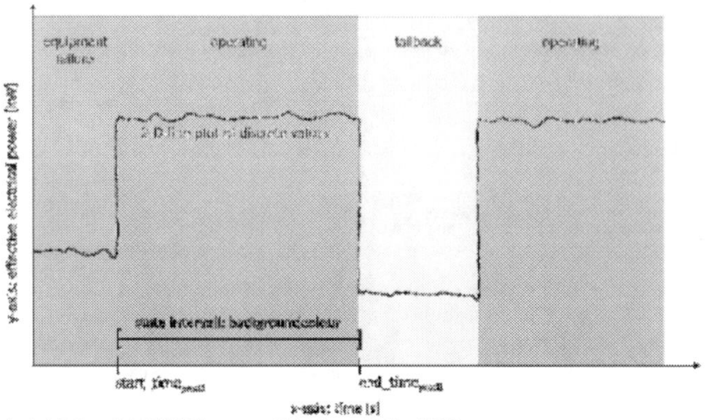

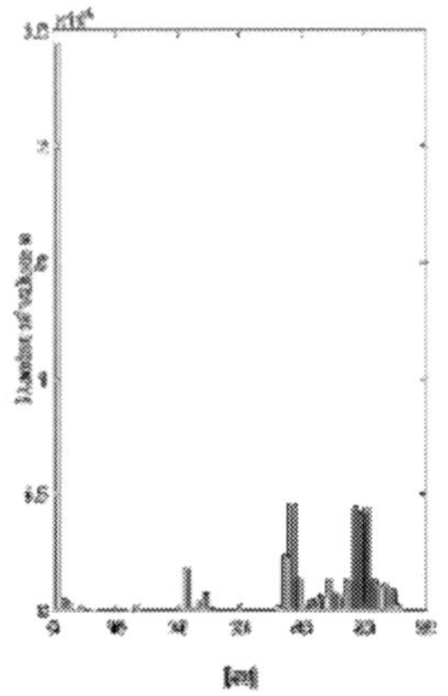

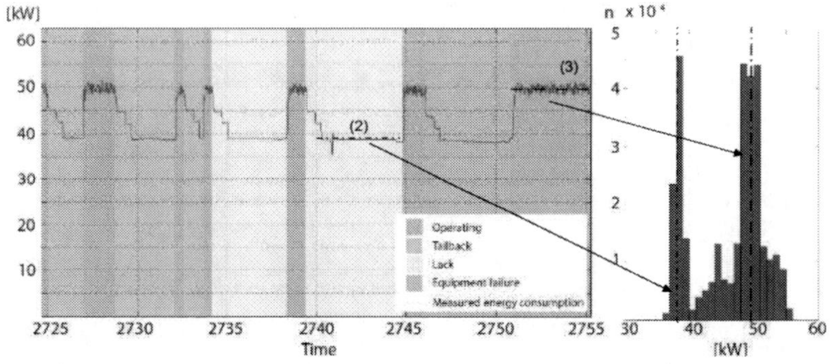

Bottle cleaning machine A: Detail of a typical consumption pattern and related operational states (left) and histogram of measured values for the operating time (time period of 14 days, right).

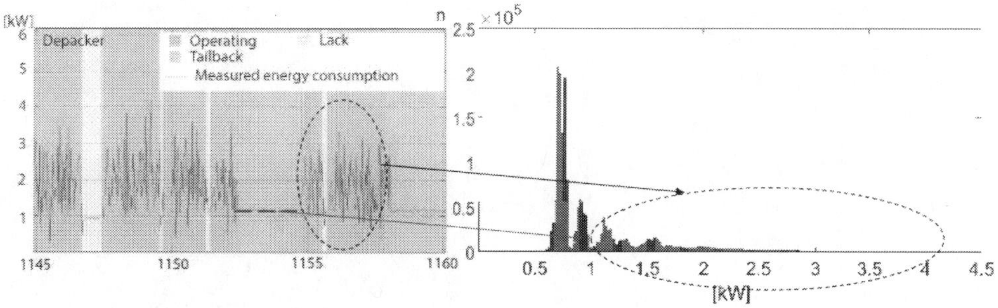

Depacker A: Detail of a typical consumption pattern and related operational states (left) and histogram of measured values for the operating time (time period of 14 days, right).

Findings from the analysis

➢ Limited database available on a high level of detail, no operational state data

- Correlation between energy consumption and operational states
- Energy demand level can be identified
- Demand level specific for products/ process parameters
- Energy consumption during non-productive times
- No definition of „standby" demands
- State based consumption behaviour for water and heat consumption

Scientific approach

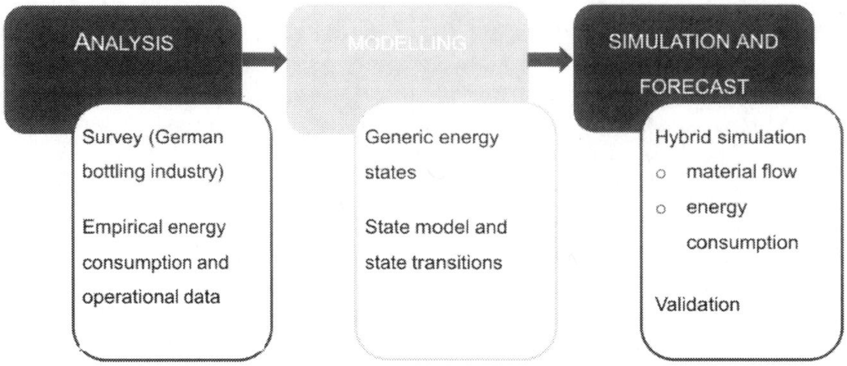

Definition and mapping of constant demand level

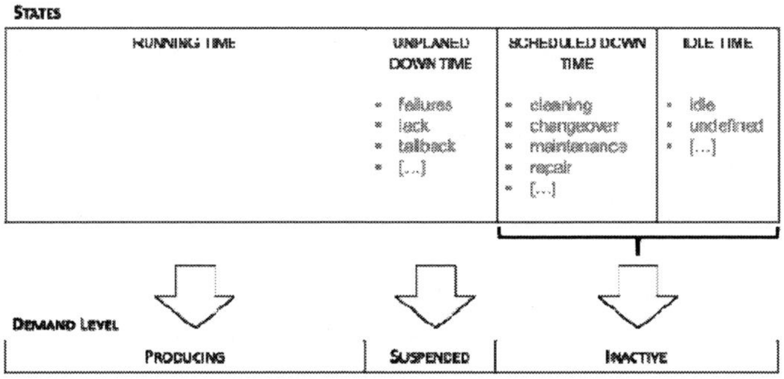

Modelling of state transitions

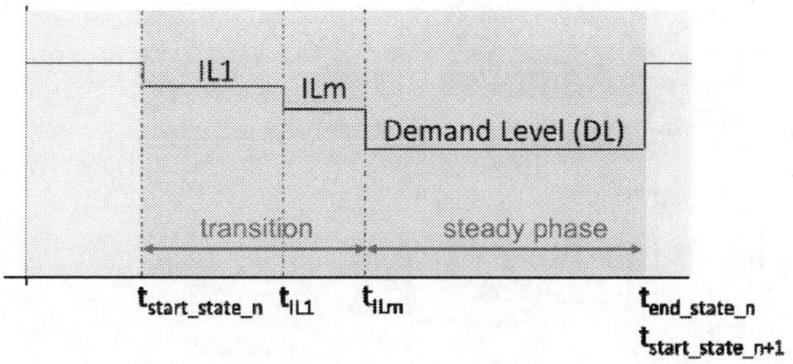

$$P_{el} = \begin{cases} IL_1 & if\ t_{start\_state\_n} \leq t < t_{IL1} \\ IL_2 & if\ t_{IL1} \leq t < t_{IL2} \\ IL_m & if\ t_{IL(m-1)} \leq t < t_{ILm} \\ DL & if\ t_{ILm} \leq t < t_{end\_state\_n} \end{cases}$$

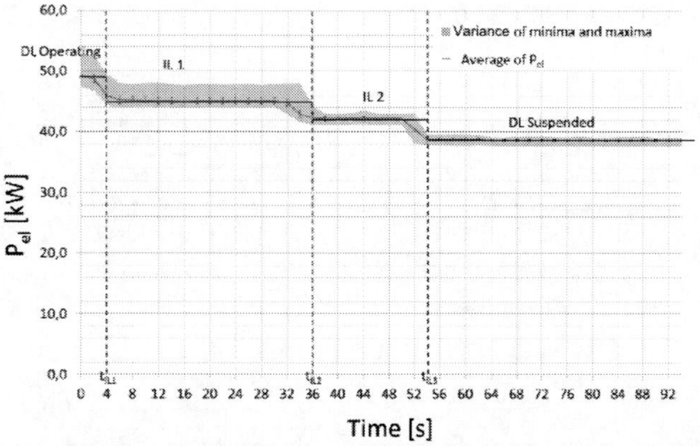

Stepwise change of the demand level of the bottle cleaning machine
(maximum variation, average value and confidence interval (α=0.05) of 20 measurements)

Scientific approach

ANALYSIS → **MODELLING** → **SIMULATION AND FORECAST**

Analysis

Survey (German bottling industry)

Empirical energy consumption and operational data

Modelling

Generic energy states

State model and state transitions

Simulation and forecast

Hybrid simulation
- material flow
- energy consumption

Validation

Implementation

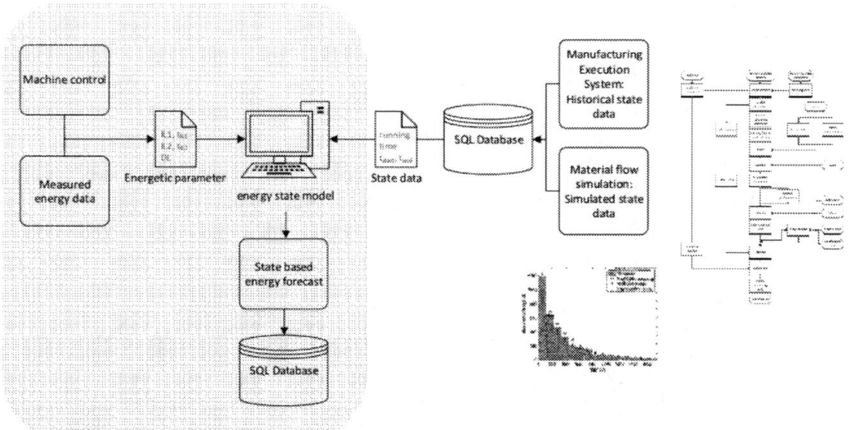

Validation criteria

- Graphical: Validation plots
- **A**verage **P**ercentage **D**eviation (1)
- **T**heil's **I**nequality **C**oefficient (2)
- Credibility (3)

$$\mathbf{APD} = \frac{1}{n}\sum_{i=1}^{n}\frac{(y_i - z_i)}{z_i} \quad (1)$$

$$\boldsymbol{TIC} = \frac{\sqrt{\sum_{i=1}^{n}(y_i - z_i)^2}}{\sqrt{\sum_{i=1}^{n} y_i^2} + \sqrt{\sum_{i=1}^{n} z_i^2}} \quad (2)$$

$$\boldsymbol{\theta} = \frac{0{,}40 - \boldsymbol{TIC}}{0{,}40} \quad (3)$$

y_i: discrete simulated value , z_i: discrete measured value

Validation: bottle cleaning machine

Example of model parameters

	Incative		*Suspended*		*Production*	
	P [kW]	t [s]	P [kW]	t [s]	P [kW]	t [s]
IL1	7,2	0	49,78	4,9	38,9	4,5
IL2	7,2	0	44,5	34,8	48,5	10,2
IL3	7,2	0	42,1	53,2	49,3	21,4
Steady state	7,2	--	37,8	--	50	--

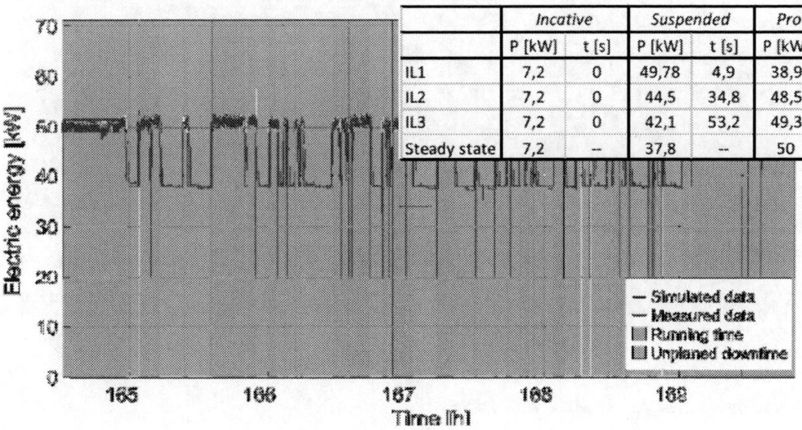

Validation: filler

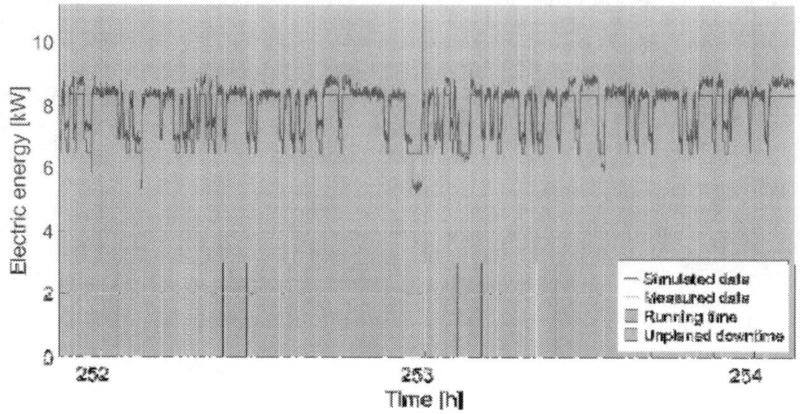

Validation: labeler

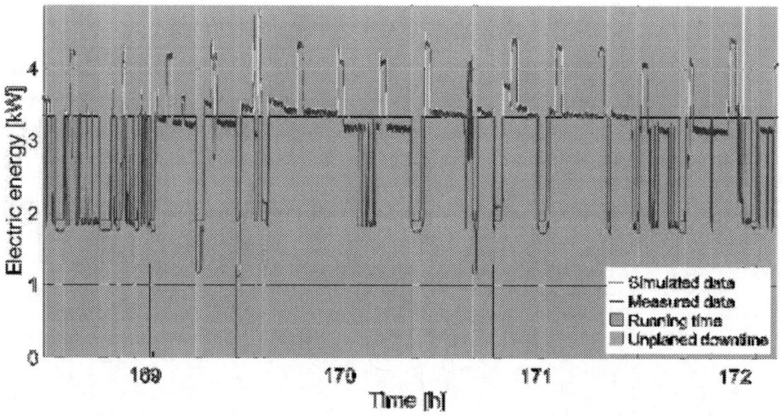

Statistical validation results

aggregat	TIC	APD [%]	measured [kWh]	simulated [kWh]	total deviation [%]	credibility	acceptability
bottle cleaning machine	0.17	60	4769	4699	-2	0.58	yes
	0.15	32	6008	6045	1	0.63	
	0.14	67	8579	8350	-3	0.65	
filler	0.13	21	1373	1473	7	0.68	yes
	0.12	17	1541	1547	< 1	0.70	
	0.09	14	1836	1779	-3	0.78	
labeler	0.19	40	352	360	2	0.53	yes
	0.16	41	392	399	2	0.60	
	0.14	22	539	524	-3	0.65	

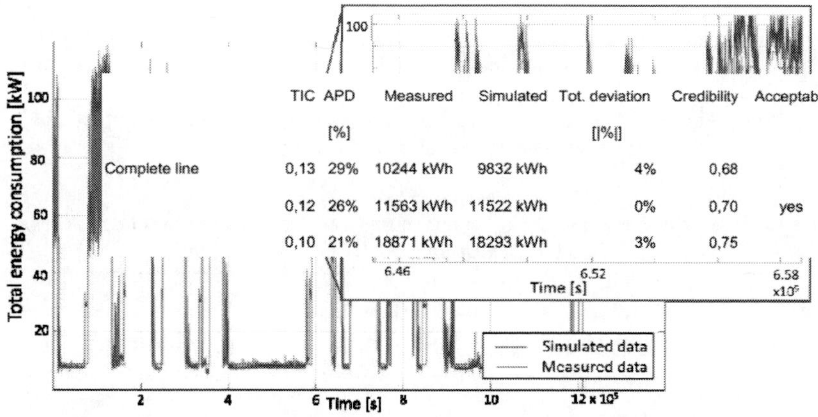

	TIC	APD [%]	Measured	Simulated	Tot. deviation [|%|]	Credibility	Acceptability
Complete line	0,13	29%	10244 kWh	9832 kWh	4%	0,68	yes
	0,12	26%	11563 kWh	11522 kWh	0%	0,70	
	0,10	21%	18871 kWh	18293 kWh	3%	0,75	

Total deviation for the total line with a 95%-confidence interval: [0,3; 4,4] for three validation runs of each fourteen days

Summary of the modelling

- DEMAND LEVEL
  - PRODUCING
  - SUSPENDED
  - INACTIVE
- INTERMEDIATE LEVEL IL and time t_{IL}
- OPERATIONAL STATE

} Forecast of Energy Demand

- Influence of availability of the line →Decrease of energy demand
- Influence of parameter chances (e.g. use of more efficient parts)
- Influence of machine control

VDMA Einheitsblatt 8751 (Draft Version) VDMA

„Einheitliche Spezifikation und Messung des Energie- und Medienverbrauchs von Abfüll- und Verpackungsmaschinen"

German Engineering Association

"Comparable specification and determination of the energy and media demand of filling and packaging machines"

- Production
- High Level Standby
- Low Level Standby
- Inactive
- Optional machine-based demand level (e.g. heating up)

HyMoME

Hybrid Modelling and Simulation of the Energy and Media demand in the beverage industry
Hybride Modellierung und Prognose des Medien- und Energiebedarfs in der Getränkeindustrie (HyMo_ME)

- Founded by the Bayerische Forschungsstiftung
- started in March 2016
- 3 years

- Electrical energy/ heat/ water/ media/ pressured air

Any questions?

Isabel Osterroth | 36th EBC congress| Ljubljana, 17 May 2017 30

iFBM

EBC wort turbidity : Nature and impact of malting and brewing process

L47 (88)

Patrick Boivin & Sophie SCHWEBEL

iFBM

Study context

⇨ EBC Method 4.5.1 : Extract of malt : Congress mash

⇨ This procedure is used for numerous malt quality parameters

4.5.1

Extract of Malt: Congress Mash

- Saccharification rate
- Soluble proteins
- pH
- Colour
- EBC wort
- Viscosity
- Free amino nitrogen content
- Odour
- Turbidity
- Extract of malt

iFBM

Material and methods

⇨ **EBC Method 4.5.1** (50 g malt fine grind + 200 ml water + 100 ml water at 70 °C)

⇨ **Turbidity measurement at 25°** (HACH 2100 AN)

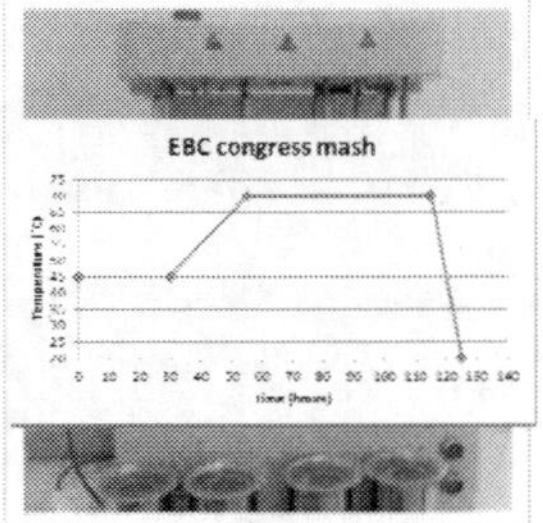

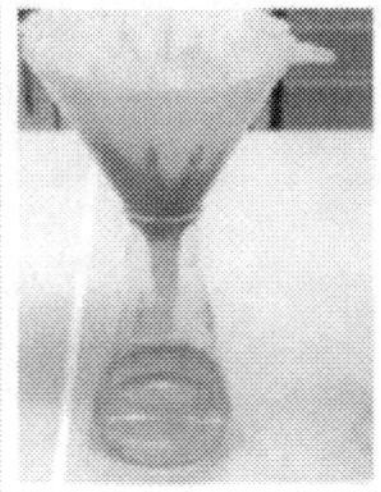

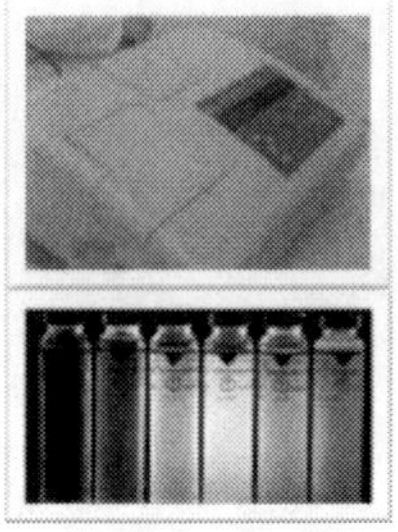

MILLING ⟶ BREWING ⟶ FILTRATION ⟶ TURBIDITY

iFBM

Study context

Is the EBC wort turbidity tightly linked to the haze of beer ?

⇨ Which compounds are responsible for this EBC wort turbidity ?

⇨ How can we manage their evolution during malting and brewing process ?

iFBM

Presentation Overview

1. Impact of malting process on EBC wort turbidity
2. Identification of compounds responsible for EBC wort turbidity
3. Impact of mashing diagram on EBC wort turbidity

Biochemical changes in barley during modification

iFBM

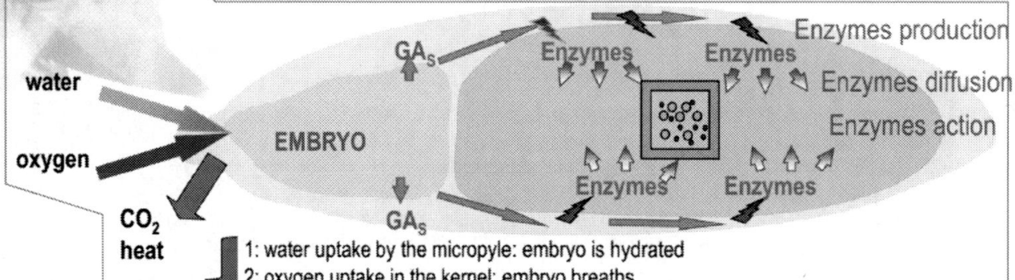

Steeping step
- 1: water uptake by the micropyle: embryo is hydrated
- 2: oxygen uptake in the kernel: embryo breaths
- 3: hydration of endosperm

Germination step
- 4: embryo products gibberellins that migrate to aleurone layer
- 5: gibberellins activate aleurone layer that will produce hydrolytic enzymes
- 6: hydrolytic enzymes migrate to starchy endosperm
- 7: β-glucanases hydrolyses endosperm layer (kernel becomes friable)
- 8: proteases act on protein: production of soluble proteins and FAN

6

Turbidity of wort & databases

- ⇨ **Databases exploitation**
  - ⇨ CBMO: French malting barley qualification
    - 1 barley variety > 1 malt batch
    - 20 parameters on malt

Turbidity of wort & databases

- ⇨ **CBMO Database**
  - ⇨ PCA with 14 malt parameters :

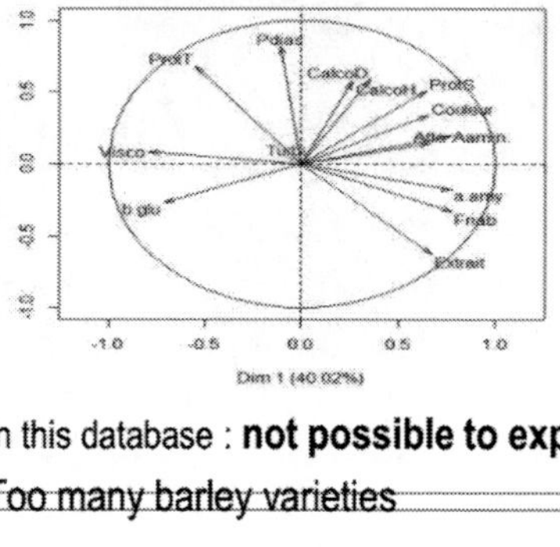

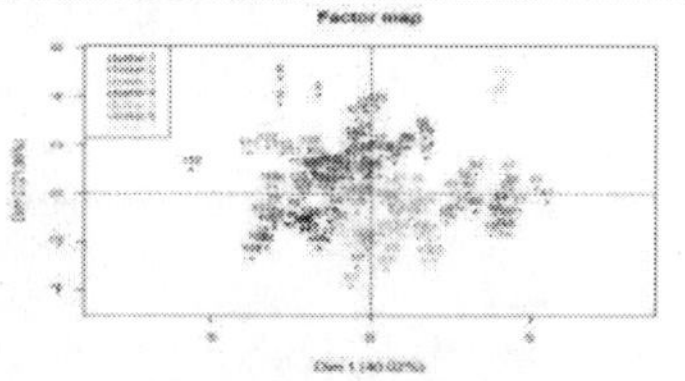

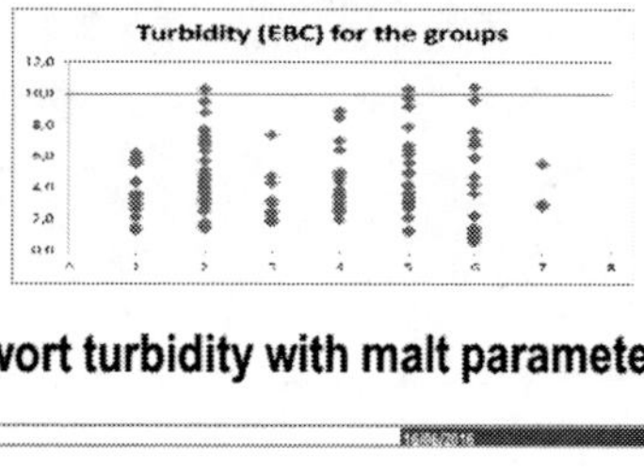

⇨ In this database : **not possible to explain wort turbidity with malt parameters**

⇨ Too many barley varieties

iFBM

Turbidity of wort & databases

⇨ **CBMO Database**

⇨PCA for SEBASTIAN

	Turb
Turb	1.00
Hum	-0.07
Extra	-0.13
Coul	-0.13
pH	0.55
Prot	-0.22
ProtS	-0.38
Kol	-0.26
diast	-0.34
visc	0.43
Fria	-0.28
gluca	0.43
Att	-0.20
Calco	-0.13
CalcoH	0.00
amy	-0.33
Azote	-0.40
Polyp	-0.26
FilTb	-0.04

⇨PCA for ESTEREL

	Turb
Turb	1.00
Hum	-0.12
Extra	0.18
Coup	0.40
pH	-0.08
Prot	-0.09
ProtS	-0.05
Kol	-0.01
diast	-0.26
visc	0.08
Fria	-0.15
gluca	0.30
Att	-0.17
Calco	-0.05
CalcoH	[illegible]
amy	[illegible]
FAN	-0.03
Polyp	-0.04
FilTb	-0.40

⇨PCA for VANESSA

	Turb
Turb	1.00
Hum	0.13
Extra	-0.03
Coul	0.24
pH	-0.31
Prot	-0.21
ProtS	-0.14
Kol	0.02
diast	-0.24
visc	0.22
Fria	-0.14
gluca	0.46
Att	-0.24
Calco	-0.23
CalcoH	-0.29
amy	0.30
Azote	-0.12
Polyp	-0.10

	SEBASTIAN	ESTEREL	VANESSA
Correlation (+)	pH, β-glucans	Color, (β-glucans)	β-glucans, (α - amylase)
Correlation (-)		α- amylase	pH

iFBM

Impact of malting process

⇨ **Methodology :** Micromalting of barley according to several process experimental conditions (~600g)

2-row spring barley varieties
- A
- B
- C
- D

4 steeping diagrams
- **D1** : 10 / 20 / 7 (37h)
- **D2** : 10 / 17 / 3 (30h)
- **D3** : 8 / 15 / 3 (26h)
- **D4** : 6 / 8 / 5 / 7 / 3 / 2 (31h)

Germination conditions
- 4 or 5 days
- Addition of Giberrellic acid

Kilning conditions
- CTPS like curing (up to 85°C)
- CTPS like curing (70°C max)

Cylindro-conical vessel Saladin box Single desk kiln

600 Kg IFBM MALTING PLANT

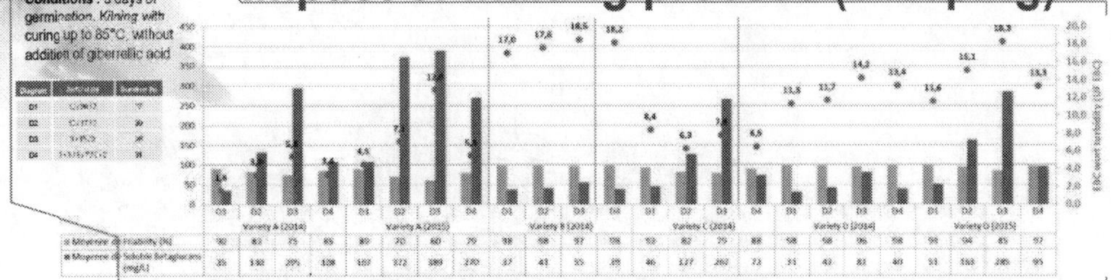

iFBM

Impact of malting process (Steeping)

- ⇨EBC wort turbidity depends on barley variety
- ⇨Effect of steeping process diagram
  - ⇨Shorter steeping diagram (D3) > bad endosperm hydration > lower modification > ➚ turbidity
  - ⇨Longer or optimized steeping > good endosperm hydration > good modification (D4): ➘ turbidity

Hydration of endosperm

iFBM

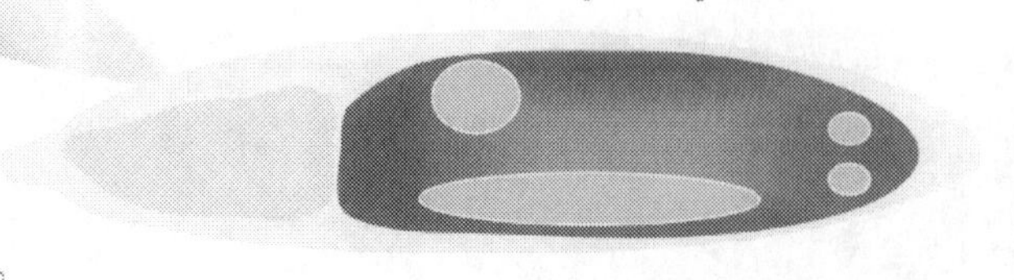

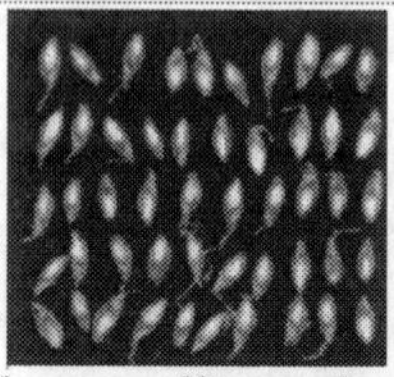

low quality malt

high quality malt

Steeping is a key step to produce a good malt

iFBM

Impact of malting process (Germination)

Conditions : D2, Kilning with curing up to 85°C

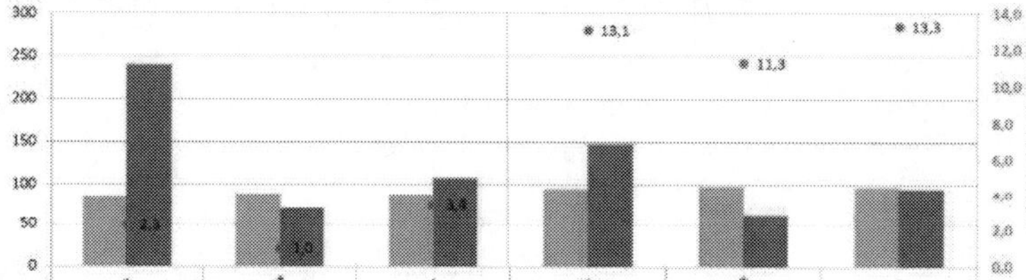

	Variety A (2014)			Variety D (2015)		
Giberellic acid (0,5 ppm)	-	+	-	-	+	-
Germination duration (days)	4		5	4		5
Moyenne de Friability (%)	82	89	85	95	98	97
Moyenne de Soluble Betaglucans (mg/L)	240	69	108	148	61	95

⇨No impact of **germination duration** on turbidity (but ↘ betaglucans content)

⇨Addition of **giberellic acid** seems to trigger the decrease of turbidity (improve modification) more significantly

Impact of Gas on wort turbidity and others malt parameters

iFBM

600 Kg IFBM MALTING PLANT

Malt analysis	malt 6RH without Ga_3	malt 6RH with Ga_3
Moisture (%)	4.7	4.4
Fine grind extract (% D.M.)	79.8	81.2
Color (EBC)	3.3	3.8
Boiled Color K.Z. (EBC)	5.0	6.9
Total protein (% d.m.)	10.00	9.9
Soluble soluble (% d.m.)	3.57	4.83
Friability (% flour)	86	96
Calcofluor % modification	98	99
Calcofluor %homogénéity	87	96
β-glucans (mg/l of wort)	169	38
diastasic power (WK m.s.)	330	360
α-amylase (UD 20°C m.s.)	34	55
Viscosity (mPa.s)	1.53	1.46
Free amino nitrogen (mg/l)	116	184
Fermentability (%)	81.5	84.3
Turbidity (EBC)	4.9	0.9

Biochemical changes in barley during modification

iFBM

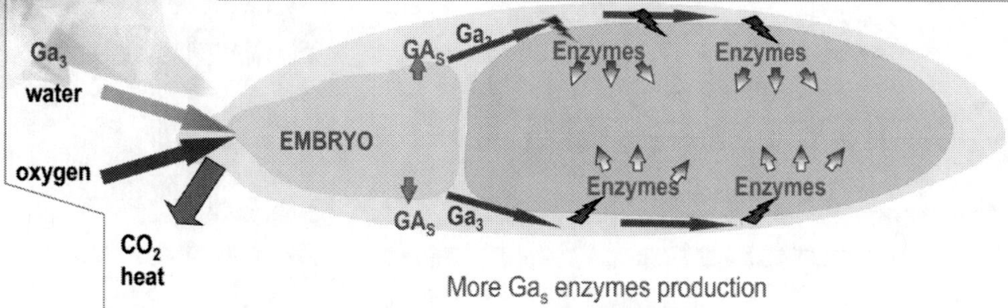

More Ga_s enzymes production
β-glucanases
α-amylases
proteases

16

Endosperm modification

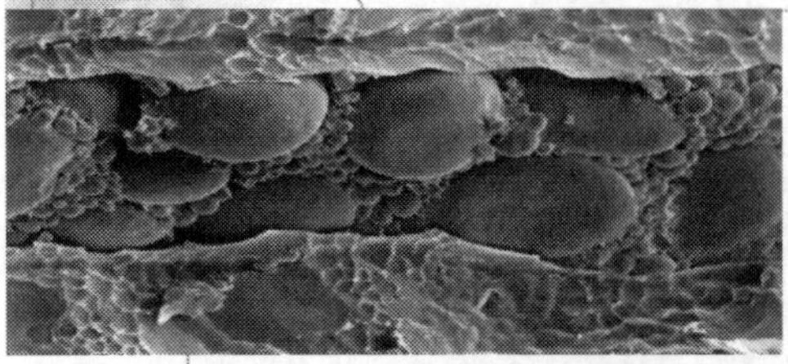

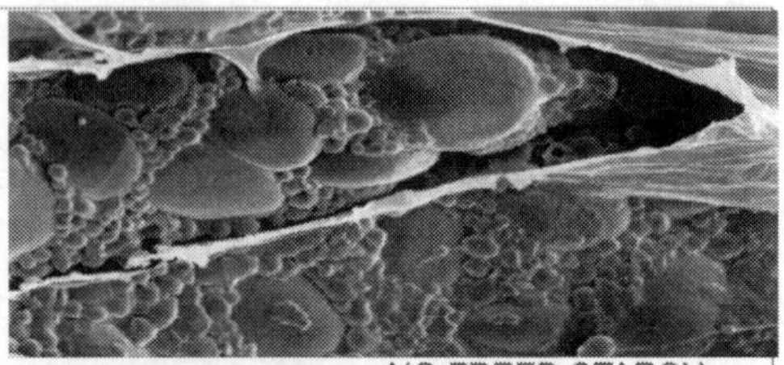

STARCH TRAPPED
& HARD KERNEL

NO FREED STARCH
& BAD MODIFICATION

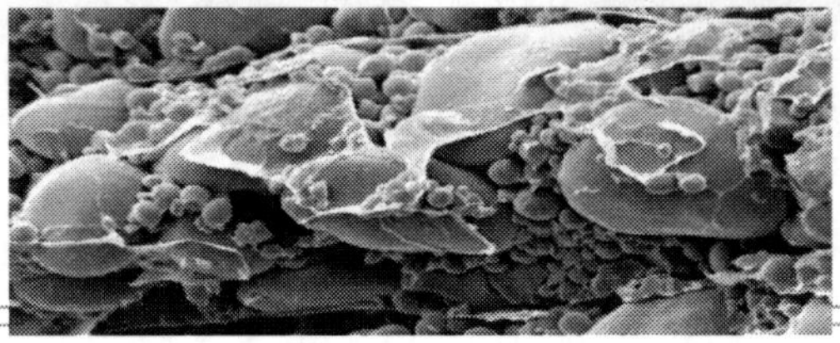

FRED STARCH
& GOOD MODIFICATION

17

Impact of malting process (Kilning)

Conditions : D2, *Kilning* with curing at
to 70°C
to 85°C

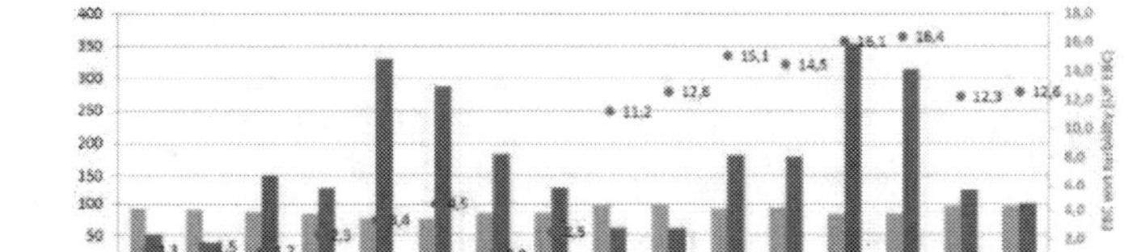

	70°C	CTPS	70°C	CTPS	70°C	CTPS	70°C	CTPS	70°C	CTPS	70°C	CTPS	70°C	CTPS	70°C	CTPS
	D1		D2		D3		D4		D1		D2		D3		D4	
	Variety A (2014)								Variety D (2015)							
Moyenne de Friability (%)	93	91	87	84	78	76	86	83	99	98	92	94	83	84	97	97
Moyenne de Soluble Betaglucans (mg/L)	52	39	151	131	331	288	183	131	62	61	181	179	353	314	123	101
Moyenne de Turbidity	1,3	1,5	1,2	2,3	3,4	4,5	0,9	2,5	11,2	12,6	15,1	14,5	16,1	16,4	12,3	12,6

⇨No significant impact of **curing temperature on turbidity**

Kilning: Enzymes activities

iFBM

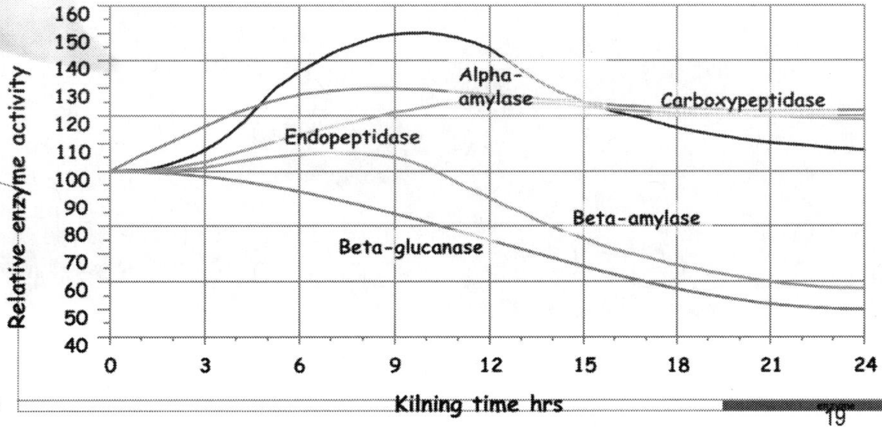

19

Impact of malting process

iFBM

- ⇨ **EBC wort turbidity can be very different according to barley variety**
- ⇨ **<u>EBC wort turbidity is tightly linked to grain modification</u>**
  - ⇨During the malting process, EBC wort turbidity is influenced by :
    - the steeping diagram: hydration of endosperm
    - the addition of giberellic acid during the germination step by improving some enzymes production
  - ⇨Effect on β-glucans to keep in mind

iFBM

Identification of compounds

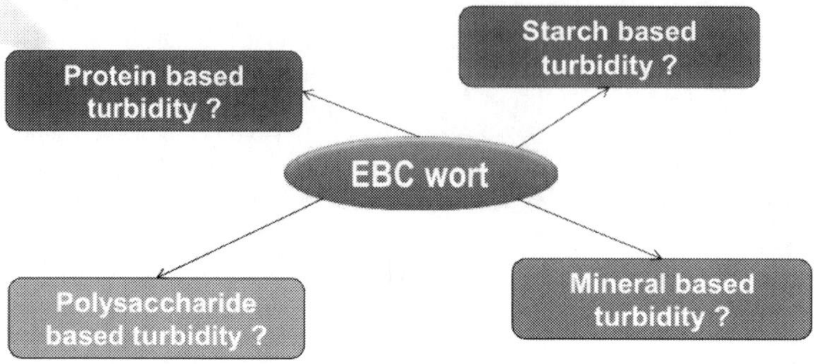

iFBM

Identification of compounds

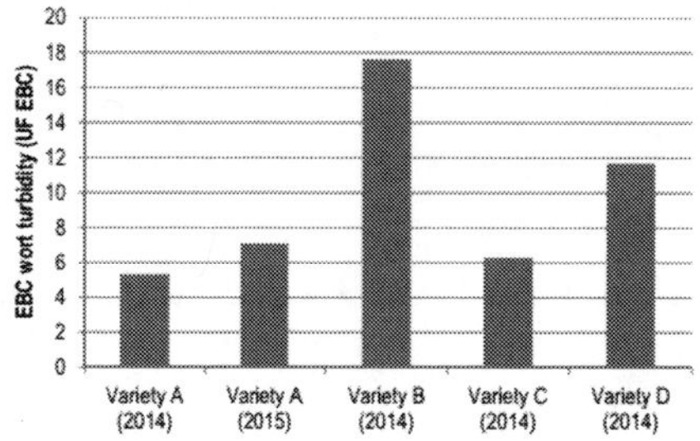

⇨ **Turbidity of EBC wort** depends a lot on barley variety

⇨ To identifiy haze compounds, different barley varieties were used

Turbidity of wort

iFBM

⇨ **Centrifugation to collect « haze »**

Malt

EBC congress mash

EBC wort

Centrifugation

supernatant → Analyzes

Pellet fraction → Analyzes

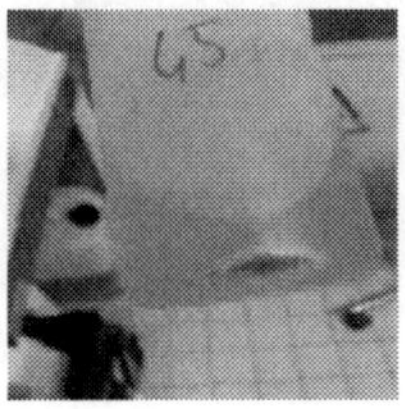

Electrophoresis of pellets

iFBM

Diagram	[illegible] (h)	Duration (h)
D1	10 / 20 / 7	37
D2	10 / 17 / 3	30
D3	8 / 15 / 3	26
D4	[illegible]	31

Diagram	D1	D1	D1	D2	D2	D3	D3	D4	D4
Giberellic acid (ppm)	0	0	0.5	0	0	0	0	0	0
Germination (days)	4	5	4	4	5	4	4	4	4
Kilning	CTPS	CTPS	CTPS	CTPS	CTPS	CTPS	70°C	CTPS	70°C

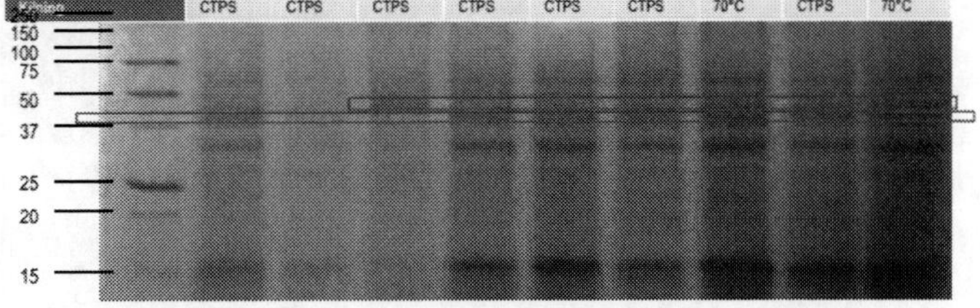

⇨ Protein profile appears not to be modified with malting diagram or germination times

⇨ **Protein A (~ 40 kDa) seems to be present in pellet (= Z ?)**

⇨ **Protein B (~45 kDa) modified with giberellic acid**

iFBM

Identification of haze compounds

- ⇨ **Arabinoxylan and mineral no impact**
- ⇨ **β-glucan difficult to conclude**
- ⇨ **Protein profile appears not to be modified**

iFBM

Identification of compounds

⇨ **Methodology** to determine a potential starch based turbidity:

EBC Wort

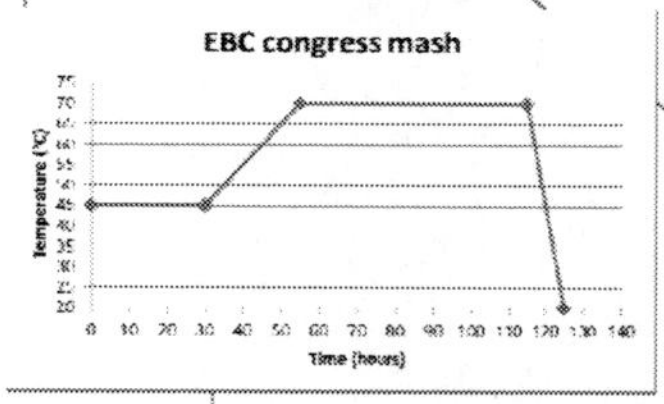

Turbidity measurement

starch

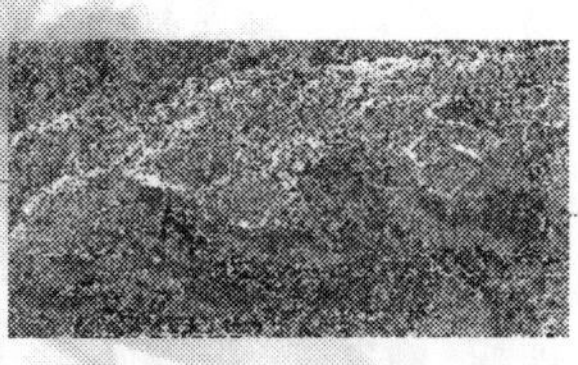

STARCH

↓

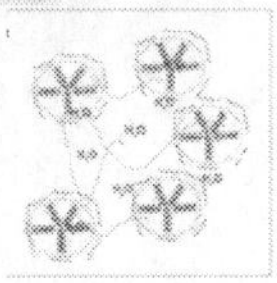

GELATINIZATION OF BIG GRANULES: 62-65°C

↓

FERMENTABLE SUGARS STEP : 63-65°C ← β-amylase, Limit dextrinase

↓

DEXTRINS STEP : 72-74°C ← α-amylase

27

Identification of compounds

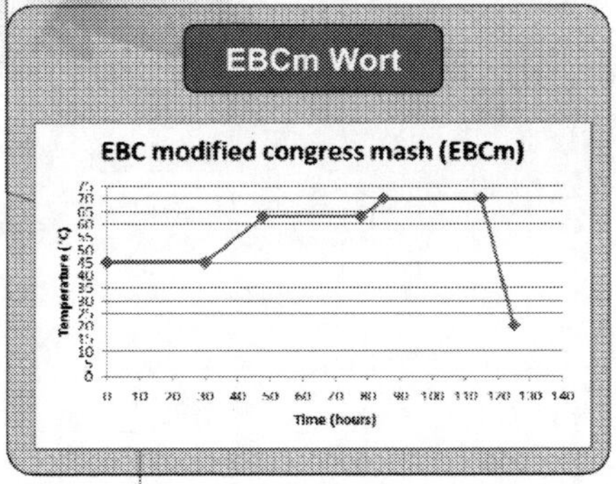

By adding a rest at 63°C, the ß-amylase and limit dextrinase activities which degrade starch is more efficient !!

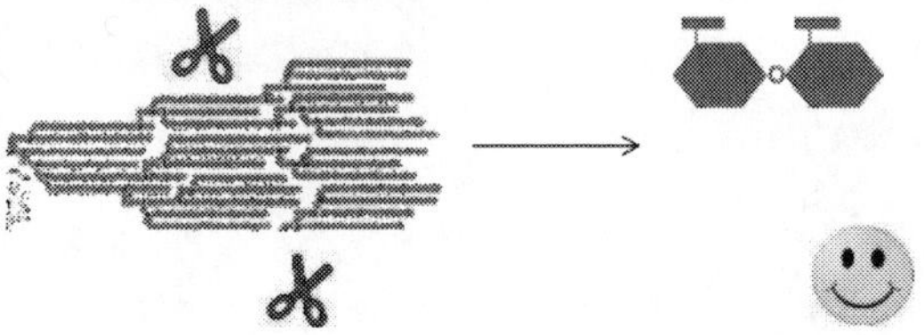

iFBM

Identification of compounds

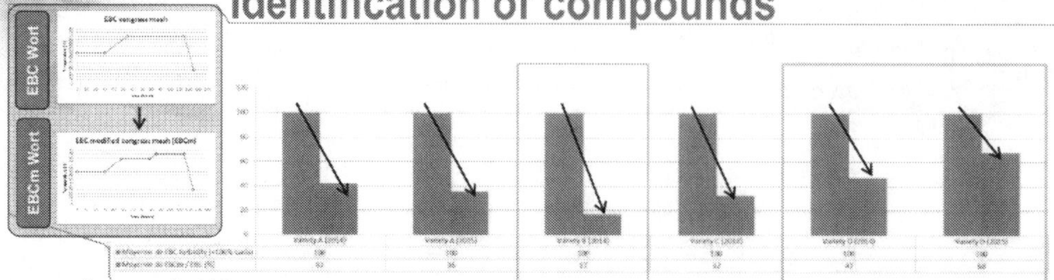

Decrease of turbidity from EBC wort to EBCm	58%	64%	82%	67%	53%	32%

⇨ **Strong reduction of turbidity with the rest at 63°C**

⇨ The EBC wort turbidity is partly starch-based for these samples

⇨ **Variable according to the variety**

iFBM

Gelatinization temperature

CROP	Gelatinization Temperature °C
2014	61-63
2015	62-65

Identification of compounds

⇨ **Methodology** to determine a potential starch based turbidity:

EBC Wort | EBCm Wort | EBCa Wort

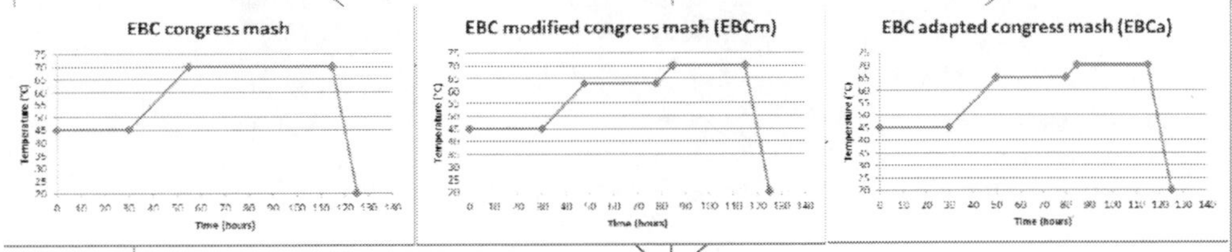

Turbidity measurement

Identification of compounds

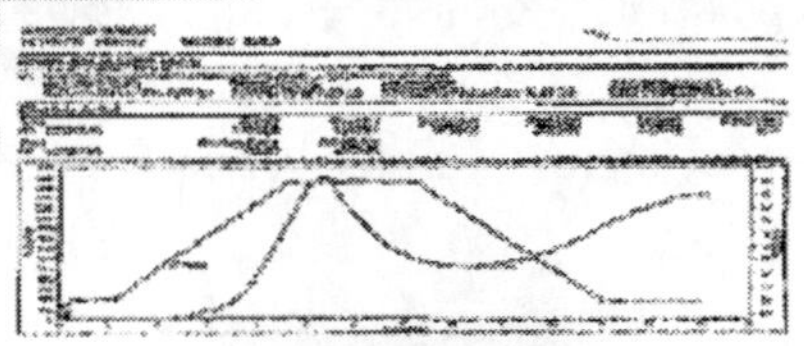

We assessed the gelatinisation temperature (RVA, Rapid Visco Analyser) of each sample to be sure that all the big starch granules are gelatinazed

iFBM

Identification of compounds

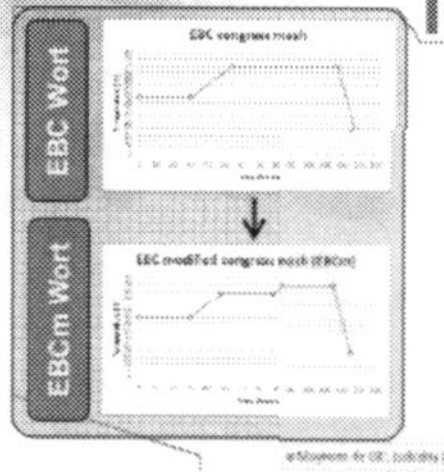

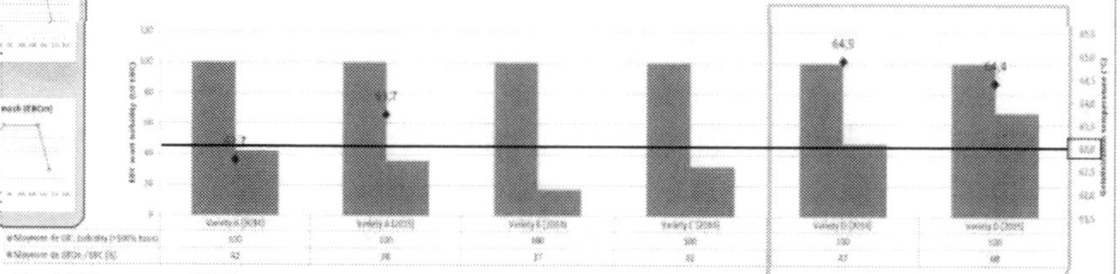

	Variety A (2014)	Variety A (2015)	Variety B	Variety C	Variety D (2014)	Variety D (2015)
Decrease of turbidity	58%	64%	82%	67%	53%	32%
Gelatinization temperature	62,7°C	63,7°C			64,9°C	64,4°C

⇨ **Hypothesis :** the higher gelatinization from the variety D (2014 and 2015) could explain the less efficient amylolyse

⇨ **Decision :** Adapt the additional rest at 65°C instead of 63°C

iFBM

Identification of compounds

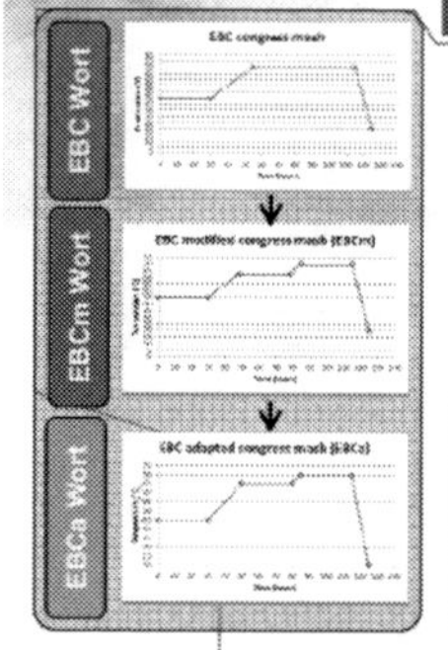

⇨ **A temperature above the gelatinisation temperature allows a decrease of turbidity**

⇨ the EBC wort turbidity is not improved for the variety D (2015)

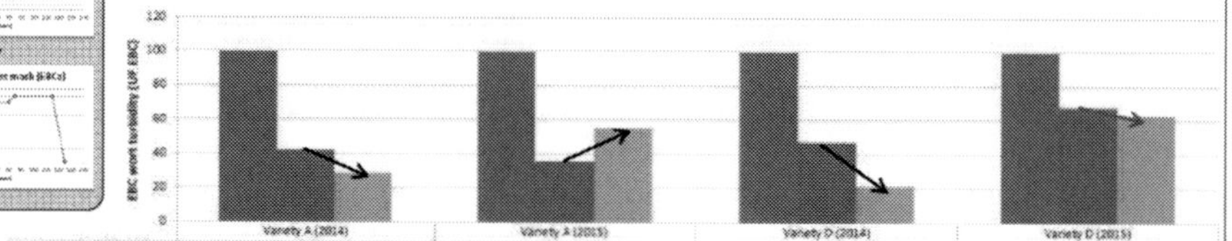

	Variety A (2014)	Variety A (2015)	Variety D (2014)	Variety D (2015)
Moyenne de EBC turbidity (=100% basis)	100	100	100	100
Moyenne de EBCm / EBC (%)	42	36	47	68
Moyenne de EBCa / EBC (%)	28	55	21	63
Decrease of turbidity from EBCm to EBCa	13%	↗ 19%	26%	5%

Optimum temperature and pH of amylolytic malt enzymes

iFBM

Amylolytic enzymes	pH	Temperature °C	
	optimum	optimum	inactivation
α-amylase	5.6-5.8	70-72	75
β-amylase	5.4-5.6	62-64	68
Dextrinase limite	5.4	60-65	68

Mashing regime

iFBM

⇨ **EBC Mashing regime should be optimized**

⇨For big starch granules gelatinization

⇨Optimum temperature and pH for amylolytic enzymes actions

iFBM

Identification of compounds

PCA correlation coefficients

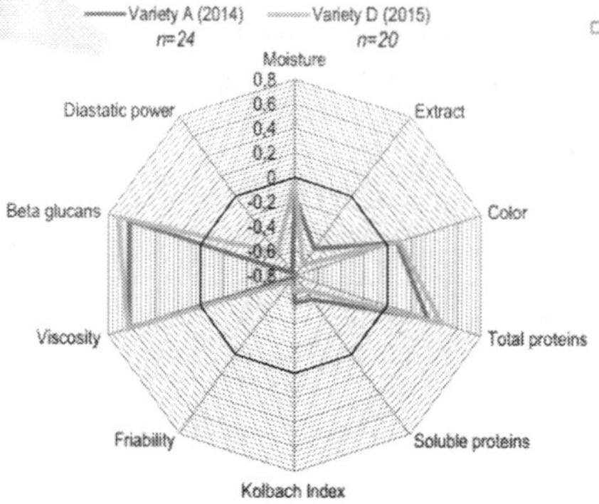

- ⇨ **The principal component analysis (PCA) shows that**
  - ⇨ Diastatic power is strongly negatively correlated to the turbidity of the **variety A (2014)**
  - ⇨ Turbidity from **variety D (2015)** seems to be explained mostly by the presence of proteins and beta glucans

iFBM

Identification of compounds

- ⇨ **To conclude**
  - ⇨ **Compounds** involved in the EBC wort turbidity **may not be similar according to barley variety** (protein- or starch-based…)
    - Differences between varieties A and D
  - ⇨ Presence of starch may explain part of the EBC wort turbidity as amylolytic activities are not well promoted in this type of diagram
    - BUT in some cases, this **starch-based turbidity could be removed by using an adapted mashing diagram** (rest around 63°C)

iFBM

CORRELATION BETWEEN EBC CONGRESS WORT HAZE AND BEER PROCESS & QUALITY

iFBM

Semi-industrial wort and beer production

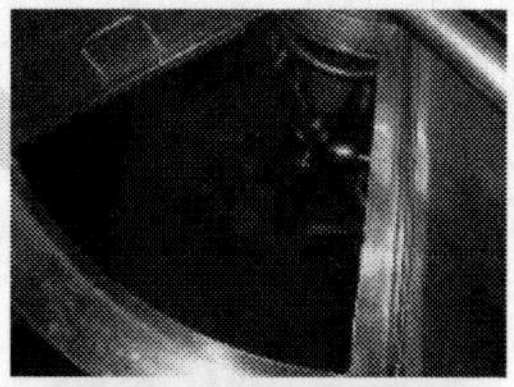

20 HL Brewing plant

iFBM

Turbidity of wort

⇨ **Information of IFBM data-base :**

⇨ Pilot trials from CBMO

- 1 malt batch => 1 brew => 1 beer
- 150 malt batches > 150 beers
- Mashing regime

Malt grits = 300 kg
Water = 950 kg
pH = 5.4+/- 0.1

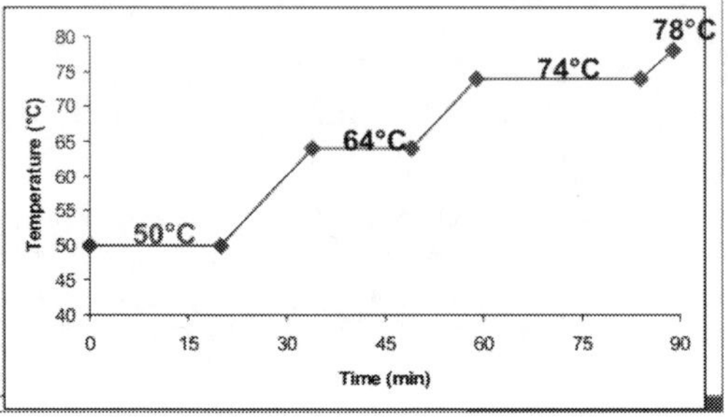

iFBM

Congress wort turbidity and final haze of beer

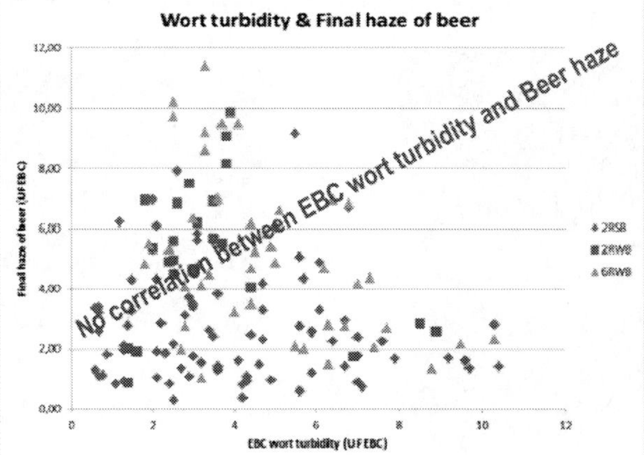

iFBM

wort turbidity and beer process

⇨ No relation with mash filtration

⇨ No relation with cold wort turbidity

⇨ No relation with beer filtration

iFBM

Take home message

⇨ **Turbidity analysis on EBC wort is not a good indicator to predict the turbidity of an industrial wort and beer**

⇨ **If still this parameter on some brewers specifications**

- ⇨ Impact of barley variety :
  - Nature of turbidity can be different according to variety
- ⇨ Malting process can influence EBC wort turbidity
  - Poor modification of the barley leads to turbidity
    - Optimization of steeping regime
    - Ga_3 if allowed by national regulation & brewers

iFBM

Take home message

Modification of mashing regime and mash filtration to be closed to industrial mashing regime

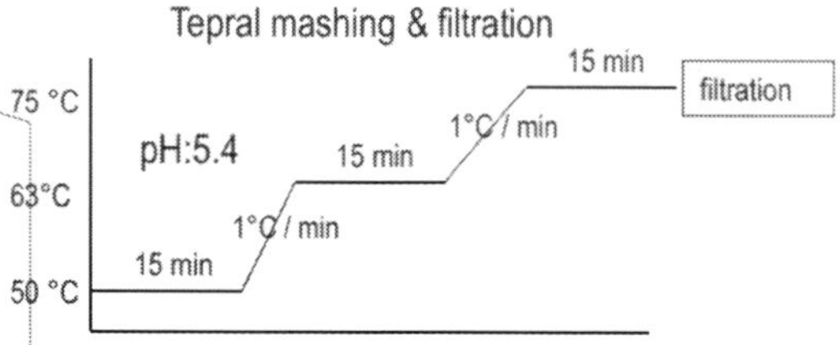

filtration under pressure
sparging with water at 75°c

iFBM

Thank you for your attention !

with the support of :

MALTEURS DE FRANCE

BRASSEURS DE FRANCE

Technische Universität München

TUM School of Life Sciences Weihenstephan

Lehrstuhl für Brau- und Getränketechnologie
Univ.-Prof. Dr.-Ing. Thomas Becker

Using pulsatile target jets to clean spent grain loaded filter cloths

36th CONGRESS EBC 2017 EUROPEAN BREWERY CONVENTION 14-18 MAY LJUBLJANA SLOVENIA

Werner, R.
Geier, U.
Becker, T.

Ljubljana, 17. May 2017

Introduction - motivation

Reduction of possible risks of non optimized cleaning
Cross-contamination, less product-safety, longer system down-times, decreased filter performance

Cleaning of rough surfaces and complex residues
Heterogeneous soil, "hard-to-clean areas", insufficient cleaning efficiency, over-dimensioned cleaning concepts

Enhancement of filter cloth cleaning
Material, construction, pore size, residue removal, area of application

Fluid mechanics of wash jets - interaction - forces
Adhesion, cohesion, wash jet profile, adhesive forces, physio-chemical effects, fouling, backflow

Economical and ecological issues
Detergent reduction, shorter cleaning intervals, less waste water

Introduction - mash filter systems

System characteristics
- Filter press
- Membrane-techniques
- Large filtration surfaces
- Increased brew sequences
- Unmalted grain possible
- Poorer malt qualities possible

Cleaning difficulty
- Complex filter cloth
- Processing limits due to filter cloth material
- After cake discharge, residues remain in filter gauze

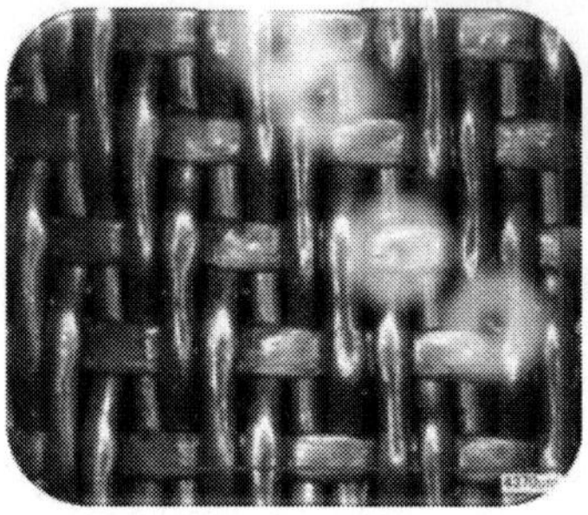

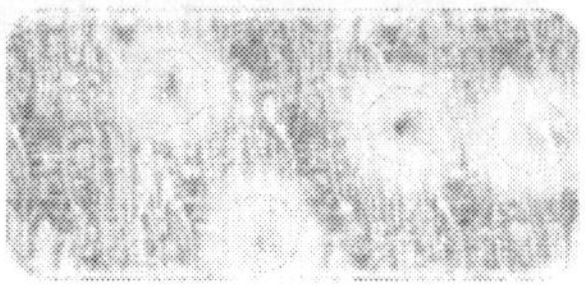

Introduction - possibilities of cleaning mash filter cloth

Why focus on jet cleaning?

Cleaning parameters
- Chemical
- Mechanical
- Temperature
- Time

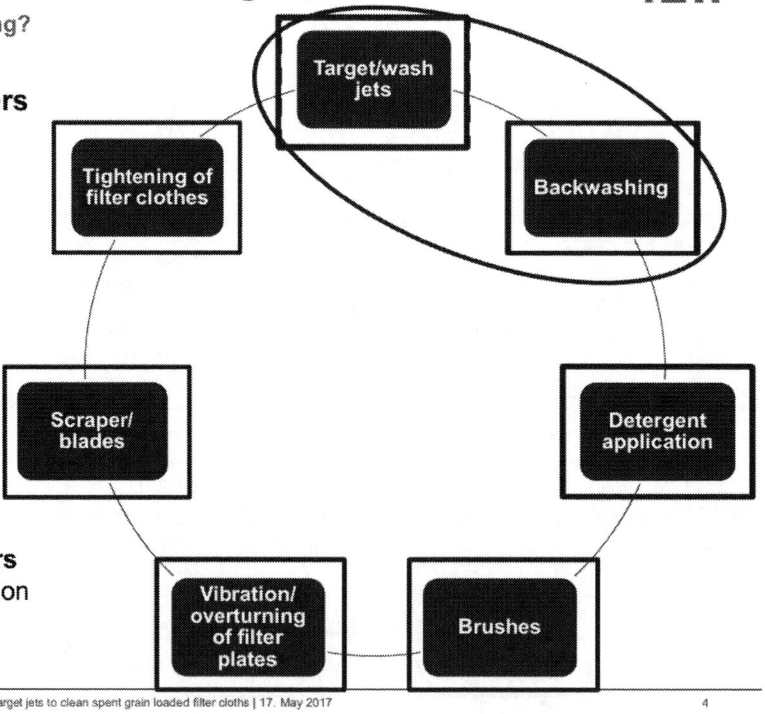

Additional parameters
- Type of contamination
- Surface condition

Theory: wash jet – fluid mechanical considerations

TUM

Wash jet can be divided into:

- Free jet
- Impinging jet
- Transport stream

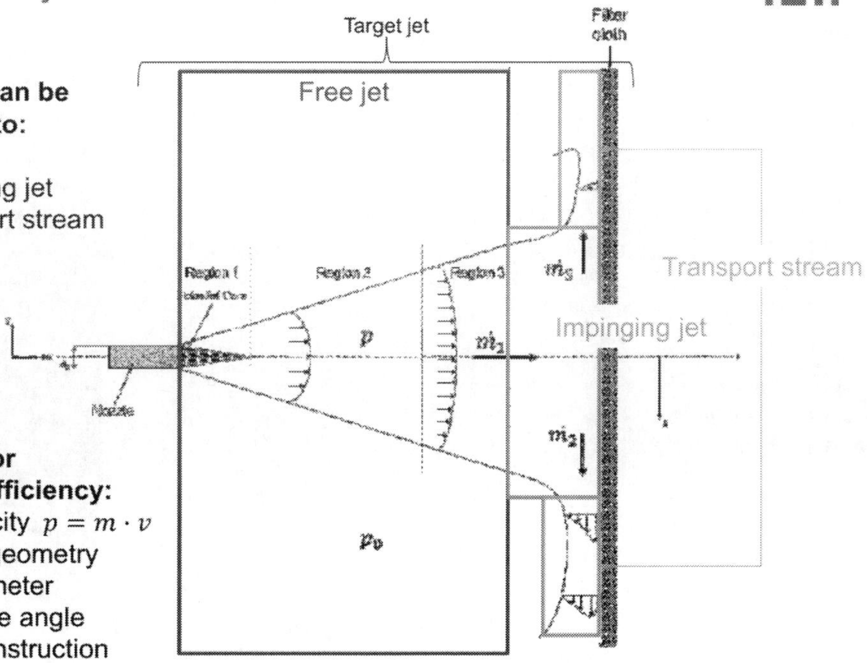

Decisive for cleaning efficiency:

- Jet velocity $p = m \cdot v$
- Nozzle geometry and diameter
- Incidence angle
- Filter construction

Theory: free jet – regions and calculation

TUM

Region 1:

- Potential core
- Length: from nozzle to potential core apex
- All parameters stay constant

Region 2:

- More droplets resulting
- Velocity decreases
- Gaussian stream profile

Region 3:

- Diffused droplet region

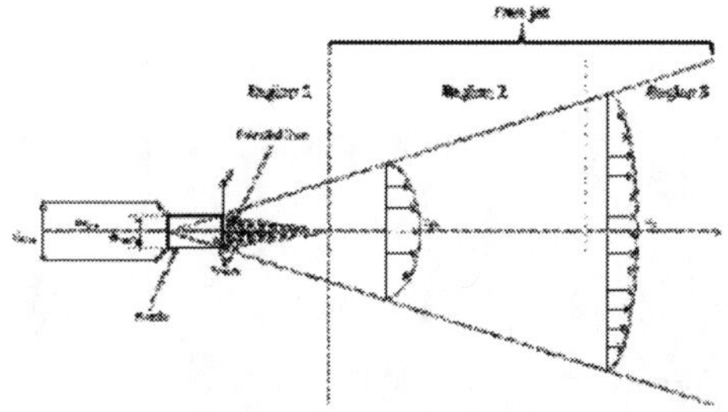

<u>Velocity region 1</u>

$$v_{Nozzle} = \frac{d_{Pipe}^2 \cdot v_{Pipe}}{d_{Nozzle}^2}$$

<u>Velocity region 2</u>

$$v_{C,2} = \frac{v_{Nozzle} \cdot c_{Core} \cdot d_{Nozzle}}{x}$$

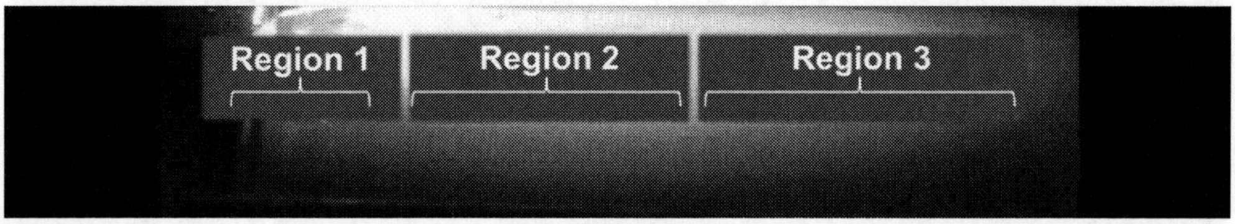

Abbreviations: d = diameter; v = velocity; c= dimensionless potential core length; x = coordinate

Theory: impinging jet and transport stream

Impinging jet:
- All entering and leaving streams are balanced
- Navier-Stokes and Euler
- Integral form describes all influences ➔ forces are determinable

$$\int_{V_c} \frac{d\rho v_j}{dt} dV + \int_{A_i} \rho v_i v_j dA_i = -\int_{A_j} p dA_j - \int_{A_i} \tau_{ij} dA_i + \int_{V_c} \rho g_j dV + \sum F_j$$

Transport stream:
Impinging jet is deviated into two streams (here two)
➔ „Backflow“

$$\varepsilon_{Fluid} = \frac{\dot{m}_2}{\dot{m}_1}$$

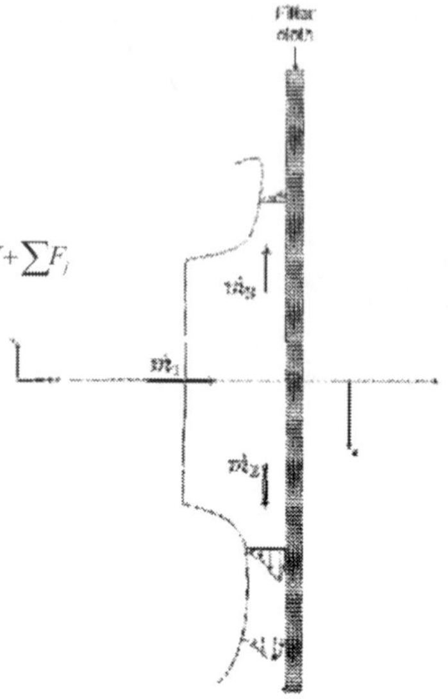

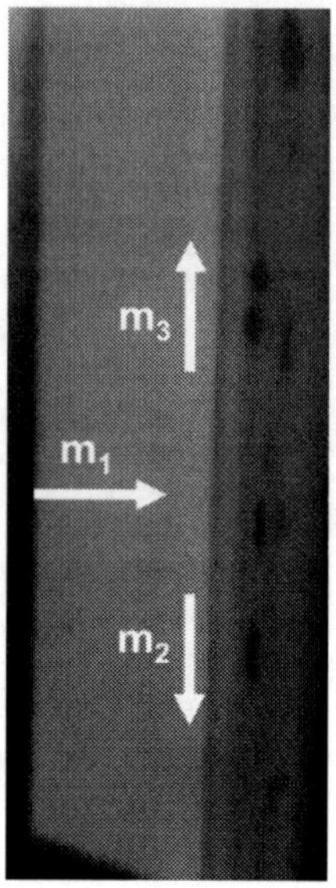

Abbreviations: ρ = density; A = surface; F = force; v = velocity; t = time; V = volume; τ = wall shear stress; g = gravitational acceleration; x = coordinate 1; y = coordinate 2

Possible process design of cleaning jets

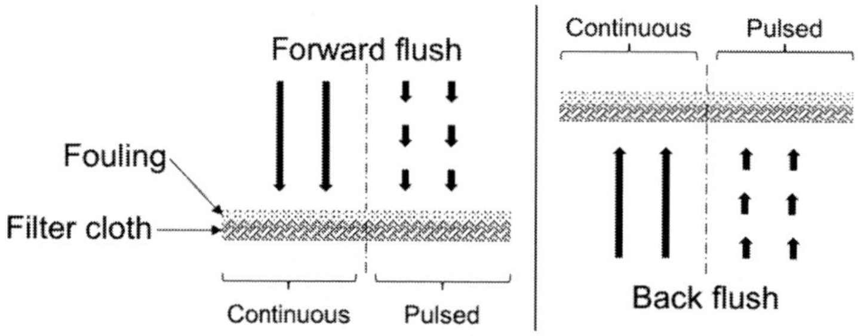

Combination of both designs as ideal solution?

• Direct contact with contamination • Full impulse discharge • Potentially soil is pressed into gauze pores	• Only indirect contact with residues • Decelerated stream due to cushioning gauze • If reached, soil is removed away from fabric

Method & material

Cleaning setup via stream cell

- Laboratory scale
- Undyed contamination ➔ new method
- Pulsatile jets via solenoid valve
- Diameter nozzle: 2 mm
- Variable distances nozzle to filter
- Nozzle angle variable (43 ° and 70 °)
- Cleaning fluid: $H2O$, 20 °C

Residue analysis via digital microscopy

- Defined mash residues
- Static image acquisition
- Magnification: 20 – 5000
- Sharp depth image
- 3D-function
- Image based analysis
- Cleaning degree G (as decisive parameter)

Source: Keyence

$$G[-] = 1 - \frac{A_{After}}{A_{Before}}$$

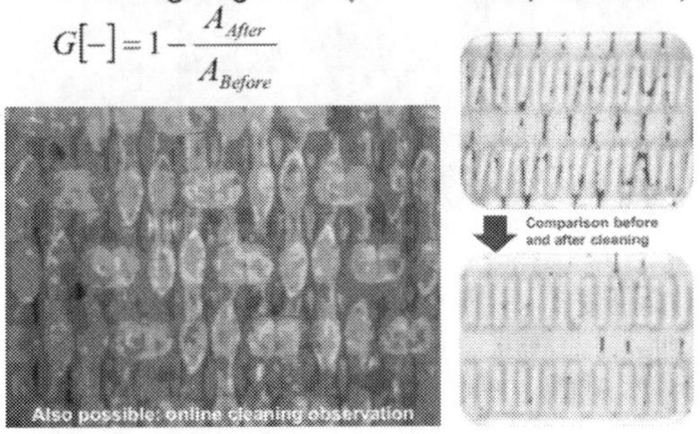

Results: pause length variation

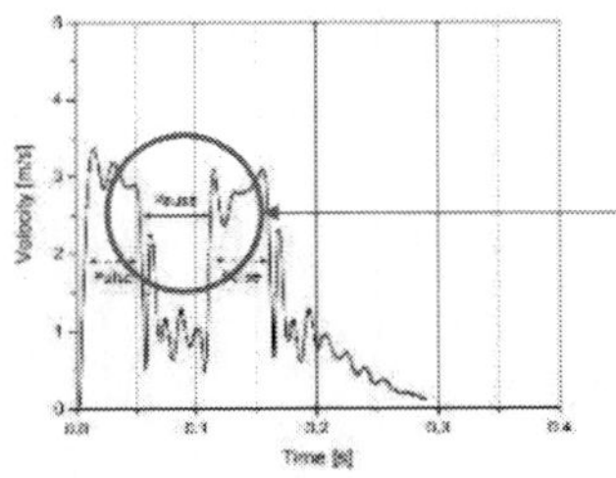

$$f_{Pulse} = \frac{1}{t_{Pulse} + t_{Pause}}$$

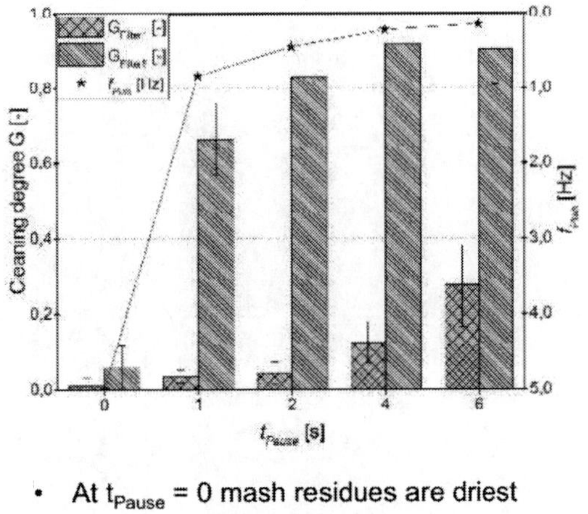

- Results show: pulse pause decisive (due to flow off of prior pulse)
- Frequency depends on pause length
- Extensive pauses are uneconomic!
- At t_{Pause} = 0 mash residues are driest
- Filter 5 is easier cleanable significantly
- Calculated frequency correlates with results

Results: jet velocity and incidence angle variation

TUM

Incidence angle 43 °

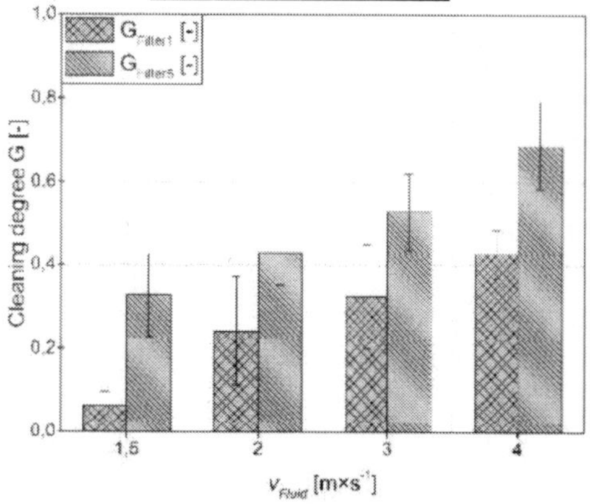

Incidence angle 70 °

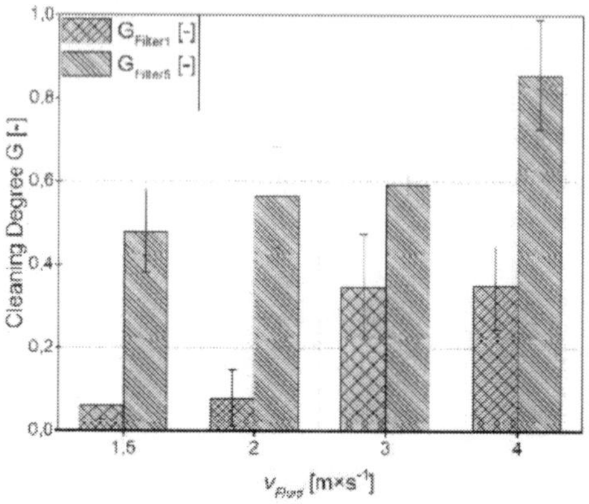

Code	Product	Mesh aperture	Material
Filter 1	05-1001-K 120 (Sefar)	80 & 100 µm	polypropylene
Filter 5	03-1001-SK 066 (Sefar)	50 µm	Polyamide 6.6

- Filter 5 easier cleanable due to smoother surface
- Higher stream velocities result in higher cleaning degrees
- $t_{Pause} = t_{Pulse} = 100$ ms

Difference between sample filters

TUM

Filter type	Top view	Lateral view	Surface roughness & construction/gauze
Filter 1 • 05-1001-K 120 (Sefar) • 80 & 100 µm • Polypropylene • Plain weave	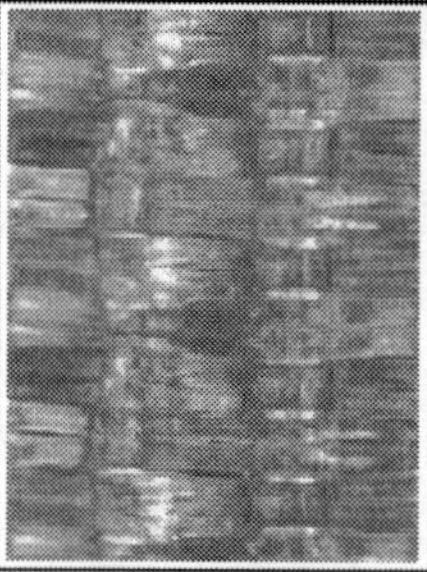	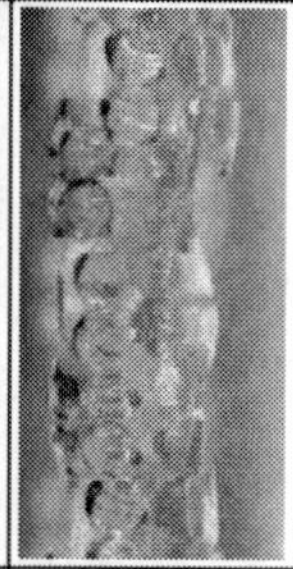	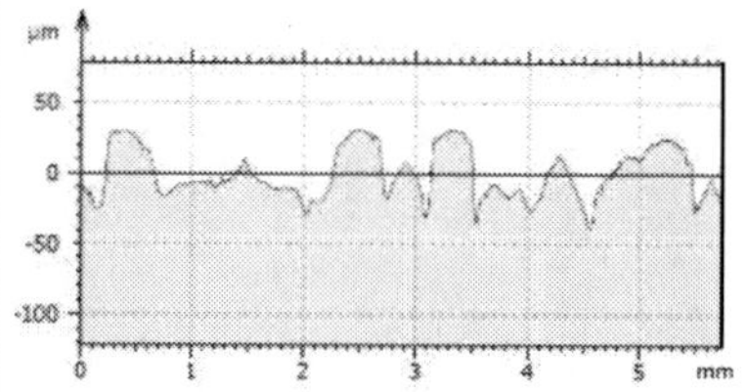
Filter 5 • 03-1001-SK 066 (Sefar) • 50 µm • Polyamide 6.6 • Twill weave	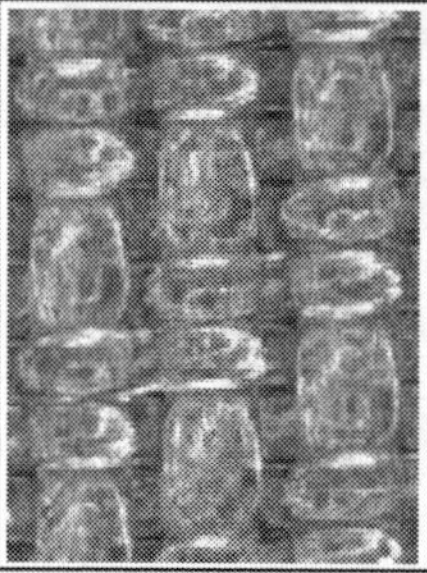	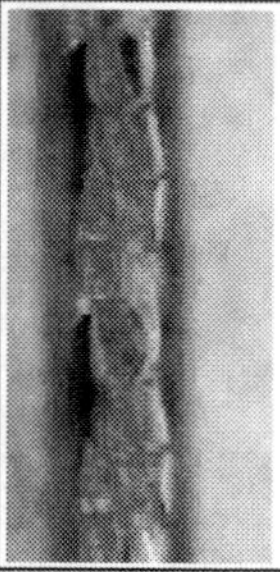	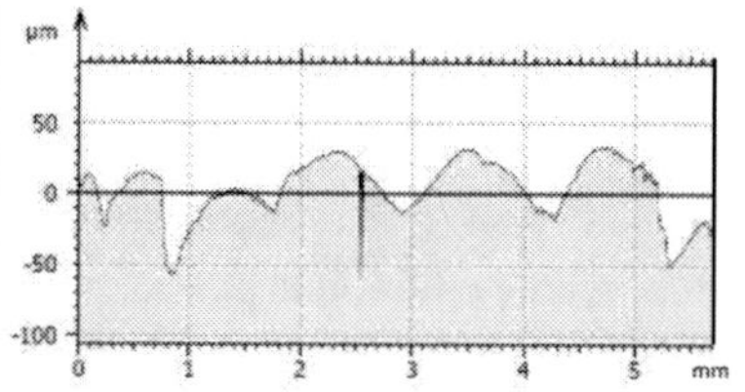

Results: pulsatile vs. continuous forward flush

Incidence angle 43 °

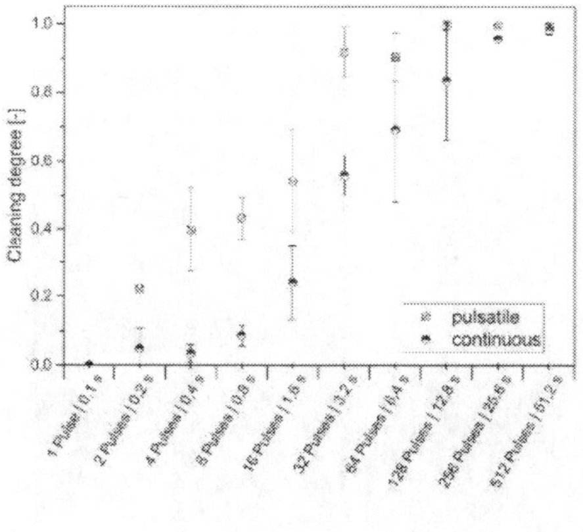

Incidence angle 70 °

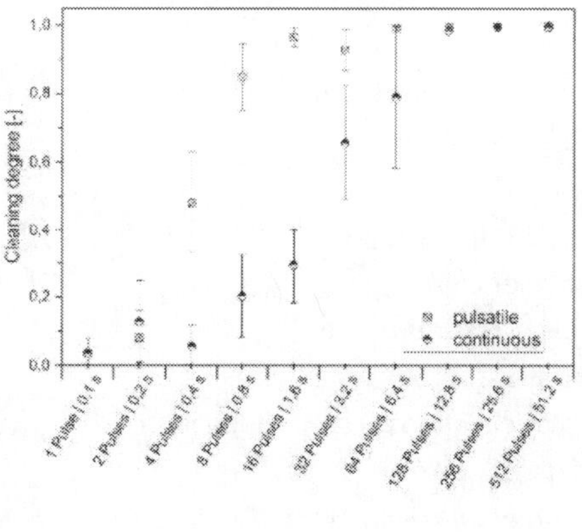

- Filter 1: 05-1001-K 120 (Sefar) - aperture size: 80 & 100 µm - polypropylene
- Pulsatile jets result in higher cleaning degrees
- Incidence angle of 70 ° is more effective due to higher impulses on residues

Characterization of pulsatile jets

Fluid mechanical enhancement
- Increased wall shear stress
- Impulse conservation
- Dryer filter surfaces enhance following impulse

Disadvantages:
- Additional installations necessary
- Potentially higher equipment costs

Advantages (Economic & Ecologic):
- Reduced detergents
- Shorter cleaning intervals
- Higher cleaning degrees

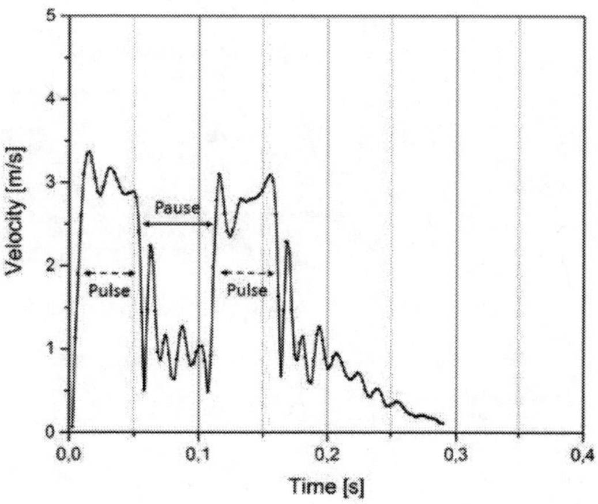

Conclusions and outlook

TUM

- Filter cloth cleaning depends on…
  - Incidence angle
  - Filter construction and material
  - Cleaning velocity
- Pulsatile jets are advantageous due to …
  - higher efficiency
  - smaller cleaning intervals
  - Reduced detergent
  - Easy adjustability
- Next investigations: comparison with backflush cleaning
- Insertion of image analysis and pulsatile jet cleaning in demand-oriented cleaning units for filter presses

General:
- Temperature
- Chemistry/detergent
- Cleaning time
- Mechanical effects
- Soil composition
- Surface

Filter media parameter:
- Material
- Construction
- Finish
- Product composition
- Detergent
- Costs (systems, detergents, life span of cloths)
- Technical requirements

Wash jet:
- Velocity
- Process type (pulsatile, continuous)
- Incidence angle
- Fluid pressure
- Nozzle geometry
- Nozzle distance
- Surrounding fluid
- Backflow

Detergent cleaning:
- Type and amount of soil
- Detergent type
- Concentration
- Surface properties
- Surface Material
- Economy and costs
- Ecologic aspects

TUM

Thank you ...

... for your attention

... a project of **industrial community research (IGF)**

sponsored by / via

iGF Bundesministerium für Wirtschaft und Energie AiF IVLV

The IGF Project (AiF 17607 N) of the IVLV is supported via AiF within the program for promoting the Industrial Collective Research (IGF) of the German Ministry of Economic Affairs and Energy (BMWi), based on a resolution of the German Parliament.

Institute of Brewing and Beverage Technology
TUM School of Life and Food Science Weihenstephan
Technical University of Munich
Weihenstephaner Steig 20
D-85354 Freising

Tel.: +49 8161 71 3261
Fax: +49 8161 71 3883
E-Mail: verwaltung@bgt.wzw.tum.de
lbgt.wzw.tum.de

Appendix

Theory – mash/spent grain as complex "residue contamination"

- Mash or spent grain residues exists of diverse compounds
- Composition varies ➔ contamination are „unpredictable"
- Different soil types are distinguishable due to water solubility

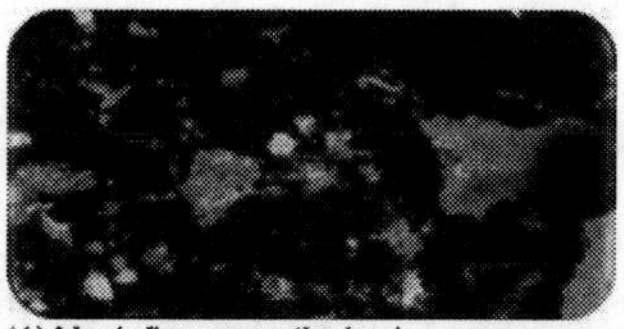
(1) Mash fluorescently dyed

Water solubility	Example
Soluble	Acids, salts, low-molecular carbohydrates
Acid-soluble	Fur, boiler scale
Swellable	Proteins, higher-molecular carbohydrates
Emulsifiable	Fats, oils
Suspendable	Particles

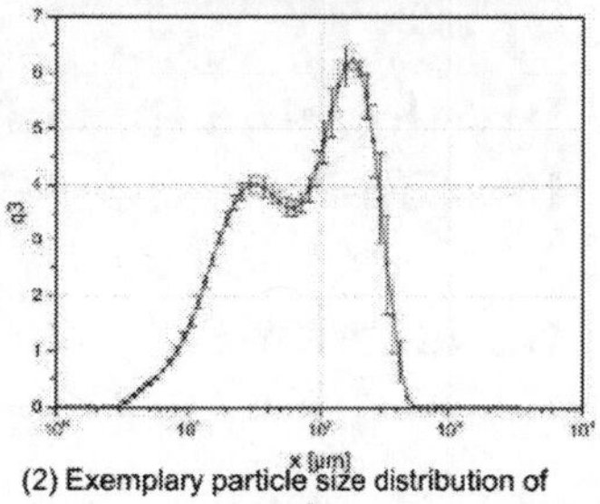

(2) Exemplary particle size distribution of mash

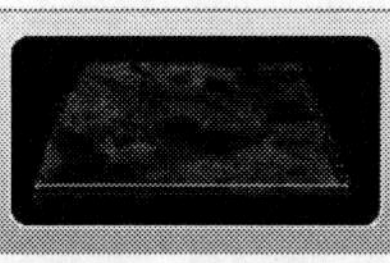

Definition of „fouling"
Permanent deposits of molecules, colloids and/or particles on membranes/clothes or their pores.

Dry Hopping And Its Effects On Beer Bitterness, The IBU Test, Beer Foam, And pH

By Dr. John Paul Maye

Technical Director

Hopsteiner

Topics

Hopsteiner

Major and Minor Hop Acids In Hops

Humulinones & Their Concentrations In Hops and Hop Pellets and Dry Hopped Beers

Dry Hopping Experiments: Effects on Beer Bitterness, the IBU Test, Beer Foam, and pH

Conclusion

Major and Minor Hop Acids In Hop Acids

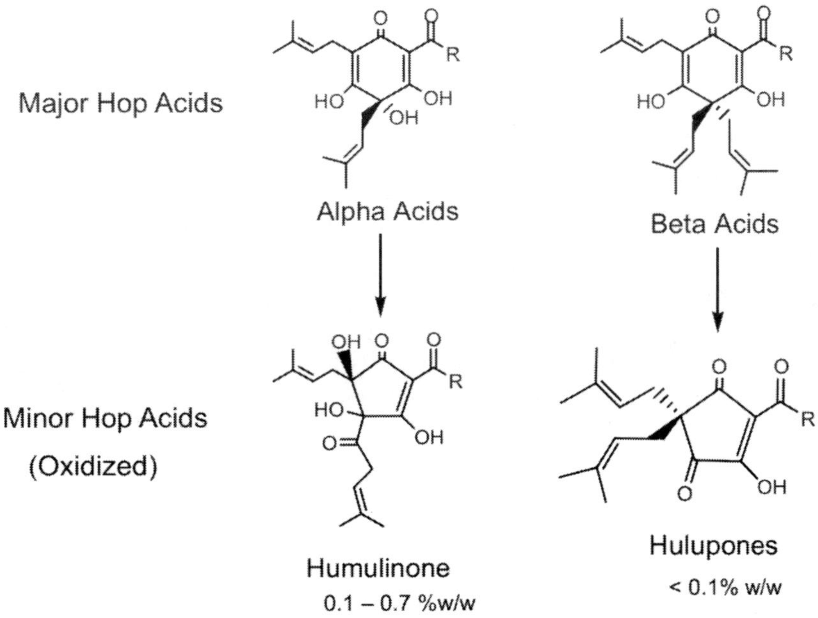

HPLC Trace of Hops

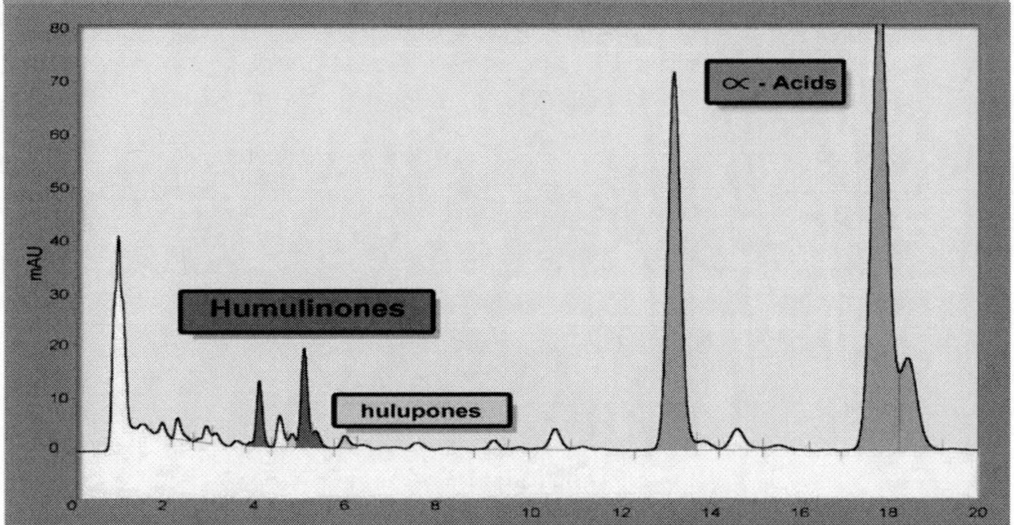

Preparation of Dicyclohexylamine Humulinone
A HPLC Calibration Standard For Humulinones

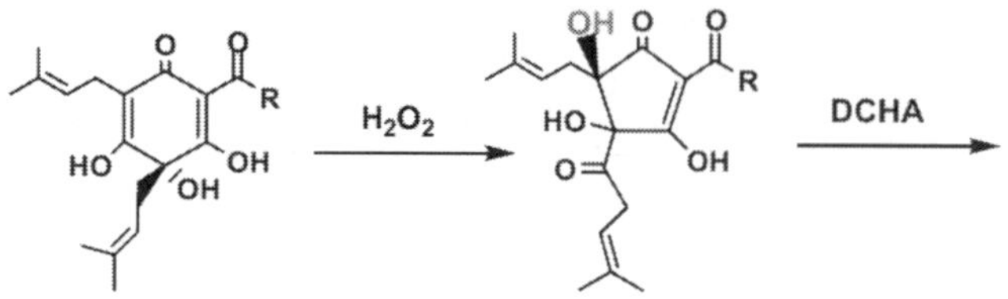

DCHA-Humulinone

64.5% Humulinone
(Organic Salt)

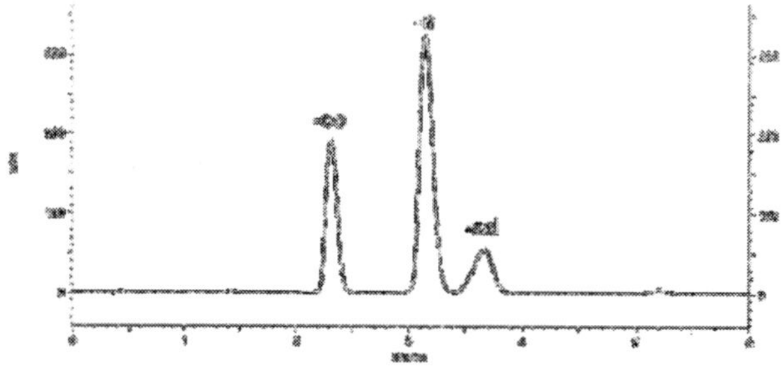

Humulinone Formation in Zeus Hop Pellets vs Temp

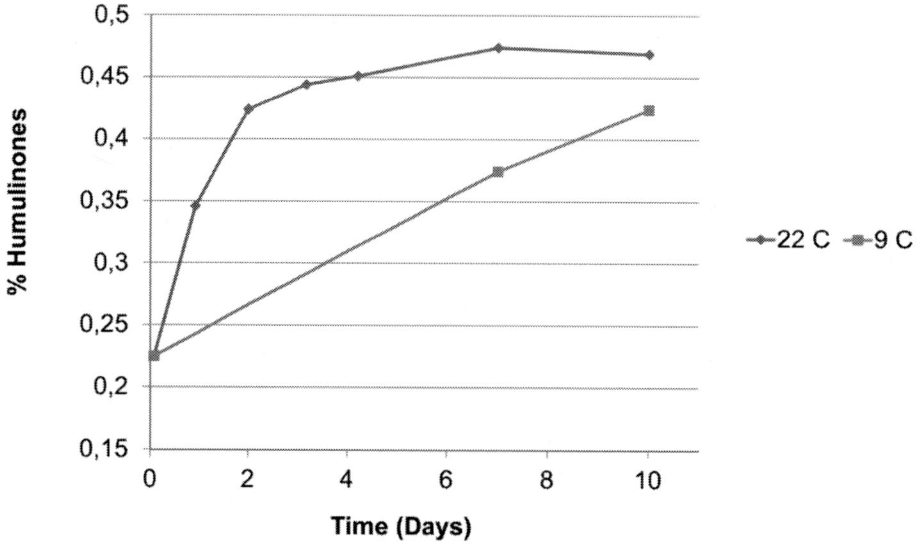

Hopsteiner

Humulinone Concentration vs HSI

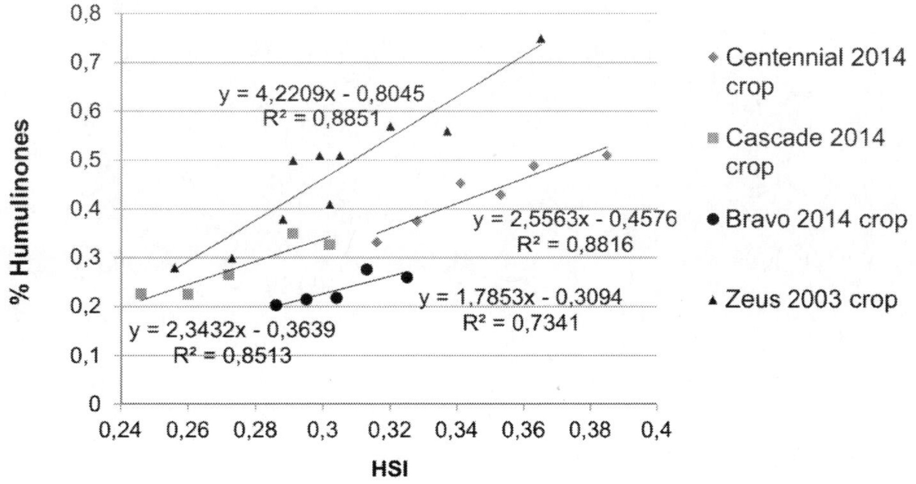

HPLC Trace of Hop Acids In Beer

Hopsteiner

Humulinone Concentration In Commercial Beers Labeled IPA's

We analyzed 30 commercial IPA's by HPLC and found the humulinone concentration to range from 3 ppm to 33 ppm.

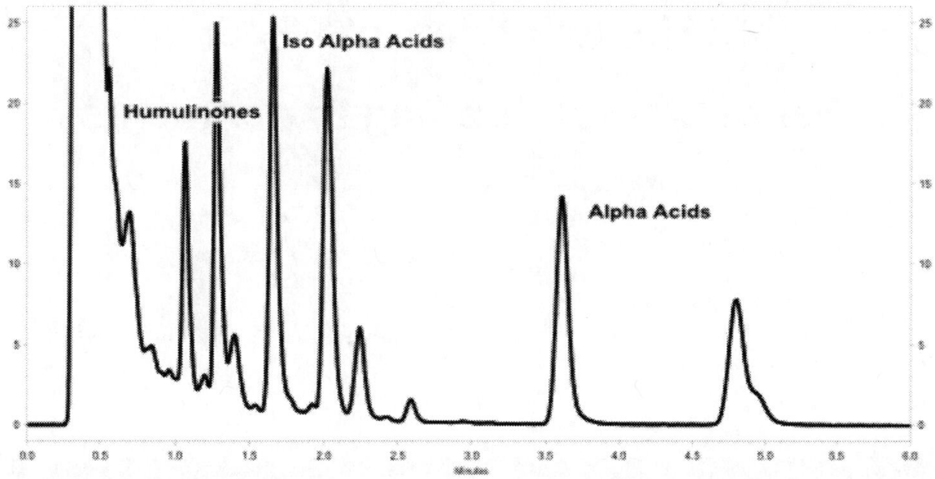

Hopsteiner

Humulinones Are Extremely Similar In Molecular Structure to Iso-Alpha Acids

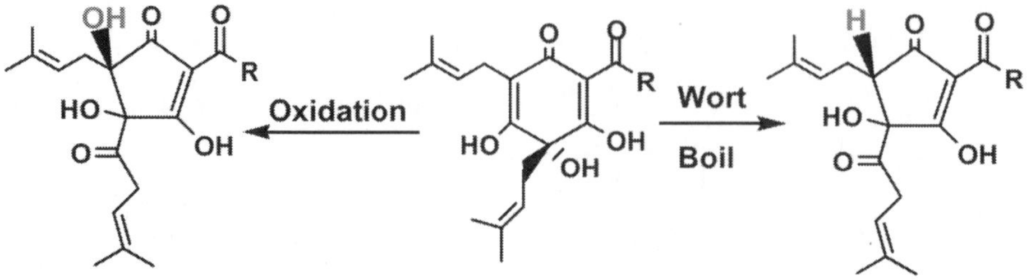

Humulinone — 1 ppm ~ 0.66 IBU

Alpha Acids

Iso-alpha Acids — 1 ppm ~ 1 IBU

Humulinones are more polar than isoalpha acids and therefore should be more beer soluble than isoalpha acids. Also humulinones have been reported to be 66% as bitter as isoalpha acids.

Hopsteiner

To Better Understand The Incorporation Of Humulinones Into Beer A Series of Dry Hopping Experiments Were Performed

Hopsteiner

Experiment #1 : Humulinone Utilization

Conditions:

Centennial hop pellets assayed 0.35 w/w% Humulinone the day of dry hopping.

Dose Rate: 0, 0.5, 1.0, and 2.0 lbs/barrel

Hop pellets were simply dumped on top of the beer.

Beer Type: Low IBU and High IBU

Temperature of dry hopping, 16 °C

Contact Time: 5 Days

Hopsteiner

Centennial hop pellets assaying 0.35% Humulinone were added to a low and high IBU beer and dry hopped for 5 days

Sample	lbs pellets Barrel of beer	ppm of Humulinone in beer	% Utilization Humulinone	ppm of Iso-a-acid in beer
Low IBU Beer (8.6)	0	0.8	-	8.6
	0.5	8	98	8.1
	1.0	14	91	7.9
	2.0	28	88	7.5
High IBU Beer (48)	0	1	-	48
	0.5	8	98	39
	1.0	14	91	35
	2.0	27	87	30

*Does this loss in isoalpha acids effect the beers bitterness?

Hopsteiner

Question #2:

How fast does humulinone dissolve into beer?

How fast are isoalpha acids lost?

What else dissolved into beer when one dry hops?

How do these changes effect beer bitterness?

Hopsteiner

Experiment #2 : What Dissolves Into Beer and How Fast?

Cascade Hop Pellets

Compound	Concentration, %
Alpha acids (ICE-3)	5.6
Beta acids (ICE-3)	5.8
Humulinones (DCHA humulinones, 65.9%)	0.26
Xanthohumol (99.7% pure, in-house std)	0.26
Hulupones (DCHA hulupones, 67.0%)	0.05

54 IBU All Malt Beer

Dry Hop Dose Rate: 0, 0.5, 1.0, 2.0 lbs per barrel

Contact Time: 1, 2, 5 days

Dry Hop Temperature: 16 °C

All Results Reported by HPLC

Day 1

Hop Acid Concentration After 24 hours of Dry Hopping

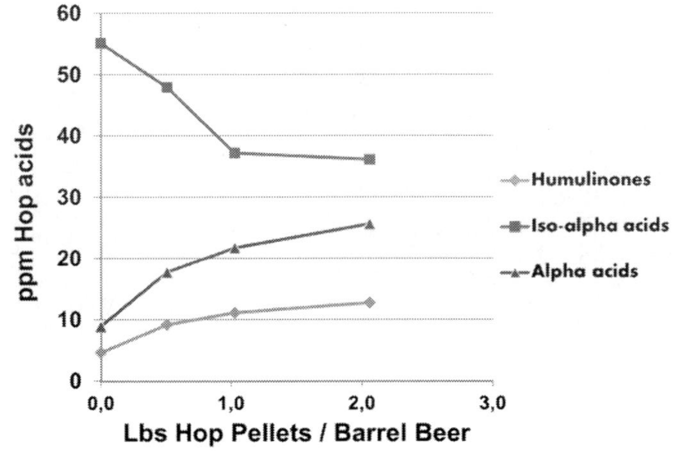

<u>1 lb/barrel (Day 1)</u>

IAA 37 ppm
AA 21 ppm
Hum 11 ppm

Day 2

Hop Acid Concentration After 2 Days of Dry Hopping

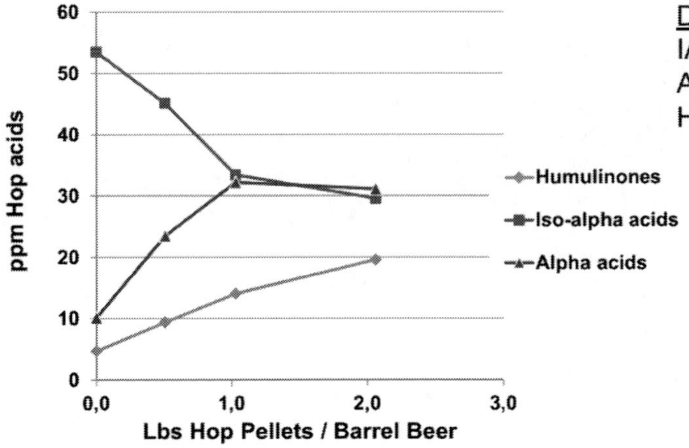

1 lb/barrel

Day 2	(Day 1)
IAA 32 ppm	(37 ppm)
AA 31 ppm	(21 ppm)
Hum 15 ppm	(11 ppm)

Hop Acid Concentration After 5 Days of Dry Hopping

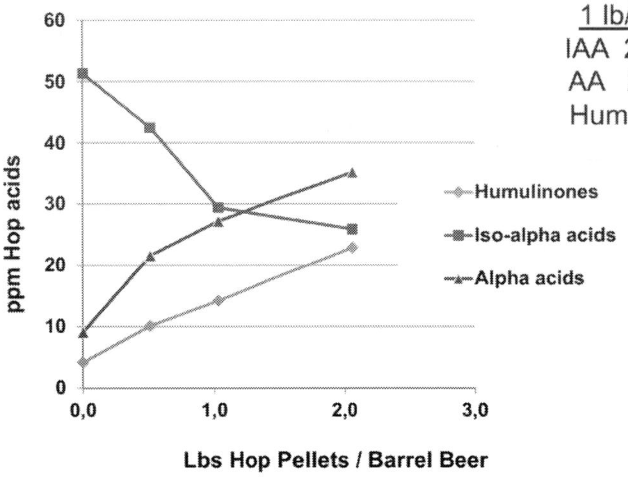

Day 5

1 lb/barrel		(Day 2)	(Day 1)
IAA	29 ppm	(32 ppm)	(37 ppm)
AA	28 ppm	(31 ppm)	(21 ppm)
Hum	15ppm	(15 ppm)	(11 ppm)

HPLC Analysis of Beers Dry Hopped for 5 Days with Cascade Hop Pellets

Lbs Hops per BBL	PPM Iso-α-acid	PPM Alpha Acids	PPM Hum	PPM XN	% Utilization Iso-α	% Utilization α-acid	% Utilization Hum	% Utilization XN
0	51.3	9	4.2	0.24	-	-	-	-
0.5	42.5	21.6	10	0.62	- 17%	11%	100%	12%
1.0	29.4	27	14	0.79	- 43%	8%	97%	8%
2.0	25.9	35	23	0.79	- 49%	6%	90%	4%

Hum = Humulinone & XN = Xanthohumol

Note: No Beta Acids or hulupones were seen in any of the dry hopped beers.

Hopsteiner

How Does This Change In Hop Acid Composition Effect Beer Bitterness?

Lbs Hops per BBL	PPM of Iso-a-acid	PPM of Humulinone	PPM Alpha Acids	Calculated Bitterness*
0	51.3	4.2	9	55.0
0.5	42.5	10	21.6	51.2
1.0	29.4	14	27	41.3
2.0	25.9	23	35	44.6

*** Calculated Bitterness = ppm IAA + (0.66 x ppm Humulinone) + (0.1 x ppm Alpha Acids)**

Hopsteiner

Experiment #3: Extreme Dry Hopping

What Happens when you dry hop with 3, 4 or 6 lbs of hops per barrel of beer?

<u>Conditions:</u>

Cascade hop pellets assayed 5.7% alpha acids, 5.5% Beta Acids and 0. 23w/w% Humulinone the day of dry hopping.

Dose Rate: 0, 1, 2, 3, 4, and 6 lbs/barrel

Hop pellets were simply dumped on top of the beer.

Beer Type: 42 ppm Isoalpha Acid Beer by HPLC (High IBU)

Temperature of dry hopping, 16 ^{o}C

Contact Time: 3 Days

Extreme Dry Hopping With Cascade Hop Pellets

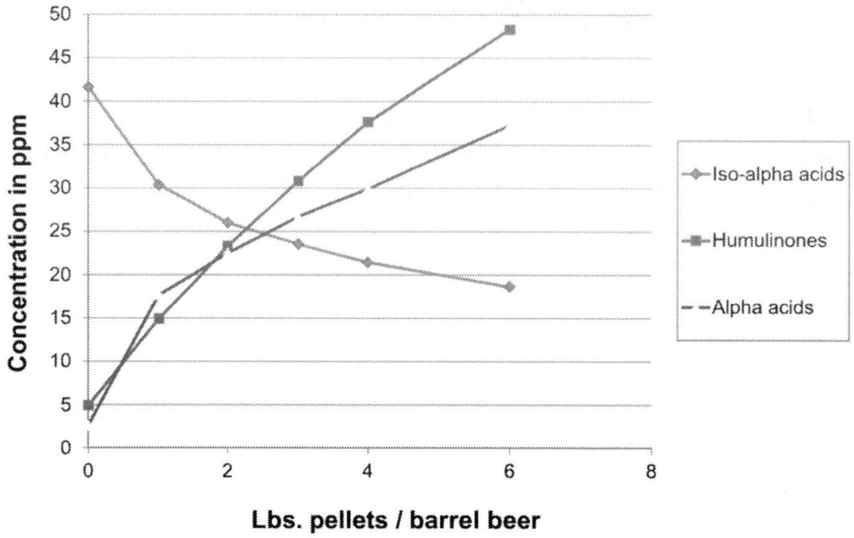

Spectro IBU vs Calculated Bitterenss

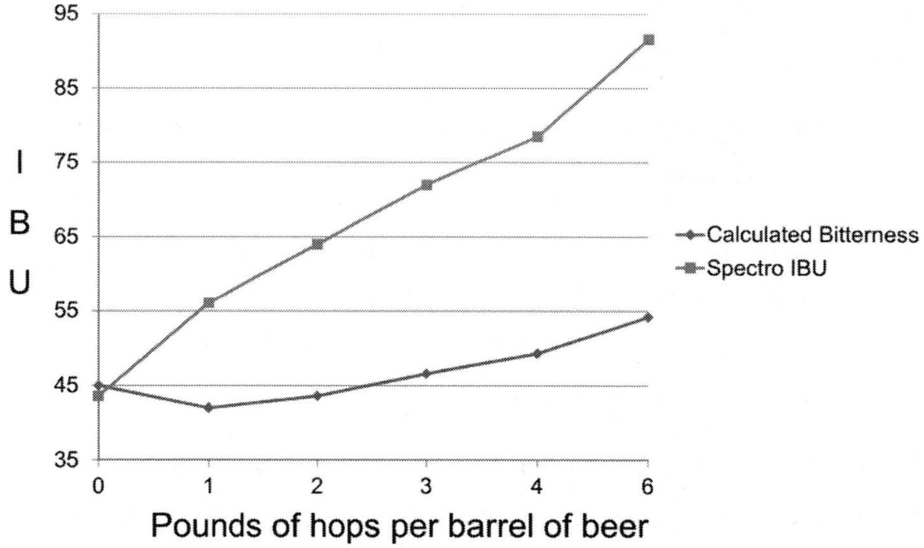

Calculated Bitterness = ppm IAA + 0.66xppm Hum + 0.1xppm AA

Hopsteiner

Why Dry Hopping Effects the International Bitterness Units Test

Hopsteiner

The IBU Test Method Was Developed to Measure Only Isoalpha Acids In Beer

The IBU Method was developed to measure isoalpha acids in beer and did not take into account dry hopped beers that contains high concentrations of humulinones and alpha acids (and polyphenols).

Add 10 mL beer + 20 mL isooctane + 1 drop Octyl alcohol + 1 mL 3N HCl to 50 mL centrifuge tube, shake 15 minutes, centrifuge, transfer upper clear isooctane layer to cuvette, read absorbance at 275 nm.

$$BU = absorbance_{275} \times 50$$

UV Spectra Under IBU Conditions

Hopsteiner

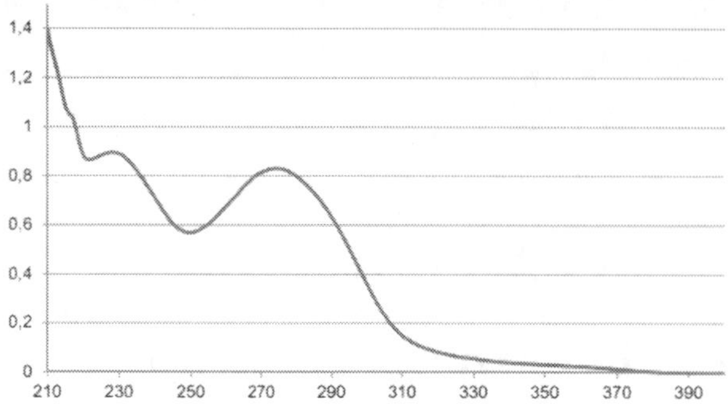

IBU Spectra of Control Beer

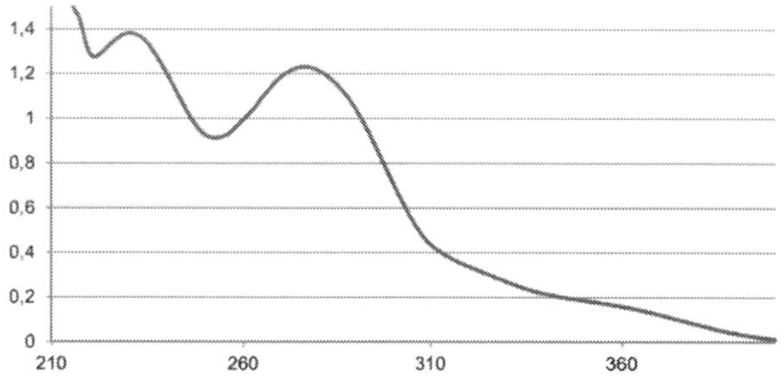

IBU Spectra of Dry Hopped Beer 2 lb/BBL Cascade

IBU spectra of IAA, HUM, and AA – All Absorb at 275 nm

Hopsteiner

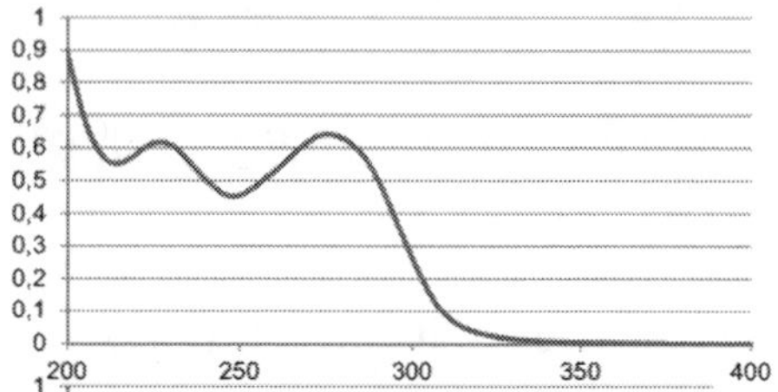

IBU Spectra of Isoalpha Acids

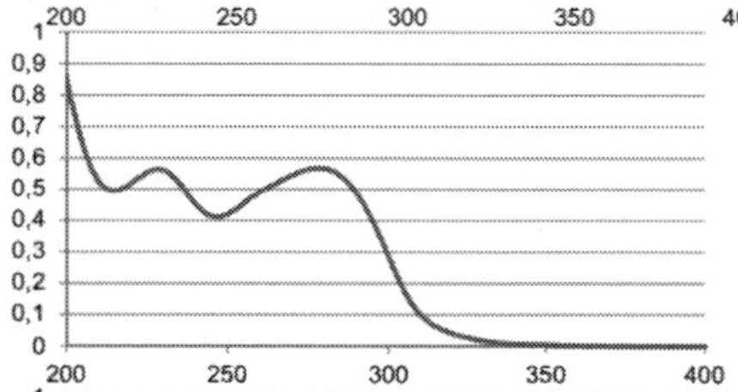

IBU Spectra of Humulinones

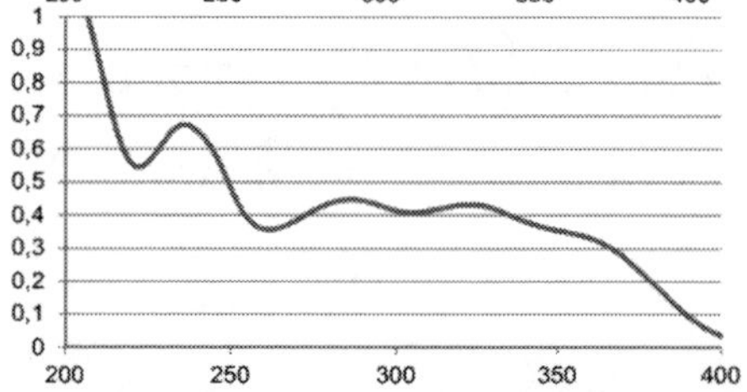

IBU Spectra of Alpha Acids

Hopsteiner

Determining The IBU Response Factor For Isoalpha Acids, Alpha Acids and Humulinones in Beer

Sample	ppm Isoalpha Acid	ppm Alpha Acids	ppm Humulinone	IBU	Δ IBU / Δ Hop Acid
32 ppm Isoalpha acid beer	32	-	-	27	0.7
32 ppm Isoalpha Acid beer + 19 ppm isoalpha acids	51	-	-	40	0.7
32 ppm IAA beer + 13 ppm Alpha Acids	32	13	-	35	0.62
32 ppm IAA beer+ 13 ppm Humulinone	32	-	13	34	0.54

Hopsteiner

The Reason The IBU Test Method Doesn't Work For Heavily Dry Hopped Beers is Because Isoalpha acids, Alpha Acids, and Humulinones Absorbs Light at 275 nm Differently And Because Each Hop Acid Has A Different Bitterness Intensity You Can Not Correlate IBU to Beer Bitterness.

	IBU Response Factor	Relative Bitterness
1 ppm Iso-Alpha Acids	0.70 IBU	1
1 ppm of Alpha Acids	0.62 IBU	0.1
1 ppm of Humulinone	0.54 IBU	0.66

Hopsteiner

Dry Hopping's & Its Effect on Beer Foam Using the Nibem Foam Stability Tester

Dose Rate vs Nibem Foam Stability

Hopsteiner

Dry Hopped 3 days at 16 °C With Cascade Hops

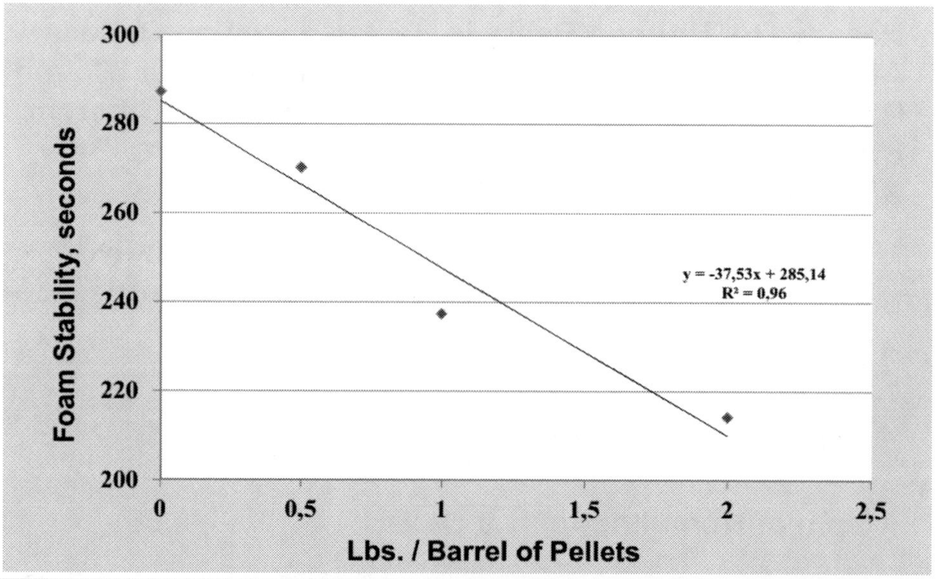

Contact Time vs Foam

Hopsteiner

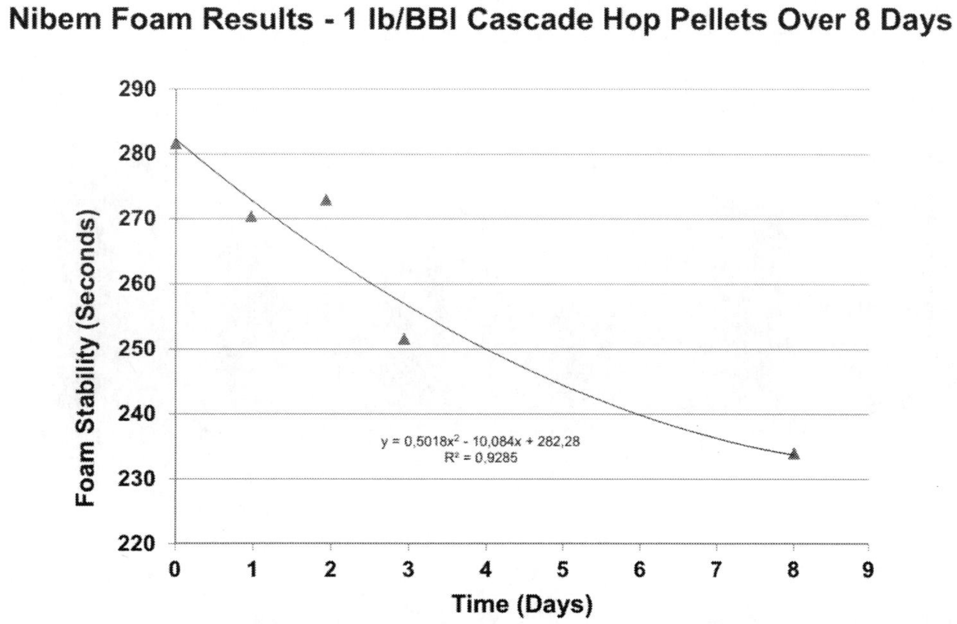

Hop Acid Composition vs Foam

Hopsteiner

Is The Reduction of Beer Foam Due to The Change In Hop Acid Composition That Occurs With Dry Hopping ?

Sample	ppm isoalpha acids	ppm alpha acids	ppm humulinone
Control	48	0	0
Control + 1lb/bbl Cascade hop pellets	30	19	17

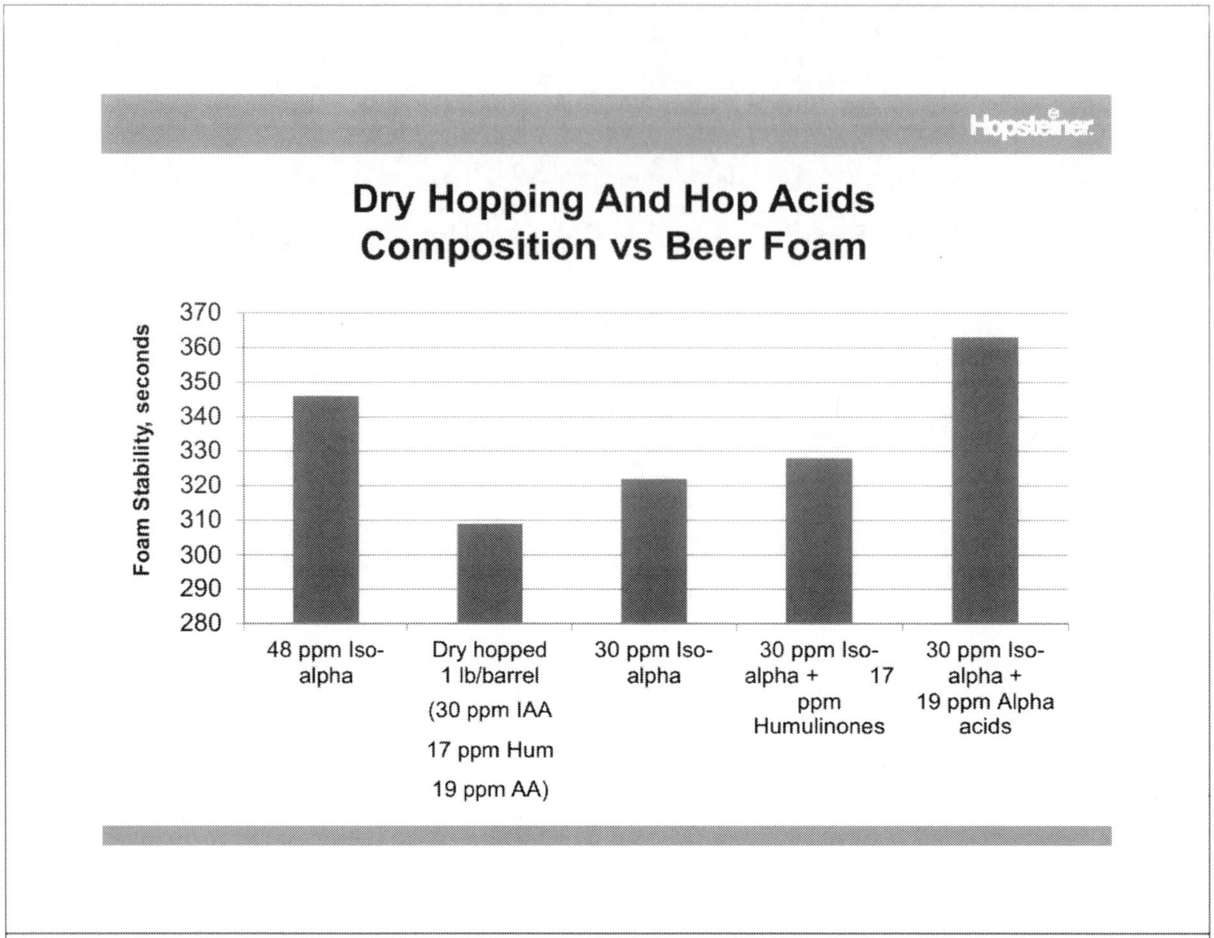
Hopsteiner
Dry Hopping And Hop Acids Composition vs Beer Foam
Foam Stability, seconds
370
360
350
340
330
320
310
300
290
280
48 ppm Iso-alpha
Dry hopped 1 lb/barrel (30 ppm IAA 17 ppm Hum 19 ppm AA)
30 ppm Iso-alpha
30 ppm Iso-alpha + 17 ppm Humulinones
30 ppm Iso-alpha + 19 ppm Alpha acids

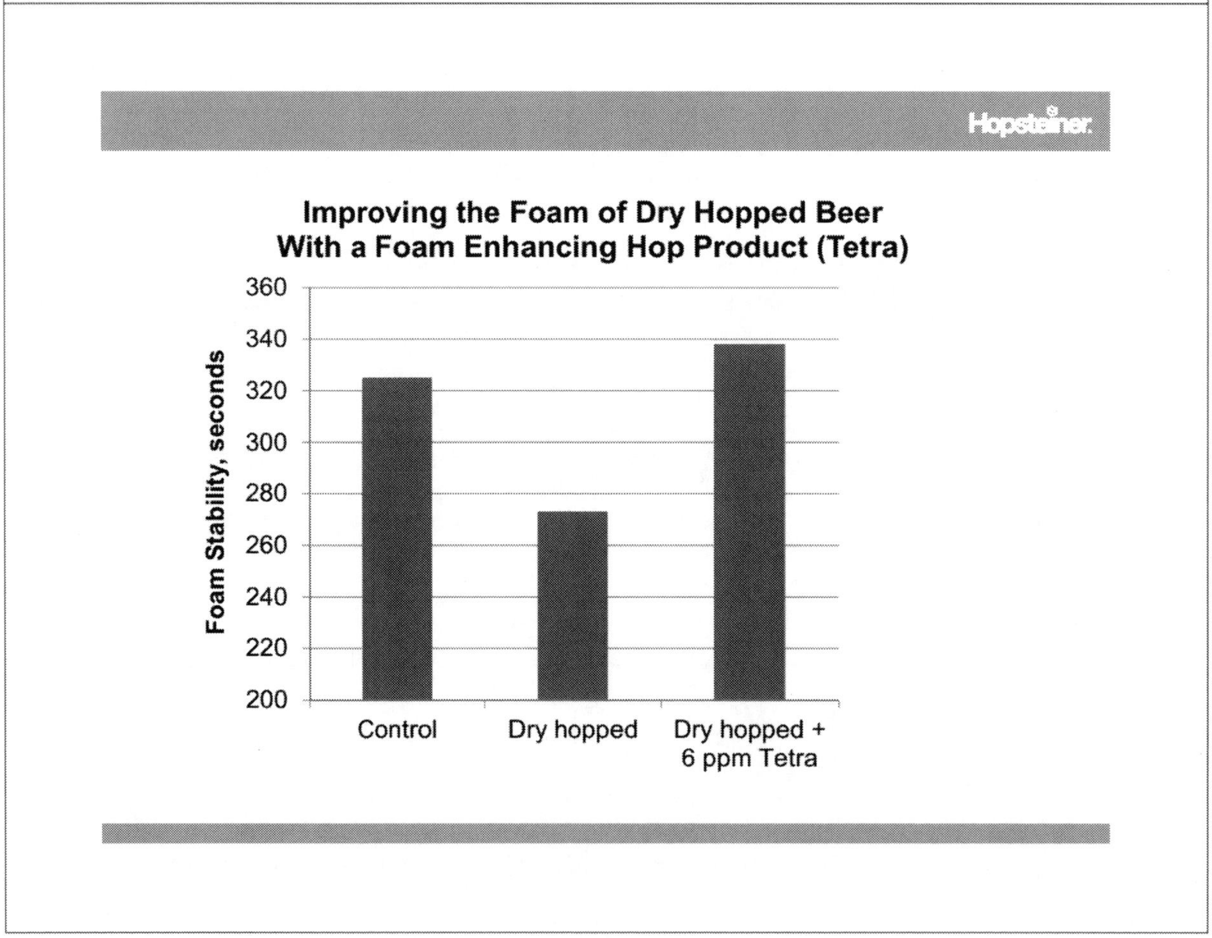
Hopsteiner
Improving the Foam of Dry Hopped Beer With a Foam Enhancing Hop Product (Tetra)
Foam Stability, seconds
360
340
320
300
280
260
240
220
200
Control
Dry hopped
Dry hopped + 6 ppm Tetra

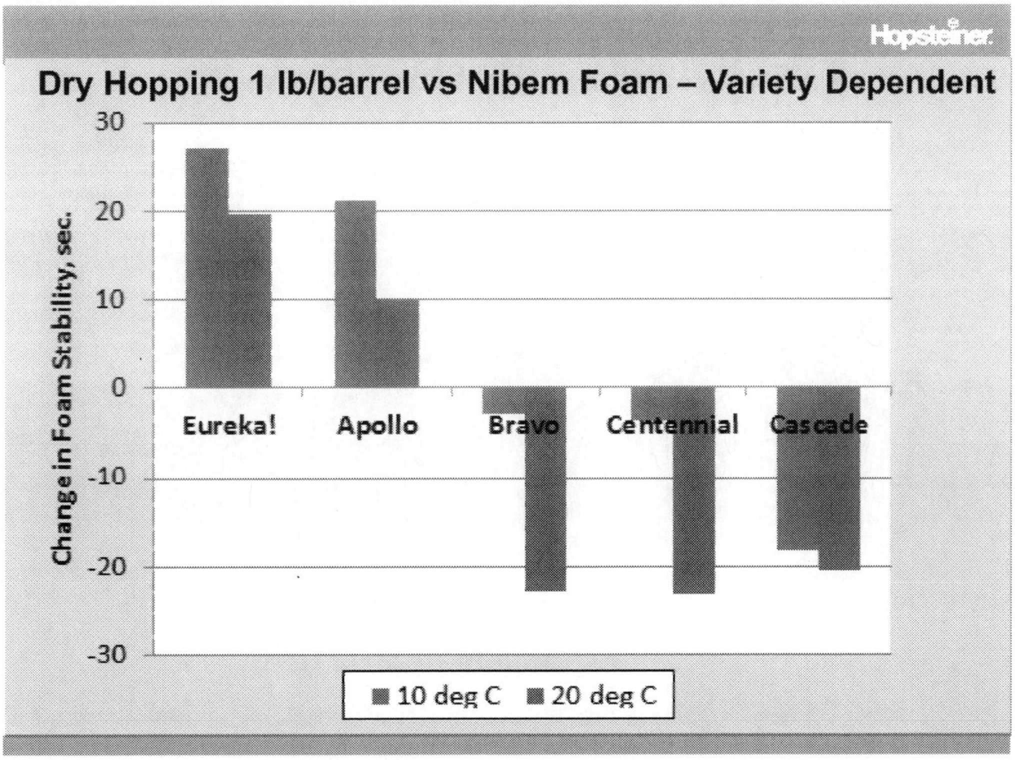
Hopsteiner
Dry Hopping 1 lb/barrel vs Nibem Foam – Variety Dependent
Change in Foam Stability, sec.
30
20
10
0
-10
-20
-30
Eureka!
Apollo
Bravo
Centennial
Cascade
10 deg C
20 deg C

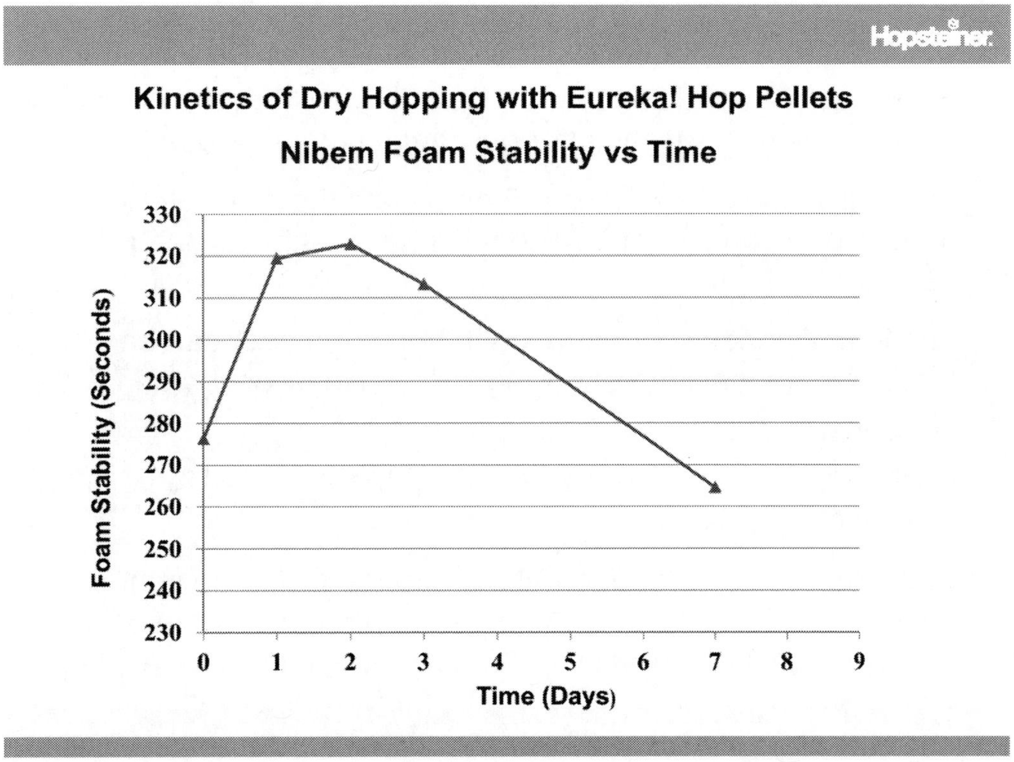
Hopsteiner
Kinetics of Dry Hopping with Eureka! Hop Pellets
Nibem Foam Stability vs Time
Foam Stability (Seconds)
330
320
310
300
290
280
270
260
250
240
230
0
1
2
3
4
5
6
7
8
9
Time (Days)

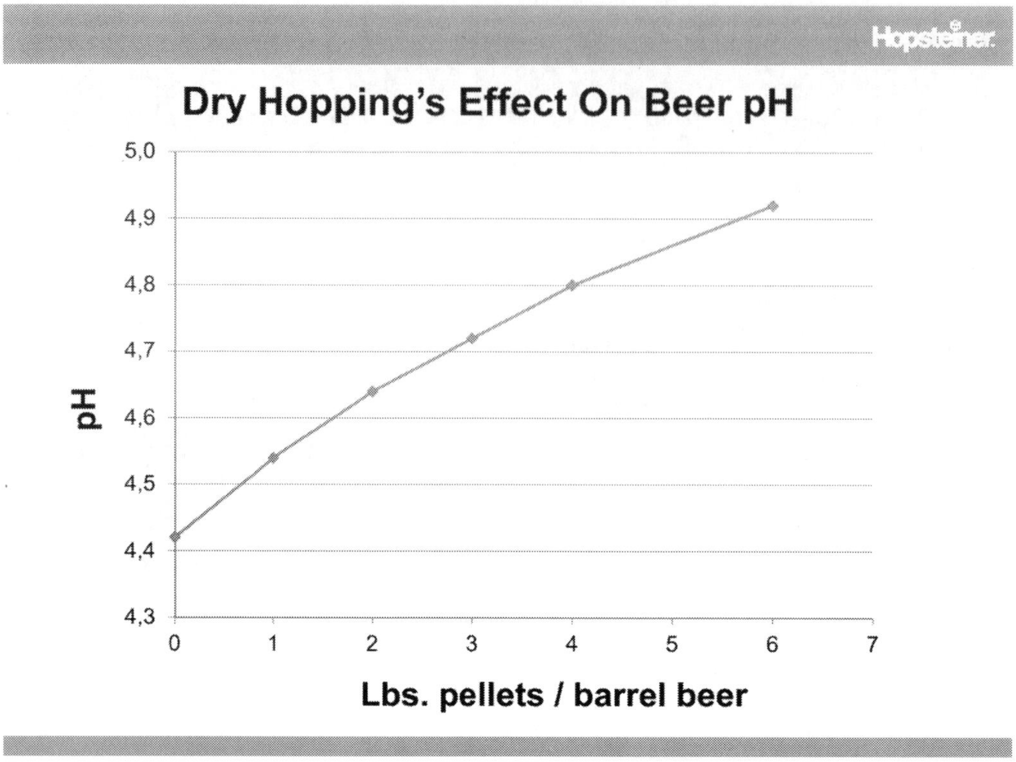

Conclusion

Hops contain small concentrations of humulinones and their concentration increases after pelleting and then stops.

The higher the HSI in hops or hop pellets the higher the humulinone concentration and this relationship is variety dependent.

The humulinones concentration in hops and beer can be measured very accurately using HPLC. Need Calibration Std

Humulinones are more polar than isoalpha acids and are extremely soluble in beer and are reported to be 66% as bitter as isoalpha acids.

Beers that are dry hopped will experience a change in hop acid composition. That is, there will be a reduction in isoalpha acids and an incorporation of humulinones and alpha acids.

Hopsteiner

Dry hopping a low IBU beer (20 IBU) will remove very little isoalpha acids but add low bitter humulinones and alpha acids and can make a beer more bitter.

Dry hopping a high IBU beer will remove a considerable amount of isoalpha acids and can make a beer less bitter when dosing 2 lbs hop per bbl beer or less. However due to the high solubility of humulinones, dry hopping any beer with 3 lbs hops/bbl or more can make a beer more bitter.

Because the IBU Test Method was developed to measure only isoalpha acids in beer, dry hopped beers that contain humulinone and alpha acids and polyphenols will give IBU test results that don't correlate with percieved bitterness. That is, the sensorial bitterness will be significantly lower than the IBU test result would suggest.

Hopsteiner

HPLC can be used to accurately measure the various hop acids in dry hopped beer. By using that data one can calculate a beers bitterness which should correlate more closely with the perceived bitterness.

Dry hopping can reduce beer foam however this appears to be variety dependent. Also, extended dry hop times can reduce beer foam even further. Foam enhancing hop products like Tetra can be used to enhance the foam of beers that are negatively impacted due to dry hopping.

Because beta acids are extremely non-polar we saw little to no beta acids in any of the dry hopped beers we tested and very small quantities if any of hulupones.

Dry hopping increases a beers pH by about 0.14 pH units per lbs of hops per barrel dosed and that increases in pH can increase a beer's bitterness by ~ 2 to 3 IBU.

Hopsteiner

Acknowledgements

Bob Smith & Jeremy Leker

Hopsteiner

&

Thank You For Your Kind Attention

36TH CONGRESS EBC 2017 EUROPEAN BREWERY CONVENTION 14-18 MAY LJUBLJANA SLOVENIA

Advanced Yeast Hunting

Hutzler M., Meier-Dörnberg T., Wagner S., Zarnkow M., Jacob F.

The Fate of Yeast During the Ages

Isolation of potential starter material that is historically related to brewing

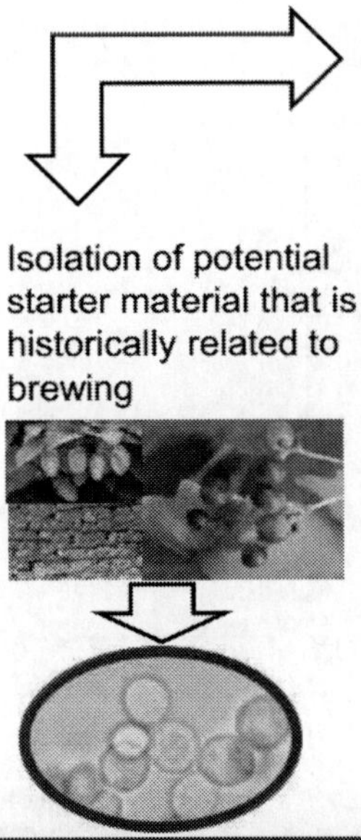

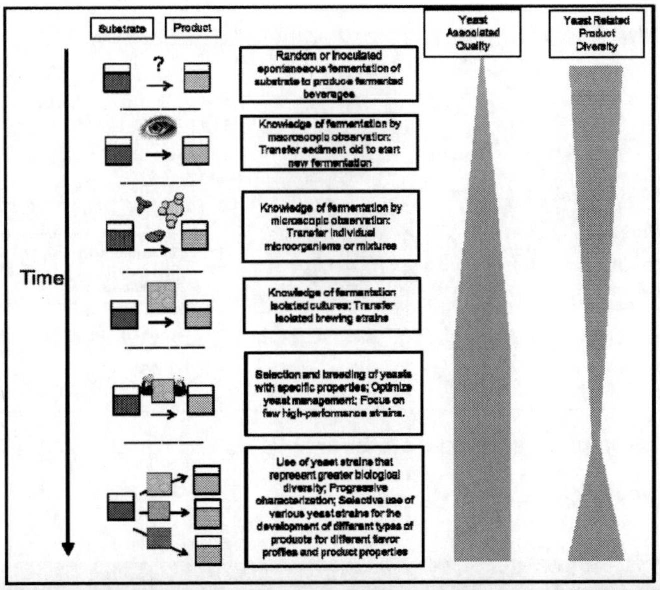

Hutzler M., Jacob F. (2015): Hefe im Wandel der Zeit, Brauindustrie, Juli 2015 (2015), Sonderausgabe 100 Jahre Brauindustrie, 76-80.

Outline

- Yeast species in a modern brewing environment
- Aims

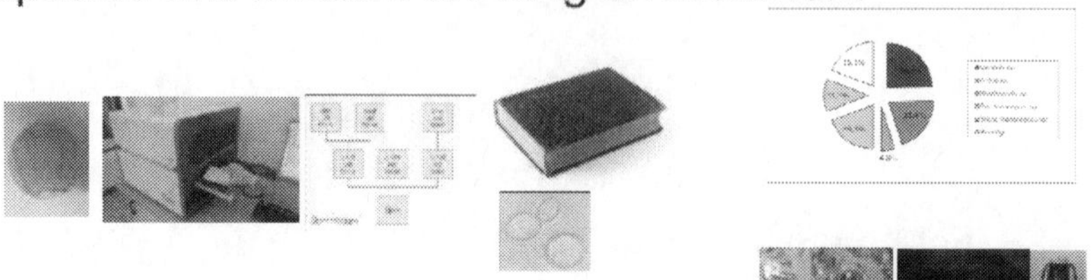

- Sampling targets, subjects, environments

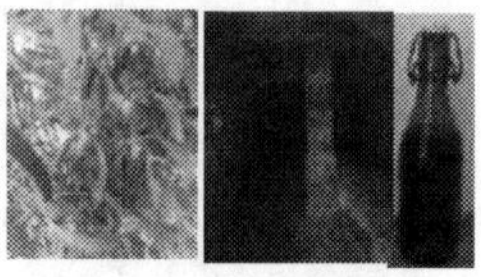

- Hunting tools and strategies and characterisation work flows (open or focused)
- Pre-screening test / detailed strain profiling

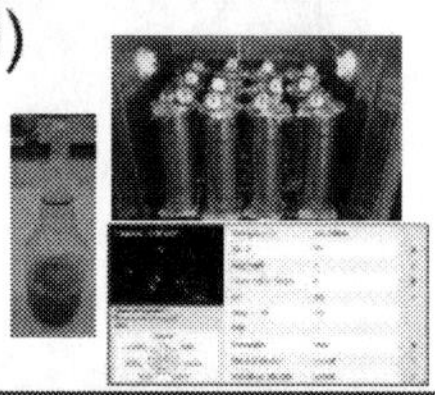

Yeast in a modern brewing ecosystem

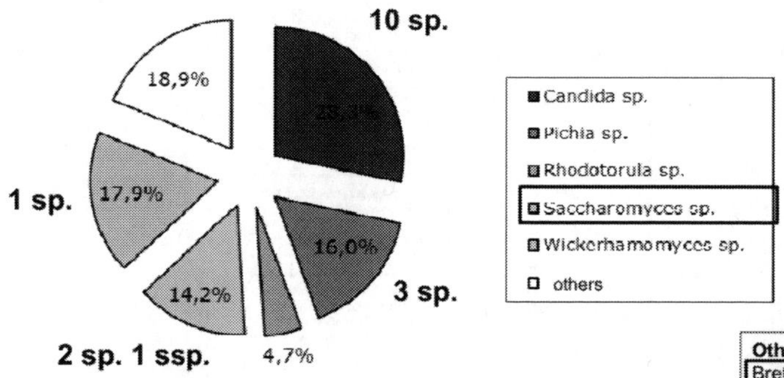

Other genera:
Brettanomyces/Dekkera
Cryptococcus
Debaryomyces
Issatchenkia
Kregervanrija
Schizosaccharomyces
Torulaspora
Williopsis

102 samples isolated from beer

→ **29 sp. 1 ssp.**

Yeast in a modern brewing ecosystem

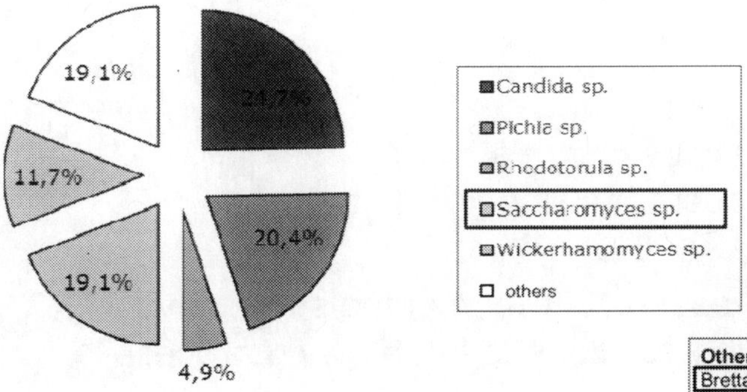

Other genera:
Brettanomyces/Dekkera
Cryptococcus
Debaryomyces
Issatchenkia
Kregervanrija
Schizosaccharomyces
Torulaspora
Williopsis
etc.

162 samples (known sampling points)

→ **42 sp. 1 ssp.**

Mathias Hutzler | Slide 5 | Forschungszentrum Weihenstephan für Brau- und Lebensmittelqualität

Brewing yeast species –starter cultures

Fermentation Type/ Flocculation		Beer Type	Genus/ Species	
bottom-fermenting	Strong flocculation / Low flocculation	Lager, Pilsener, Export, bottom-fermented special beers, bottom-fermented low alcohol beer, etc.	*S. pastorianus* (var. *carlsbergensis*) (*Saccharomyces* hybrids)	
top-fermenting	Low flocculation	German Wheat Beer, Ale, Stout, Koelsch, Alt, Belgian special beer styles (Witbeer, Trapist Beer), African indigenous beer styles etc.	*S. cerevisiae* (*Saccharomyces* hybrids)	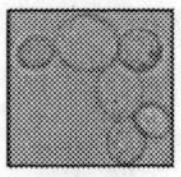
-	-	Non- and low alcohol beer styles	*Saccharomycodes ludwigii*	
-	-	Berliner Weiße, Belgian special beer styles (e. g. Lambic)	*Brettanomyces bruxellensis*	
-	-	African indigenous beer styles	*Schizosaccharomyces pombe*	
-	-	Special beer styles with fruity character	*Torulaspora delbrueckii*	

Huztler M, Koob J, Riedls R, *et al.* 2015. *Yeast identification and characterization.* In: *Brewing microbiology. Editor Hill A.* London: Woodhead Publishing, 65-104.

Mathias Hutzler | Slide 6 | Forschungszentrum Weihenstephan für Brau- und Lebensmittelqualität

Aim of Advanced Yeast Hunting Project

- Develop tailored methods to quickly identify yeast species from unknown samples

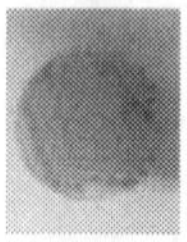

- Find old brewing strains and ancestors/ relatives to modern brewing yeast strains

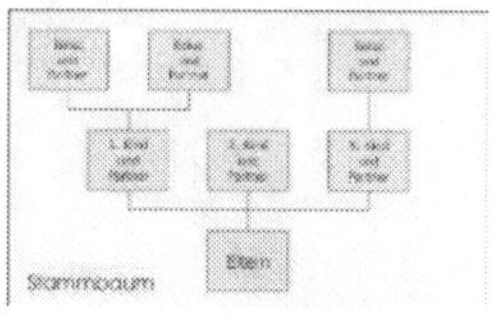

- Find yeast on carrier materials that have historical relations to the brewing process

- Find new *Saccharomyces* and non-*Saccharoymces* brewing strains

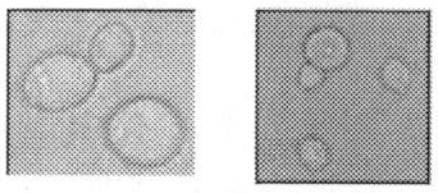

Potential Starter Material (historical relation)

Sumerian Prayer-Goddess Ninkasi

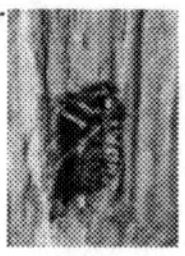

Insect

Ingredient, Beer Clarification

Hop

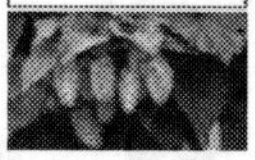

Barrel production

Tree bark

General fermentation starter

Fruit

other ingredients, raw materials,

Sallaberger W.(2012):Bierbrauen in Versen: Eine neue Edition und Interpretation der Ninkasi -Hymne", in Sabine Ecklin/Catherine Mittermayer (Ed.), Altorientalische Studien zu Ehren von Pascal Attinger. Orbis Biblicus et Orientalis 256, Fribourg: Academic Press, Göttingen(Vandenhoeck & Ruprecht 2012), 291–328.

Meussdoerfer F., Zarnkow M. (2014): Das Bier: Eine Geschichte von Hopfen und Malz, Verlag C. H. Beck, Munich.

Environmental Samples (Groups)

FRUIT 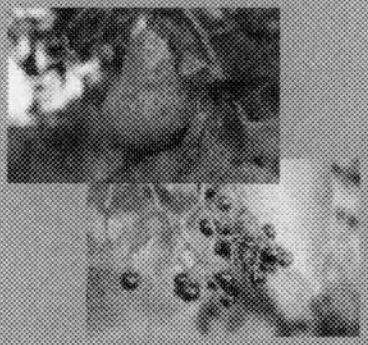	Apple Pear Blackberry Strawberry Rose hip Raspberry Elder Plum Grape Wild cherry Sloe	MISTLETOE 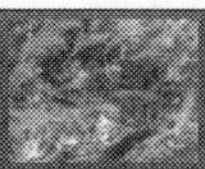	Stems, Berrys, different varieties
		OAK Bark 	Oak tree bark Beech tree bark Rowan tree bark
		HOP  	Hop Wild hop
		INSECT 	Fruit fly Earwig Harvestman Wasp

Environmental Samples (Examples)

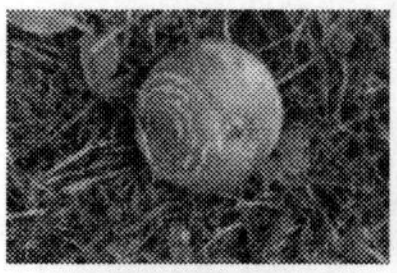
Apple

Pear

Blueberry

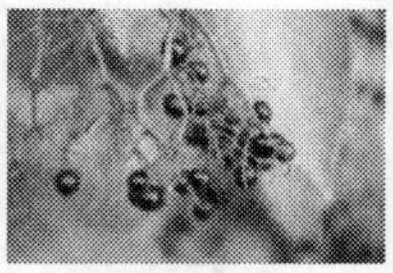
Elder

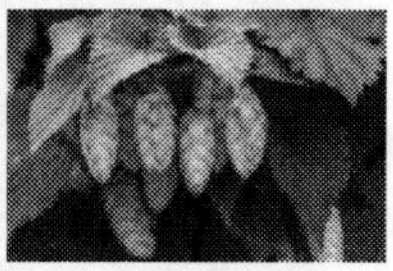
Hop

Plum

Rose hip

Grape

Bark

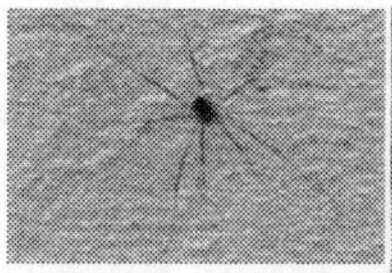
Harvestman

Wasp

Brewing cellar samples / old bottles

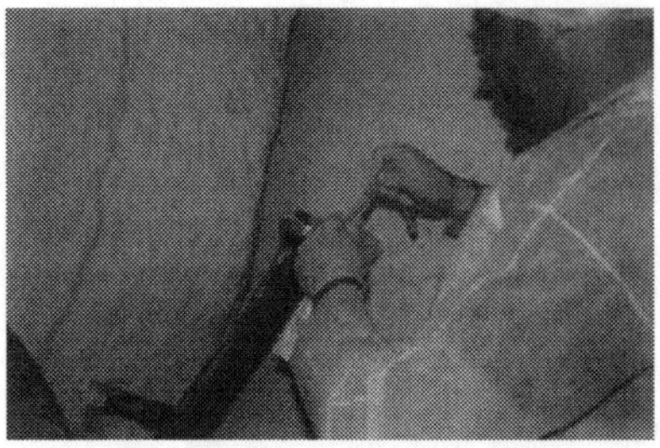

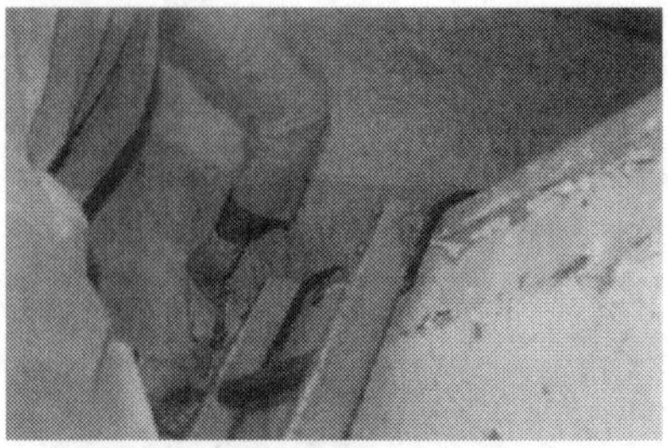

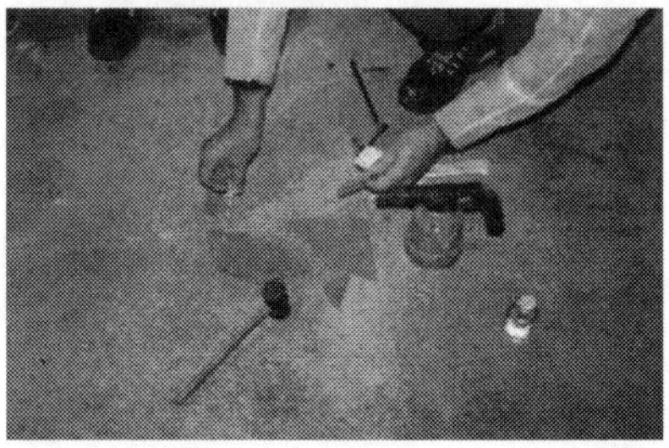

Mathias Hutzler Slide 11 Forschungszentrum Weihenstephan für Brau- und Lebensmittelqualität

Brewing cellar samples / old bottles

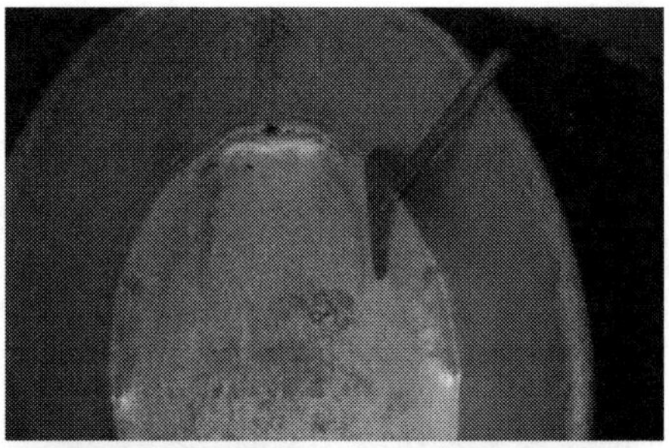

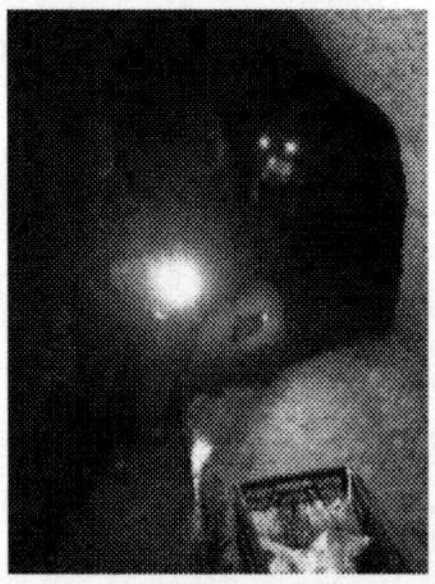

Mathias Hutzler Slide 12 Forschungszentrum Weihenstephan für Brau- und Lebensmittelqualität

Brewing cellar samples / old bottles

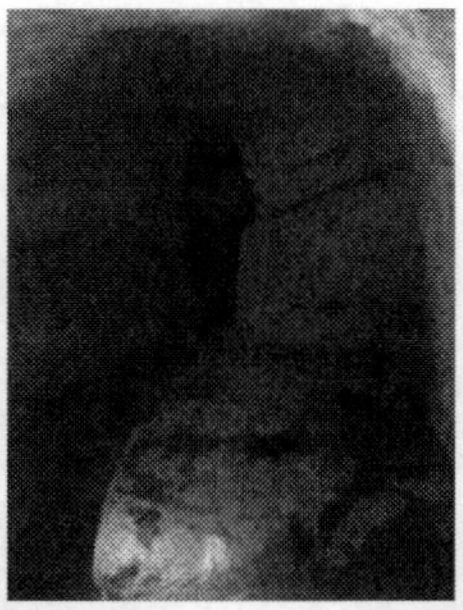
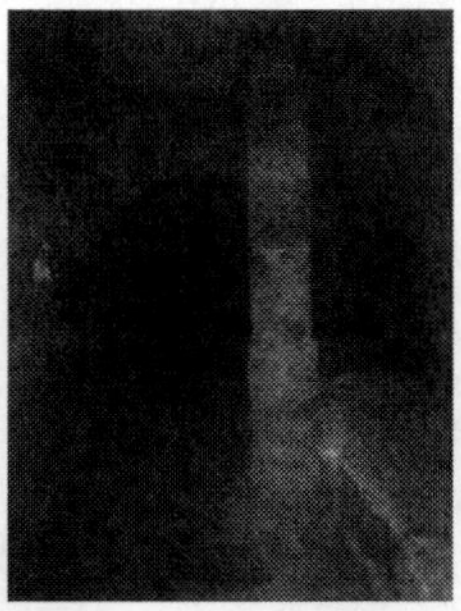

Mathias Hutzler | Slide 13 | Forschungszentrum Weihenstephan für Brau- und Lebensmittelqualität

Hunting tools and strategies

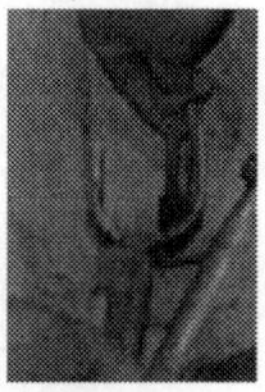

sample (bottle liquid, fruit, bark, air, sediment)

sampling under "sterile" conditions
(sterile bag, flamable knife, swab, forceps, pipette, air sampler, sterile tubes)

open system

wort + tetracycline
+ fermentation lock
(optional pre-incubation)

single colony isolation
universal medium

focused system (specific)

YM + tetracycline
+ low pH + alcohol
+ fermentation lock

specific real-time
PCR → "fishing"
for specific species
(e. g. *S. cer.*)

Mathias Hutzler | Slide 14 | Forschungszentrum Weihenstephan für Brau- und Lebensmittelqualität

Hunting tools and strategies

open system

single colony isolation
universal medium

Library based ident.
rRNA gene
sequencing
MALDI-TOF
FTIR

species
identification

Hit No.	Hit Quality	Sample Name
1	0.294697	Saccharomyces cerevisiae [illegible]
2	0.414482	Saccharomyces cerevisiae WSYC [illegible]
3	0.503783	Saccharomyces cerevisiae [illegible]
4	[illegible]	Saccharomyces cerevisiae [illegible]
5	[illegible]	Saccharomyces cerevisiae WSYC [illegible]

focused system (specific)

specific real-time
PCR → "fishing"
for specific species
(e. g. *S. cer.*)

S. cer.
COXII

Sample selection
e. g. 3 of 21 *S. cer.* +

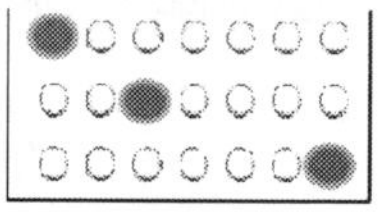

PCR confirmation

single colony

Hunting tools and strategies

open system ⟷ **focused system (specific)**

combination

Challenges
mould growth
many colonies
labouriuous
Broad yeast spectrum

Challenges
only isolation of specific species
sample selection
specificity can be determined

Results focused system

Positive real-time PCR results for *Saccharomyces cerevisiae* In 60 environmental samples in 2015

		RT-system					
Group	Sample	Sce ITS	Sce COX II	Spast	Sbp	Seub	Sdia
Tree Bark	Oak Tree Bark	+	+	-	-	-	-
Fruit	Apple 4A	+	+	-	-	-	-
	Grape blue A	+	+	-	-	-	-
	Grape blue B	+	+	-	-	-	-
Hop	12	+	+	-	-	-	-
	14	+	+	-	-	-	-

Hop 12

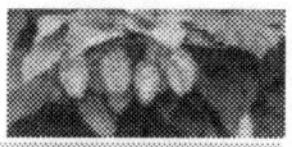

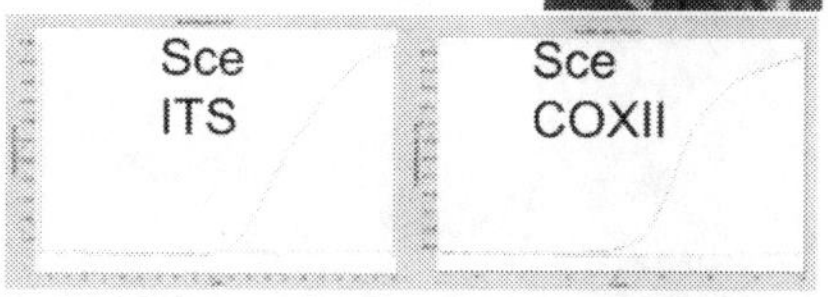

Ct-Value: 22,46 Ct-Value: 23,03

Apple 4A

Ct-Value: 23,04 Ct-Value: 22,56

Mathias Hutzler Slide 17 Forschungszentrum Weihenstephan für Brau- und Lebensmittelqualität

Results open system (ITS1-5.8S-ITS2 sequencing)

		Region	Place/ sampling type	ITS1-5.8S-ITS2 sequencing
cellar	#830 A	Kulmbach	Träffnerkeller wood	*Cryptococcus terreus*
	#832 A	Kulmbach	Träffnerkeller sediment	***Debaryomyces hansenii***
	#832 B	Kulmbach	Träffnerkeller sediment	***Candida norvegica***
	#832 C	Kulmbach	Träffnerkeller sediment	***Candida sake***
	#832 C	Kulmbach	Träffnerkeller sediment	***Torulaspora delbrueckii***
	#837 A	Kulmbach	Träffnerkeller swab	***Cyberlindnera sp.***
	#837 B	Kulmbach	Träffnerkeller swab	***Cyberlindnera sp.***
	#837 C	Kulmbach	Träffnerkeller swab	***Kazachstania servazii***
	#837 D	Kulmbach	Träffnerkeller swab	***Torulaspora delbrueckii***
	#839 A	Kulmbach	Träffnerkeller swab	***Candida sake***
	#843 A	Kulmbach	7 old bottles wood	*Metschnikowia sp.*
	#800 A	Schönbrunn	Langkeller wood	***Candida norvegica***
	#809 A	Selb	Friedhofstrasse swab	***Candida norvegica***
	#811 A	Selb	Jahnstrasse 15 swab	***Debaryomyces hansenii***
	#813 A	Selb	Ludwigstrasse 1 wood	***Debaryomyces hansenii***
	#817	Wunsiedel	WUN 10 wood	***Debaryomyces hansenii***
	#883 A	Moosburg	Paddle 1	***Debaryomyces hansenii***
	#885 A	Moosburg	Paddle 2	***Debaryomyces hansenii***
	#887 A	Moosburg	Wall 2	***Debaryomyces hansenii***
	#888 A	Moosburg	Yeast cellar (2 swabs)	***Debaryomyces hansenii***
	#1204 A	Eggenfelden	Swab sediment	***Zygosaccharomyces rouxii***
	#1207	Eggenfelden	Swab sediment	***Pichia membranifaciens***
	#1209	Eggenfelden	Swab sediment	***Candida boidinii***
old bottles	#861 A	Weihenstephan	Old bottle	***Sacch. bayanus/ pastorianus***
	#861 B	Weihenstephan	Old bottle	***Sacch. bayanus/ pastorianus***
	#861 C	Weihenstephan	Old bottle	***Sacch. bayanus/ pastorianus***
	#861 D	Weihenstephan	Old bottle	***Sacch. bayanus/ pastorianus***
	#861	Weihenstephan	Old bottle	***Sacch. bayanus/ pastorianus***
	#1215	unknown	Old bottle (Blonde)	***Nauvomozyma castellii***
	#1215	unknown	Old bottle	***Brettanomyces anomalus***
	#1216	unknown	Old bottle	***Brettanomyces anomalus***

selected results 2016/ 2017

Mathias Hutzler Slide 18 Forschungszentrum Weihenstephan für Brau- und Lebensmittelqualität

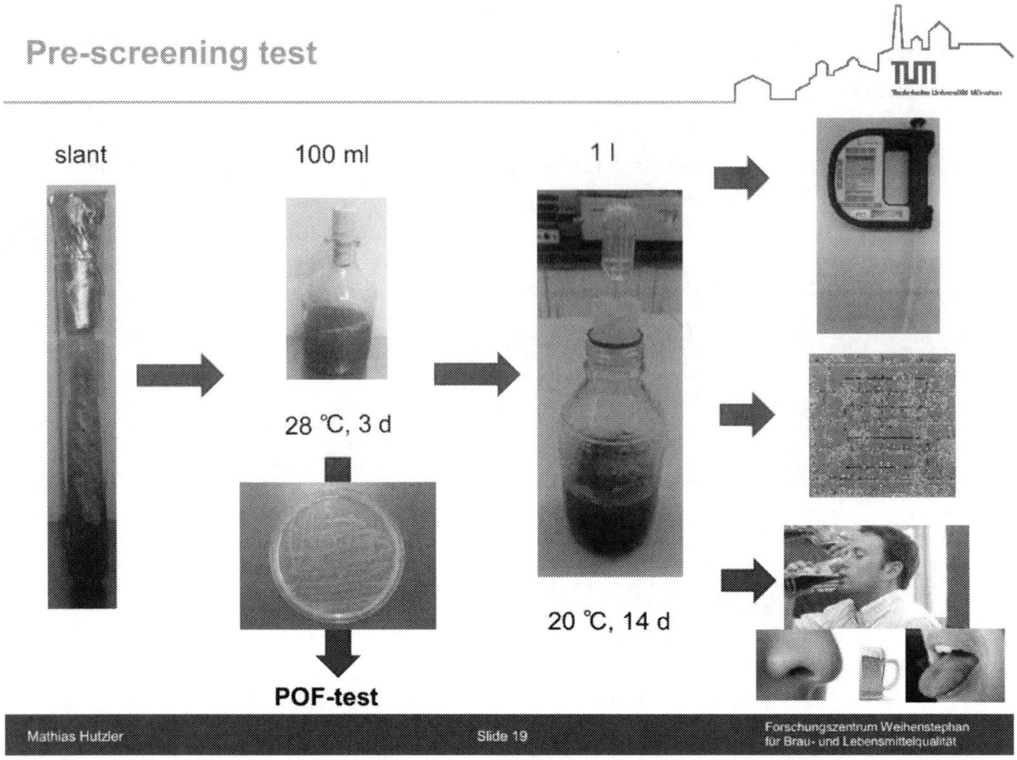
Pre-screening test
Technische Universität München
slant
100 ml
1 l
28 °C, 3 d
20 °C, 14 d
POF-test
Mathias Hutzler
Slide 19
Forschungszentrum Weihenstephan
für Brau- und Lebensmittelqualität

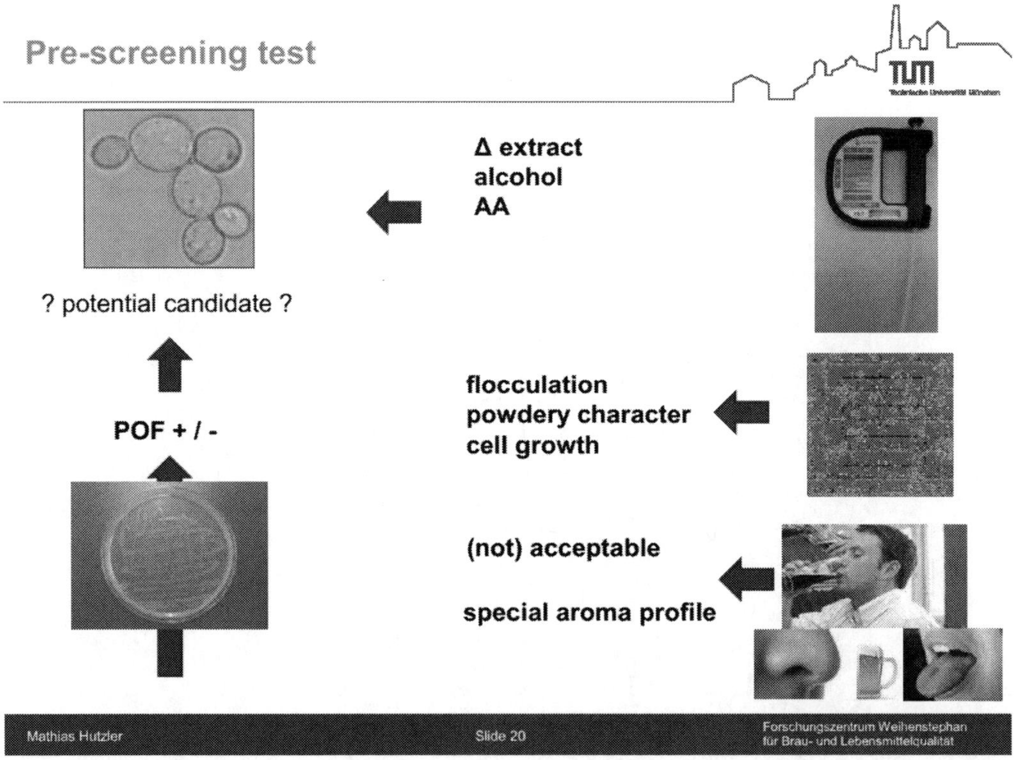
Pre-screening test
Technische Universität München
Δ extract
alcohol
AA
? potential candidate ?
POF + / -
flocculation
powdery character
cell growth
(not) acceptable
special aroma profile
Mathias Hutzler
Slide 20
Forschungszentrum Weihenstephan
für Brau- und Lebensmittelqualität

Pre-screening test - aim ale yeast

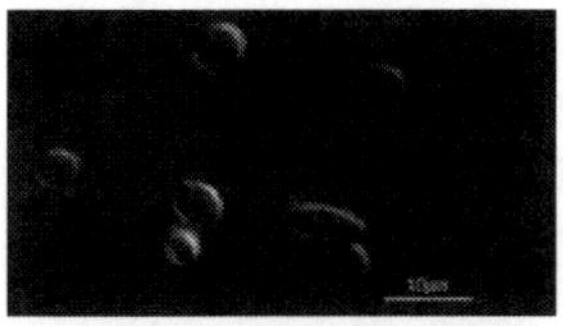

Δ extract > 8 °Plato
alcohol > 4 % [v/v]
AA > 70 %

potential candidate ale yeast *yes*

POF *negative*

flocculation *medium*
powdery character *strong*
cell growth *moderate*

acceptable yes

special aroma profile *yes*

Pre-screening test – results (7 strains)

Sample	Identification	POF-Test: POF (YM+FA)	POF-Test: POF (YM+CA)	Fermentation: Alc. [% v]	Fermentation: E [m]	Fermentation: AA [%]	Fermentation: pH	Sensory tasting: Acceptable	Sensory tasting: Description aroma attributes
Cellar sample VF2	*S. bayanus/ pastorianus*	++	+	3.23	5.0	55.8	4.5	+	spicy, slightly clove, fruity, sweet
Cellar sample VF (4)	*S. bayanus/ pastorianus*	+	-	3.2	5.6	52.7	4.5	+	sweet, citrus, fruity
Mistle berry (6)	*S. pastorianus*	-	-	5.67	2.1	84.5	4.6	+	clear taste, neutral
Oak bark # 600	*S. paradoxus*	-	-	1.33	10.2	20.0	4.6	+/-	wort like, slightly apple
Oak bark 2015	*S. cerevisiae*	++	+	4.07	4.9	62	4.5	+	vanilla, honey, floral, spicy, clove, malty, dry
Aple 4A 2015	*S. cerevisiae*	++	+	4.2	3.3	71,3	4.4	+	apple, clove
Old bottle # 861A	*S.bayanus/ pastorianus*	+	-	5.55	4.1	72,5	4.4	+	plum, dry, clear taste

Pre-screening test – results (17 strains)

Sample	Identification	POF-Test: POF (YM+FA)	POF-Test: POF (YM+CA)	Fermentation: Alc [% v]	Fermentation: E [m]	Fermentation: AA [%]	Fermentation: pH	Sensory tasting: Accept-able	Sensory tasting: Description aroma attributes
Cellar sample VF2	*S. bayanus /pastorianus*	++	+	3.23	5.02	55.8	4.52	+	spicy, slightly clove, fruity, sweet
Cellar sample VF (4)	*S. bayanus /pastorianus*	+	-	3.2	5.6	52.7	4.53	+	sweet, citrus, fruity
Mistle berry (6)	*S. pastorianus*	-	-	5.67	2.05	84.5	4.57	+	clear taste, neutral
Oak bark # 600	*S. paradoxus*	-	-	1.33	10.22	20	4.59	+/-	wort like, slightly apple
Oak bark # 606 B-2	*S.paradoxus*	+	-	1.18	9.28	19.7	4.60	+/-	wort like
Oak bark 2015	*S.cerevisiae*	++	+	4.07	4.88	62.0	4.52	+	vanilla, honey, floral, spicy, clove, malty, dry
Hops 12 2015	*S.cerevisiae*	++	+	4.13	4.46	64.4	4.41	+	floral, black berry, clove
Aple 4A 2015	*S.cerevisiae*	++	+	4.2	3.32	71.3	4.40	+	apple, clove
Old bottle # 861 A	*S.bayanus /pastorianus*	+	-	5.55	4.09	72.5	4.45	+	plum, dry, clear tate
Old bottle # 861 B	*S.bayanus /pastorianus*	+	-	4.73	3.4	73.2	4.41	+	fruity, clear taste, dry
Old bottle # 861 C	*S.bayanus /pastorianus*	++	+	5.87	4.32	72.6	4.54	+	clear taste, neutral
Old bottle # 861 D	*S.bayanus /pastorianus*	++	+	4.97	3.65	72.8	4.43	+	fruity, vanilla, clove, honey
Old bottle # 861	*S.bayanus /pastorianus*	+	-	5.56	4.05	72.8	4.42	+	fruity, floral
Sherry barrel TUM 545	*S.cerevisiae*	+	-	6.45	1.61	88.8	4.37	+	vinous, grany smith apple, fruity
Sherry barrel TUM 546	*S.cerevisiae*	+	-	5.33	1.89	84.8	4.33	+	passion fruit, peach, maracuja, black current
Sherry barrel TUM 547	*S.cerevisiae*	-	+	2.38	6.45	35.8	4.45	+	malty, tropical fruity, sweet
Sherry barrel TUM 548	*S.cerevisiae*	+	-	3.68	4.84	59.8	4.33	+	berry, apple, pier, plum, raisin, citrus

Detailed strain profiling

Isothermic fermentation in in special 2 l fermenters

Detailed strain profiling

Genetic profiling

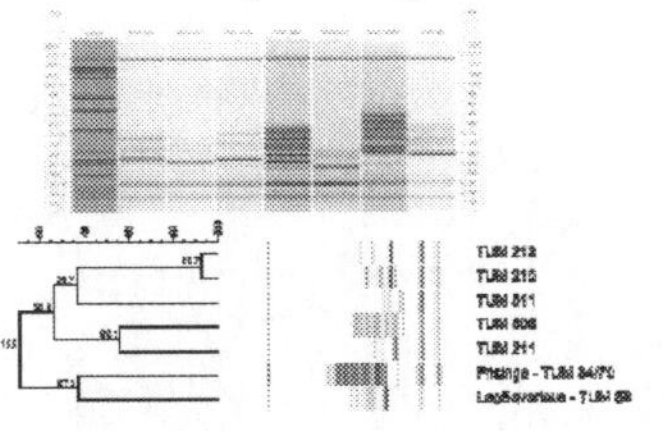

SO2

	SO₂ concentration of the finished beers (mg/l)				
	TUM 210	TUM 211	TUM 213	TUM 506	TUM 511
SO_2	2.63 ± 1.64	2.60 ± 0.98	2.63 ± 1.521	2.23 ± 1.02	0.50 ± 0.00

POF

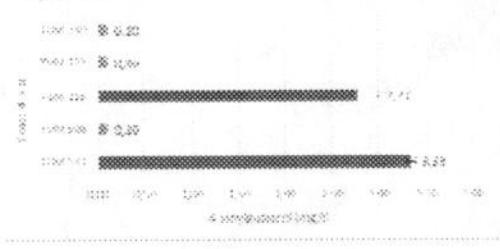

Sugar utilisation

	Sugar utilization (%)				
	TUM 210	TUM 211	TUM 213	TUM 506	TUM 511
Glucose	99.16 ± 0.05	98.55 ± 0.47	99.13 ± 0.00	98.23 ± 0.12	99.42 ± 0.12
Fructose	99.35 ± 0.21	98.57 ± 0.21	98.83 ± 0.00	96.05 ± 0.36	98.05 ± 0.00
Sucrose	98.51 ± 0.48	98.51 ± 0.48	98.21 ± 0.00	99.11 ± 0.00	95.54 ± 0.00
Maltose	98.89 ± 1.23	87.23 ± 0.82	99.39 ± 0.07	98.48 ± 0.93	99.28 ± 0.09
Maltotriose	75.01 ± 8.39	26.66 ± 0.26	38.96 ± 0.46	59.28 ± 0.81	83.91 ± 0.71

pH drop

	pH value decrease during primary fermentation				
	TUM 210	TUM 211	TUM 213	TUM 506	TUM 511
0 h	5.2	5.2	5.2	5.2	5.2
24 h	4.6	4.5	4.5	4.6	4.6
48 h	4.4	4.4	4.5	4.5	4.4
72 h	4.4	4.4	4.5	4.5	4.4
96 h	4.4	4.4	4.5	4.5	4.4
After primary fermentation	4.4	4.4	4.5	4.5	4.4
After maturation	4.5	4.5	4.5	4.6	4.4
Final Beer (after lagering)	4.6 ± 0.01	4.7 ± 0.01	4.6 ± 0.01	4.7 ± 0.01	4.4 ± 0.01
ΔpH	0.6	0.5	0.6	0.5	0.8

Sensorial analysis

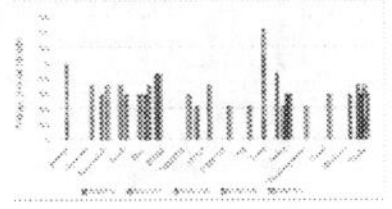

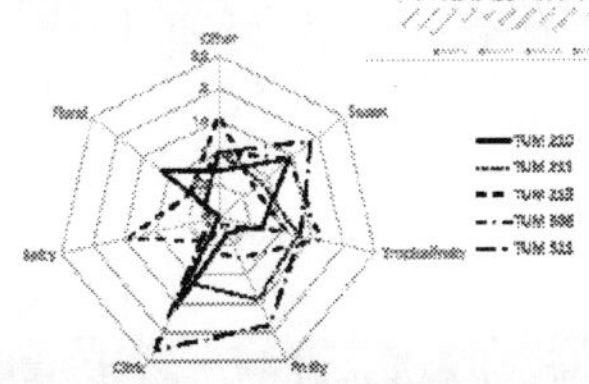

Detailed strain profiling

FAN – amino acid uptake

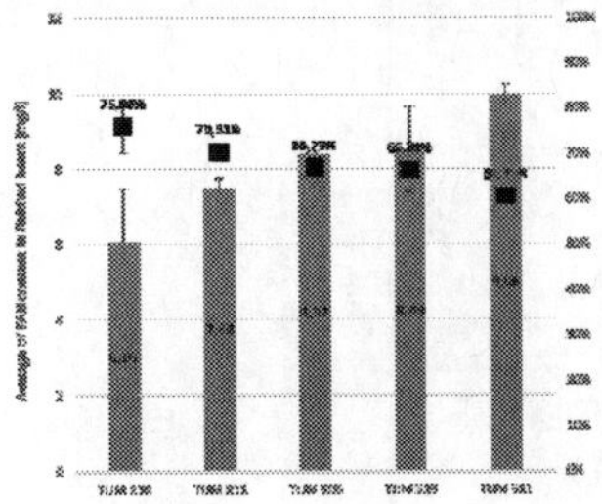

fermentation performance

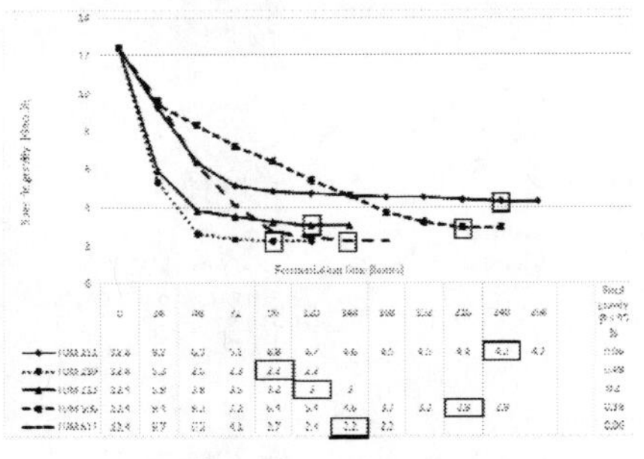

yeast in suspension

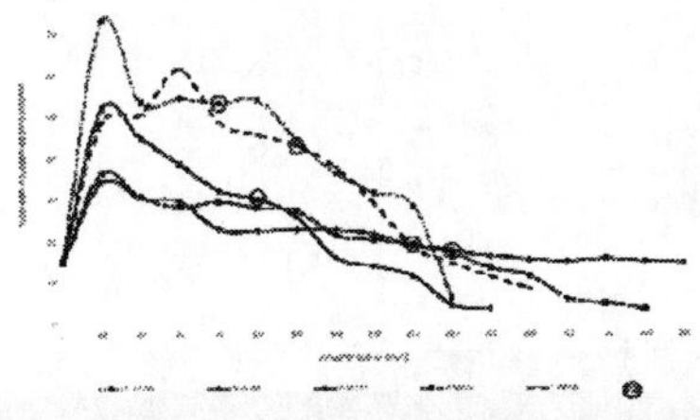

fermentation-by-products

	Fermentation by-products (mg/l)					
	TUM 210	TUM 211	TUM 213	TUM 506	TUM 511	Threshold (Lager beers)*
Isoamyl acetate	2.40 ± 0.33	2.03 ± 0.05	4.77 ± 0.36	1.53 ± 0.11	2.93 ± 0.05	1.6
Ethyl acetate	31.97 ± 3.76	37.93 ± 0.11	51.57 ± 3.35	32.57 ± 1.31	54.40 ± 0.61	30
∑ Ester (E)	34.37 ± 4.09	39.97 ± 0.14	56.33 ± 3.86	34.10 ± 1.36	57.33 ± 0.65	n.v
n-Propanol	27.30 ± 3.53	18.30 ± 0.16	26.00 ± 1.49	20.67 ± 1.01	20.77 ± 0.53	800
i-Butanol	31.77 ± 4.05	16.20 ± 0.37	13.80 ± 0.96	20.90 ± 1.40	13.13 ± 0.28	200
Amyl alcohols	100.67 ± 5.65	59.27 ± 1.32	75.97 ± 7.30	88.50 ± 5.90	74.97 ± 1.24	70
∑ Higher alcohols (HE)	159.73 ± 12.31	93.77 ± 1.83	115.77 ± 9.05	130.07 ± 8.12	108.87 ± 1.23	n.v.
4-Vinylguajacol	0.10 ± 0.00	0.10 ± 0.00	2.77 ± 0.23	0.10 ± 0.00	3.33 ± 0.05	0.3
Diacetyl	0.09 ± 0.01	0.10 ± 0.01	0.07 ± 0.00	0.12 ± 0.01	0.06 ± 0.00	0.15
2,3-Pentandione	0.02 ± 0.01	0.02 ± 0.00	0.02 ± 0.01	0.02 ± 0.00	0.01 ± 0.00	0.9
∑ Vicinal diketones	0.11 ± 0.01	0.12 ± 0.01	0.09 ± 0.01	0.14 ± 0.01	0.07 ± 0.00	n.v.
Acetaldehyde	2.53 ± 0.93	4.60 ± 0.48	5.60 ± 1.81	5.93 ± 0.93	4.03 ± 0.46	25
Ratio (∑E : ∑HE)	1 : 4.65	1 : 2.35	1 : 2.06	1 : 5.40	1 : 1.90	

* (Meilgaard 1975)

Detailed strain profiling

Tropicus - TUM 506®

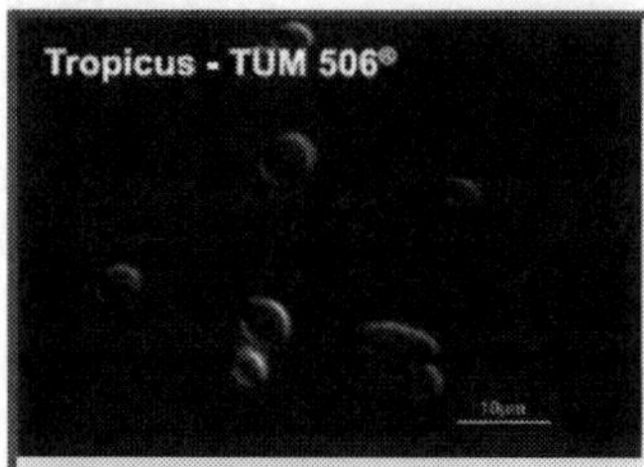

Flavor impression:
Flavor notes of tropical fruits

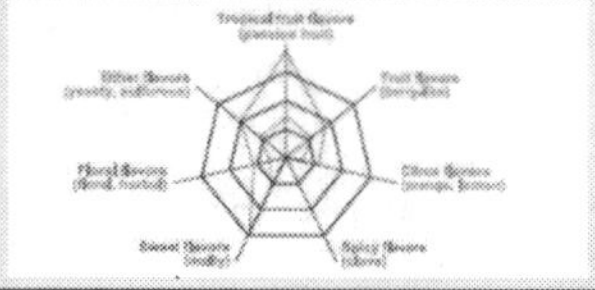

Beer type (TF)	: Ale, Kölsch	
AA (%)	: 77	●
SO_2 (mg/l)	: 2	●
Fermentation rate (d)	: 9	●
pH	: 4.6	●
Ratio E : HA	: 1:5	
POF	: -	
Flocculation	: weak	●
Diacetyl reduction	: normal	●
Maltotriose utilization	: normal	●

BF ≙ bottom/top-fermenting ● strong/fast ● normal
Flavor impression ≙ overall sensory description E ≙ ester

● weak/slow — Overall flavor impression — (individual flavor attributes)
HA ≙ higher alcohols POF ≙ phenolic off flavor AA ≙ apparent attenuation

Mathias Hutzler Slide 27 Forschungszentrum Weihenstephan für Brau- und Lebensmittelqualität

Yeast strain overview / poster

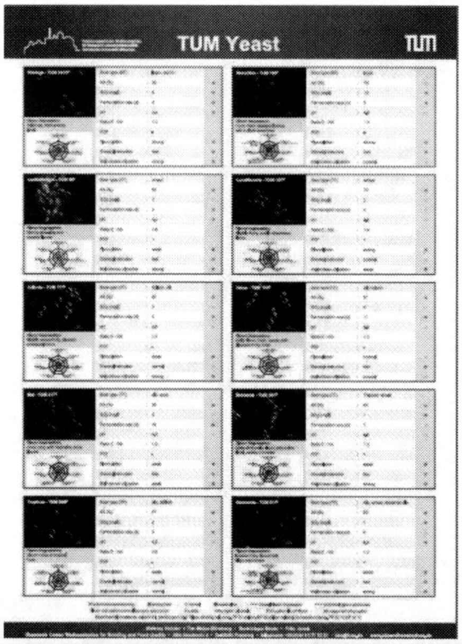

Mathias Hutzler Slide 28 Forschungszentrum Weihenstephan für Brau- und Lebensmittelqualität

Outlook

- Apply hunting/ screening methods to select *Saccharomyces* wild types from brewing related and non-related environments

- Apply hunting/ screening methods to select Non-*Saccharomyces* wild types from brewing related and non-related environments
- Compare new isolates with modern brewing strains (ancestry)

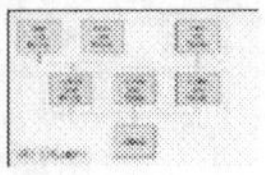

- Find new potential brewing strains -> brewing trials

Mathias Hutzler | Slide 29 | Forschungszentrum Weihenstephan für Brau- und Lebensmittelqualität

Thank you:

Max Michel
Margit Grammer
Franz Meußdoerffer
Frank-Jürgen Methner
(TU Berlin)

Dr.-Ing. Mathias Hutzler
Alte Akademie 3
85354 Freising-Weihenstephan
Telefon: +49 (0) 8161 / 71- 3100
Telefax: +49 (0) 8161 / 71- 4181
E-Mail: m.hutzer@tum.de
www.blq-weihenstephan.de

Mathias Hutzler | Slide 30 | Forschungszentrum Weihenstephan für Brau- und Lebensmittelqualität

How to analyze dry hopped beer?

Hopsteiner

Martin Biendl (Hopsteiner Germany)

How to analyze dry hopped beer?

- *Which hop-derived compounds are relevant to beer flavour?*
  - ✓ *Taste (bitterness): Non volatile compounds*
  - ✓ *Odour (aroma): Volatile compounds*
- *Add on to existing (or development of new) analysis methods?*
  - ✓ *Non volatile compounds: HPLC (liquid chromatography)*
  - ✓ *Volatile compounds: GC (gas chromatography)*
- *Goal: Inclusion in Analytica-EBC (International Methods)*

Which compounds are relevant to beer flavour?

- Verified hop-derived key bitter compounds

- ✓ *Iso-alpha-acids / Reduced iso-alpha-acids (Rho, Tetra, Hexa)*
- ✓ *Alpha-acids*
- ✓ *Hulupones* ***(Haseleu et al., 2009)***
- ✓ *Humulinones* ***(Algazalli and Shellhammer, 2016)***
- ✓ *Hard resin fraction* ***(Dresel et al., 2015):***
  - *Prenylflavonoids (e.g. xanthohumol, isoxanthohumol)*
  - *Glycosides of quercetin, kaempferol, multifidol*

Which compounds are relevant to beer flavour?

- Verified hop-derived key aroma compounds

- ✓ *Linalool* ***(Lermusieau et al., 2001; Fritsch et al., 2005)***
- ✓ *Myrcene* ***(Fritsch et al., 2005)***
- ✓ *Geraniol* ***(Kishimoto et al., 2006; Steinhaus, 2007)***
- ✓ *Esters, e.g. 2-methylbutylisobutanoate* ***(Takoi et al., 2009)***
- ✓ *Thiols, e.g. 4-mercapto-4-methyl-2-pentanone (4-MMP) and a few more compounds with sulfanyl group* ***(Steinhaus, 2007; Kishimoto et al., 2008; Takoi et al., 2009; Gros et al., 2011)***

Verified key flavour compounds transfered into beer clearly above their threshold concentrations

Beer type	Bitter resins	Bitter polyphenols	Aroma compounds
Non dry-hopped	(Reduced) Iso-Alpha-Acids	-	Linalool

Verified key flavour compounds transfered into beer clearly above their threshold concentrations

Beer type	Bitter resins	Bitter polyphenols	Aroma compounds
Non dry-hopped	(Reduced) Iso-Alpha-Acids	-	Linalool
Dry hopped	+ Alpha-Acids Humulinones Hulupones Xanthohumol	Glycosides of: Quercetin, Kaempferol, Multifidol	+ Myrcene Geraniol Several Esters, Thiols

Add on to existing EBC methods or development of new methods?

- ✓ *Bitter resins: HPLC-UV (EBC 9.47) ?*
- ✓ *Polyphenols: HPLC-UV or HPLC-MS ?*
- ✓ *Aroma compounds*
  - *Headspace GC-MS ?*
  - *For thiols:*

    more sophisticated sample preparation followed by GC-MS/MS – see oral presentation of Martin Steinhaus (L24)

How to analyze dry hopped beer?

- ✓ *Bitter resins: HPLC-UV (EBC 9.47)?*

Analytica-EBC - Method 9.47: Determination of iso-alpha- and reduced iso-alpha-acids in beer

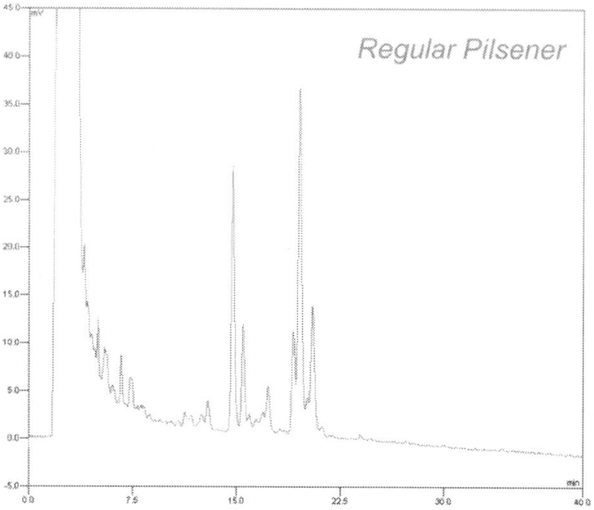

Beer sample preparation:
Dilution with methanol (1:1)
Direct injection into HPLC after separation of the precipitate

HPLC conditions:
Column: Zorbax Eclipse C8
Detection: UV (270 nm)
Eluent A: Methanol
Eluent B: 30 % Acetonitrile / 70 % Citric acid (1%)

HPLC programme:
0-5 min: 85 % Solvent B
5-45 min: 85-20 % Solvent B
45-50 min: 20-85 % Solvent B
50-60 min: 85 % Solvent B

9

Dry hopped beer according to method EBC 9.47

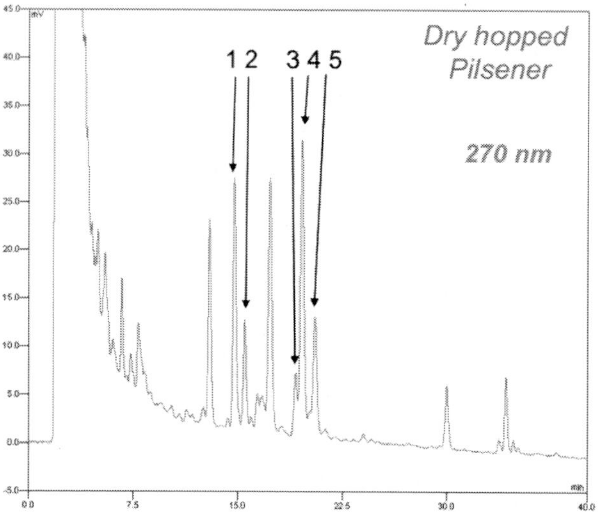

1: cis-co-isohumulone
2: trans-co-isohumulone
3: cis-ad-isohumulone
4: cis-n-isohumulone
5: trans-ad/n-isohumulones

10

Dry hopped beer according to method EBC 9.47

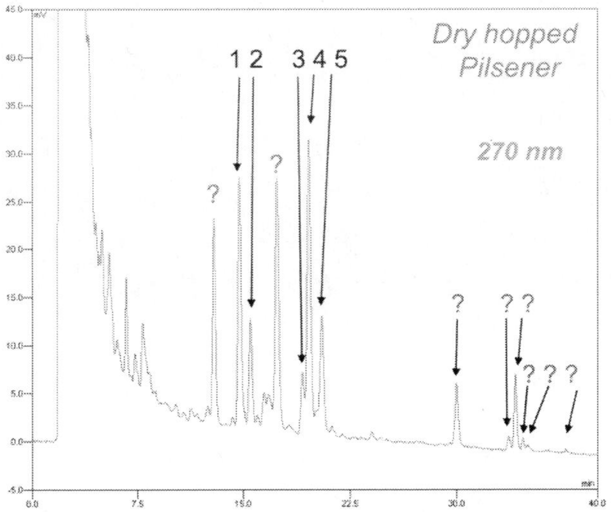

1: cis-co-isohumulone
2: trans-co-isohumulone
3: cis-ad-isohumulone
4: cis-n-isohumulone
5: trans-ad/n-isohumulones

? alpha-acids (humulones)
? humulinones
? hulupones
? xanthohumol

Correct peak assignments were finally possible with suitable pure standards

11

Composition of different beers according to method EBC 9.47 (mg/l)

Pale Ale	Iso-Alpha	Alpha	Humu-linones	Hulu-pones	Xantho-humol	Isoxantho-humol
*Regular**	*31.5*	*0.9*	*0.5*	*0.3*	*< 0.1*	*1.0*

** Kettle hopping only (Hallertau Tradition pellet type 90)*

Composition of different beers according to method EBC 9.47 (mg/l)

Pale Ale	Iso-Alpha	Alpha	Humu-linones	Hulu-pones	Xantho-humol	Isoxantho-humol
*Regular**	31.5	0.9	0.5	0.3	< 0.1	1.0
*+ Dry hopping***	33.4	11.7	7.2	1.8	0.2	1.2

** Kettle hopping only (Hallertau Tradition pellet type 90)*

*** Additional dry hopping with 500 g / hl (Bravo pellet type 90)*

Composition of different beers according to method EBC 9.47 (mg/l)

Beer	Iso-Alpha	Alpha	Humu-linones	Hulu-pones	Xantho-humol	Isoxantho-humol
*Ale**	20.1	15.7	10.8	1.6	0.8	0.4
*Stout**	12.5	23.2	7.9	2.2	8.6	1.2

** Commercial dry hopped beers*

Dry hopped Stout + alpha-acids standard

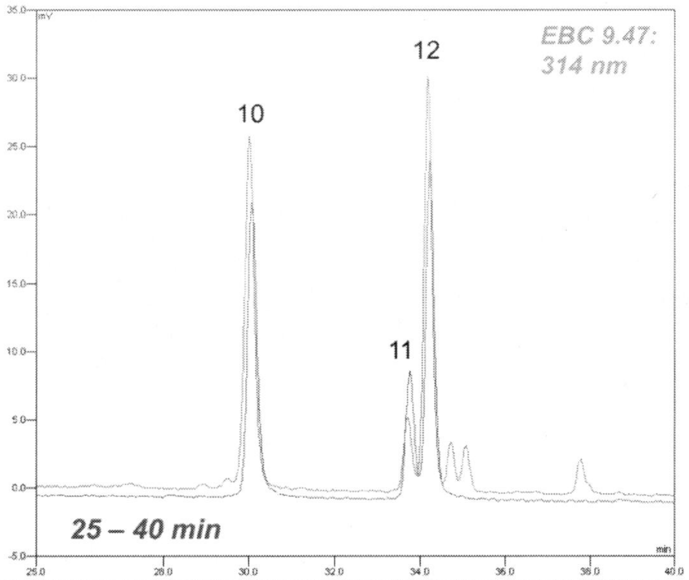

10: co-humulone

11: ad-humulone

12: n-humulone

15

Stout + xanthohumol and beta-acids standard

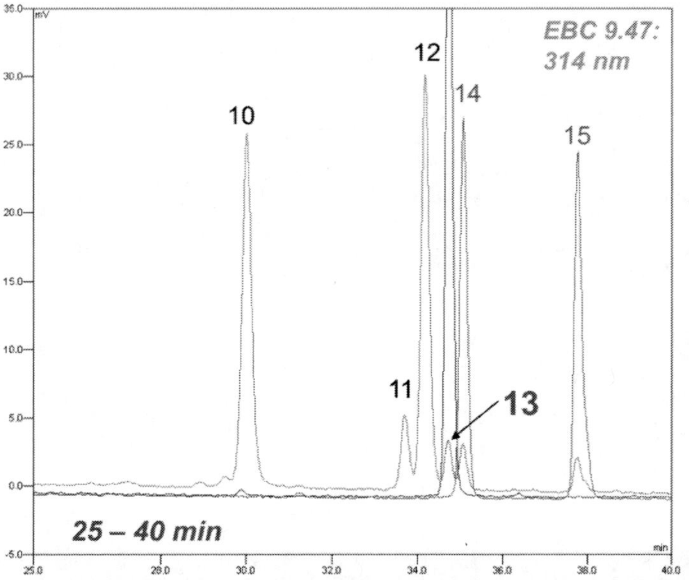

10: co-humulone

11: ad-humulone

12: n-humulone

13: xanthohumol

14: co-lupulone

15: n/ad-lupulone

16

Analytica-EBC - Method 9.47: Feasible range of bitter resins to be covered

Iso-alpha- and reduced iso-alpha-acids (270 nm)

Humulinones (270 nm)

Alpha-acids and Hulupones (314 nm)

Xanthohumol (370 nm)

Isoxanthohumol (290 nm)

Next step: International collaborative trial (on new parameters)

17

Task of International Hop Standard Committee

Preparation and release of suitable calibration standards:

*Humulinone standard (DCHA-Humulinone)**

*Hulupone standard (DCHA-Hulupone)**

*Alpha-acids standards (DCHA-Alpha)***

Pure Xanthohumol and Isoxanthohumol

According production processes and stability data published by

** Maye et al., 2016 and **Maye and Leker, 2014*

18

How to analyze dry hopped beer?

- ✓ *Bitter resins: HPLC-UV (EBC 9.47)* +
- ✓ *Polyphenols: HPLC-MS ?*

Polyphenols in beer coming exclusively from hops

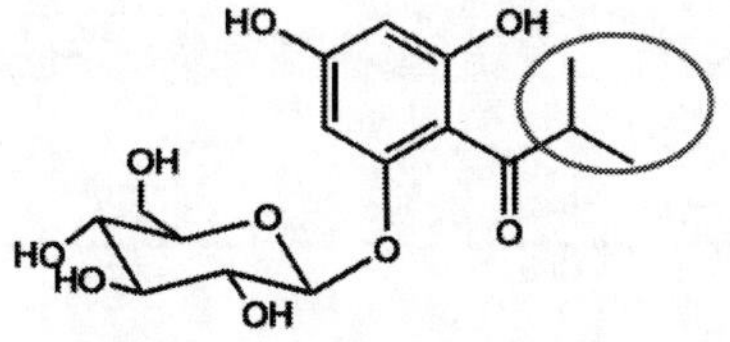

Co-multifidol glucoside

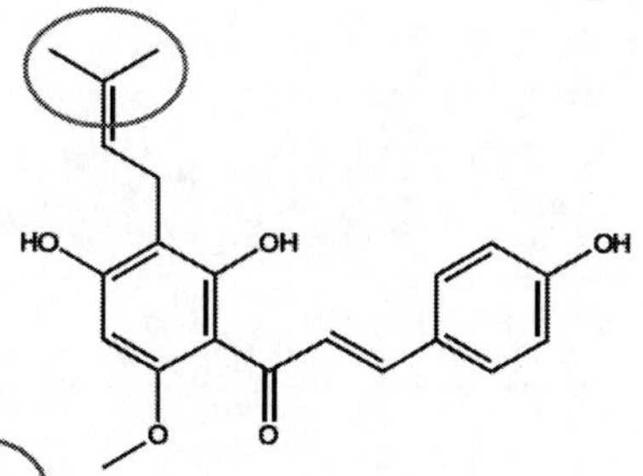

Xanthohumol

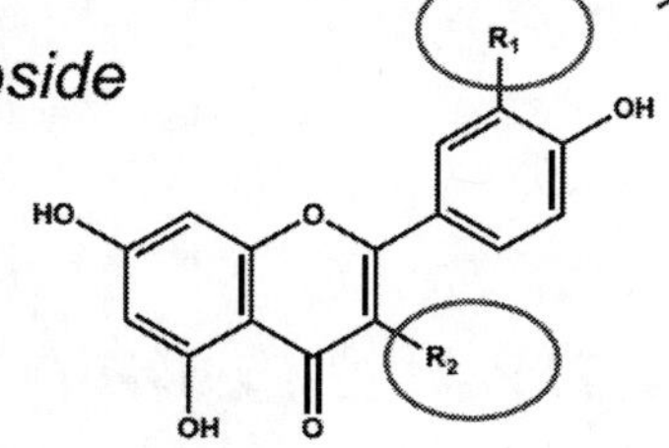

Kaempferol-3-O-glucoside	$R_1 = H$	R_2 = *Glucose*
Quercetin-3-O-glucoside	$R_1 = OH$	R_2 = *Glucose*

20

LC-MS/MS analysis of hop polyphenols in beer

Our method:

HPLC (Shimadzu) gradient method (methanol/water/formic acid)

HPLC column: Synergi Fusion-RP (Phenomenex), 50 x 2.0 mm

MS/MS (Sciex) detection: negative ion mode, scheduled MRM

Internal calibration standards: Dicamba, Bentazon, Nicarbazin

Direct injection of beer sample (after dilution with HPLC solvent)

Detailed method presented in poster P008 (Christina Schmidt)

21

LC-MS/MS analysis of hop polyphenols in beer

Pale Ale	Kaempferol glucoside	Quercetin glucoside	Multifidol glucoside	Xantho-humol
Regular*	1.1	1.3	2.0	0

* Kettle hopping only (Hallertau Tradition pellet type 90)

22

LC-MS/MS analysis of hop polyphenols in beer

Concentration increase by dry hopping (mg/l):

Pale Ale	Kaempferol glucoside	Quercetin glucoside	Multifidol glucoside	Xantho-humol
*Regular**	1.1	1.3	2.0	0
+ Bravo**	1.7	3.3	3.7	0.7

** Kettle hopping only (Hallertau Tradition pellet type 90)*

*** Additional dry hopping with 500 g / hl (Bravo pellet type 90)*

23

LC-MS/MS analysis of hop polyphenols in beer

Concentration increase by dry hopping (mg/l):

Pale Ale	Kaempferol glucoside	Quercetin glucoside	Multifidol glucoside	Xantho-humol
*Regular**	1.1	1.3	2.0	0
+ Bravo**	1.7	3.3	3.7	0.7
Threshold	0.5	0.9	1.8	3.5

** Kettle hopping only (Hallertau Tradition pellet type 90)*

*** Additional dry hopping with 500 g / hl (Bravo pellet type 90)*

24

How to analyze dry hopped beer?

- ✓ *Bitter resins: HPLC-UV (EBC 9.47)* +
- ✓ *Polyphenols: HPLC-MS/MS (negative ion mode)* +
- ✓ *Aroma compounds (GC)*
- – *Headspace GC-MS ?*

Our equipment for hop aroma analysis of beer

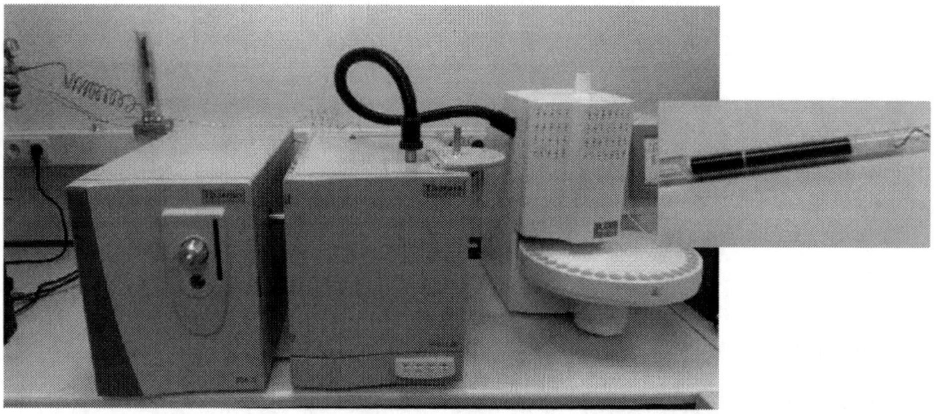

Focus GC with DSQ II	***Turbo Matrix^{TM}***
Mass Spectrometer	*Headspace-Trap Sampler*
(Thermo Scientific)	***(Perkin Elmer)***

26

Analysis of various dry hopped beers on hop aroma compounds according to HS-Trap GC-MS

Significant differences	Minor differences	Not detectable
*linalool**	*2-nonanone*	*e,z-1,3,5-undecatriene*
*geraniol**	*2-decanone*	*α- and β-pinene*
citronellol	*2-undecanone*	*propyl-2-methylbutanoate*
terpineol	*2-dodecanone*	*methyl-2-methylbutanoate*
*myrcene**	*2-tridecanone*	*ethylisobutanoate*
α-humulene	*β-damascenone*	*ethyl-2-methylbutanoate*
β-caryophyllene	*β-limonene*	*methylhexanoate*
β-farnesene	*methylnonanoate*	*methylheptanoate*
*2-methylbutylisobutanoate**	*methyldecanoate*	*methyloctanoate*
isobutylisobutanoate	*ethyldodecanoate*	*3-methylbutylpropanoate*

** Key aroma compounds as known from various papers*

27

Determination of key aroma compounds according to HS-Trap GC-MS (ppb)

Pale Ale	Myrcene	Linalool	Geraniol	2-Methylbutyl-isobutanoate
Regular	15	20	n.d.*	< 5

**n.d. = not detected*

28

Hop Analysis - New challenges by craft beer

Key hop aroma compounds in different beers (ppb):

Pale Ale	Myrcene	Linalool	Geraniol	2-Methylbutyl-isobutanoate
Regular	15	20	n.d.	< 5
Dry hopped 300 g/hl	293	329	56	73

Significant increase of key aroma compounds after dry hopping

29

Hop Analysis - New challenges by craft beer

Key hop aroma compounds in different beers (ppb):

Pale Ale	Myrcene	Linalool	Geraniol	2-Methylbutyl-isobutanoate
Regular	15	20	n.d.	< 5
Dry hopped 300 g/hl	293	329	56	73
Dry hopped 500 g/hl	513	485	77	109

Increase is correlated with amount of hops used for dry hopping

30

International (EBC/ASBC) collaborative pre-trial for analysis of beer according to HS-Trap GC-MS

Based on our published method (Schmidt and Biendl, 2016)

There are more than 10 brewery labs using this equipment

7 in the US, 4 in Germany were finally prepared to participate

- *4 beer samples (cans) were sent out to them in February*

Tested parameters:

- *Myrcene, linalool, geraniol, 2-methylbutylisobutanoate*

Collection of results not yet completed

31

Results from ring tests on hop aroma compounds

AHA (Arbeitsgruppe Hopfenanalyse) ring test (spring 2017):

US IPA* ppb	Myrcene	Linalool	Geraniol	2-Methyl-butylisobut.
AHA range**	42 - 590	175 - 452	390 - 1588	60 - 172

* same beer sample as analysed in the EBC/ASBC pre-trial, too

** 5 AHA labs using all different Headspace (HS) methods:

HS-SPME FID or GC-MS, HS-Trap GC-MS, HS GC-MS/MS

32

Results from ring tests on hop aroma compounds

AHA (Arbeitsgruppe Hopfenanalyse) ring test (spring 2017):

US IPA ppb	Myrcene	Linalool	Geraniol	2-Methyl-butylisobut.
*AHA range**	42 - 590	175 - 452	390 - 1588	60 - 172
*Hopsteiner***	571	452	390	145

** different Headspace GC methods*

*** Headspace-Trap GC-MS (average of duplicate analysis)*

33

Results from ring tests on hop aroma compounds

AHA ring test (spring 2017) vs. EBC/ASBC collaborative pre-trial:

US IPA ppb	Myrcene	Linalool	Geraniol	2-Methyl-butylisobut.
AHA	42 - 590	175 - 452	390 - 1588	60 - 172
*Hopsteiner**	571	452	390	145
*Bitburg**	649	380	460	137

** Headspace-Trap GC-MS (average of duplicate analysis)*

34

How to analyze dry hopped beer?

- ✓ Bitter resins: HPLC-UV (EBC 9.47) +
- ✓ Polyphenols: HPLC-MS/MS (negative ion mode) +
- ✓ Aroma compounds
  - For terpenes, terpene alcohols, esters: HS-Trap GC-MS +
  - For thiols: more sophisticated method (see Steinhaus L24)

Lab scale evaluation method for the determination of the dry hopping potential of different varieties

- ✓ Extraction solvent:

  50 ml Ethanol / 950 ml water buffered with citric acid to pH 4.5

- ✓ Dosage: 2.5 g milled hops (or pellets) / 1 l extraction solvent
- ✓ Extraction parameters: 24 hours / 20°C / moderate agitation
- ✓ Sample preparation before analysis: filtration

36

Analysis of filtrated aqueous solution

- ✓ *HPLC-UV (EBC 9.47)*
- ✓ *HPLC-MS/MS (negative ion mode)*
- ✓ *Headspace-Trap GC-MS*

Results from evaluation method for dry hopping

Hop Compound	Method	Apollo	Threshold
Alpha-Acids/ppm	*EBC 9.47*	*14.5*	*+/-10*
Humulinones/ppm	*EBC 9.47*	*28.2*	*+/-10*
Hulupones/ppm	*EBC 9.47*	*4.9*	*+/-10*

Results from evaluation method for dry hopping

Hop Compound	Method	Apollo	Threshold
Alpha-Acids/ppm	*EBC 9.47*	*14.5*	*+/-10*
Humulinones/ppm	*EBC 9.47*	*28.2*	*+/-10*
Hulupones/ppm	*EBC 9.47*	*4.9*	*+/-10*
Xanthohumol/ppm	*HPLC-MS/MS*	*0.4*	*+/-3*
Kaempferol-glucoside/ppm	*HPLC-MS/MS*	*0.4*	*+/-0.5*
Quercetin-glucoside/ppm	*HPLC-MS/MS*	*0.8*	*+/-1*
Multifidol-glucoside/ppm	*HPLC-MS/MS*	*1.6*	*+/-2*

39

Results from evaluation method for dry h[illegible]

Outlook: Correlation with sensory impressions?

Hop Compound	Method	Ap[illegible]	[illegible]shold
Alpha-Acids/ppm	*EBC 9.47*	[illegible]	*+/-10*
Humulinones/ppm	*EBC 9*[illegible]	[illegible]*.2*	*+/-10*
Hulupones/ppm	*E*[illegible]	*4.9*	*+/-10*
Xanthohumol/ppm	[illegible]*MS*	*0.4*	*+/-3*
Kaempferol-glucoside/p[illegible]	[illegible]*MS/MS*	*0.4*	*+/-0.5*
Quercetin-glucosid[illegible]	[illegible]*LC-MS/MS*	*0.8*	*+/-1*
Multifidol-gluc[illegible]	*HPLC-MS/MS*	*1.6*	*+/-2*
Myr[illegible]	*HS-Trap GC-MS*	*137*	*+/-100*
2-Meth[illegible]*tan./ppb*	*HS-Trap GC-MS*	*176*	*+/-100*
[illegible]*ol/ppb*	*HS-Trap GC-MS*	*44*	*+/-10*
[illegible]*alool/ppb*	*HS-Trap GC-MS*	*97*	*+/-10*
4-MMP/ppt	*Steinhaus (L24)*	*11.5*	*+/-2*

Many thanks to

- *Annika Lagemann, Stefan Hanke (Bitburg Brewery)*
- *Martin Steinhaus, Klaas Reglitz (DFA, Freising)*
- *Christina Schmidt, Timo Lambertsen (Hopsteiner, Mainburg)*
- *John P. Maye, Jeremy Leker, Bob Smith (Hopsteiner, Yakima)*

You for your attention

41

Solving the challenges of natural compressible filter aids for processing precoats on candle filters

EBC Congress 2017,
17th May 2017
Ljubljana, Slovenia

Dr. Jörg Zacharias,
Dr. Ralph Schneid

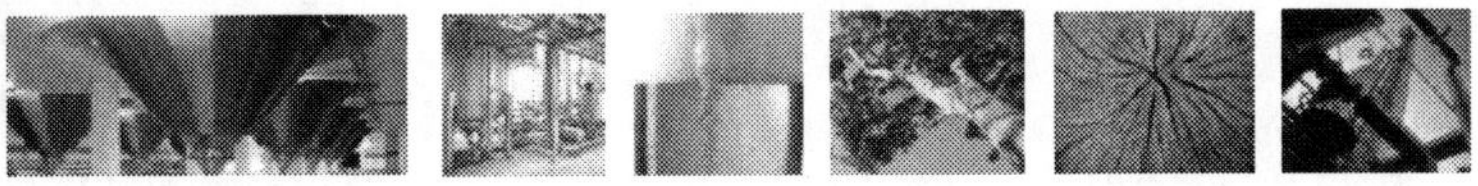

Author: Dr. J. Zacharias
Date: 16.05.2017

KRONES

Abstract

For the production of bright beer the precoat filtration is still a prevalent process. Alternative filter aids are still a considerable interest.

The authors spread their efforts with viscose fibres in previous publications.

→ But all these had in common that still some points had to be solved. Especially up scaling and stable production were difficult.

KRONES

Overview

1. Introduction
2. Development of viscose fibres for beer filtration
3. Precoat filtration with viscose fibres
4. Conclusion

KRONES

Introduction

KRONES

5 | KRONES | CRD Process Technologies

Global objective of the project: Viscose fibre as filter aid

Objective:

- Combine the advantages of the different filtration systems
- Develop a filter aid which
  - does not generate dust
  - is suitable for cost-neutral or profitable disposal
- Suitability for all existing precoat filter systems

Parameter	Precoat filtration
Quality of the filtrate	+
Flexibility	+
Quality of unfiltrate	+
Maintenance costs	+
Water consumption	+
Current consumption	+
Filter aids	+
Disposal of aids	+

KRONES

6 | KRONES | CRD Process Technologies

Viscose: Regenerate fibre on the base of cellulose

Viscose specialty fibres

- The base material for the production of viscose fibres is cellulose
  - Cellulose is dissolved in lye as a cellulose derivative and forms a honey-like textile pulp
  - When the textile pulp is extruded through a spinneret in an acid bath, the textile pulp is neutralized, the pure cellulose falls out and the viscose filaments are formed
  - The filaments are stretched, washed, dried and cut to fibres of any desired length
- The chemical and physical characteristics of the viscose fibres are tailored to customer requirements and are of a consistent quality

→ Viscose is no natural cellulose
→ It is a man-made fibre
→ It is a regenerate fibre on the base of cellulose

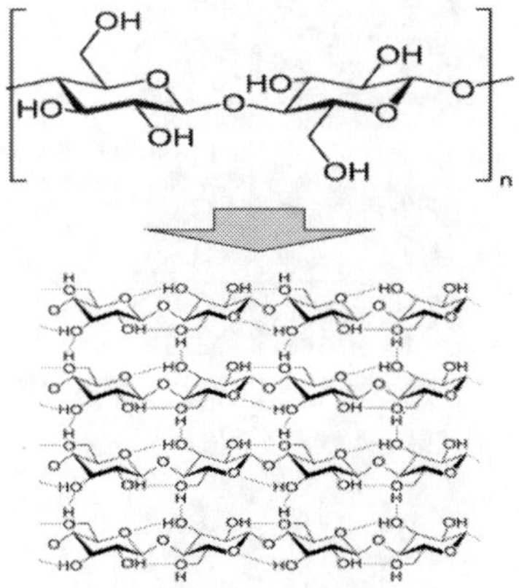

(Source: http://de.wikipedia.org/wiki/Zellstoff)

→ This offers unlimited possibilities for tailor-made shapes & functions

KRONES

Approach: Viscose fibres as filter aid for precoat filtration

Kieselguhr

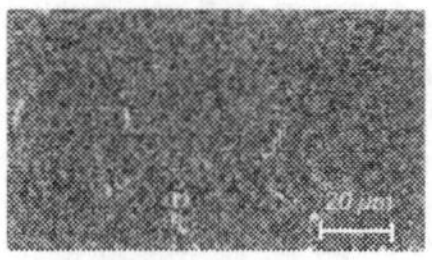

Image: Jasmin Sieger, https://de.wikipedia.org/wiki/Kieselgur

Diatoms

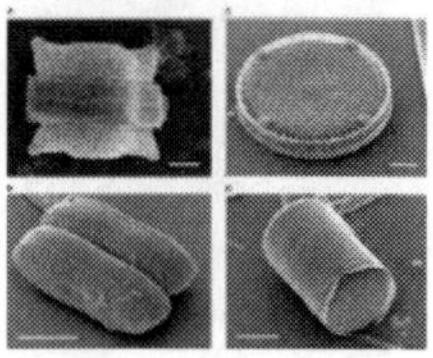

Images: Mary Ann Tiffany, San Diego State University. – Bradbury J. Nature's Nanotechnologists: Unveiling the Secrets of Diatoms. PLoS Biol 2(10)2004: e306. doi:10.1371/journal.pbio.0020306

Viscose fibres

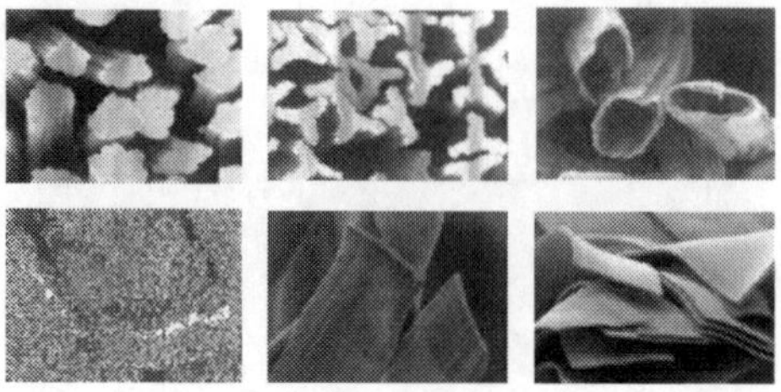

- Shape of cross section
- Surface structure

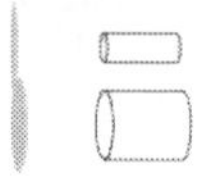

- Fibre diameter
  6,5 – 48 µm
  (0,5 – 27 dtex)

- Fibre length
  0,3 – 12 mm
  20 – 110 mm

Kelheim Fibres

KRONES

Natural cellulose ←→ Viscose fibres

(„cotton")

(Quelle: http://de.wikipedia.org/wiki/Baumwollfaser)

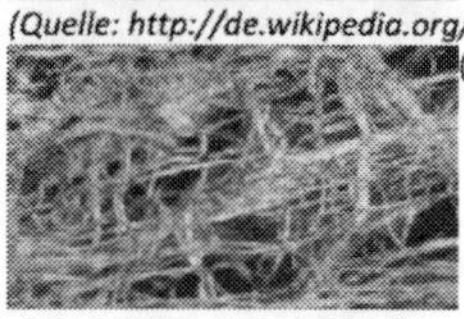

(„cellulose")

(Quelle: http://de.wikipedia.org/wiki/Zellstoff)

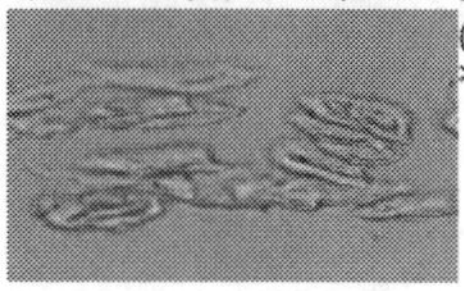

(„cellulose")
>cross section<

(Source: Kelheim Fibres)

(Source: Kelheim Fibres)

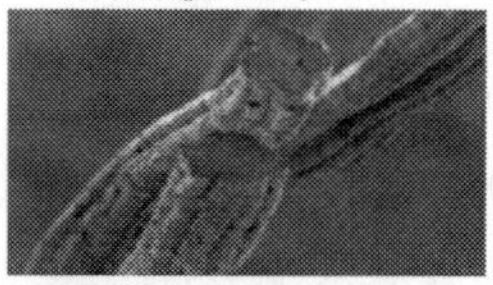

(„functionalized round fibre: ion exchanger fibre")

(„flat, plain surface")

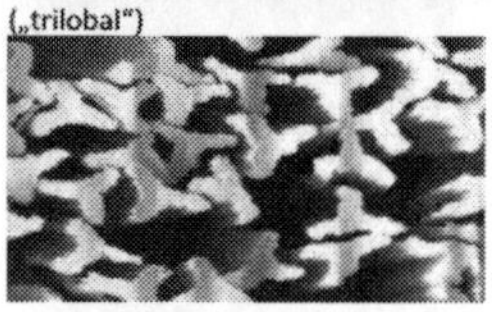

(„trilobal")

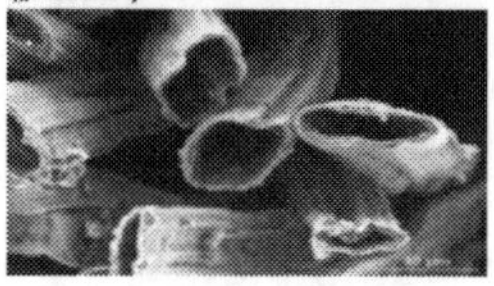

(„hollow")

(„round"),

Kelheim Fibres

KRONES

Development of viscose fibres for beer filtration

KRONES

Sedimentation ↔ precoat formation ↔ filtration

Sedimentation

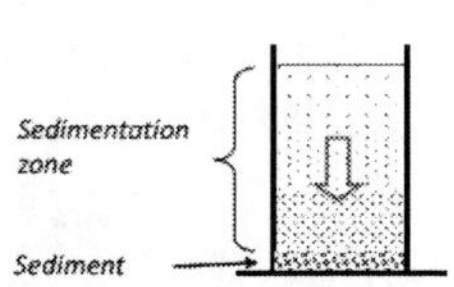

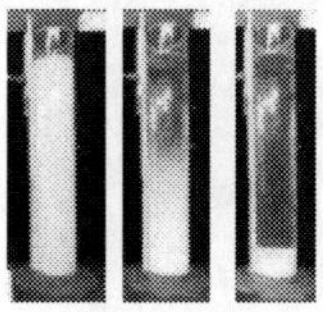

- Sedimentation parameters
- Particle / fibre parameters

⇒ Sediment formation and structure

⇒ Tool for initial estimate of filter cake structure

$$Porosity\ \varepsilon = \frac{V_{Pores}}{V_{Sediment}}$$

Precoat Formation

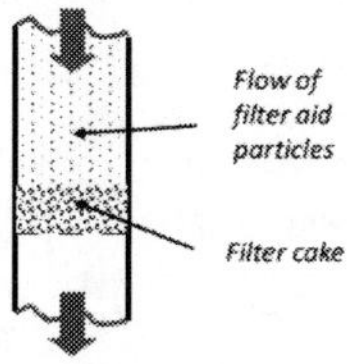

- Particle / fibre parameters
- Process parameters

⇒ Filter cake formation and structure

⇒ Filtration properties

⇒ Quality of filtrate (clear beer)

KRONES

Sedimentation, cake formation – effect of fibre geometry

Sedimentation of viscose fibre dispersions
Short cut (0,3 – 0,9 mm fibre length)

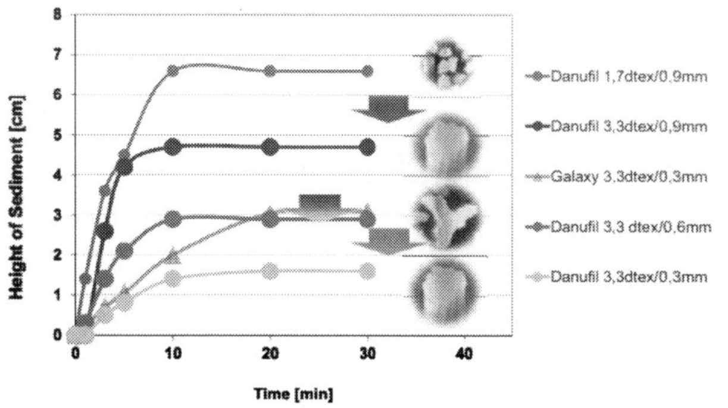

Densitiy of cake increases (porosity falls) with:

- falling Fibre length
- rising fibre diameter
- circular fibre cross section

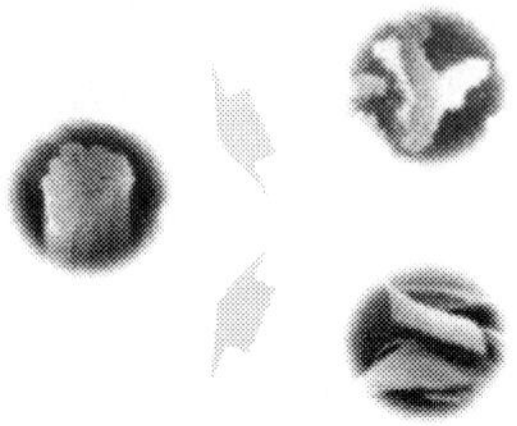

KRONES

Porosity of sediment, filtration – precision cut

Precision cut (flock), 100 µm fibre length

Sedimentation – Porosity $\varepsilon = \frac{V_{Pores\,(total)}}{V_{Filter\,Cake}}$

12 µm (wet)

Titre 0,9 dtex (dry fibre)
Diameter 9 µm
1,5 dtex (wet fibre)
⇒ Porosity (Sediment) ≈ 81 %

Titre 17 dtex (dry fibre)
Diameter 38 µm
29 dtex (wet fibre)
⇒ Porosity (Sediment) ≈ 64 %

Kieselguhr
⇒ Porosity (Sediment) ≈ 82-85 %

Filtration – pressure drop and efficiency

Variation of fibre diameter, shape of cross section and fibre mixtures

⇒ Porosity of filter cake similar or even lower than of filter kieselguhr but pore structure of cake too coarse

⇒ Very low pressure drop

⇒ Very low filtration efficiency (particle retention)

⇒ Even with shorter required fibres

KRONES

Fibre length distribution – precision cut and flock

Precision Cut (Guillotine)

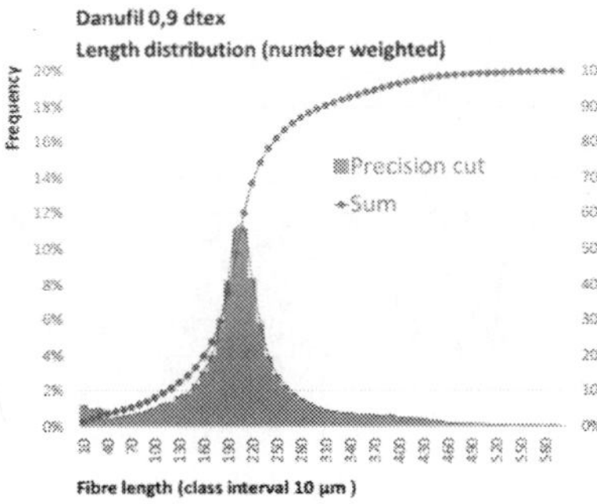

$\bar{L} \approx$ 209 µm (Number average)
$\bar{L} \approx$ 248 µm (Mass average)
(N = 60050)

Cutting Mill (Sieve Classifier)

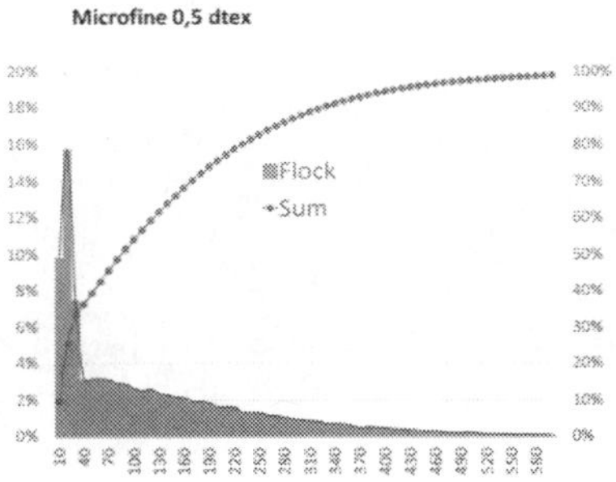

$\bar{L} \approx$ 133 µm (Number average)
$\bar{L} \approx$ 292 µm (Mass average)
(N = 60132)

Cutting Mill (Air Classifier)

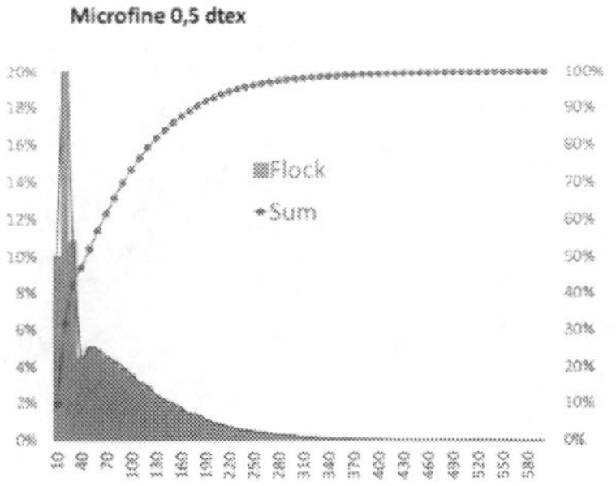

$\bar{L} \approx$ 74 µm (Number average)
$\bar{L} \approx$ 156 µm (Mass average)
(N = 100216)

KRONES

Microscopy – precision cut and flock

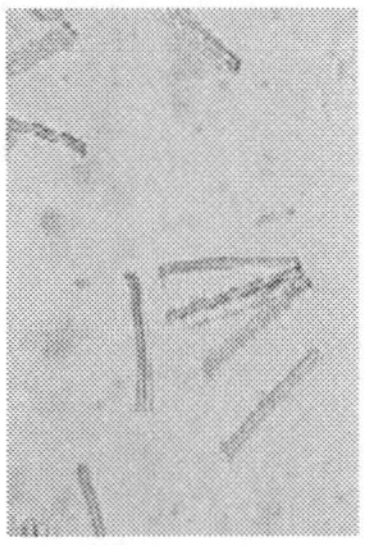

Precision cut

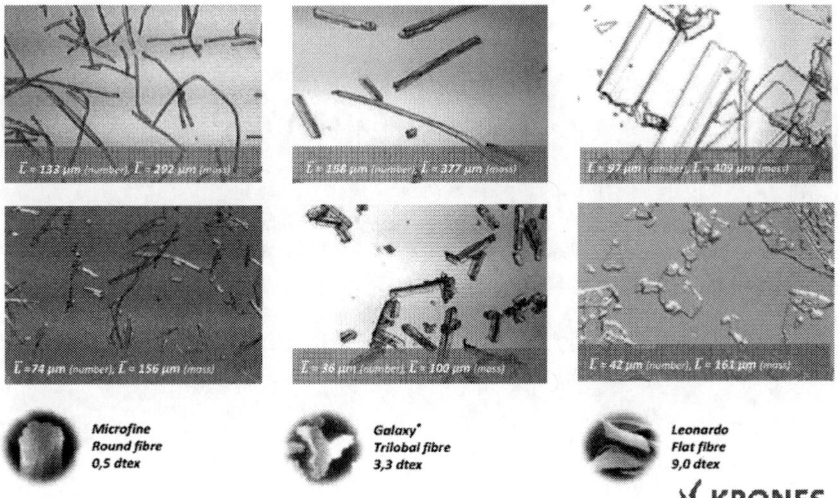

Danufil® Round fibre 1,7 dtex

Microfine Round fibre 0,5 dtex

Galaxy® Trilobal fibre 3,3 dtex

Leonardo Flat fibre 9,0 dtex

KRONES

Conclusion for the half way – development of viscose fibres

- Sedimentation of fibres
  ⇒ Estimation of filter cake formation
- Fibre length
  ⇒ Control of cake porosity
  ⇒ Short fibres deliver compact cake
- Fibre diameter and shape of fibre cross-section
  ⇒ Control of specific fibre surface
  ⇒ Control of total fibre surface in cake
  ⇒ Unavoidable simultaneous effect on porosity of cake
- A general quantitative estimation of cake porosity based on form parameters of fibres was not possible
- Hydraulic diameter of sediments and filter cake
  ⇒ Simple working model for pore size estimation (not shown here)
  ⇒ High filtration efficiency with short hydraulic diameters (low porosity of filter cake and high specific surface area of fibre)

KRONES

Precoat filtration with viscose fibres
-
Lab scale

KRONES

4 steps upscale to optimise viscose fibres

1.) Lab testing **2.) Lab test rig** **3.) Pilot plant** **4.) Industrial scale plant**

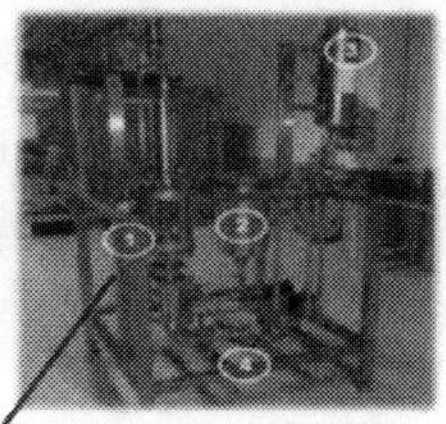

→ Special filter cake evaluation chamber
developed to analyze the shape and the functionality related to their filtration behavior.

KRONES

Definitions

- Permeability

$$\beta = \frac{V_L * h_k * \eta_L}{A * \Delta p * t} = v * \eta_L * \frac{h_k}{\Delta p}$$

β =	Durchlässigkeit in Darcy-Einheiten
v =	Filtergeschwindigkeit
V_L =	Filtratvolumen
h_K =	Höhe des Filterkuchens
η_L =	Viskosität des filtrierten Mediums (Wasser 20°C η = 1)
A =	Filterfläche (26,42 cm²)
Δp =	Druckdifferenz
t =	Durchflusszeit für V

- Compressibility

$$\frac{\beta_0}{\beta} = \left(\frac{p_S}{p_0}\right)^n$$

- $n \to 0$ Sehr schwach kompressibler bzw. inkompressibler Kuchen:
- $0<n<1$ „Mäßig kompressibler Kuchen"
- $n>1$ „Hoch kompressibler Kuchen"

- Porosity

$$\varepsilon = \frac{\rho_{Trocken;Einzelfaser} - \rho_{Nass;Kuchen}}{\rho_{Trocken;Einzelfaser}}$$

- Filtration

Evaluation based on curves:
- Turbidity 25°/90°
- Pressure difference across filter
- Filter performance

KRONES

Impact grinding and length on permeability

$$\beta = \frac{V_L * h_k * \eta_L}{A * \Delta p * t} = v * \eta_L * \frac{h_k}{\Delta p}$$

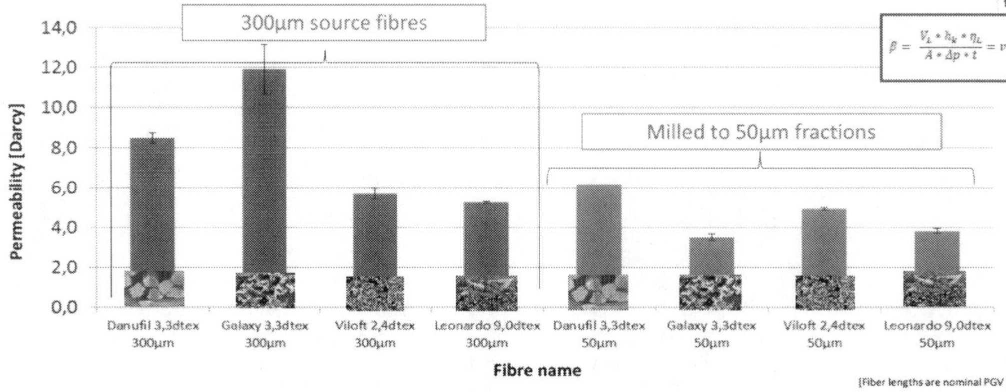

[Fiber lengths are nominal PGV priorities]

- All milled fibres have 50 microns with a lower permeability as their 300 µm source fibres.
- Danufil 50 µm shows greater permeability than the 300 µm source fibres Viloft and Leonardo.
- The Galaxy fraction has the lowest permeability.
- **<u>BUT: All the Darcy values are not in the area of Kieselguhr (0,2 to ~1,0)).</u>**

... Do we need that for the fibres??

What is missing? → The dependency on the cake flow velocity !!!

KRONES

Compressibility, functional phenomenon of viscose fibres

Analysis of a known issue of alternative filter aids

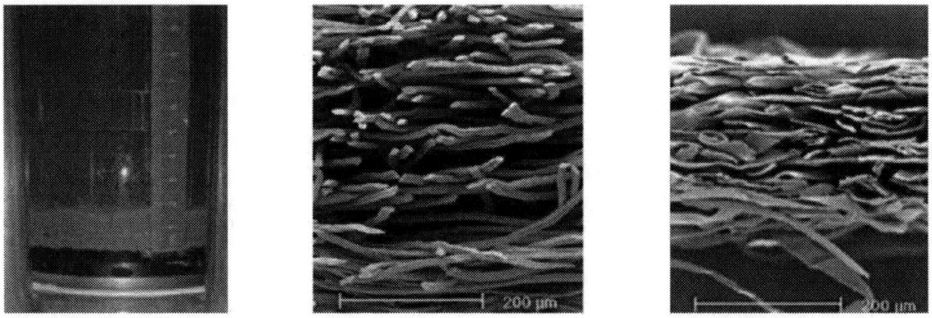

Figures: Filter cake in different compression levels

- It is <u>possible to influence</u> the compressibility of the filter cake <u>and to adjust</u>.
- <u>The compressibility is therefore part of the filtration strategy.</u>

KRONES

Exemplary Results of Compression Tests: Cake Height Development

Bandwidth in exemplary form factor and variant titles

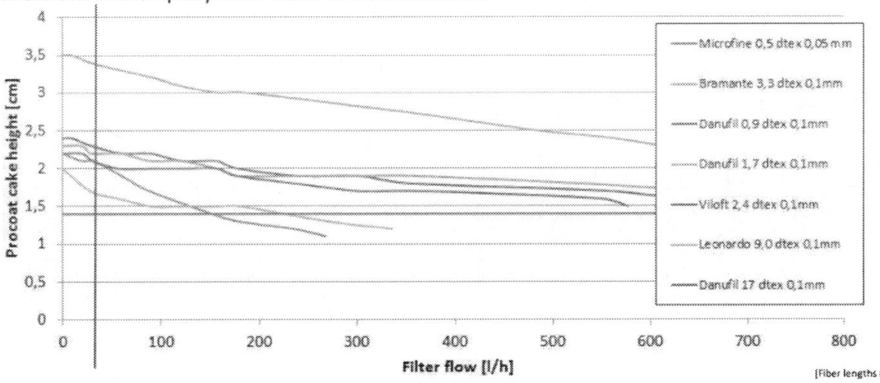

[Fiber lengths are nominal PGV priorities]

➢ Danufil 17 dtex shows no compression

KRONES

Exemplary Results of Analysis of the Influence of Form Factor and Length

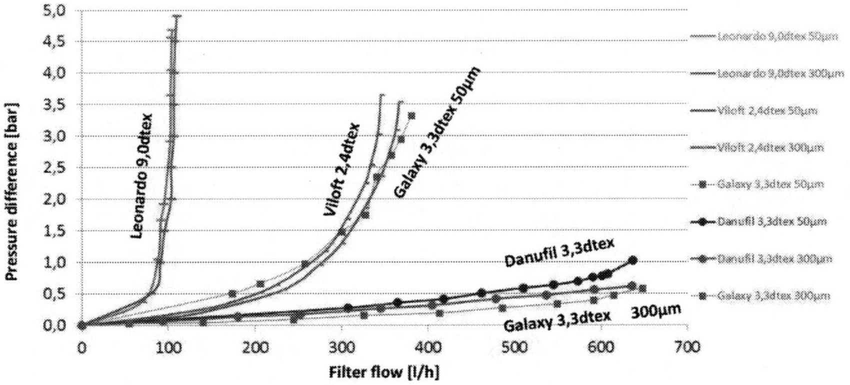

$$\frac{\beta_0}{\beta} = \left(\frac{p_s}{p_0}\right)^n$$

- Danufil, Microfine: the higher the titer the lower the pressure difference
- Leonardo: strikingly rapid increase of the pressure difference
- The compressibility is largely dominated by titer and the form factor.

[Fiber lengths are nominal PGV priorities]

➢ The "most" milled fibres 50 µm achieve the same pressure differences as their output fibers 300 µm
➢ ... except the Galaxy fibres

KRONES

Conclusion of the basic filtration test rows - ... on the two-thirds way

As a result it was shown that:

- The permeability and porosity does not lead directly to comparable correlations.
- Small titer and short lengths show better turbidity values.
- The compressibility must be taken into account.
- The form factor is important for the pressure buildup.

- Expectation: **Compressibility drives permeability!** ... and starts filtration !!!

Upscale to a pilot plant with selected fibres is possible!

KRONES

Precoat filtration with viscose fibres

-

Pilot plant trails

A) The candle

KRONES

Upscale: Beer filtration on the pilot plant → Benchmark: Turbidity

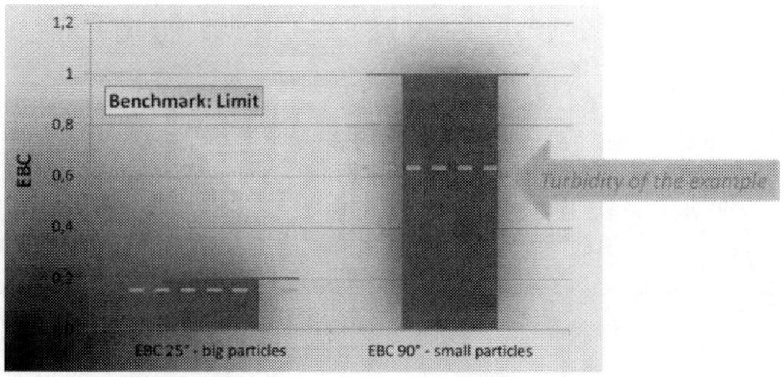

- Comparable retention rates / cut-offs need to be achieved !

KRONES

The twin-flow principle

- Bypass flow (typical parameter for kieselguhr)
  - A recommended bypass flow can be 12% of the nominal filter
  - The bypass flow is adapted to the respective filter aid (FA)

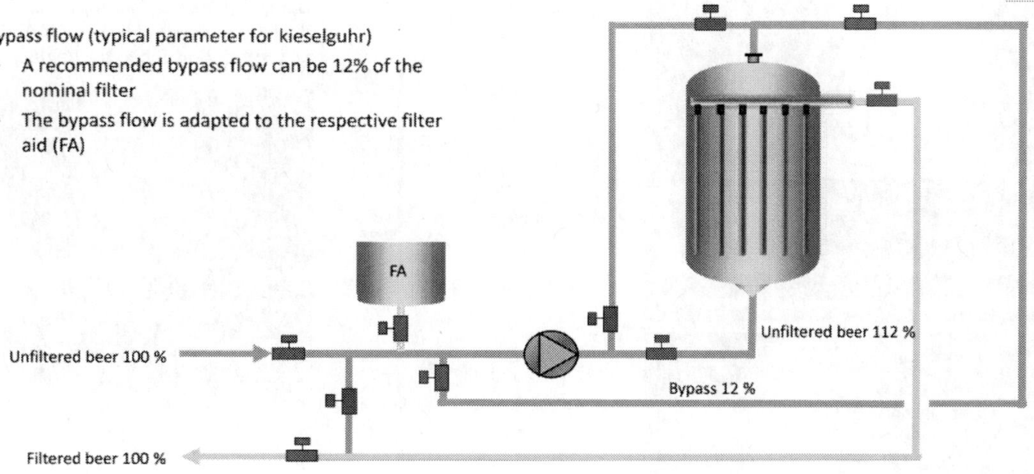

KRONES

The Venturi effect

- How is the precoat growing?

→ Radial OR growth from top to bottom in several layers?

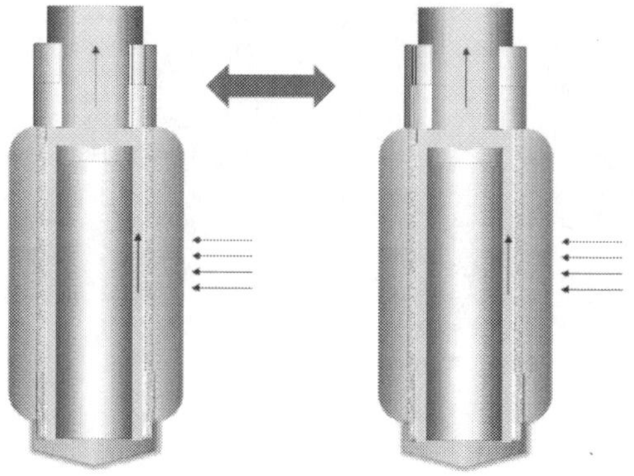

Radial precoating "Growing"precoating

→ For fibres the "growing" precoating is found in the prevailing cases !

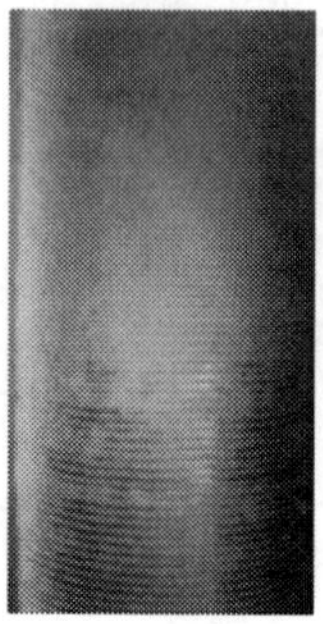

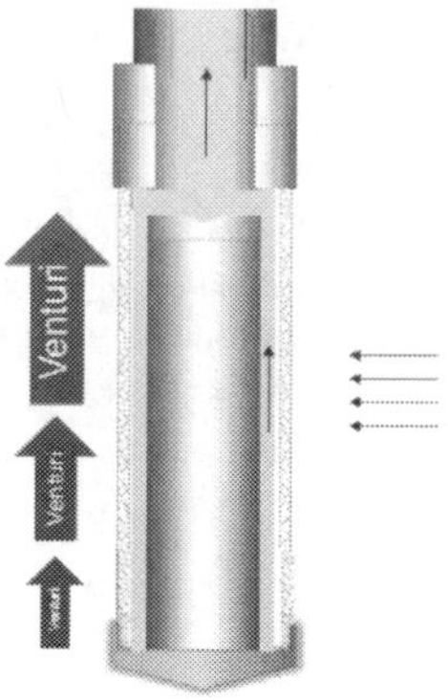

KRONES

Result: From unevenness in growth leads to a uniform, cohesive layer

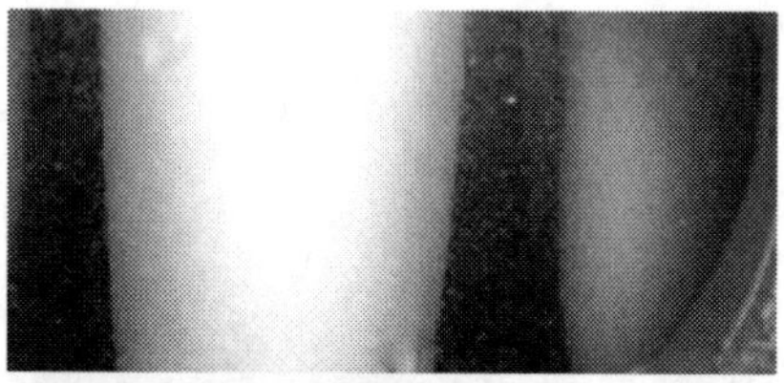

- Conical top to bottom structure

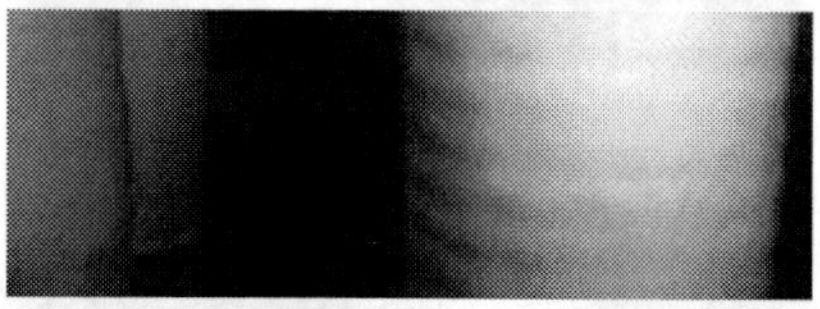

- Fluffy wave structure

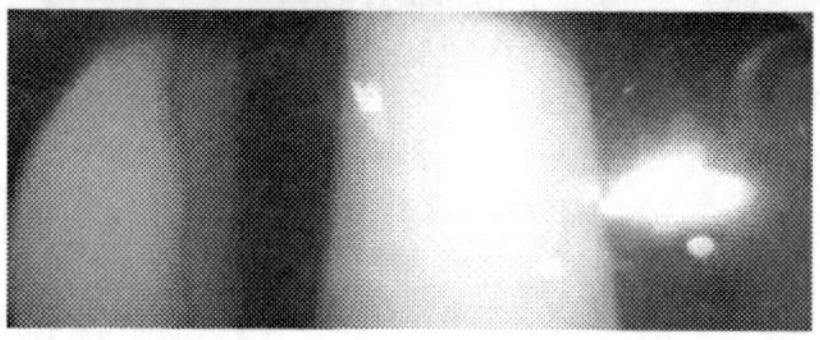

- Pear shape

- Optimisation of the flow inside and outside the candles needs to be done to achieve a uniform suction due to the Venturi effect

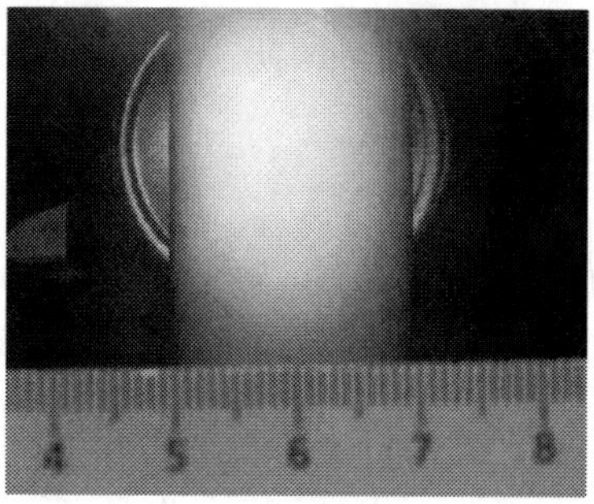

KRONES

Precoat filtration with viscose fibres
-
Pilot plant trails

B) The process

KRONES

The turbidity adjusting pressure control for viscose fibers

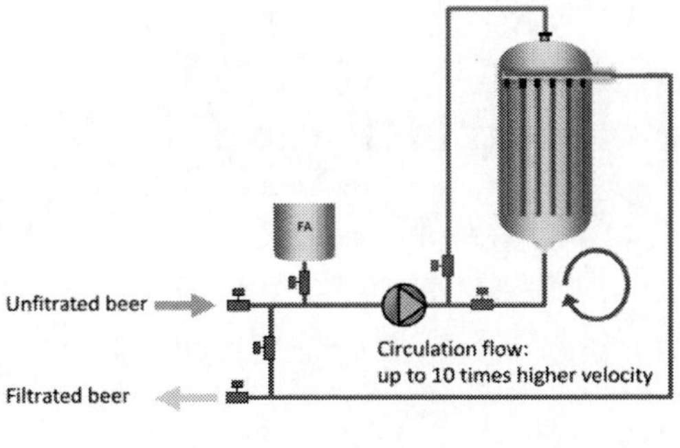

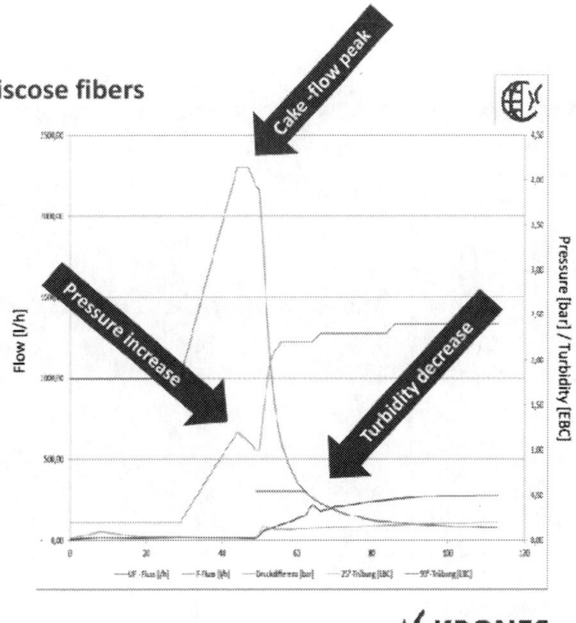

KRONES

Minimising the peak for long term filtrations

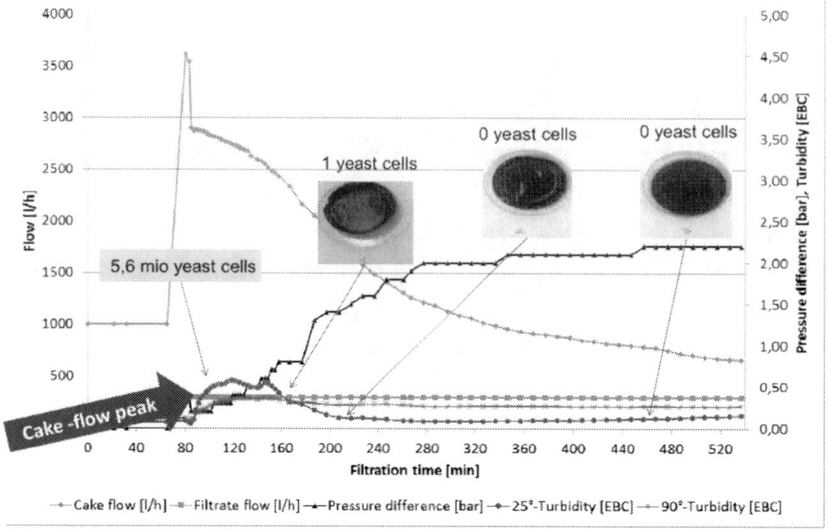

KRONES

Technological results of the practical filtration

KRONES

33 | KRONES | CRD Process Technologies

Results beer analyses pilot trials 05/2017 (over all analysis to demonstrate suitability)

Analysis	unfiltered Beer	Kieselguhr-Filtrate	Viscose-Filtrate
Original Extract [%]	11.70	11.35	10.77
Real Extract [% w/w]	4.12	3.93	3.75
Alcohol [% vol.]	4.98	4.86	4.59
Degree of Attenuation app. [% w/w]	80.3	81.0	80.7
Colour [EBC]	7.0	6.6	6.6
pH-Value [-]	4.33	4.33	4.40
Viscosity [mPa*s] (12%)	1.557	1.537	1.541
Bitterunits [BE]	18	16	16
Total Nitrogen [mg/L] (12%)	944	914	916
$MgSO_4$-perceptible N [mg/L] (12%)	219	183	196
FAN [mg/L]	139	130	128
Total-Polyphenols [mg/L]	264	217	223
Anthocyanogene [mg/L]	70	62	63
β-Glucan [mg/L]	245	234	220
SO_2 [mg/L]	6.6	5.5	4.6
Photometric Iodine Reaction [dE]	0.41	0.23	0.24
Head Retention NIBEM 30 [s]	--	293	301
Turbidity 90°/ 25° [EBC] 20°C	> 99	0.19 / 0.10	0.25 / 0.09
Iron [µg/L]	10	156	27
Copper [µg/L]	67	125	71

*due to pilot plant equipment slight dilution with the dosing

➢ **Why this big table of data?**

Technische Universität Berlin

KRONES

34 | KRONES | CRD Process Technologies

Results beer analyses pilot trials 05/2017 (*calculated in comparison to 12% original extract)

Analysis	unfiltered Beer	←*	Kieselguhr-Filtrate	←*	Viscose-Filtrate	←*
➡ Original Extract [%]	11.70	12	11.35	12	10.77	12
Real Extract [% w/w]	4.12	4.23	3.93	4.16	3.75	4.18
Alcohol [% vol.]	4.98	5.11	4.86	5.14	4.59	5.11
Degree of Attenuation app. [% w/w]	80.3	80.3	81.0	81.0	80.7	80.7
➡ Colour [EBC]	7.0	7.2	6.6	7.0	6.6	7.4
pH-Value [-]	4.33	4.33	4.33	4.33	4.40	4.40
Viscosity [mPa*s] (12%)	1.557	1.557	1.537	1.537	1.541	1.541
Bitterunits [BE]	18	19	16	17	16	18
Total Nitrogen [mg/L] (12%)	944	944	914	914	916	916
$MgSO_4$-perceptible N [mg/L] (12%)	219	219	183	183	196	196
FAN [mg/L]	139	143	130	138	128	143
Total-Polyphenols [mg/L]	264	271	217	229	223	249
Anthocyanogene [mg/L]	70	72	62	66	63	70
β-Glucan [mg/L]	245	251	234	249	220	245
➡ SO_2 [mg/L]	6.6	6.8	5.5	5.8	4.6	5.2
Photometric Iodine Reaction [dE]	0.41	0.42	0.23	0.25	0.24	0.27
➡ Head Retention NIBEM 30 [s]	--	--	293	293	301	301
➡ Turbidity 90°/ 25° [EBC] 20°C	> 99	> 99	0.19 / 0.10	0.20 / 0.11	0.25 / 0.09	0.28/ 0.10
➡ Iron [µg/L]	10	10	156	162	27	30
➡ Copper [µg/L]	67	69	125	130	71	79

*due to pilot plant equipment slight dilution with the dosing

➢ **To show: There is no influence on the general technological parameters.**

➢ **And, having positive influence on the oxidation.**

Technische Universität Berlin

KRONES

ESR-Measurements

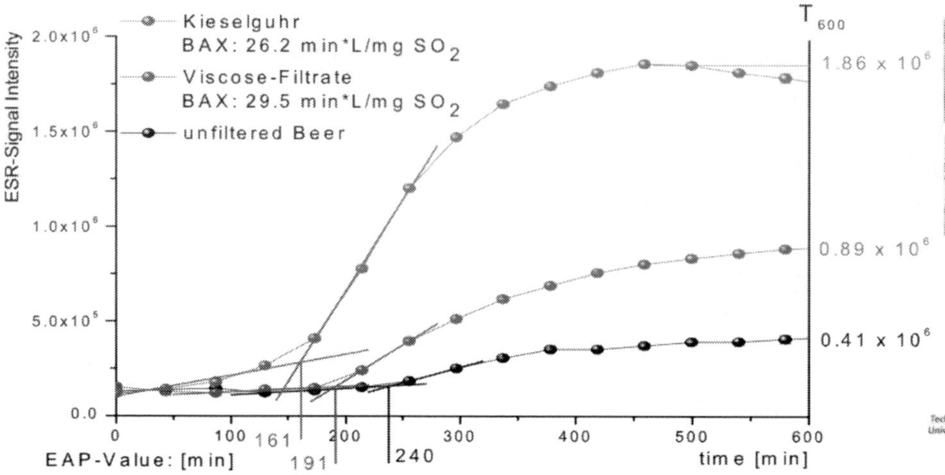

➢ **No negative influence of viscose fibres !**

KRONES

Conclusion

KRONES

Conclusion

- After the modification of the candles, compressible filter aids could be precoated almost without problems.
- An increase of the cake flow resulted in a suitable cake diameter and the compression became visible.
- From this point onwards, attempts to retain the filter cake with beer or a yeast suspension were carried out.
- The discrepancy of minimal pressure loss was solved with a higher cake flux.
- Control of turbidity in consideration of compression and porosity of the precoat is possible.
- In the end, this yields a solution in avoiding instable filtration results.
- Technological parameters at least comparable to kieselguhr are reached, with less oxidation potential.
- AND: Viscose fibres are completely biodegradable and also usable in biogas plants

KRONES

BITBURGER BRAUGRUPPE
STARKE MARKEN

Improved HS Trap GC-MS analysis of hop aroma compounds in dry hopped beer

Annika Lagemann, Dominique Dixius, Stefan Hanke, Georg Stettner

Bitburger Braugruppe GmbH. Römermauer 3. 54634 Bitburg/Germany e-mail: annika.lagemann@bitburger-braugruppe.de

Introduction

To give beer its bitter taste and to prolong its shelf life, brewers have added hop for a long time. Lately, beer brewers have put more attention to the flavor potential.

Dry hopping can increase aroma components of hop oil in beer, which are substantially composed of hydrocarbons (monoterpene, sesquiterpene), oxygenated compounds (terpenic alcohols, sesquiterpenic alcohols) and sulphur containing compounds (thioester, sulfide). To analyze the hop flavor in dry hopped beer quantitatively, a recently published HS GC-MS method (Schmidt & Biendl 2016) has been used and optimized within this study. Approaches for optimizing were focused on Headspace Trap conditions.

Within this project, 126 volatile substances with origin from hops could be identified in dry hopped beer and 18 of them were chosen for their aroma activity and determined in a single run method.

A first analysis of dry-hopped beers does not clearly show correlations between quantitative amounts of singular hop oil components and the intensity and specification of the hop flavor in beer but rather point to the importance of the composition of hop oil and the interaction of those components. Following steps focused on optimal conditions of dry hopping. In particular, quantitative differences of hop aroma compounds based on varied contact times were examined in order to find a potential possibility to accelerate the brewing process.

Method and Results

Beer was pretreated by the addition of NaOH, ethanol concentration was set to 5.2 %Vol, NaCl and THF were added, and spiked with the internal standard butanol. Beer analysis was performed using a Clarus 680 gas chromatograph coupled to a Clarus SQ 8 T mass spectrometer (Perkin Elmer, Rodgau, Germany). The TurboMatrix HS-40 Trap (Air Monitoring Trap M041-3628) was used as a HS sampler (sampling parameters are given in table 1). A Optima 1701 MS capillary column (60 m length x 0.25 mm i.d.; film thickness, 0.25 µm) was used for chromatographic separation. Helium served as carrier gas with a column head pressure of 200 kPa controlled by the HS-Trap sampler. The GC temperature program was from 45 °C (held for 2.0 min) up to 100 °C at a rate of 3 °C/min (held for 0 min), then up to 200 °C at a rate of 5 °C/min (held for 0 min) and further up to 270 °C at a rate of 40 °C/min (held for 10 min). The MS transfer line was set to 220 °C and the ion source temperature was 250 °C. The mass spectrometer was operated in selected ion monitoring (SIM) mode using electron ionization (70 eV). The analytes were detected in time windows. The identification was done on the basis of their retention times and the mass spectrum in comparison to the standard compounds.

Table 1 HS-trap sampling conditions for hop aroma compounds

parameter		value	
		Schmidt & Biendl, 2016	Bitburger, 2017
	temperatures		
oven		85°C	85°C
needle		90°C	100°C
transfer line		130°C	130°C
trap low		40°C	40°C
trap high		280°C	280°C
	pressures		
column		150 kPa	200 kPa
vial		240 kPa	240 kPa
desorption		150 kPa	120 kPa
	times		
thermostat		45 min	45 min
dry purge		8 min	*4 min*
heating hold / trap hold		15 min	25 min
desorb		0.6 min	0.2 min
pressurization		1.0 min	1.0 min
trap load ("decay")		1.6 min	*0.9 min*
# cycles		2 cycle	2 cycle
	options		
dry purge		On	On
outlet split		On	On

Optimization contains mainly adjustments of pressures and times of HS-Trap conditions, which lead to an increased resolution and intensity of analytes. E. g. lowering the trap load time effected improved chromatographic resolution of terpenic alcohols (sign.1). Reducing the dry purge time increased the intensity significantly because of the simultaneously occurred loss of ethanol and terpenic alcohols during dry time.

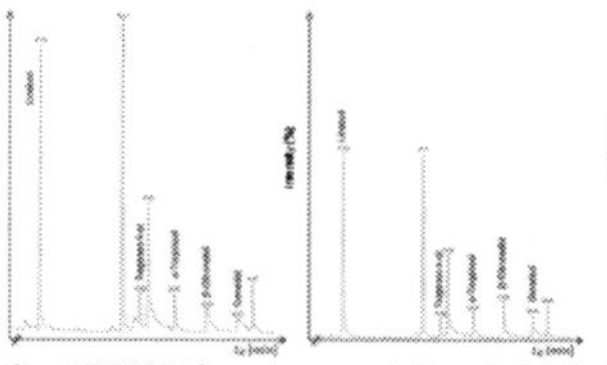

Sign. 1 Full-SCAN-Chromatogramm-part of terpenic alcohols: left: Trap Load of 1.6 min; right: Trap Load of 0.9 min

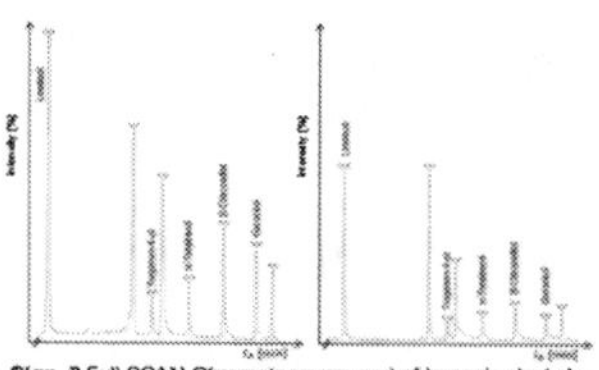

Sign. 2 Full-SCAN-Chromatogramm-part of terpenic alcohols: left: dry purge of 1 min; right: dry purge of 8 min

As a first step, we analyzed 26 different beer samples to get an impression of concentration differences, and a potential correlation between the aroma compounds and sensory tests. Results are shown in table 2 as range next to one example, which is a dry-hopped pale ale (Craftwerk Tangerine Dream).

Table 2 Amounts [µg/L] of hop aroma compounds of 26 different beer samples and Craftwerk Tangerine Dream Pale Ale

compound	*range [µg/L]* Min	Max	Craftwerk Pale Ale Tangerine Dream
Ethylisobutanoate	3,8	309,1	31
Isopentylacetate	173,7	4634,9	1688
Isobutylisobutanoate	0,5	56,8	57
Camphen	0,0	0,3	0,2
β-Myrcene	0,3	409,3	287
D-Limonen	0,1	1,6	1,6
Ethylhexanoate	114,7	488,8	149
Isopentylisobutanoate	0,1	33,9	12
2-Methylbutylisobutanoate	1,2	193,4	85
p-Cymene	0,0	0,2	0,1
Linalool	5,4	462,8	94
Terpinen-4-ol	0,4	10,4	2,5
α-Terpineol	3,4	88,9	24
β-Citronellol	0,9	98,6	9,6
Geraniol	11,6	363,2	86
β-Caryophyllen	0,0	13,8	6,2
β-Farnesen	0,0	6,8	0,8
α-Humulen	0,0	86,2	43

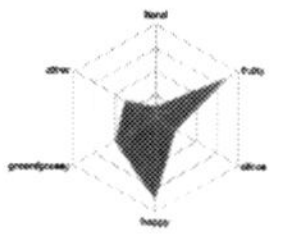

Sign. 3 Sensory profile of Craftwerk pale ale Tangerine Dream

A correlation between sensory attributes and the concentration of individual aroma compounds is visible but not clear. Sensory evaluation was done by a trained taste panel internally. The tested pale ale showed a high concentration of the fruity ester isobutylisobutanoate (table 2) which correlated with the fruity impression of the sample (sign. 3). Other aroma active substances which are discussed for the hoppy flavor (myrcene) were determined in a medium value of 287 µg/L, and the beer sample was described as strong noticeable hoppy. These findings pointed out the known importance of the interaction of aroma compounds.

The other focus of our study was checking the influence of contact time of hops during dry hopping. Therefore we used an unfiltrated, separated Pilsener beer with no salient hoppy flavor. Approaches were done in semi-technical 6 hL scale at -1.5°C. Hop addition (307 g/hL corresponds to about 2 mL hop oil/hL) was done in two different ways: with 2 h pumping after addition and without pumping into the tanks. The hop was the variety "Hallertauer" from harvest 2015 grown in the Hallertau (alpha acids content EBC 7.7 HPLC: 2.4%).

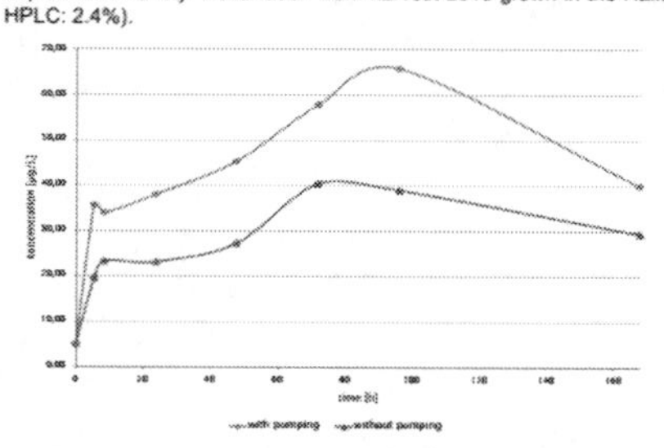

Sign. 4 Course of linalool concentration depending on contact time of hops

As shown in sign. 4 results revealed, that higher concentrations of aroma compounds were detected when pumping the beer for two hours after the hop addition. Concentrations of all detected aroma compounds increased compared to the approach without pumping The increase of linalool concentration averaged about 58%. Furthermore the course of concentration of nearly half the substances showed a maximum level of transferred aroma components between 50 and 80 hours.

Conclusion and Outlook

The optimization of Headspace-Trap-autosampler conditions in combination with the selection of different ion masses for detection resulted in a significantly higher intensity and improved chromatographic resolution of the aroma compounds´ signals. Additional changes in sample preparation lead to a single run method for 18 substances.

Further steps will be the quantification of dry hopped beers with varied hops contact times between 50 and 80 hours to verify our findings. Parallel sensory tests of these beers should be performed to show the influence of different concentrations depending on contact times of hops. Furthermore investigation of sensory and analytical differences of fresh and aged sample beer samples should be performed.

References

Schmidt, C. & Biendl, M.:Headspace Trap GC-MS Analysis of Hop Aroma Compounds in Beer. Brewing Science, pp. 9-15, 69, 2016

Aberl, A. & Coelhan, M.: Determination of Volatile Compounds in Different Hop Varieties by Headspace-Trap GC/MS - In Comparison with Conventional Hop Essential Oil Analysis. J. Agric. Food Chem. 2785-2792, 60, 2012

36th EBC Congress, Ljubeljana, Slovenia, May 14 – 18. May 2017

Improved PCR detection of beer spoiling bacteria –

optimised enrichment with NBB and innovative lateral flow detection

P004 Agnes J. Huber,
Döhler GmbH, Darmstadt, Germany

Introduction

PCR-based detection methods for beer spoiling bacteria have gained importance in microbiological quality control since PCR permits a fast and reliable detection. However, cultural enrichment is still essential to detect trace contaminations. (see Fig. 1)

We aimed to develop a fast and universal enrichment broth with an optimised nutrient composition for beer spoiling bacteria compatible with the most common PCR detection kits.

A novel PCR detection, followed by an easy visualization using a lateral flow system similar to a pregnancy test, was also evaluated in combination with the new broth to prove compatibility.

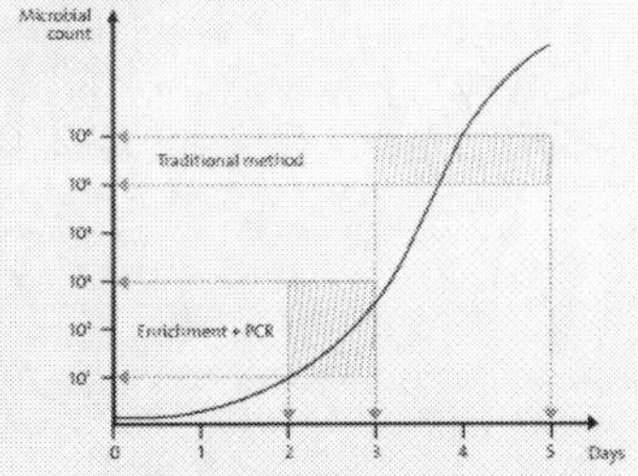

Fig. 1: Microbial growth curve for comparing the sensitivity of traditional cultural methods and PCR.
Traditional culture media need approx. 10^5 - 10^6 microorganisms/ml for optical detection. In comparison PCR detection is possible with 10^1 - 10^3 microorganisms. For microbial trace contaminations, an enrichment step is still necessary for successful PCR detection.

Enrichment with NBB®-PCR

Detection sensitivity

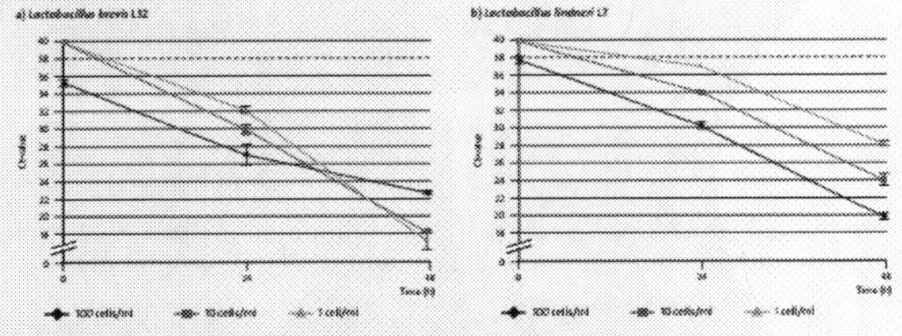

Fig. 2: Enrichment of microbiological trace contamination with NBB®-PCR Broth – real-time PCR analysis
a) C_t values of *Lactobacillus brevis* L32 and b) C_t-values of *Lactobacillus lindneri* L7 with 1; 10; 100 cells/ml inoculated in 15 ml NBB®-PCR Broth. For this, the cells were counted microscopically. Sampling took place after 0 h, 24 h, and 48 h of anaerobic incubation. The real-time PCR detection was done with a 1 ml sample, which was analysed according to manufacturer specification with the GeneDisc® Beer Spoilage Bacteria Kit (Pall Corporation). Data points are mean values of a threefold determination; the standard deviation is indicated. A C_t value of >38 means that the detection of beer spoiling bacteria is not reliable (dotted line).

- Bacteria concentrations of 100 cells/ml are detected at the beginning of the experiment – results at 0 h
- After 24 h of incubation in NBB®-PCR, bacteria are detected in each tested cell concentration
- After 48 h of incubation, bacteria reached C_t values of at least 30 even with a low initial cell count

▶ **Enrichment of trace contaminations with NBB®-PCR Broth for a reliable detection is possible within 48 to 72 h even for slow-growing bacteria like *L. lindneri*.**

Compatibility with common, commercially available PCR kits

	L. brevis BLQ6	*L. lindneri* L7	*P. damnosus* BLQ12	*P. frisingensis* DU2	Unspiked NBB®-PCR
foodproof® Beer Screening Kit (BIOTECON Diagnostics GmbH)	+	+	+	+	–
First Beer PCR Kit / L. lindneri PCR / P. damnosus PCR (GEN-IAL® GmbH)	+	+	+	+	–
GeneDisc® Beer Spoilage Bacteria (Pall Corporation)	+	+	+	+	–
PCR detection Kit B™ LP Screening & PCR detection kit B™ Pectinatus sp. (PIKA Weihenstephan GmbH)	+	–*	+	+	–

+ = C_t-value < 38; - = C_t-value >38 (not detectable); *Growth of bacteria was determined using a different detection kit

Table 1: Real-time PCR detection with commercially available detection kits in combination with NBB®-PCR
10 ml NBB®-PCR Broth were spiked with 100 cells/ml of *Lactobacillus brevis, Lactobacillus lindneri, Pediococcus damnosus* and 1000 cells/ml of *Pectinatus frisingensis* respectively. The cells were counted microscopically and the PCR analysis was performed according to the manufacturers instructions after 48 h of anaerobic incubation. Data points are mean values from a threefold determination. These external validation tests were performed at the Research Center Weihenstephan for Brewing and Food Quality (Technical University Munich)

- All real-time PCR systems succeeded in finding the respective beer spoiling bacteria after 48 h
- No false positives or PCR inhibition were observed

▶ **NBB®-PCR Broth is compatible with the most common PCR detection kits for beer spoiling bacteria.**

Methods

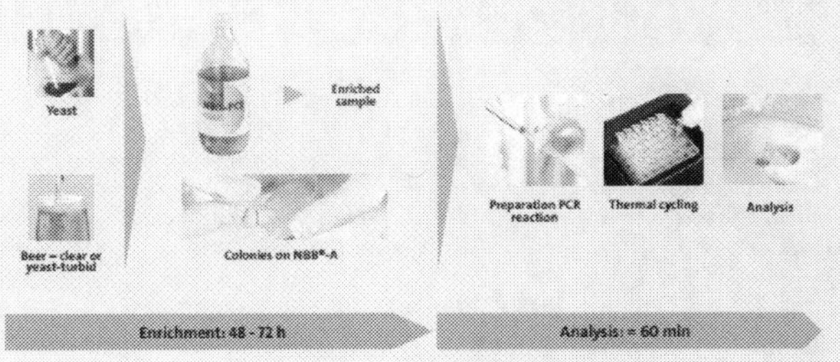

Enrichment of different sample types with NBB®-PCR:

- Yeast samples: 9 ml of NBB®-PCR Broth are added to 1 ml of yeast.
- Clear beer samples (e.g. pilsner): Filtered through a membrane, incubated in 10 ml of NBB®-PCR Broth.
- Non-filterable samples (e.g. wheat beer): Mixed with NBB®-PCR Broth, 1:1 ratio.

All samples are incubated anaerobically for 48 - 72 h at 28 - 30°C and analysed with a PCR detection kit for beer spoiling bacteria.

For PCR analysis with DMD® Screen and Ident Kits, no purification step is needed:

- Add 2 µl sample to prepared PCR mastermix
- 50 min PCR reaction in thermal cycler
- Transfer 2µl of PCR mixture below blue area of the detection strip
- Visible result after 5 min assay buffer migration

Lateral Flow Detection with DMD® Screen and Ident Kits

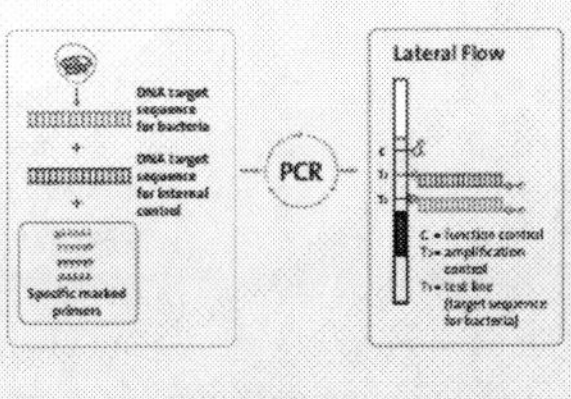

Fig. 3: The technology behind the DMD® Screen and Ident Kits
During **PCR**, the DNA of a target microorganism – here shown in blue – is specifically multiplied with the help of an enzyme called polymerase (amplification). The internal control DNA used to show if the reaction was inhibited or not – here shown in red – is amplified as well. The amplified DNA is then made visible on the **lateral flow** using antibodies labeled with gold particles.

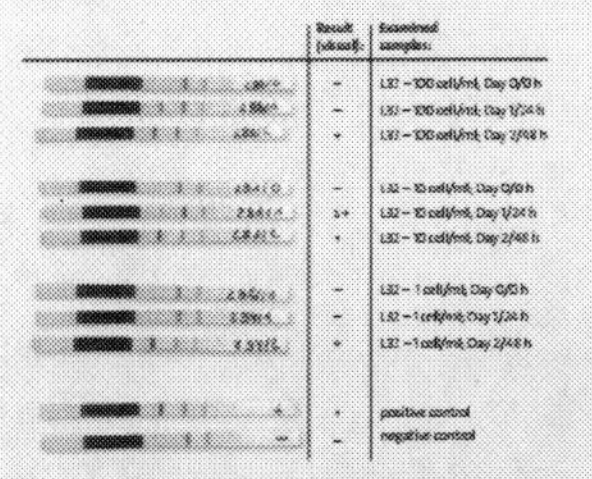

Fig. 4: Enrichment of microbiological trace contaminations with NBB®-PCR Broth – lateral flow analysis with DMD® Screen and Ident Kit
NBB®-PCR was inoculated with 1; 10; 100 cells/ml as described in Fig. 2. Samples were collected after 0 h, 24 h, and 48 h of anaerobic incubation, 2 µl of each were analysed with DMD® Lactobacillus Pediococcus Screen Kit and DMD® Universal Detection Module (threefold determination, data only shown for *L. brevis* L32).

- After 24 h of incubation in NBB®-PCR, bacteria are detected in the tubes with cell concentrations of 100 cells/ml and 10 cells/ml for *L. brevis* and 100 cells/ml for *L. lindneri*
- After 48 h incubation, reliable detection of a low initial cell count – 1 cell/ml – is possible with DMD® Lactobacillus Pediococcus Screen Kit
- Negative control and amplification controls perform as expected – no false positives or PCR inhibition are observed

▶ **Reliable detection of trace contaminations with DMD® Screen and Ident Kits is possible after enrichment with NBB®-PCR Broth.**

Conclusion

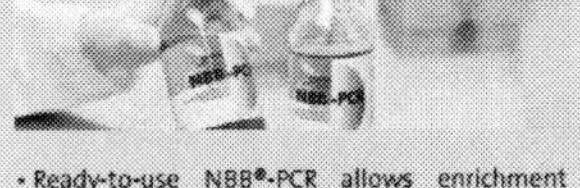

- Ready-to-use NBB®-PCR allows enrichment of beer spoiling bacteria within 48-72 h from all sample types arising in a brewery
- It is compatible with the most common, commercially available real-time PCR kits for beer spoiling bacteria
- DMD® Screen and Ident Kits are easy to use: no need for DNA purification and require minimal hands-on time
- Simple visual analysis of PCR products on a test strip allows detection without sophisticated know-how
- The use of an internal amplification control provides safe and reliable results

Author's contact information: Agnes Huber, R&D Teamleader, Doehler GmbH, e-mail: agnes.huber@doehler.com

KU LEUVEN

FLAVOUR PLUS

Volatile fingerprinting of coriander and orange peel in view of improved flavour-consistency of specialty beers and distilled spirits

Filip VAN OPSTAELE, Gert De Rouck, Valérie Winne, Guido Aerts, and Luc De Cooman

INTRODUCTION

Particular flavouring herbs play a crucial role in the unique sensory properties of specialty beers. Variability in intrinsic aromatic quality of the herbs (due to origin, climate conditions, processing, storage, ...) represents however a serious problem in view of consistent flavour quality of the end product. In this respect, a methodology based on HS-SPME GC-MS/O is used for (1) fast and accurate quality control of botanicals used in brewing and distilling, and (2) improved control of aromatisation based on key-impact odorants.

MATERIALS AND METHODS

Raw materials: coriander (*Coriandrum sativum* L; origin: Ukraine, Italy, Romania and orange peel (*Aurantii dulcis pericarpium tot*; origin: Ghana, Senegal, Turkey); **Alcohol extracts**: 4g ground coriander or orange peel in 40 mL Ethanol/water (75/25% v/v), 7 days; **Analytical profiling** of ground coriander/orange peel and alcohol extracts via HS-SPME GC-MS; **Determination of key-odorants** in coriander and orange peel via HS-SPME GC-MS/O; **Sensory evaluations** of alcohol extracts and beer aromatised with alcohol extracts via descriptive and triangular tests

DISCRIMINATION OF CORIANDER AND ORANGE PEEL VIA HS-SPME GC-MS

The bi-plots of PCA based on the total composition of the volatile fraction (variables) of coriander and orange peel samples (objects) demonstrate the separation of the samples according to their origin/aromatic characteristics. This approach allows to evaluate the aromatic quality of incoming samples. Clearly, breweries receive coriander or orange peel samples with different aromatic properties (see e.g. Brewery 3 (Brew3), incoming coriander C1 (Brew3-C1) vs incoming coriander C2 (Brew3-C2)).

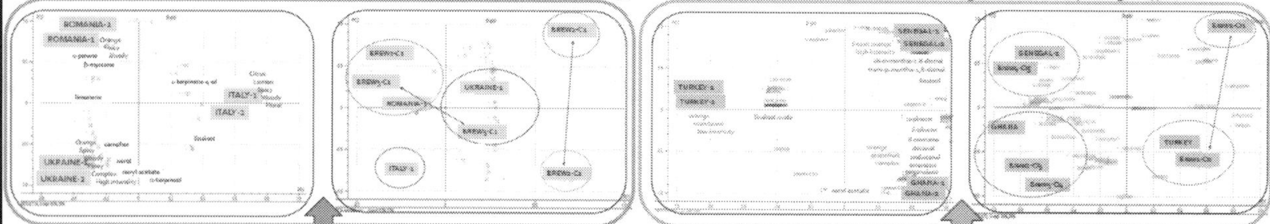

DETERMINATION OF KEY-IMPACT COMPOUNDS

Linalool

= Predominant key-odorant in coriander (55-70% of total volatiles). Level is used for standardised aromatisations (equal linalool levels).

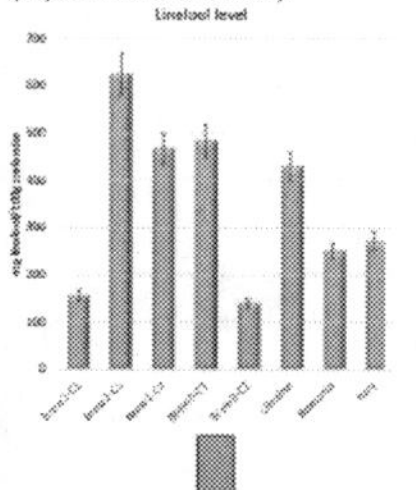

CORIANDER

- α-pinene
- β-myrcene
- limonene
- linalool
- camphor
- nerol
- neryl acetate
- α-terpineol
- α-terpinen-4-ol

Key-impact compounds for odour/aroma

MONITORING incoming samples
- absolute level
- relative composition

AROMATIC PROPERTIES

On the basis of additions for **aromatisation** in brewing/distilling practice for **improved aroma-consistency**

ORANGE PEEL

- α-pinene
- β-pinene
- β-myrcene
- limonene
- terpinolene
- linalool
- decanal
- undecanal
- neryl acetate
- α-terpineol
- linalool oxide
- *cis*-p-mentha-2,8-dienol
- *trans*-p-mentha-2,8-dienol

Limonene

= predominant key-odorant in orange peel (>90% of total volatiles). Level used for standardised aromatisations (equal limonene levels).

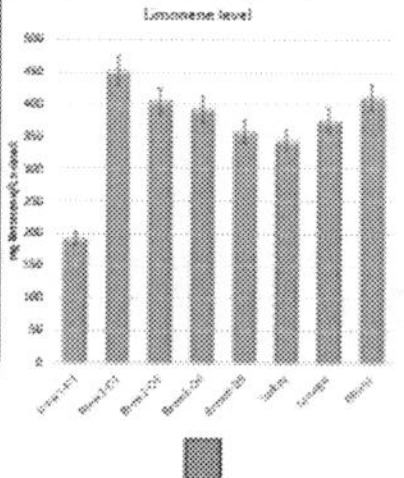

SENSORY EVALUATIONS CORIANDER

Triangular tests – Alcohol extracts (10% v/v) – FRESH SAMPLES	Correct answ.(out of 12) (α=sign.lev.)
No standardisation (equal vol.)	
Ukraine vs Italy	11 (α=0.001)
Ukraine vs Romania	9 (α=0.05)
Italy vs Romania	10 (α=0.01)
Ukraine vs Brew3-C1	5 (α=0.3)
Brew2-C1 vs Brew2-C2	11 (α=0.001)
Brew3-C2 vs Romania	6 (α=0.20)
Standardised extracts (equal linalool)	
Ukraine vs Italy	9 (α=0.05)
Ukraine vs Romania	6 (α=0.20)
Italy vs Romania	9 (α=0.05)

Alcohol extracts (fresh coriander)

°aromatic differences between different origins/brew samples
°differences less pronounced with standardised extracts (i.e. equal linalool level)

Alcohol extracts (stored coriander, 12months, 20°C)

°decrease of aroma intensity, no off-flavours via GC-MS/O
°via standardisation (linalool level): aroma fresh ≈ aged

Beer aromatised with alcohol extract (fresh and stored)

°addition level determines whether origins are discriminated
°standardised additions: differences less pronounced
°addition: equal linalool level: no difference fresh/stored

Beer aromatisation with alcohol extract – FRESH	Correct answ. (/12) (α=sign.lev.)
Addtion rate: 250 μL extract	
Ukraine vs Romania	5 (α=0.3)
Addition rate: 500 μL extract	
Ukraine vs Romania	10 (α=0.01)
Addition rate: (standardised addition)	
Ukraine vs Romania	6 (α=0.20)
Ukraine vs Italy	8 (α=0.10)
Beer aromatisation with alcohol extract - STORED	
Addition rate: 500 μL extract	
Ukraine 0 months vs 12 months	10 (α=0.01)
Addition rate: standardised addition	
Ukraine 0 months vs 12 months	5 (α=0.3)

CONCLUSIONS Volatile fingerprinting of botanicals via HS-SPME GC-MS is a fast and powerful tool for evaluating the aromatic quality of incoming samples. When combining the total analytical passports with data derived from key-impact odorants for each type of botanical, brewers and distillers dispose of a powerful tool for improved control of aromatisation of the final product (beer/spirit). The high potential of the proposed methodology in view of improved flavour-consistency was demonstrated via (advanced) aromatisation of final beer with alcohol extracts and will be further investigated via conventional aromatisation in real brewing practice.

Acknowledgements: Flanders Innovations Entrepreneurship (Vlaio) is thanked for financial support of the TETRA-project 150155 'Consistent production of alcoholic beverages aromatised with herbs'

KU Leuven, Technology Campus Ghent| Faculty of Engineering Technology | Dept. Microbial and Molecular Systems (M2S) |
Cluster Bio-engineering Technology (CBeT) | Lab of Enzyme, Fermentation, and Brewing Technology (EFBT) | Gebroeders De Smetstraat 1, B-9000 Ghent, Belgium |
E filip.vanopstaele@kuleuven.be

LC-MS/MS analysis of hop flavonoids in dry-hopped beers

Christina Schmidt and Martin Biendl

Hopsteiner - Hallertauer Hopfenveredelungsgesellschaft mbH (HHV), Mainburg, Germany

Introduction

The spectrum of bitter substances in hops includes α- and β-acids but also prenylflavonoids and glucopyranosides (**Figure 1**). Multifidol glucosides were isolated from hops in 2005, for the first time (**1**). These compounds are intermediate products of the biosynthesis of α- and β-acids. Both multifidol glucosides and prenylflavonoids (e.g. xanthohumol) are typical for hops. The composition of quercetin and kaempferol glucosides in hops is suitable to differentiate hop varieties (**2**). New findings about key bitter compounds in hops and their contribution to the bitter profile of beer were introduced and discussed in literature recently (**3**,**4**).

Whereas HPLC-UV analysis of selected bitter substances like α-acids or humulinones (oxidized α-acids) in dry-hopped beers is feasible, the analysis of hop flavonoids like multifidol, kaempferol and quercetin glucosides requires the use of HPLC-MS/MS technique. For this reason, the present study was executed to investigate a suitable HPLC-MS/MS method for monitoring hop flavonoids in dry-hopped beers.

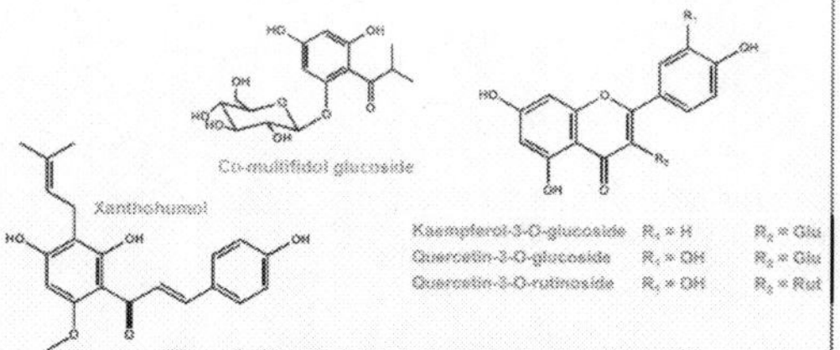

Figure 1. Chemical structure of selected hop flavonoids

Materials and methods

	hop product	hop variety	amount for dry hopping
Trial 1:			
Beer No.1	**pellets type 90**	**Apollo**	**500 g/hl**
Beer No.2	**pellets type 90**	**Bravo**	**500 g/hl**
Beer No.3	**pellets type 90**	**Calypso**	**500 g/hl**
Beer No.4	**pellets type 90**	**Denali**	**500 g/hl**
Beer No.5	**pellets type 90**	**Lemondrop**	**500 g/hl**
Trial 2:			
Pale Ale No.1	pellets type 90	Lemondrop	500 g/hl
Pale Ale No.2	pellets type 90	Bravo	500 g/hl

Table 1. Dry hopping conditions for both brewing trials.

Analyses of degassed beer samples without further clean-up procedures were performed after addition of internal standards (dicamba, bentazon and nicarbazin) and dilution by LC-MS/MS. The HPLC system (Shimadzu Corporation, Kyoto, Japan) was coupled with a mass spectrometer (SCIEX, Darmstadt, Germany) running in the negative ion mode. The quantitation was done using the scheduled MRM mode of the instrument with the fragmentation parameters optimized prior to analysis. Data processing and integration was performed by using Analyst software and MultiQuant (SCIEX, Darmstadt, Germany). For chromatography, an analytical 50 x 2.0 mm Synergi 4µ Fusion-RP 80A column (Phenomenex, Aschaffenburg, Germany) served as the stationary phase. Quantitation was done by external calibration in a range between 1 and 500 ng/ml.

Beers tested in this study were from 2 different brewing trials. For the first trial, the basic beer (Pale Ale) contains alpha extract only (**5**). This initial beer was then split into different parts and each was dry-hopped using one of the five varieties shown in table 1. For the second trial, another Pale Ale basic beer was split too and dry-hopped with two different hop varieties (**Table 1**). Hallertau Tradition (pellets type 90) has been used for the Pale Ale basic beer (trial 2) during wort boiling.

Results

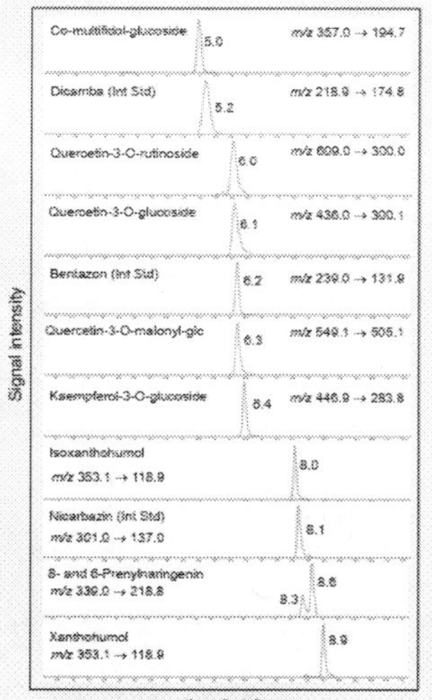

Figure 2. LC-MS/MS chromatograms of hop flavonoids in dry-hopped beers.

The developed HPLC-MS/MS method allows the identification and quantitation of 9 hop flavonoids in dry-hopped beers using a single LC-MS/MS run with selective mass transitions as given in Figure 2.

The quantitative data for both brewing trials are summarized in Table 2. The basic beer of trial 1 was prepared using alpha extract only for bittering. None of the substances tested was detected by this method in the beer. The results for beers No.1-5 show the influence of dry hopping on the hop flavonoid composition. The highest amount for co-multifidol glucoside was observed in beer No. 2 with hop variety Bravo. The lowest amount was detected in beer No. 4, dry-hopped with the hop variety Denali. Beer No. 2 (with Bravo) showed the highest amounts for all substances tested. 6- and 8-prenylnaringenin were not detected in the dry-hopped beers.

The analysis of Pale Ale basic beer (trial 2) gives quantitative data for 7 of 9 hop flavonoids tested. The results show the entry of these substances to the beer during wort boiling already. The impact of dry hopping on the hop flavonoids pattern in beer is obvious after subtraction of the results from the Pale Ale basic beer.

For the ratio of quercetin and kaempferol glucosides in beer No. 5 and Pale Ale No. 1, both dry-hopped with the hop variety Lemondrop, it could be observed that the amounts of these 2 compounds were equal in these 2 beers whereas the concentration of quercetin glucoside was higher in comparison to kaempferol glucoside in all other beers. The composition of quercetin and kaempferol glucosides is genetically determined and therefore depend on the variety (**2**).

Table 2. Amounts of hop flavonoids in dry-hopped beers in mg/L (both brewing trials).

	Trial1						Trial 2		
Compound	**Basic beer**	**Beer No.1**	**Beer No. 2**	**Beer No. 3**	**Beer No. 4**	**Beer No. 5**	Basic beer (Pale ale)	Pale Ale No.1	Pale Ale No.2
Co-multifidol glc	n.d.	1.5	3.3	1.2	0.4	1.6	2.0	3.5	3.7
Quercetin-3-O-glc	n.d.	0.7	1.7	0.6	0.5	0.4	1.3	2.2	3.3
Quercetin-3-O-rutinoside	n.d.	0.5	0.5	0.3	0.2	0.3	0.5	0.5	1.0
Quercetin-3-O-malonyl-glc	n.d.	0.3	1.2	1.3	0.4	0.6	1.3	1.4	3.4
Kaempferol-3-O-glc	n.d.	0.2	0.4	0.2	0.1	0.5	1.1	2.2	1.7
Isoxanthohumol	n.d.	0.1	0.1	0.1	0.1	0.1	1.3	0.7	0.9
Xanthohumol	n.d.	0.1	0.2	0.2	0.1	n.d.	n.d.	0.3	0.7
6-Prenylnaringenin*	n.d.	n.d.	n.d.	n.d.	n.d.	n.d.	0.1	0.1	0.1
8-Prenylnaringenin	n.d.	n.d.	n.d.	n.d.	n.d.	n.d.	n.d.	n.d.	n.d.

*6-Prenylnaringenin was quantified with 8-Prenylnaringenin

To evaluate the influence of bitter compounds on the overall bitter profile of a dry-hopped beer, the knowledge of so called flavour thresholds is necessary.

Dresel et al. published in 2015 human recognition threshold concentrations of taste compounds found in hops (**3**). The flavour threshold for co-multifidol glucoside is 1.8 mg/L, for quercetin glucoside 0.9 mg/L, and for kaempferol glucoside 0.5 mg/L (**3**). The concentrations of these 3 bitter compounds in the beers from trial 2 exceed the flavour threshold clearly and contribute to the overall bitter profiles of these dry-hopped beers. Also the amounts of co-multifidol glucoside and quercetin glucoside in beer No. 2 from trial 1 exceed the flavour thresholds clearly.

Conclusions

The developed HPLC-MS/MS method allows the identification and quantitation of selected hop flavonoids in dry-hopped beers using a single LC-MS/MS run. Dry-hopped beers produced with different hop varieties showed significant differences in their hop flavonoids pattern. The contribution of these bitter substances to the overall bitter profile of beer could be evaluated using the quantitative data and the flavour thresholds known from literature.

References

1. Bohr, G.; Gerhäuser, C.; Knauft, J.; Zapp, J.; Becker, H.: Anti-inflammatory Acylphloroglucinol Derivatives from Hops (*Humulus lupulus*). J. Nat. Prod., **2005**, 68, 1545-1548.
2. Kammhuber, K: Differentiation of the World Hop Collection by Means of the Low Weight Molecular Polyphenols. BrewingScience, **2012**, 65, 16-23.
3. Dresel, M.; Dunkel, A.; Hofmann, T.: Sensomics Analysis of Key Bitter Compounds in the Hard Resin of Hops (*Humulus lupulus* L.) and Their Contribution to the Bitter Profile of Pilsner-Type Beer. J. Agric. Food Chem., **2015**, 63, 3402-3418.
4. Dresel, M.; Vogt, C.; Dunkel, A.; Hofmann, T.: The Bitter Chemodiversity of Hops (*Humulus lupulus* L.). J. Agric. Food Chem., **2016**, 64, 7789-7799.
5. Product data sheet "Alpha Extract" (https://hopsteiner.com/wp-content/uploads/2016/02/SS_51_07_Alpha_Extract.pdf)

P009 Chemical studies on bitter acid oxides derived from stored hops (*Humulus lupulus* L.)

KIRIN

Yoshimasa Taniguchi[1], Yasuko Matsukura[1], Kazutoshi Shindo[2], Takahiro Yamazaki[3], Mikio Katayama[3]
[1]Kirin Company Limited, Central Laboratories for Key Technologies, [2]Japan Women's University,
[3]Kirin Company Limited, Research Laboratories for Health Science & Food Technologies

Introduction

Hop-derived bitter acids, such as α-, β- and iso-α-acids, affect beer quality and display various physiological effects. They readily oxidize; however, chemical information on bitter acid oxides has been limited and their effects on beer quality or potential health benefits are not well understood. Here, we summarize our recent studies on bitter acid oxides.

Bitter acid oxides in stored hops

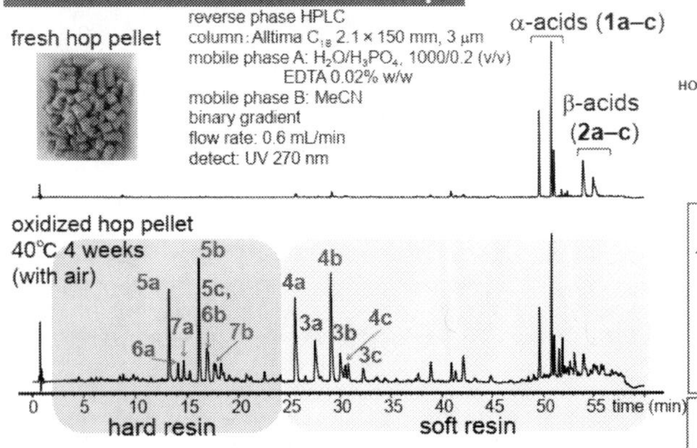

concentration changes in hops stored at 40°C (with air)

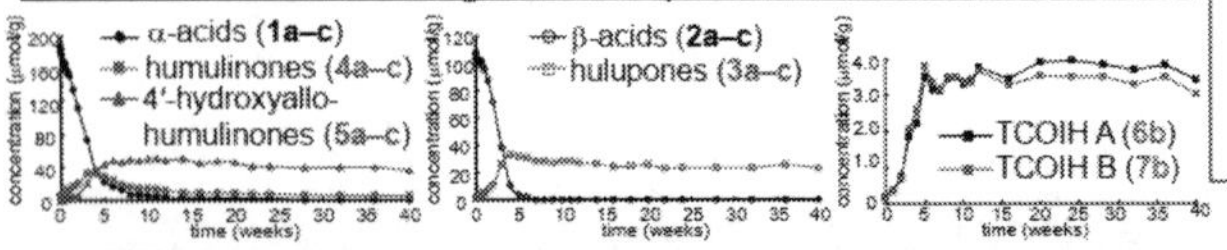

oxidation pathway

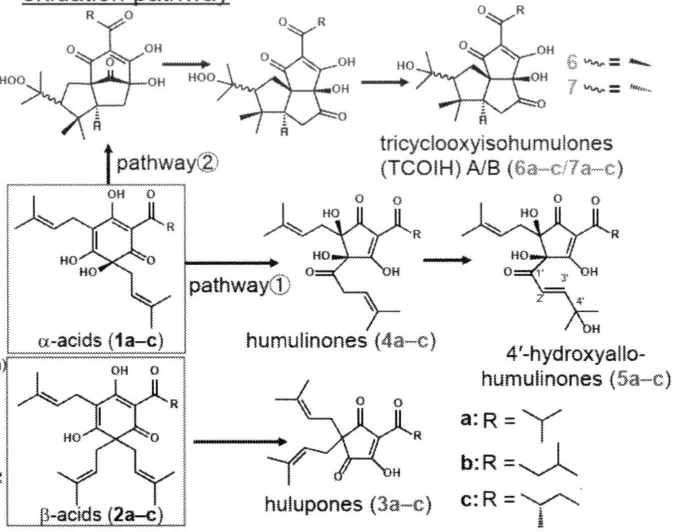

J. Agric. Food Chem. **2013**;61:3121-3130. Taniguchi *et al.*
J. Nat. Prod. **2014**;77:1252-1261. Taniguchi *et al.*

Chemical changes of bitter acid oxides (hard resin) during word boiling and beer aging

Bitter acid oxides in hard resin were investigated for their compositional changes in experimental models of wort boiling and beer aging. **6** and **7** are stable, whereas **5** is readily transformed.

wort boiling model

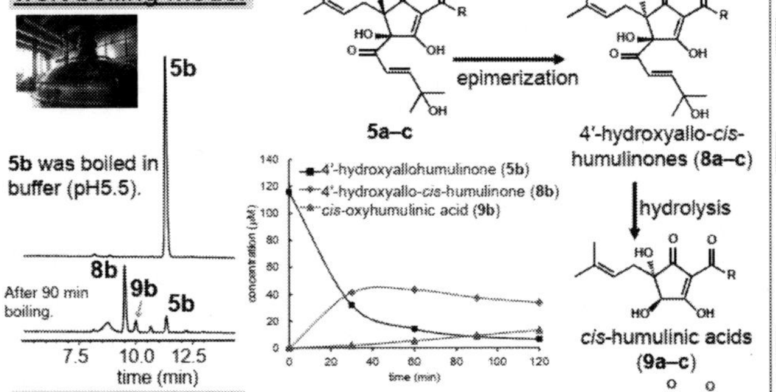

beer aging model

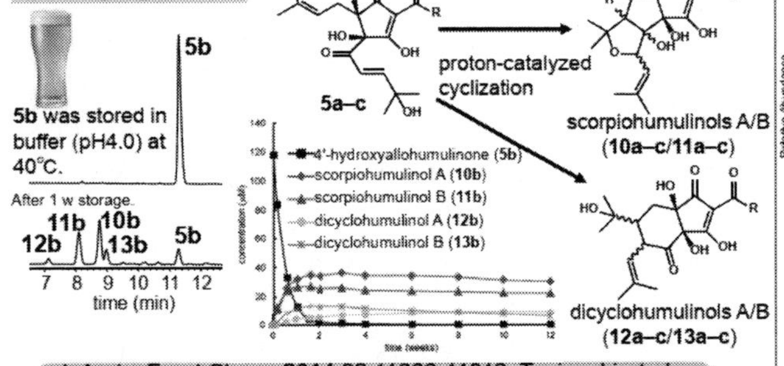

J. Agric. Food Chem. **2014**;62:11602-11612. Taniguchi *et al.*
J. Agric. Food Chem. **2015**;63:10181-10191. Taniguchi *et al.*

Preparative method of the hop bitter acid oxide fraction

hop pellet
- oxidized completely (e.g. 60°C 120 h under air)

oxidized hop pellet
- extraction with water (oxidized hard resin could be recovered)

water extract
- heating to transform unstable **5** into rather stable **8–13** (e.g. 90°C 4 h) (optionally)
- extraction with CH_2Cl_2

bitter acid oxide fraction

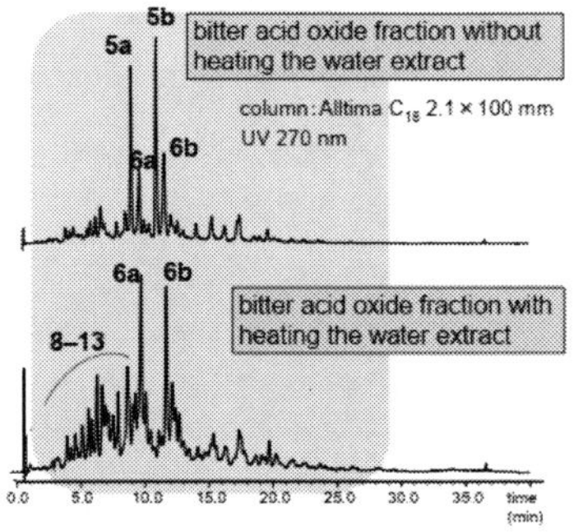

MS analyses confirmed that the fraction consisted of bitter acid oxides possessing a common partial structure (β-tricarbonyl moiety).

high resolution MS analysis

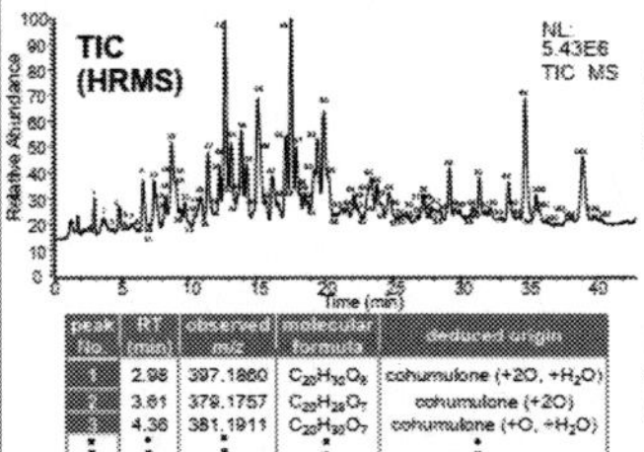

peak No.	RT (min)	observed m/z	molecular formula	deduced origin
1	2.98	397.1860	$C_{20}H_{30}O_8$	cohumulone (+2O, +H_2O)
2	3.61	379.1757	$C_{20}H_{28}O_7$	cohumulone (+2O)
3	4.36	381.1911	$C_{20}H_{30}O_7$	cohumulone (+O, +H_2O)
⋮	⋮	⋮	⋮	⋮
107	40.26	377.1964	$C_{21}H_{30}O_6$	*n*- or adohumulone (+O)

All of the detected peaks show molecular formula of the bitter acid oxides.

MS/MS analysis (precursor ion scan)

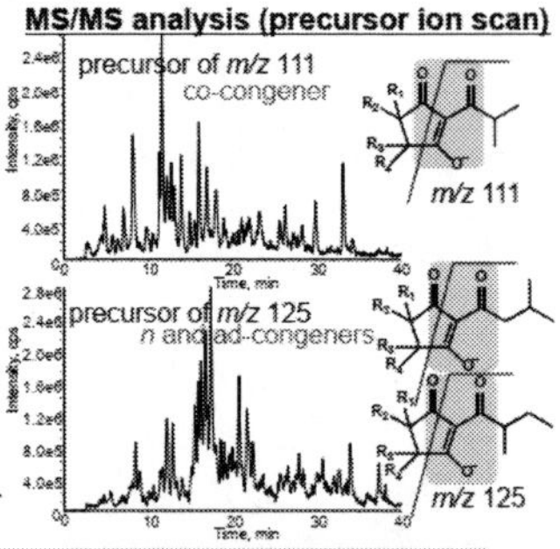

Biosci. Biotechnol. Biochem. **2015**;79:1684-1694. Taniguchi *et al.*

Laboratory of Food Chemistry and Biochemistry
Leuven Food Science and Nutrition Research Centre

KU LEUVEN

Fonds Baillet Latour

Studying the conversion of malt to fermentable sugars during mashing by GC and HPAEC-PAD analysis

N.A. Langenaeken[1], C. De Schepper[1], D. De Schutter[2], C.M. Courtin[1]

[1] KU Leuven, Laboratory of Food Chemistry and Biochemistry and Leuven Food Science and Nutrition Research Centre (LFoRCe), 3001 Leuven, Belgium, [2] Fund Baillet Latour, 3000 Leuven, Belgium
Contact: niels.langenaeken@kuleuven.be

Introduction

From a carbohydrate point of view, the efficiency of barley-to-beer conversion can be assessed as the potential to convert fermentable carbohydrates into alcohol and is mainly determined by the efficacy of enzymes during malting, mashing and fermentation. In former times, this extract yield was evaluated based on density measurements. However, vital information on fermentable sugars is often overlooked.

The aim of this study is to establish a new approach that allows the comparison of different brewing techniques or installations on the production of fermentable sugars.

Here, we use the combination of GC and HPAEC-PAD carbohydrate analyses to calculate the fermentable extract yield. **Analytical challenges** that were addressed include the potential interference of high enzyme activities or high sugar contents that are known to affect conventional measurements.

Set-up

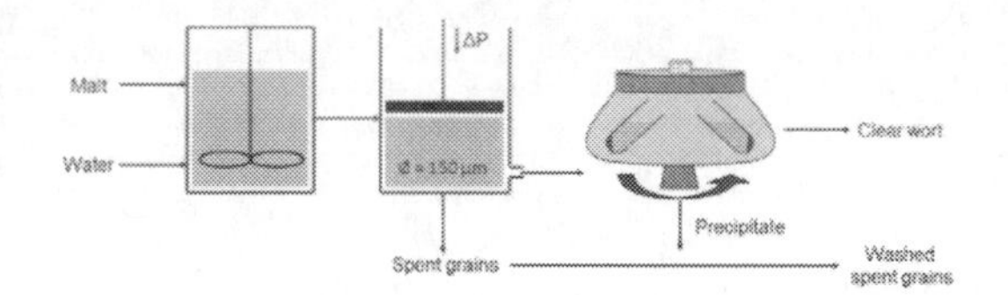

Figure 1: Lab-scale mashing and wort separation set-up in order to perform reproducible brewing trials. After mashing, wort and spent grains are separated by use of a press with a mesh size of 150 µm. The turbid wort is then centrifuged, which results in a clear wort. The spent grains are washed together with the precipitate to remove the remaining fermentable extract.

Table 1: The hydrolysis factor takes water into account that is incorporated during the hydrolysis of oligomers, to express the level of fermentable sugars on a monomer equivalent basis. This value depends on the degree of polymerization and the molecular weight of the respective oligomer.

Starch and dextrins	Hydrolysis factor
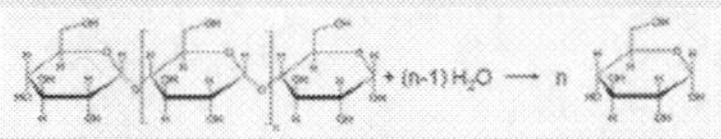	0.90
Maltotriose	
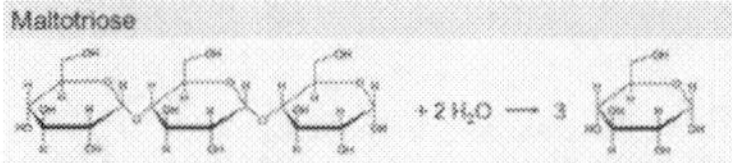	0.93
Maltose	
	0.95
Glucose	
	1.00

Materials & Methods

Mashing

- Triplicate mashing procedure was performed with 50 g barley malt and 200 mL deionized water
- Mash was modified with 75 mg $CaCl_2$ and 0.3 mL of a 0.5 M H_2SO_4-solution[1]
- Figure 1 gives a schematic overview of the process
- Resting times were at 45°C, 62°C, 72°C for 30 min and 78°C for 10 min

Analysis of total monosaccharide composition after hydrolysis (Englyst method - GC)

- Dry samples and 1:100 diluted wort samples were hydrolyzed with trifluoroacetic acid (2N)
- Reduction with $NaBH_4$ and acetylation with acetic acid anhydride was performed[2]
- The alditol acetates were separated with GC as described by Courtin *et al.* (2000).
- The total monomeric glucose content is the highest theoretical yield of fermentable glucose, maltose and maltotriose and is here referred to as the **glucose potential**.

Enzyme inactivation in malt prior to fermentable sugar analysis

- Milled malt was boiled for 2 h in a 80% (v/v) aqueous ethanol solution under constant reflux.
- The solvent was removed by a rotary evaporator and the sample dried overnight at 50°C[3].

Analysis of fermentable sugars after mashing (HPAEC-PAD)

- Dry samples and 1:5 diluted wort samples were diluted to 5 µg/mL after addition of internal standard.
- Glucose, fructose, sucrose, maltose and maltotriose are quantified.
- For maltose, additional calibration samples with concentrations up to 40 µg/mL were used.
- Samples were separated with HPAEC-PAD.[4,5]
- Fermentable sugars were converted to **monomer equivalents** using the hydrolysis factors (Table 1).

Results & Discussion

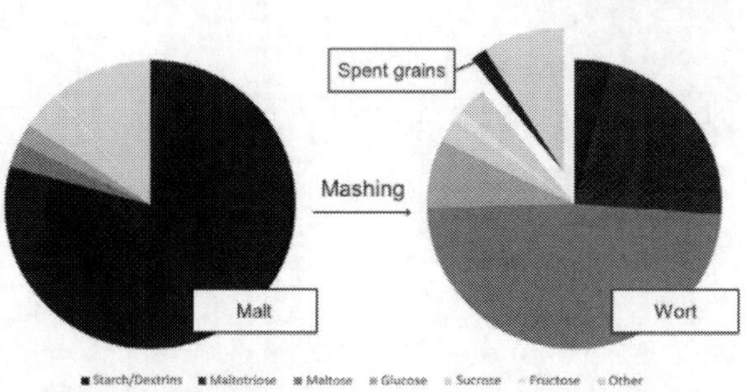

Figure 2: Carbohydrate mass balance of the lab scale mashing unit. All carbohydrates are expressed as monomer equivalents to make it possible to compare before and after hydrolysis. Starch or dextrin content is calculated as the glucose potential excluding the fermentable sugars.

Figure 2 shows the carbohydrate composition of malt and its conversion to wort and spent grains. Predominantly maltose and maltotriose are formed during mashing, while sucrose is partially hydrolyzed to glucose and fructose. The residual fraction comprises cellulose, arabinoxylan, β-glucan and protein.

This lab-scale mashing process can be evaluated based on two critical values, being the fermentable extract yield based on the glucose potential (FEY-GP) and the fermentable extract yield based on the malt mass (FEY-M).

$$FEY - GP = \frac{Fermentable\ sugars\ in\ wort\ (HPAEC - PAD)}{Glucose\ potential\ in\ malt\ (GC)} = 93.0\%$$

$$FEY - M = \frac{Fermentable\ sugars\ in\ wort\ (HPAEC - PAD)}{malt\ mass\ (dm)} = 79.3\%$$

This lab-scale mashing unit has a FEY-GP of 93.0%, which means that 7.0% of the carbohydrates that could be converted to fermentable sugars is not fermentable. This value is independent of the starch content of malt and only reviews the mashing process. Both values can be compared with values of other systems with different external factors, e.g. with the addition of exogenous enzymes.

This leads to the **conclusion** that the comparison of mashing processes by means of the FEY-GP allows a better understanding of the conversion to fermentable sugars and is therefore inevitable for brewhouse optimization.

References: [1] Li, Y., Lu, J., & Gu, G. (2005). Control of arabinoxylan solubilization and hydrolysis in mashing. Food Chemistry 90, 1-2, 101-108. [2] Courtin, C. M., Van den Broeck, H., & Delcour, J. A. (2000). Determination of reducing end sugar residues in oligo- and polysaccharides by gas-liquid chromatography. Journal of Chromatography A 866, 97-104. [3] Scott, R. W. (1972). The viscosity of worts in relation to their content of β-glucan. Journal of the Institute of Brewing 78, 2, 179-186. [4] Rohrer, J. (ed). Analysis of carbohydrates by high-performance anion-exchange chromatography with pulsed amperometric detection (HPAE-PAD). [5] Rohrer, J. (ed). Optimal settings for pulsed amperometric detection of carbohydrates using the Dionex ED40 electrochemical detector.

Acknowledgements: N.A. Langenaeken thanks the Baillet Latour Fund (FBL, 3000 Leuven, Belgium) for financial support.

Application of a central composite design for optimization of an HS-SPME-GC-MS method for determination of volatiles in beer

UNIVERSITY OF CHEMISTRY AND TECHNOLOGY PRAGUE

Nešpor Jakub*, Karabín Marcel, Hanko Vojtěch, Dostálek Pavel

University of Chemistry and Technology, Prague, Department of Biotechnology, Technická 5, CZ-166 28 Prague 6, Czech Republic

*Corresponding author. Tel.: +420 220444036; e-mail address: nesporj@vscht.cz

KEYWORDS: Beer, central composite design, HS-SPME, GC-MS, volatile compounds

INTRODUCTION:

Solid-phase microextraction (SPME) combined with gas chromatography and mass spectrometry (GC-MS) is a rapid and precise method that has been applied to the analysis of many different aromatic compounds in beer and other alcoholic beverages [1, 2, 3]. However, any chromatographic system is affected by many different parameters and optimization of the method is a time consuming process, but essential for establishing optimal conditions for the correct quantification of analytes.

Setting optimal conditions for the extraction step is essential for maximizing the analytical response, thus ensuring their correct quantification. The goal of any analytical extraction method is to establish conditions for optimal isolation, with minimal risk of false results. Usually, establishing optimum conditions is based on performing a large number of experiments, the number of which increases with number and range of optimized conditions. For this purpose, many different experimental designs using Responsible Surface Methods (RSM) are available [4]. One of them is CCD (Central Composite Design), which works with a reduced number of laboratory experiments to describe the system as a whole [4, 5, 6, 7]. The number of necessary experiments of CCD is based on the number of monitored parameters (conditions) and can be calculated from the equation, $2^N + 2 \times N + n_0$, where (N) is the number of variables (optimized parameters). For example determination of four variables at three levels will require $2^4 + 2 \times 4 + 1 = 25$ laboratory experiments. Furthermore it is also recommended to measured the center point (n_0) in more than one repetition (2 – 5 repetitions) in order to verify measurement accuracy, especially by optimization of the chromatographic system. The CCD model, in combination with ANOVA statistical evaluation, represents a universal tool for optimization of chromatographic systems.

PRINCIPLE OF OPTIMIZATION:

- HS-SPME coupled with GC-MS was optimized
- Tested SPME fibers:
  - 50/30 µm DVB/CAR/PDMS
  - 75 µm CAR/PDMS
  - 65 µm PDMS/DVB
  - 85 µm PA
- Conditions of the HS-SPME method were optimized using a CCD; α = 2.000
  - Time of extraction (25 – 45 min)
  - Temperature of extraction (30 – 50 °C)
  - The effect of added NaCl (15 – 26% w/v)
  - The influence of pH (3 – 5)
- Equation of mathematical dependence for parameters:

$$Y = \beta_0 + \sum_{i=1}^{4} \beta_i x_i + \sum_{i,j=1 (i \neq j)}^{6} \beta_{ij} x_i x_j + \sum_{i=1}^{4} \beta_{ii} x_i^2 \quad (1)$$

Selection of parameters → Matrix of analysis → Real results → Mathematical equations → Model data → Application of new conditions

RESULTS:

- 29 analyses was performed (Table 1)
- Results were interpreted as total peak areas of higher alcohols, esters and organic acids
- From the experimental data, predictive values were counted (Table 1) via mathematical equation (1)

Table 1: Experimental conditions (designed by CCD), the sum of the peak areas of analytes and their predicted surfaces

Run	NaCl	Temperature of extraction	Time of extraction	pH	Sum of peak area (Y)	
	(g)	(°C)	(min)		Observed	Predicted
1	1.75	35	30	4.00	6054926444	6141462601
2	2.25	35	30	4.00	5611307809	6064702629
3	1.75	45	30	4.00	6679712516	6931853439
4	2.25	45	30	4.00	7465930506	6865986152
5	1.75	35	40	4.00	5752011797	5609449526
6	2.25	35	40	4.00	7685591806	5539687789
7	1.75	45	40	4.00	6959182948	6379267323
8	2.25	45	40	4.00	7447908246	6320398272
9	1.75	35	30	5.54	7041780241	6187049846
10	2.25	35	30	5.54	4956057328	6068932150
11	1.75	45	30	5.54	7046327690	6980618667
12	2.25	45	30	5.54	7776546458	6873383658
13	1.75	35	40	5.54	5203868496	5802042025
14	2.25	35	40	5.54	4502471482	5490922564
15	1.75	45	40	5.54	6233549319	6375037806
16	2.25	45	40	4.77	4613210703	6274811032
17	1.50	40	35	4.77	4813402519	6303133575
18	2.50	40	35	4.77	4706486587	6126146828
19	2.00	30	35	4.77	4642841601	5452387501
20	2.00	50	35	4.77	5845287843	7026566807
21	2.00	40	25	4.77	4117592924	5657560235
22	2.00	40	45	4.77	5623307296	6788155936
23	2.00	40	35	3.23	5726554102	6249478427
24	2.00	40	35	6.31	5154203276	6249478431
25	2.00	40	35	4.77	3964673298	5324208585
26	2.00	40	35	4.77	4386887185	5324208585
27	2.00	40	35	4.77	4129843918	5324208585
28	2.00	40	35	4.77	4183026829	5324208585
29	2.00	40	35	4.77	4596613677	5324208585

DISCUSION:

- The model and experimental data were compared to determine the applicability of the CCD. The correlation coefficient between real and model data was 0.645, which is sufficient for level of significance of p = 0.01
- Each parameter was evaluated by ANOVA to determine the level of significance, which was set to p = 0.05. Major impact had temperature (Pareto diagram)
- Joint effect of parameters (3D graphs) implies from mathematical dependence (2)

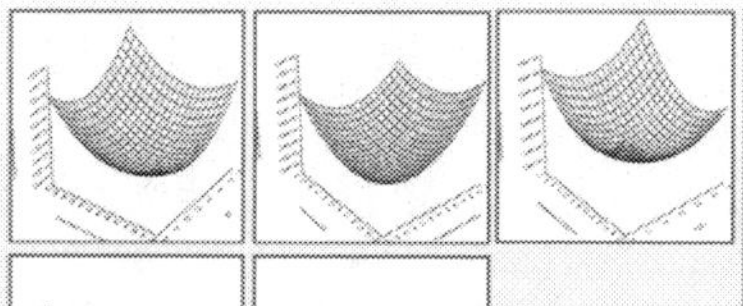

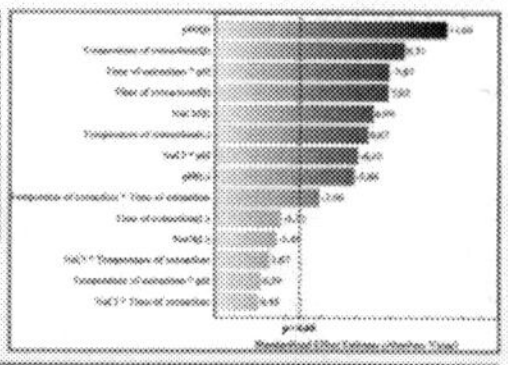

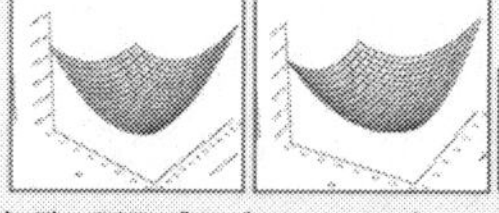

$$\begin{aligned} Y = {} & 5622107057 - 44246687\, x_1 + 393569826\, x_2 - 282648925\, x_3 \\ & - 297898472\, x_4 + 2723171\, x_1 x_2 + 1749559\, x_1 x_3 - 10339431\, x_1 x_4 \\ & - 5143260\, x_2 x_3 + 794496\, x_2 x_4 - 13248686\, x_3 x_4 + 222607904\, x_1^2 \\ & + 228804642\, x_2^2 + 224662375\, x_3^2 + 231317461\, x_4^2 \end{aligned} \quad (2)$$

- The RSD of each compound was computed from ten replications and did not exceed 15% (Table 2)

Table 2: Linear regression data and intra-day and inter-day precision of volatiles in beer

Analytes	Linear regression data		LOD	LOQ	Intra-day precision	Inter-day precision
	Linear range (mg/l)	r^2	(mg/l)	(mg/l)	(%)	(%)
Ethyl acetate	11.352-33.395	0.9947	0.1190	0.5778	7.2	5.7
n-Propyl acetate	0.019-0.055	0.9796	0.0032	0.0121	6.0	6.1
Butanoic acid, ethyl ester	0.018-0.054	0.9850	0.0031	0.0142	5.2	7.8
Isobutanol	10.489-30.789	0.9966	0.2042	1.2379	8.4	10.0
Isoamyl acetate	1.213-3.568	0.9583	0.0089	0.0542	7.9	10.5
Amyl alcohol / Active amyl alcohol	10.445-30.728	0.9952	0.0345	0.2093	8.2	8.2
Hexanoic acid, ethyl ester	0.115-0.337	0.9795	0.0015	0.0091	6.9	8.2
Octanoic acid, ethyl ester	0.107-0.315	0.9840	0.0002	0.0004	9.8	11.7
Linalool	0.098-0.287	0.9976	0.0018	0.0048	8.5	10.0
Decanoic acid, ethyl ester	0.104-0.307	0.9826	0.0003	0.0008	11.2	12.8
2-Furanmethanol	10.906-32.083	0.9665	0.2595	0.7052	8.2	10.7
Acetic acid, 2-phenylethyl ester	0.119-0.349	0.9779	0.1072	0.0105	6.3	6.4
Hexanoic acid	1.037-3.052	0.9978	0.0058	0.2293	6.2	10.0
2-Phenylethanol	14.183-41.724	0.9559	0.0496	0.0903	5.3	10.8
Tetradecanoic acid, ethyl ester	0.107-0.316	0.9680	0.0017	0.0030	4.1	6.6
Octanoic acid	9.300-27.993	0.9933	0.0283	0.0915	11.2	11.4
Hexadecanoic acid, ethyl ester	0.094-0.278	0.9563	0.0134	0.0244	4.1	5.0
n-Decanoic acid	0.943-2.774	0.8765	0.0425	0.0775	4.7	4.8

CONCLUSIONS:

- The HS-SPME-CG-MS method was optimized in two effective steps.
  - ✓ CAR/PDMS fiber was selected
  - ✓ Extraction conditions were set at: extraction temperature = 50 °C; extraction time = 45 min; NaCl = 26% w/v; pH = 3.3
  - ✓ Calibration curves were performed under these conditions and precision of the method was verified
  - ✓ The optimized method was used for determination of 19 volatile substances in 6 Czech lagers obtained from mid-scale breweries (data not shown)

REFERENCES:

[1] [illegible]

[2] [illegible]

[3] [illegible]

[4] [illegible]

[5] [illegible]

[6] [illegible]

[7] [illegible]

Acknowledgement: This work was supported by TACR "Centre for Innovative Use and Strengthening of Competitiveness of Czech Brewery Raw Materials and Products" – TE02000177.

Comparative study of accelerated shelf-life tests on physical ageing of beverages containing weighted flavor emulsions

Sören Rossmann, Verena Wolff, Ulrich Müller, Jan Schneider
North Rhine Westfalia Institute of Food Technology ILT.NRW, Lemgo, Germany

ILT.NRW Institute of Food Technology — 36th Congress EBC 2017 European Brewery Convention, Ljubljana, Slovenia

Hochschule Ostwestfalen-Lippe
University of Applied Sciences

GeGenstrom emulgieren — Sponsored by the Federal Ministry of Education and Research — Forschung an Fachhochschulen

Motivation

Emulsion based beverages

Beverage emulsions are an important product for the preparation of increasingly popular beer-based mixed drinks and soft drinks in order to incorporate water insoluble colors or flavors (flavor emulsion), to generate turbidity (cloud emulsion) or both combined [1]. Compared to other oil-in-water food emulsions, beverage emulsions are a unique class of type, as they are highly concentrated semi-finished products and after being processed into drinks finally consumed considerably diluted [2].

Physical stability problems

Despite the use of stabilizing ingredients and intensive mechanical forces like high-pressure homogenization, physical instability phenomena are a great issue:

Creaming ("Ringing") ***Turbidity Loss***
Sedimentation ***Inhomogeneity***

The reason is the inherent instability of beverage emulsions as generally known from macroemulsions. In addition, governmental regulations restrict the application of stabilizers.

Visible ringing of an orange soft drink

Accelerating Shelf-life Tests (ASLT)

New formulations always have to be analyzed in terms of physical stability beside chemical and sensory evaluation leading to time-consuming development processes. The industry consequently employs ASLTs to accelerate these changes. In difference to beer a systematic approach to evaluate the effect of artificial storage conditions for rapid tests in emulsion based beverages is not available yet. Therefore, different tests were compared in regard to physical changes of beverages with orange oil emulsion.

Experimental Setup

Enhancing physical instabilities by integrating artificial „product defects":

A) Reference Emulsion	B) Reduced Ester Gum	C) Low Homog. Pressure	D) Reduced Emulsifier
12 % Emulsifier (OSA-starch)	12 % Emulsifier (OSA-starch)	12 % Emulsifier (OSA-starch)	***6 % Emulsifier*** (OSA-starch)
6 % Orange oil	10 % Orange oil	6 % Orange oil	6 % Orange oil
6 % Weighting agent	***2 % Weighting Agent***	6 % Weighting agent	6 % Weighting agent
HP-homogenization: 2x 300/50 bar	HP-homogenization: 2x 300/50 bar	*HP-homogenization: 1x 250/50 bar*	HP-homogenization: 2x 300/50 bar
x_{90}= 736 nm	x_{90}= 620 nm	x_{90}= 5710 nm	x_{90}= 785 nm

Emulsions A to D were diluted to Ready-to-drink beverages for stability analyzing:
0,12 % emulsion diluted with beverage base (10 °Bx sucrose, pH 3,2)

50 days storage under accelerating storage conditions:
(Comparative real-time storage: 180 d at 20 °C)

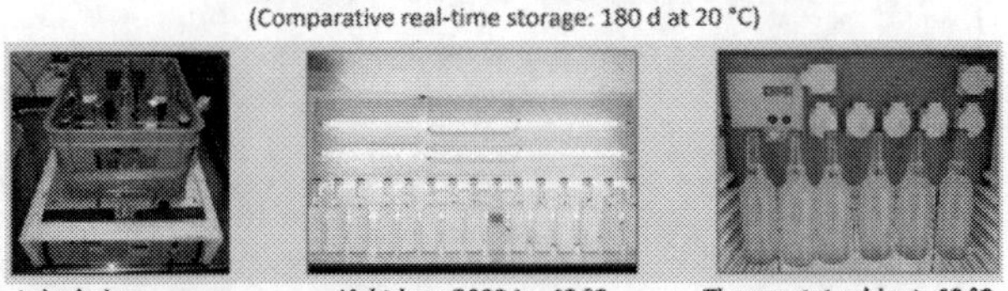

Lab-shaker 12 h·d^{-1} at120 min^{-1} — **Light box: 5000 Lx, 40 °C** 6500 K daylight lamps — **Thermostat cabinet: 40 °C** dark

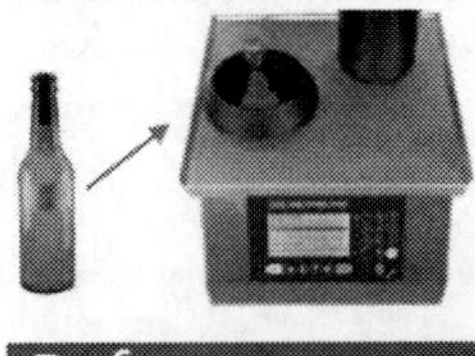

Physical stability was analyzed with:

- Turbidity measurement (*Optek*): direct analyzing of scattered light in the bottle
- Particle size measurement (*Malvern*): laser diffraction
- Calculation of instability index:

$$\text{Instability index} = 1 - \frac{\text{turbidity } \tau(t)}{\text{initial turbidiy } \tau_0}$$

Results

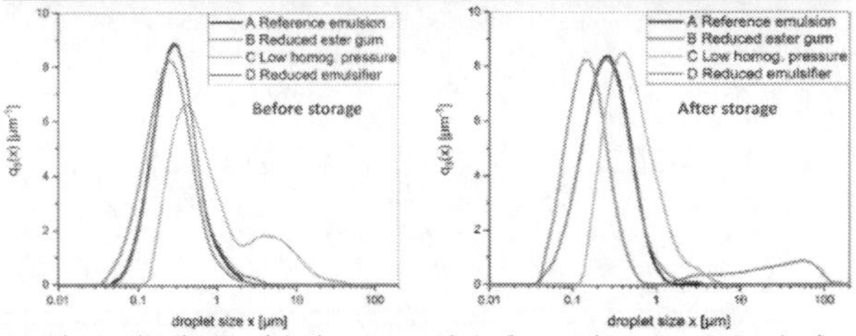

Particle size distribution of the beverages with the four emulsions A to D directly after preparation (left) and after storage under accelerating conditions (right).

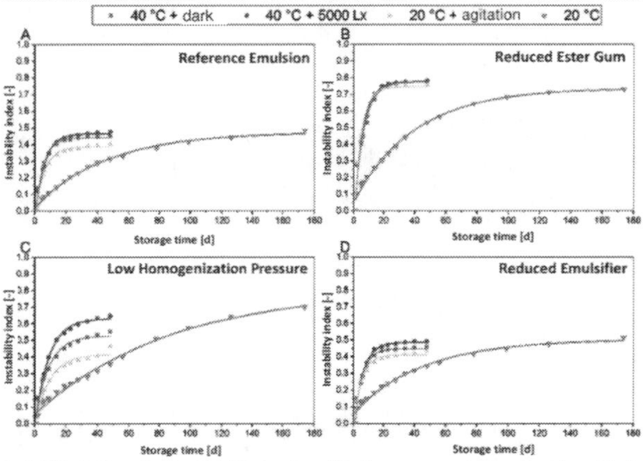

Instability indices showing the deterioration of the beverages containing emulsions A to D (different "product defects") over a storage period of 50 days (accelerated) and 180 days (real-time storage), respectively. To compare the efficiency of the ASLTs, exponential functions were fitted to the measured values, which allow calculation of the **accelerating factors** of the ASLTs against the real-time storage.

Accelerating factors of the ASLTs

	Agitation 20 °C, 120 min^{-1}	*Light box* 40 °C, 5000 Lx	*Thermostat cabinet,* 40 °C	Instability-mechanism [1,4]
A) Reference emulsion	8.2	9.1	7.9	*(Diffusion)*
B) Reduced Ester Gum	7.7	5.8	6.5	Ostwald Ripening
C) Low Homog. Pressure	11.0	9.1	9.6	Creaming
D) Reduced Emulsifier	9.2	8.7	7.9	*(Diffusion)*

Conclusion

The examined ASLTs are capable of accelerating the physical deterioration of emulsion-based beverages remarkably. Accelerating factors of 5.8 up to 11.0 were achieved regardless of the predominant instability mechanism. Surprisingly, the periodic agitation without elevated temperatures reveal the highest acceleration. Unfortunately, agitation exhibits large deviations in the resulting extent of instability compared to real-time storage. Considering all intrinsic deficiencies of the emulsion beverages, the warm storage with additional illuminance shows the deterioration most similar to the real-time storage. Since the differences in time are not extreme this test condition reveals as most promising.

It is thus obvious that the results cannot be easily transferred to other emulsion formulations. Nevertheless, for practical purposes the results of the presented work can serve as a basis for the development and assessment of accelerated shelf-life tests.

References

[1] Piorkowski, D. T. and McClements, D. J. (2014): Beverage emulsions: Recent developments in formulation, production, and applications, Food Hydrocolloids, 42.

[2] Tan, C.-T.: Beverage Emulsions, in Friberg, S. E. and Larsson, K. (Eds.)(1997): Food emulsions, 3. ed., rev. and expanded, Dekker, New York.

[3] Reiner, S. J.; Reineccius, G. A. and Peppard, T. L. (2010): A Comparison of the Stability of Beverage Cloud Emulsions Formulated with Different Gum Acacia- and Starch-Based Emulsifiers, Journal of Food Science, 75,5.

[4] Lim, S. S.; Baik, M. Y.; Decker, E. A.; Henson, L.; Michael Popplewell, L.; McClements, D. J. (2011): Stabilization of orange oil-in-water emulsions: A new role for ester gum as an Ostwald ripening inhibitor, Food Chemistry, 128,4.

Contact

ILT.NRW
Beverage Technology
Sören Rossmann
soeren.rossmann@hs-owl.de

Acknowledgements. This work was funded under the program FHprofUnt (FKZ: 03FH037PX3) by the Federal Ministry of Education and Research (BMBF) of the German Government.

www.ilt-nrw.de

36th CONGRESS EBC SLOVENIA 2017 14-18 MAY LJUBLJANA EUROPEAN BREWERY CONVENTION

Sensory and analytical changes in beer-mix-beverages during shelf life

H. Klein, R. Steingruber, C. Forster
Brau Union Österreich AG, 4020 Linz, Austria

Description of Topic:

Beer-mix beverages, in particular Radlers have been increasingly successful in recent years. Due to the composition and specific ingredients of beer-mix beverages the known mechanisms and control strategies of beer staling have only limited relevance for these products.

A broad range of Radlers in different pack types representing diverse flavours has been investigated by physical-chemical methods, gas and liquid chromatography as well as descriptive sensory analysis over the course of the shelf life.

Material and Methods:

Radlers with different natural complex flavours were investigated at different stages during shelf life: fresh, 2 months, 4 months and 6 months (end of BBD). All samples were stored dark at ambient temperature (23 ± 1,5 °C).

Tab. 1: Investigated beer-mix beverages

Sample	Type	O.G. [% m/m]	Alc. [% vol]	Beer ratio [%]	Juice / aroma in non-beer part
1	Lime	9,6	1,9	40	lime, lemon
2	Lemon	9,8	1,8	50	lemon, lime, acerola
3	Herbs	9,8	1,8	40	orange, apple, herbs
4	Lemon	9,8	1,7	40	lemon, orange, apple

Analytical parameters for investigation were chosen with respect to methods relevant for beer quality and beer staling as well as quality control of lemonades and fruit juices.

Tab. 2: Investigated analytical parameters

Parameter	Method	Parameter	Method
Extract, alcohol	EBC 9.4	Carbohydrates	HPLC-RI
Colour	EBC 9.6	Ascorbic acid	Photometric
pH	EBC 9.35	Total sulfite	DTNB
Bitterness	EBC 9.8	Furfural	HPLC-UV
Carbon dioxide	Volume expansion	5-HMF	HPLC-UV
Turbidity	Nephelometric 90°	FAN	NOPA
Total acids	End-point titration pH 8,1	Strecker aldehydes	SPME GC-ECD (PFBOH)
Citric acid	HPLC-UV	Aroma compounds	GC-FID / GC-MS

In addition to the analytical characterization of the products sensory descriptive profiling was performed on a 1 to 5 intensity scale.

Sensory Results:

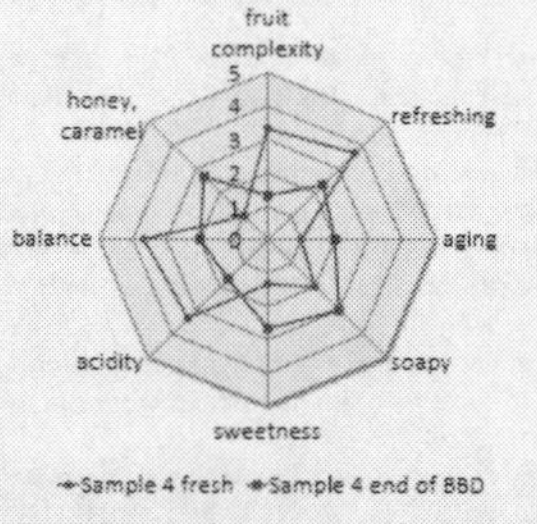

Fig. 1: Sensory changes in sample 4

In general, sensory changes of beer mix beverages during shelf-life are characterized by a loss of:

- complexity of fruity notes
- acidity
- refreshment
- overall balance

Typical beer staling sensory attributes such as papery, bready or catty are not predominant. On the contrary, honey-sweet, caramel and soapy notes appear.

Analytical Results:

The change of analytical parameters during shelf life as illustrated in table 3 is quite consistent in the Radlers investigated. Different to beer, there is no increase of colour and a significant decrease of pH by 0,1. Concentrations of furfural and 5-HMF is approx. 2 times higher than in lager beer, supposedly due to the presence of high concentrations of carbohydrates in beer-mix beverages.

Tab. 3: Change of analytical parameters during shelf life

Parameter	Change	Parameter	Change
Extract, alcohol	stable	Carbohydrates	stable / sucrose inversion
Colour	no significant increase	Ascorbic acid	slight decrease
pH	significant decrease	Total sulfite	slight decrease
Bitterness	stable	Furfural	strong increase
Carbon dioxide	depending on pack type	5-HMF	strong increase
Turbidity	strong sedimentation	FAN	stable
Total acids	stable	Strecker aldehydes	strong increase
Citric acid	stable	Aroma compounds	see table 4

Specific volatile substances like limonene, terpineols and linalool originating from fruit juice compounds could be identified as markers for the organoleptic deterioration of the product. Moreover, α-pinene and ß-pinene very found to be highly thermo-sensitive and disappeared within a short period of time after a moderate heat load. Reaction rates of volatiles differed significantly in dependence of the pack type such as glass, or PET.

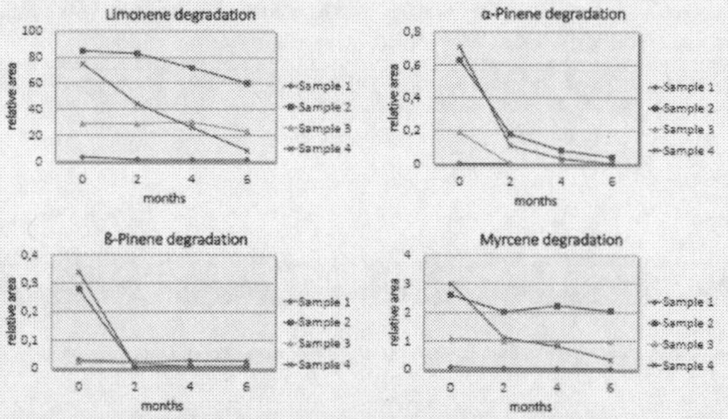

Fig. 2: Change of selected aroma compounds during shelf life

The degradation of limonene as the main aroma component of citrus fruits is strongly pH, temperature and oxygen dependent.

Tab. 4: Change of aroma compounds during shelf life in Radlers

Decrease of aroma compounds % at end of BBD compared to fresh product		Formation and increase of aroma compounds
Limonene	10-80%	Terpineoles (α-, ß-, γ-)
Linalool	50-80%	1,4-Cineol
α-Pinene	0-10%	1,8-Cineol (Eucalyptol)
ß-Pinene	0-10%	Terpinolen
Myrcene	10-90%	1-Terpinenol

Conclusion:

The sensory loss of fruity aromas in beer-mix beverages can be explained by the degradation of key character aroma compounds like limonene. The typical known beer staling flavours are not predominant in aged Radlers. Most likely, the complex fruity flavours deriving from the non-beer part exert a masking effect.

BRAU UNION ÖSTERREICH

BARTH-HAAS GROUP

THE LANGUAGE OF HOPS - HOW TO IMPLEMENT AND APPLY A COMMON DESCRIPTION OF HOP FLAVOURS

Dr. Elisabeth Wiesen[1], Georg Drexler[1], Mark Zunkel[1], Sebastian Hinz[1], Alicia Muñoz Insa[1], Victor Algazzali[2], Tim Kostelecky[2], Dr. Christina Schönberger[1],
1. Joh. Barth & Sohn GmbH & Co. KG, Nuremberg, Germany; 2. John I. Haas Inc., Yakima/Washington D.C., USA

WHY DO WE NEED A UNIFORM TASTING SCHEME?

Hops are the soul of beer. In the last 10 years, hop usage in the brewing industry has changed tremendously. Until then, a subtle to moderate hop aroma achieved with kettle additions was standard. Now with the importance of dry hopping and very high amounts of addition for hop aroma in the brewing process in general, the need for a sensory hop language is obvious. With the help of flavourists and sommeliers, we have developed a uniform tasting scheme specifically for hops and hoppy beers that works with 12 aroma categories and identifying specific aroma attributes. This tasting scheme's result is a defined and comparable aroma profile for the relevant hop variety or beer. This tasting scheme is meant to be a standard language in the brewing industry for hops so that an easier comparison of hop varieties and hoppy beers in a sensory context is possible.

WHAT DOES AROMA MEAN?

Aroma refers to all volatile components of a food or beverage, which are in interaction with our olfactory system (especially Bulbus olfactorius). The aroma and flavour perception is complemented by retronasal aroma release in the oral cavity. The processing of sensory data is highly complex. The stimuli forwarded to our brain influence various regions in the brain. *

AROMA CLASSIFICATION

Aroma components produce sensory impressions which are referred to as aroma. Not only the relevant stimuli but also these sensations can be classified. Both of these factors make it difficult because of the pure abundance of aromas, which is why attempts have so far been unsatisfactory. One of the early formulations was completed by Linnaeus (1756) following a classification according to the names of plants. All formulations have been working with 4 to 44 categories. In 1968, Harper established a characterising system for the food industry based on 44 different categories. Later on industry and institutes have selected about 160 descriptors that were used more frequently out of a pool of 800. The Harper's Scale was published in 1985 in the Atlas of Odor Character Profiles and and is today seen as the standard. **

SO WHAT IS REALLY THE CORRECT NUMBER OF CATEGORIES?

The BARTH-HAAS GROUP presents one possible solution with 12 different categories:

AROMA CATEGORIES AND EXAMPLES OF THE RELEVANT HOP VARIETIES

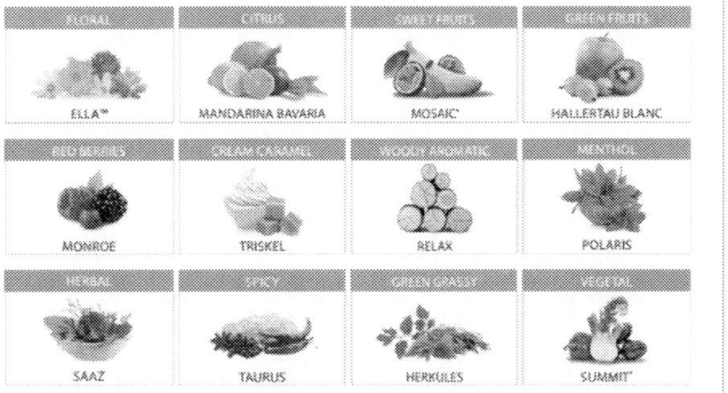

BARTH-HAAS GROUP UNIFORM TASTING SCHEME

CATEGORY	SPECIFIC ATTRIBUTES	INTENSITY (1 – 10)
FLORAL	Elderflower, camomile blossom, lily of the valley, jasmine, apple blossom, rose, geranium, carnation, lily, lilac, lavender	
CITRUS	Grapefruit, orange, lemon, lime, bergamot, lemon grass, ginger, tangerine	
SWEET FRUITS	Banana, watermelon, honeydewmelon, peach, apricot, passionfruit, lychee, dried fruit, plum, pineapple, cherry, kiwi, mango, guava	
GREEN FRUITS	Pear, apple, quince, gooseberry, white wine grape	
RED BERRIES	Cassis, blueberry, raspberry, blackberry, strawberry, redcurrant, black currant, wild strawberry, cranberry	
CREAM CARAMEL	Butter, chocolate, yoghurt, honey, cream, caramel, toffee, coffee, tonka bean, vanilla	
WOODY AROMATIC	Tobacco, cognac, barrique, leather, woodruff, incense, myrrh, resin, cedar, pine, earth	
MENTHOL	Mint, balm, camphor, menthol, wine yeast	
HERBAL	Lovage, marjoram, tarragon, dill, parsley, basil, fennel, coriander, rosemary, thyme, green tea, black tea, mate tea, sage	
SPICY	Pepper, chilli, curry, juniper, aniseed, liquorice, fennel seeds, clove, cinnamon, gingerbread, coriander seeds, nutmeg	
GREEN GRASSY	Green grassy, tomato leaves, green pepper, fresh cut grass, hay, nettle, cucumber	
VEGETAL	Celery root, celery stock, leek, onion, artichoke, garlic, wild garlic	

FOR BEER	RATING
AROMA INTENSITY (1-10)	
AROMA QUALITY (1-10)	
BITTER INTENSITY (1-10)	
BITTER QUALITY (1-10)	
HARMONY (1-10)	
BODY AND MOUTHFEEL (1-10)	
PREFERENCE	
ESTIMATED BITTERNESS (IBU)	

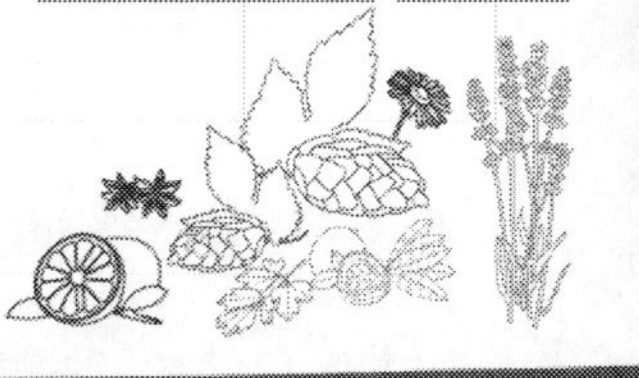

HOW TO IMPLEMENT THIS TASTING SCHEME?

Implementing a new tasting scheme takes time to train for hop flavor based on these twelve categories. Whole cone and pellet hops are the first step in familiarizing a tasting panel with the categories. This requires rubbing whole cone hops or using a pestle and mortar for pellets in releasing the aroma effectively. After evaluating multiple hop varieties, a cold extraction (tea) is created to simulate dry-hopping dosing 2 g of hops per 200 ml of water at 20°C. The next stage is using the same amount of hops to a base beer to achieve a quick impression of how the hops will taste in the beer matrix. These three methods can be completed in a short amount of time and are quite effective for training purposes.

* Jelen, H. (2011), Food flavors. Chemical, sensory and technological properties, CRC Press. ** Thiemer, E. T. (2012), Fragrance Chemistry, Academic Press.

For information about a professional training on this scheme and how to use it please contact our Technical Sales Support team at team_tss@johbarth.de

A member of BARTH-HAAS GROUP

FIRST IDENTIFICATION OF CYSTEINYLATED AND GLUTATHIONYLATED PRECURSORS OF 3-MERCAPTOHEXAN-1-OL IN BARLEY.

WHAT IS THE INFLUENCE OF MALTING PROCESS?

Aurélie Roland*, Stéphane Delpech, Florence Reillon, Clément Viel and Laurent Dagan

Nyseos, 53 Rue Claude François, Parc 2000, 34080 Montpellier, France
aurelie@nyseos.fr - +33 6 08 64 86 40

INTRODUCTION

Barley (*Hordeum vulgare*) is the major source for brewing malts, which constitutes the single most important raw material for beer production. Mainly composed by starch (more than 65 % of dry weight), it also contains proteins, cell wall polysaccharides and in a smaller amount, fat and minerals. Kishimoto and co-workers [1] hypothesized that thiol precursors might occur in malt since unhopped beers contained 3MH (3-mercaptohexan-1-ol). In 2016, Roland *et al.* [2] demonstrated for the first time the occurrence of both cysteinylated and glutathionylated precursors of 3MH in malts and the influence of enzymatic activities on such molecules.

To go deeper in thiol precursor research, we performed a new study focused on barley :

1. to identify cysteinylated and glutathionylated precursors of both 3MH and 4-mercapto-4methylpentan-2-one (4MMP) in barley
2. to investigate the impact of malting process on thiol precursors

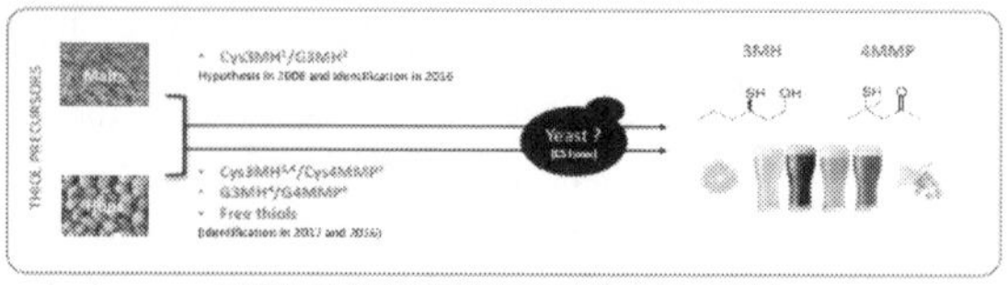

Fig 1: Biogenesis of 3MH and 4MMP during brewing (alcoholic fermentation)

MATERIAL AND METHODS

1. Samples

Six row barley samples (5 spring and 1 winter varieties) and their corresponding malts were gifted by Institut Français de la Brasserie et de la Malterie to be studied.

2. Extraction procedure

Freshly ground flours of barley and malts (1 g) were extracted by a mixture of water/MeOH (50/50, volume) during 1 hour at room temperature. After centrifugation (12000 rpm; 15 min), samples were aliquoted (800 µL), concentrated to dryness and stored at -20°C prior to be analyzed.

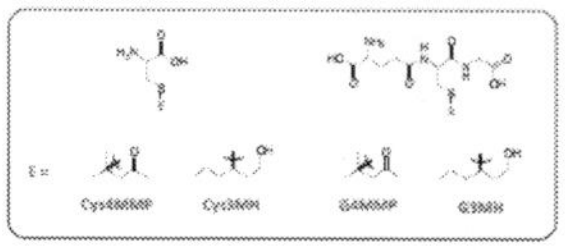

Fig 2: Cysteinylated and glutathionylated precursors of 3MH and 4MMP

3. Analysis of thiol precursors by UPLC-MS/MS

Extracts were suspended in a synthetic must matrix containing deuterated thiol precursors (Cys3MH-d_2 at 111.4 µg/L, Cys4MMP-d_6 at 22.7 µg/L, G3MH-d_2 at 205.6 µg/L and G4MMP-d_6 at 22.4 µg/L). Analysis was performed by UPLC-MS/MS as already published [4]. Briefly, samples were directly injected on a reverse phase separation column (1.9 µm; 2.1 x 100 mm). Mobile phases consisted in (A) water and (B) acetonitrile both acidified with 0.1 % of formic acid. The gradient began at 0 % of B during 30 sec, increased to 2 % of B for 1 min and then to 11 % of B in 7 min. System was then flushed with 99 % of B during 2 min before being reconditioned with 100 % of A for 1.5 min. Detection was carried out under multiple reaction monitoring mode (MRM) using the most specific and sensitive transitions for each precursor

RESULTS AND DISCUSSION

1. Identification of thiol precursors in barley

Thiol precursors were synthesized according to published methods [5,6] and fully characterized ^{1}H NMR using trimethylsilylpropionic acid as internal standard.
Cysteinylated and glutathionylated conjugates were considered as identified in the samples when meeting the following criteria specified in the commission decision 2002/657/EC 36:
- The relative retention time of an analyte in a sample and a calibration solution has to be within the ± 5 %.
- The presence of four identification points (parent ion and three daughter ions).
- The relative ion intensities have to comply with the ± 30 % permitted tolerances.

In practice, a mixture of pure synthetic standards in water was analyzed by UPLC-MS/MS and the retention times and MRM spectra were compared with those obtained from natural extract of malts.
Cys3MH and G3MH were formally identified for the first time in all the samples (Fig 3) whereas we did not detect any trace of 4MMP precursors under our conditions. As for hops and malts, 3MH precursors seemed to be more ubiquitous than 4MMP ones in barley.

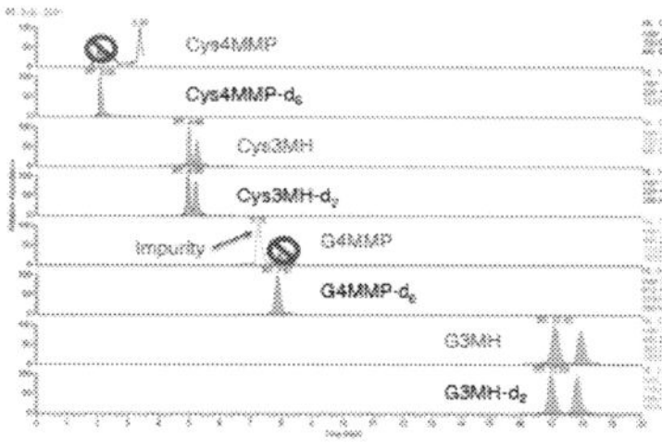

Fig 3: Chromatograms of thiol precursors in barley samples

2. Quantification of thiol precursors in barley

The 3MH precursors ranged from 1 to 2 ppb for Cys3MH and from 2 to 59 ppb for **G3MH** (Fig 4), **this latter representing more than 90% of 3MH thiol precursors.**

Cys3MH amounts did not change whatever the considered sample whereas the G3MH concentration could be twice higher in spring barley 4 than other ones. G3MH seemed to be a powerful marker to characterize the 3MH thiol precursors in barley.

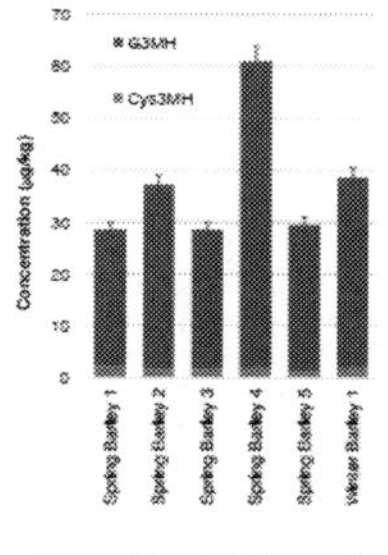

Fig 4: Quantification of Cys3MH and G3MH in barley samples

As recently demonstrated, both G3MH and Cys3MH were present in hops [3,4] and malts [2] at significant amounts. Using these published data, we can compare the thiol precursors origins from raw materials (Fig 5) and hypothesize their evolution during brewing. We can assume that less than 3% of thiol precursors contained in malts could lead to more than 64 % of 3MH in final beers.

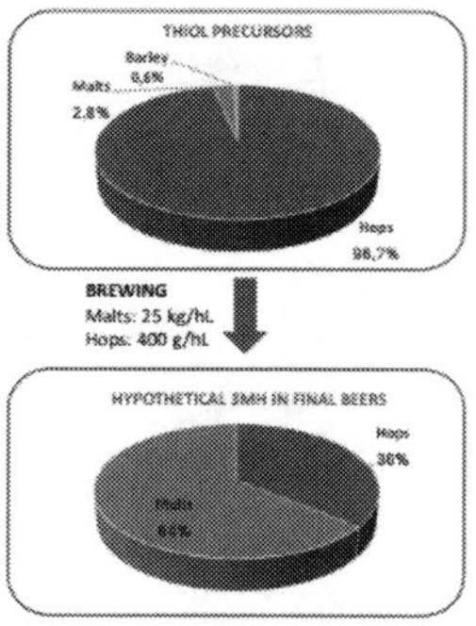

Fig 5: Thiol precursor evolution during brewing

3. Influence of malting process

After malting, the Cys3MH amounts did not change whereas the G3MH ones were impacted considerably (Fig 5). **Overall, G3MH concentrations was higher in malts than in barley** (Fig 6) **suggesting an *in-situ* formation during malting.** Indeed, *(E)*-2-hexenal [7] and glutathione were present (until 128 ppm for GSH) in barley and could react to produce the corresponding conjugate that could be reduce by specific enzymatic activities from malt.

By contrast, decreased G3MH levels were found only in spring barley malt 1, consolidating our previous findings [2] concerning either the enzymatic degradation of G3MH into Cys3MH or an effect of malting temperature on these molecules.

These preliminary results exhibited the importance of malting process to drive thiol precursors amount in raw material.

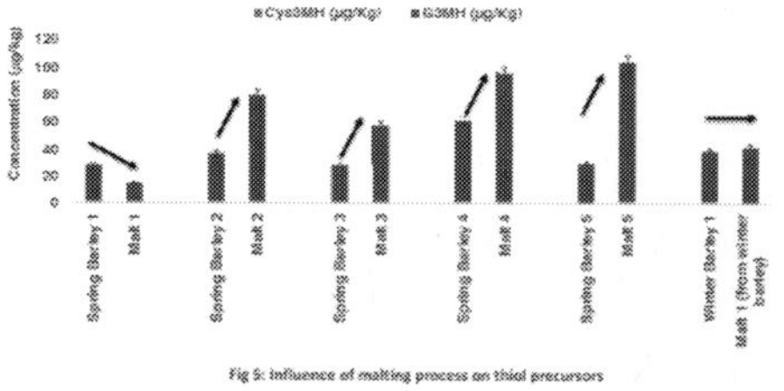

Fig 5: Influence of malting process on thiol precursors

CONCLUSION

Up to now, malts were only considered as a sugar source for the fermentation and not as an aroma reservoir. For the first time, this study counterbalanced this point of view by identifying both cysteinylated and glutathionylated precursors of 3MH in barleys. Malting seemed to be a crucial step to modify the thiol precursor levels from barley and so, to drive the raw material quality.
This study opened avenue to optimize brewing process by using complementary raw materials (malts, hops).

REFERENCES

[1]. Kishimoto *et al.*, J. Am. Soc. Brew. Chem., 2008, 66, 192.
[2]. Roland *et al.*, Poster in *World Brewing Congress*, Denver (USA), 2016.
[3]. Gros *et al.*, J. Agric. Food Chem., 2012, 60, 32.
[4]. Roland *et al.*, Flavour Fragr. J., 2016, 31, 455.
[5]. Fedrizzi *et al.*, J. Agric. Food Chem., 2009, 57, 991.
[6]. Roland *et al.*, J. Chrom. A., 2010, 1217, 1626.
[7]. Cramer *et al.*, J. Agric. Food Chem., 2005, 53, 7526.

ACKNOWLEDGMENTS

We acknowledged Marc Schmidt from Institut Français des Boissons, de la Brasserie et de la Malterie for providing the samples of barleys and malts.

SHIMADZU

Investigation of mycotoxins in different beers with several clean-up techniques using the Mycotoxin Screening System

Carola Schultz[1]; Philipp Jochems[1], Robert Ludwig[1], Gesa Schad[1]
[1]Shimadzu Europa GmbH, Duisburg, Germany

1. Introduction

Mycotoxins are secondary metabolites, produced by fungi and were found to be toxic and carcinogenic [1]. Therefore, quantification in food and beverages is crucially important, in order to ensure the health of humans and animals. This poster presents simple and fast methods for sample preparation and analysis of several mycotoxins in different matrices using the Shimadzu "Mycotoxin Screening System". It is a package with ready-to-use methods and reports to screen mycotoxins for European maximum residue levels. This work focuses on determination of Aflatoxin B1, B2 (AFB1, AFB2), G1, G2 (AFG1, AFG2), Deoxynivalenol (DON), Nivalenol (NIV) and Ochratoxin A (OTA), as these mycotoxins can be produced by fungal infestation during or after harvest of grain and can therefore end up in processed products such as beer, brewed from malt.

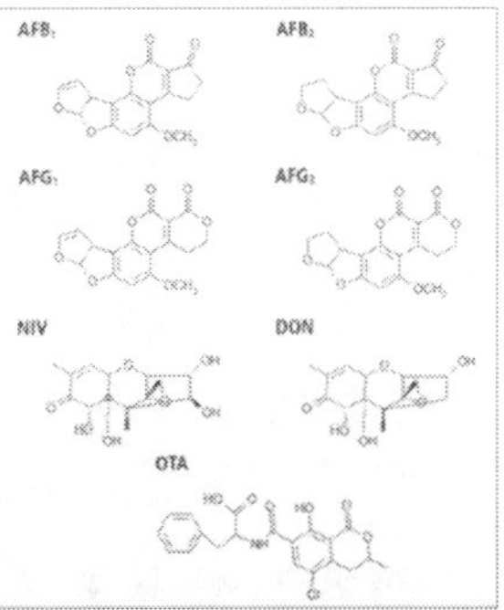

Figure 1: Structures of mycotoxins

All analytes of interest could be quantified with LOQs at or below the European maximum residue levels (EUMRL) (Table 1). Highly sensitive detection of all analytes was achieved by the combination of fluorescence and photodiode array (PDA) detectors.

Table 1: Different mycotoxin regulations in EU, US and Japan [2,3]

Target compound	Examples of targeted product	EU	US (Codex)	Japan
Aflatoxin B1,B2,G1,G2 (AFB1,B2,G1,G2)	Grain, Grain product (Flour etc.)	Total 4~15 µg/kg AFB1 2~12 µg/kg	Total 10~15 µg/kg	Total 10 µg/kg
Ochratoxin A (OTA)	Wheat, Wheat flour	2~10 µg/kg	5 µg/kg	Not regulated
Deoxynivalenol (DON)	Wheat, Wheat flour	500~1750 µg/kg	1000 µg/kg	1100 µg/kg (prov. std. value)
Nivalenol (NIV)	Wheat, Wheat flour		Not regulated	

2. Materials and Methods

2-1. Sample pretreatment with three different solid phase extraction cartridges

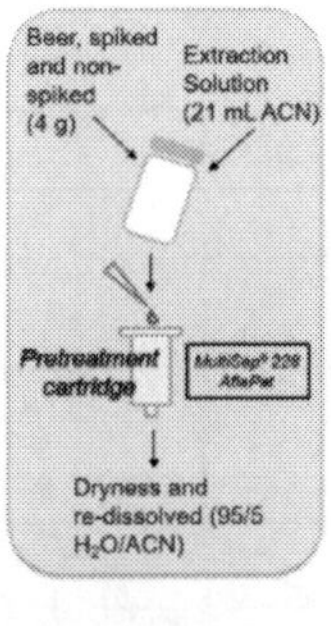

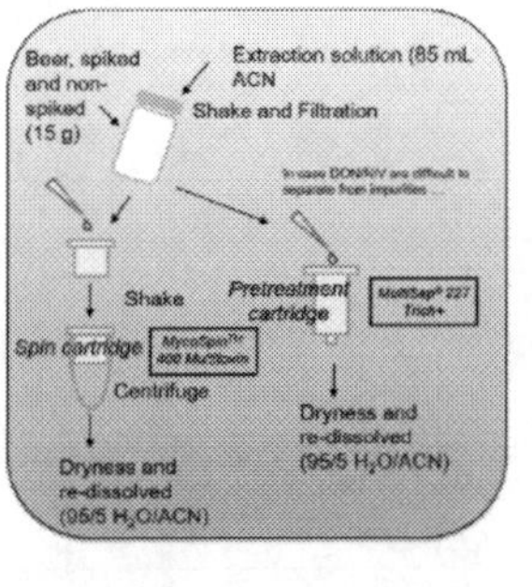

2-2. Analytical conditions

For analysis, a Shimadzu i-series system with integrated photo diode array (LC-2040C 3D) and additional fluorescence detector (Rf-20AXs) was used (figure 2). The detailed analytical conditions used are shown in table 2.

Figure 2: Shimadzu Mycotoxin Screening System

Table 2: Analytical conditions for analysis of mycotoxins

System:	LC-2040C 3D (Shimadzu Corporation, Japan)
Column:	Shim-pack GIST C18 3.0 mm x 75 mm, 2 µm (Shimadzu Corporation, Japan)
Mobile Phase A:	20 mmol/L NaH_2PO_4 in H_2O, pH 2.5
Mobile Phase B:	Acetonitrile
Mobile Phase C:	Methanol
Gradient:	5 %B to 35 %B; 15 %C in 10 min
Injection volume:	10 µL
Oven Temp.:	55 °C

For Research Use Only. Not for use in diagnostic procedures.

3. Results

All mycotoxins were detected with fluorescence or PDA detection. For NIV and DON, the clean-up cartridge MultiSep 227 was used. Both were clearly identified in the spiked beer sample using PDA detection (figure 3).

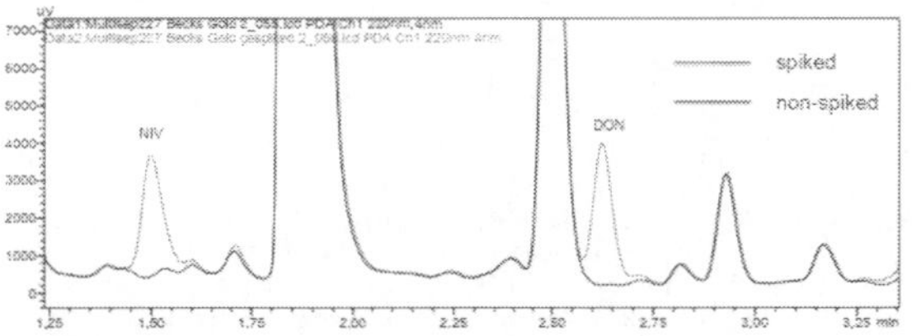

Figure 3: PDA Chromatogram of spiked and non-spiked beer sample

A chromatogram (fluorescence detection) of a spiked and a non-spiked beer sample can be found in figure 4. By comparison of the unaltered beer samples, a difference in the two batches was identified. The chromatogram obtained from batch 2 shows a slightly higher mycotoxin content than the sample of batch 1 (figure 5). However, the concentrations are still below the EU regulations and therefore acceptable. The clean-up cartridge MultiSep 228 was used.

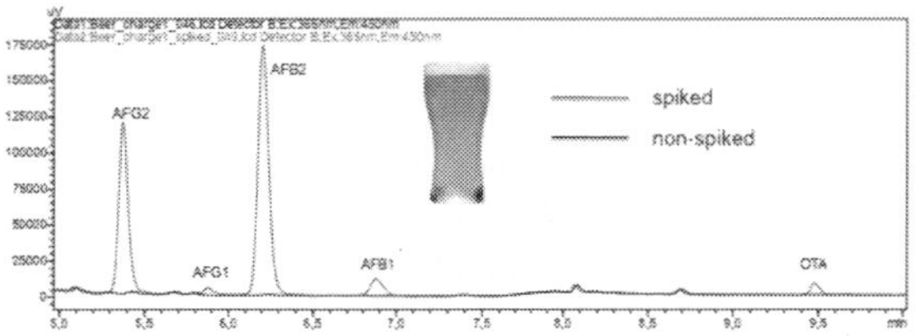

Figure 4: Chromatogram of the analysis of a spiked and a non-spiked beer sample

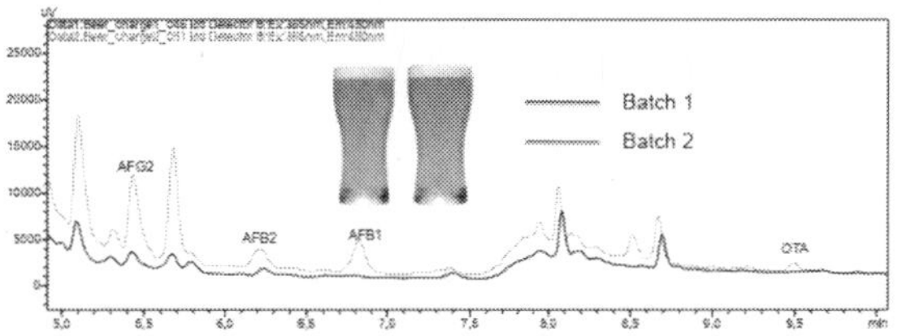

Figure 5: Comparison of samples of two different batches of the same beer brand

For quantification of the mycotoxins, calibration curves were produced in the range of 2.5 - 25 µg/L. These calibration curves all showed good linearity with **R^2 values ≥ 0.97** (figure 6).

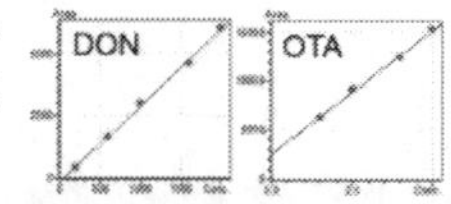

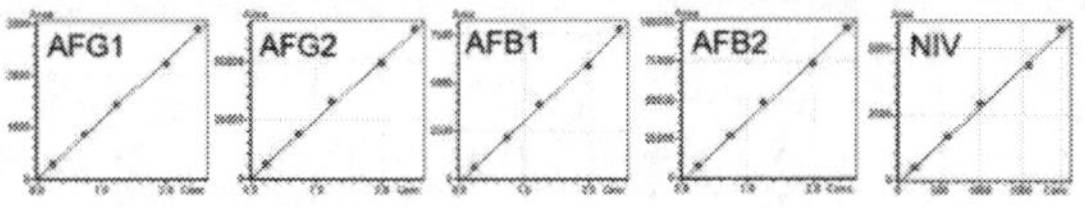

Figure 6: Calibration curves for mycotoxines of interest in the range of 2.5 – 25 µg/L

4. Conclusion

In this study, the applicability of the "Mycotoxin Screening System" from pretreatment to analysis and final reporting of analytical results for the simultaneous analysis of 7 mycotoxins in beer was shown. AFB1, AFB2, AFG1, AFG2, OTA, DON and NIV which are most commonly tested in malt products were extracted and analyzed in spiked and non-spiked beer samples. The sample preparation technique as well as the measurement were proven to be easy and fast. The beer samples analyzed contained mycotoxin concentrations below the EU regulations. The system offers a fast, safe and easy method for the analysis of mycotoxins in beverages such as beer.

References

[1] European Food Safety Authority (http://www.efsa.europa.eu/de/topics/topic/aflatoxins)
[2] EU: Commission Regulation (EC) No 165/2010 of 26 December 2010 amending Regulation (EC) No 1881/2006. Setting maximum levels for certain contaminants in foodstuffs as regards aflatoxins.
[3] EU: Commission Regulation (EC) No 105/2010 of 5 February 2010 amending Regulation (EC) No 1881/2006. Setting maximum levels for certain contaminants in foodstuffs as regards ochratoxin A.

Open your eyes: In microbiological results some important information may be hidden

Marta Orive Camprubí , Núria Feliu Besora, Jaume Lluís Tartera, Benet Fité Luís

Mahou S.A, Quality/R+D+i, Lleida, Spain

EBC 2017 SLOVENIA

MAHOU SAN MIGUEL

1. INTRODUCTION

It is well known that microbiology analyses are no Mathematics. Traditionally, decisions are made based on microbiological analytical results. Usually the risk evaluation of spoilage of beer that is released to the market, is based on colony count or presence/absence of spoilers. Growth in broth tends to be quicker than on solid media but the latter may be preferred, when filtration is possible, because they allow following the evolution by colony counts. Anyway, results from traditional analyses may only give partial information of the real spectrum of microorganisms present in the sample.

Bias may exist because methods could favour one group of microorganisms while hindering the detection of others. Do broth or solid media give a wide view of the spectrum of spoilers present in a beer sample? Another reason for bias in traditional methods is the existence of spoilage bacteria that may be Viable But Non-Culturable (VNBC) and / or be hard to culture (Oliver 2005; Suzuky *et al.* 2008; Deng *et al.*, 2014; Matoulková and Kubizniaková, 2015).

The availability of techniques which are non-dependant on culture, like Real Time Polymerase Chain Reaction (RT-PCR), allows for the detection and identification of spoilers present in the sample even if they are hard to culture or VNBC. At the moment, RT-PCR seems to be a good choice because obligate and/or potential spoilers are rapidly detected and identified in one single reaction.

Breweries will use one or two methods for detection of spoilers and a choice must be made between a wide range of methods available in the market: detection by means of culture in broth or on solid media after membrane filtration, forcing tests, or combining different methods, and even RT-PCR, ...

In this poster, brewery spoiled samples from a pilot plant, where hygiene conditions were forced to be under-standard, are evaluated. Samples are analysed using different traditional methods and RT-PCR was used for identification. Focus was placed on Lactic acid bacteria, either grown in traditional methods or applied directly to forcing test spoiled samples.

It is shown that a different picture arises when looking at the results coming from the different methods and, consequently, the perception of risk derived from them is completely different.

Some obligate spoilers may be hidden when looking at microbiological results coming from different methods. Are microbiologists and brewers aware of such important hidden risks?

With this study, we aim at helping microbiologists and brewers to open their eyes when interpreting microbiology results before any decision is made.

2. EXPERIMENTAL DESIGN

In this study, lactic acid bacteria recovered from beer by different methods were identified to species level and the obtained results were compared.

Detection:

Beer samples were incubated at 27ºC as a Forcing test until macroscopic alteration was detected or up to 45 days (Internal method).

In parallel, beer samples were analysed using a broth medium and samples were incubated until macroscopic growth was detected (EBC method 4.3.1).

Samples were also analysed by membrane filtration (MF) with incubation onto two different solid media (A and B) for beer spoilers (EBC method 4.3.2).

Identification:

RT-PCR (TM Detection kit B LP Identification, Pika-Weihenstephan) was used for species identification.

Samples either from Forcing test or from broth with visible growth, were further analysed by direct RT-PCR and species identified. Moreover, 0.1 ml of spoiled beer was inoculated onto two solid media A and B and in broth C. If necessary, purification and isolation were done. Finally microorganisms grown were identified by RT-PCR again.

When colonies were grown on solid media from membrane filtration, purification and isolation were done, if necessary. Finally they were identified by RT-PCR.

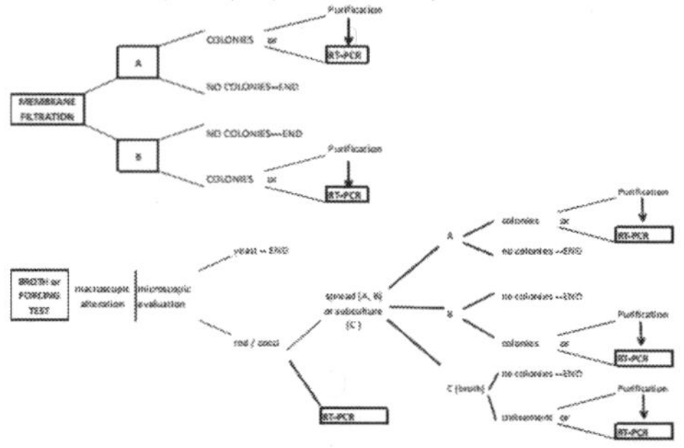

3. RESULTS

Detection + identification results from parallel traditional methods corresponding to 6 samples, which presented visible alteration in forcing test, are shown in the following table.

Direct results coming from Forcing test, broth, and MF are included in red rectangles and are compared. The other columns show results of additional subcultures and identifications that allowed to build up a "complete picture" of the spoilers present in the original samples.

Next a few examples are discussed in order to present biased results and different degrees of risk perception of spoilage.

If the decision in the brewery was to use only membrane filtration on solid media (either A or B): in sample 1, A or B, the paradox would be that no spoiler would be detected and then the risk of spoilage would be considered to be inexistent. Additionally, in samples 2 A and 3 A, a potential spoiler would be detected but no obligate spoiler would be present. However, in both samples both in Forcing test and in the complete picture, obligate spoilers were present.

If only broth culture was the choice, in samples 1, 2, 3, 5 and 6, bias exists as the identified species in broth are different from those identified in the Forcing test. But these species were all obligate spoilers, and so, the risk evaluation of spoilage would be similar in both cases.

A special case is sample 4, where no culture method could detect *P. damnosus* that was detected in the Forcing test and identified with RT-PCR. In this case, *P. damnosus* grew on media only when the beer sample came direct from the forcing test.

It has to be noticed that Forcing test did not give a complete picture and so, there was a biased result as well. As an example, no *L. brevis* was detected in the Forcing test while in the complete picture it is found in most of the samples.

	FORCING TEST				BROTH MEDIA				MF		COMPLETE PICTURE
	When +	A	B	C	When +	A	B	C	A	B	
1	L.lindneri L.coryniformis	no detection	L.coryniformis	L.brevis	L.lindneri L.coryniformis	no detection	no detection	L.lindneri L.coryniformis	no detection	no detection	L.lindneri L.coryniformis L.brevis
2	L.collinoides L.lindneri	L.brevis	L.brevis	L.brevis L.collinoides L.lindneri P.damnosus	L.collinoides L.lindneri L.brevis L.coryniformis	L.brevis	L.brevis	L.brevis L.collinoides	L.collinoides L.coryniformis	L.brevis L.coryniformis	L.collinoides L.lindneri L.brevis P.damnosus L.coryniformis
3	L.collinoides L.lindneri	no detection	L.collinoides L.lindneri P.damnosus	L.collinoides	L.brevis	L.collinoides	L.brevis L.collinoides	L.brevis L.collinoides	L.collinoides L.plantarum	L.brevis	L.collinoides L.lindneri P.damnosus L.brevis L.plantarum
4	P.damnosus	P.damnosus	P.damnosus	P.damnosus	no detection	no detection	no detection	no detection	no detection	no detection	P.damnosus
5	L.collinoides P.damnosus	no detection	L.collinoides P.damnosus	L.collinoides P.damnosus	L.brevis	L.brevis	L.brevis	L.brevis P.damnosus	no detection	L.brevis L.collinoides	L.collinoides P.damnosus L.brevis
6	L.lindneri P.damnosus	no detection	L.lindneri P.damnosus	L.lindneri P.damnosus	L.brevis	L.brevis	L.brevis	L.brevis L.coryniformis	L.collinoides L.brevis	L.brevis	L.lindneri P.damnosus L.brevis L.coryniformis L.collinoides

4.CONCLUSIONS

If only MF and solid media for colony counts are preferred, medium B shows better detection of spoilers than A in this study. In any case, in order to have a more complete picture of the spoilers present in a sample, would it be interesting to have a parallel Forcing test or broth culture? Examples: Sample 1 A or B or 2A and 3 A.

If only broth is chosen, no colony counting is possible but for a complete picture, should there be a parallel Forcing test? Example: Sample 4.

A brewery could consider to use only a Forcing test combined with RT-PCR for detection. As seen in Results, these combined methods do not give the complete picture. Example: Samples 1, 2, 3, 5 and 6. If it was the choice, should not it be complemented with a broth culture or a FM culture on solid medium for a more complete picture?

Less risk is perceived if a potential spoiler (as *L. collinoides*) is detected. But an obligate spoiler (*L. brevis*) could be hidden there. Risk perception of spoilage is very different. Exemples 2A and 3 A.

Another question that comes out from this experience is if it is better to go for group detection (spoiler detection or lactic acid detection) rather than species detection? It is open for discussion.

We have collected a valuable set of results that we hope will help users of microbiological results to open their eyes and think in a wider manner before risk evaluation is finally made.

We would like very much to know your experience and discuss it with you so that we could also learn more about the complexity of microbial results, as well.

5. REFERENCES

Deng, Y., Liu, J., Li, H., Tu, J., Fang, H., Chen, J., Qian, F. (2014). An improved plate culture procedure for the rapid detection of beer-spoilage lactic acid bacteria. Journal Institute of Brewing, 120:127-132

European Brewery Convention (2005). Section 4.3 Detection of Contaminants in beer,. EBC Analytica Microbiologica, pp 209-211, 211-212, 215-217, Fachverlag Hans Carl, Nürnberg

Matoulková, D. And Kubizniaková, P. (2015) Lactic Acid Bacteria and Cultivation Methods of their detection-Part I. Kvasny Prum., 2015, 61: 76-88

Oliver, K.D. (2005). The viable but nonculturable state in bacteria. J.Microbiol. 43: 93-100

Suzuki, K., Asano, S., Iljima, K., Kuriyama, H., Kitagawa, Y. (2008) Development of detection medium for hard-to-culture beer spoilage lactic acid bacteria. J.Appl.Microbiol, 104:1458-1470

TM Detection kit B LP Identification (for identification of beer spoiling bacteria) Cat.No 2001-37. PIKA Weihenstephan GmbH , Germany

COMPARISON OF RAPID METHODS FOR DETECTION OF BEER SPOILERS IN LOW ALCOHOL BEER MIXES

Núria Feliu Besora, Marta Orive Camprubí, Jaume Lluís Tartera, Benet Fité Luís
Mahou S.A, Quality/R+D+i, Lleida, Spain

EBC SLOVENIA

MAHOU SAN MIGUEL

1. Introduction

In low or free alcohol beer mixes the rapid microbiology becomes a need because the microbiological risks are higher than in traditional beer. If these products are packed in PET or kegs containers the risk is even higher. For these type of products Mahou San Miguel has a control plan based on traditional methods that need 9 days to give the results (3 days of product pre-incubation + 6 days for reference method results for *D.anomala*, slowest spoilage yeast in our products to grow on media). Different rapid methods were evaluated to reduce that time to results and a chart with the main characteristics of the best available techniques was built according to Mahou San Miguel needs. Two different techniques were selected and validated for several Mahou San Miguel products.

Characteristics	Tecnichal principle	Range of microorganisms capable to detect	Detection level	Compatible range of products	time to results after 2-3 days of pre-incubation	samples to process in parallel
BACTIFLOW® 	Enzymatic reaction + fluorescence staining + flow citomety detection	- Viable - viable non cultivable	1MO/muestra	Filterable and non filterable products	2-3h	24
MILLIFLEX QUANTUM® 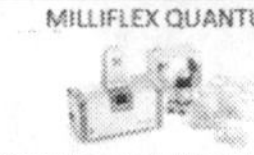	Enzymatic reaction + fluorescence staining + digital camara detection	viable	1MO/volum	Filterable products	Between 1/2 and 1/3 of traditional method	1

Table 1. Main characteristics for the 2 rapid methods selected by Mahou San Miguel to reduce the time for results of their low or free alcohol beer mixes products.

2. Materials and methods

Products: A (Lager beer 5.5ABV + clear lemonade with preservatives), B (Lager beer 5.5ABV + clear lemonade without preservatives), C (Lager beer 5.4ABV + hazy lemonade with preservatives), D (Lager beer 5.4ABV + hazy lemonade without preservatives), E (0.0ABV beer + lemon juice with preservatives), F(0.0ABV beer + lemon juice without preservatives).

Yeast strains: *Sacharomyces cerevisiae, Saccharomyces pastorianus; Sporodiaboluls salmonicolor; Zygosacharomyces bailii; Dekkera anomala; Dekkera bruxellensis* as potential spoilage yeast for the products to be tested.

Reference method: to analyse the spoilage yeast in this type of products we have used the traditional method of membrane filtration (MF) using WLNM (EBC method 4.3.2)

Validation procedure for rapid method: Before using a rapid method this should be validated as equivalent to the reference method. To validate each technique we have followed the following sequence:

1. Compatibility of the samples with the technique: is there any inhibitor factor? Is there any auto-fluorescent signal that could interfere the result?
2. Is it necessary to adapt the protocol/methodology provided for the supplier? check if the methodology available could be applied to our products
3. Qualitative validation of the technique following the ISO 16140:2003 criteria: efficacy, relative detection level, relative specification, relative sensitivity were determined

3. Bactiflow system validation (BF)

This technique detects microorganisms which need to be grown during a pre-incubation time in the product.

1 microorganism/bottle → **2-3 days pre-incubation at 27ºC + analysis in BF (1.5ml)** → positive signal in BF **> 100microorganism/ml**

The slowest spoilage microorganism for each studied product is known from previous studies (data not published). We have checked that 1 microorganism/bottle had altered the product. The objective was to detect the slowest microorganism in each case in less than 9 days (total time required for results using the reference method).

Validation procedure:

1. Products and microorganisms are compatible with BF methodology
2. Several adaptations were needed in the protocol proposed by the supplier.

Products+ slowest spoilage microorganism	Theoretical: 3d pre-incubation+1.5ml analyzed	Longest incubation: Up to 8 d pre-incubation + 1.5ml analyzed	IMPROVED METHOD: Up to 8 days pre-incubation + 100ml analyzed	CONCLUSION
A + Z.bailii	No signal	No signal	No signal	Alternative rapid method needed
B+ S. pasturianus	No signal	No signal	No signal	Alternative rapid method needed
C + D.anomala	No signal	No signal	Signal detected after 6 days of pre-incubation	Validation of BF according ISO
D + S.pastorianus	No signal	No signal	No signal	Alternative rapid method needed
E + D.anomala	No signal	No signal	Signal detected after 6 days of pre-incubation	Not analysed
F + D.anomala	No signal	No signal	Signal detected after 6 days of pre-incubation	Not analysed

Table 2. Adaptation of Bactiflow protocol

3. Qualitative validation: Only product C was validated in BF versus *D.anomala* and *D.bruxellensis*. 10 samples without microorganisms, 10 spiked at desired detection level (<10 cel/bottle) and 10 spiked above detection level (20 cel/bottle) were prepared and analyzed in parallel using the reference method and the Bactiflow system..

Validation results:

Relative efficacy	98%
Relative detection level	4-13 microorganisms/bottle
Relative specification	100%
Relative sensitivity	97%

Table 3. Validation results according ISO 16140

4. Milliflex Quantum system validation (MFQ)

This technique detects micro-colonies before they become visible to the eye.
We needed to know the slowest spoilage microorganism to grow in reference media.
For our products it was *D.anomala*. The objective was to detect *D.anomala* in less than 6 days (time to grow on of the reference media) using Milliflex Quantum system.

Validation procedure:

1. Products and microorganisms are compatible with MFQ methodology
2. No adaptation of the protocol was needed.
   Micro-colonies of *D.anomala* were detected after 72h.

Days	Flow for reference method	Flow for MFQ in Mahou
0	Membrane filtration	Membrane filtration
		Incubation 27°C
3 (72h)	Incubation 27°C	Staining + MFQ Results (micro-colonies)
		Incubation 27°C
6	Results	Results on petri dishes

Table 4. Reference method protocol vs. MFQ

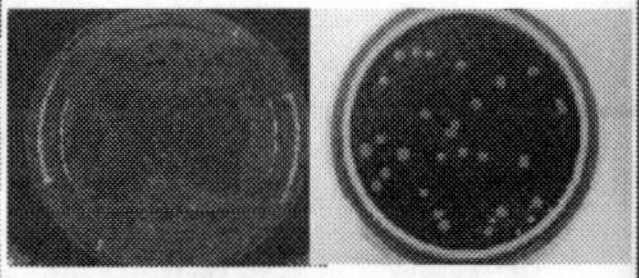

Figure 1 micro-colonies (72h) vs. colonies (6days)

3. Qualitative validation was carried out using 3 different products (A, C, E) combined with both *D.anomala* and *D.bruxellensis* and also 1 product with *Z.bailii*.. For each combination 6 samples without microorganisms, 6 spiked at detection level desired (<10cel/bottle) and 6 spiked above detection level (20 cel/bottle) were prepared and analyzed in parallel using the reference method and the Milliflex Quantum system.

Validation results:

Relative efficacy	100%
Relative detection level	4 CFU / 100 mL
Relative specification	100%
Relative sensitivity	100%

Table 5. Validation results according ISO 16140

5. Conclusions

- Both methods are considered equivalent to the reference method .
- In Bactiflow system: an important adaptation was needed.
  - 3 products of 6 had potential to be analysed using that methodology.
  - 6 days were needed for results
- Milliflex Quantum was a method very close to the reference method
  - could be used in all of the products
  - 3 days were needed for results after 3 days of pre-incubation
  - Further identification of the microorganism s is possible.
- The control plan using Milliflex Quantum as an alternative to the reference method is designed in 6 days (instead of 9 days using the reference method):

3days product pre-incubation + 3days for results (micro-colonies)

6. References

European Brewery Convention (2005). Section 4.3 Detection of Contaminants in beer,. *EBC Analytica Microbiologica*, pp 211-212, Fachverlag Hans Carl, Nürnberg

Hutzler., M., Wellhoener, U., Tenge, C. and Geiger.E., (2008) Beer mixed beverages: dangerous spoilage yeasts, susceptible beverages. *Brauwelt International*, 206-211.

ISO 16140:2003. Microbiology of food and animal feeding stuffs - Protocol for the validation of alternative methods - Amendment 1

Valton, N., Caron, A., Vicente, M. and Chollet, R. (2012) Beverages: Rapid detection of microorganisms. *Filtration+Separation*, May/June pp 22-24

Walker S. (2010). Bri_Quantum report http://www.bri-advantage.com/instrumentcms/pdf/QuantumReport.pdf (accessed 25/10/2014)

DEVELOPMENT OF A NEW METHOD FOR CLASSIFICATION OF NON-VOLATILE NITROSO COMPOUNDS IN BEER USING GC WITH CHEMILUMINESCENCE DETECTION

Tomáš Vrzal[1,2], Dagmar Matoulková[1], Jana Olšovská[1]

e-mail: tomas.vrzal@beerresearch.cz

[1] Research Institute of Brewing and Malting, PLC (RIBM), Prague, CZ. [2] Department of Analytical Chemistry, Faculty of Science, Charles University, Prague, CZ.

AIM OF STUDY

- develop of a method for classification of non-volatile nitroso compounds in beer by chemiluminescence detection and chemometric evaluation
- distinguish chromatographic peaks of nitroso and interfering compounds
- make the first step in identification of unknown non-volatile nitroso compounds in beer
- demonstration of method performance on a microbially contaminated beer sample

INTRODUCTION

Nitroso compounds, specifically carcinogenic N-Nitrosamines, are well-known contaminating compounds in beer. These compounds can be classified as volatile and non-volatile. These days, the main source of a higher amount of total N-Nitrosamines (ATNC) in beer is microbial contamination. Volatile N-Nitrosamines concentrations in beer are very low, even after microbial contamination, e.g. N-Nitrosodimethylamine concentration is usually about 0.2 μg/kg. Based on our previous study, only about 1 % of ATNC in beer is formed by volatile N-Nitrosamines, non-volatile ones representing the vast majority of total ATNC in beer. A routine and reliable analytical method for determination of non-volatile N-Nitrosamines doesn't currently exist because the structure of these compounds has not yet been elucidated either.

Nitrosamine specific chemiluminescence detector (NCD) is not selective enough and some false positive peaks are common in the chromatogram. These peaks could be falsely identified as nitroso compounds. This probem could be solved with CTR filter, which traps interfering pyrolytic products before they enter the reaction chamber, but the main disadvantage is the significant signal decrease of nitroso compounds leading to sensitivity decrease (see. Fig 1) and the signals from nitro and nitro compounds are not distinguished. Therefore the classical NCD is blind in searching for non-volatile constituents of ATNC.

For these reasons, there is requirement for a method which will be able to find unknown non-volatile nitroso compounds as individual chromatographic peaks for subsequent detailed structural elucidation. Based on this requirement we developed a new method for classification and detection of nitroso compounds by pyrolytic profiling NCD. This method uses changes in peak signal depending on pyrolytic temperature (see Fig. 2) and classification of peaks into one of compound classes by multivariate chemometric method.

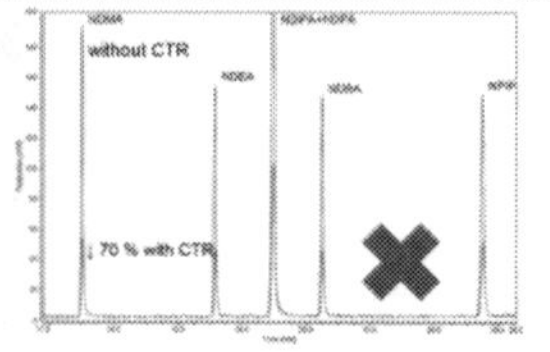

Figure 1: N-nitrosamine signal decrease with CTR filter.

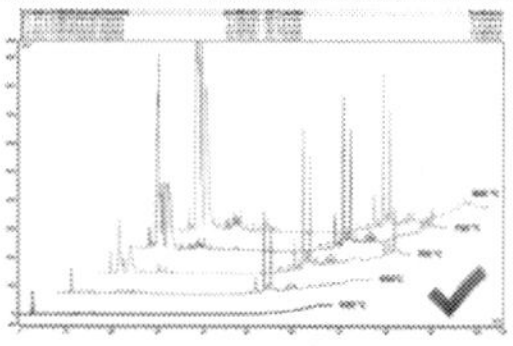

Figure 2: GC-TEA chromatogram of beer sample at different pyrolytic temperatures.

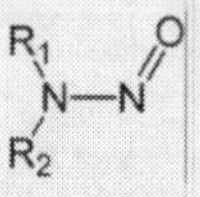

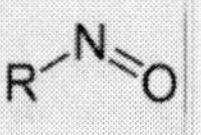

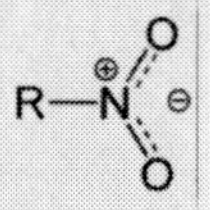

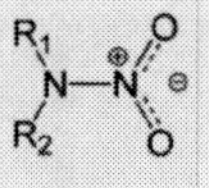

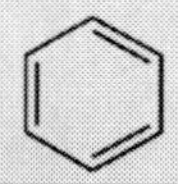

METHODS

Chemicals:

Pure substances used for construction of clasification function by discriminant analysis:
(at least two different concentrations of each)

- N-Nitrosamines – aliphatic and aromatic
- C-Nitroso compounds – aliphatic, aromatic and combined with nitro group
- interferences – aliphatic and aromatic nitro compounds, non-nitroso aromatic compounds and substances with unsaturated hydrocarbon chain
- N-Nitramines – aliphatic (prepared by oxidation of N-nitrosamines)

Data processing:

Relative peak areas (related to the highest area across the pyrolytic profile) and pairwise ratio between these areas were used as independent variables for discriminant analysis. Data set was randomly divided into training and test set (1:1). Forward selection was used for choosing the most discriminating variables for final discriminant function ($\alpha = 0.05$). The training set was used to construct discriminant function and its performance was checked by the test set as well as leave-one-out validation method.

Beer sample preparation:

- 6 ml beer sample + 1 ml 0.2 mol/l ammonium sulphamate in 0.2 mol/l H_2SO_4 (nitrite scavenger)
- precipitation by acetonitrile (beer:acetonitrile, 1:4, v/v)
- distillation of acetonitrile from supernatant (vacuum evaporator)
- sample clean-up with SPE C-18
- sample evaporation to dryness (vacuum evaporator)
- reconstitution of residues and derivatization by BSTFA:TMCS (99:1, v/v) solution

GC-NCD analysis:

GC-NDC analysis was carried out by Thermo Trace 1310 gas chromatograph equiped with Ellutia 820 TEA chemiluminescence detector. Samples were measured at pyrolytic temperatures 500, 650, 700, 750 and 800° C to obtain pyrolytic profiles.

The final method procedure is schematically described in Fig.3.

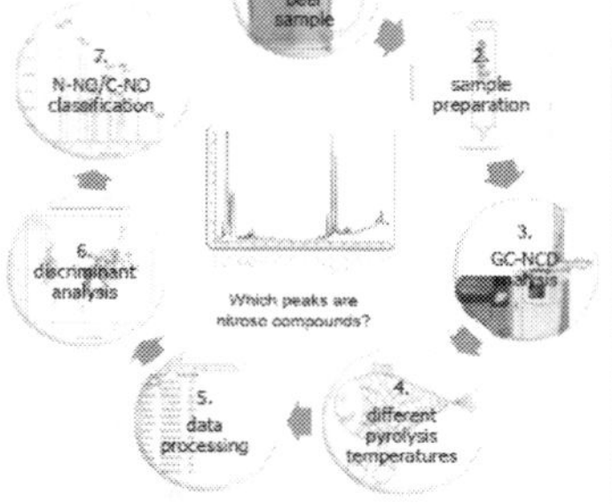

Figure 3: Method flowchart

RESULTS

Table 1: Sensitivity comparison for different compound classes (pyrolysis at 750 °C).

	nitroso		interferences	
	N-NO	C-NO	R-NO_2	aromatic compounds
rel. sensitivity (%)	100.00	24.84	198,1	0.13

Table 2: Confusion matrix obtained by leave-one-out cross-validation

Class	Correct (%)	C-Nitroso	N-Nitroso	interferences	C-Nitroso/nitro combination	N-Nitro
C-Nitroso	76.92	13	0	0	0	3
N-Nitroso	100.00	0	35	0	0	0
Interferences	97.82	0	0	42	1	0
C-nitroso/nitro combination	100.00	0	0	0	6	0
N-Nitro	100.00	0	0	0	0	3
Total	96.12					

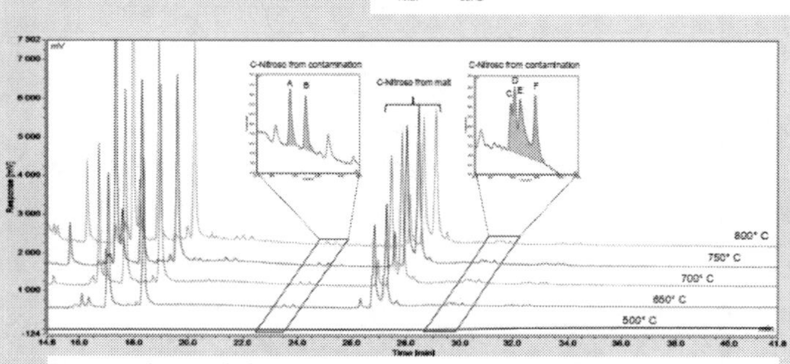

Figure 4: GC-NCD chromatogram of artificially bacterially contaminated (Escherichia coli) beer sample at different pyrolytic temperatures. Nitroso compounds produced by contamination are highlighted at pyrolytic temperature 750° C.

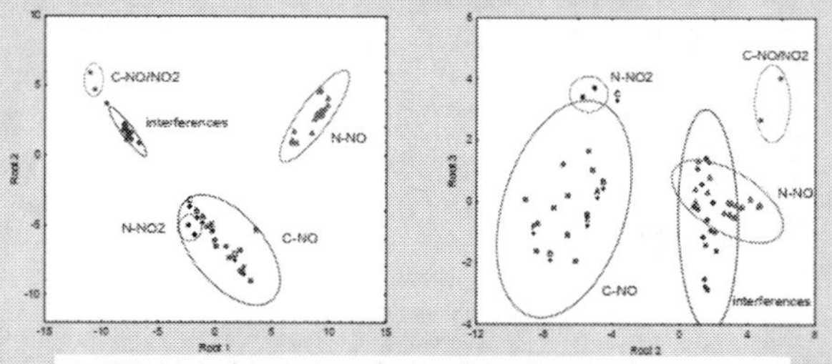

Figure 5: Discriminant analysis canonical scores of pyrolytic profiles from different classes of compounds. Highlighted peaks from Fig 4 are marked as black crosses and classified.

CONCLUSION

A method for classification of NCD chromatographic peaks for non-targeted nitroso compound analysis with good performance was developed. This method is able to detect and classify nitroso compounds into C- and N-Nitroso group and distinguish them from other interfering compounds. It is the first and significant step in our understanding of unknown non-volatile nitroso compounds in beer. The method was demonstrated on artificially microbially contaminated beer sample. The great majority of nitroso compounds detected was classified into C-Nitroso group.

The method is also usable for malt samples or other food matrices.

References:

- Čulík et al., Non-Volatile N-Nitrosamines in Brewing Industry Part I. Arising and methods of estimation. Kvasny Prum. 58 (1), 6 – 11, 2012
- Hansen et al., Characterization of Pyrolysis Conditions and Interference by Other Compounds in the Chemiluminescence Detection of Nitrosamines. Analytical Chemistry 51, 1526 – 1528, 1979.
- Smith, Nitrate Reduction and N-Nitrosation in Brewing, J. Inst. Brew. 100, 347 – 355, 1994.

Acknowledgement

This study was supported by the project of Ministry of Education Youth and Sports of the Czech Republic No. LO 1312.

Pesticide residue analysis in Hops: analysts' nightmare or a unique opportunity to advance

Martin Dušek, Vladimíra Jandovská, Jana Olšovská
Research Institute for Brewing and Malting, Prague, Czech Republic
e-mail: dusek@beerresearch.cz

AIM OF STUDY

- To develop the sample preparation procedure based on QuEChERS approach for determination pesticide residues in Hops
- To optimize the sample clean-up method based on dispersive or cartridge solid phase extraction (SPE) to overcome the problem of signal suppression.

INTRODUCTION

Hops is considered the green gold among the brewery raw materials and the cost of this key ingredient is related to the yield per plant. Agricultural pesticide application makes it possible to reach good yields, reduce the losses during storage and improve the quality of the product. In fact the hop plant belongs to crops with **intensive chemical protection**.

The range of problems that can arise in growing the hop plant includes:

- **bacterial diseases** (*Crown gall*)
- **fungal diseases** (*Downy mildew, Verticillium wilt, Powdery mildew*)
- **virus diseases** (*Mosaic virus, Hop latent virus*)
- **insects and mites invasion** (*California prionus beetle, Japanese beetle, Potato leafhopper, Rose chafer, Hop aphid, Two-spotted spider mite*).

The residues of pesticide active ingredients can persist in the crop for a long time and can potentially endanger consumer's health. This is the reason why the **maximum residue levels** (MRLs) of pesticide residues in hops were established.

DESIGN OF EXPERIMENTS

1. **Composition of SPE sorbent mixtures** (PSA, GCB, C18) *vs.* **Matrix effects**
2. **Composition of SPE sorbent mixtures** *vs.* **Recovery of pesticides**
3. **Cartridge SPE** *vs.* **Dispersive SPE**
4. **Sample dilution for overcoming or reducing of matrix effects**

METHODS and EXPERIMENTS

List of pesticide interests

	Compound	MRL (ppm)*		Compound	MRL (ppm)*		Compound	MRL (ppm)*
1.	Acephate	0.05	17.	Imazalil	0.1	33.	Pyraclostrobin	15
2.	Acetamipride	0.05	18.	Imidacloprid	10	34.	Pyridaben	10
3.	Abamectin	0.1	19.	Malathion	0.02	35.	Quinoxyfen	2
4.	Azoxystrobin	30	20.	Malaoxon		36.	Spirodiclofen	40
5.	Bifenthrin	10	21.	Mandipropamid	90	37.	Spirotetramat	15
6.	Boscalid	60	22.	Mepanipyrim	0.05	38.	Spiroxamine	0.1
7.	Carbendazim	0.1	23.	Metalaxyl	10	39.	Tebuconazole	40
8.	Chlorpyrifos	0.1	24.	Myclobutanyl	5	40.	Tebufenpyrad	0.5
9.	Clothianidin	0.05	25.	Oxadiazon	0.05	41.	Thiabendazole	0.1
10.	Cyazofamid	20	26.	Penconazole	0.5	42.	Thiamethoxam	0.1
11.	Cymoxanil	2	27.	Pendimethalin	0.05	43.	Triadimefon	10
12.	Dimethomorph	80	28.	Pirimicarb	0.05	44.	Triadimenol	
13.	Etoxazole	15	29.	Propamocarb	0.05	45.	Trifloxystrobin	40
14.	Fenpyroximate	10	30.	Propargite	0.05	46.	Triflumizole	0.1
15.	Flonicamid	2	31.	Propiconazole	0.05			
16.	Hexythiazox	20	32.	Pymetrozine	15			

*Regulation (EC) No. 396/2005 of the European Parliament and of the Council on maximum residue levels of pesticides in or on food and feed of plant and animal origin and amending Council Directive 91/414/EEC.

Sample preparation

- Unbuffered original QuEChERS procedure [1,2]
- Dispersive SPE (PSA, C18, Z-Sep – zirconia based sorbent)
- Cartridge SPE (PSA)
- 1 g Saaz variety of organically grown hops spiked at the level 500 ppb (50 µg/L)
- 46 pesticides + 3 x ISTD (TPP, thiamethoxam-*d3* and azoxystrobin-*d4*)

HR/AM LC-MS/MS (Q Exactive hybrid quadrupole-orbitrap MS)

- PRM: (Resolution: 17,500; Isolation window: 1.4 *m/z*), two HR-MRM transition per analyte
- Column: Atlantis T3 3µm, 2.1x100 mm (Waters), MF: *(A)* 2 mM ammonium formate + 0.1% FA, *(B)* MeOH; flow rate = 340 µL/min; T=40°C; inject volume = 2 µL

FINDINGS and RESULTS

Optimizing sample preparation is essential to reduce matrix components and thus to improve reproducibility and accuracy of the analysis. In this study, an extraction of pesticides from the hops matrix was developed and optimized using QuEChERS based method. Although the QuEChERS approach has yielded increased speed and sensitivity, co-extracted interferences in difficult matrices remained as obstacles. In addition to advantages such as the use of less solvent and materials, QuEChERS employs a novel, quick and more versatile dispersive solid phase extraction clean-up (dSPE). The selection of appropriate dSPE sorbent(s) or their combination is critical for elimination of matrix effects of analytes and has to be done individually in accordance with the specific composition of these matrices. The hop cone basically consists of the hop leaves and is made up of the outer structure and the fine yellow resinous powder known as lupuline glands (resin glands). Although a multitude of chemicals are present in lupuline glands, the three main ones are soft and hard resins, also called α-bitter acids (*humulones*) and β-bitter acids (*lupulones*), along with several different types of essential oils (*e.g. myrcene, humulone* and *caryophyllene*). The average content of total resins varieties from 15 to 30wt% and the amount of essential oils is 0.5-3wt% [3]. These compounds and especially resins are, in comparison to other acetonitrile extractable compounds such as polyphenols, waxes, chlorophyll, and monosaccharides, the main source of adverse effects from co-extracted matrix. Thus, it is obvious that the first step in optimization of the sample clean-up has to be based on removing bitter acids by application of appropriate dSPE sorbent.

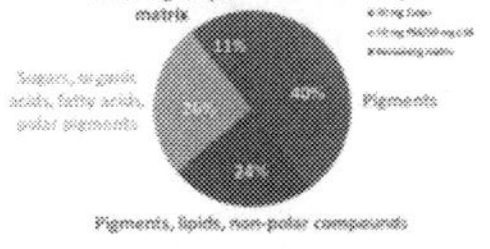

FIGURE 1. The amount (%) and type of co-extracted compounds possible to remove by application of various dSPE sorbents.

1-2. Dispersive SPE (dSPE) - sorbent(s) selection

Experimental: dispersive SPE: 1) 100mg PSA, 2) 150mg PSA, 3) 100mg PSA/25 mg C18, 4) 100mg PSA/25 mg C18/25 mg Z-Sep, 5) 10mg GCB/50 mg PSA/60 mg Z-Sep+; 1 mL of QuEChERS extract, centrifuged for 2 min.

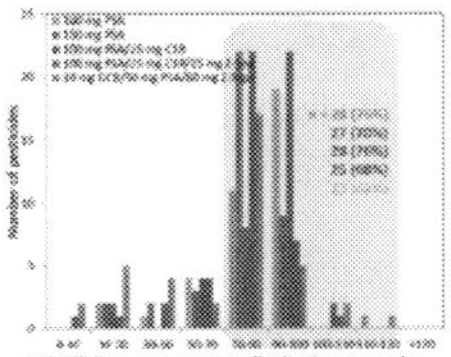

FIGURE 2. Percent recovery distribution comparing various dSPE sorbent blends.

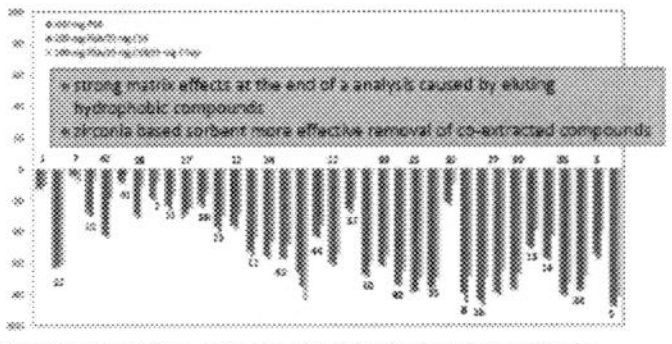

FIGURE 3. Matrix effects, expressed as %ME-100, for three dSPE sorbents in retention time order.

3. Cartridge SPE vs. Dispersive SPE

PROS: - 36 from 38 pesticides have recovery between 80-120%

CONS: - cost of SPE cartridge
- solvent evaporation step has to be involved into the protocol
- c-SPE: combination c-SPE protocol with other sorbents (C18) is clumsy

Adopted QuEChERS procedure for hops samples

- **Weight 1 g of minced hops in 50 mL extraction vessel**
- **Add 10 mL water and wait 30 min**
- **Add 10 mL Acetonitrile + ISTDs**
- Shake
- **Add partitioning/Buffer Salts**
  - 4 g $MgSO_4$
  - 1 g NaCl
- Shake / Centrifuge (4,500 rpm, 7 min)
- **d-SPE: 1 mL aliquot and mix with 150 mg $MgSO_4$ + 100 mg PSA + 25 mg C18 + 25 mg Z-Sep**
- Shake / Centrifuge (13,000 rpm, 2 min)
- **Re-acidify Extract** (10 µL formic acid 5% (v/v) per mL)
- **LC-MS**

4. Effect of Sample Dilution on Matrix Effects

The matrix effects could be compensate using extensive extract dilution as a valuable tool to overcome the problem of signal suppression.

Experimental: Undiluted d-SPE (100 mg PSA/mL) extract spiked at concentration level (500 ppb, 50 µg/L) and dilutions of **1X** (25 µg/L), and **1/4X** (10 µg/L).

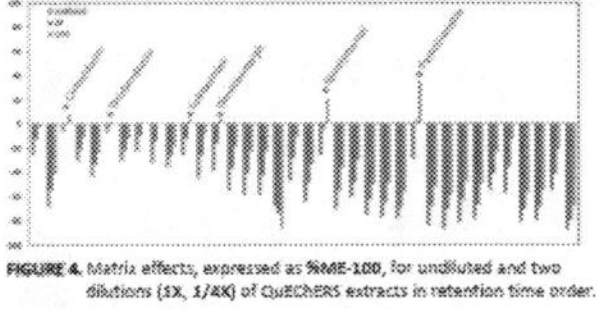

FIGURE 4. Matrix effects, expressed as %ME-100, for undiluted and two dilutions (1X, 1/4X) of QuEChERS extracts in retention time order.

CONCLUSION

- The additional sample clean-up by dispersive or cartridge SPE is sample has to be involved in sample preparation procedure.
- The using of PSA sorbent is essential to and ensure good data quality.
- The SPE method, using combination of PSA, C18 and Z-Sep as dispersive sorbents, minimize adequate the matrix effects of hops and delivered acceptable recoveries for most of the pesticides.
- The using of other tested sorbents must carefully balance minimization of matrix effects with analyte recovery.
- Sample dilution is also powerful tool for elimination matrix effects. However, higher dilution then 1/4X will cause that LOD for some pesticides to fall below their MRL.

Literature
[1] Anastadiades M. et al.: J. AOAC Int. 86, 412-431 (2003).
[2] Lehotay S. J. et al.: J AOAC Int. 88, 615-29 (2005).
[3] Almaguer C. et al.: J. Inst. Brew. 120, 289–582 (2014).

ACKNOWLEDGEMENT
This study was supported by the project TE02000177 "Centre for Innovative Use and Strengthening of Competitiveness of Czech Brewery Raw Materials and Products" of the Technology Agency of the Czech Republic.

36TH CONGRESS EBC SLOVENIA 2017 EUROPEAN BREWERY CONVENTION 14-18 MAY LJUBLJANA

Are different hop pellet properties for dry hopping and hop dosing in the brewhouse justified?

Adrian Forster* and **Florian Schüll**, HVG Hopfenverwertungsgenossenschaft e.G., Wolnzach, D
Andreas Gahr and Richard Lenz, Hopfenveredlung St. Johann GmbH, Train, D

INTRODUCTION

What does **pelletization** mean?
In a pellet press, hop powder is forced through holes in a die. The degree of pressing is primarily dependent on two parameters:

- Length of the channel through which the powder is pressed; the longer the channel, the greater the compression
- Friction in the channel, which varies with compounds in the product which "lubricate"; this is moisture content and other liquid media such as resins (alpha) and oil

Some characteristics of hop pellets:

- Usual diameter 6 mm, length 5 – 12 mm
- Bulk density increased from 135 kg/m³ in baled hops to approx. 500 kg/m³ in pellets
- Easy packaging in laminated foils without any oxygen diffusion (inert atmosphere)
- Free flowing product
- Transportable in screws, elevators, pneumatic systems
- Easy weighing
- Therefore suitable for automatic dosing systems in a brewery
- Lupulin membrane is crushed => substances dissolve faster

PELLET QUALITY AND STABILITY

The more stable or harder a pellet the easier all processes like transport, dosing or weighing will be. Therefore many brewers prefer stable pellets and complain often about fine particles. However in order to achieve this, higher pressure needs to be applied to squeeze the powder through the holes of a die. Thereby heat is generated by the friction in the channel increasing the temperature of the product too.

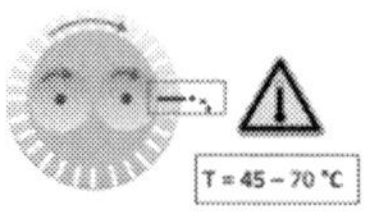

This is the critical point during pelletizing process! Higher temperature means lower quality!
The temperature of the pellets must be controlled and the product has to be cooled after pelletizing. Each brewery can formulate its requirements for pellet stability consequently and should be aware of the fact:
The more stable the pellets, the higher the temperature during the pelletization process.
Pelletization involves finding a "proper" compromise between stability and temperature.

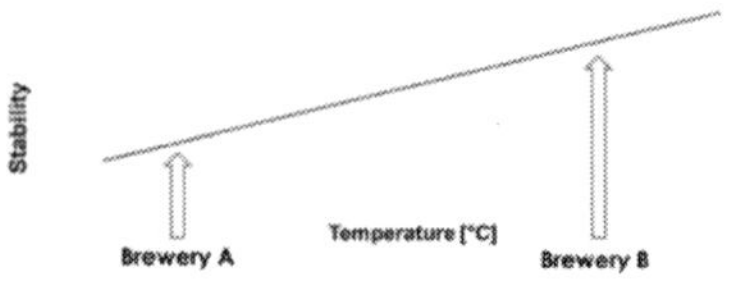

Figure 1: Basic correlation between pellet stability and pelletizing temperature depending on brewery specifications

Indicators of damage to the hops through pelletization:

- Differences in the sensory impression of the powder and pellets which can be tested sensorily by a direct comparison of powder and pellets or via a hop tea (hops brewed in hot water)
- A measurable loss in α-acid content, indicating a significant damage by oxidation

Breweries often demand a certain pellet stability, which is specified as the amount of loose powder per package. They need to be aware that lower amounts of fine particles and high pellet stability is associated with high pelletization temperatures and product damage.
The pelleting temperature can be part of a specification. But the design of a pellet dosing unit in a brewery should fulfill the demand for soft and high quality pellets anyway.

SPECIAL DEMANDS ON PELLETS FOR DRY HOPPING ?

	Brewhouse	Tank (dry hopping)
Temperature	80 – 100 °C	0 – 20 °C
Movement	Intensive (boiling)	almost no
Dissolution of pellets	quickly	(rather) slowly
Automatic dosing = conveying mechanically	common	uncommon
Loose hop powder disturbes	yes (conveying / vapour stack)	no
Aroma of pellets	important	particularly sensitive

Table 1 : Typical differences in application of pellets

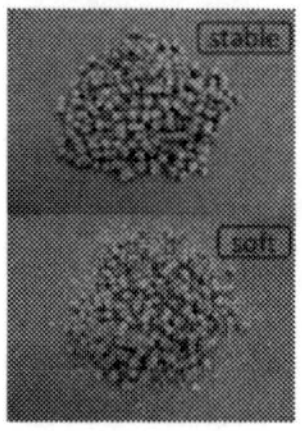

Figure 2 : Stable and soft pellets

Interesting differences between pellets for kettle and dry hopping:

- Pellets for dry hopping dissolve slowly in cold beer. Therefore they should be pressed as soft as possible for a better disintegration.
- **"Dry Hop Pellets"** do not need a specific stability because there are fully automated systems scarcely. Fine particles do not disturb in most cases.
- When dosed into a tank without a specific device pellet particles are supposed to sediment slowly for better extraction.
- Flavour quality is even more important in dry hop pellets; lowest possible pelletizing temperatures are on strict demand.

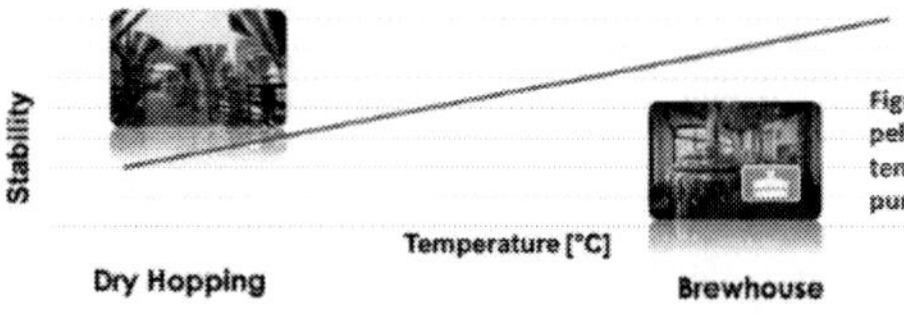

Figure 3: Correlation between pellet stability and pelletizing temperature for different purposes

Possibilities to describe pellet properties:

- Temperatures during pelletization
- Measures for stability:
  - Hardness by Kahl = disintegration force [kg]
  - Bulk density [kg/l]
  - Ratio of fine particles in a package [% rel.]
- Disintegration of pellets at 20°C:
  - Visual observation in a glass cylinder (estimation of time)
  - Sedimentation in a glass cylinder (time)
- Solubilisation of hop substances:
  in a 0.5 L glass bottle (1 g / 0.5 L for 30 min),
  like alpha, IBU, polyphenols, linalool [4]

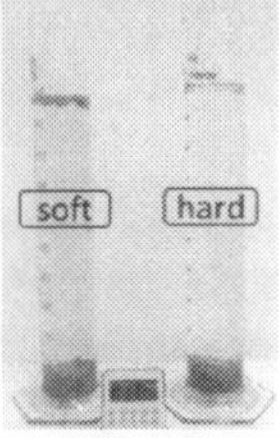

Figure 3: Disintegration of soft and hard pellets

ROLE OF PARTICLE SIZE FOR DRY HOP PELLETS

There are two major possibilities for dry hopping with pellets:

- External dosing systems like "Hopgun" or "DryHopnik"; optimal particles size for those not discussed here.
- Direct dosage of hop pellets into the tank; discussed here.

The particle size of the dissolved pellets has an influence on the sedimentation. The bigger the particles the faster they sediment. The extraction of valuable substances is better, when particles float in a tank for a longer period. There is a possibility to influence the particle size of hop powder prior to pelletization ➡ **Type 45 process: milling and sieving at -30°C.**

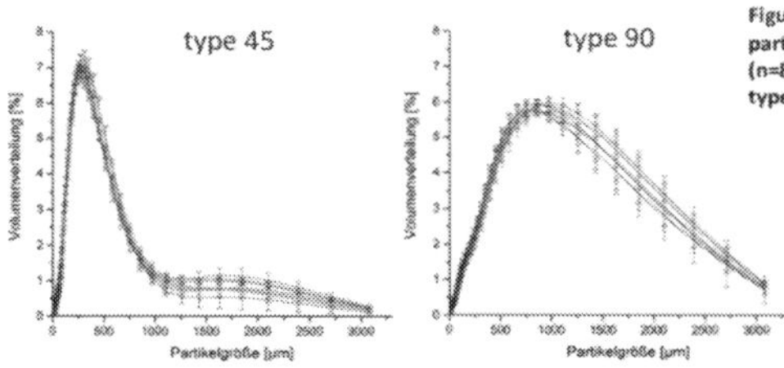

Figure 4: Difference of particle size distribution (n=8) of type 45 and type 90 hop powder

Some authors have recently proposed to achieve a smaller particle size by using specific rollers in a pellet mill [3]. However this involves higher temperatures of the powder even before pressing and consequently higher pellet temperatures.

	unit	kettle	dry
Pellet temperature	°C	54	48
Hardness acc. Kahl	kg	4	1
Bulk density	kg/l	530	485
Ratio of fine particles	% rel.	2	10
Disintegration	min	5	2
Sedimentation (30cm)	sec	5	60*
Solubilisation of α-acids	mg/l	1.5	4.0

Table 2: Characteristic properties of pellets for brewhouse use (kettle) and dry hopping (dry) respectively.

* Pellets and part of pellets floating for longer time (Figure 3).

SUMMARY

General concerns regarding pellet quality:

- Brewers often prefer pellets with a high stability without fine powder particles. Mostly there is no rational reason for this prejudice besides the fact that hard pellets are more advantageous in some dosing systems.
- Stable pellets are inevitably produced in a pellet mill at higher temperatures which has a negative impact on the product quality.
- The brewer has to accept and find a compromise between hard pellets pressed at higher temperatures and soft pellets with a higher ratio of fines that require a lower pelletization temperature.

For dry hopping in a tank the specifications on pellets are different to those on pellets dosed in the brewhouse:

- They should be softer for a faster disintegration time, which also allows a pelletization at reduced temperatures and thus a production of a high aroma quality pellet.
- A smaller particle size is also an advantage when dosing pellets into a tank. They sediment slower and enable a better extraction.

An ideal dry hop pellet with these properties (soft pelletization / lower particle size) can be produced in a production plant for lupulin enriched hop pellets at -30°C which allows for a more gentle production process.

REFERENCES

1) EBC Manual of Good Practice, Hops and Hop Products, Fachverlag Hans Carl, Nürnberg 1997, ISBN 3-418-00758-9, pp. 56-60
2) Biendl M., Engelhard B., Forster A., Gahr A., Lutz A., Mitter W., Schmidt R. and Schönberger C.: Hops – Their Cultivation, Composition and Usage; Fachverlag Hans Carl, 09/2014, ISBN: 978-3-418-00823-3, pp. 159-161
3) Engstle J., Kuhn M., Kohles M., Briesen H. and Först P.: Disintegration of Hop Pellets during Dry Hopping, Brewing Science 69, 2016, 123-127
4) Krottenthaler M.: Factors influencing the transfer rate of hop aroma compounds during dry hopping of lager beers, Oral Presentation, EBC Congress, Glasgow, 2011

* adrian.forster@gmx.de

FTIR characterisation of sorbent-polyphenol interactions involved in beer stabilization

UNIVERSITY OF CHEMISTRY AND TECHNOLOGY PRAGUE

Karabín* M., Jelínek L., Nešpor J., Dostálek P.

University of Chemistry and Technology, Prague, Department of Biotechnology, Technická 5, CZ-166 28, Prague 6, Czech Republic.
*e-mail: karabinm@vscht.cz

AIM OF THE WORK

The aim of this work is to evaluate the activity of individual functional groups of phenolic compounds involved in binding with sorption stabilizers used for colloidal stabilization of beer

INTRODUCTION

The colloidal stability of beer is one of the most complex and not fully elucidated issues the brewing industry is facing these days. The disturbances in colloidal equilibrium lead to the formation of small particles that can grow over time to a size that can be observed by the naked eye (1-10 µm). Whereas the majority of customers perceive fresh beer as a clear sparkling beverage, the appearance of this non-biological haze is perceived by these consumers as a major defect.

The exact mechanism of haze formation is still the subject of scientific investigation. The most accepted theory was introduced by Siebert et al. in the second half of the 1990s. In his work he suggested that the reduction in the colloidal stability of beer is caused, almost exclusively, by the precipitation reactions of the oxidized forms of polyphenols (especially flavan-3-ols) with the fraction of malt polypeptides rich in proline. The importance of oxidation and its influence on the rate of the haze-formation process was confirmed in the 2003 by Leemans et al. A critical prerequisite for the ability of polyphenols to form colloidal haze is the presence of two or more sufficiently distant hydroxyl groups. Polyphenols that do not meet this requirement inhibit the whole process because they are unable to link protein molecules to larger structures.

Modern methods for the prevention of colloidal turbidity (stabilization) are therefore currently based almost exclusively on the removal of haze-active precursors (generally during the beer filtration and packaging), usually by binding to a suitable sorbent. The removal of polyphenols by PVPP is the most frequently used technique, although polyamide-based products (BEERPAP® - polyamide layer on kieselguhr core) have also been tested for this purpose in recent years. While this process is effective and well-researched from the point of view of technology, only a little is known about the precise mechanism and the types of bonds involved. This knowledge should allow selective removal of haze-active polyphenol compounds while preserving polyphenols with antioxidant properties would help maintain the sensory stability of the product during storage.

Siebert, K. J.; Carrasco, A.; Lynn, P. Y. Formation of protein-polyphenol haze in beverages. J. Agric. Food Chem. 1996, 44 (8), 1997–2005.
Siebert, K. J.; Troukhanova, N. V.; Lynn, P. Y. Nature of polyphenol-protein interactions. J. Agric. Food Chem. 1996, 44 (1), 80–85.
Leemans, Ch.; Pellaud, J.; Melotte, L.; Dupire, S. Opportunities for beer haze prediction: A new tool assess beer colloidal stability. In Eur. Brew. Conv. Proc. 29th Congress, 29th EBC Congress, Dublin 2003; pp 904–914.

FTIR ANALYSIS

The analysis was performed using a FTIR spectrometer NICOLET 6700 equipped with DTGS and MCT-A detectors (measuring range 7500 - 350 cm^{-1})

500 ml of solution of phenolic substances (catechin, epicatechin, protocatechuic acid, vanilic acid and ferulic acid) in 5% (v/v) ethanol with an approximate concentration of 0.1 g/l (pH=4.3, treated by hydrochloric acid) was mixed with sorbent (2.5 g of PVPP or 3 g of BEERPAP®) and the suspension was stirred for 10 minutes on a magnetic stirrer. The sorbent was filtered off, rinsed with 200 ml of distilled water and the moisture was removed by lyophilization. 0.1 g of this dry polyphenol-sorbent complex or same amount of pure polyphenol and pure sorbent were analyzed by FTIR.

RESULTS AND DISCUSSION

FTIR analysis is based on the comparison of the infrared spectra of the adsorbent, the adsorbate and the absorbent-adsorbate complex. Due to the characteristic shifts of the peaks corresponding to the individual functional groups of polyphenols and sorbents, it is possible to determine not only which of them is involved in the formation of the bond but also the type of this bond.

The FTIR spectrum of pure PVPP (Fig. 1) is characteristic by appearance of wide peak at 1648 cm^{-1} (vibrations of the carbonyl group, adjacent to the tertiary amine). The analogous peak (at 1631 cm^{-1}) was obtained also for the carbonyl group adjacent to the secondary amine belonging to the polyamide 6 (Fig. 2).

Spectra of the polyphenol-sorbent complex are characterized by the shift of the PVPP carbonyl band (1648 → for PVPP; 1631→ for polyamide). This clearly indicates that all analyzed polyphenols are capable of binding to this sorbent. In the case of polyamide 6 sorbent, these shifts are not nearly as intense as for PVPP, which means that the binding of polyphenols to polyamide 6 is slightly weaker, and in addition to the relatively strong hydrogen bonds, the weak physical bonds such as hydrophobic interactions can play a significant role here.

Table I : Polyphenol-PVPP interactions

Spectrum	Peak shift (cm^{-1})	Assignment
PVPP	1648→?	Polyphenol-sorbent bonding
Catechin	1281→1282 1236→1225 1181→1176 1108→1099 1046→1028	Interaction between C3´, C4´, C5 a C7 hydroxyls and carbonyl group of PVPP
	1142→1142 1077→1077	No interaction between C3 hydroxyl and carbonyl group of PVPP
Epicatechin	1289→1279 1224→1196 1068→1064 1044→1026 1016→1010	Interaction of C3, C3´, C4´, C5 and C7 hydroxyls with carbonyl group of PVPP
	1289→1289 1140→1140 1093→1093	No interaction between C3 hydroxyl and carbonyl group of PVPP
Protocatechuic acid	1054→1005 1119→1110	Interaction between C3 and C4 hydroxyls and carbonyl group of PVPP
	1277→1258 1246→1219	Interaction of C1 carboxyl group with carbonyl group of PVPP
Vanillic acid	1166→1157	Interaction between C4 hydroxyl and carbonyl group of PVPP
	1231→1198	Interaction between C3 methoxy- group and carbonyl group of PVPP
	1277→1282 1236→1223	Interaction between C1 carboxyl group and carbonyl group of PVPP
Ferulic acid	1282→1282 1034→1018	Interaction of C4 hydroxyl with carbonyl group of PVPP
	1266→1246 1231→1224	Interaction of carboxyl group with carbonyl group of PVPP
	1164→1164	No interaction between C3 methoxyl group and carbonyl group of PVPP

PVPP:

- ✓ In the case of flavan-3-ols, significant shifts corresponding to the hydroxyl groups attached to aromatic ring carbons were found
- ✓ Heterocyclic ring carbon (C3) hydroxyl does not participate in PVPP-polyphenol binding.
  - hydroxyls on aromatic cycle are more reactive than the hydroxyls attached to the heterocycle.
  - outwardly directed aromatic ring hydroxyls are more accessible and the hydroxyl on carbon C3 is also partially shielded by structure of the aromatic ring (Fig. 3)
- ✓ Polyphenols containing vicinal hydroxyl groups on C3 and C4 carbons are capable of binding to PVPP.
- ✓ In addition, analyzed phenolic acids are able to interact with PVPP also via the carboxyl group, and vanillic acid also through the unspecified weak non-covalent interaction in which the methoxy group is involved.
- ✓ Based on the sorbent and sorbate structures, it can be assumed that this interaction is not a hydrogen bond, since the methoxy group lacks hydrogen which is bound to sufficiently electronegative atom.
- ✓ The same type of interaction has not been demonstrated for ferulic acid, probably due to the chain length.

POLYAMIDE 6:

- ✓ The binding of polyphenols to the polyamide sorbent is relatively weak
  - the intensity of the shifts of individual bands was only about half compared to PVPP,
  - all four hydroxyl groups, bonded on aromatic rings, involved in the interaction, while the C3 substituent remains passive.
- ✓ For protocatechuic acid have not been documented any shifts indicating the contribution of hydroxyl groups to binding to the sorbent.
- ✓ Relatively strong shift confirms that this phenolic acid binds to the polyamide sorbent almost exclusively via carboxyl group.
- ✓ For vanillinic acid, shifts of the carboxyl, C4 hydroxyl and methoxy group were detectable.
- ✓ Only the carboxyl group of major malt polyphenol - ferulic acid was largely involved in the bonding to the polyamide sorbent. The remaining two substituents (the C3 methoxy group and the C4 hydroxy group) did not participate in the interaction.

Table II : Polyphenol-polyamide interactions

Assignment	Peak shift (cm^{-1})	Spectrum
Polyphenol-sorbent bonding	1631→?	Polyamide 6
Interaction between C3´, C4´, C5 a C7 hydroxyls and carbonyl group of polyamide 6	1281→1276 1236→1226 1100→1096 1077→1068 1048→1040	Catechin
No interaction between C3 hydroxyl and carbonyl group of polyamide 6	1042→1042 1077→1077	
Interaction of C3, C3´, C4´, C5 a C7 hydroxyls with carbonyl group of polyamide 6	1289→1279 1224→1216 1062→1061 1044→1036 1016→1010	Epicatechin
No interaction between C3 hydroxyl and carbonyl group of polyamide 6	1289→1288 1140→1140 1093→1088	
No interaction between C3 a C4 hydroxyls and carbonyl group of polyamide 6	1094→1094 1119→1119	Protocatechuic acid
Interaction of C1 carboxyl group with carbonyl group of polyamide 6	1277→1268 1249→1238	
Interaction between C4 hydroxyl and carbonyl group of polyamide 6	1165→1128	Vanillic acid
Interaction between C3 methoxy- group and carbonyl group of polyamide 6	1231→1102	
Interaction between C1 carboxyl group and carbonyl group of polyamide 6	1277→1281 1238→1228	
No interaction of C4 hydroxyl with carbonyl group of polyamide 6	1282→1282 1034→1034	Ferulic acid
Interaction of carboxyl group with carbonyl group of polyamide 6	1266→1248 1231→1225	
No interaction between C3 methoxyl group and carbonyl group of polyamide 6	1164→1164	

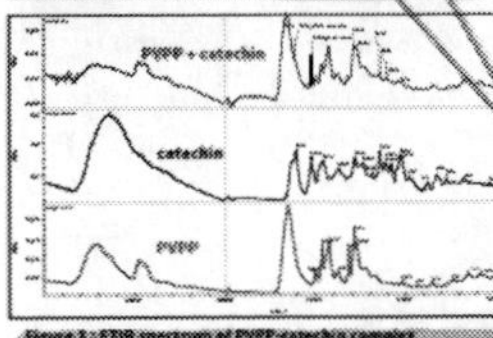

Figure 1: FTIR spectrum of PVPP-catechin complex

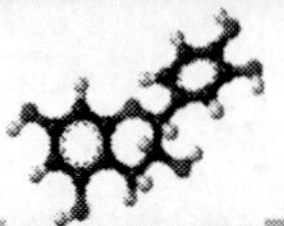

Figure 3: 3D-structure of catechin

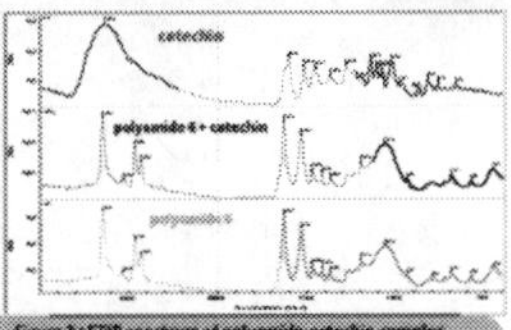

Figure 2: FTIR spectrum of polyamide-catechin complex

CONCLUSIONS

In this work was tested the ability of both traditional (PVPP) and newly developed polyamide (BEERPAP®) sorbents to bind selected polyphenolic substances. The types of bonds were described using the FTIR method for the obtained polyphenol-sorbent complexes and importance of the individual functional groups in polyphenolic molecules was defined.

We found that, in addition to the expectable interactions of the hydroxyl groups, also carboxyl group of phenolic acids and, by so far unidentified interaction, also methoxy- moieties participate in the stabilization effect. Lower absolute efficiency of polyamide sorbent could be compensated by its ability to preferentially bind larger flavonoid structures and thus increase the colloidal stability of beer without losing antioxidant activity.

Acknowledgement: This work was supported by TACR "Centre for Innovative Use and Strengthening of Competitiveness of Czech Brewery Raw Materials and Products" - TE02000177.

Characterization of beer haze using AF4

Pieter Stam, Jozé Mutsaers, Angela de Bruine - Paulus, Wilbert Heijne* and Erwin Kaal

DSM Biotechnology Center, part of DSM Food Specialties B.V., Delft, The Netherlands

*DSM Food Specialties, Delft, The Netherlands

Wilbert.Heijne@dsm.com

Introduction

Crystal clear beer and colloidal beer haze stability are important quality parameters of many beer types.
Beer haze is in most cases unwanted and consists of a very complex mixture of macromolecules, including polysaccharides, polyphenols and proteins.

Better understanding of the nature of the beer haze is essential to prevent or resolve it. For this, advanced analytical methods enable to assess the complex structure of haze. Recently, the applicability of Asymmetrical Flow Field-Flow Fractionation (AF4) was investigated for beer haze characterization.

In this study, the development of a novel analytical method for (native) beer haze profiling based on AF4-UV-FLR-MALLS-dRI (ultraviolet, fluorescence, differential refraction-index and multi angle laser light scattering detector) is described. Preliminary results are shown of several randomly chosen (lager type) beers from the market. AF4 reduces sample preparation to a minimum and the obtained beer haze profiles contain valuable information on haze composition.

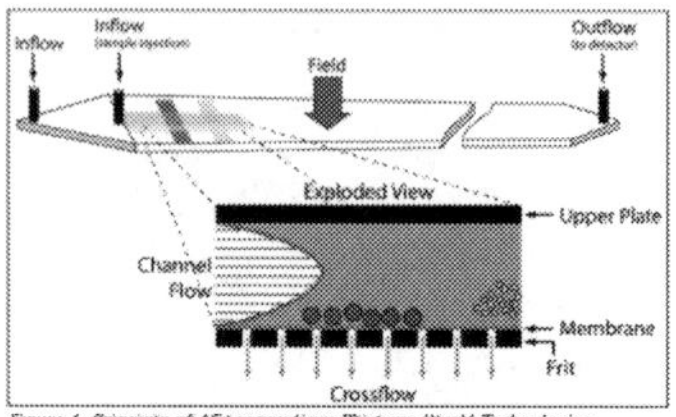

Figure 1. Principle of AF4 separations [Picture: Wyatt Technologies, http://www.wyatt.com/images/graphics/theory/fff/fff-crossflow.png]

Experimental

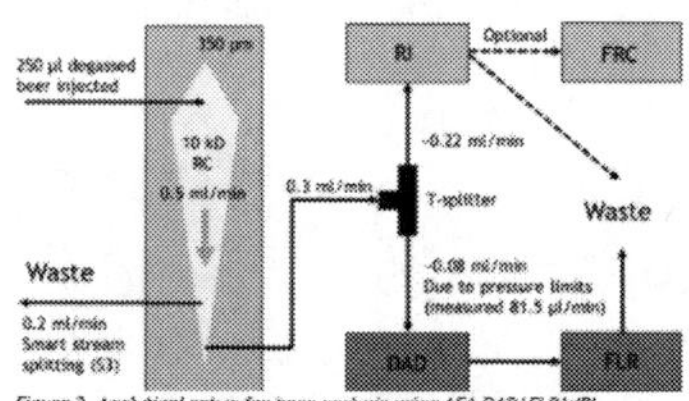

Figure 2. Analytical setup for beer analysis using AF4-DAD/FLR/dRI

Table 1. Detector settings

	Detector settings	Specificity
UV	280 nm	Proteins + polyphenols
UV	320 nm	Polyphenols
DAD	UV/VIS spectrum 240-800 nm	Any compound with UV absorption in the 240-800 nm range
FLR	Ex278, Em338 nm	Proteins + probably some polyphenols
FLR	Ex280, Em305 nm	Polyphenol like
FLR	Ex280, Em440 nm	Polyphenols
FLR	Ex320, Em450 nm	Polyphenols
dRI	Refractive index	Polysaccharides after corrections
MALLS	19 angles	Molecular size properties

Results

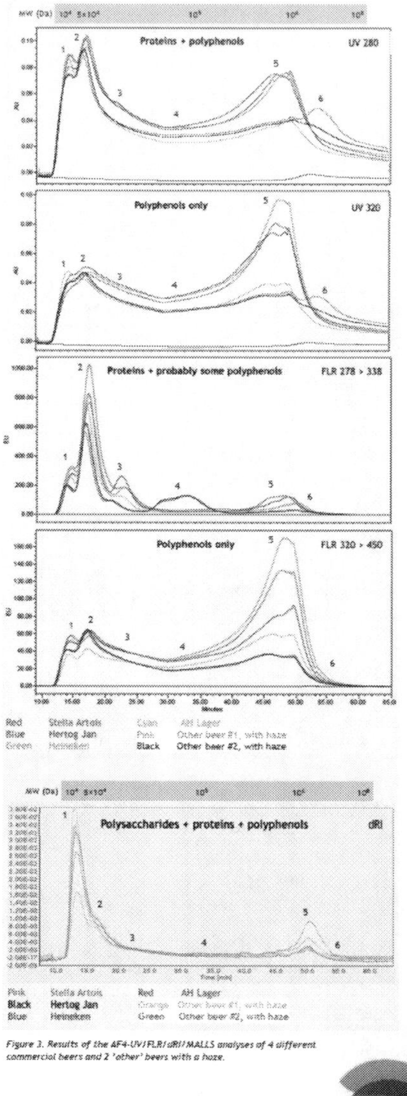

Figure 3. Results of the AF4-UV/FLR/dRI/MALLS analyses of 4 different commercial beers and 2 'other' beers with a haze.

Conclusions

- This AF4 method developed for beer shows to be very promising for characterization of macromolecules in beer.
- Allows quick and easy sample preparation: the sample is directly injected, after degassing the beer.
- Interesting information can be obtained from the results; 6 different regions can be recognized in the AF4 profiles.

Table 2. Characterization of the AF4 profiles.

Region	Information
1	Rich in polysaccharides, ~5×10^4 Da (cut-off of the AF4 membrane)
2	Rich in proteins, 10^4-10^5 Da
3	Rich in proteins, ~10^5 Da
4	Flat part in the fractograms of the lager beers. A protein rich peak was detected in the 2 'other' beers, ~10^5 Da
5	Rich in polyphenols, ~10^6 Da
6	Ultrahigh MW part, 10^6 - 10^8 Da, only detected in 'other' beers, with haze.

- The AF4 method allows for easy fractionation of the samples, for further characterization with different techniques: NMR, proteins-, polyphenols-, and sugar analyses.
- Next steps focus on correlating observed particular beer hazes to specific signals in the spectrum

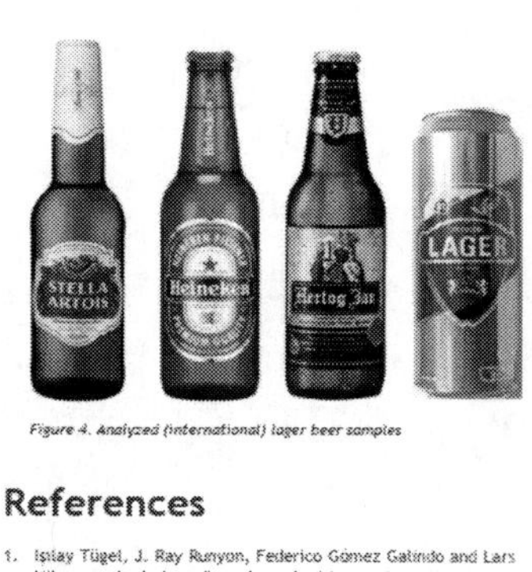

Figure 4. Analyzed (international) lager beer samples

References

1. Işılay Tügel, J. Ray Runyon, Federico Gómez Galindo and Lars Nilsson, Analysis of polysaccharide and proteinaceous macromolecules in beer using asymmetrical flow field-flow fractionation, Institute of Brewing & Distilling, (wileyonlinelibrary.com) DOI 10.1002/jib.195

2. Karl J. Siebert. Haze formation in beverages, *LWT* 39, 2006, p. 987-994

3. Paola Quifer-Rada, et al. A comprehensive characterisation of beer polyphenols by high resolution mass spectrometry (LC-ESI-LTQ-Orbitrap-MS), *Food Chemistry*, 2015, Volume 169, p. 336-343

HEALTH · NUTRITION · MATERIALS

DSM
BRIGHT SCIENCE. BRIGHTER LIVING.

hochschule mannheim

Particle Size Characterization of Precipitated Non-Spherical Protein Particles

Karlsruhe Institute of Technology

J. Ross-Jones[1,2], T. Teumer[1,3], F. Garcia[1], M.J. Krause[2], M. Rädle[1], F-J. Methner[3], H. Nirschl[2]

[1] Institute of Process Control and Innovative Energy Conversion, University of Applied Sciences Mannheim, Germany
[2] Karlsruhe Institute of Technology, Institute for Mechanical Process Engineering and Mechanics, Karlsruhe, Germany
[3] Institute of Food Technology and Food Chemistry, Brewing and Beverage Technology, Technical University of Berlin, Germany

INTRODUCTION

Motivation / Background: In most beers, producers strive to minimize haze to maximize visual appeal. To detect the formation of particulates, a measurement system for sub-micron particles is required. To facilitate the construction of a characterization system for non-spherical protein particles, a simulation software was written. The software simulates a monochromatic light source and subsequent light scattering from particles 50 to 200 nm in diameter. The results are compared to Mie-theory calculations [1] for larger particle sizes.

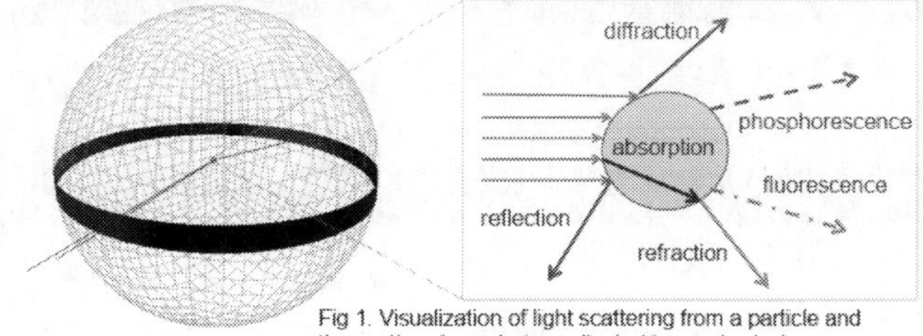

Fig 1. Visualization of light scattering from a particle and the scattered rays being collected by a spherical sensor

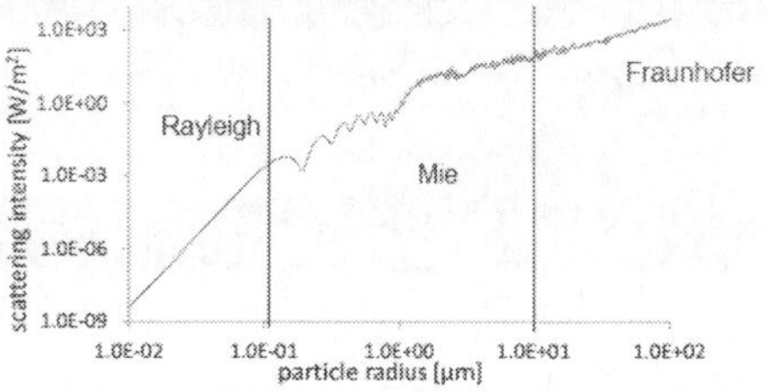

Fig 2. Scattering intensity in Mie and Rayleigh regions for sub micrometer particles

APPROACH

A ray tracing algorithm using wave theory was developed to simulate particle light scattering, describing light interference, reflection and diffraction. The light is modeled as 10^{10} rays. In the case of small particles, the rays scatter and reflect when they collide with the simulated particle. A spherical shaped sensor accumulates these scatter and reflected waves. The intensity and wavelength of the wave are determined by the surrounding medium. Millions of rays are simulated to capture the system behavior.

Simulation results are shown for spherical and non-spherical particles as well as the effect of changing the incident angle to the particle. These results are used to design a sensor probe for non-spherical sub-micrometer particles.

$$I(r,t) = A_0 e^{-2\pi \cdot \alpha \cdot \frac{d}{\lambda}} \cos(k \cdot r - wt)$$

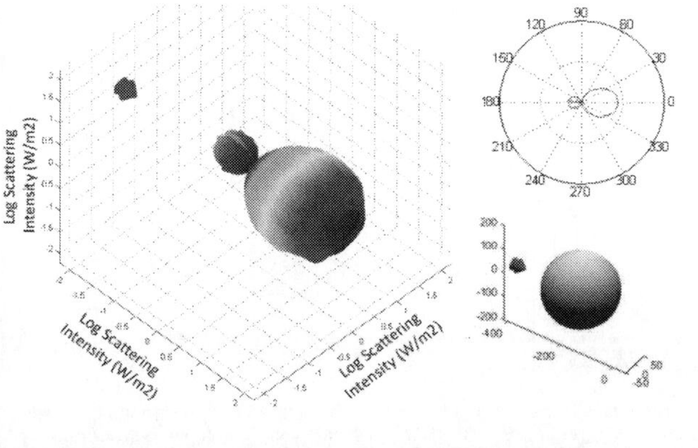

Fig 3. A spherical particle with a 150 nm radius scatters light mostly forward and back, however little light scatters orthogonal to the light source

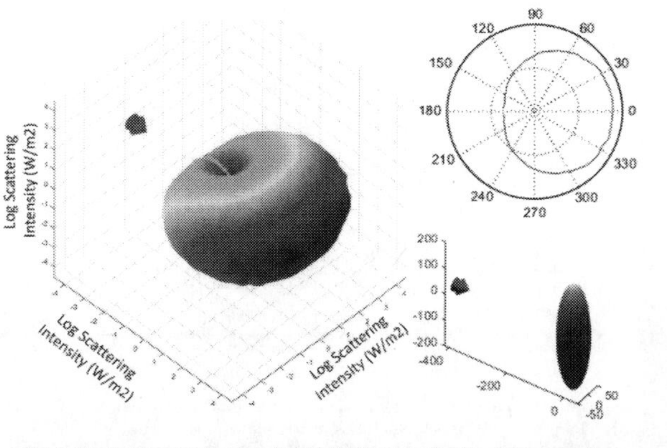

Fig 4. An elliptical particle with a height of 200 nm and width of 50 nm scatters light only orthogonally to its major axis

RESULTS / CONCLUSSION

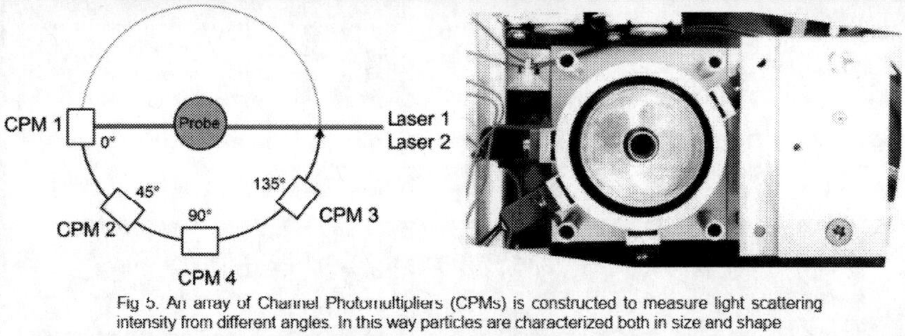

Fig 5. An array of Channel Photomultipliers (CPMs) is constructed to measure light scattering intensity from different angles. In this way particles are characterized both in size and shape

This method allows **determining** the **size of particles** as small as 250 nm as well as the detection of slow growing particles. With this forthcoming knowledge toward precipitation reactions in beer, **improved quality measurement** equipment for the brewery industry can be developed. The lack of accurate information about imaginary and real components of the refractive indices, as well as non-spherical and non-homogeneous particles, is a known challenge for laser diffraction methods. In a defined environment, like beer, where polyphenols and proteins naturally agglomerate, however, this can be turned into an advantage. It is proposed that **through inverse calculation** of light scattering together with possible refractive indices, **conclusions about the particle** nature can be made.

BIBLIOGRAPHY

[1] Calculation basis: Dr. Warren J. Wiscombe (wiscombe@climate.gsfc.nasa.gov), NASA Goddard Space Flight Center, Code 913, Greenbelt, MD 20771

Contact: Jesse Ross-Jones, j.ross-jones@hs-mannheim.de
EBC 2017, European Brewery Convention, May 14 – 18, 2017, Ljubljana, Slovenia

KU LEUVEN

FLAVOUR PLUS

SIMULATION OF BEER TRANSPORT: IMPACT OF VIBRATION ON BEER QUALITY AND STABILITY

Barbara JASKULA-GOIRIS, B. De Causmaecker, A. Paternoster, G. De Rouck, J. De Clippeleer, J. Braet, G. Aerts, L. De Cooman

INTRODUCTION

Beer transport often occurs under suboptimal conditions. Especially when transporting overseas, beer is stored for a prolonged period before actual consumption. During this time, the fresh beer flavour deteriorates, resulting in the occurrence of off-flavours and a significant decrease in drinkability. Therefore, in this study we investigated the influence of temperature and vibrations on beer quality and flavour stability under lab conditions.

FACTORS INFLUENCING BEER QUALITY DURING TRANSPORT AND STORAGE

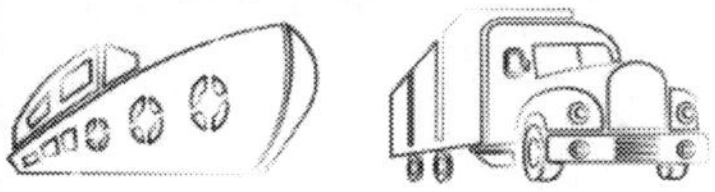

HUMIDITY VIBRATIONS

TIME BEER QUALITY LIGHT

TEMPERATURE

MATERIALS AND METHODS

Fresh beers (stored at 0°C), beers after transport simulation (30 days at 30°C) with or without vibrations (1.7 Hz), and forced aged samples (60 days at 30°C) were analyzed using advanced analytical tools for aldehyde profiling (GC-MS), bitterness (UPLC), turbidity (haze meter), total package oxygen, etc.

A beer kept at 0°C (fresh beer)
B beer kept at 30°C for 60 days
C beer kept at 30°C for 30 days without vibrations
D beer kept at 30°C for 30 days without vibrations and subsequently at 30°C for 60 days
E beer kept at 30°C for 30 days with vibrations
F beer kept at 30°C for 30 days with vibrations and subsequently at 30°C for 60 days

RESULTS

OXYGEN CONENT

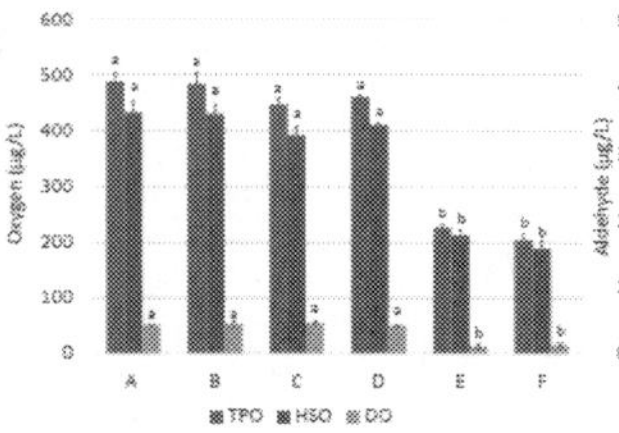

ALDEHYDE PROFILE

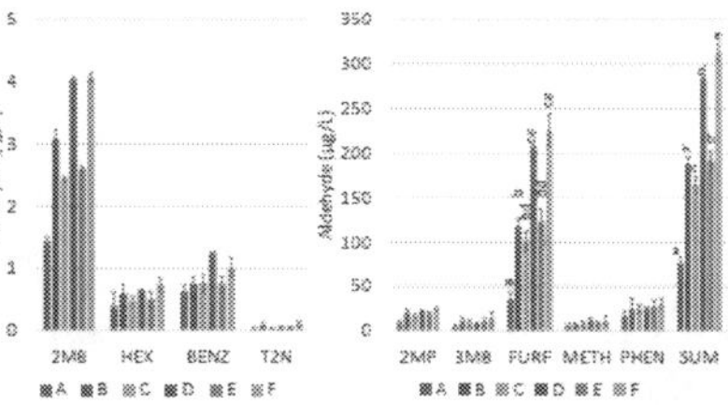

BITTERNESS PROFILE

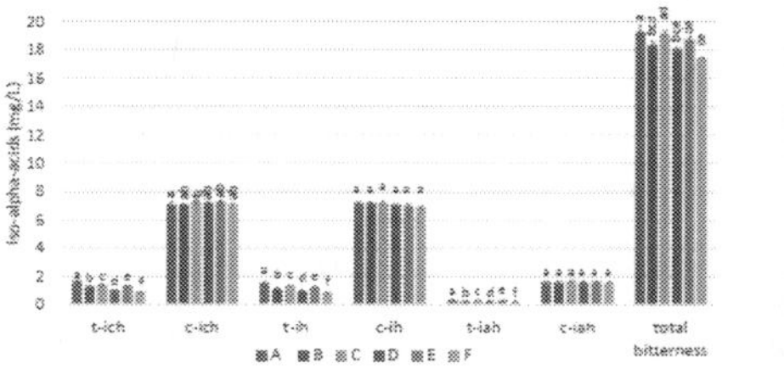

Component identification: TPO: total package oxygen; HSO: headspace oxygen; DO: dissolved oxygen; 2MP: 2-methylpropanal; 2MB: 2-methylbutanal; 3MB: 3-methylbutanal; HEX: hexanal; FURF: furfural; METH: methional; BENZ: benzaldehyde; PHEN: phenylacetaldehyde; T2N: *trans*-2-nonenal; SUM: sum aldehyde markers; t-ich: *trans*-isocohumulone; c-ich: *cis*-isocohumulone; t-ih: *trans*-isohumulone; c-ih: *cis*-isohumulone; t-iah: *trans*-isoadhumulone; c-iah: *cis*-isoadhumulone.
Bars from the same parameter with different letters are significantly different (α=0.05).

Results obtained under laboratory conditions show that vibration of the beer results in an oxygen decrease. Increased temperature has a negative effect on beer quality (i.e. higher aldehydes levels, increased colour and decrease in iso-α-acids content), especially when combined with vibrations. A relatively high loss of *trans*-iso-α-acids can be observed after beer vibration and storage up to 60 days at 30°C (beer F). Furfural, a heat load indicator, increases most during ageing. Furthermore, vibrations have a negative influence on beer colloidal stability, especially during prolonged ageing.

COLOUR FORMATION

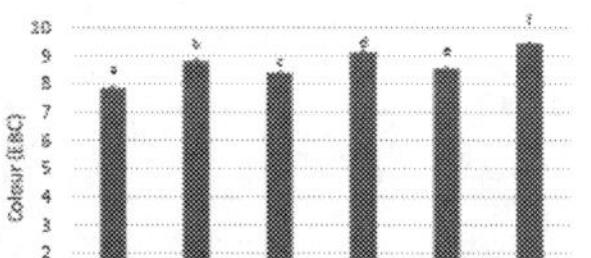

HAZE FORMATION

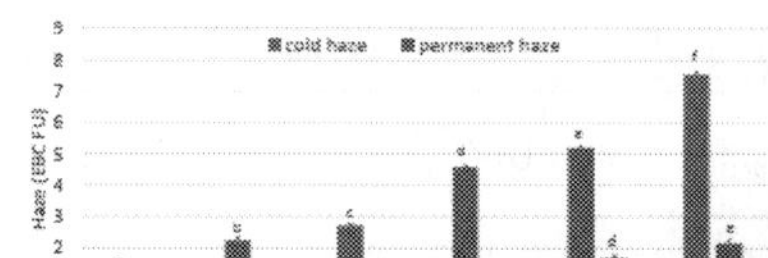

SENSORY EVALUATION

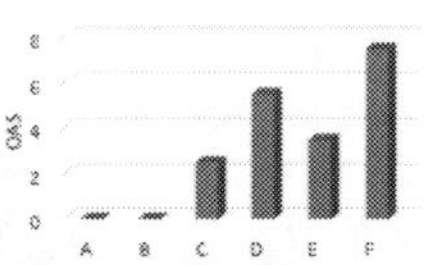

Overall-ageing-score (OAS) based on an 8-point scale (in-house procedure) 0: fresh, oxidized flavour not detectable; 2: very weakly aged; 4: weakly aged; 6: clearly aged; 8: strongly aged, undrinkable.

Sensory evaluation shows a clear effect of vibrations in combination with increased temperature on the appearance of off-flavours. Flavour changes are more pronounced in the beer samples E and F (with vibrations) when comparing with beer samples C and D (without vibrations).

CONCLUSION

Vibrations in combination with high temperature have a negative impact on beer quality, leading to a rise in total aldehyde content, increased colour and a decrease in bitterness. Moreover, combination of high temperature and vibration showed pronounced negative effect on the colloidal stability of beer. Cooled transport and better beer traceability is necessary for preservation of freshness and a reduced return of export beers.

ACKNOWLEDGEMENT

The authors thank the Flanders Innovation & Entrepreneurship, Brussels, Belgium and Belgian Brewers and Maltsters for the financial support.

KU Leuven Technology Campus Ghent, Faculty of Engineering Technology, Department of Microbial and Molecular Systems (M2S), Cluster for Bioengineering Technology (CBeT), Lab of Enzyme, Fermentation and Brewing Technology, Gebroeders De Smetstraat 1, 9000, Ghent, Belgium
E: Barbara.jaskulagoiris@kuleuven.be

Potential effect of "chit" malt on beer foam stability

Stefano Buiatti[1], Stefano Bertoli[2] and Paolo Passaghe[1]

[1]*University of Udine, Dept. of Agricultural, Food, Animal and Environmental Sciences, Udine, Italy;*
[2]*Baladin, Independent Italian Farm Brewery, Piozzo (CN), Italy*

Corresponding author: stefano.buiatti@uniud.it

36TH CONGRESS EBC 2017 14-18 MAY LJUBLJANA EUROPEAN BREWERY CONVENTION SLOVENIA

INTRODUCTION

One of the most important properties customers observe in a beer is the foam; considering consumer preferences, it was found that they are inclined to prefer a beer with stable and persistent foam (Iimure *et al.*, 2008). Foam tightness depends on the presence or absence of some molecules. The main group of molecules are polypeptides, derived from barley malt, and α-acids coming from hops. The goal of this work is to look for a correlation between foam stability and positive foam proteins (<10 KDa, > 30 KDa) and to investigate the possibility to increase their content using an alternative raw material (Chit malt) in the formulation (Evans and Bamforth, 2009; Bamforth, 2011). The Chit malt (Fig. 1) is basically raw barley that has been barely malted.

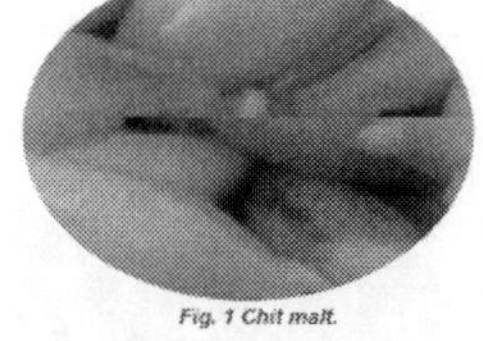

Fig. 1 Chit malt.

METHODS

Tab. 1 Chemical analyses of beer samples.

Beer samples code	1	2	3	4	5
Chit malt (%)	40	30	20	10	0
Pils malt (%)	60	70	80	90	100
Alcohol (% v/v)	5.3	5.4	5.6	5.2	5.6
Plato (°P)	11.7	11.8	12.0	11.6	12.1
Real degree of fermentation (RDF %)	70.4	70.2	70.5	70.0	71.3
Bitterness (IBU)	53	47	50	51	50

At the pilot plant of University of Udine (capacity of 200 L) five craft beers were brewed using different amount of Chit malt (10, 20, 30, 40%). This malt has a content of protein fractions potentially higher than a standard base malt. The process conditions were kept constant for all the brews in order to have standardized beer samples in terms of alcohol content and bitterness (Tab. 1). 100% Pils malt was used to produce reference beer (beer 5). Alcohol (% vol/vol), Plato (°P), Real Degree of Fermentation (RDF, %) were determined by the Alcolyzer Beer Analyzing System (Anton Paar). The bitterness of the five beers was measured according to European Brewery Convention (EBC) method 9.8. The Pils and Chit malts were purchased from Weyermann Specialty Malting Company (Bamberg, Germany), the dry ale yeasts (Safale S-04) and hops (Hallertau Perle, in pellets) from P.A.B. (Udine, Italy).

CE–UV-VIS analysis was performed on a 7100 capillary electrophoresis coupled to a UV-VIS detector (Agilent Technologies). The separation was performed in a capillary with an inner diameter of 50 µm and a total length of 100 cm working in normal polarity (Nunnally *et al.*, 2006; Tian *et al.*, 2014). Beer samples and polypeptide standards (Fig. 2) were prepared and used immediately. They were prepared following the protocol in the ProteomelabTM SDS-MW analyses (Beckman Coulter kit, MA, USA). The Bicinchoninic Acid Protein Assay was performed following the protocol in the Beckman Coulter Kit BCA purchased from Sigma-Aldrich (Milan, Italy). Foam stability was measured with a NIBEM foam stability tester (supplied by Haffmans). The NIBEM values quoted are for foam collapse over 30 mm (EBC method 9.42).

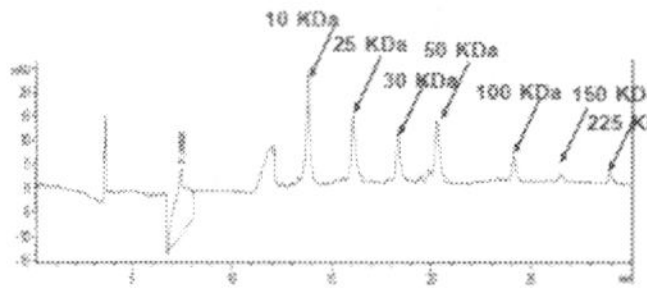

Fig. 2 Representative electropherograms of the molecular weight standards (CE analysis).

RESULTS...

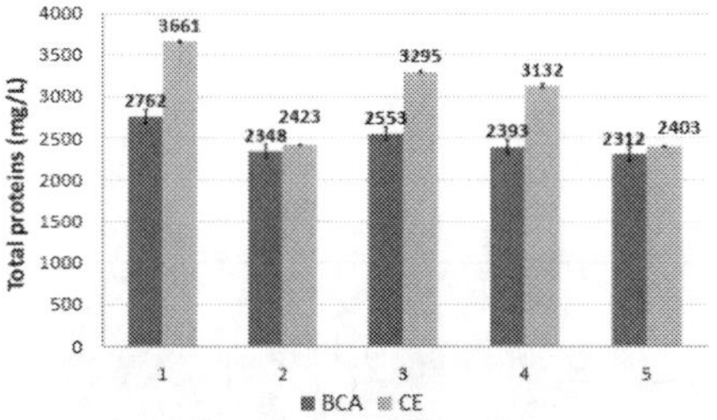

Fig. 3 Total proteins (mg/L) (BCA and CE analysis) of the five beers. The results are reported as mean values (n = 3). The error bars represent the standard deviation (SD).

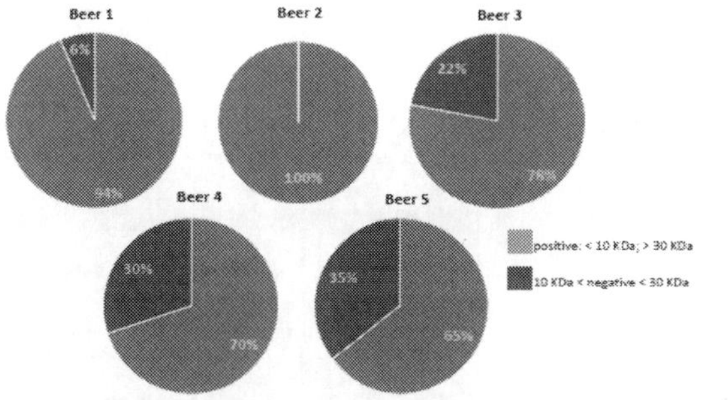

Fig. 4 Polypeptides content (positive and negative) as a percentage of total protein content (CE analysis) of the five beers.

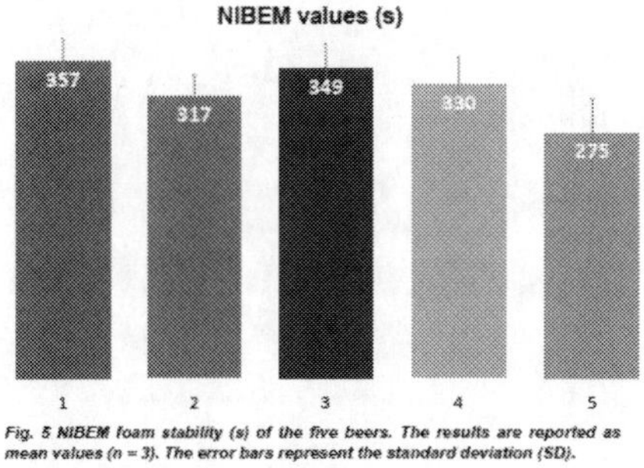

Fig. 5 NIBEM foam stability (s) of the five beers. The results are reported as mean values (n = 3). The error bars represent the standard deviation (SD).

... and DISCUSSION

The CE results confirm those obtained with the BCA assay (Fig. 3); beers produced with Chit malt (beers 1, 2, 3 and 4) showed higher values in terms both the BCA assay and CE analyses compared to the reference beer 5; the sample 2 revealed lower protein content and at the same time a lower foam stability in comparison to the samples 1, 3 and 4 (Figures 3 and 5). However, the beer produced with 30% of Chit malt provided higher values with BCA, CE and NIBEM analyses than reference beer 5 (Figures 3, 4 and 5). The data confirm a direct relationship between the content of positive protein fractions and the persistence of foam; our results suggest that Chit malt could increase the foam active polypeptides and at the same time reduce the amount of negative polypeptides (haze-active compounds) (Fig. 4). So based on the observation that beers brewed with a low modified malt contained less haze-active proteins than the one brewed with 100% of a standard modified malt, it is tempting to speculate that the use in the formulation of under modified malt could be a viable alternative to traditional techniques adopted to improve foam and beer colloidal stability.

REFERENCES

Bamforth, C.W., The non-biological instability of beer. *J. Inst. Brew.* 2011, 117, 488-497.
Evans, D.E. and Bamforth, C.W., Beer foam: achieving a suitable head. In: Beer: A Quality Perspective, C. W. Bamforth, Academic Press: Burlington MA. 2009, 1-60.
Iimure, T., Takoi, K., Kaneko, T., Kihara, M., Hayashi, K., Ito, K., Sato, K. and Takeda, K., Novel prediction methods of beer foam stability using protein Z, barley dimeric alpha-amylase inhibitor-1 (BDAI-1) and yeast thioredoxin. *J. Agric. Food Chem.* 2008, 56, 8664-8671.
Nunnally, B., Park, S.S., Patel, K., Hong, M., Zhang, X., Wang, S.-X., Rener, B., Reed-Bogan, A., Salas-Solano, O., Lau, W., Girard, M., Carnegie, H., Garcia-Cañas, V., Cheng, K.C., Zeng, M., Ruesch, M., Frazier, R., Jochheim, C., Natarajan, K., Jessop, K., Saeed, M., Moffatt, F., Madren, S., Thiam, S. and Altria, K. A series of collaborations between various pharmaceutical companies and regulatory authorities concerning the analysis of biomolecules using capillary electrophoresis. *Chromatographia*. 2006, 64, 359-368.
Tian, M., Zhang, J., Mohamed, A.C., Han, Y., Guo, L. and Yang, L., Efficient capillary electrophoresis separation and determination of free amino acids in beer samples. *Electrophoresis.* 2014, 35, 577-584.

Spent Hop Materials Rich in Xantohumol: An Overlooked Foam Stability Enhancer

UNIVERSITY OF CHEMISTRY AND TECHNOLOGY PRAGUE

Jelínek L.*, Hanko V., Nešpor J., Karabín M., and Dostálek P.
University of Chemistry and Technology, Prague, Department of Biotechnology, Technická 5, CZ-166 28 Prague 6, Czech Republic
Corresponding author: Tel.: +420 220 444 427, Email address: lukas.jelinek@vscht.cz

The aim of this work is to introduce possibilities of using several spent hop materials (residues after CO_2 extraction) as natural enhancers of beer foam.

Introduction

Foam stability is the key factor in the assessment of beer quality. It is well known that the stability of beer foam is positively influenced by hydrophobic proteins, isohumulones, some metallic ions and polysaccharides. Our team recently found that a small dose of vegetative hop residues after CO_2 extraction of high-alpha varieties can significantly increase the stability of wort foam. Additionally, we demonstrated, that foam stability strongly and positively correlated with content of xanthohumol in spent hop materials. It is generally known, that the physico-chemical properties of xanthohumol are similar to α-bitter acids and therefore it can be assumed that isoxanthohumol, i.e. the isomeric product of xanthohumol formed during hop boiling (see in Fig. 1), will be involved in foam stability vie an enhancing mechanism similar to isohumulones. In this work, we tested the foam enhancing properties of spent hop materials after CO_2 extraction of hop varieties Taurus and Vital (see in Fig. 2 and 3), which were chosen for their high content of xanthohumol. Additionally we tested also the hop residue after production of lupulin pellets T45 (variety SAAZ), which was characterised by low xanthohumol content. In the future, the results of this research could significantly contribute to the development of a natural and beer-friendly foam enhancing product.

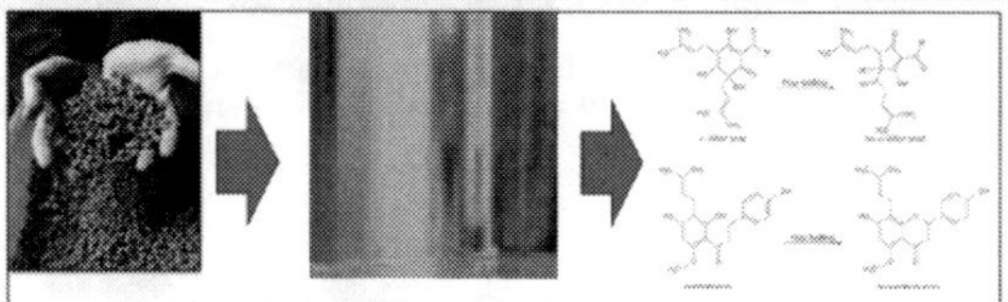

Fig. 1 Isomerisation of α-bitter acids and xanthohumol during hop boiling

Fig. 2 Pelletized residue after CO_2 extraction (Taurus)

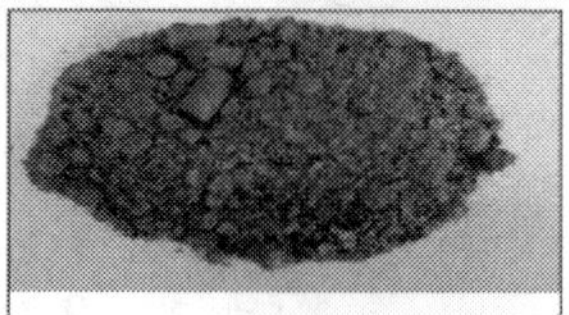

Fig. 3 Powdery residue after CO_2 extraction (Vital)

Materials and method

Hop material: Residue after CO2 extraction (Taurus); residue after CO2 extraction (Vital) and spent hop after production of lupulin pellets T45 (Saaz) →**HPLC-DAD analyses of bitter acids and polyphenols (see in Tab. I)**
Micro-brewing trials: The pale sweet wort concentrate (0.36 kg) was added 2 L of water and the mixture was heated to boiling. The CO_2 hop extract (0.38 g; 36 % of α-bitter acids) and dose of spent hop materials (0; 1; 2.5 and 5 g/L) were added at the beginning of boiling (duration = 90 min). The hoped wort was cooled and fermented (7 days) in 5 L plastic casks at 10 °C. The pitching rate was 1 g of the dried lager yeast / L of cooled wort. Additionally, the 4 samples of green beer adjusted with xanthohumol (0.3; 0.6; 0.8 and 1.4 mg/L of green beer) were prepared.
Beer analyses: Determination of xanthohumol and isoxanthohumol concentrations by HPLC-DAD; determination of foam stability (sec.) by FSA-COMFORT device (1-CUBE, CR)

Tab. I Concentration of α-bitter acids, xanthohumol and isoxanthohumol in spent hop materials

Hop material	α-bitter acids (% w/w)	Xanthohumol (mg/kg)	Isoxanthohumol (mg/kg)
Residue after CO_2 extraction (Taurus)	0.8	10630	290
Residue after CO_2 extraction (Vital)	0.6	8670	160
Spent hop after production of pellets T45	0.5	380	50

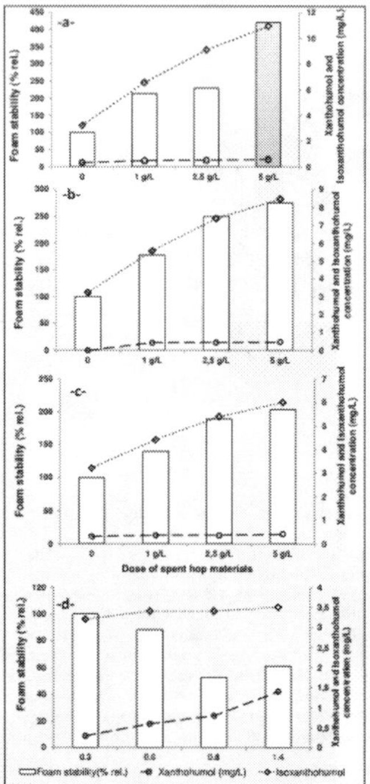

Fig. 4 Foam stability, xanthohumol and isoxanthohumol concentration of green beers prepared from different doses of residue after CO_2 extraction-Taurus **(a)**; residue after CO_2 extraction-Vital **(b)**, spent hop after production of pellets T45-Saaz **(c)** and green beer adjusted with different dose of xanthohumol **(d)**

Results

- As can be seen on Fig. 4 a-c and 5, the foam stability of green beers was positively influenced by addition of spent hop materials to wort
- The most promising results were obtained in the case of hop residues after CO_2 extraction of high alpha varieties Vital and Taurus (foam stability increase by 40-320 % rel.)
- The lowest tested dose (1 g/L) seems to be optimal for brewing uses (increase the foam stability by more than 50 %)
- As can be seen in Fig. 6, the foam stability strongly and positively correlated (σ=0.93) with the concentration of isoxanthohumol in green beers, therefore we can assume, that this compound plays an important role in stabilisation of beer foam
- The foam stability dropped with the increasing concentration of xanthohumol in beer (see in Fig. 4d), therefore it is very important to boil the spent hop materials at least 60 min. for a good conversion of xanthohumol to isoxanthohumol.

Fig. 5 Stability of beer foam 5 minutes after pouring into a glass. The control beer (left) and beer prepare from residue after CO_2 extraction of Taurus – 1g/L of wort (right)

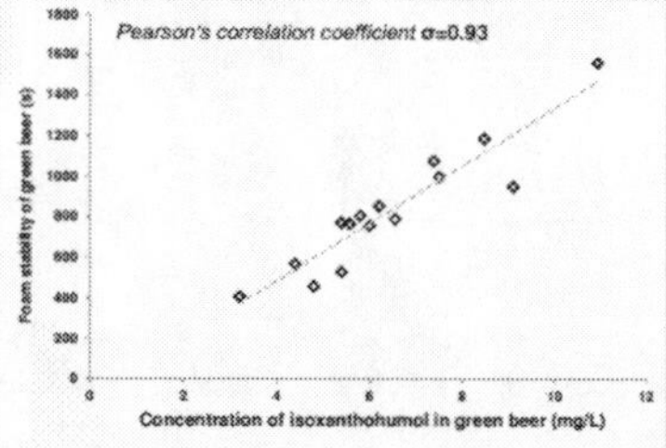

Fig. 6 Pearson's correlation between the foam stability and isoxanthohumol concentration in green beer

Conclusions

The vegetative residues after CO_2 extraction of high alpha varieties Taurus and Vital seem to have great potential for enhancing the foam stability of beer. The key attribute of these materials is their xanthohumol content. This compound isomerised during hop boiling and newly formed isoxanthohumol can extend the time of foam stability, probably by the a similar mechanism as observed in isohumulones. Interestingly, it was proved, that a high level of xanthohumol itself significantly reduce the foam stability.
Until now, spent hops has been an overlooked material with minimal brewing value. In this work, we introduce one of its brewing uses, which can contribute to improving the quality of beer.

Acknowledgment

This work was supported by TACR „Centre for Innovative Use and Strengthening of Competitiveness of Czech Brewery Raw Materials and Products" TE02000177.

Foam standard

P031

P. Kosin[1], T. Branyik[2], J. Savel[1], F. Ulmann[2]; petr.kosin@budvar.cz

[1]Budejovicky Budvar, n.p., Ceske Budejovice, Czech Republic; [2]University of Chemistry and Technology Prague, Prague, Czech Republic

Foam standard can be used for two purposes:

- as a study material for beer foam studies
- as a reference material for metrological control in QC lab

Current recipes for foam standards are deficient in foam negative components. Only foam positive components (proteins, bitter acids) in buffer and ethanol are being used. Fatty acids should be included for realistic foam structure and stability.

Our optimized recipe for foam standard comprise of:

- buffer (pH 4.6, 9 g/L KH_2PO_4 in distilled water)
- ethanol (5 % vol.)
- bitter acids (19 mg/L isomerized alpha acids)
- protein (395 mg/L bovine serum albumin - BSA)
- fatty acid (10 mg/L linoleic acid - LA)
- CO_2 (5 g/L when NIBEM or pouring test is needed)

Advantages of fatty acids in the recipe:

- Structure smooth as real beer foam (fig. 1) and realistic foam stability (fig. 2)

Fig. 1. Optimization of foam composition: Foams with the same foam stability, but different structure: A without fatty acid, 140 mg/L bovine serum albumin (BSA), B 10 mg/L linoleic acid (LA), 395 mg/L BSA, C 16 mg/L LA, 650 mg/L BSA. The more fatty acids, the smoother was the foam structure. Foam standard B resembled beer foam the most. All samples also contained buffer pH 4,6, ethanol 5 % vol. and isomerized alpha acids 19 mg/L.

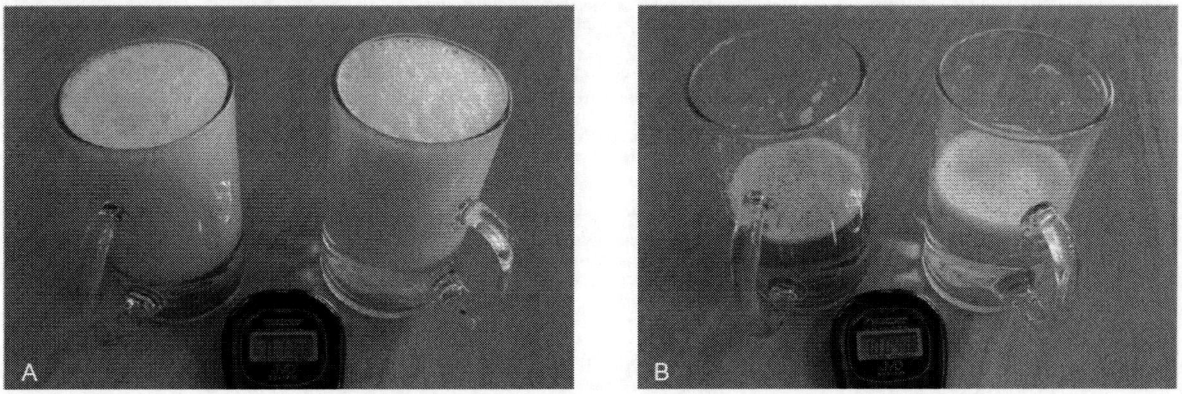

Fig. 2 Foam created by pouring beer (left glass on each picture) and model solution containing fatty acids and CO_2. Foam after pouring (A) has realistic structure, similar stability and appearance of last foam layer after breakdown (B).

BITBURGER BRAUGRUPPE

STARKE MARKEN

The Use of the Hop Variety Polaris as Bittering Hop

Stefan Hanke, Georg Stettner

Bitburger Braugruppe GmbH. Römermauer 3. 54634 Bitburg/Germany e-mail: Stefan.Hanke@bitburger-braugruppe.de

Introduction and Agronomical Properties

In 2015 the hop variety Polaris (PA) was categorized as dual purpose variety. When looking to the acreage development in the Hallertau (Fig. 1) it can be seen, that the acreage for bitter varieties is dominated by Herkules (HS) whose acreage increases every year. The varieties Taurus (TU) and Magnum (HM) have reduced acreage every year (1). Due to its agronomical properties (2) the share of Polaris increases since its release and though it is a question of time when Polaris joins the Top3 bitter varieties in the Hallertau. Focus of these trials was to compare the bitter quality and impression of Polaris compared to the varieties Herkules, Magnum and Taurus. As seen in the acreage development Polaris is getting more interest because of its ability to give good yields particularly on heavy soils where Polaris is superior compared to Herkules or other varieties. The α-concentration of Polaris was always the highest since 2012 compared to other bitter varieties (3). Although Polaris has only a medium yield potential the combination with high α-content makes this variety interesting for good alpha yields at certain soil conditions.

Table 1

Official α-Content per Year and Variety

Variety	2012	2013	2014	2015	2016	5 year average
Polaris	20.0	18.6	19.5	17.7	21.3	19.4
Magnum	14.3	12.6	13.0	12.6	14.3	13.4
Taurus	17.0	15.9	17.4	12.9	17.6	16.2
Herkules	17.1	16.5	17.5	15.1	17.3	16.7

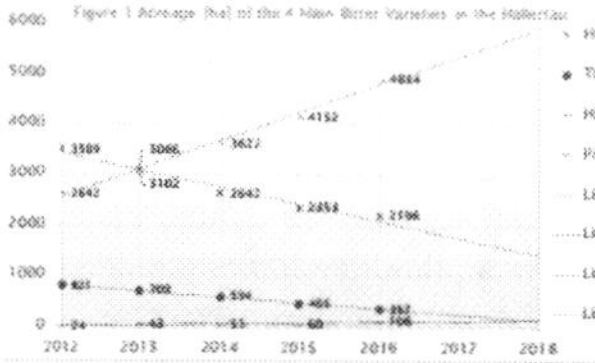

Figure 1 Acreage (ha) of the 4 Main Bitter Varieties in the Hallertau

Table 2

Agronomical Properties of the 4 Main Bitter Varieties

	Yield	Quality			Resistance towards			
	kg/ha	a-Acids (%)	Cohumulone (%)	Vert. Wilt (mild)	Downy Mildew	Powdery Mildew	Spider Mite	Hop Aphid
Magnum	2000	11-16	21 - 29	++	+	- - -	-	- - -
Taurus	2000	14 - 19	20 - 25	+	-	- - -	- -	- -
Polaris	2200	18 - 23	17 - 22	+ +	-	+/-	-	- -
Herkules	2700	14 - 19	32 - 38	+ +	-	+/-	-	- -

++ good to very good; + good; +/- medium; - poor; -- poor to very poor; — very poor

Experimental Setup

All brews were performed in duplicate in a semi-technical 15 hl scale. The base wort was an all malt wort made from 100 % pale pilsen type malt mashed with a two mash decoction regime. Mash separation was done in a lauter tun and boiling time was 75 min with low pressure boiling and internal calandria. Trub separation was performed in a whirlpool. Wort was cooled to 10°C and fermented with a lager strain. Maturation was done at 11°C until diacetyl <0.1 ppm. After 2 weeks of cold storage (-1.5°C) the beer was filtered in a kieselgur frame filter without stabilisation. Hop addition was done at the beginning of wort boiling according to HPLC α-content of the varieties (9 g α/hl). The hops were harvested in 2015, grown in the Hallertau. Hop pellets were provided by Hopsteiner.

Sensory evaluation was done by an internal trained taste panel. The samples were presented as triangular test. When the tasters were right in the triangular test, subsequently a descriptive testing was performed. Ageing was performed at 27°C over 2 and 4 months. All analytics were performed according the standards of MEBAK and EBC.

Results

Figure 2 Overall Bitter Intensity

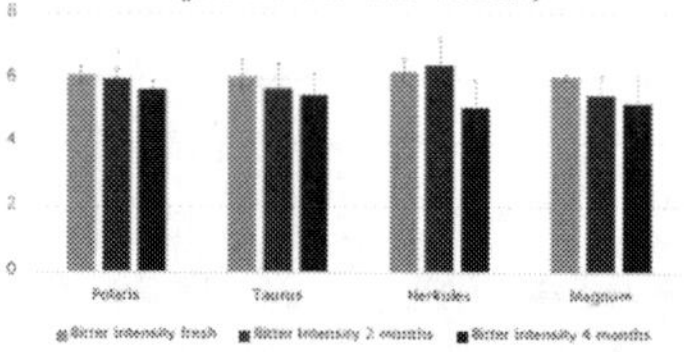

Figure 3 Bitterness Intensity Aftertaste

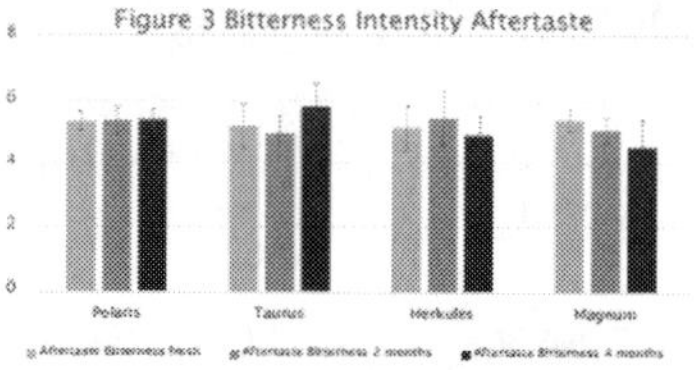

The hop analysis showed the α-potential of the variety Polaris. Due to heat and dry stress 2015 was a year with very low α- contents. Even under these stressful conditions PA showed the highest α-concentrations (+19 % compared to HS).

The analytical data from the beers are shown in table 4. As it can be seen there are no significant differences between the beers. The IBUs of the TU-beers were a bit lower. TU showed in these trials the lowest utilisation rate.

In the triangular tastings only three significant different combinations were obtained but not confirmed in the duplicate trials (Table 5) and also not confirmed with ageing. The descriptive evaluation showed no significances.

Table 4 Analytical Data (mean) of the Beers

	Polaris	Taurus	Herkules	Magnum
Apparent Extract [%]	2.53	2.60	2.54	2.58
Apparent Attenuation Degree [%]	77.6	76.9	77.4	77.2
ABV [%]	4.6	4.6	4.6	4.6
Original Gravity [%]	11.2	11.2	11.2	11.3
pH-value [-]	4.47	4.46	4.49	4.52
IBU [EBC]	30	28	30	30
Foam (Nibem) [s]	269	270	271	271
Iso-α-Acids HPLC [mg/l]	29.4	28.2	30.3	30.1
α-Acids HPLC [mg/l]	0.5	0.4	0.5	0.5
Total Humulones HPLC [mg/l]	30	29	31	31

Table 3 Hop Analysis

	Polaris	Taurus	Herkules	Magnum
α-Acids EBC 7.5 [%]	17,95	11,53	15,29	12,44
α-Acids HPLC EBC 7.7 [%]	16,04	9,87	13,43	10,92
Co-Humulon [%]	23,1	23,2	32,9	24,5
β-Acids HPLC EBC 7.7 [%]	4,10	3,43	3,63	5,90

Table 5 Correct Answers in the Triangular Tests

		fresh	2 months	4 months
PA vs TU	I	5/15	4/13	6/16
	II	2/12	7/15	3/13
PA vs HS	I	5/11	4/13	4/12
	II	7/13*	4/15	6/15
PA vs HM	I	5/11	5/15	8/11**
	II	5/10*	6/15	5/14

Figure 2 showed the overall intensity rating of the bitterness over the ageing period. For PA, TU and HM it can be seen that the bitterness intensity decreased slightly over the 4 month, HS showed after 2 months of ageing a slightly bitter peak. No significant differences between the four varieties were observed. An important quality attribute for German Pilsner style beer is the intensity of the aftertaste which should not be to intense. Figure 3 shows the behaviour of the four varieties with ageing. PA showed a stable bitter aftertaste while TU had the tendency to increase and HM had the tendency to decrease the with ageing.

Conclusion

Due to pest tolerances, very high alpha contents and medium to high yields the acreage of the variety Polaris is increasing every year. Established varieties like Magnum and Taurus loose acreage. In these trials the bitterness of PA was compared to the well established bitter varieties (HS, TU, HM). The results could not show significant differences between Polaris and Herkules, Taurus or Magnum, when used as a bittering hop. The bitter profile showed also no significant differences. The variety Polaris could be used as a replacement for established bitter varieties. An impact on flavour, when added at the begin of wort boiling, is negligible.

References

(1) Seigner, E., Lutz, A.: Data provided by Bay. Landesanstalt für Landwirtschaft, Institut für Pflanzenbau und Pflanzenzüchtung, Hopfenforschungszentrum Hüll

(2) Seigner, E., Lutz, A.: Data provided by Bay. Landesanstalt für Landwirtschaft, Institut für Pflanzenbau und Pflanzenzüchtung, Hopfenforschungszentrum Hüll

(3) Anonym: Alpha-Säurewerte von Hopfen der Ernte 2016 und aktualisiert mehrjährige Durchschnittswerte. In Brauwelt 156 (2016) No. 43, S. 1250

What a robot can tell you about beer quality? Implementing the RoboBEER to assess beer quality based on sensory descriptors

Claudia Gonzalez Viejo [1], Sigfredo Fuentes [1*], Kate Howell[1], Damir Torrico[1] and Frank R. Dunshea[1]

[1]University of Melbourne, School of Agriculture and Food; Faculty of Veterinary and Agricultural Sciences, VIC 3010, Australia
*Corresponding author: sfuentes@unimelb.edu.au

Introduction

Beer quality is defined by consumers from its sensory characteristics, being the visual descriptors such as foamability and color the most important. However, other attributes such as mouthfeel, tastes, aromas and flavors, which are linked to foam-related parameters are also important to monitor. Traditional methods to assess beer sensory descriptors involve the training of panelists to rate the intensity levels of each attribute such as in the quantitative descriptive analysis (QDA®). However, these methods are still considered subjective, time-consuming and costly to conduct routinely.

Materials and Methods

A total of 22 different beer samples were analyzed in triplicates for foamability (total lifetime of foam, lifetime of foam, maximum volume of foam, foam drainage), bubble size distribution (small, medium and large), color (CieLab* and RGB), alcohol, temperature and carbon dioxide (CO2) using a robotic pourer RoboBEER (Fig. 1) constructed with Lego® blocks (The Lego Group, Billund, Denmark), open source sensors and Arduino® boards (Arduino, Italy). All parameters were analyzed using computer vision algorithms in Matlab® ver. 2016a (Mathworks Inc., Matick, MA. USA). Furthermore, a double blind QDA® session was conducted for the same beer samples with 12 trained panelists to rate ten sensory descriptors in beer such as levels of yeast, grains and hops aromas, bitter, sour and sweet tastes, hops flavor, and viscosity, astringency and carbonation mouthfeel. An artificial neural networks (ANN) regression model using the Matlab Neural Network Toolbox™ 7 (Mathworks Inc., Matick, MA. USA) was developed with the 15 color and foam-related parameters obtained using the RoboBEER as inputs to predict the intensity levels of the ten sensory descriptors assessed in the sensory session. To create this model, three hidden neurons (Fig. 2) and a Levenberg-Marquardt training algorithm were used along with a random data division with 70% (n = 46) used for training, a mean squared error performance algorithm with 15% (n = 10) for validation and a default derivative function with 15% (n = 10) for testing.

Fig. 1 RoboBEER constructed with Lego® blocks, coupled with remote sensors and controlled using Arduino® boards, showing the sensors results in real-time and the pouring of a beer sample being recorded using an iPhone 5S.

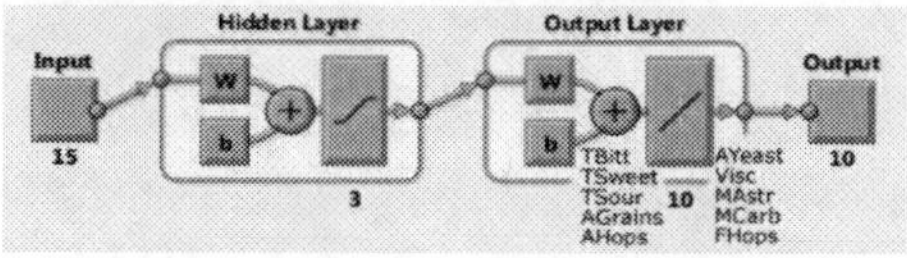

Fig. 2 Two layer feedforward network diagram with sigmoid functions using three hidden neurons and ten outputs for regression models. A total of 15 inputs from RoboBEER for foamability, bubble size distribution, color in two scales, alcohol and CO2, along with ten targets from sensory descriptors were used to create the ANN model.

Results

The ANN model developed had high correlation and determination coefficients ($R = 0.91$; $R^2 = 0.83$) for the prediction of the intensity levels of ten sensory descriptors (Fig. 2 and 3). The same results were obtained after several retraining attempts and using both normalized and non-normalized data.

Furthermore, the error histogram of the model presented a normal distribution. As shown in Table 1, the correlation (R) and determination coefficients (R2) for all training, validation and testing stages were high ($R \geq 0.87$ and $R2 \geq 0.76$), and the mean squared errors (MSE) were low (<1.6). The overall model presented a slope of 0.98 and was statistically significant (p-value < 0.0001). This shows that the model obtained is highly accurate and reliable.

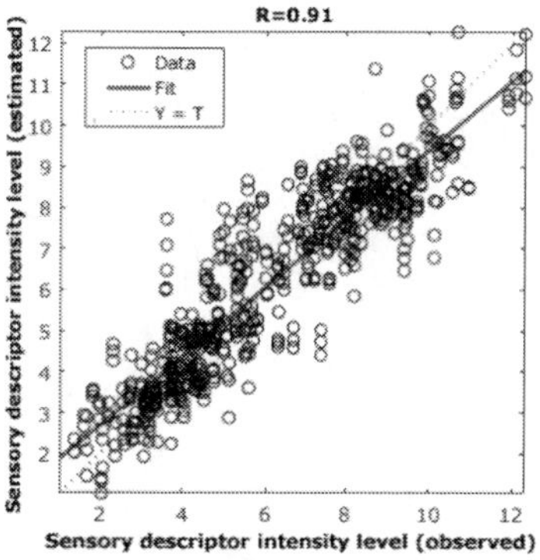

Fig. 3 Artificial neural networks (ANN) overall model showing the correlation coefficient (R) in which x-axis represents the estimated and y-axis the observed sensory descriptors intensity levels

Table 1. Statistical results from the ANN model showing the three stages and overall values for number of samples, means squared error (MSE), correlation (R) and determination (R^2) coefficients.

Stage	Samples	MSE	R	R^2
Training	46	0.98	0.92	0.85
Validation	10	1.55	0.87	0.76
Testing	10	1.00	0.93	0.86
Overall	66	0.90	0.91	0.83

Conclusions

RoboBEER showed to be an effective, accurate, less time consuming and more objective tool that, along with machine learning algorithms, can be used as a novel method to assess beer quality based on the color and foam-related parameters, and the intensity levels of its main visual, aroma, taste, flavour and mouthfeel descriptors. Therefore, RoboBEER can potentially be used in the industry as part of the quality control at the end of the production line.

References

Gonzalez Viejo, C., Fuentes, S., Li, G., Collmann, R., Condé, B., & Torrico, D. (2016). Development of a robotic pourer constructed with ubiquitous materials, open hardware and sensors to assess beer foam quality using computer vision and pattern recognition algorithms: RoboBEER. Food Research International, 89, 504-513.

DEVELOPMENT OF ELECTRONIC BEER NOSE

56th CONGRESS EBC 2017 EUROPEAN BREWERY CONVENTION 14-18 MAY LJUBLJANA SLOVENIA
BEER QUALITY – SENSORY
NEW PRODUCT DEVELOPMENT
BREWING SCIENCE AND RESEARCH

Pinto L.T.[1,2], Reitenbach A.[1,2]

[1]Universidade Federal de Santa Catarina, EQA, Florianópolis, Brazil
[2]Federal University of Santa Catarina, EQA, Florianópolis, Brazil

INTRODUCTION & OBJECTIVES

Beer's flavor and aroma vary due to a wide range of influences which can occur at every single stage of its production and storage. Beer is made of over one thousand components which can contribute to the product's flavor, providing it with pleasant or, in several cases, unpleasant aromas, due to the presence of what is called off-flavor compounds.

This project aims at developing an electronic nose which is able to evaluate and detect undesirable compounds in beer samples, as a new proposal to the aroma's quality control.

WHAT IS AN ELECTRONIC NOSE

An electronic nose is an instrument which comprises an array of chemical sensors with partial specificity combined with an appropriate recognition system for recognizing simple or complex odors (Gardner & Bartlett, 1994), by mimicking the mode of action of the human nose (Wiśniewska, 2017).

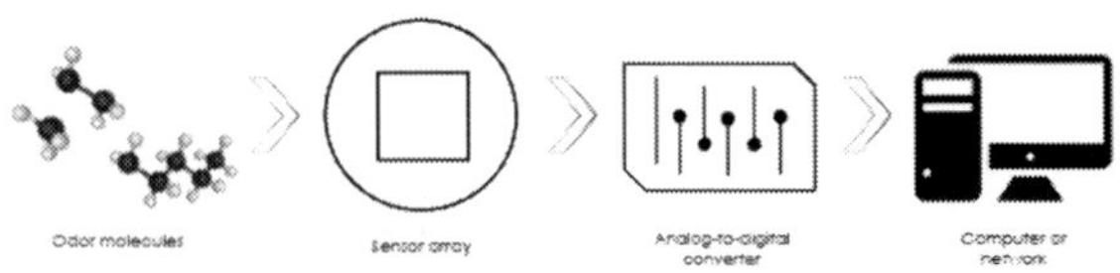

Figure 1. Odor molecules reach the sensor array and this chemical information is converted and analyzed using a network.

The sensor array captures odor molecules that comprise a single volatile organic compound (VOC) or a group these substances (Fig. 1). An analog-to-digital converter transforms chemical signals into electrical signal. Signals are transmitted to a computer or neural network. Then, as the signals are compared to previously established standards, data is processed and the odors are identified.

NEURAL NETWORKS

A neural network is a tool designed to solve problems using artificial intelligence. Its main algorithmic is inspired to some issues of the functioning of the human brain (Varnamkhasti et al., 2011).

A comparison between a natural neuron and an artificial neuron is shown in Figure 2A. The pathway that stimuli follow until a response is obtained (that is, an odor is identified) is shown in Figure 2B.

INPUT (1) The stimulus that happens in the dendrites of a natural neuron can be compared to the input stimulus in the set when odor is being analyzed. The stimuli correspond to a vector $x = [x_1; x_2; \ldots; x_n]$.

BODY (2,3) Decisions about the stimuli occur based on the information stored in the neuron. The artificial body consists of functions. The weighted sum of the inputs x_i by their weights w_i (linear output), which is then sent to a function f_x.

OUTPUT (4) Artificial neurons provide an output activation y (response) for each type of stimulus as in an axon of a natural neuron, which sends a response to other axon terminals, forming a network.

(A)

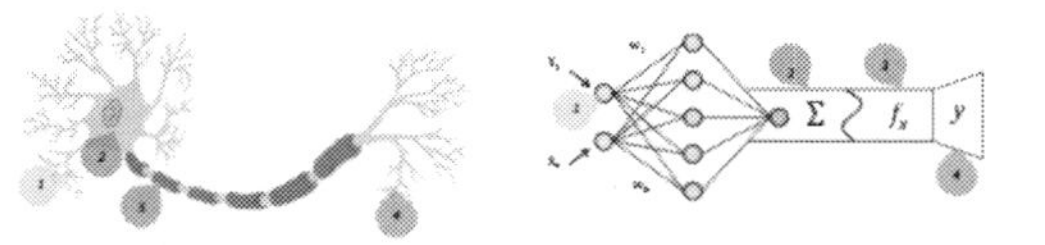

(B)

Figure 2. (A) Comparison between a natural neuron and an artificial one. (B) Pathway that stimuli follow until a response is obtained (odor is identified).

CONSTRUCTION OF THE ELECTRONIC NOSE

The electronic nose was built using different gas sensors and a temperature and humidity sensor connected to the protoboard (Fig. 3B, white board), which is connected to the Arduino (Fig. 3B, data acquisition board, blue one). Beer samples are properly prepared and transferred to test tubes. The sensor array is placed in order to establish contact with the sample vapors. They are able to detect and quantify VOCs.

As a way of assessing the success of an electronic nose combined with a neural network data can be compared to two groups of panels, one consisting of untrained people (no previous knowledge of aroma compounds in beer) and a trained one (Table 1). Success percentage using the apparatus and a PPN reached 86%, considerably higher than the percentage using the other methods.

METHOD	SUCCESS %
Untrained panel	28%
Trained panel	47%
Electronic nose + PNN	86%

Table 1. Success percentage for 3 methods used in beer sensorial analysis.

(A) (B)

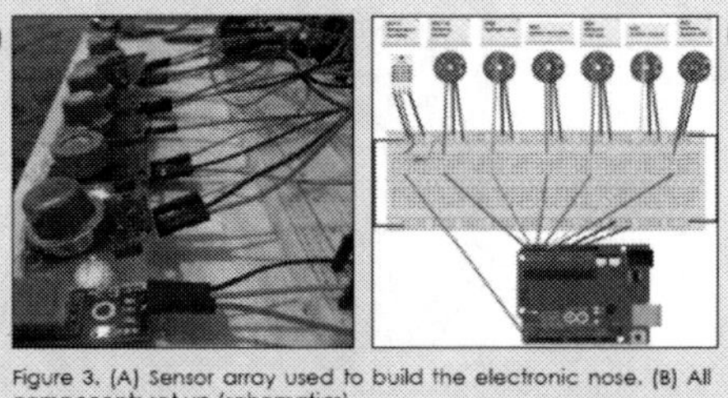

Figure 3. (A) Sensor array used to build the electronic nose. (B) All components set up (schematics)

Using bootstrap technic, a classification of the compounds detected by the different sensors was made (Fig. 4). As the success percentage demonstrates, the use of the electronic nose associated to the PNN network is a very interesting method and represents a real gain in reliability of the sensorial beer analysis.

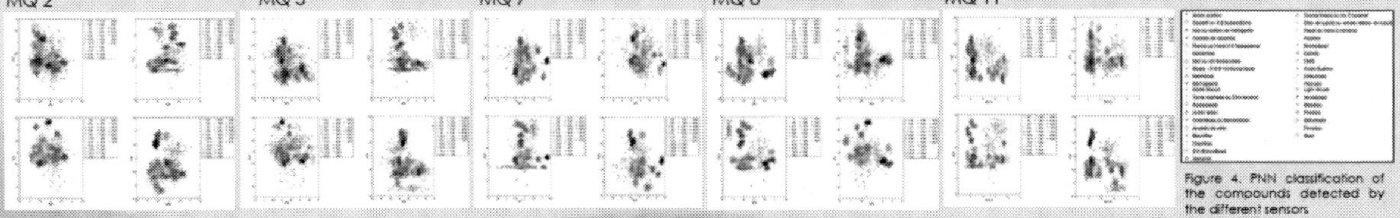

Figure 4. PNN classification of the compounds detected by the different sensors

SOME REFERENCES

GARDNER, J.; BARTLETT, P. A brief history of electronic noses. **Sensors and Actuators B**, 18-19 (1994) 211-220.

VARNAMKHASTI, I.I. Ghasemi et al. Potential application of electronic nose technology in brewery. In: **Trends in Food Science & Technology**, v. 22, p. 165-174, 2011

WIŚNIEWSKA, P.et al. Comparison of an Electronic Nose Based on Ultrafast Gas Chromatography, Comprehensive Two-Dimensional Gas Chromatography, and Sensory Evaluation for an Analysis of Type of Whisky. **Journal of Chemistry** v. 2017 (2017)

Relevance of Oxygen for the Formation of Strecker Aldehydes During Beer Production and Storage

36th CONGRESS EBC 2017 EUROPEAN BREWERY CONVENTION 14–18 MAY LJUBLJANA SLOVENIA

Philip C. Wietstock, Thomas Kunz, Frank-Jürgen Methner

Technische Universität Berlin, Institute of Food Technology and Food Chemistry, Chair of Brewing Science, Germany

Introduction & Beer Staling Mechanism

The stale character of food is a quality issue for many producers and distributors. It is caused by certain carbonyl compounds emerging during storage and, once they exceed their flavor thresholds, producing distinct off-flavors. Despite decades of research, flavor instability is still a major problem for many breweries and pathways are still not clear.

A potential role of free radicals in beer staling in beer staling was initially investigated by Bamforth and Parsons, and their harmful character was further verified by others.

Free Radical Formation:

i. Activation of molecular oxygen
ii. Formation of Reactive Oxygen Species (ROS)
iii. Hydroxyl (˙OH) and hydroxyethyl radical (˙EtOH) formation in the so-called Fenton reaction; transition metal ions such as iron and copper act as catalysts
iv. In beer, potential 'recycling' of metal ions by reducing substances

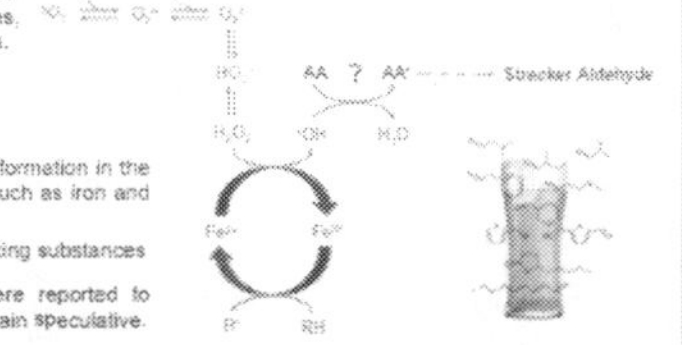

While oxygen exposure and oxidative conditions were reported to promote beer staling, the reasons for these findings remain speculative.

Aim of the Study:

Toward the objective to deploy and further elucidate beer staling and off-flavor development, different factors which potentially affect beer staling were tested in model solutions and in finished beer. Particularly, it was to be tested if ROS are involved in the formation of Strecker aldehydes during beer storage and wort production.

Research Justification & Approach

Initial Situation/Research justification

- (Oxidative) beer deterioration reactions during storage yielding off-flavors are far from being understood
- Particularly Strecker aldehydes respond to oxygen exposure
- No reasonable explanation in brewing-related literature

Hypothesis

- Reaction Oxygen Species (ROS) such as the hydroxyl radical and the hydroxyethyl radical are involved in the formation of Strecker aldehydes during beer production and storage

1st step: Trials in buffered model systems at beer-like conditions

- Oxidative degradation of amino acids by Fenton-type reactions

2nd step: Elucidating relevance during beer production/storage

- Beer storage trials using Response Surface Methodology (RSM)
- Labeling experiments to test de novo formation of aldehydes during beer storage
- Effects of oxygen exposure during lab-scale wort production

Analytical methods

Staling aldehyde concentrations in beers and worts were measured by solvent-assisted flavor evaporation (SAFE)-GC/MS using selected ion monitoring (SIM) mode. Diethyl ether was used to extract beers. Pentanal was used as internal standard (IS).

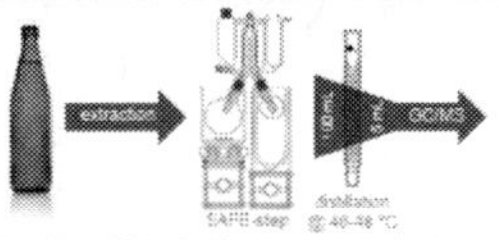

Table: Retention times and m/z ratios used for quantitating carbonyl compounds by GC/MS.

Substance	t_R [min]	m/z
2-methylbutanal	10.15	58
3-methylbutanal	10.63	58
pentanal (IS)	12.96	58
2-furfural	19.05	96
heptanal	21.26	70
methional	21.50	104
benzaldehyd	23.03	106
benzaldehyde-d^5	23.03	111
octanal	23.76	84
phenylacetaldehyde	24.79	91
phenylacetaldehyde-d^5	24.79	96
ethyl nicotinate	27.61	106
γ-nonalactone	29.95	85

Hydroxy radical (˙OH) and hydroxyethyl radical (˙EtOH) levels were measured using a Bruker® BioSpin X-band electron spin resonance spectrometer with α-(4-pyridyl-1-oxide)-N-tert.-butylnitrone (POBN) serving as a spin trap.

Iron concentration in beer was quantitated using an iCAP 6200 inductively coupled plasma-optical emission spectroscopy (ICP-OES) system fitted with a CID 86 detector and argon as the carrier gas.

Extract (2.9.2.3), color (2.12.2), pH (2.13), free amino nitrogen (2.6.4.1.1), total polyphenols (2.16.1), and bitter units (2.17.1) in beers and/or worts were analyzed according to MEBAK.

Total packaged oxygen in beers was measured according to MEBAK (2.28.1.1.2) using a DIGOX 6.1 apparatus (Dr. Thiedig, Berlin, Germany).

Response Surface Methodology (RSM)

High gain of knowledge, low effort:

- ✓ Evaluation of every variable's effect on the response.
- ✓ It is possible to observe the interaction effect of the independent parameters on the response.
- ✓ Regression function valid for complete design space allows calculation/prediction of values.

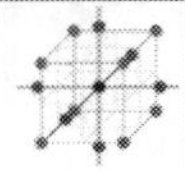

Procedure for RSM analysis:

1. Determination of the independent parameters and their levels.
2. Selection of the experimental design and the prediction and verification of the model equation.
3. Use of backward elimination to identify important variables and deplete insignificant factors from the model.
4. Examine the fitted model plot, interaction plots, and ANOVA statistics (R^2, adjusted R^2, lack-of-fit test, etc.).
5. Obtain the response surface plot and contour plot of the response as a function of the independent parameters.

Results: Trials in Buffered Model Solutions

A buffered model system (0.1 mM acetate buffer, pH 4.3, 5.0 % v/v ethanol) was used to study the potential involvement of ROS in the formation of Strecker aldehydes from their parent amino acids.

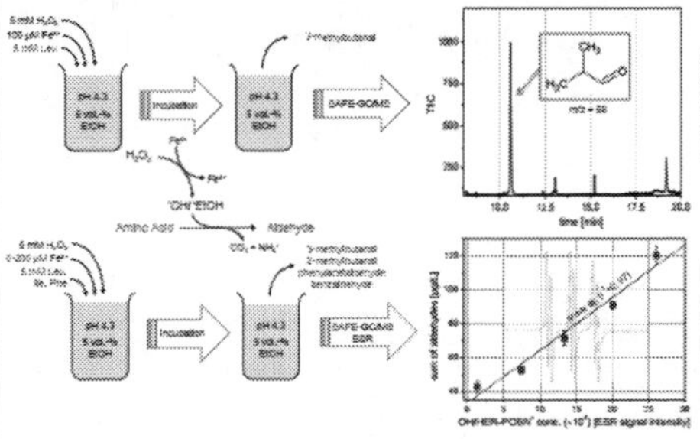

Oxidation of amino acids by the Fenton reagents H_2O_2 and Fe^{2+} yields Strecker aldehydes

Conc. of ˙OH/˙EtOH correlates positively with amount of Strecker aldehyde levels formed

➡ ˙OH/˙EtOH are involved in Strecker aldehyde formation!

Results: Beer Storage Trials Using RSM

Response surface methodology (RSM) and a central composite rotatable (CCRD) design was used to assess the relevance of different factors and potential involvement of ROS in the formation of Strecker aldehydes during beer storage. A commercially available filtered lager-type beer (11.5 °P, 4.9 % v/v alcohol, pH 4.41, 32 bitter units, 48 µg/L iron) available in 500 mL bottles served as the base beer for all trials. Labelling experiments with Phe-d^5 were used to investigate the oxygen-dependent *de novo* formation of Strecker aldehydes during storage.

Beer storage trial 1:

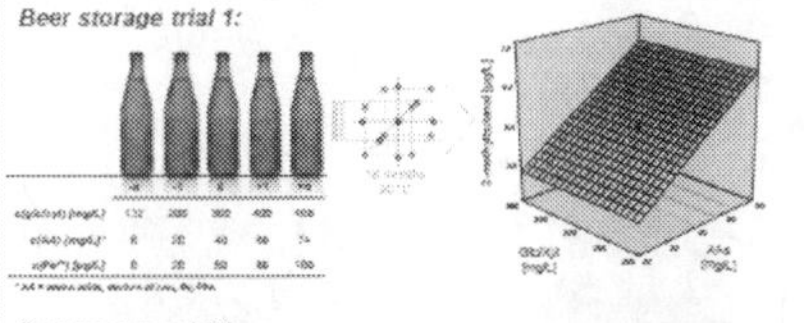

Table: F-values of aldehydes (responses). Only significant responses where p < 0.05 are shown.

Responses	Glc/Xyl	AAs
3-methylbutanal	not significant	20.21
2-methylbutanal	not significant	101.1
phenylacetaldehyde	not significant	5.23
2-furfural	4.83	not significant

Beer storage trial 2:

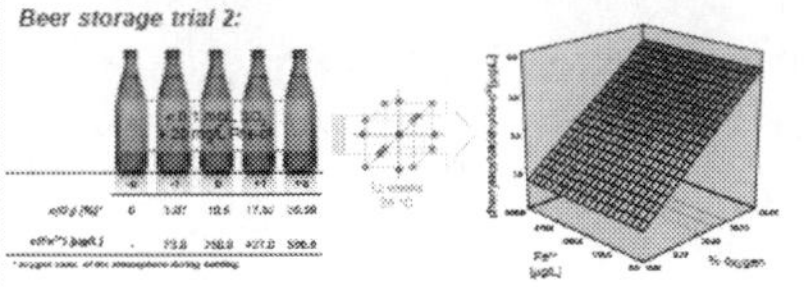

Table: F-values of aldehydes (responses). Only significant responses where p < 0.05 are shown.

Responses	O_2	Fe^{2+}
3-methylbutanal	93.75	3.33
2-methylbutanal	62.35	not significant
methional	50.71	not significant
phenylacetaldehyde	78.58	not significant
benzaldehyde	144.24	not significant
phenylacetaldehyde-d^5	97.54	not significant

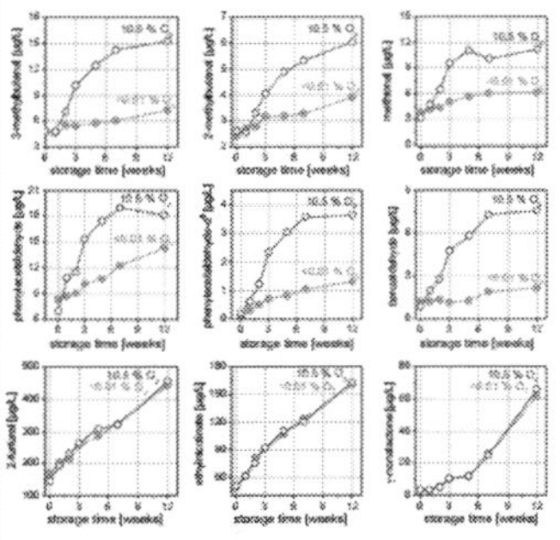

- *De novo* formation of phenylacetaldehyde-d^5 during storage in dependency of oxygen concentration.
- Amino acid concentration and oxygen concentration are significant ($p < 0.001$) for Strecker aldehyde formation.
- Fe^{2+} concentration was insignificant for all aldehydes with the exception of 3-methylbutanal where it was kept in the model ($p = 0.0879$).
- 2-Furfural, ethylnicotinate, and gamma-nonalactone were not affected by oxygen or Fe^{2+} concentration during bottling; however, 2-furfural responded to Glc/Xyl concentration
- Aldehydes from fatty acid oxidation (heptanal, octanal) were not affected by any of the factors tested (data not shown).

➡ ROS are involved in the *de novo* formation of Strecker aldehydes during beer storage

The role of Fe^{2+} in beer staling:

It is noteworthy to mention that in all storage trials, Fe^{2+} additions had no effect on the amounts of Strecker aldehydes found at the end of storage, even though Fe^{2+} acts as a catalyst in the Fenton reaction system! Two assumptions can serve as an explanation for this behavior:

- In bottled beer, oxygen is always the Fenton reactions rate limiting step.
- Fe^{2+}, once it's oxidized to Fe^{3+}, it is rapidly reduced again to its lower valence state by reducing substances present in beer and is thus available again for its catalytic action in the Fenton reaction system again.

Two test these assumptions, 10 mg/L of EDTA (a good iron chelator) was added to beers and beers were stored for 12 weeks at 28 °C.

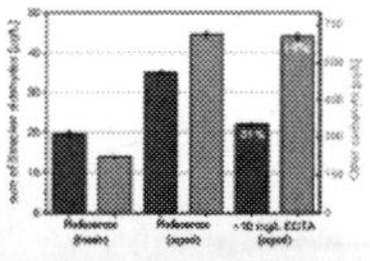

➡ EDTA additions reduced the concentration of Strecker aldehydes formed! Iron must therefore be involved in the formation reaction!

Results: Impact of Oxygen During Mashing on Aldehyde Formation

The influence of oxygen during mashing and mash separation was investigated by producing worts from 100 % Pilsner malt either under nitrogen atmosphere or under air atmosphere in a glove box.

- Strecker aldehydes are significantly lowered when oxygen is omitted during mashing.
- All other carbonyls are unaffected.

➡ Potential involvement of amino acid oxidation during wort production!

Table: Influence of oxygen during mashing. Bold values are significantly different at p < 0.05.

		Mashing	
		under O_2 atmosphere	under N_2 atmosphere
Wort analysis			
extract	[% wt./wt.]	14.0 ± 0.26	13.9 ± 0.23
pH	[-]	5.91 ± 0.01	5.90 ± 0.01
color	[° EBC]	4.6 ± 0.0	4.1 ± 0.1
FAN	[mg/L]	209.7 ± 5.6	216.3 ± 5.86
total polyphenols	[mg/L]	148.5 ± 2.3	188.8 ± 8.8
Carbonyl compounds			
3-methylbutanal	[µg/L]	75.1 ± 5.2	36.8 ± 3.4
2-methylbutanal	[µg/L]	29.7 ± 1.6	20.3 ± 1.7
methional	[µg/L]	38.7 ± 2.3	16.2 ± 1.1
phenylethanal	[µg/L]	47.0 ± 2.8	14.4 ± 0.9
benzaldehyde	[µg/L]	7.1 ± 0.4	8.8 ± 0.8
2-furfural	[µg/L]	81.5 ± 3.4	78.2 ± 7.6
γ-nonalacton	[µg/L]	1.9 ± 0.2	1.8 ± 0.2

Conclusions and Outcomes of this Study

The beer staling mechanism was expanded by a thus far unknown pathway involving hydroxyl and hydroxyethyl radical attack on amino acids thereby forming the corresponding Strecker aldehydes.

This study therefore helps to further clarify food staling mechanisms, particularly, as related to the oxidative degradation of amino acids and concomitant formation of Strecker aldehydes during wort production and beer storage.

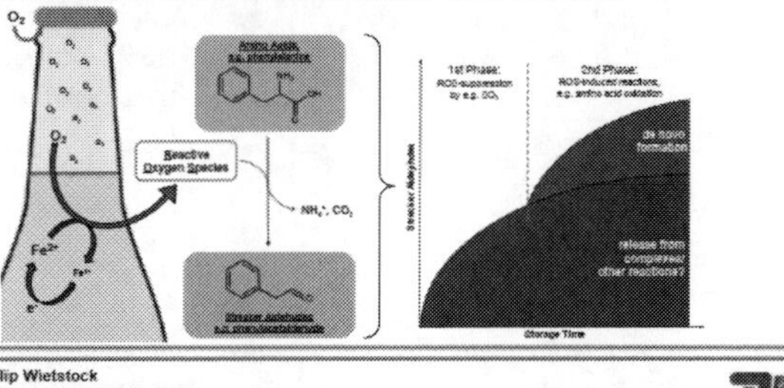

Contact: Philip Wietstock
Email: philip.wietstock@tu-berlin.de
Phone: +4930-314 27505

Occurrence of biogenic amines in craft corn beers and commercial barley beers

*Villarreal-Sanchez L.A.**, *Romero-Medina M.A.*, *Verde-Calvo J.R.*, *Escalona-Buendía H.B.*
Metropolitan Autonomous University, Biotechnology, Mexico, Mexico
***lavs@xanum.uam.mx**

Casa abierta al tiempo
UNIVERSIDAD AUTÓNOMA
METROPOLITANA

36TH CONGRESS EBC 2017 EUROPEAN BREWERY CONVENTION 14-18 MAY LJUBLJANA SLOVENIA

Keywords: Biogenic amines, corn beer, RP-HPLC-DAD

INTRODUCTION

Tyramine (TYR) and putrescine (PUT) are two of the most important biogenic amines (BA) found in fermented beverages and foods such as cheese, sausage, vegetables, fish, beer and wine, inter alia. Large amounts of this compounds ingested might produce toxicological effects and sensory disturbances. There are no reports about the occurrence of BA in beers made 100% corn malt, thus this work was carried out in order to know the presence of BA in beers made of blue and red corn as raw material.

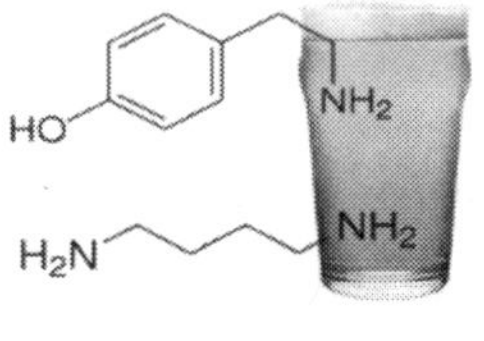

MATERIAL & METHOD

Samples[1]

Six craft beers were made by duplicate in a pilot plant (20L) of the Metropolitan Autonomous University followed a typical brewing method and Ale fermentation where all the conditions (T°C, time, hop and yeast quantity) remained constant, the type of malt was the only different factor used in beer formulations.

INGREDIENTS:

- Two varieties of corn red & blue Chalqueño breeds from Milpa Alta, Mexico City.
- *Saccharomyces cerevisae* Safale US-05 (Fermentis Division of S. I. Lesaffre, Marcq-en-Baroeul, France)

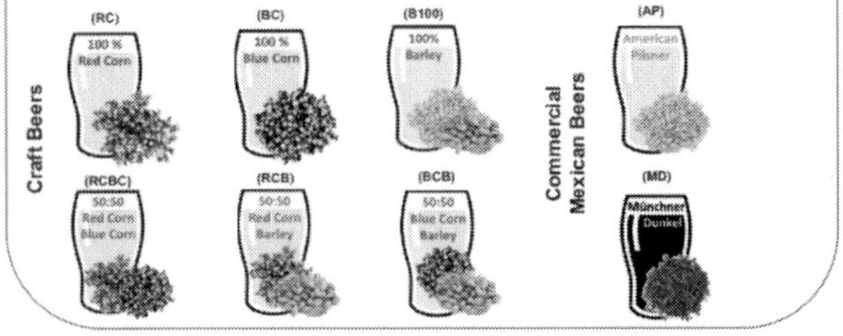

Derivatization[2]

1.75 mL Borate buffer solution 1M (pH=9)
750 µL Methanol
1 mL Sample of beer
30 µL Diethyl ethoxymethylenemalonate (DEEMM)

Ultrasound bath (30 min) → Heat (60 min, 70 °C)

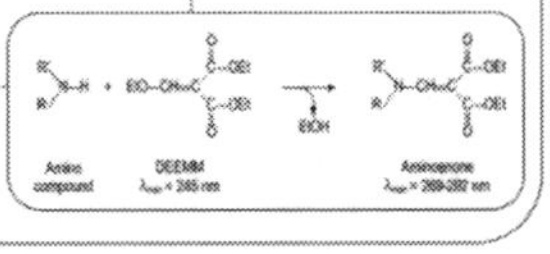

BA determination (RP-HPLC-DAD)[2]

Quantitation by external standard method

Chromatograph
HPLC Agilent® T1100 Series

Column
Agilent® ZORBAX Eclipse XDB – C18 3.5 µm (50 mm x 4.6 mm)

Detector
Diode Array (280 nm)

RESULTS

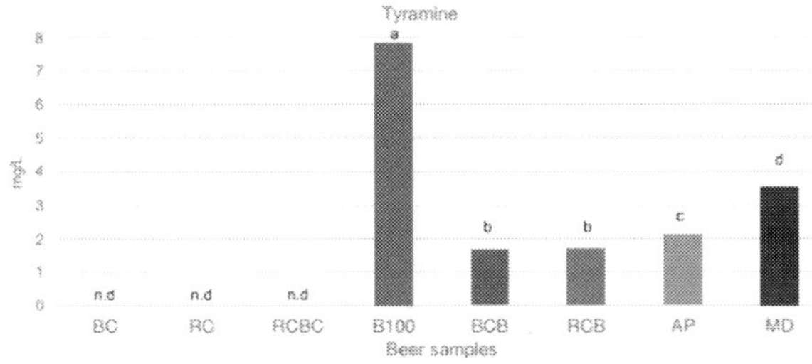

Fig. 1. Concentration of tyramine in beers
n.d: *No Detected. Different letters indicate significant differences among the samples for type of malt, based on Tukey's test ($p \leq 0.05$).). Standard error <0.1*

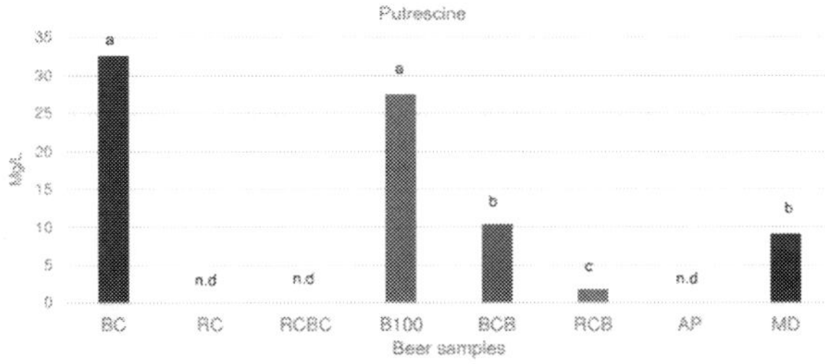

Fig. 2. Concentration of putrescine in beers
n.d: *No Detected. Different letters indicate significant differences among the samples for type of malt, based on Tukey's test ($p \leq 0.05$).*

In brewing, the types of amines are dependent on the raw materials in the beverage, as well as the method of brewing, and any microbial contamination that may have occurred during the brewing process or during storage[3].

Results in this study suggest that red and blue corn do not influence the tyramine occurrence. Otherwise, barley beers (craft and commercials) showed the highest concentration of tyramine. The highest amount of putrescine was found in blue corn beer followed by barley beer. Within the commercial beers, putrescine was not detected in American-Pilsner beer but it was in the Münchner-Dunkel one.

CONCLUSION

Tyramine and putrescine concentrations in all beers analyzed are not hazardous to health people, however, putrescine levels could cause undesirable sensory effects. This work was carried out in order to continue the investigations on biogenic amines in beers and lead off the continuous research on corn beer quality.

BIBLIOGRAPHY

1. Verde Calvo, José Ramón, Héctor Bernardo Escalona Buendía, Nicte Nataly Cruz Rodriguez, and María Angélica Romero Medina. 2014. Proceso de elaboración de cerveza a base de maíz. MX/a/2014/009140,issued2014.
2. Gómez-Alonso, S., Hermosín-Gutierrez, I., & García-Romero, E. (2007). Simultaneous HPLC Analysis of Biogenic Amines , Amino Acids , and Ammonium Ion as Aminoenone Derivatives in Wine and Beer Samples. *Journal of Agricultural and Food Chemistry, 55*(13), 608–613.
3. Kalač, P., & Krízek, M. (2003). A Review of Biogenic Amines and Polyamines in Beer. *Journal of the Institute of Brewing, 109*(2), 123–128.

Impact of dry hopping techniques on xanthohumol content of beer

Júlio C. Machado Jr.[1], Miguel A. Faria[1], Armando Maia[1,2], Caterina Monalisa[1], Isabel M P.L.V.O. Ferreira[1]

INTRODUCTION

Beer is the alcoholic beverage most consumed in the world. In recent years were produced almost 2 billion hectoliters of beers per year. In the brewing process, hop (*Humulus lubpulus L.*) is one of the main ingredients, together with malt, water and yeast. Hydrophobic compounds from hop cones present bitterness and characteristic hoppy aroma. In the recent scenario, the microbreweries are a new market trend that increase and motivate for a development of knowledge in hops and hops utilization.

Dry hopping techniques are being used by breweries to highlight aroma characteristic of hops in beers. Besides the volatiles fraction the dry hopping techniques also increase the extraction of other compounds like xanthohumol. Xanthohumol is a prenylchalcone which occurs almost exclusively in hops and have a multiple bioactive properties, in particular antioxidant, anti-inflammatory, anticancer, immunomodulatory and antimicrobial. However, studies about yours extraction are scarce.

OBJECTIVES

This study evaluated the extraction of xanthohumol in dry hopping techniques using three varieties of hops in beer with different conditions.

MATERIALS AND METHODS

American-Style Pale Ale beer Base
(5 Kg of Pale Ale and 0.5 Kg of Caramel – Crystal)

Mashing
60 min at 62°C → 10 min at 72°C → **Filtration** 78°C → **Wort boiling** 60 min at 100°C + 42 g of Summit hop → **Fermentation** 7 days at 20°C → **Maturation** 21 days at 4°C

Dry hopping techniques

Used Hops
American hops Chinook (CHI)
English hops East Kent Golding (EKG)
German hops Tettnanger (TET)

Hops Concentrations:
0.7 g/ L, 1.4 g/ L and 2.8 g /L

Dry hopping Time:
21 days, 10 days and 5 days

Analytical Procedure

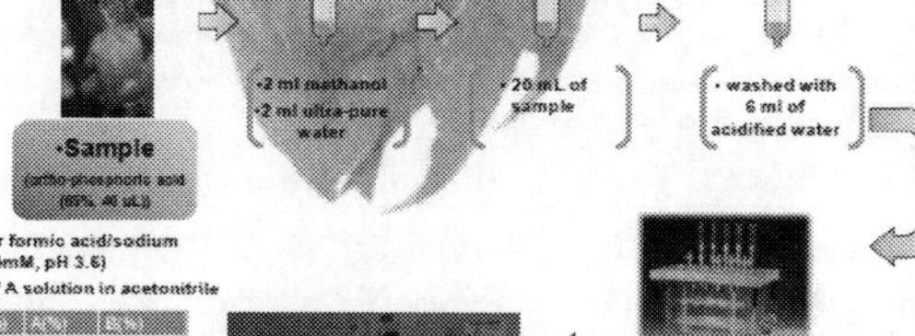

•(A) buffer formic acid/sodium format (25mM, pH 3.6)
•(B) 2% of A solution in acetonitrile

Time(min)	A(%)	B(%)
0-3	65	35
23	25	75
40-45	0	100
50-56	65	35

DAD detection

RESULTS

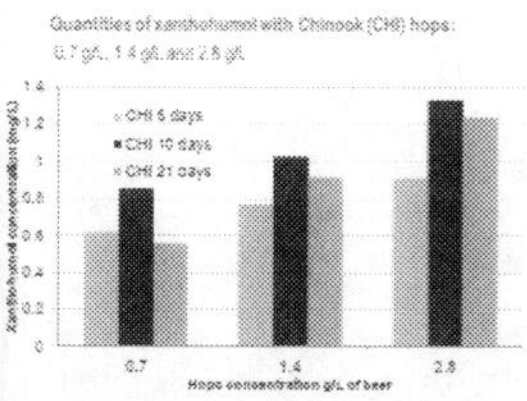

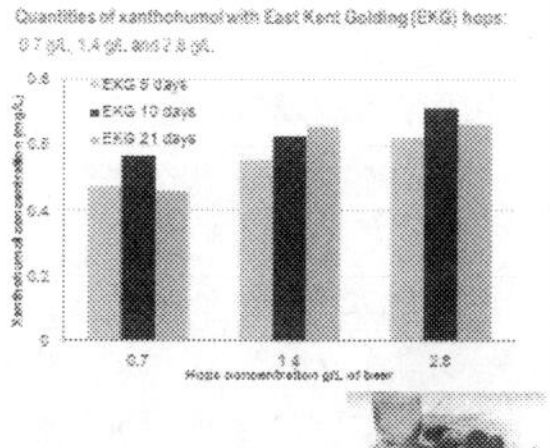

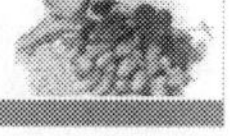

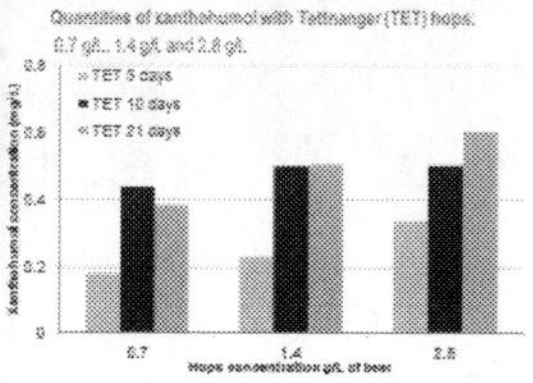

In the comparison between different hops at the same concentration and time of exposition (dry hopping), in all cases, the beers with Chinook demonstrated higher levels of xanthohumol and the beers with Tettnanger presented the lowers values.

The total amount of xanthohumol quantified in hops presented a statistical differences among CHI and the others tested hops, having 0.50% (± 0.04%), that is 0,5g of xanthohumol in 100g of hops. The other hops present 0.29% (± 0.02%) in EKG and 0.26% (± 0.01%) in TET.

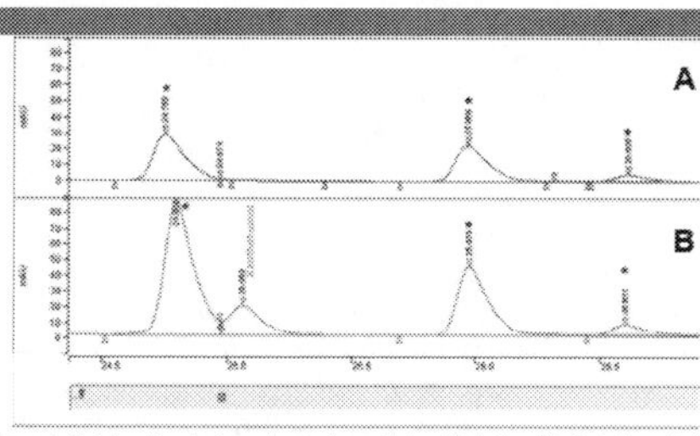

The Control did not present detectable amounts of xanthohumol.

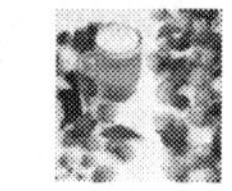

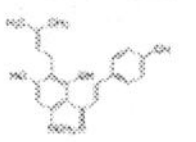

Figure 1: Typical section of chromatogram for xanthohumol detection. (A) Control sample (B) Sample. * different alpha-acids present in sample.

With regard of variations among different hops concentrations used it was demonstrated a tendency which how much more hops were added, more xanthohumol were present in beers. However when calculated the recuperation indices, doing a relation of quantity of xanthohumol in beer with the initial quantities of xanthohumol in each hop, was observed a best extractions in lower concentrations. In general, extracted 20.9% (± 4.8%) when applied 0.7 g/L of hops, 14.0% (± 2.0%) in beers with 1.4 g/L of hops and 8.1% (± 1.4%) in beer with 2.8 g/L of hops.

Concerning to times of dry hopping in maturation 10 days as the better time for a higher extraction of xanthohumol. It was well pronounced when beers was maturated with 0.7 g/ L of CHI, the final concentration of xanthohumol were 0.62 (± 0.03) mg/ L to 5 days, 0.85 (± 0.02) mg/ L in 10 days and 0.55 (± 0.06) mg/ L in 21 days.

CONCLUSION

In general the lower value of xanthohumol was 0.18 (± 0.14) mg/ L in beer matured 5 days with Tettnanger and the higher value of 1.33 (± 0.03) mg/ L in beer matured 10 days with 2.8 g/ L of Chinook.

The beers with Chinook demonstrated higher levels of xanthohumol and the beers with Tettnanger presented the lowers values. For higher time of dry hoping the reduction of xanthohumol could be explained by a degradation in these conditions.

Acknowledgements
This work received financial support from the European Union (FEDER funds [illegible]) and National Funds (FCT/MEC, Fundação para a Ciência e Tecnologia and Ministério da Educação e Ciência) under the Partnership Agreement PT2020 [illegible]. One of the authors (A. Melo) wishes to thank the Fundação Ciência Tecnologia grant [illegible]. To all financing sources the authors are greatly indebted.

Evaluation of α-acids content of beer produced by dry hopping techniques

Julio Cesar Machado Junior, Miguel A. Faria, Armindo Melo, Isabel M.P.L.V.O. Ferreira
LAQV/REQUIMTE, Laboratório de Bromatologia e Hidrologia, Departamento de Ciências Químicas, Faculdade de Farmácia, Universidade do Porto, 4051-401 Porto, Portugal

INTRODUCTION

The bitter α-acids from the hops plants, *Humulus lupulus* L., are recognized as bioactive compounds with proved beneficial health properties. Scientific evidence has been accumulated concerning the cancer preventive potential of hop compounds due to its antioxidant, apoptosis-inducing, anti-estrogenic, anti-angiogenic, anti-inflammatory and antiproliferative activities, as well as, modulation of carcinogens metabolism[1]. The different modes of hop addition influence directly beer composition since at high temperatures the α-acids isomerize to iso-α-acids, consequently, α-acids are lost when hops are added during the hot phase of beer production. However, enhancement of α-acids occurs by the addition of hop at cold stages of brewing process (named dry hopping techniques).

OBJECTIVES

Current work is focused on evaluate the impact of dry hopping techniques on α-acids content.

METHODS

Brewing process

According the mashing procedure of 60 min at 62°C, 10 min at 72°C and heated to 78°C for filtration, 20 L of an American-Style Pale Ale type beer was produce from 5.5 Kg of barley malt (5 Kg of Pale Ale and 0.5 Kg of Caramel – Crystal). At the beginning of wort boiling was added 42 g of Summit hop. Wort was boiled for 60 min at 100°C, fermented for 7 days at 20°C and maturated for 21 days at 4°C.

Maturation and dry hopping techniques

Three hop types at 3 different concentrations (0.7; 1.4 and 2.8 g/L) were added at 3 different points during 21 days maturation. Chinook (CHI) East Kent Goldings (EKG) and Tettnanger (TET) hops. The control was a sample without hopping sample.

	0.7 g/L	1.4 g/ L	2.8 g/ L
All 21 days			
Last 10 days			
Last 5 days			

Sample preparation[2]

After maturation time, beers were centrifuged for 5 min at 3.000 RPM to separate the hops and stored at 4°C until the analysis.

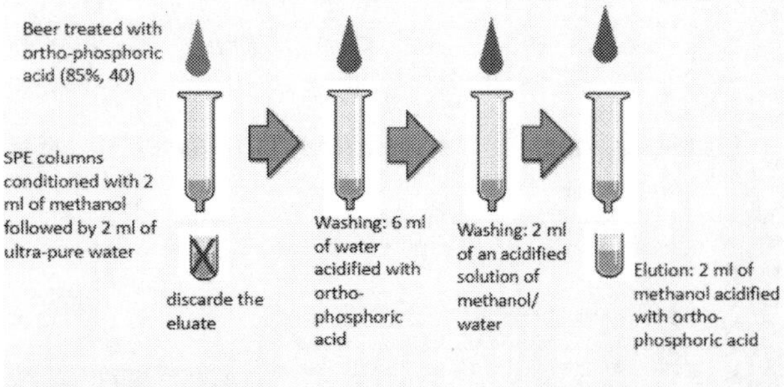

Liquid Chromatography with Ultra-Violet Diode Array Detector (HPLC-DAD) to analyze α-acids[3]

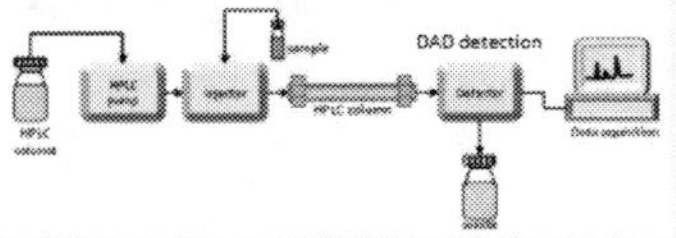

Time (min)	A (%)	B (%)
0-3	65	35
23	25	75
40-45	0	100
50-56	65	35

(A) buffer formic acid/sodium format (25mM, pH 3.6)
(B) 2% of A solution in acetonitrile

RESULTS

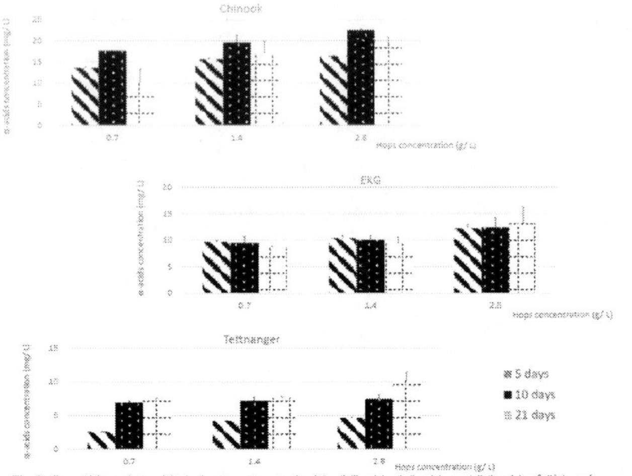

Fig 1. Quantities of α-acids in beer maturated with 0.7 g/ L, 1.4 g/ L and 2.8 g/ L of Chinook, EKG and Tettnanger hops in 5; 10 and 21 days.

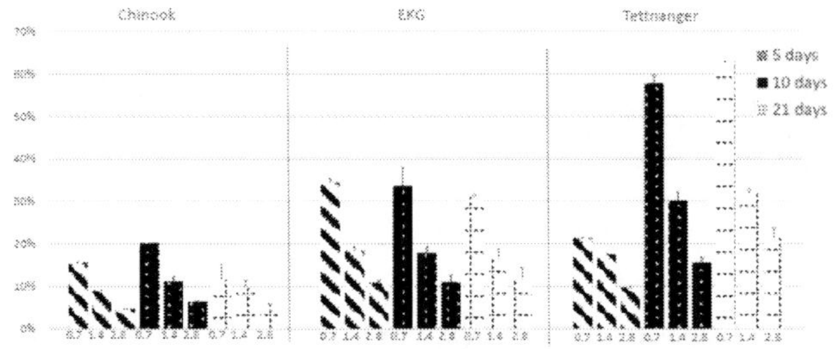

Fig 2. Recovery indices. Percentage of α-acids extracted in beer maturated with 0.7 g/ L, 1.4 g/ L and 2.8 g/ L of Chinook, EKG and Tettnanger hops in 5; 10 and 21 days.

RELEVANT FINDINGS

- When compared at the same concentration and time of exposition, in all cases, the beers with CHI presented higher levels of α-acids, and the beers with TET presented the lowest values, what is expected since the contents of α-acids of hops pellets was 12.5% (± 1.0%) in CHI, 4.0% (± 0.2%) in EKG and 1.7% (± 0.1%) in TET.
- In beers, total of α-acids ranged between 2.5 mg/ L (beers with 0.7 g/ L of TET, added 5 days) and 22.5 mg/L (beers with 2.8 g/ L of CHI, added 10 days).
- In the most cases, hop addition 10 days before finish maturation promoted the highest α-acids content in all beers.

CONCLUSION

Results demonstrated a general tendency indicating a positive correlation among the quantity of hops added and the level of α-acids present in beers. However when recovery indices were calculated (the ratio between the quantity of α-acids in beer and the initial quantities of α-acids in each hop) the highest extractions were observed at lower concentrations.

REFERENCES

1. Karabín, M., Hudcová, T., Jelínek, L., & Dostálek, P. (2016). Biologically Active Compounds from Hops and Prospects for Their Use. Comprehensive Reviews in Food Science and Food Safety, 15, 542-567.
2. ASBC (1968) α-acids na β-acids in hops and hop extracts by HPLC (international Method).
3. Prencipe, F. P., Brighenti, V., Rodolfi, M., Mongelli, A., dall'Asta, C., Ganino, T., Bruni, R., Pellati, F. (2014). Development of a new high-performance liquid chromatography method with diode array and electrospray ionization-mass spectrometry detection for the metabolite fingerprinting of bioactive compounds in Humulus lupulus L. J Chromatogr A, 1349, 50-59.

Further use of PVPP regeneration caustic for CIP applications – A significant step to enhance an environmental friendly beer production

Dipl.-Braumeister Philipp Zeuschner, Dr.-Ing. Deniz Bilge, Dr.-Ing. Roland Pahl, Research and Teaching Institute for Brewing in Berlin (VLB), Research insitute for Beer and Beverage Production (FIBGP)

VLB BERLIN

Motivation

In beer stabilization plants that work with regenerable PVPP in a recycling process, hot sodium hydroxide is used to remove the polyphenols from the PVPP after beer stabilization. During this process the used caustic turns deep black and is regularly discarded directly after the process. Some breweries apply a nanofiltration to the spent caustic to separate the diluted polyphenolic substances [1-3], which can be seen as a useful but comparatively complex and cumbersome issue. However, the discoloration originating from the dissolved polyphenols should theoretically not affect the dirt dispensing capacity of the caustic. Hence, VLB Berlin investigated the possibilities of a direct re-use of the blackened caustic for CIP purposes.

Figure 1: Preparation of comparable soiled plates

1. Properties of PVPP regeneration caustic

- 11 samples from 5 different German breweries were investigated
- Results varied depending strongly on sampling time and brewery

Attribute	Values
NaOH concentration [% w/w]	0.9 - 2.5
pH value	12.5 - 13.8
Conductivity [mS/cm]	46.7 - 121.9
Surface tension [mN/m]	44 - 55
COD [mg/l]	2053 - 10500
Total polyphenols [mg/l]	136 - 4190

- ➢ Alkalinity and surface tension are not significantly affected by regeneration
- ➢ Alkalinity and electric conductivity show linear correlation
- ➢ Accumulation of polyphenols in the caustic and thus increase of chemical oxygen demand by regeneration time
- ➢ Mixing phases with degassed water at the beginning and at the end of the process → dilution of NaOH concentration
- ➢ Adjustment of concentration by addition of fresh NaOH might be necessary

BUT

- ➢ **Anyway, all water and all sodium hydroxide used for the regeneration process can be collected and re-used**

Figure 2: Trial desing – laboratory scale

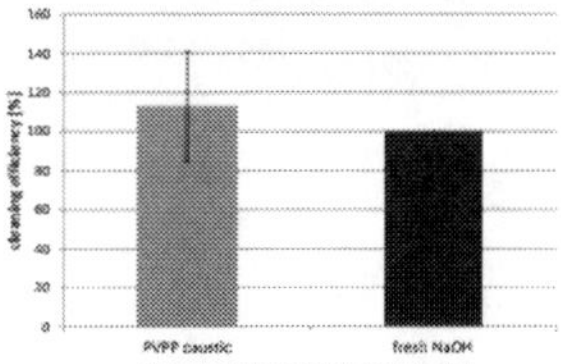

Figure 3: Cleaning results, laboratory trials average values and standard deviation, n = 5

2. Cleaning trials – laboratory scale

- Preparation of comparable soiled plates by hanging up stainless steel plates in fermentation vessel of VLB pilot plant from pitching to hosing → same deposits on plates as on tank wall (fig. 1)
- Fixation of soiled plates in bench-top CIP model for a defined flushing with the respective cleaning medium in recirculation (fig. 2)
- Cleaning efficiency was evaluated by time for complete (optical) dirt removal
- To compare results between different batches both cleaning media were testet always on one plate next to each other → cleaning efficiency of fresh caustic set to 100 %
- ➢ The results (fig. 3) show that there are **no significant differences in cleaning efficiency between fresh NaOH and spent caustic from PVPP regeneration**

3. Material compatibility

- Brewery relevant materials (stainless steel, gaskets, hoses, tiles) were submerged in PVPP regeneration caustic in different conditions (hot, cold, open or closed system) over partially extreme periods

Material	System	Temperature	Period	Result after cold water rinsing
Stainless steel	closed	20°C	15 weeks	no permanent discoloration
	open	20°C	7 days	no permanent discoloration
		40°C	3 hours	no permanent discoloration
		80°C	3 hours	no permanent discoloration
CIIR	closed	20°C	15 weeks	no permanent discoloration
	open	20°C	7 days	shadow on phase boundary
		80°C	3 hours	no permanent discoloration
NBR	closed	20°C	15 weeks	no permanent discoloration
	open	20°C	7 days	shadow on phase boundary
		80°C	3 hours	no permanent discoloration
PTFE	closed	20°C	15 weeks	no permanent discoloration
	open	20°C	7 days	shadow on phase boundary
		80°C	3 hours	no permanent discoloration
PVC	closed	20°C	15 weeks	no permanent discoloration
Tiles	open	20°C	7 days	no permanent discoloration

- ➢ The only permanent discoloration was observed after intense dry up on the tested rubber materials
- ➢ All other discolorations could be removed easily by rinsing with cold water
- ➢ **Negative interactions with materials in brewery relevant conditions can be excluded**

4. Cleaning trials – pilot scale

- Comparative cleaning trials carried out in 0.5 hl CCTs and 2.5 hl open fermentation vessels of VLB pilot plant
- Cleaning principle: lost recirculation with static sprayballs (25 l/min·m) with and without 2 minutes pre-rinse
- Optical evaluation with UV spotlight
- Evaluation of rinsing water
- ➢ **No significant differences in cleaning time between fresh and spent caustic**. Pre-rinse brings no time advantage (fig. 4).
- ➢ **Fresh water quality is reached after 2 minutes of final rinse for both cleaners** (fig. 5)
- ➢ **At hot cleaning in the brewhouse no conspicuous observations were made. Cleaning result equal to fresh caustic.**
- ➢ **No interactions with cleaning additives could be determined**

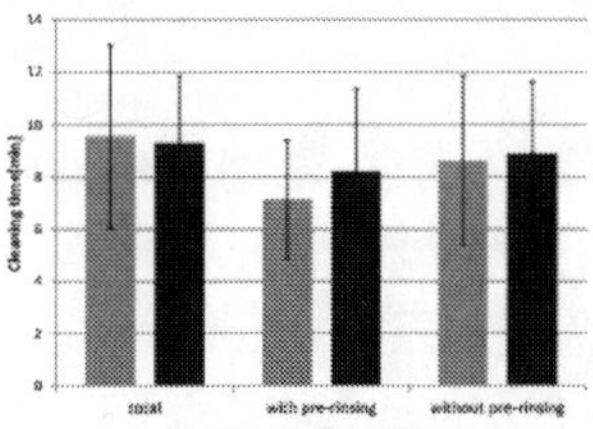

Figure 4: Cleaning results, pilot trials average values and standard deviation, n (total) = 18

Supported by:
Federal Ministry for Economic Affairs and Energy
on the basis of a decision by the German Bundestag

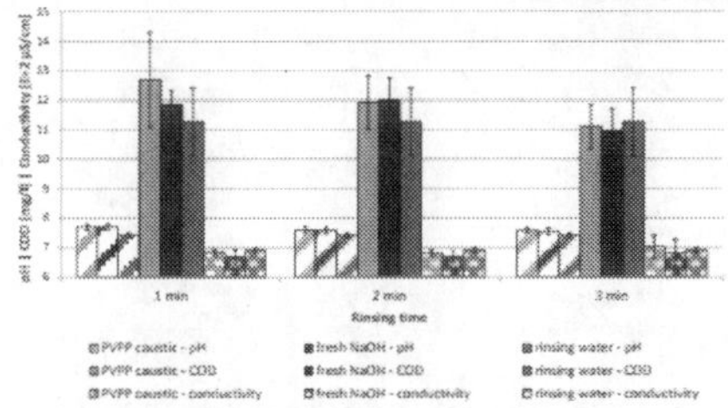

Figure 5: Investigation of final rinsing water, pilot trials average values and standard deviation, n = 9

5. Cleaning trials – field studies

- Comparative cleaning trials in medium sized brewery in Berlin
- Lost recirculation cleaning of 30 hl dished bottomed fermentation tanks with static sprayball
- Cleaning program of brewery: 2 min pre-rinse, 20 min caustic recirculation, 3 min intermediate rinse, 20 min acidic cleaning, 30 sec final rinse
- Optical comparison after every celaning step including UV spotlight
- Evaluation of rinsing water
- ➢ **Cleaning result with regeneration caustic equal to fresh caustic** (fig. 6)
- ➢ **Fresh water quality is reached after 2 minutes of final rinse for both cleaners** (fig. 7), polyphenols below detection limit

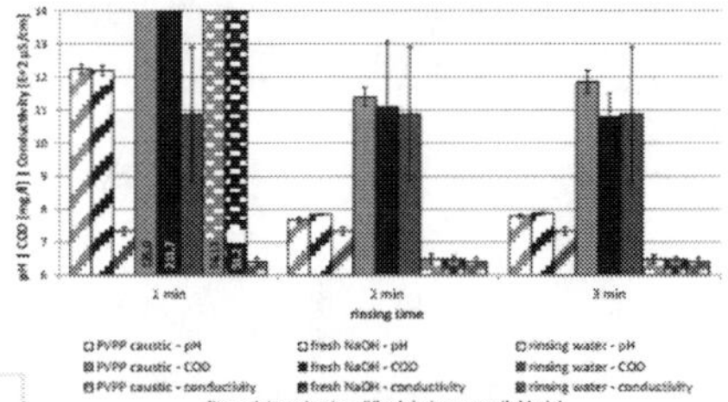

Figure 6: Investigation of final rinsing water, field trials average values and standard deviation, n = 6

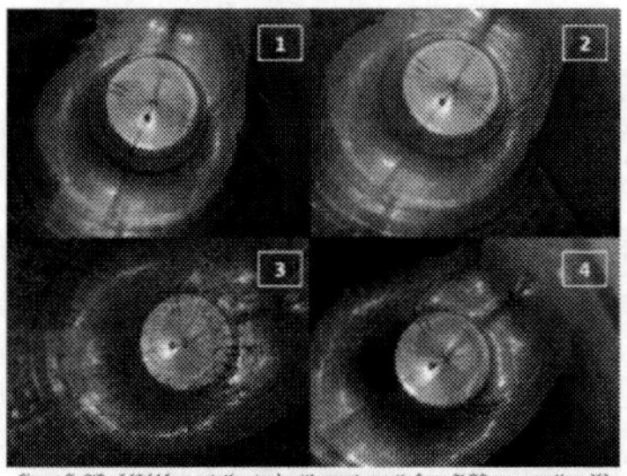

Figure 7: CIP of 30 hl fermentation tank with spent caustic from PVPP regeneration: (1) before CIP, (2) after 2 minutes pre-rinse, (3) after 20 minutes caustic recirculation, (4) after 3 minutes intermediate rinse

Conclusion

- According to the here presented investigation of cleaning behavior and side-effects of spent caustic from PVPP regeneration, fresh NaOH can be substituted with, as long as lost CIP is applied
- By the addition of the necessary piping network, pumps, eventually an additional buffer tank as well as some programming significant savings in water, waste water and chemicals can be realized, implementing this new approach of caustic re-use
- Hence, another step towards environmentally friendly beer production would be done

Literature:

[1] Chmiel, H., Fähnrich, A., Janke, H.D., Kaschek, M., Kluth, J., Mavrov, V., Rogener, F.: „Kosteneinsparpotenziale durch Wasser- und Laugenmanagement", Brauwelt 1/2 (1997), 16-26

[2] Linehan, R., Grennan, D.: Spent caustic recovery Dundalk Brewery, Liquid effluent reduction using membrane technology, 2008

[3] Götz, G., Geißen, S.-U., Ahrens, A., Reimann, S.: „Adjustment of the wastewater matrix for optimization of membrane systems applied for water reuse in breweries", Journal of Membrane Science 465 (2014), 68-77

Contact: VLB Berlin / FIBGP / Philipp Zeuschner / Seestr. 13 / 13353 Berlin / Germany / Email: zeuschner@vlb-berlin.org

Valorisation of brewer's spent yeast

Z.E. Martins[1], O. Pinho[1,2], I.M.P.L.V.O. Ferreira[1]
[1]LAQV/REQUIMTE/ Department of Chemical Sciences, Laboratory of Bromatology and Hydrology, Pharmacy Faculty of University of Porto, Portugal;
[2]Faculdade de Ciências da Nutrição e Alimentação da Universidade do Porto, 4200-465 Porto, Portugal
(Email: isabel.ferreira@ff.up.pt, +351 220428642)

INTRODUCTION

Beer is one of the most popular and most highly consumed alcoholic beverages in the world. Although many types of beer exist, the large majority may be classified into two major groups, ale and lager, depending on the yeast used and the fermentation conditions. Traditionally, the fermentation for ale beer is carried out using a "top-fermenting" yeast (e.g. *Saccharomyces cerevisiae*) at temperatures from 16 °C to 24 °C. Fermentation is followed by a short period of aging, or none at all. In contrast, lager beers are traditionally produced by using "bottom-fermenting" yeast strains (e.g. *Saccharomyces pastorianus*) at lower temperatures, from 8 °C to 15 °C. In this type o beer, fermentation is followed by a long (often a few weeks) low-temperature period of maturation (aging, "lagering").

Brewing industry is a huge global business, nevertheless beer production inevitably involves generation of by-products. Alternative uses for perishable brewing wastes are paramount, not only for environmental reasons but also because beer production peak occurs when there is less demand for animal feed.

Brewer's spent yeast (BSY) is the second major brewing by-product, but has received little attention as a marketable commodity, and its disposal is often an environmental problem, being mostly used for animal feed. However, it can be a valuable source of dietary fibre, especially β-glucans. The production of a functional food ingredient from BSY that can contribute to indirect income generation and circle economy.

OBJECTIVES

This study aimed to obtain fibre rich extracts from Ale and Lager BSY and characterize their fibre composition. Moreover, sensory, colour, and crumb structure properties of bread fortified with Lager BSY extract was studied.

MATERIALS AND METHODS

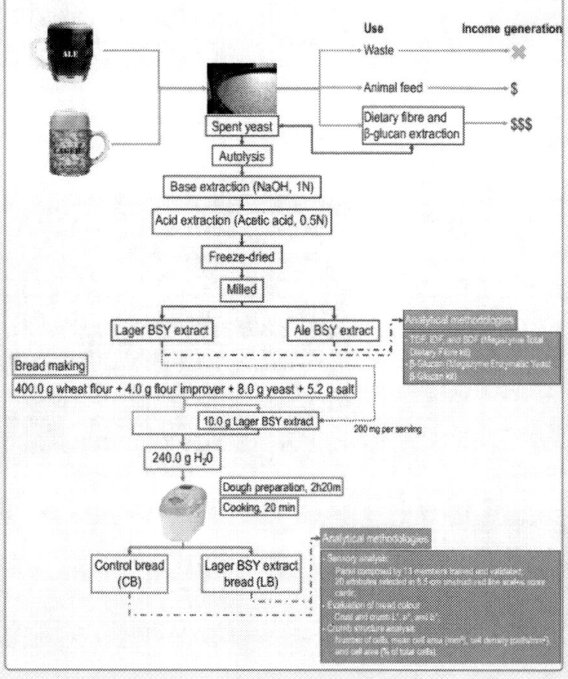

RESULTS

BSY extract analysis

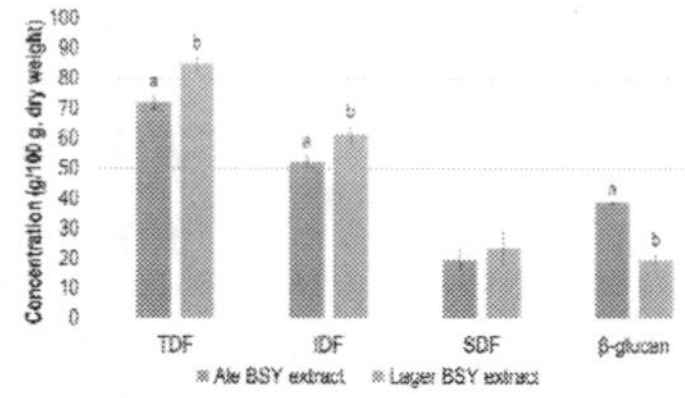

Figure 1: Dietary fibre composition of each extract (g/100 g), on dry weight basis.
BSY, Brewer's spent yeast; IDF, Insoluble dietary fibre; TDF, Total dietary fibre; SDF, Soluble dietary fibre.

- Significant differences ($p < 0.05$) were found for every dietary fibre, except SDF;
- IDF fraction represents 72% of TDF in both BSY extracts;
- Lager BSY extract has the highest content of TDF and IDF;
- Ale BSY extract has the highest content of β-glucan.

Bread analysis

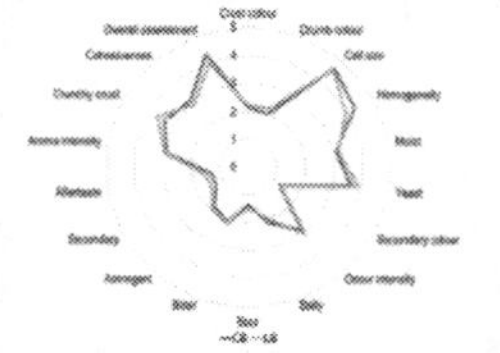

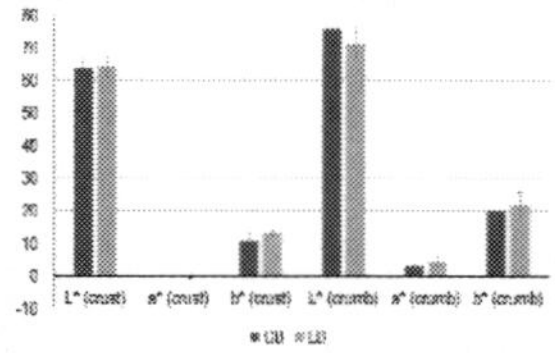

Figure 2: Overall bread descriptors for each type of bread. Figure 3: Values for color parameters for each type of bread.

- No significant differences ($p > 0.05$) were found between control bread (CB) and Lager BSY extract bread (LB) for sensory profile and colour;
- Crust colour: both bread types had a greenish (negative a* values) and yellowish colours (positive b* values);
- Crust colour: both bread types had a reddish (positive a* values) and yellowish colours (positive b* values);

Table 1: Values for crumb image analysis for each type of bread.

Bread	Number of cells	Mean area (mm²)	Cell density (cells/mm²)	Cell area (% of total cells)			
				0.2 ≤ CA (mm²)	0.2 < CA ≤ 3.0 (mm²)	3.0 < CA ≤ 10.0 (mm²)	CA > 10.0 (mm²)
CB	529 (308–569)	6.09 (4.65–8.22)	88.00 (38.92–156.57)	58.50 ± 2.50	30.70 (27.51–34.94)	4.09 ± 1.25	5.16 (4.92–8.12)
LB	487 (359–677)	7.00 (4.98–9.65)	69.55 (37.30–135.90)	57.65 ± 2.50	33.55 (28.48–35.43)	4.63 ± 1.42	5.39 (3.40–6.98)
p	ns	ns	ns	ns	ns	ns	ns

- No significant differences ($p > 0.05$) were found between control bread (CB) and Lager BSY extract bread (LB) for crumb morphological features;
- Cell area distribution, more than 80% of cells in every bread formulation had an area below 3.0 mm²;
- Data gathered from image analysis provided information that would not be possible to obtain from sensory data.

CONCLUSION

Lager and Ale BSY extracts are rich sources of dietary fibre, including β-glucan. Also, breads with 200 mg yeast β-glucan per serving were successfully obtained without major impact on bread characteristics and with sensory acceptability. Therefore, Lager and Ale BSY have the potential to be used in food industry, not only as a functional ingredient but also as a profitable way to optimize indirect income generation.

Acknowledgements
[illegible]

36th European Brewery Convention 2017

ILT NRW
INSTITUTE OF FOOD TECHNOLOGY

Study on the implementation of a biocatalytic power-to-methan reactor in a brewery

Timo Broeker, Marc Hoffarth, Klaus Heikrodt, Jan Schneider
Northrhine Westfalia Institute of Food Technology ILT.NRW, Lemgo, Germany

36th CONGRESS EBC 2017 EUROPEAN BREWERY CONVENTION 14-18 MAY LJUBLJANA SLOVENIA

Hochschule Ostwestfalen-Lippe
University of Applied Sciences

P052

bioCONNECT

Federal Ministry of Education and Research
FORSCHUNG AN FACHHOCHSCHULEN

Motivation

Environmental concerns drive industrial development towards efficiency. Not seldom this leads to advanced fields of activity, for example through adding value to a side stream. Beside spent grains, breweries produce a major side stream: CO_2. Present utilization like liquefaction consumes further energy and offers a solution only for a small percentage of the overall emission produced in breweries. On the other hand, power-to-gas and methanisation of the electrolytic produced hydrogen with CO_2 are becoming an intensively discussed technology to contribute to the transformation of the energy system. In order to benefit from that technology, there is a demand for cheap and pure CO_2 sources, which makes breweries interesting for implementation. Further, storage capacities for carbon dioxide are on site as well as technology and competences to run biotechnological processes.

An innovative lab scale reactor for the biocatalytic methanisation has been build up and attached to the alcoholic fermentation. Besides the investigation of the dynamic response of the biological archaea process, the implementation into a local brewery has been investigated. Experimental and published results have been used to study several possibilities to realise a biocatalytic P2G concept in a brewery.

Methanisation of CO_2 and H_2

Following reaction equation shows the conversion of CO2 and hydrogen to methane:

$$CO_2 + 4H_2 \longrightarrow CH_4 + 2H_2O\,(l) \qquad \Delta H_R = -165\ \mathrm{kJ\,mol^{-1}}$$

An alternative to chemical methanisation (Sabatier process) are microorganisms (archaea), which metabolize CO_2 to CH_4 under anaerobic conditions. The advantages of this biological over chemical process are moderate process parameters.

Tab. 1: Process parameters of chemical and biological methanisation

Process	Temperature ϑ	Pressure p	needed quality of CO_2 flow
chemical	400 °C	20 bar	high (catalysts)
biological	50 °C to 80 °C	< 10 bar	low

Germany's P2G Potential

The German target to produce 80 % electricity from renewables by 2050 will lead to approx. 153 GWh installed power, while the demand decreases down to 74 GWh. In order to balance the resulting fluctuation DLR* and Fraunhofer predict the need of about 25 TWh storage capacities. A possible storage technology is power-to-gas (P2G). Because of the lack of an adequate hydrogen infrastructure, P2G-hydrogen can be converted with carbon dioxide into methane. Netto electricity exports of Germany were 88 TWh in 2015. 50 % of this energy could be used for P2G to balance fluctuating energy production. This energy can be converted via electrolysis into **8.8 bn m³ H_2** and afterwards via methanisation into **2.2 bn m³** respectively **25 TWh CH_4**. That demands **2.2 bn m³ CO_2**. The cheapest and purest CO_2 source are breweries and wine- or champagne producers with CO_2 excess from alcoholic fermentation.

Fig 2: Production of beer, wine, sparkling wine/champagne* and bioethanol in Germany (2014) and resulting CO2 missions from fermentation (*data based on sales) and their share of overall CO2 demand for methanisation in the 2050 scenario.*

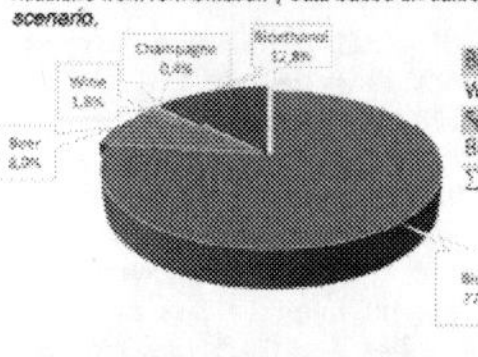

	Production (Germany)	CO_2 production
Beer	95 600 000 hl a⁻¹	312 000 t a⁻¹
Wine	9 202 000 hl a⁻¹	69 000 t a⁻¹
Sparkling wine / Champagne	3 174 000 hl a⁻¹	16 000 t a⁻¹
Bioethanol	726 000 t a⁻¹	693 000 t a⁻¹
Σ		1 090 000 t a⁻¹

Those 1 090 000 t a⁻¹ CO_2 can be converted to about 400 000 t a⁻¹ CH_4 which are about 5.6 TWh of storable energy.

References

[1] Federal Statistical O ce. Absatz von Bier - Fachserie 14 Reihe 9.2.1 - July 2015, Wiesbaden, article no. 2140921151074
[2] Federal Statistical O ce. Wein -Fachserie 3 Reihe 3.2.2 - 2014, Wiesbaden, article no.2030322147004
[3] Federal Statistical Oce. Schaumweinsteuer- und Zwischenerzeugnissteuerstatistik -Fachserie 14 Reihe 9.5 - 2014, Wiesbaden, article no. 2140950147004
[4] German Bioethanol Association. Marktdaten 2015, July 2016;
[5] Federal Ministry for Economic Afairs and Energy. Energiedaten 2015, April 2015
[6] Federal Network Agency. Biogas- Monitoringbericht 2015, August 2016

Experimental Approach

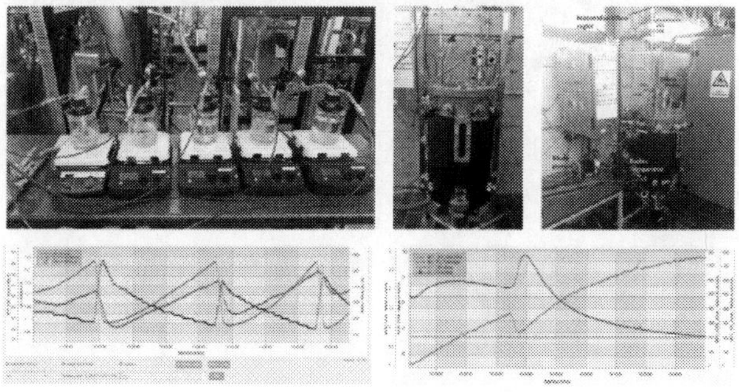

Fig 3. Top: Pictures of the experimental setup, 500 mL batch reactors (start pressure = 1,5 bar), 20 L continues reactor. Left: Display of the pH value, CH_4 concentration and redox potential during steady state as an example of optimisation process. Fluctuation of the system is due to the pH regulation resulting in change of the redox potential and lower methane yield. Right: Example of system run-up until 98 % CH_4 concentration.

Eq.1. Systems Volume-time yield

$$VTY = \frac{gasyield \cdot flow\ CO_2}{reactorvolume} = \frac{0.98\,\frac{CH_4}{CO_2} \cdot 2 \cdot 10^{-4}\,\frac{Nm^3}{min}\,CO_2 \cdot 60\,\frac{min}{h} \cdot 24\,\frac{h}{d}}{0.020\ m^3} = 14.1\ Nm^3 m^{-3} d^{-1}$$

Model Brewery

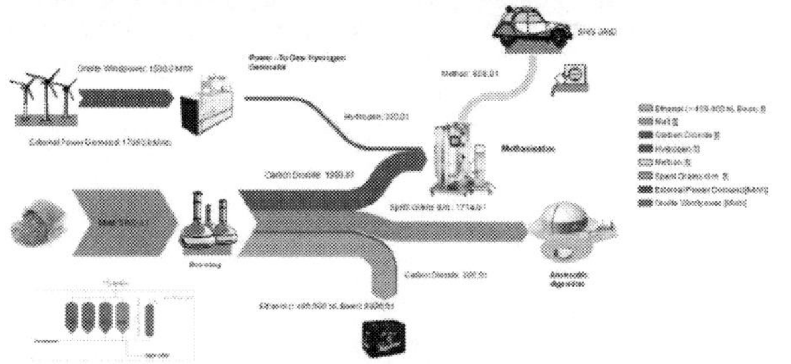

The brewery in the model is a mid-scale brewery located in Westfalia, Germany with an output of approx. 400 000 hL beer. About 1800 t of carbondioxid are used for inerting the tanks and further release to the atmosphere. An onsite wind-power plant generates 1500 MWh of 18560 MWh needed for methanisation of 1800 t carbondioxid. That results in 656 t of methane, which can be injected into the natural gas grid or used onsite for co-generation of process energy.

Conclusion

Objectives of the experimental part were optimization of agitation and phase transition as well as operation start and shut down sequences. For implementation, objectives like technical feasibility and on-site synergies, on-site power production for maximum capacity utilisation, power market instruments and coupling of existing electric consumers for the high power demand (reduced tax for electricity consumption) were investigated.

Markt instruments of interest are: (i) selling of net balancing services, (ii) ex-trade (spot contracts, day-ahead, etc.), (iii) onsite process energy co-generation .

The models showed that, depending on upcoming legislative framework, biocatalytic methanisation could work well using carbon dioxide source from alcoholic fermentation. Further modelling will be carried out and published .

Contact

ILT.NRW
Integrated Biorefinery Concepts
Timo Broeker
timo.broeker@hs-owl.de

www.ilt-nrw.de

Wet oxidation of spent grains from breweries for the production of fermentable sugars

Jan Schneider, Timo Broeker, Marc Hoffarth
North Rhine Westfalia Institute of Food Technology

ILT NRW — INSTITUTE OF FOOD TECHNOLOGY

36th CONGRESS EBC 2017 EUROPEAN BREWERY CONVENTION 14-18 MAY LJUBLJANA SLOVENIA

Fossil based economy is sooner or later going to be substituted by bio based economy. Therefore, drop-in carbohydrates from renewables are desired by the chemical industry, often as a substrate for biotechnological applications. Sidestreams from agriculture and food processing are potential sources for such a purpose. A major sidestream at industrial scale is brewer's spent grain (BSG), which constitutes at least 80 - 85 % of the organic by-products in breweries.

Experimental Set-up

Besides the usability of the process to release fermentable sugars, the feasibility of the wet oxidation and the developed direct cell method were of interest in this study. The glucose yield, the amount of degraded lignin and the consumed concentration of oxidizing agents where investigated. Finally, the submerse oxidation was compared with a direct treatment of BSG and other substrates in the electrolysis reactor DCT (direct cell treatment).

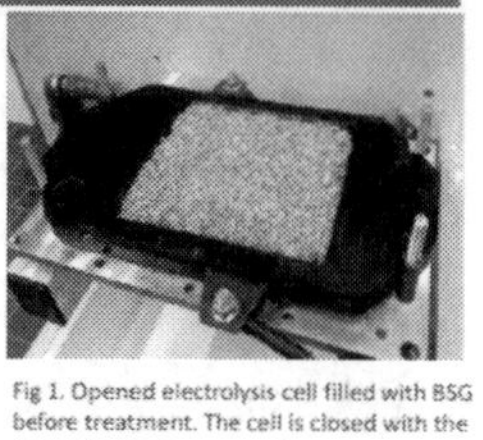

Fig 1. Opened electrolysis cell filled with BSG before treatment. The cell is closed with the upper electrode.

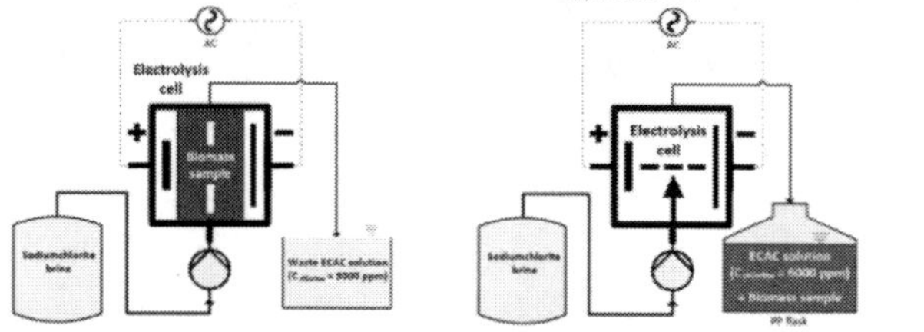

Fig. 2. Electrolysis reactor manufactured by aquagroup AG, Weiden, and the two applied reaction techniques. Left – for direct cell treatment (DCT) of the biomass Sodium brine was pumped through the cell, in which it reacted and active chlorine was produced directly in the biomass. The waste solution from DCT was collected and the chorine content was analysed in the waste solution. Right – The brine was electrolysed without cell filling and the chlorine solution used in an external vessel for the pretreatment.

General Approach

Since there are no reported investigations on the oxidative pretreatment of BSG, forming of inhibitors from the high amount of contained protein or similar problems were taken into account. For this reason, other second generation cellulose from an agricultural sidestream substrates were used for comparison: wheat straw and maize silage.

Tab. 1. Cellulose, Lignin and Dry matter content (after sample preparation) of substrates (BSG - brewer's spent grain, WS - wheat straw, MS - maize silage).

	Cellulose (%wt d.m.)	Lignin (%wt d.m.)	D.M. (g/100 g)
BSG	21,8	22.5	90.2
WS	38,6	24.3	91.7
MS	19.2	19.9	92.3

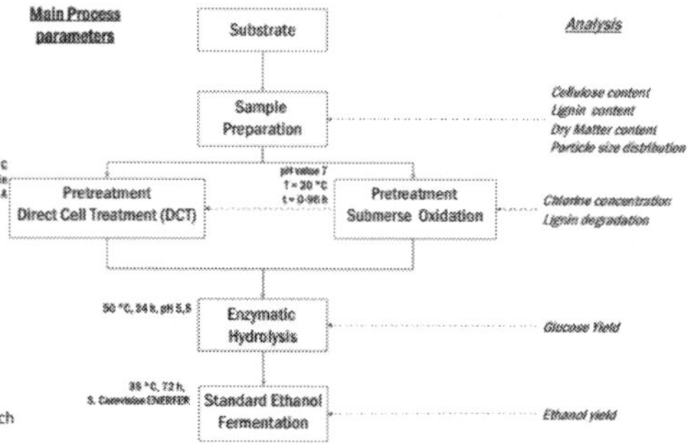

Fig 3. General Approach

Glucose Yield

$$Y_{glucose/biomass} = \frac{c\ glucose\ in\ g/100\ g}{\chi biomass\ in\ g\ d.m./100\ g}$$

Lignin degradation

$$\Delta\chi_{lignin} = \chi_{ligning,0} - \chi_{ligning,f}$$

Oxidation rate

$$c(t) = c_\infty + c^* \cdot e^{-k \cdot t}$$

The ***coefficient*** $c^* = c_0 - c_\infty$ represents the consumption of chlorine. In order to find appropriate processing conditions and to verify the delignification, the specific consumption was observed as the ratio of c* versus the dry matter content of the biomass suspension $x_{d.m.}$:

$$AC_{sp.\ biomass} = \frac{c^*}{x_{d.m.}}$$

Chlorine consumption rate k indicates particularly the temperature and mass transfer influencing sizes as the particle sizes and the hydrodynamic conditions. Thus, this figure gives an information about the different investigated reactions methods.

Results

Before comparing different reaction techniques and substrates, the process parameters pH value and temperature were investigated with 39 settings (Tab.2). The target figure was the ethanol concentration after standardised enzymatic hydrolysis and fermentation.

Tab.2. Influence of temperature and pH-value settings in submerse wet oxidation pretreatment WS (initial chlorine concentration 6000 ppm) on the ethanol concentration (as mean value x and range). Fermentation was done directly in the mash without further separation.

pH value	Temperature	Ethanol concentration	Range	Sample size
(-)	(°C)	(% w/w)	(% w/w)	(-)
5	4	0,21	0,04	3
5	20	0,22	0,02	3
5	50	0,25	0,08	4
7	4	0,28	0,07	5
7	20	0,27	0,08	7
7	50	0,24	0,02	3
9	4	0,20	0,09	5
9	20	0,22	0,01	3
9	50	0,23	0,05	5

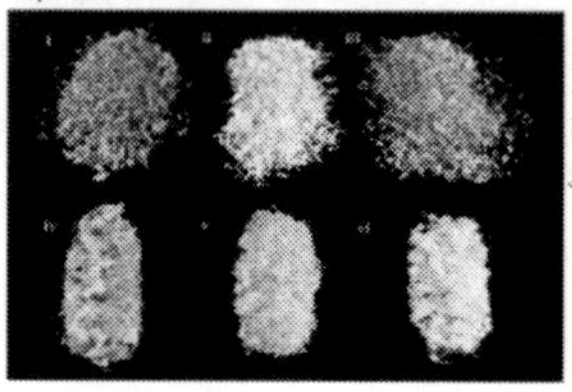

Fig 4. Bleaching effect and the visible grade of released cellulose of active ECA chlorine on BSG and wheat straw. i: untreated spent grain; ii: BSG 120 h; iii: BSG DCT 0,5 h, iv: WS DCT 30 min; v: WS 6 h; vi: WS 120 h

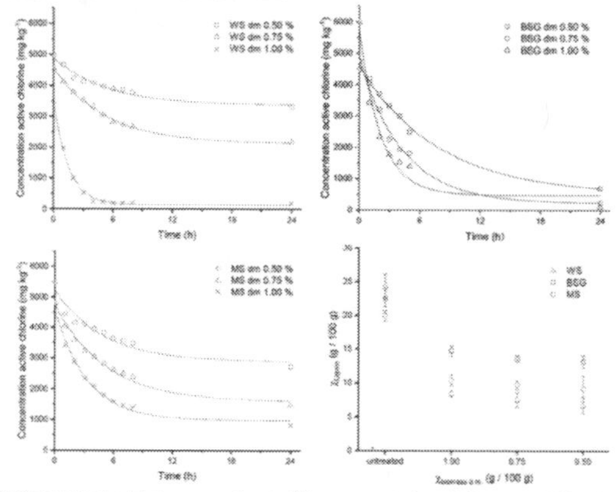

Fig 5. Kinetics of active chlorine consumptions in different concentrations of wheat straw (WS), maize silage (MS) and brewer's spent grain (BSG) by submerse ECAC-treatment at pH value 7 and 20 °C. The correlated Lignin content (lower right) of the untreated sample and after 24 h indicates the effect of the oxidization.

	$\chi_{d.m.}$ (g/100 g)	c* (mg/kg)	k (h^{-1})	R^2_k (-)	$AC_{sp.\ biom.}$ (mg/g)	$\Delta\chi_{Lignin}$ (g/100 g)	$AC_{sp.lignin}$ (mg/g)
WS	0.50	1517	0.190	0.956	303	17.7	8.6
	0.75	2427	0.194	0.997	324	16.8	14.4
	1.00	3404	0.680	0.995	340	16.0	21.3
MS	0.50	2330	0.207	0.944	466	10.7	21.8
	0.75	3145	0.195	0.981	419	10.3	30.7
	1.00	3600	0.306	0.989	360	9.5	37.8
BSG	0.50	4080	0.128	0.993	816	9.4	43.4
	0.75	4621	0.237	0.985	616	8.9	51.9
	1.00	5311	0.478	0.959	531	7.8	67.8

Complete delignification at low concentrations of oxidative chemicals is possible. The feasibility of the delignification of brewer's spent grain and other lignocellulosic material in a direct cell treatment reactor as well as in a submerse treatment could be shown. The process condition during oxidation at 20 °C and pH 7.0 are suitable. The active chlorine concentration and the biomass content in the treated mash must be individually adjusted in order to obtain both, a rapid process and a high extent of lignin degradation. The direct cell treatment accelerates the glucose yielding process compared with the immersion in oxidation agent. However, the consumption of oxidation agent is higher. The process works well for BSG to provide glucose, but straw performs remarkably better likely due to less oxidation agent consuming non-lignin compounds as protein. In contrast to the good results in delignification and saccharification, the fermentation to ethanol delivered poor conversion efficiency. While in studies with an acidic hydrolysis pretreatment yields of 11 % EtOH (d.m.) for WS (80 % of max possible yield) and 10 % for BSG have been reached, the fermentation of oxidised samples in this study resulted in maximum of 3.7 % EtOH yield, possibly because of residues of the oxidation agents which may affect the yeast activity.

Acknowledgements:
The Project KF2402603BN3 was funded under the program Central Innovation Programme for SMEs (ZIM) by the Federal Ministry of Economics and Technology based on a decision of the German Bundestag.

www.ilt-nrw.de

Contact:
The corresponding author marc.hoffarth@hs-owl.de

Anita Van Landschoot[1,2], Anneleen Decloedt[1,2], Sylvie Vandoorne[1,2], Dana Vanderputten[2]
[1]Ghent University, Faculty of Bioscience Engineering, 1 Valentin Vaerwyckweg, B-9000 Ghent, Belgium
[2]University College Ghent, Faculty of Nature and Technology, 1-Valentin Vaerwyckweg, B-9000 Ghent, Belgium

EFFECT OF WORT FILTRATION TYPE ON BLOND FINAL BEER: INDUSTRIAL CASE STUDY

After the mashing process the spent grains have to be separated from the wort. Either a lauter tun or membrane filtration can be used.
Many brewers are shifting from the traditional lauter tun to mash filtration mainly because the latter technique recovers more extract from the grains.

A Belgian craft brewer replaced his lauter tun by a wort filter Meura 2001 Junior (2016)
→ Impact on the wort and beer when using the same infusion mashing process?

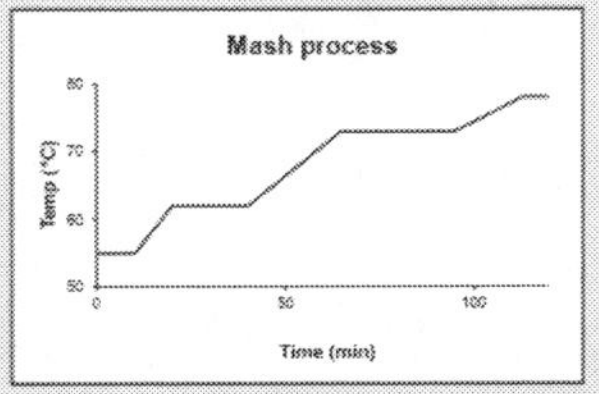

Colour of the blond beers 2-11 EBC ↑

Explanation:
Free Amino Nitrogen (FAN) in the wort (>300 ppm) ↑
→ Maillard reaction during cooking ↑

Colour problem solved by adapting the mashing process
→ skip the protease rest at 55 °C and directly mash in at 62 °C

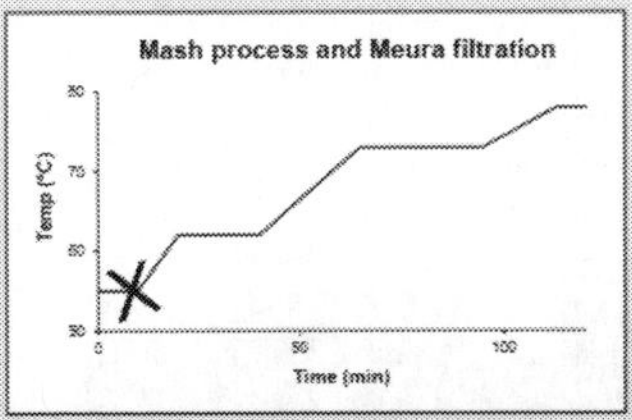

Results of the final full malt wort and blond beer

Wort: pH 5.5-5.6; colour 6-9
Sugar profile of the wort: ~ 67% fermentable sugars (~ 8% glucose, < 0.2% fructose, ~ 4% sucrose, ~ 43% maltose, ~ 12% maltotriose)

Characteristics blond beer: pH 4.2-4.3, colour 10-12, 7.4-7.5 Alcohol V/V%, Eo 16.2-16.4 °P, Er 5.2-5.3 °P, RDF ~ 68%
Taste of the blond beers: OK
Additionally: Beers produced with the adapted mash process and Meura filtration have a better foam stability

Conclusions

Changing type of wort filtration can effect the characteristics of the beer produced. A small adaptation to the mashing process can resolve the occuring problem and in parallel increase foam stability.

Contact
Anita.vanlandschoot@ugent.be
www.ugent.be/bw/en
Universiteit Gent
@ugent
Ghent University

Fastest Mash and Beer Filtration

Enzyme Solution Development from Lab to Large Scale Brewery Performance

Ana Bela Pinheiro, Wilbert Heijne*, Tiago Freitas, Claudia Teixeira, Hans Jansen*, Marcel Jansen* and Jozé Mutsaers*

Unicer Bebidas, S.A., Porto, Portugal

**DSM Biotechnology Center, DSM Food Specialties B.V., Delft, The Netherlands*

Contact: Wilbert.Heijne@dsm.com

Introduction

For efficient and sustainable beer production, brewers want to maximize starch yield and the filtrations of mash and beer should be as fast and smooth as possible, and enzymes are essential for this.

Besides Beta glucanases, proteases and amylases, it was recently shown that xylanases contribute to efficient filtration of wort and beer.

We developed a solution with classical enzymes for the maximization of starch, as well as for the fastest mash and beer filtration. On lab scale, filtration time, yield, beta glucan levels and wort viscosity were found to be outstanding with the solution Filtrase NL Fast.

The relevance for large industrial scale brewing is shown here with wort and beer production at the brewery of Unicer. Mash and beer filtration and yield demonstrated a strong performance correlating to the lab work.

This solution enables brewers to optimize wort and beer filtration maximize the output.

Not only beta-glucans

Established enzyme solutions for mash filtration aim primarily at beta-glucan degradation.

However, based on the raw material composition (Table1) also degradation of hemi-cellulose substances like arabino-xylans from the malted barley or other cereal is essential to get good starch liberation, mash separation but even more important to achieve fast and flawless beer filtration at cold temperatures, where influence on viscosity of such substances becomes rate limiting.

Table 1. Barley cell wall composition

Barley cell wall composition	Endosperm	Aleurone layer
ß-(1-3,1-4) Glucans	70-75%	26%
Arabino-xylans	20%	65-70%
Cellulose	2%	2%
Other non-starch polysaccharides	2%	3%

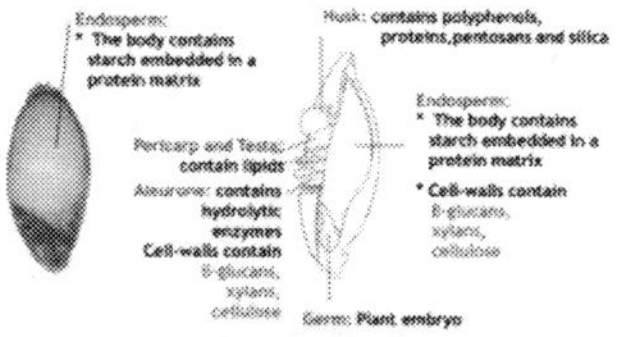

Figure 1. Barley composition

Experimental setup

Lab screening & development - Laboratory scale tests were used to screen amongst a large library of enzymes for optimal performance, for example on reduction of the viscosity of wort and beer.

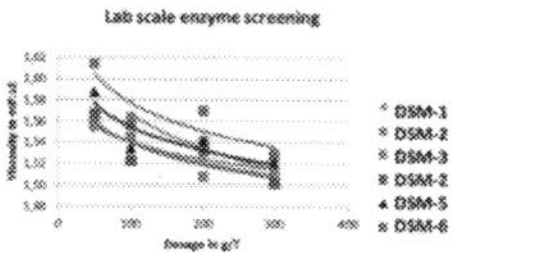

Figure 2. Enzyme screening with lab scale mashing and filtration tests

Brewing trials

Most promising enzymes from the screening were taken to pilot brewing tests.

Parameters such as beta-glucan, beer viscosity and filtration performance were used. A coefficient summarizes the success of mash or beer filtration to correct for pressure adjustment while controlling the flow of wort or beer.

Filt.Coeff = (Time for a fixed volume filtered (min.) * Max Filt. Pressure (bar))

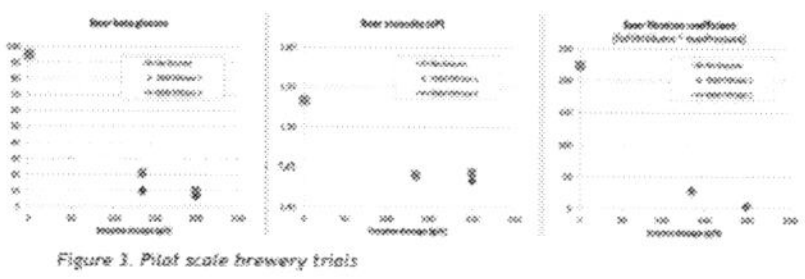

Figure 3. Pilot scale brewery trials

Large Scale Results

Large brewing trials were performed at Unicer Bebidas, Porto. Brewhouse operations as well as beer filtration with the new enzyme were tested versus the established solution, and observed to deliver faster and more efficient wort production and beer filtration.

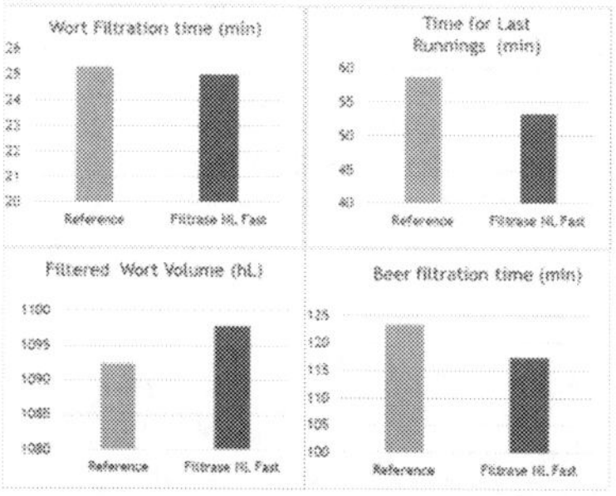

Figure 3. Large brewery trials : 1st phase trials

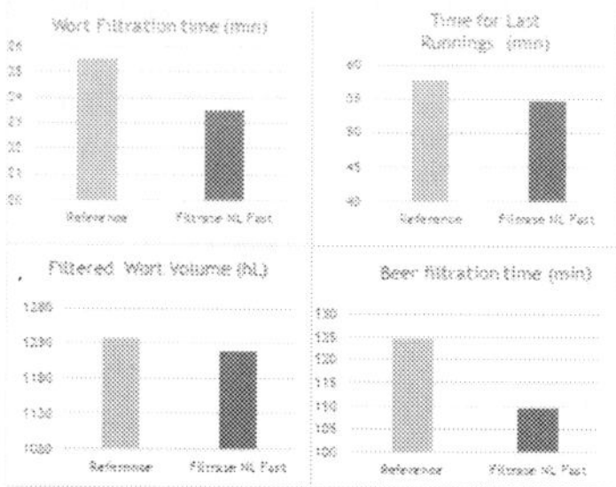

Conclusions

This work presents a solution with classical enzymes for the maximization of starch yield, and the fastest mash and beer filtration.

On lab scale, filtration time, yield, beta glucan levels and wort viscosity were found to be outstanding with the Filtrase NL Fast enzyme solution.

The relevance for large industrial scale brewing is shown here with wort and beer production at the brewery of Unicer. Mash and beer filtration as well as yield demonstrated a strong performance correlating to the lab work.

Taken together, the present study shows that this solution enables brewers to optimize wort and beer filtration and maximize the output of the raw materials, thus allowing the sustainable beer brewing.

Acknowledgements

The authors wish to thank several people, institutes and partners that were instrumental in developing present work:

-Pilot brewery at Technology Campus Ghent, KU Leuven. Laboratory of Enzyme, Fermentation and Brewing; Gert de Rouck, Brecht de Causmaecker and Guido Aerts

- Many large industrial breweries world wide that were important in large scale performance testing and validation of this enzyme solution to optimal mash and beer filtration

unicer

DSM
BRIGHT SCIENCE. BRIGHTER LIVING.

SZENT ISTVÁN UNIVERSITY
FACULTY OF FOOD SCIENCE

Studying fouling in crossflow microfiltration of rough beer

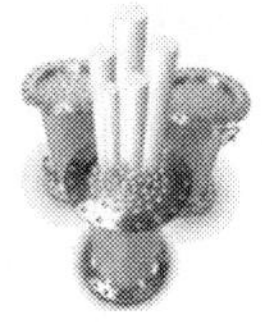

Áron VARGA, Edit MÁRKI
Szent István University, Faculty of Food Science,
Department of Food Engineering
mr.aron.varga@gmail.com

Introduction

The clarification of rough beer is important, because of improving product quality, efficiency, environmental regulations and sustainability. However, one of the main problems of ordinary application of crossflow microfiltration for clarification of rough beer is fouling mechanisms. Membrane fouling is a process whereby a solution or a particle (e.g. yeast and colloidal particles) is deposited on a membrane surface and in membrane pores so that the membrane's performance is degraded. One of the hydrodynamic techniques for fouling reduction is increase of the turbulence intensity with use of static mixer in the flow channel.

Aims

Our aims were the following: study that the usage of static mixer has effect on membrane fouling and analytical parameters throughout laboratory filtration of rough beer, develop a membrane cleaning method.

Materials and Methods

1. Standardized lager beer was brewed for the filtration investigations at the Department of Brewing and Distilling in pilot scale
2. 2^p type full factorial experimental design (*Table 1*) was used:
   - The three factors were the following: static mixer, transmembrane pressure and recirculation flow rate
   - Flux was considered as a response
3. A membrane cleaning method was developed
4. Analytical measurements, technical calculations and statistical analysis were performed

Table 1.: Combination of different operating parameters for filtrations

Filtration code	Static mixer	Transmembrane pressure (bar)	Recirculation flow rate (l/h)
m1	No	0.4	50
m2	Yes	0.4	50
m3	No	1.2	50
m4	Yes	1.2	50
m5	No	0.4	200
m6	Yes	0.4	200
m7	No	1.2	200
m8	Yes	1.2	200
m9	No	0.8	125
m10	Yes	0.8	125

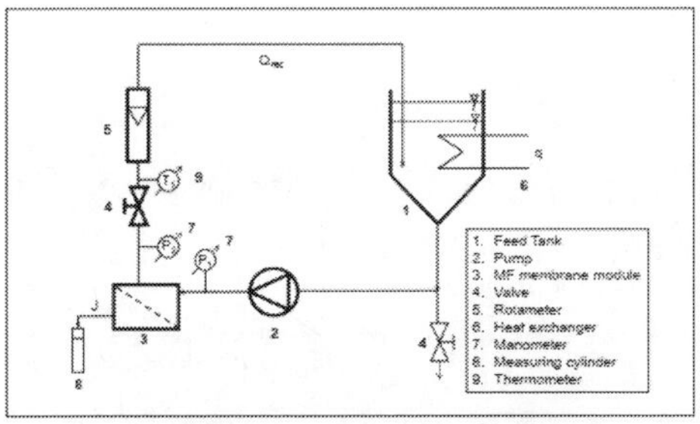

Figure 1.: Schematic flow diagram of beer membrane filtration

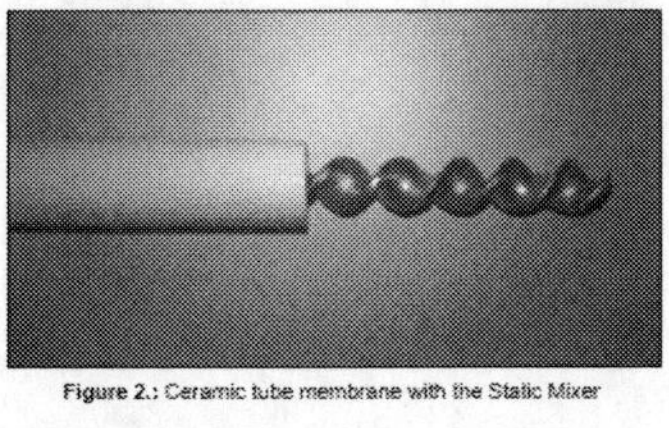

Figure 2.: Ceramic tube membrane with the Static Mixer

Results and Discussion

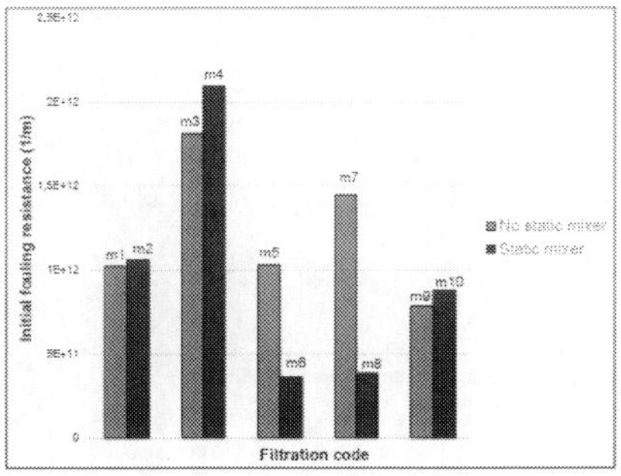

Figure 3.: Initial fouling resistances of different filtrations

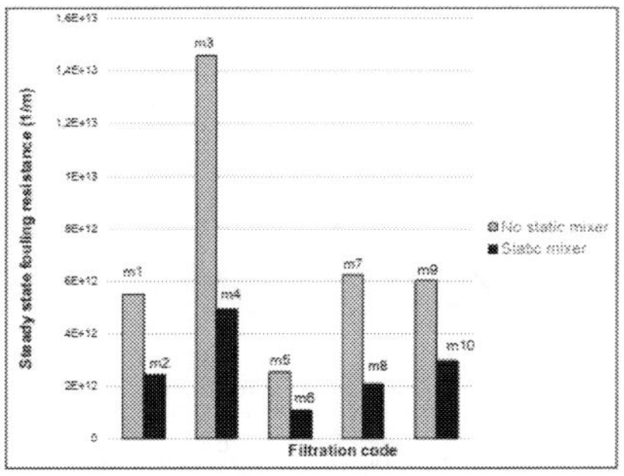

Figure 4.: Steady state fouling resistances of different filtrations

In many cases it has been proven that application of static mixer has immense effect on decrease of initial fouling resistance (e.g. 73% decrease) and steady state fouling resistance (e.g. 67% decrease) values.

Table 2.: Results of analytical measurements of rough beer and two different filtrations

Parameter	Rough beer	Filtered beer m3	Filtered beer m4
Alcohol (V/V %)	4.6	3.4	4.3
Real extract (w/w %)	4.48	3.24	3.89
Bitterness (IBU)	18	13	15
Colour (EBC)	6.53	4.23	5.70
pH	4.547	4.651	4.645
Turbidity (EBC)	2.50	0.25	0.25

As it can be seen in the *Table 2* the values of alcohol content, extract, bitterness, colour and turbidity decreased in both cases compared to the values of rough beer, however the pH values increased. Analytical parameters of *m4* permeate are better than analytical parameters of *m3* permeate.

Conclusion

According to the experimental results, it is shown that static mixer has positive effect on fouling resistance (initial and steady state) values and results of analytical measurements.

References

Ambrosi, A., N.S.M. Cardozo, and I.C. Tessaro. 2014. "Membrane Separation Processes for the Beer Industry: A Review and State of the Art." Food and Bioprocess Technology 7 (4): 921–36. doi:10.1007/s11947-014-1275-0.

Gaspar, I., A. Koris, Z. Bertalan, and G. Vatai. 2011. "Comparison of Ceramic Capillary Membrane and Ceramic Tubular Membrane with Inserted Static Mixer." Chemical Papers 65 (5): 596–602. doi:10.2478/s11696-011-0045-y.

Acknowledgements

We would like to say thanks to Szent István University, Faculty of Food Science, Department of Brewing and Distilling; Szent István University, Faculty of Food Science, Department of Food Preservation and Szent István University, Faculty of Horticultural Science, Department of Biometrics and Agricultural Informatics for supporting the investigations, measurements and statistical analysis.

Filtration and Stabilization Optimization and its Effect on Quality Relevant Parameters such as EAP

Zimmermann T.[1], Dr. Dong J.[2], Dr. Zhang J.[3]

[1] BASF South East Asia Pte Ltd. Vice Director Regional Technical & Application Management, Nutrition & Health, Singapore | [2] Tsingtao Brewery Co. Ltd. Vice President, R&D Center, Qingdao, China | [3] BASF (China) Company Ltd. Human Nutrition, Shanghai, China

ABSTRACT

Filtration and stabilization changes were carried out at Tsingtao No. 3 Brewery in Qingdao, China. Before and during the implementation phase, beer samples were taken and accordingly analysed.

The main emphasis of running the process optimization was on sensory and stability improvement of finished beer as well as on cost optimization combined with sustainability. The usage of e.g. electron spin resonance (ESR) spectroscopy was an important tool for measuring free radical content during beer production process and to assess the anti-oxidative characteristics of individual compounds. Besides standard beer analysis and sensory profiling, metal content change was measured.

All analytical data show a significant improvement in respect to ageing. Sensory evaluation also confirmed the analytical findings. The implemented change at Tsingtao No. 3 Brewery had a significant and most important positive contribution to overall beer stability.

Source: Tsingtao Brewery Co. Ltd.

INTRODUCTION

With support of the R&D Center of Tsingtao Brewery Co. Ltd. and BASF, Tsingtao's No. 3 Brewery in Qingdao, China decided to optimize the existing filtration and beer stabilization process. The installed kieselguhr (KG) filter line was retrofitted, so Crosspure® could be used for the process optimization project.

Crosspure® consists of two polymers: Polystyrene (70%) & PVPP (30%). It offers a 2:1 solution - filtration & stabilisation within one filter; reduced overall filtration/stabilization costs; and no undesirable heavy metal uptake.

PROCESS DESCRIPTION

After a proper site inspection, the future filter lay-out was designed and feasibility study conducted. Tsingtao No. 3 Brewery also decided to switch from manual to automatic beer filtration process. The filter retrofitting could be completed within one month. During that time, automation system, regeneration unit and Crosspure® dosing tank were installed.

MATERIAL & METHOD

- Same unfiltered Tsingtao beer was filtered in parallel, either with kieselguhr & PVPP or with Crosspure® alone.
- Packaged beers were taken and analyzed according to international methods such as EBC Analytica & MEBAK.
- Predicting and enhancing the oxidative stability of beer at an early stage of beer production is a topic of great interest for major brewing groups such as Tsingtao Brewery.
- Electron spin resonance (ESR) spectroscopy is an important tool for measuring free radical content during beer production processes and to assess the antioxidative characteristics of individual compounds.
- Determination of Endogenous Antioxidative Potential (EAP) of beer via ESR is achieved by forced oxidizing the sample, typically at 60°C and atmospheric conditions, and measuring its course of radical formation over a specific time. The lag-time during which the propagation of free radicals are absorbed or suppressed by the system can be directly correlated with its EAP and therefore with its expectant shelf-life.
- Influence of beer matrix on oxidative stability is affected by pH-value, metallic ions, Maillard reaction products, reductons, etc.

RESULTS & DISCUSSION

The following tests were conducted:

1. Beer Analysis
2. Turbidity
3. Oxidative Stability of Beer
4. Sensory Analysis

1 Beer Analysis

ANALYSIS	UNIT	KG + PVPP	CROSSPURE®
Original extract	%	10.84	10.91
Alcohol	Vol. %	4.89	4.93
Degree of attenuation app.	%	85.7	85.8
pH value		4.33	4.33
Color	EBC	6.9	6.9
Bitter units	BU	10.5	11.8
Iron	mg/L	0.075	0.019
Copper	mg/L	0.083	0.060
Tannoids	mg PVP/L	15.6	0
Sensitive proteins	EBC	0.12	0.03
Polyphenols, T125	mL	14.23	>60
Proteins, P40	mL	47.7	52.25
Total polyphenols	mg/L	130.38	82.00

Table 1: For most tested parameters, beer analysis data do not show significant differences. The major changes are metals content and all polyphenols related parameters such as tannoids, T125 and total polyphenols.

2 Turbidity

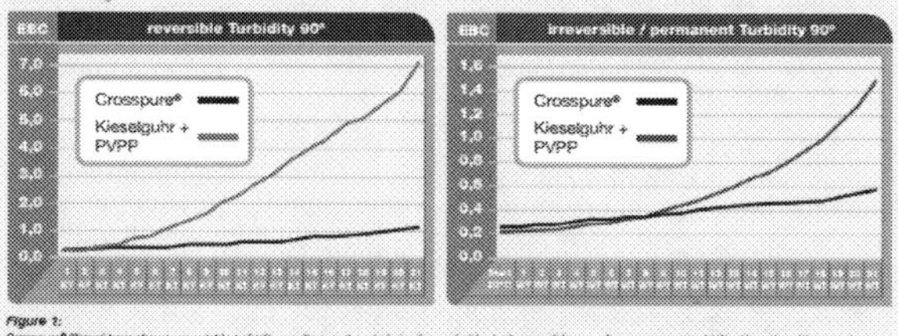

Figure 1: Crosspure® filtered beer shows very stable turbidity results over the whole testing period for both reversible as well as permanent turbidity. More than 20 warm days could be achieved and still the critical 2.0 EBC mark was not reached.

3 Oxidative Stability of Beer (ESR, EAP)

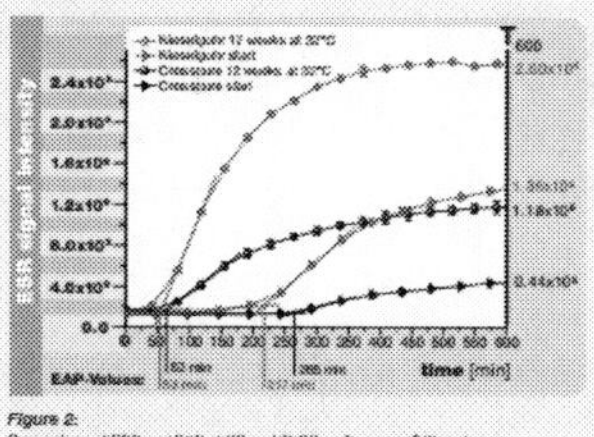

Figure 2: Comparison of ESR and EAP of KG and PVPP vs Crosspure® filtered beers showed excellent results for the Crosspure® filtered beer.

4 Sensory Analysis

ANALYSIS	KG + PVPP		CROSSPURE®	
	fresh	forced aged	fresh	forced aged
Odor	**4.4**	**4.0**	**4.4**	**4.1**
Flavor	**4.3**	**3.5**	**4.3**	**4.0**
Palate Fullness	**4.0**	**3.8**	**4.2**	**3.9**
Fizziness	**4.3**	**3.3**	**4.6**	**3.5**
Quality of Bitterness	**4.2**	**3.6**	**4.2**	**3.8**

Table 2: In case of sensory evaluation, the situation is a bit different. Whereas the fresh beers resulted in similar evaluation results, the preference of forced aged beer tends to Crosspure® filtered. Beer was stored at 32°C for 12 weeks.

CONCLUSION

The filter line optimization project at Chinese Tsingtao No. 3 Brewery was realized in Q4, 2016. In general the sensory and stability of finished beer could be improved. Moreover the filter line retrofitting resulted in lower production costs, hence overall target of filter line optimization was fully achieved. The change from single use filter aid to regenerative Crosspure® also underpins Tsingtao's sustainability efforts.

REFERENCES

- EBC / MEBAK analytics
- EBC Congress 2011 in Glasgow, poster of Thomas Kunz
- EBC Congress 2013 in Luxemberg, poster of Thomas Zimmermann

ACKNOWLEDGEMENT

BASF would like to thank Tsingtao Brewery Co. Ltd. and Technische Universität Berlin for their kind support and co-operation.

BASF
We create chemistry

P058 The 36th EBC Congress in Ljubljana, Slovenia 14-18 May, 2017 SUNTORY

Effect of harvest time on health-promoting activities of hop polyphenols

Takako Inui [a,*], Koharu Okumura [b], Hiroo Matsui [c], Takahiro Hosoya [b], and Shigenori Kumazawa [b]

[a] Suntory Beer Ltd., Beer Development Department, Osaka, Japan
[b] Department of Food and Nutritional Sciences, University of Shizuoka, Shizuoka, Japan
[c] Suntory Global Innovation Center Ltd., Research Institute, Kyoto, Japan
* Corresponding author: E-mail address: takako_inui@suntory.co.jp (T. Inui)

Objective

Our previous study [1] confirmed that harvest time of Saaz hop affects aroma characteristics. Similarly, we studied the effect of Saaz hop harvest time on health benefits and elucidated the relationship between health-promoting activities and hop polyphenol (HPP) profiles.

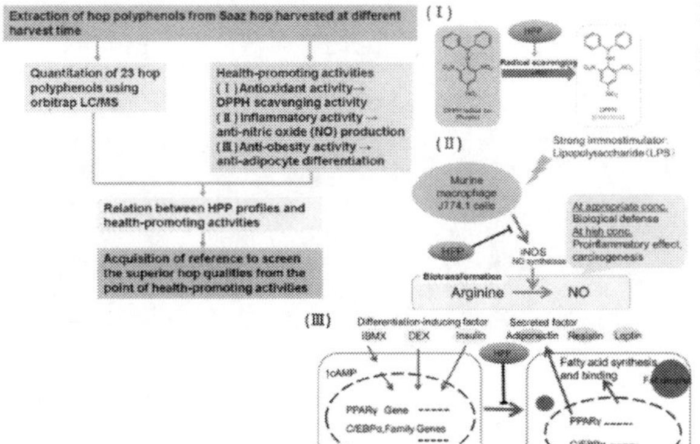

Murine pre-adipocyte 3T3-L1 cells Adipocyte

Materials and methods

Hop raw materials:

Cultivation trial was implemented between 2011 and 2014 at two farms north of Saaz region with similar soil types located close to each other. Plant age was between 10 and 20 years. All plants were uninfected. Harvest time is the number of days from flowering to harvest in each cultivation trial. Table 1 shows the pruning date and harvest time of each hop samples.

Table 1. Experimental design of cultivation trial

Harvest year	2011				2012				2013				2014			
Farm	Rybnany								Steknik							
Pruning date	April 15				April 16				April 20				April 15			
Harvest time (days)	22	29	50	57	23	32	51	60	26	37	53	61	29	40	57	64

Extractions of hop polyphenols:

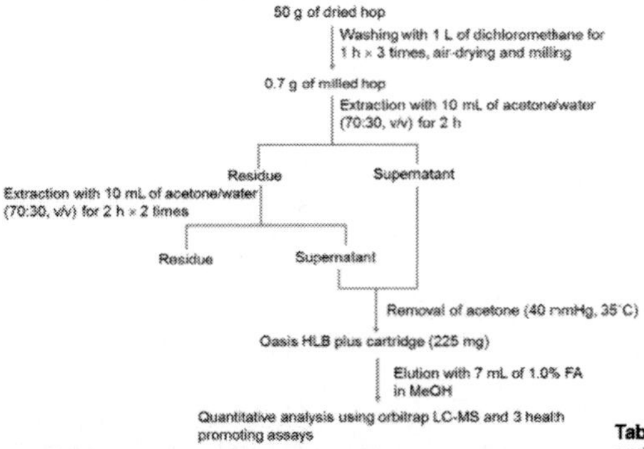

Quantitative analysis of hop polyphenols using orbitrap LC-MS:

Column: Phenomenex Luna C18(2) (5-μm, 4.6 × 150 mm)
Flow rate: 700 μL/min
Solvent: A) H_2O (0.1% FA), B) MeCN (0.1% FA),
B; 2%-98% (30 min)

Quantitated 23 polyphenols:

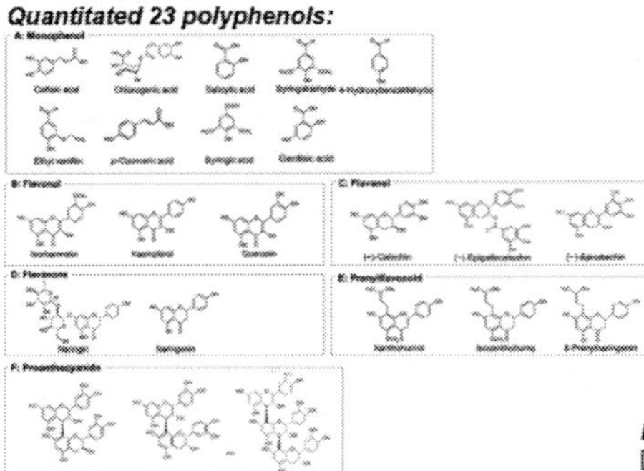

Results

Fig. 1. Health-promoting activities of polyphenols extracted from hops (HPP) harvested at different times over 4 years per unit weight of HPP.

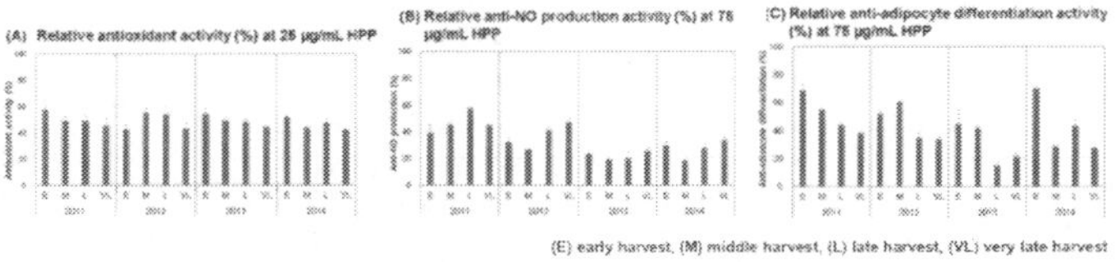

(E) early harvest, (M) middle harvest, (L) late harvest, (VL) very late harvest

Fig. 2. Eight and five polyphenols indicating significant increases and decreases per unit weight of HPP (μmol/kg) and total content of HPP (mg/kg).

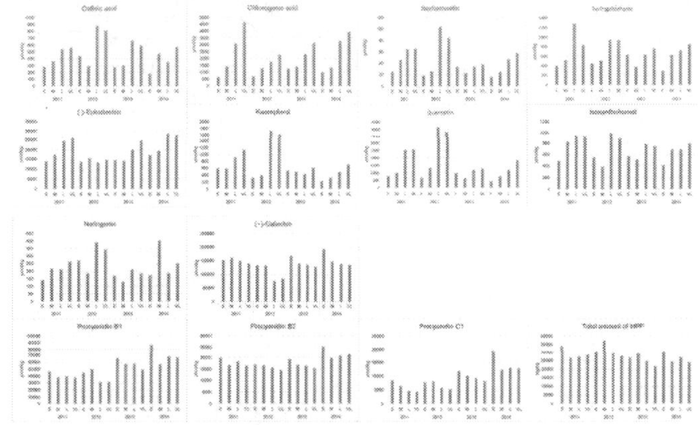

(E) early harvest, (M) middle harvest, (L) late harvest, (VL) very late harvest

Table 2. Half-maximal inhibitory concentration (IC_{50}) for inhibitory activities in three assays, mean concentration (μM) of each polyphenol (PP), and total polyphenol contents in active concentrations (25 and 75 μg/mL) of hop polyphenols (HPP) using standard reagents.

	IC_{50} (μM) for antioxidant activity	IC_{50} (μM) for anti-NO production	IC_{50} (μM) for anti-adipocyte differentiation	Mean concentration (μM) of each PP in active concentration of HPP	
				HPP = 25 μg/mL	HPP = 75 μg/mL
Caffeic acid	23.7	-	-	0.012	0.036
Chlorogenic acid	18.5	-	-	0.053	0.158
Syringic acid	33.3	-	-	0.766	2.299
Gentisic acid	11.6	-	-	0.007	0.020
Isorhamnetin	23.7	-	56.8	0.001	0.003
Kaempferol	16.0	38.4	NA*	0.017	0.052
Quercetin	13.1	41.3	90.2	0.035	0.105
(+)-Catechin	12.9	-	-	3.474	10.421
(−)-Epigallocatechin	10.7	-	NA*	0.012	0.035
(−)-Epicatechin	11.6	-	-	0.481	1.442
Xanthohumol	-	11.7	NA*	0.190	0.569
Isoxanthohumol	-	27.5	NA*	0.018	0.053
8-Prenylnaringenin	-	37.4	76.4	0.011	0.034
Procyanidin B1	12.0	-	-	1.301	3.904
Procyanidin B2	8.6	-	-	0.454	1.361
Procyanidin C1	18.4	-	-	0.232	0.696
Total content				7.185	21.554
					*: Non-active

Table 3. Multivariate correlation between content of 23 hop polyphenols (HPP) and total polyphenol (PP) content in each hop, antioxidant activity, anti-NO production, and anti-adipocyte activity.

Compounds	Antioxidant activity at 25 μg/mL HPP	Anti-NO production at 75 μg/mL HPP	Anti-adipocyte differentiation at 75 μg/mL HPP
Caffeic acid	-0.4143	0.3421	-0.7090
Chlorogenic acid	-0.5008	0.2845	-0.5541
Salicylic acid	-0.0448	0.1703	-0.1707
Syringaldehyde	-0.3056	0.5926	-0.5075
4-Hydroxybenzaldehyde	0.0586	0.7855	0.1438
Ethyl vanillin	-0.1627	-0.3637	-0.3703
p-Coumaric acid	-0.0890	0.1486	-0.1678
Syringic acid	-0.2009	0.5692	-0.3231
Gentisic acid	-0.1806	-0.4980	-0.3668
Isorhamnetine	-0.1310	0.8385	-0.3662
Kaempferol	-0.0608	0.8295	-0.2780
Quercetin	-0.1087	0.8238	-0.3327
(+)-Catechin	0.2262	-0.2213	0.4736
(−)-Epigallocatechin	0.0658	0.0205	0.3113
(−)-Epicatechin	-0.4852	0.1084	-0.4114
Naringin	0.1896	-0.5865	0.1966
Naringenin	-0.4043	0.1910	-0.4137
Xanthohumol	-0.4561	0.0504	-0.7763
Isoxanthohumol	-0.4071	0.5832	-0.6043
8-Prenylnaringenin	-0.4057	-0.1010	-0.6956
Procyanidin B1	0.0766	-0.6161	0.1572
Procyanidin B2	0.1562	-0.1851	0.4257
Procyanidin C1	0.0703	-0.5894	0.2101
Total polyphenol content	0.6492	0.1394	0.7807

Conclusion

☞Strong effects of low concentrations of HPP were observed in three assays.
☞ A significant increase in anti-adipocyte differentiation activity per unit weight of HPP was observed in early harvested samples.
☞ A significant difference in anti-NO production activity per unit weight of HPP was observed among different harvest years.
☞ HPP significantly increased in early harvested samples.
☞Present findings reveal that differences in health-promoting activities of hop are dependent on different polyphenols other than 23 polyphenols. Moreover, the synergistic effect of some polyphenols might also contribute to the observed differences in their health promoting activities.

References:

[1] Inui, T., Matsui, H., Hosoya, T., Kumazawa, S., Fukui, N. and Oka, K.: Effect of harvest time and pruning date on aroma characteristics of hop teas and related compounds of Saaz hops. J. Am. Soc. Brew. Chem., 74, 231–241 (2016).

36TH CONGRESS EBC SLOVENIA 2017 EUROPEAN BREWERY CONVENTION 14-18 MAY LJUBLJANA

Comparison of different dosage criteria when using aroma hops for late hopping

Florian Schüll and Adrian Forster, HVG Hopfenverwertungsgenossenschaft e.G., D-Wolnzach,
Andreas Gahr, Hopfenveredlung St. Johann GmbH, D-Train

HVG — Your German Hopportunity!

INTRODUCTION

Aroma hops are often used in late additions (end of boil/whirlpool) to create a distinctive hop aroma in beer. Without consideration of the brewing technology there are at least the following problematic issues regarding a comparable aroma intensity:

- Usage of different crop years of the same variety
- Substitution between established varieties
- Evaluation of breedings and comparison with the existing variety portfolio

It is still being discussed which criterion works best to guarantee a comparable and reproducible aroma intensity.

Possible dosing criteria are:

- Dosing of a defined quantity (g/hl or g/l) – *Common criterion*
- Dosing according to α-acids in hops (g/hl or mg/l) – *Common*
- Dosing according to total hop oil (ml/hl) – *Less common*
- Dosing according to Linalool content in hops (mg/hl or µg/l) – *Uncommon*

A comparison of the same variety but two different crop years shows that the dosage of hops acc. to quantity and alpha acids results in different aroma intensities. Only a dosage acc. to hop oil is a reasonable criterion [1, pp.279]. Also Linalool content is proposed as a suitable criterion in [2].

[1] Hops - Their Cultivation, Composition and Usage; Fachverlag Hans Carl, 09/2014,
[2] Mitter W., Jährliche Schwankungen im Hopfen – Korrekturmöglichkeiten im Sudhaus, BRAUWELT 14, 2008, Seite 372-373

THOUGHTS ON AROMA CAPACITY

1) Comparison of the linalool content in beer, using the two varieties *Perle* and *Saphir*. Dosage acc. to QUANTITY, ALPHA ACIDS and HOP OIL:

HOPS	Perle	Saphir
Alpha Acids [%-w/w]	7.4	4.1
Total Oil [ml/100g]	1.3	1.1
Linalool [mg/100g]	4	10

Values taken from [1, pp.147]

BEER	Quantity [100g/hl]		Alpha [5g/hl]		Hop Oil [1ml/hl]	
	Perle	Saphir	Perle	Saphir	Perle	Saphir
Dosage of Linalool [µg/l]	40	100	27	122	31	91
Linalool* in beer [µg/l]	12	30	8	37	9	27
Hop Flavour	NO	YES	NO	YES	NO	YES

* Linalool: Transfer Rate 30 %; Flavour Threshold 20 µg/l

In all cases, the objective to get a beer with distinctive hoppy flavour can be achieved with *Saphir*, but not with *Perle*. The Aroma Capacity of *Saphir* is higher than that of *Perle*.

2) Impact of the hop oil composition on the aroma capacity explained with the example MYRCENE and LINALOOL when 100g/hl are dosed

	Myrcene	Linalool
Quantity in hops [mg/100g]	1000	10
Flavor treshold in beer [µg/l]	>150	>10
Transfer rate in beer [%-rel]	<1	≈30
Quantity in beer [µg/l]	<100	30
Aroma Capacity	NO	YES

It is evident, that Myrcene beside of the high quantity in hops has no impact to a hop aroma in beer in comparison to Linalool. All other Mono- and Sesquiterpenes behave similar like Myrcene: High quantity in hops but very low solubility in beer and flavour thresholds of more than 100 µg/l.

About 80-90% of the total hop oil, the Hydrocarbon-fraction (= Mono- and Sesquiterpenes), has no contribution to the aroma in late hopped beer.

The much better soluble part of hop oil in beer is the oxygen fraction (10-20%).

This fraction consists of the following group of components:

- Carboxylic acid esters (flavour threshold varies from approx. 10 to >100 µg/l beer) like 2-Methylbutyl-2-Methylpropanoate
- Ketones like 2-Undecanon
- Monoterpene alcohols like Linalool, Geraniol, Nerol (flavour threshold 10 to 40 µg/l in beer)
- Sesquiterpene alcohols like Eudesmol, Farnesol
- Sesquiterpene oxides originating from β-Caryophyllene and Humulene

Considering the much better solubility in beer our hypothesis is, that the oxygen fraction contributes strongly to the aroma capacity of a hop. With a first set of brewing trials we investigated whether the oxygen fraction would be a suitable dosage criterion for a better prediction of the hop aroma intensity in beer.

MATERIAL & METHODS

The used hop varieties for these test series were two breeding lines of the Hop Research Center Huell (89/25 and 96/24), their genetic mothers (*Spalter* resp. *Tettnanger*) and additionally the aroma variety *Saphir*. Analytical details of the hop varieties and the brewing setup are shown in poster N°64 of this session. The experimental setup to evaluate different dosage criteria regarding their resulting aroma intensity in beer consisted of 3 brews per variety. The tested criteria were:

- Dosing of a defined quantity of pellets (500 g/hl)
- Dosing according to the total hop oil content of pellets (6 ml/hl)
- Dosing according to the oxygen fraction content of pellets (300 mg/hl)

HOP DOSAGES OF BREWING TRIALS

Calculated dosages of quantity, hop oil, oxygen fraction and linalool depending on the dosage criterion (hops dosed as Pellets T90):

Dosage Criteria	Pellet Quantity 500 g/hl			Hop Oil 6 ml/hl			Oxygen fraction 300 mg/hl		
Dosed Quantities	Oil	Oxygen Fraction	Linalool	Quantity	Oxygen Fraction	Linalool	Quantity	Oil	Linalool
89/25	8.5	645	85	353	455	60	188	3.4	38
96/24	8.8	475	60	343	326	41	204	3.0	24
Spalter	3.5	355	30	857	609	51	275	1.9	17
Tettnanger	3.0	270	25	1000	540	50	536	3.2	21
Saphir	4.8	610	35	632	771	44	254	2.8	20
Max : Min	2.9	2.3	3.4	2.9	2.4	1.5	2.9	2.1	2.2

RESULTS OF BREWING TRIALS

Since Linalool is considered to be an indicator for hoppy impressions in beer and esters contribute to the fruity aroma, the evaluation of the conducted brewing trials focused on Linalool and 2-Methylbutyl-2-Methylpropanoate. Sensory testing of the beers was limited to the attribute "intensity of hop aroma" (smell and taste, scale 0 to 10).

Linalool and 2-Methylbutyl-2-Methylpropanoate contents in beer:

Dosage Criteria	Values in beer [µg/l]	89/25	96/24	Spalter	Tettnanger	Saphir	Max : Min
Pellet Quantity 500 g/hl	Linalool	541	385	170	138	215	3.9
	2-methylbutyl-2-methylpropanoate	43	4	4	3	24	21.5
Oil 6 ml/hl	Linalool	322	228	222	201	207	1.6
	2-methylbutyl-2-methylpropanoate	168	18	28	7	135	24.0
Oxygen Fraction 300 mg/hl	Linalool	230	135	89	107	115	2.6
	2-methylbutyl-2-methylpropanoate	27	8	9	8	9	3.4

- The ratio Max. to Min. of Linalool was largest when dosing acc. to quantity, became lower with dosing acc. to oxygen fraction and was lowest for dosing acc. to hop oil.
- The ratio Max. to Min. of the ester was considerably higher than for Linalool and nearly the same for the dosage by quantity and oil. The dosage acc. to the oxygen fraction reduced this ratio from of over 20 to 3.4. This is a first indicator that this type of dosage might be useful.

Sensorial Intensity of Hop Aroma (0 to 10, 10 is highest)

Dosage Criteria	Line 89/25	Line 96/24	Spalter	Tettnanger	Saphir	Max -Min
Pellet Quantity 500 g/hl	7.7	6.0	5.8	4.5	6.2	3.2
Oil 6 ml/hl	8.1	6.2	5.7	5.5	5.8	2.6
Oxygen Fraction 300 mg/hl	7.4	6.5	6.3	6.0	6.7	1.4

- Since the aroma capacity of the used hop varieties varied in a wide range, the biggest difference between the intensity of hop aroma was observed when the same quantity of hops was dosed. This finding was expected.
- But the difference became lower with a dosage acc. to the criterion hop oil and was the lowest when hops were dosed acc. to the oxygen fraction.

SUMMARY

The dosage of aroma hops for late hopping in the brewhouse can be based on several criteria. It is known that a dosage acc. to alpha acids does not fit with the goal of a comparable intensity of hop aroma in beer. In a first series of experiments, 3 other dosage criteria (quantity, hop oil and oxygen fraction of hop oil) were compared by means of 5 hop varieties. Under consideration of the evaluation of two hop aroma components and the sensorial intensity of hop aroma in all beers, the dosage acc. to the oxygen fraction of hop oil proves to be a criterion which results in low analytical and also sensorial fluctuations. This result supports our hypothesis that the oxygen fraction is an appropriate criterion for a better prediction of hop aroma intensity in beer.

P60 Identification of geranic acid contributing to varietal aroma in SORACHI ACE and synergy with other hop-derived flavour compounds

Ljubljana, Slovenija
May, 2017

Atsushi Tanigawa[1], Ayako Sanekata[2], Kiyoshi Takoi[1], Koji Takazumi[2], Ichiro Matsumoto[1], Yasuyuki Nakayama[1]

[1]Product & Technology Innovation Department, SAPPORO BREWERIES LTD.
[2]Frontier Laboratories for Value Creation, SAPPORO HOLDINGS LTD.

Corresponding author: Atsushi.Tanigawa@sapporobeer.co.jp

SAPPORO

Abstract

'SORACHI ACE' hop was first grown in Japan by SAPPORO BREWEIES LTD. in 1984. Recently, this hop has been used as a 'flavour hop' to give beer a unique characteristic aroma , for example, woody, pine-like, citrus, dill-like and lemon grass-like. We investigated the specific flavour compounds derived from 'SORACHI ACE' hop by solid phase micro-extraction-gas chromatography-mass spectrometry (SPME-GC-MS) and gas chromatography- olfactometry (GC-O) analysis. We showed that one of the unique volatile compounds comprising SORACHI ACE was geranic acid, and the highest level of geranic acid was found in this hop. In addition, as a result of model sensory evaluations, we reported that geranic acid itself had little flavour, but that geranic acid could enhance the flavour intensities of linalool and geraniol in spite of its very low flavour intensity. In this study, we showed that the unique varietal aroma of SORACHI ACE beer arose from synergy between geranic acid and other hop-derived flavour compounds.

< Breeding process of SORACHI ACE >

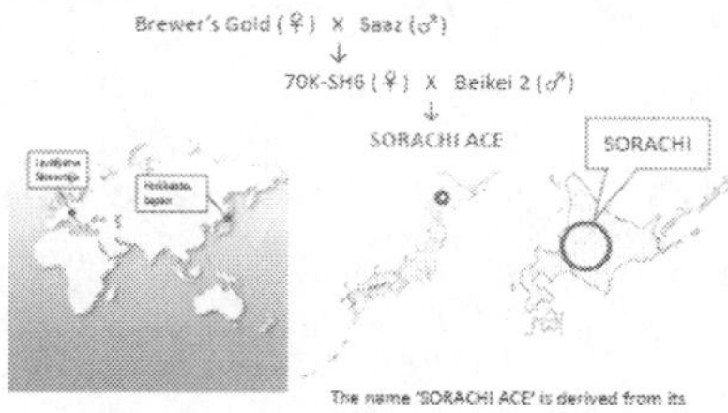

The name 'SORACHI ACE' is derived from its birth place: 'SORACHI' in Hokkaido, Japan.

< Unique aroma of SORACHI ACE beer >

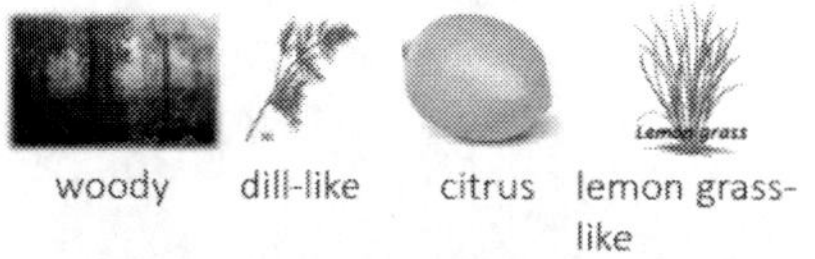

< Specificity of geranic acid >

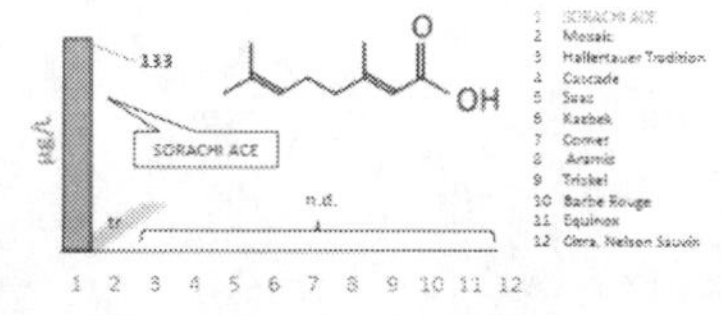

Fig.1 Concentration of geranic acid in test beers made with various hops
tr : trace, <25 μg/L

< Sensory evaluation 1. Triangular Test >

Two different sets of flavour model solutions (Model 1 and Model 2) were prepared by spiking each key compound at a three times higher concentration than in 'SORACHI ACE' beer.

In Model 1, the spiked geranic acid could not be recognized at a significant level.
However, the beer containing geranic acid could be detected in Model 2 under coexistence with linalool and geraniol.
It is thought that geranic acid could enhance the flavor intensities of linalool and geraniol.

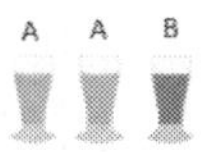

Table 1 Triangular test involving 9 panelists

	Sample A	Sample B	Correct answers /total answers
Model 1	Pilsner beer	Pilsner beer + Geranic acid	2/9
Model 2	Pilsner beer + Linalool & Geraniol	Pilsner beer + Linalool & Geraniol + Geranic acid	7/9 ($p < 0.05$)

< Sensory evaluation 2. Addition Test >

We brewed the test beer with SORACHI ACE, analyzed and selected 16 key hop compounds. Next, we carried out the sensory evaluation by addition tests to evaluate the synergy between geranic acid and other hop-derived flavour compounds.

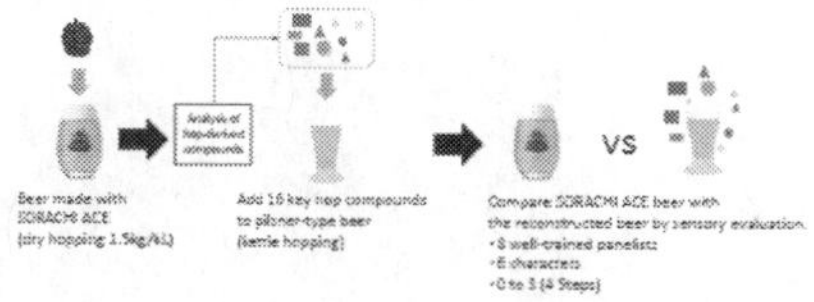

The beer with only fatty acids (including geranic acid) could not simulate flavour profiles of SORACHI ACE beer (Fig.2 A). On the other hand, the beer with 16 key compounds could reproduce flavour profiles of SORACHI ACE beer, especially the flowery, fruity and lemon-like flavour (Fig.2 B).

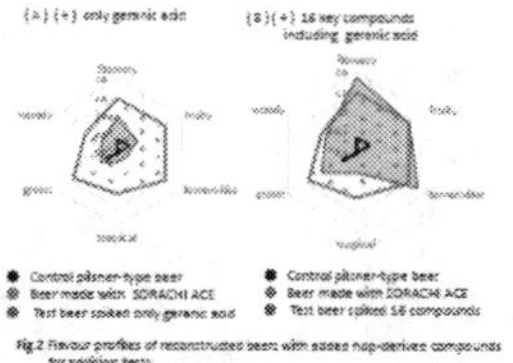

Fig.2 Flavour profiles of reconstructed beer with added hop-derived compounds for addition tests

Table 2 Concentrations of 16 key flavour compounds in SORACHI ACE beer

		(μg/L)
Fatty acids	Geranic acid	138
	Nonanoic acid	5
Alcohols	Geraniol	56
	Nerol	50
	Linalool	97
	α-Terpineol	20
	β-Citronellol	15
Aldehydes	Geranial	1.7
	Neral	0.6
Esters	Methyl geranate	1.2
	Methyl neralate	0.01
Hydrocarbons	Limonene	0.2
	Terpinolene	0.08
	α-Terpinene	0.3
	γ-Terpinene	0.1
	Myrcene	10.1

✓ SORACI ACE contains a higher level of geranic acid than other hop varieties.
✓ Geranic acid itself had little flavour, but geranic acid could enhance the flavour intensities of linalool and geraniol.
✓ Test beer spiked 16 key flavour compounds could simulate the flavour profiles of the beer made with SORACHI ACE, especially flowery, fruity and lemon-like characters.

What kind of flavour character is offered by geranic acid ?

< Sensory evaluation 3. Omission Test >

For omission test, six different model samples were prepared.
Sample A comprised all 16 key compounds as a control simulating SORACHI ACE beer.
Sample B to F comprised all key compounds except one group of compounds as tests.

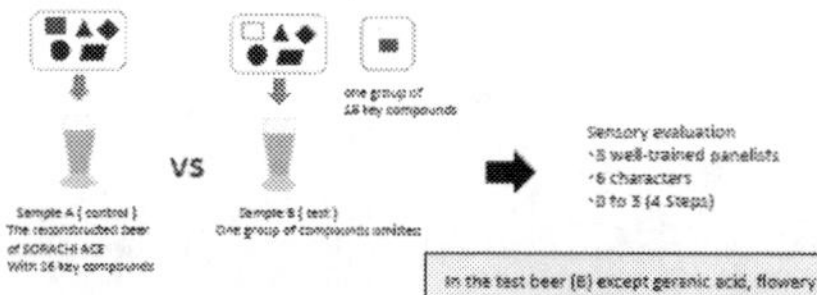

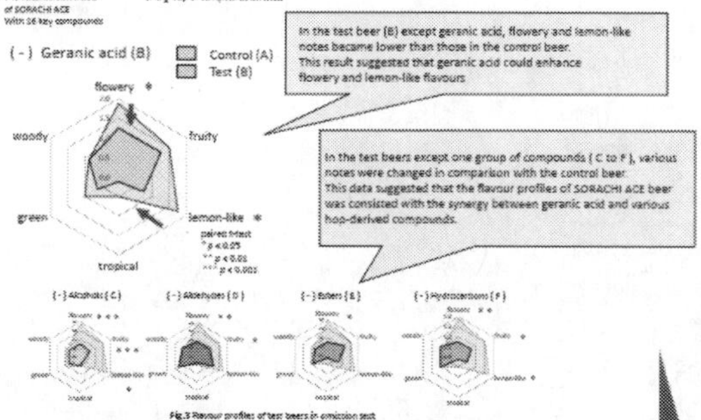

Fig.3 Flavour profiles of test beers in omission test

Table 3 Average scores of test beers in omission test (Concentrations of 5 groups were showed in Table 2)

		flowery	fruity	lemon-like	tropical	green	woody
A	Control (all compounds)	1.9	1.5	1.8	0.5	1.0	0.9
B	(-) Fatty acids (including geranic acid)	1.3 *	1.3	0.8 *	0.1	0.6	0.9
C	(-) Alcohols	0.8 ***	0.6 ***	0.3 *	0.3	0.3	0.3
D	(-) Aldehydes	2.0 **	1.5 *	1.1	0.4	1.3	1.0
E	(-) Esters	0.3 *	1.1	0.8	0.1	1.5	0.3
F	(-) Hydrocarbons	0.8 **	0.3 *	0.3 *	0.1	1.0	1.0

How to form the specific flavour characters of SORACHI ACE beer?

< Comparison of hopping conditions >

We brewed the test beers with different hop-addition timings (1.5 kg/ kL SORACHI ACE) . Of 3 test beers, dry-hopped beer showed the strongest lemon-like character in sensory evaluation.
However, geranic acid remained even in kettle-hopped beer in amounts as large as in late-hopped and dry-hopped beers.

We suggested that the reason why dry-hopped beer showed more lemon-like notes rather than late-hopped beer was not the difference of the amount of geranic acid but the differences of the amount of other hop-derived flavour compounds (Fig. 6).

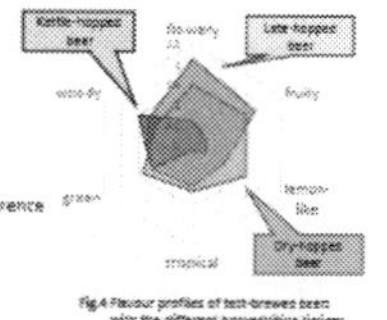

Fig.4 Flavour profiles of test-brewed beers with the different hop-addition timings

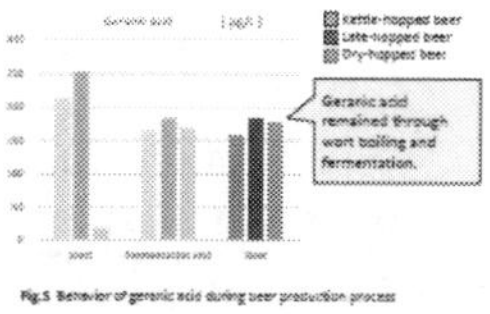

Fig.5 Behavior of geranic acid during beer production process

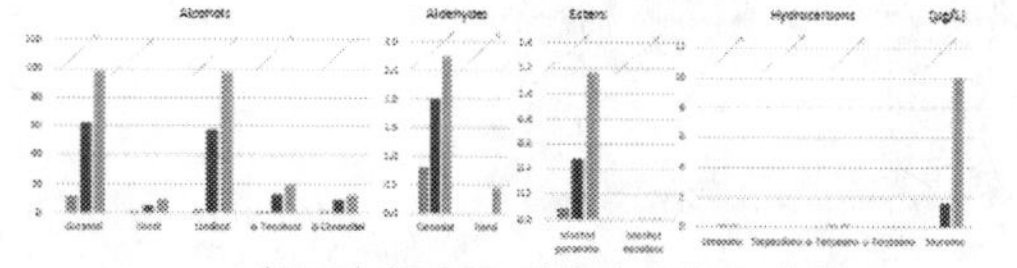

Fig.6 Concentration of hop-derived compounds in finished beers with different hop-addition timings

Can SORACHI ACE change flavour profiles on blending with other flavour hops ?

< Synergy between SORACHI ACE and other hop varieties >

We brewed the test beers with blended SORACHI ACE (late hopping 1.0kg/kL) and other flavour hops (dry hopping 0.5kg/kL).

In Cascade, Kazbek and Hallertau Blanc, the blend-hopped beers showed less characteristic flavour profiles than single-hopped SORACHI ACE beer.
However, hop-blended beers made with Citra or Mosaic showed more characteristic flavour profiles than single-hopped beer with SORACHI ACE.

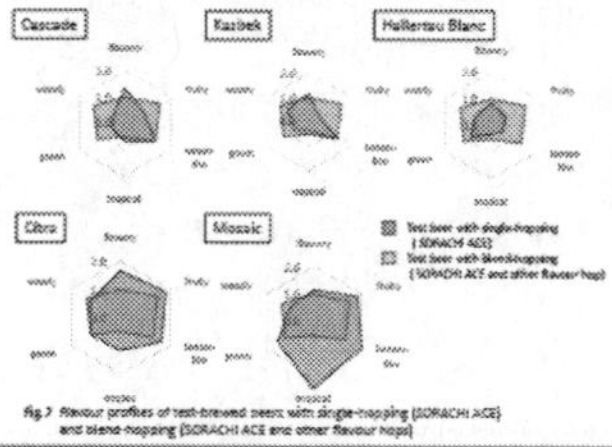

Fig.7 Flavour profiles of test-brewed beers with single-hopping (SORACHI ACE) and blend-hopping (SORACHI ACE and other flavour hops)

We brewed test beers made with only Mosaic(dry hopping 0.5kg/kL) and with blending SORACHI ACE (late hoppng 1.0kg/kL) and Mosaic (dry hopping 0.5kg/kL) .

In the test-brewed beer made with SORACHI ACE and Mosaic, lemon-like and green notes became higher than those in the beer made with only Mosaic. The flavor profiles of the beer blended SORACHI ACE and Mosaic was changed from that of the beer made with only SORACHI ACE(Fig.7) or with only Mosaic(Fig.8).

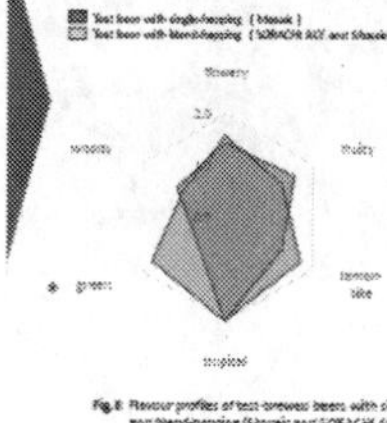

Fig.8 Flavour profiles of test-brewed beers with single-hopping (Mosaic) and blend-hopping (Mosaic and SORACHI ACE)

Table 4 The amounts of hop-derived compounds including some thiols in two test beers

	Late hopping (1.0kg/kL)	-	SORACHI ACE
	Dry hopping (0.5kg/kL)	Mosaic	Mosaic
Fatty acids (μg/L)	Geranic acid	[illegible]	31
	[illegible]	[illegible]	[illegible]
Alcohols (μg/L)	Geraniol	25.7	38.6
	Nerol	5.8	7.0
	Linalool	36.0	74.5
	α-Terpineol	2.8	8.3
	β-Citronellol	10.5	36.3
Aldehydes (μg/L)	Geranial	0.7	0.8
	Neral	0.8	0.8
Esters (μg/L)	Methyl geranate	20.3	27.6
	Methyl neralate	3.0	3.0
Hydrocarbons (μg/L)	Myrcene	1.36	4.01
	Limonene	0.03	0.77
	Terpinolene	0.06	0.07
	α-Terpinene	0.06	0.30
	γ-Terpinene	0.01	0.05
Thiols (ng/L)	[illegible]	13.3	16.5
	[illegible]	30	36
	[illegible]	238	272

Conclusion

✓ Geranic acid could enhance the flavour intensities of other hop-derived flavour compounds and affect especially flowery and lemon-like notes.
✓ The unique aroma of SORACHI ACE beer was consisted with the synergy between geranic acid and other hop-derived compounds.
✓ Dry-hopped SORACHI ACE beer showed the strongest lemon-like flavour, depending on not the difference of the amount of geranic acid but the differences of the amount of other hop-derived flavour compounds.
✓ The beer made by blending SORACHI ACE and Mosaic had higher lemon-like and green notes than the beer made with only Mosaic.

Recent hop varieties from Slovenian breeding program

Čerenak A., Radišek S., Košir I.J., Kolenc Z., Oset Luskar M.

Slovenian Institute of Hop Research and Brewing

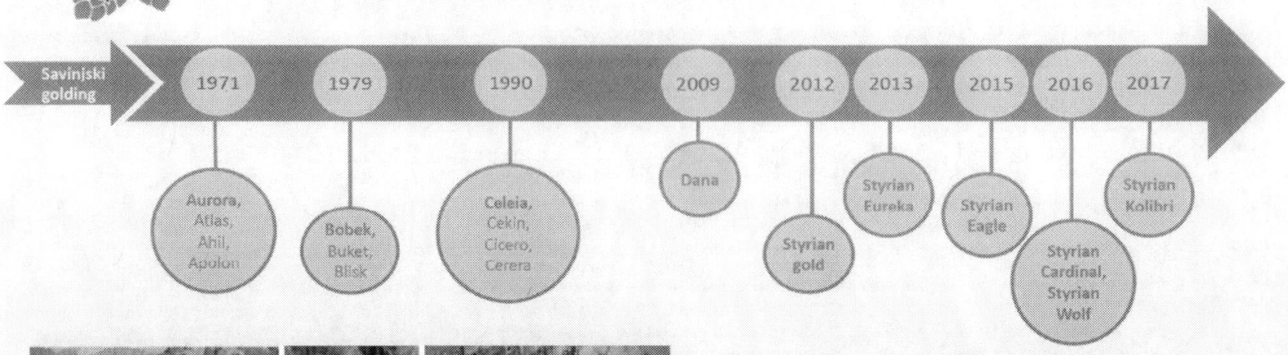

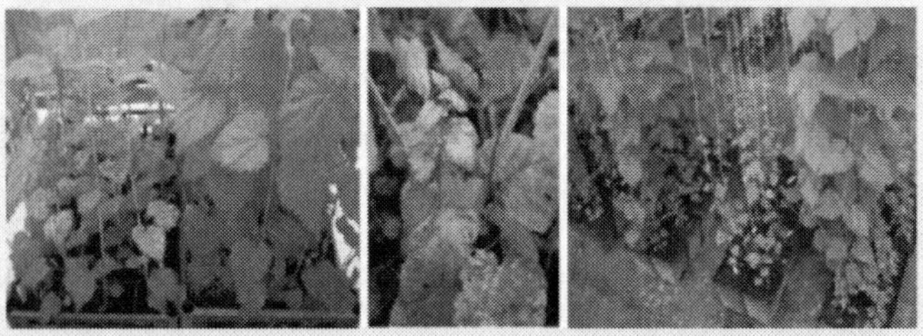

Selection of seedlings to powdery mildew, downey mildew and Verticillium wilt in hop disease resistance breeding.

Selection methods are concentrated on improved quantity and quality of the yield and resistance against diseases and pests.

The Slovenian hop industry relies on own varieties, mainly **Celeia, Aurora, Savinjski golding** and **Bobek**, which cover together with new varieties 99 % of crop production.

Slovenian varieties are growing on 99 % of Slovenian hop fields

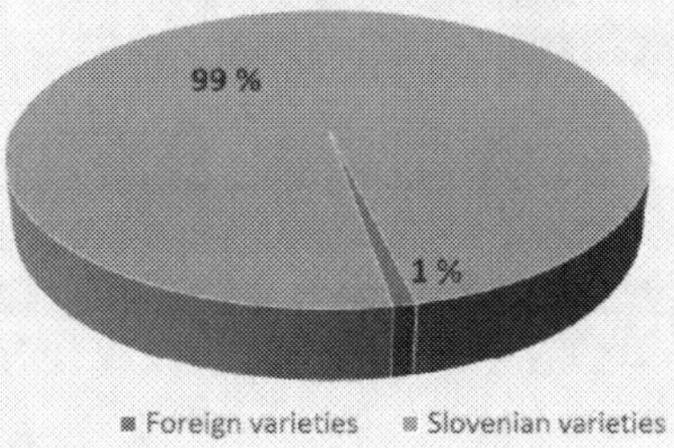

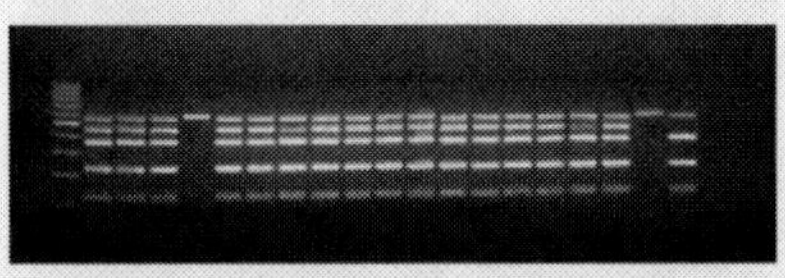

Sex determination with multiplex molecular markers is included in breeding program.

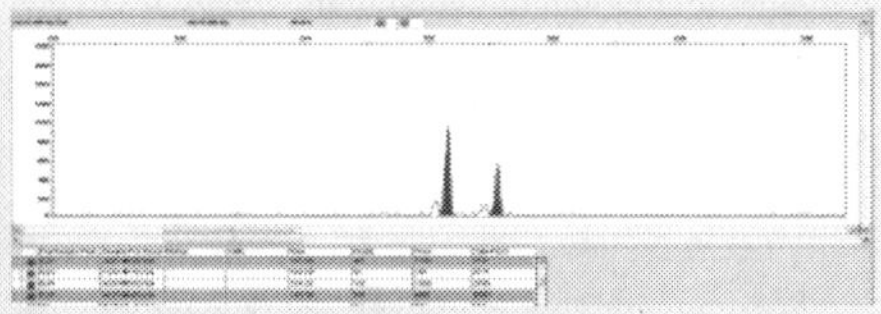

Hop variety fingerprinting with a fluorescent labelled microsatellite markers.

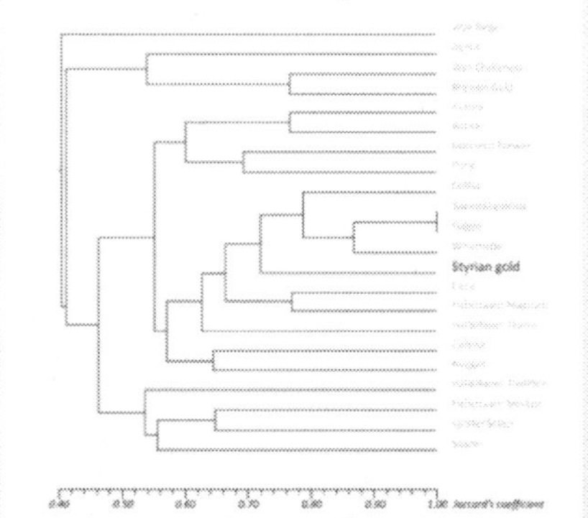

Genetic relationship based on analysis of molecular marker.

Hop cones of flavour variety Styrian Wolf.

Without neglecting the classical breeding objectives, new hop varieties with fruity, citrusy and floral aroma compositions have been bred for a strong, differentiating hop-derived aroma and flavor notes. In various brewing trials using both traditional and dry hopping techniques, various top and bottom fermented beers revealed unique aroma and flavor impressions imparted by new Styrian varieties.

Experimental brewery and laboratory for agrochemistry at SIHRB used for brewing evaluation of hop varieties.

In recent years, 6 hop varieties (**Styrian Gold, Styrian Eureka, Styrian Eagle, Styrian Cardinal, Styrian Wolf** and **Styrian Kolibri**) have been released.

Speeding up hop breeding through in vitro seed germination and precocious sex determination

UNIVERSITÀ DI PARMA

C. Liberatore, G. Mattion, M. Rodolfi, T. Ganino. B. Chiancone

Department of Food and Drug, University of Parma

Introduction

To keep innovative the craft beer sector, it is necessary to deepen knowledge on existing hop biodiversity, but also to renew the varietal panorama. Traditional hop breeding techniques are time consuming, due the seasonality of seed germination and the impossibility of precocious sex plant identification. Biotechnologies are a valid tool to overcome these problems.

In this research, in order to accelerate the hop breeding process, seed germination was stimulated using different Gibberellic acid (GA_3) concentrations, to imbibe seeds before the *in vitro* culture and added to the culture medium; moreover, markers linked with plant sex were used to precociously identify male and female seedlings.

In vitro hop seed germination

Seeds were isolated from Columbus cones, collected in Manaro sul Panaro (Italy).

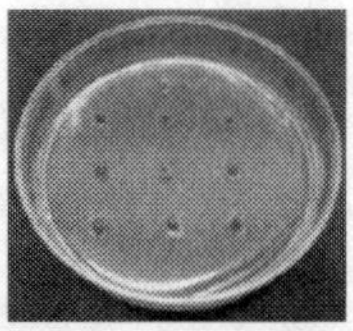

Sterilized seeds were placed in 60 mm Petri dishes (100 seeds/thesis).

<u>**HF culture medium composition**</u>

Half concentration of Murashige and Skoog (MS) medium (Murashige and Skoog, 1962), MS vitamins, 30 g/l of sucrose, 8 g/l of agar; pH 5.8.

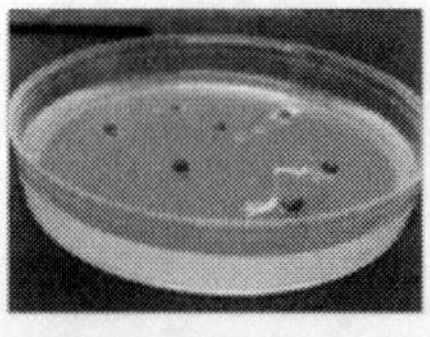

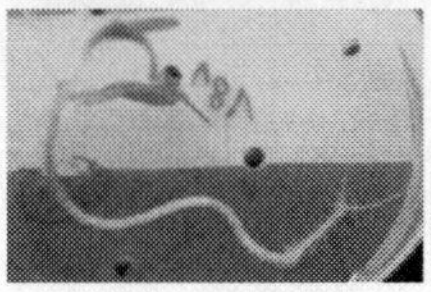

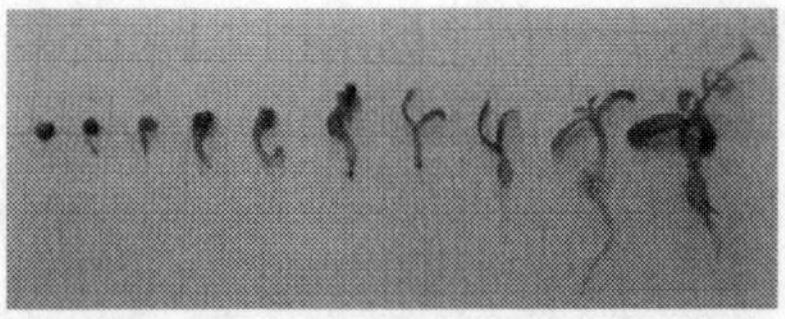

After one day in culture, seeds in culture started to germinate and carried on germinating for the following 60 days. Seedlings followed the normal development, showing the formation of the hypocotyl and the cotyledons. Well developed seedlings were moved to jars, containing 100 ml of HF medium. In the jars, seedlings continued their development forming epicotyl and the first photosynthetically active leaflets.

Influence of GA_3 on *in vitro* hop seed germination

In order to accelerate *in vitro* hop seed germination, the influence of GA_3, added in the imbibition solution, in the colture medium or in both, was evaluated.

Thesis	GA_3 concentration imbibition solution	culture medium
C/HF	-	0.0mg/l
C/0.5mg	-	0.5mg/l
C/1mg	-	1.0mg/l
IC/HF	0.0mM	0.0mg/l
IC/0.5mg	0.0mM	0.5mg/l
IC/1mg	0.0mM	1.0mg/l
I0.5mM/HF	0.5mM	0.0mg/l
I0.5mM/0.5mg	0.5mM	0.5mg/l
I0.5mM/1mg	0.5mM	1.0mg/l
I1mM/HF	1.0mM	0.0mg/l
I1mM/0.5mg	1.0mM	0.5mg/l
I1mM/1mg	1.0mM	1.0mg/l

C: no imbibition; **IC**: 24h imbibition in sterile distilled water; **I0.5mM**: 24h imbibition in 0.5mM GA_3 solution; **I1.0mM**: 24h imbibition in 1.0mM GA_3 solution; **0.0mg/l**: no GA_3 added in the culture medium; **0.5mg/l**: 0.5mg/l GA_3 added in the culture medium; **1.0mg/l**: 1.0mg/l GA_3 added in the culture medium.

Influence of culture medium composition on *in vitro* hop seed germination, after 60 dd of culture.

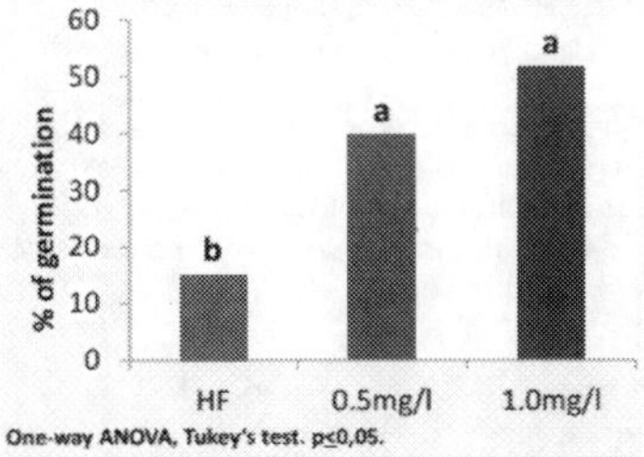

One-way ANOVA, Tukey's test. p≤0,05.

The presence of GA_3 in the culture medium induced a statistically significant increasing of the *in vitro* germination percentage.

Influence of GA_3 on *in vitro* germination percentage (FGP) and on mean germination time (MTG) of hop seeds, after 60 dd of culture.

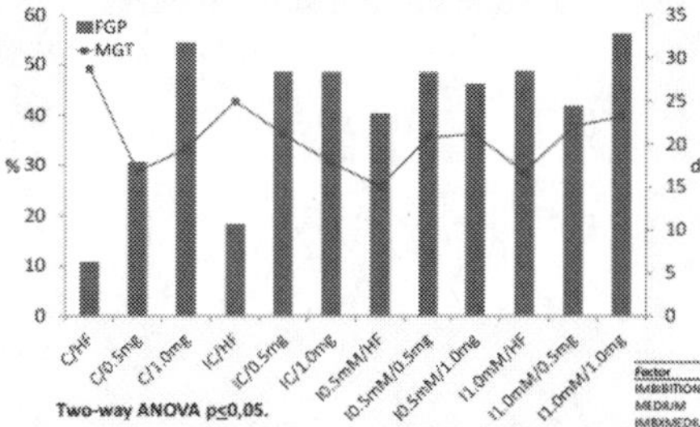

Two-way ANOVA p≤0,05.

Factor	p-Value
IMBIBITION (IMB)	0.344
MEDIUM	0.000
IMBXMEDIUM	0.339

Two-way ANOVA shows that "Medium" is the only significant factor.

MGT is high in seeds non imbibed and cultured on HF medium, whilst the presence of GA_3, alternatively in the imbibition solution or in the culture medium, reduces, on the average, the time needed for germination.

Germination trends of hop seed during 60 days of culture

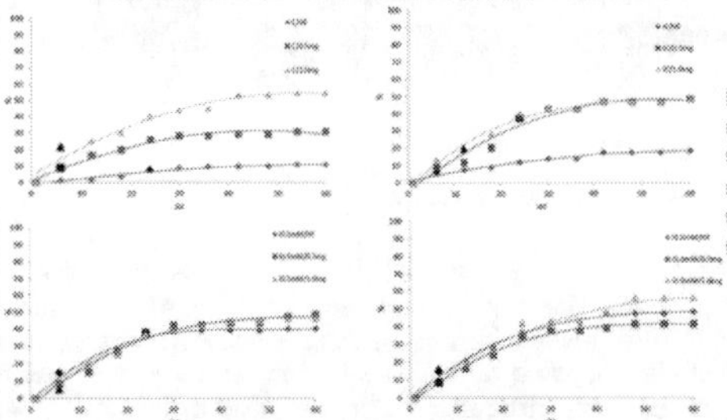

Peak value (black dots in the graphics) stands within the first 8 days for most thesis.

Seeds not imbibed (a) or imbibed for 24h in water (b) showed different trends depending on the medium in which they were cultured; different behavior was observed for seeds imbibed in GA_3 solutions (0.5 and 1.0mM): seeds followed very similar trend, independently on the medium used.

Sex determination

To precociously identify male and female seedlings, DNA was extracted from acclimatized seedlings and FORO marker was used (Rodolfi et al., 2015).

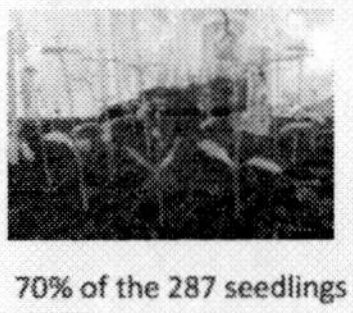

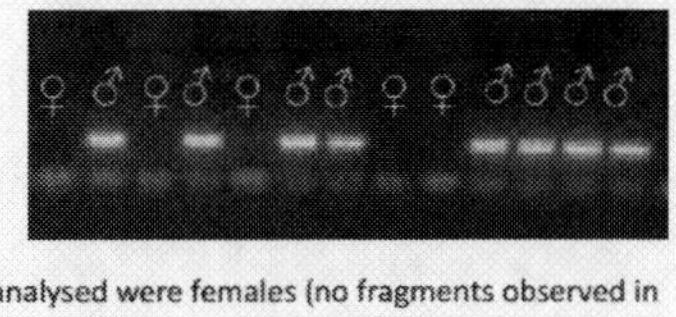

70% of the 287 seedlings analysed were females (no fragments observed in the agarose gel). Female plants will be moved to open field conditions to test their vegeto-productive responses.

Conclusion

This is the first study, in our knowledge, investigating the influence of Gibberrellic acid on *in vitro* hop seed germination.

The best treatment (**I1mM/1.0mg**) induced the **56% of seeds to germinate**, percentage greatly higher than the one reported by Suciu et al. (1977) on *in vivo* cultured seeds.

Of the 644 hop seeds germinated *in vitro*, 287 formed healthy and well developed plants, **70% of which were females**.

Further studies are needed to characterize the vegeto-productive performances of the obtained female plants in open field conditions.

References

Murashige, T., Skoog, F. (1962). A revised medium for rapid growth and bioassays with tobacco tissue cultures. Physiologia Plantarum, 15, 473-479.

Rodolfi, M., Mongelli, A., Marieschi, M., Ganino, T., Bruni, R., & Torelli, A. (2015) Optimization of PCR marker for early sex identification in Hop. In EBC-European Brewering Convention, Porto, 24-28 May 2015.

Suciu, T., Salontai, A., Muntean, L., Felecan, V., & Vaida, L. (1977). Notulae Botanicae Hortu Agrobotanici, 9, 79-84.

TotallyNaturalSolutions
aromas & flavours to inspire

Dry hopping – a new look at techniques, utilisation and economics

56TH CONGRESS EBC 2017 EUROPEAN BREWERY CONVENTION 14-18 MAY LJUBLJANA SLOVENIA

R.Marriott ray.marriott@totallynaturalsolutions.com and C.Wilson colin.wilson@totallynaturalsolutions.com

Totally Natural Solutions Ltd, East Peckham, Kent, TN12 5LG, UK

Introduction

Dry hopping of beers has seen a rapid revival as the craft brewing sector has grown and the consumer looks for beers with more diverse flavour and aroma. Traditionally dry hopping with cones, plugs or pellets has provided an adequate method of flavour introduction but higher raw material prices and beer losses through absorption onto hops has led to a re-evaluation of the options that are available to deliver dry hop character. There have been very few studies on the use of alternative techniques and products to improve aroma utilisation partly due to the difficulty in obtaining robust analytical data for aroma molecules present at ppb levels. This work reviews the available options for dry hopping and present the results of trials using a full range of products from hops to hop oil products from a single hop variety showing the relative utilisation as measured by transfer of key aroma molecules into beer, beer losses through the dry hopping process and the relative economics.

Traditional dry hopping

Traditional dry hopping is still the most prevalent method and carried out with either hop cones, pellets or plugs.

Transfer of hop aroma in traditional dry hopping has been studied in beer and model systems and the efficiency depends very much on factors such as product used (hops, pellets or plugs), contact time, hopping method (static or dynamic) and product ABV. High transfer efficiency can be achieved [1] in optimised conditions and the results of dry hopping an IPA at 5%ABV using a recirculating filter and T90 Citra pellets are shown below.

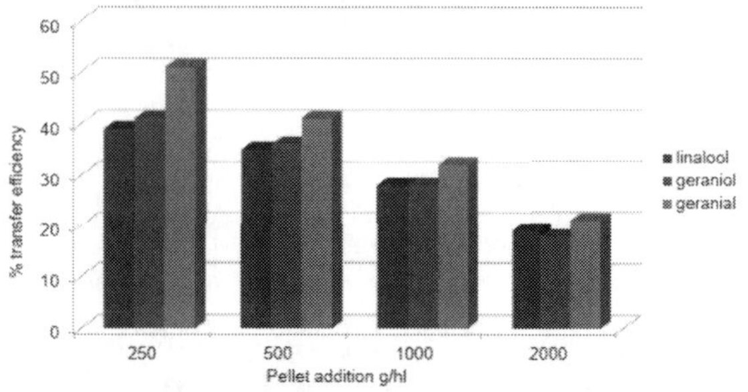

Figure 1. Transfer efficiency of monoterpene alcohols and aldehydes from "Citra" hops resulting from dry hopping at 250 - 2000g/hl T90 pellets (determined by headspace – SPME)

In craft beers hopping rates up to 2000g/hl are used by some brewers and as the hopping rate increases the transfer efficiency decreases. Concurrently beer losses increase as more liquid is absorbed by the hops and at high hopping rates this loss can be a significant economic factor. This can be mitigated to some extent by the use of a "concentrated" hop pellet produced by separation of the lupulin glands at low temperature.

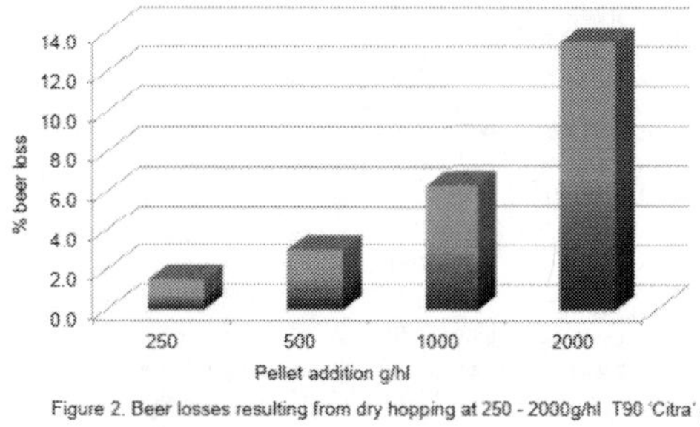

Figure 2. Beer losses resulting from dry hopping at 250 - 2000g/hl T90 'Citra' pellets

Dry hopping with extracts and oils

Hop oil products have been available since the late 1960s, but whole hop oil obtained by steam distillation of hops did not impart a satisfactory hop flavour to beer largely as a result of the labile nature of some of the essential oil components. In the 1970s the development of vacuum steam distilled oil, as a stable emulsion in water, was developed as a usable substitute for dry hops in U.K. cask-conditioned ales[2]. Selective extraction of hop oil from single hop varieties with liquid CO_2 was demonstrated in 1981[3] which was used to replace dry hopping and also found to be stable during pasteurisation.

Despite the early optimism that hop oils or extracts could replace dry hopping this was not widely adopted as these early extracts and oil preparations were not a direct substitute in terms of flavour and aroma delivery in beer.

Recent developments to produce oils extracted with cold CO_2 from single hop varieties has improved the flavour profile and these can be used at 100% transfer efficiency without beer loss. An example of this is shown below where 2g/hl oil replaces 250g/hl dry hopping with T90 pellets.

	linalool	α-terpineol	nerol	geraniol	geranial
T90 pellets	236.3	5.8	4.4	69.0	10.8
250g/hl	39%	43%	44%	41%	51%
Hop oil	303.0	6.8	5.0	84.2	10.6
2g/hl	100%	100%	100%	100%	100%

Table 1. Concentration (ppb) of monoterpene alcohols and aldehydes from the addition of "Citra" hop oil or T90 pellets

Analytically this work demonstrates that the level of key aroma molecules can be delivered by the use of suitably extracted hop oils. Although they deliver an authentic dry hop character some development work is always needed to determine the required dose rates and the point of addition and maturation time.

Conclusions

Traditional dry hopping can provide efficient transfer of hop aroma molecules to beer but this decreases markedly as dose rates increase. Beer losses also increase sharply at these elevated hopping rates. The use of extracts and oils can provide distinctive dry hop character and provided that they deliver the desired aroma profile can be more economic as beer losses are avoided and easier to use.

References

1. Bailey, B., Schönberger, C., Drexler, G., Gahr, A., Newman, R., Pöschl, M., & Geiger, E. (2009). The influence of hop harvest date on hop aroma in dry-hopped beers. *Technical Quarterly of the Master Brewers Association of the Americas*, *46*(2), 1-7

2. Laws, D. R. J., Peppard, T. L., Sharpe, F. R., and Pickett, J. A. (1978) Recent developments in imparting hop character to beer. *J. Am. Soc. Brew. Chem.* 36, 69-72

3. Scott, R. W., Theaker, P. D., Marsh, A. S., Grimmett, C. M., Laws, D. R. J., and Sharpe, F. R. (1981) Use of extracts prepared with liquid carbon dioxide as a substitute for dry hops. *J. Inst. Brew.* 87, 252-256

PO63

Description and Classification of Two "New" Huell Aroma Breeding Varieties

Andreas Gahr*, Hopfenveredlung St. Johann GmbH, Train, D
Adrian Forster and Florian Schüll, HVG Hopfenverwertungsgenossenschaft e.G., Wolnzach, D

INTRODUCTION

At present new hop varieties with unique aroma and flavour are popping up like flowers at springtime. But many of those are emerging from a rush job in order not to miss the "Flavour Hop Train". However, it must not be forgotten that there might be a need for classical aroma varieties as well in the future, no matter if quality, efficiency (aroma yield), yield (kg/ha) or disease and climate tolerance will be the driving factors.
Although a lot of new flavour varieties were released by the Hop Research Center in Huell, Bavaria, the last few years (*Polaris, Mandarina Bavaria, Huell Melon, Hallertau Blanc, Ariana* and *Callista*) there are still promising "normal" aroma breeding lines in the experimental gardens that have a number only. On two of those we want to draw your attention to: **89/002/025** and **96/001/024**.

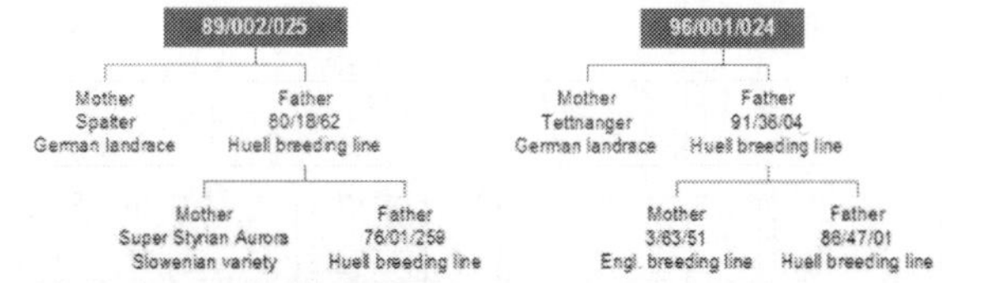

Figure 1: Pedigree of 89/002/025 and 96/001/024

HOP ANALYSES

		89/25	96/24	SSP	TTE	HSR
α-acids	% w/w	5.3	4.1	3.7	3.3	2.8
β : α		0.89	1.00	1.86	1.55	2.11
Cohumulone ratio	% rel.	22	23	26	25	14
Hop oil	ml/100g	1.70	1.75	0.70	0.60	0.95

Table 1 : Basic Hop Analyses of the two experimental varieties compared to the two mother varieties *Tettnanger* (TTE) and *Spalter* (SSP) as well as to *Saphir* (HSR), the aroma variety with the highest aroma capacity yet.

		89/25	96/24	SSP	TTE	HSR
Xathohumol	% w/w	0.42	0.44	0.46	0.32	0.38
Sum of low molecular-weight polyphenols (LM)	mg/100g	1181	1007	1777	1756	1837
Total polyphenols (TP)	% w/w	5.7	5.7	6.4	6.5	6.7
LM : TP	%	21%	18%	28%	27%	27%

Table 2 : Hop Polyphenol Analyses; LM:TP ratio = indicator for low molecular character of the polyphenols

		89/25	96/24	SSP	TTE	HSR
Sum of Monoterpenes (MT)	mg/100g	733	659	340	248	342
Sum of Sesquiterpenes (ST)	mg/100g	353	478	195	191	204
Hydrocarbon fraction (HCF)	mg/100g	1086	1137	535	438	546
Farnesen (%) of ST	mg/100g	26	3	29	29	7
Sum of 6 Esters	mg/100g	54	18	12	5	37
Linalool	mg/100g	17	12	6	5	7
Geraniol	mg/100g	2	2	6	4	2
Sum of 4 Epoxides	mg/100g	10	9	19	14	21
Oxygen fraction (OF)	mg/100g	129	95	71	54	132

Table 3 : Hop Aroma Substance Analyses

Major differences: 89/25 and 96/24 lower in total and low molecular polyphenols, but high aroma capacity (total oil, oxygen fraction, linalool).

BREWING TRIALS

- 2-hl research brewery in St. Johann, Germany
- 100% malt, infusion, 12 % original extract
- Bitter hop addition begin of boil: *Herkules* pellets (to 20 IBU)
- Late hop addition according to oil content (6 ml/hl)
- Bottom fermentation with yeast strain W34/70
- Fermentation 8°C, maturation 14°C, cold conditioning 0-1°C
- Kieselguhr filtration, low-oxygen bottling

Dosing 6 ml of oil/hl results in:		89/25	96/24	SSP	TTE	HSR	Max : Min
Amount	g/hl	353	343	857	1000	632	2.9
α-acids	mg/l	187	141	317	330	177	2.3
Oxygen fraction	mg/hl	455	328	609	540	771	2.4
Linalool	mg/hl	60	41	51	50	44	1.5

Table 4 : Dosed amounts of other criteria and ratio max/min

BEER ANALYSES

Original extract, alcohol, app. extract, degree of fermentation and pH show good reproducibility with variation coefficients of 0.8 to 1.9 % and are therefore not shown. Iso-α-acids vary moderately from 13 to 15 mg/l and the bitterness from 18 to 22 IBU.

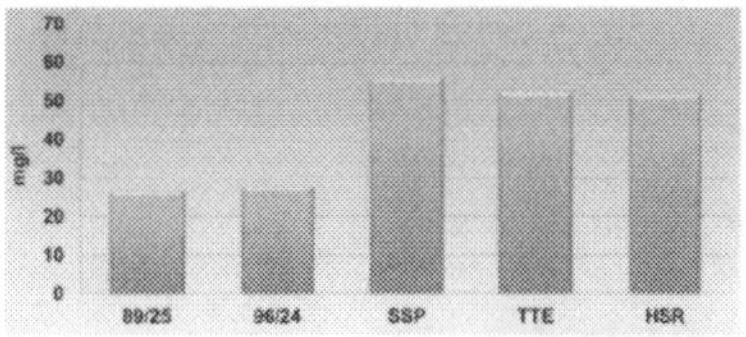

- 89/25 and 96/24 significantly lower in hop derived low molecular (LM) polyphenols mostly as a result of lower dose of hop amount (s. Tab. 2 and 4)

Figure 3 : Beer analyses (hop derived low molecular polyphenols)

- 89/25 significant higher linalool content in beer; beer values follow dose rates.

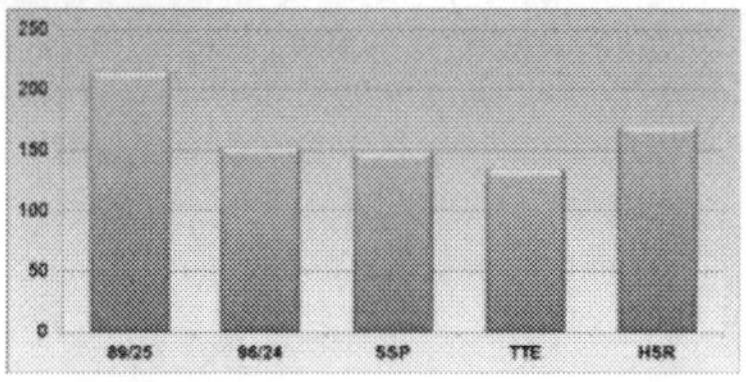

Figure 4: Beer analyses (Total Linalool); Ratio of S-Linalool = approx. 10% = const.

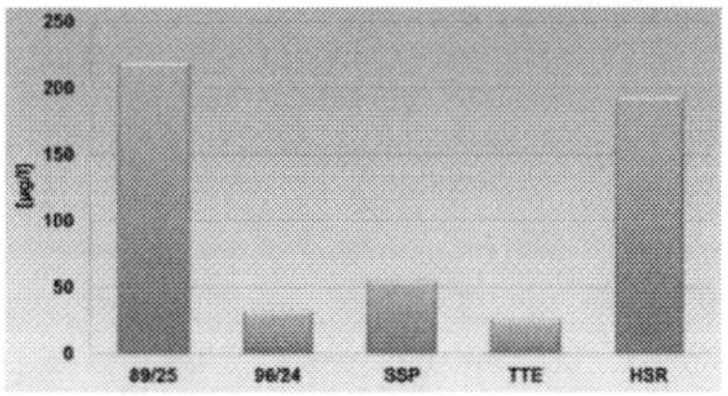

- 89/25 highest ester content in beer, HSR has high levels as well
- single components can exceed their threshold level

Figure 5: Beer analyses (Sum of 6 Esters)

SENSORY ANALYSES

Beers were tasted by the St. Johann taste panel (s. table 5, figure 6) and a small consumer panel, where 89/25 has been preferred 6:1 to TTE and 5:2 to SSP).

	89/25	96/24	SSP	TTE	HSR
DLG (average grade)	4.47	4.27	4.48	4.41	4.21
Preference	1*	3	2	3	5

- good average marks for all beers
- 89/25 sign. preferred

Table 5: Tasting Results (enhanced DLG form, St. Johann panel, n=13)

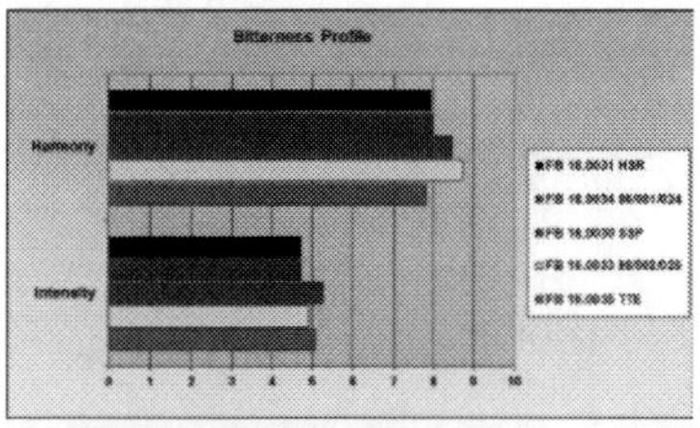

- all beers on a comparable bitterness level (intensity)
- harmony (quality) of bitterness all beers were rated as fine, 89/25 and SSP by trend better

Figure 6: Tasting Results (bitterness profile , St. Johann panel, n=13)

SUMMARY

Two Huell aroma breeding varieties, 89/002/025 and 96/001/024 are compared to their mothers *Spalter* and *Tettnanger* as well as to *Saphir*, the aroma variety with the highest aroma capacity in Germany yet.

- Both breeds have reduced low molecular polyphenolic character compared to mother varieties.
- 89/25 shows a high farnesene content, an indicator for its affiliation to the Saaz hop family.
- Especially 89/25 is very rich in aroma substances of good solubility like linalool and esters.
- A dosage according to hop oil content resulted in a reduced throw of the breeding varieties: 35% of TTE, 41% of SSP and 55% of HSR.
- The 89/25 beer showed the highest values of flavour-active compounds (linalool and esters).
- Particularly 89/25 has been rated good in the tastings, the beer has been preferred significantly.
- These results are supplemented by two more test series varying the dosage criteria
- In order to achieve a comparable hop aroma in beer the breeding lines can be dosed at significantly lower quantities.
- Due to the results HVG has decided to grow both varieties on an amplified acreage this year: 89/25 on 3 ha and 96/24 on 1 ha.

Especially 89/002/025 is a very promising candidate for a registration as a new aroma variety for late hopping purposes.

ACKNOWLEDGEMENT

Thanks to the Bayerische Landesanstalt für Landwirtschaft (LfL), Hüll, for providing the hops and information on the experimental varieties.

* andreas.gahr@hopfenveredlung.de

Varietal Comparison of Cytolysis of Barley during Malting

Xiang S. Yin, Patricia Aron, Rebecca Jennings, Paul Kramer
Rahr Malting Co. USA

Introduction

Satisfactory cytolysis of barley during malting directly affects the malt in brewhouse performance and the beer production efficiency. Despite significant amounts of research carried out in the field over the decades, knowledge on the barley factors affecting biochemical degradation of cell wall components in early stages of germination is limited.

The present study examined the varietal differences in cytolysis during germination of selected North American malting barleys. Malt samples germinated for 1-4 days under laboratory and production conditions were tested for progressive changes in the levels of β-glucans, wort viscosity, and enzyme development including β-glucanase and xylanase, followed by kernel modification studies.

Experimental

Barley samples

Malting quality CDC Copeland grown in Canada and Pinnacle from USA were collected from the 2015 and 2016 crop years.

Total Protein contents:

Vairiety	2015	2016
Pinnacle	10.3	11.8
CDC Copeland	11.4	11.4

Laboratory malting conditions

Steep: 9 hrs (wet) -15 hrs (dry) - 6 hrs (wet) - 18 hrs (dry)
Germination: 18°C, no watering, upto 4 days
Kiln: 20hrs @65°C, then 3 hrs @80°C

Production malting

Samples were generated from production batches that had been germinated for 1-4 days and put in strainer bags then kilned under the same conditions. The process parameters were optimized based on the varietal characteristics.

Results and Discussion

Laboratory malting

2016 crop CDC Copeland and Pinnacle were subject to malting under the same conditions, and showed the following outcomes:

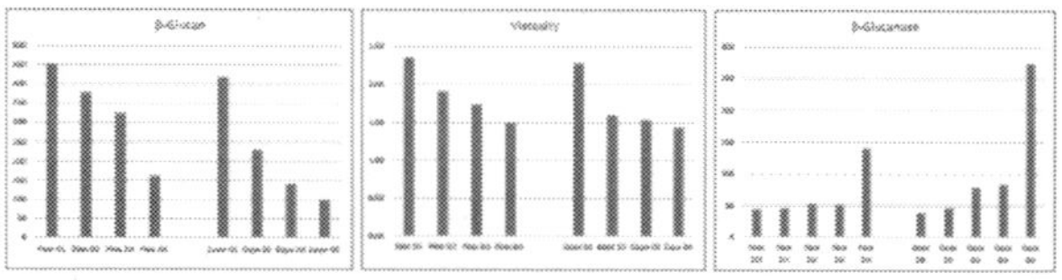

Copeland showed faster degradation of β-glucans with the first two days of germination, but Pinnacle started relatively slowly. While the large portion of modification was done in the first two days for Copeland, Pinnacle showed very gradual cytolysis as germination went on and the last day of germination was critical for bringing down its β-glucans to acceptable level.

Results and Discussion – continued

Production malting

Malt samples of 2015 crop Copeland and Pinnacle kilned in strainer bags were demonstrated to have the following properties:

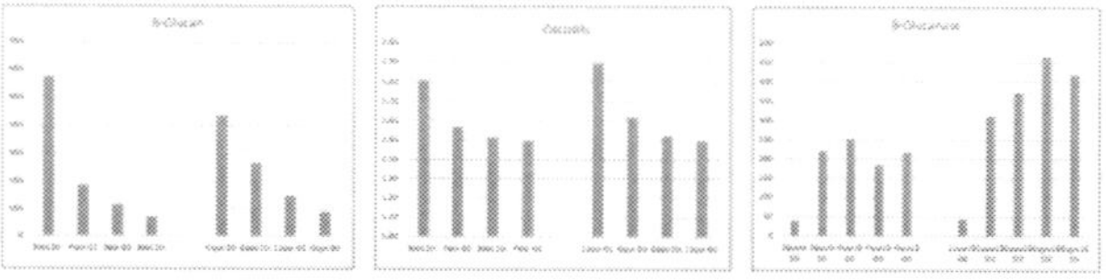

Pinnacle showed faster degradation of β-glucans within the first two days of germination under this particular production condition which was set for promoting modification of the barley. The process compensation adjustment appeared to have compensated the lower speed of modification inherent in the variety. Other possible factors may include ~1% lower protein content in the 2015 crop Pinnacle barley.

Malt modification by Microfluo Malt Analyser

The above malt samples from production conditions were subject to modification analysis as demonstrated below.

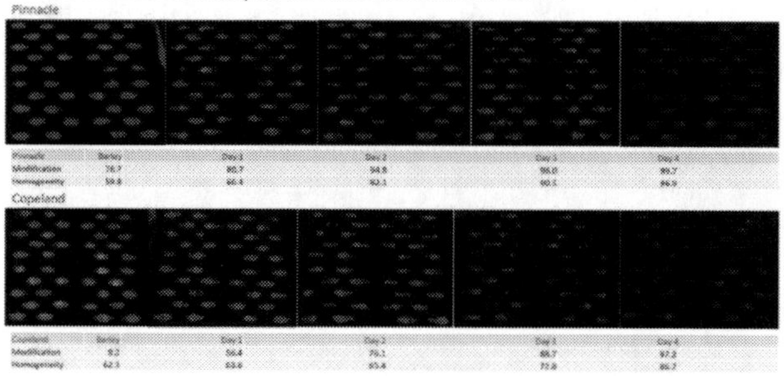

Conclusions

Varietal difference

Under the same process conditions CDC Copeland showed more active cytolysis than Pinnacle as demonstrated by the progressive reduction of β-glucans, viscosity and speed of β-glucanase development, while xylanase did not present significant escalation (results not shown).

Process and barley quality factors

With the adjusted process conditions to promote the modification on lower protein 2015 crop sampled barley, the cytolysis of Pinnacle was made comparable to or even above that of CDC Copeland.

Balance of FAN and β-glucans in malt

With barley proteins at >11% when the β-glucan is controlled at <200, the FAN level is likely >200.

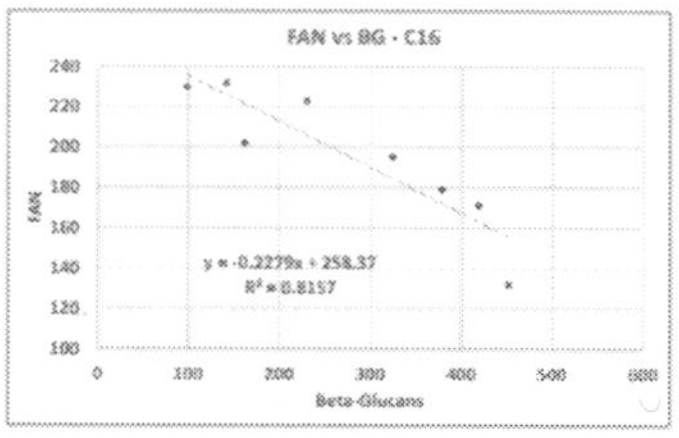

Acknowledgements

Thanks are due to Robert Monsour, Rachel LaCount and QC & Operations colleagues for the technical assistance, and to Rahr Corporation for permission to publish this research.

P-067

A Study on Advanced Technologies for DON Reduction in Malting

Patricia Aron, Xiang S. Yin, Rebecca Jennings, Paul Kramer
Rahr Malting Co. USA

Introduction

Fungal load particularly *Fusarium spp.* on barley kernels is a major cause of deoxynivalenol (DON) production on malt. In the present study, an exploration of the state-of-the-art clean technologies including treatments with ozonated water, cold plasma and pulsed light on the potential of *Fusarium* reduction on barley was carried out. The work was conducted in comparison with conventional approaches including the use of alkaline, acid, ClO_2 and sodium bisulphite, respectively. More natural ingredients such as hop beta-acids were also included in the tests.

Experimental

Barley samples:
A 2016 crop sample of Lacey variety harvested in ND was used as the base barley, which had a relatively high DON level of 3.2ppm. Other analysis were 12.2% moisture, 12.9% protein, and 97% germination capacity.

Laboratory malting conditions:
Sample: 100 g malt + 200 mL RO H2O
Steep: 16 H (wet)- 8 H (dry) – 5 min soak (wet) – 16 H (dry)
Germination: 18°C, no watering, 72 hours
Kiln: 20hrs @70°C
Clean: Remove rootlets
Treatments were executed on the raw barley or steeped barley on the first day of germination.

DON by GC-ECD:
Quantification of DON was carried out utilizing gas chromatography and µECD detection according to the ASBC Barley Method 11- A

Germination Score:
At the end of the germination before kilning, observe the number of kernels with rootlets extending over one kernel length

# grains with rootlet over 1 kernel length (%)	Germination Score
0-20	1
21-40	2
41-60	3
61-80	4
81-100	5

Results and Discussion

- The three state of the art clean technologies demonstrated to have a pronounced effect on malt DON reduction; DON reduction increased with dosage. However, barley germination vigor was also affected accordingly.
- Treatments appeared to be more effective on Day 1 malt than on the raw barley.
- The optimal range of treatment varied depending on the setup of the equipment, and further investigation of the energy level in the treatment water is required in order to scale up the experimentation.
- Under the present conditions, the ranges of optimal contact time on Day 1 malt were 15-60 mins for ozonated water, 60-90 secs for pulsed-light water, and 5-15 mins for plasma water.

Results and Discussion – continued

- All chemical treatments (B-acid, NaOH, sodium bisulphite, free-fatty acids) did show effects on DON, but also affected germination vigor. Dosages could not be readily controlled as the storage time of the barley appears to be a variant as well.

Effect of State-of-the Art Clean Technologies:

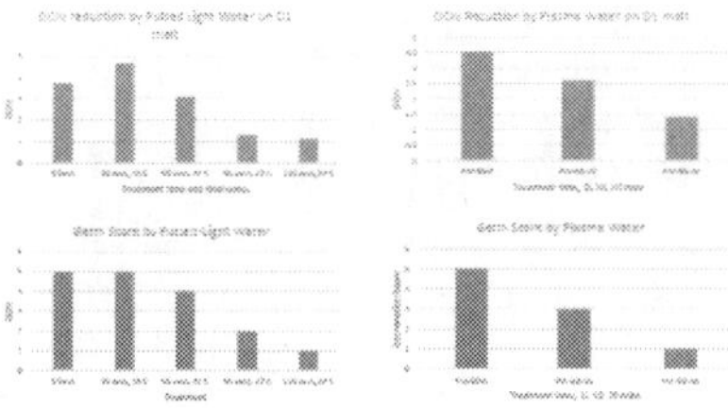

Effect of Chemical Treatments:

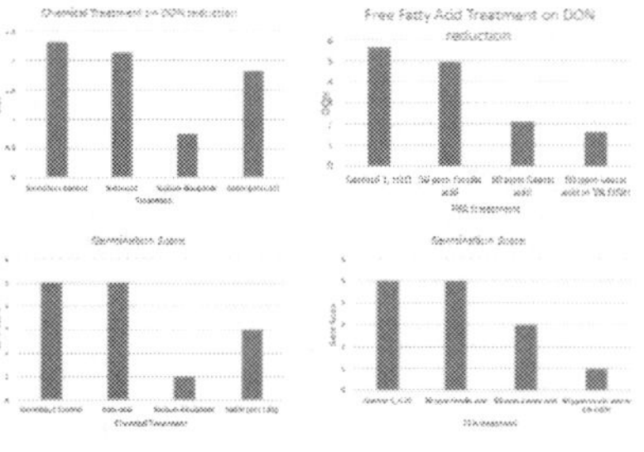

Conclusions

State-of-the Art Technologies:
Pulsed light water treatments showed a reduction in DON from 60 sec of treatments. However, treatments longer than 90 seconds negatively impact germination. Plasma water treatment on either steeped barley or Day 1 malt did significantly impact DON levels. Higher doses may be required for testing. Ozonated water treatment confirmed a positive effect on Day 1 malt with germination slightly affected (data not shown).

Chemical Treatments:
Chlorine dioxide and acid treatments lead to increased malt DON possibly due to the elimination of the inhibitors of Fusarium. NaOH did not show consistent reduction of malt DON under the testing conditions. Sodium bisulphite treatment demonstrated a reduction to malt DON and hop derived Beta-acids also demonstrated a positive effect on DON reduction. However for all effective treatments dosage amounts and time need optimization.

Plasma water treatment at concentrated dosage produced pronounced effect on both the DON and germination capacity of the malt produced.

Acknowledgements:
Thanks are due to Jigar Brown for assistance with DON analysis and Dongjie Chen, Roger Ruan and Paul Chen of the Department of Bioproducts and Biosystems Engineering, University of Minnesota and Rahr Operations colleagues for the technical assistance, and to Rahr Corporation for permission to publish this research.

Reference:
ASBC Methods of Analysis, Barley Method 11-A. Deoxynivalenol. Approved 1996, rev. 2014. American Society of Brewing Chemists, St. Paul, MN, U.S.A. doi: 10.1094/ASBCMOA-Barley-11.

Deutsche Forschungsanstalt für Lebensmittelchemie | Leibniz-Institut

Characterisation of Key Aroma Compounds in Two Special Malts by Means of Aroma Extract Dilution Analysis

DFA
Leibniz Institut
Deutsche Forschungsanstalt für Lebensmittelchemie

Leibniz-Gemeinschaft

Veronika Mall[1], Jens Voigt[2], Andreas Richter[3] and Peter Schieberle[4]

[1] German Research Center for Food Chemistry, Leibniz Institute, Freising, Germany; [2] University of Applied Sciences Trier, Trier, Germany; [3] Weyermann Specialty Malts, Bamberg, Germany; [4] Technical University of Munich, Freising, Germany

Introduction

Malt is one of the most important ingredients for the beer production, bringing in sugars, enzymes and aroma. With the uprising trend of craft brewing the varieties in beer flavours and the therefore used malts with special aroma characteristics are increasing steadily. In order to rate the aroma potential of two special malts, key aroma compounds in a roasted malt and a Belgian type malt were characterised by means of aroma extract dilution analysis (AEDA).

Isolation of Aroma-Active Compounds

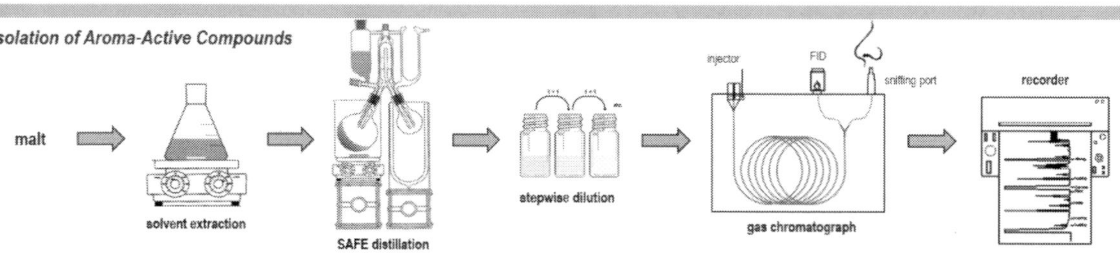

For an appropriate identification and characterisation of odorant compounds, the concept of molecular sensory science has been developed in our group in the last decades [1]. After solvent extraction of the malt, the volatile compounds were gently separated from non-volatile malt components by means of solvent-assisted flavour evaporation (SAFE) distillation. The resulting aroma distillate was then subjected to gas chromatography-olfactometry (GC-O) to locate the aroma active compounds. AEDA was performed to rate their importance.

Outcomes from AEDA of Belgian Type Malt

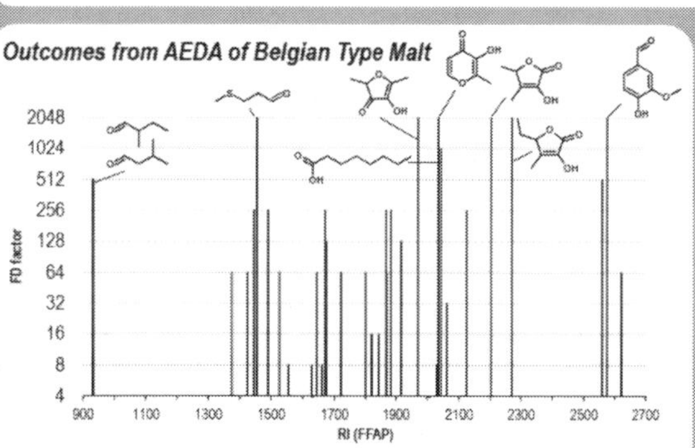

Identification was carried out by gas chromatographic and mass spectrometric methods, in comparison with reference compounds. Crucial identification parameters: retention index (on 2 columns of different polarity), odour quality and intensity, and mass spectral data (MS-EI, MS-CI).

Following this approach, a total of 34 aroma compounds with FD factors ranging from 8 to 2048 were detected. The highest FD factor of 2048 was assigned to the caramel-like 4-hydroxy-2,5-dimethyl-3(2*H*)-furanone and 3-hydroxy-2-methyl-4(4*H*)-pyranone (maltol), the seasoning like furanones, 3-hydroxy-4,5-dimethyl-2(5*H*)-furanone (sotolon) and 5-ethyl-3-hydroxy-4-methyl-2(5*H*)-furanone (abhexon), all deriving from *Maillard* reactions. Other important compounds were characterized as the vanilla-like vanillin and the *Strecker* aldehyde 3-(methylthio)-propanal (cooked potato-like).
The characteristic 2- and 3-methylbutanal with their malty aroma reached an FD factor of 512.

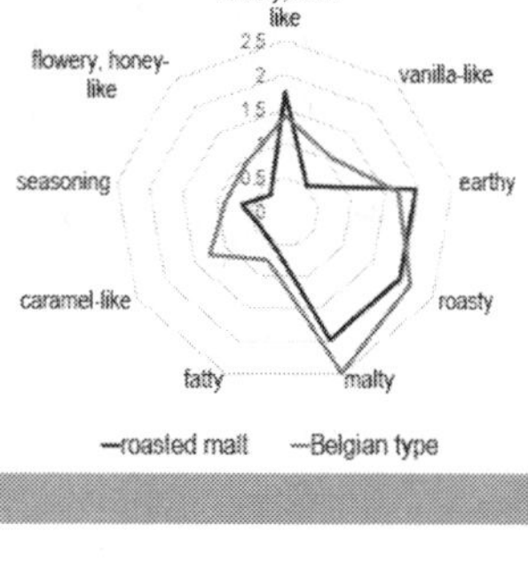

Comparison of AEDA of Belgian Type and Roasted Malt

Next to high FD factors for the typical *Maillard* reaction products, the rosted malt also featured thermally generated odorants, such as the smoky 2-methoxyphenol, 3-isopropyl-2-methoxypyrazine (earthy) or the roasty, popcorn-like 2-acetylthiazol as important aroma compounds in the AEDA. These compounds are responsible for a more smoky and earthy aroma profile. In comparison, the Belgian type malt showed a more seasoning, caramel, malty and honey-like profile.

		FD factor	
odorant	odor description	Belgian type malt	roasted malt
4-hydroxy-2,5-dimethyl-3(2*H*)-furanone	caramel-like	2048	2048
vanillin	vanilla-like	2048	2048
3-hydroxy-2-methyl-4(4*H*)-pyranone	caramel-like	2048	1024
3-hydroxy-4,5-dimethyl-2(5*H*)-furanone	seasoning	2048	1024
3-(methylthio)propanal	cooked potato-like	2048	512
5-ethyl-3-hydroxy-4-methyl-2(5*H*)-furanone	seasoning	2048	256
octanoic acid	musty	1024	512
phenylacetic acid	honey-like	512	256
3-methylbutanal	malty	512	2
2-methylbutanal	malty	512	2
2-methoxyphenol	smoky, sweet	64	2048
acetic acid	sour, vinegar-like	256	1024
2- and 3-methylbutanoic acid	sweaty, cheese-like	256	1024
2-ethyl-3,5-dimethylpyrazine	earthy	4	1024
(*E,E*)-2,4-nonadienal	fatty	4	1024
2-acetylthiazol	popcorn-like, roasty	2	1024
3-isopropyl-2-methoxypyrazine	earthy, pea-like	<1	1024
unknown compound	roasted	<1	1024
4-allyl-2-methoxyphenol	clove-like, smoky	<1	512

In general, different malting processings result in distinguishable aroma profiles and varying sets of key aroma compounds, which can lead to special aroma notes in beers brewed with these malts. However, quantitative data and the calculation of odor activity values will be necessary to map the whole potential of the special malt as aroma influencing ingredient to special beers.

[1] Schieberle, Hofmann, In: Food Flavors - Chemical, Sensory and Technological Properties, 2011, 411-437

36TH CONGRESS EBC 2017 14-18 MAY LJUBLJANA EUROPEAN BREWERY CONVENTION SLOVENIA

Comparative analysis of brewhouse efficiency, energy and materials cost in three types of microbreweries

Panagiotis Tataridis[1], Olivie Pitlinger[1], Papakonstantinou Spyros[1]

[1] Laboratory Sector of Malting & Brewing, Department of Enology & Beverage Technology, School of Food Technology & Nutrition, Technological Educational Institute of Athens, E-mail: p_tataridis@yahoo.gr, Tel. +30 210 5385547, Ag. Spyridonos, 12243, Athens, Greece.

Introduction

The rapid growth of microbreweries in recent years gave birth to an array of production facilities. But not all microbreweries are created equal. The difference in knowledge and experience of the brewers, as well as, their choices in equipment and processes, creates big disparities in their efficiency and production cost. The current study compares the data collected from three different types of microbreweries. Extract and volume production, consumption of raw materials (total water, malts, hops and yeast), cleaning agents and energy usage were monitored on a monthly basis for the period of one year. The following table summarizes the differences for each brewery.

Brewery	Product type	Beer types	Brewhouse type	Annual production in hL	Heat type	Energy source
A	Filtered, unpasteurized	Pale Ale, IPA	Mash/Lauter - Boiler/whirlpool	1.000	Steam	Electricity & Diesel
B	Non filtered unpasteurized	Pils, Brown Ale	Mash/Boiler – Lauter	1.000	Heating element	Electricity
C	Filtered pasteurized	Pils, Lager, Black Lager, Smoked Lager, Weiss, Red Ale	Mash/Boiler – Lauter - Whirlpool	15.000	Steam	Electricity & LPG

Results & Discussion

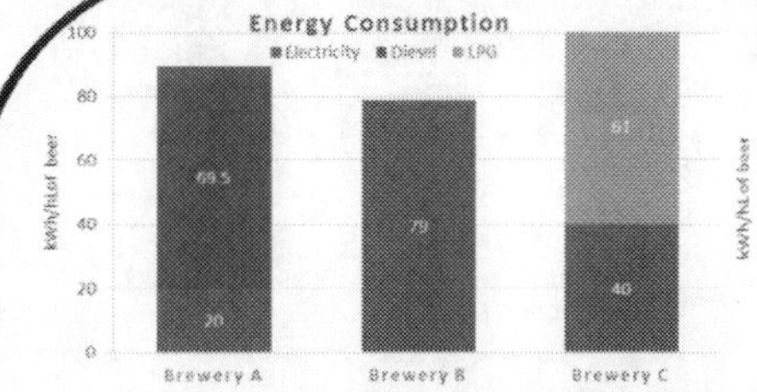

Brewery C is using 1/3 of its capacity, thus exhibiting higher energy consumption per hL of beer. The electricity consumption was extremely high at 40kWh/hL, double the optimum of its class at 20kWh/hL of beer.

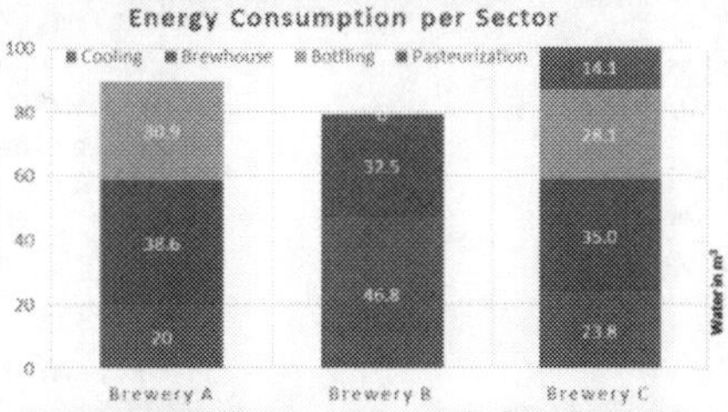

All breweries have similar brewhouse energy consumption. Brewery B has twice the cooling consumption of the others. Even though brewery B appears to have the lower energy consumption, it should be noted that it does not use energy for product stabilization and bottling.

Example of data collected from Brewery C

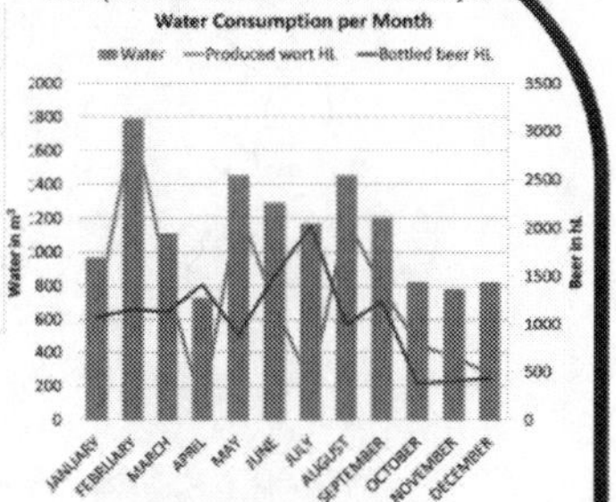

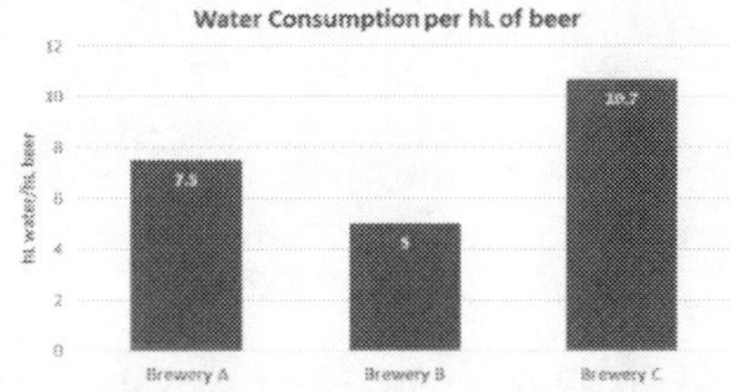

Brewery B has the lower water consumption, however it does conform to the 6hL/hL optimum as it does not use water for filtration or product stabilization. Brewery C has a consumption close to the upper limit (10.9 hL/hL, Esslinger, 2004). Brewery A has a consumption higher than the optimum, but typical for its size.

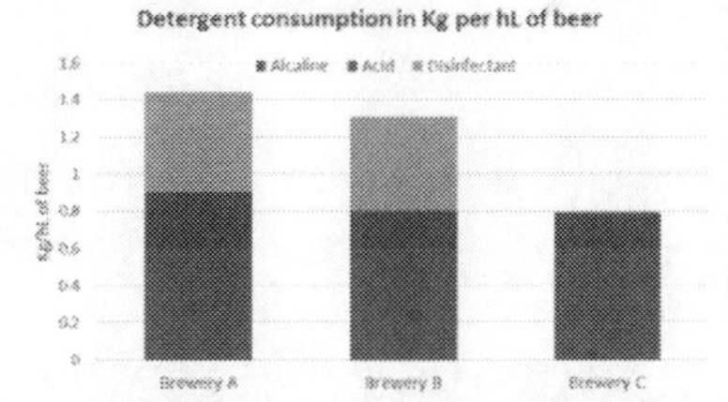

All breweries have the same consumption in caustic detergent. Brewery C uses a more expensive acid detergent with disinfectant action, as well as, an automated CIP system (dosing unit & on-line conductivity measurement). As seen in the cost diagram, all breweries have almost the same cost of detergent use per hL of beer.

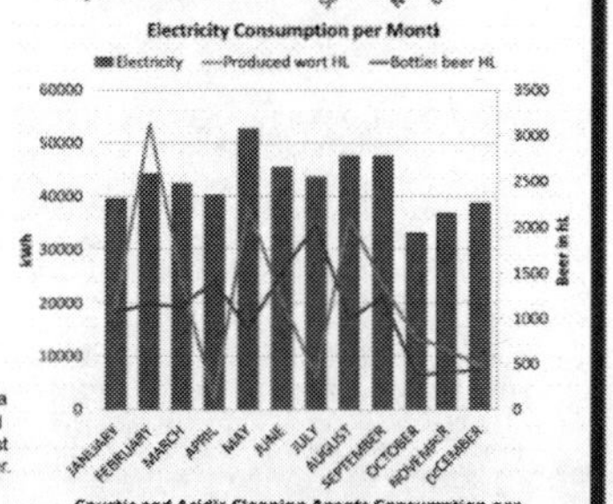

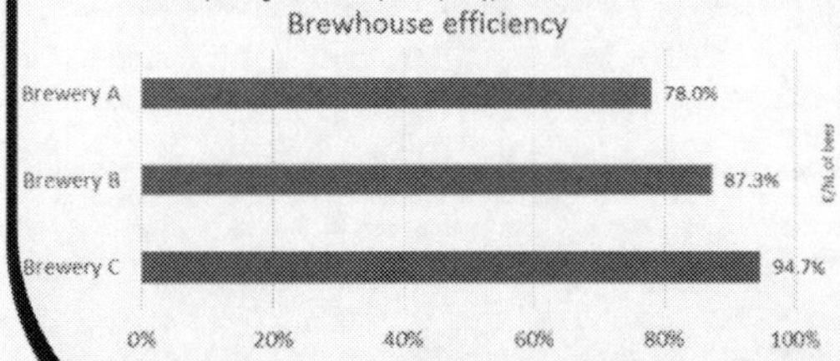

Breweries A & B have much lower efficiency than expected, both due to their unsophisticated design and brewhouses types, as well as, due to lack of rigorous process control.

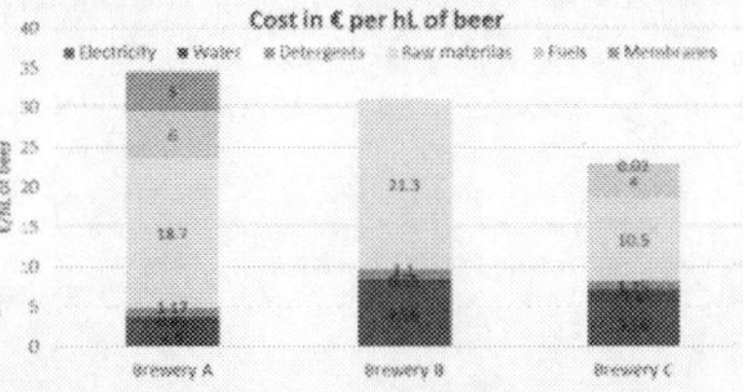

Brewery C has the lower cost because of its efficiency and bulk purchases. Brewery A has an added high cost of sterilizing membranes.

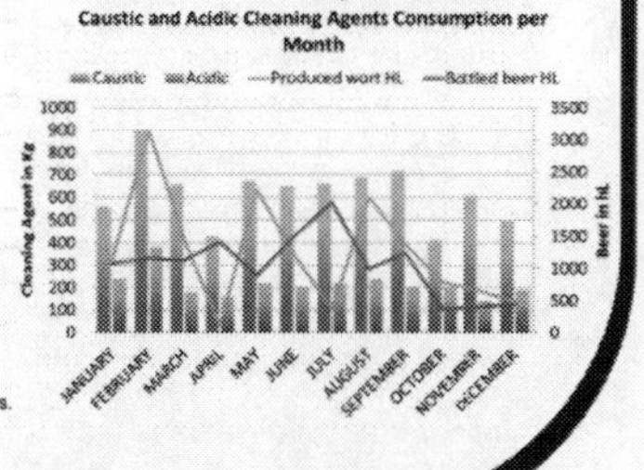

Conclusions

Results showed significant differences in various cost and efficiency indexes, as well as, mismatch of equipment capacity and process problems, commonly found in many new microbreweries. For smaller microbreweries (A & B), the most common problem is lower efficiency and higher variability due to lack of process control. Brewery C operates at a relatively low capacity of 30%, thus significantly increasing operational cost. Data evaluation led to the following cost saving opportunities and production optimization suggestions. Initial implementation has already shown positive results.

Process	Electricity	Thermal energy	Water	Raw materials	Cleaning
• Product recovery from pipes and vessels • Better process control (pH, temperature, time) to increase efficiency • In-house quality control (Brewery A & B)	• Fewer bottling runs of higher volume (All) • Optimization of the cooling system or replacement with a direct ammonia system (Brewery C) • Use of solar energy due to location (Brewery C)	• Fine tuning & longer runs for bottling to reduce the startups and the number of sterilization cycles (All) • Wort production in three continuous shifts to avoid shutting down the steam boiler, leading to fuel savings (Brewery C) • Proper use and good maintenance of the energy recovery systems (Faduko, 2 stage wort cooler) • Implementation of energy saving systems in the boiler room (economizer, superheater) (Brewery C)	• Higher volume bottling sessions to minimize sterilizations, cleaning and hot and tap water consumption (All) • Measuring the concentration of cleaning agents during the rinse steps to avoid water overconsumption • Optimization of the wort cooling system to allow for highest possible hot water recovery in the highest temperature (Breweries A & B)	• Larger purchase orders to reduce the transportation cost per kg (Brewery A & B) • Supplier contracts for better prices (All) • Re-pitch yeast (Breweries A & B) • Use of domestic malt suppliers, for lower transportation costs to achieve lower total price (All)	• Good aeration of CCT and BBT tanks before alkaline cleaning to avoid the weakening of the caustic cleaning agent (Brewery A, B) • Reuse caustic cleaning solutions for pre-cleaning (Brewery A) • Implement a two-shift schedule during bottling to avoid cleaning and sanitizing twice (Brewery A)

Chair of Food Packaging Technology
TUM School of Life Sciences Weihenstephan
Technical University of Munich

Weihenstephan Standards for brewing processes

A standardized MES interface for process areas of breweries

R. Bär, T. Voigt

Introduction

Breweries have to produce high-quality products at acceptable market prices. In order to remain competitive, they rely on effective IT assistance for their in-company processes. They are increasingly using process-orientated production management systems, so-called **Manufacturing Execution Systems** (MES). MES closes the gap between the production level (**P**rocess **C**ontrol **S**ystem) and the enterprise management level (**E**nterprise **R**esource **P**lanning) and thus enables a vertical data continuity. Exemplary tasks of MES are shown in Figure 1.

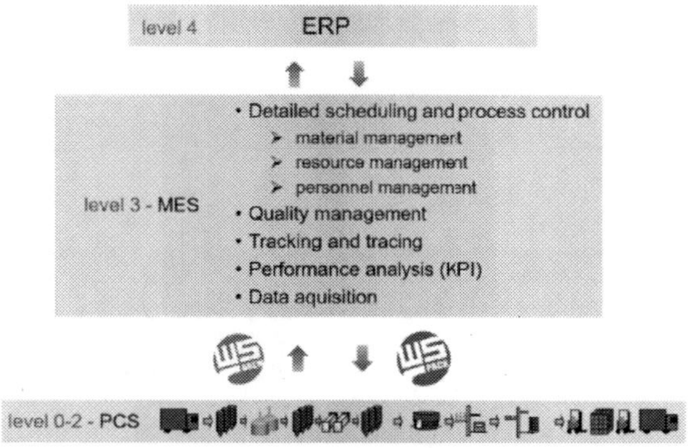

Figure 1: Automation pyramid: MES functions in breweries based on [VDI 5600] and [ANSI-ISA S-95]

For effective use of MES functions, a data exchange network between all relevant production facilities and the involved IT systems is essential. However, the connection of the machines, process control and computer systems is complex. Automation companies use different communication mechanisms. The availability of data of relevance for the MES varies. In many cases, no data are available. Therefore, time-consuming definition work and expensive engineering and programming is required to connect the different systems.

Standard definitions for the communication interfaces, especially the contents of the provided and acquired data, help to reduce integration problems and save time. This has been proved for the bottling and packing area – WS Pack (also meat processing - WS Food and baking industry – WS Bake) by the Weihenstephan Standard concept. In the mentioned application domains, Weihenstephan Standards provide detailed definitions of the most relevant production data tags needed for MES implementation. Together with recommendations about the communication protocol and explanations about data processing to gain MES information, food and beverage producers, OEMs and IT system providers use them in a uniform way.

Approach

For the process part of a brewery, standards are missing. The new MES interface will cover the most relevant process areas, in particular brewhouse, fermentation cellar, filtration and the related utilities like steam generator or compressed air supply. In order to figure out the needed data tags a bottom up approach is followed (see Figure 2):

1. MES function description
2. Modelling based upon three founding pillars
   - Equipment
   - Process
   - MES functions
3. Derivation and definition of the relevant data tags

Methods

Within the scope of a research project a representative survey in the form of expert interviews was conducted. The requirements and needs of the three industry sectors were considered.

Based on the results, relevant MES functions can be derived and an optimal MES for breweries can be described.

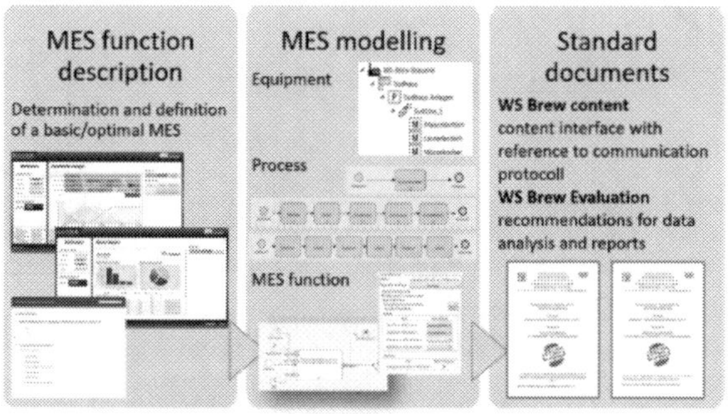

Figure 2: Used method to determine MES relevant data tags

The modelling tool is based on already existing standards. The brewery equipment is pictured in a tree diagram based on ANSI/ISA-S88 (Batch Control). The process is modeled using Business Process Modeling Notation (BPMN) and can be divided in a granularity of three levels. The process can be linked to the equipment and by using standard (mathematical) templates, KPIs like the brewhouse yield and consumptions (related e. g. on a batch) can be calculated.

First Results

In cooperation with the project partners relevant MES functions have already been worked out and are currently formulated and modeled.

Tracking and Tracing	Process- and product quality	Long-term evaluation
Key Performance Indicator (KPI)	Energy and media consumption	Capacity and states
Detailed scheduling	Information management	Maintenance

Table 1: Most relevant MES functions in a brewery

One of the most important functions of a MES is the interface between the production and the enterprise management, in order to enable a transparent and effective production. In combination with an appropriate communication protocol such as OPC UA, the MES is the leading IT system to manage and evaluate the production, where all relevant data are merged.

Project partners

Contact: **Raik Bär, M.Sc.**
raik.baer@tum.de
+49 8161 71 3580
www.lvt.wzw.tum.de
www.weihenstephaner-standards.de
Chair of Food Packaging Technology
TUM School of Life Sciences Weihenstephan
Technical University of Munich

hochschule mannheim

Investigation of a new crystallizer with included post-treatment for ion depletion of spring water for the brewery industry

Lars Erlbeck[1], Matthias Rädle[1], Thomas Kunz[2], Frank-Jürgen Methner[2]

Introduction

Guarantee of water quality is eminently important for constant high product quality (e.g. Brewery industry)

Necessary depletion of contaminations like compounds of Manganese, Iron and Nitrate as well as water hardness and TOC

- Manganese concentration responsible for taste and color of the beer
- High Iron concentration effects negative influence on yeast, thus on fermentation and taste of the beer
- Residual alkalinity changes the pH-value and enzyme activity
  → Yield loss
- During fermentation Nitrate transforms into Nitrite
  → Inhibiting impact on yeast

Ion exchanger are most commonly used having following side effects:

- Disadvantage of necessary usage of chemical regeneration compounds
- Risk of overdosage, thus discoloration (precipitation of Manganese)
- Batch process only → Double version → High investment

<u>Alternative:</u>

Crystallization of spring water with subsequent separation and post-treatment

- Lack of necessary usage of chemical compounds
- Less corrosion and scaling effects
- Continuous process → No double version

Material & Methods

Crystallization vie scrape chiller:

- Heat transfer surface to form ice crystals
- Forced conveyance by screw to transport ice to the cone
- Press section through a tapering, perforated cone

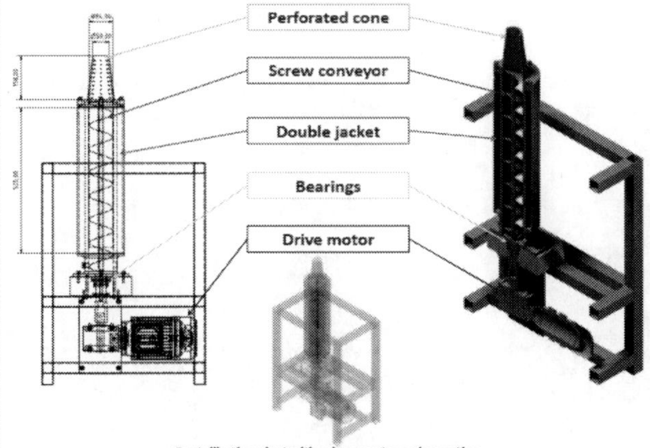

Crystallization plant with subsequent pressing section

Characteristic values of the crystallization plant

Description	Unit	Parameter
Heat transfer surface	[m²]	0.14
Drive power motor + transmission	[kW]	0.55
Torque motor + transmission	[60 Nm]	60
Max. rotational speed screw	[1/min]	62
Cooling capacity heat pump	[kW_{el}]	5.5
Heat transfer coefficient	[W/(m²*K)]	300
Area ratio of the tapering cone A_1/A_2	[-]	0.49

Results

Test parameters

Description	Unit	Parameter
Motor speed	[1/min]	40
Inlet temperature coolant	[°C]	-5
Flow rate coolant	[m³/h]	0.94
Mass flow ice	[kg/h]	0.20

Composition test solution and limits

Contamination	Limit [mg/l]	Reagent
Iron	0.2	$FeCl_3$
Manganese	0.05	$MnSO_4$
Residual alkalinity	Depends on beer type	Tap water
Nitrate	50	NH_4NO_3
TOC	-	$CO(NH_2)_2$

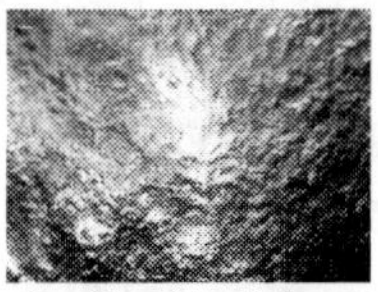

Produced ice mixture before cone
→ Lots of small, flaky ice crystals

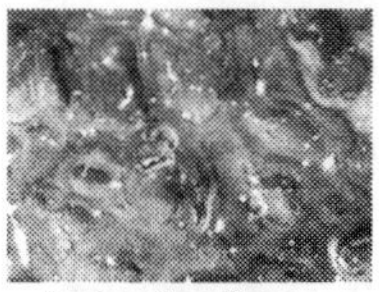

Produced ice after cone
→ Few big, three-dimensional crystals

- Significant reduction of Nitrate, Iron and Manganese
- Decrease of residual alkalinity and TOC
- Lower deviation of all limit values

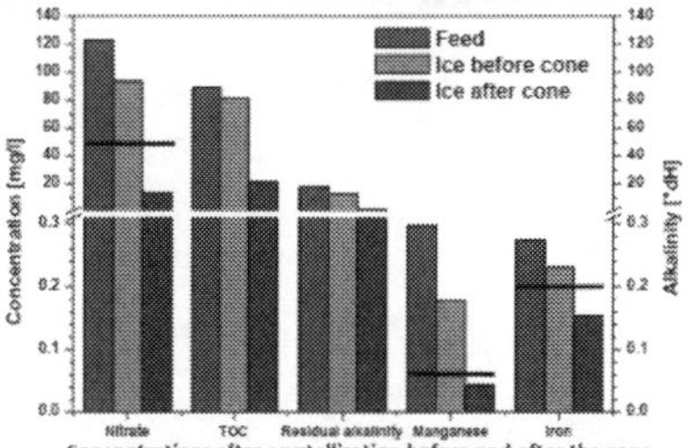

Concentrations after crystallization before and after the cone

- Exceptional increase of depletion through pressing in the cone
- Higher residual concentration of Iron due to corrosion of the crystallizer (no stainless steel)

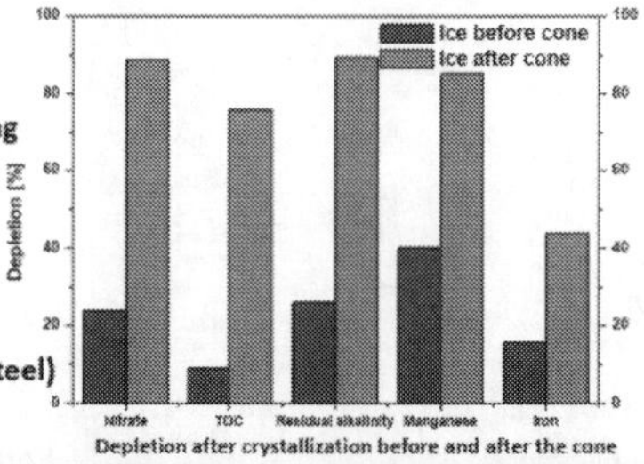

Depletion after crystallization before and after the cone

Conclusions / Outlook

- Reduction of all ion concentrations under limit values
- Achievement of required spring water quality for brewing industry
- Successful presentation of functionality of a new, innovative and alternative process

- Further optimizations to increase mass flow
- Investigations on parameter change and impact on results
- Experiments on influence on microorganisms

Kontakte

[1] Institute for Process Control and Innovative Energy Conversion, University of Applied Sciences Mannheim, Germany

[2] Chair of Brewing Science, Department of Food Technology and Food Chemistry, Technical University of Berlin

P073

Contact: Lars Erlbeck, l.erlbeck@hs-mannheim.de

European Brewery Convention, 14 – 18 May 2017, Ljubljana, Slovenia

Flash pasteurization
How long is the optimal heat-holding time/temperature for achieving the best beer quality?

R. Feilner, T. Oehmichen/KRONES AG

36TH CONGRESS
EBC
2017 14-18 MAY LJUBLJANA
EUROPEAN BREWERY CONVENTION
SLOVENIA
P076

Abstract

Flash pasteurisation of beer is an important instrument for reliable extension of shelf life. In conjunction with a suitable hygienic filler, flash pasteurisation constitutes a product-friendly alternative to tunnel or chamber pasteurisers. The haze- and taste-stabilising effect of a flash pasteuriser, on wheat beer, for example, is also gaining steadily in perceived importance. Established figures for pasteurisation units (PUs) specify the dimensions of the heat-holding sections (30 seconds) and the heat-holding temperatures (64°C – 72°C). The question here arises of whether these traditional parameters – using the analytical and technological options now available – actually constitute the optimum. The following article shows results of a comparison 30 seconds heat holding tube and a shorter one with 6 seconds. The following poster shows the influence to heat-, oxygen- and aging indicators and allows a detailed comparison between the processes and the beer quality.

This paper describes a heat-holding process that enables haze-stabilised, microbiologically stable wheat beers and at the same time also filtered beers to be produced in better quality using a system with reduced exposure to heat.

The crucial constituent of this process is a dramatic shortening of the heat-holding section and a concomitant increase in the pasteurisation temperature. It is crucial, particularly for the filtered beers, to fix a maximum temperature that, in contrast to a desired haze in the unfiltered beers, does not cause any change in the clear-filtered beers.

Materials and Methods

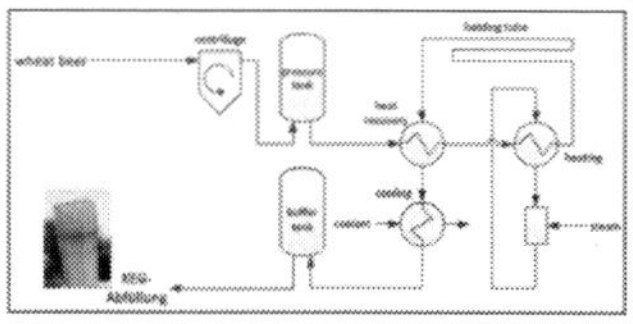

Beer type	Time [s]	Tempera-ture [°C]	PU [-]
Wheat beer	30	80	380
	6	83.2	220

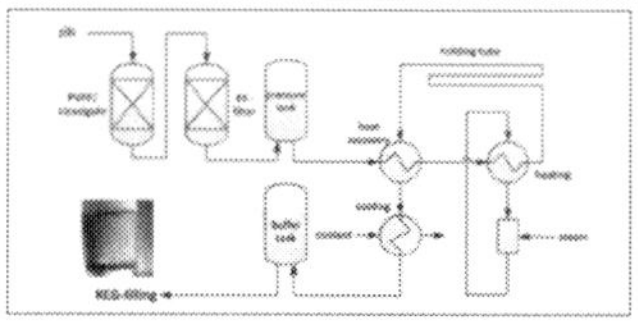

Beer type	Time [s]	Tempera-ture [°C]	PU [-]
Pils	30	73.6	45
	6	78.4	45

Profess comparision between 30 and 6 seconds heat holding time

Results

Dwell time simulations

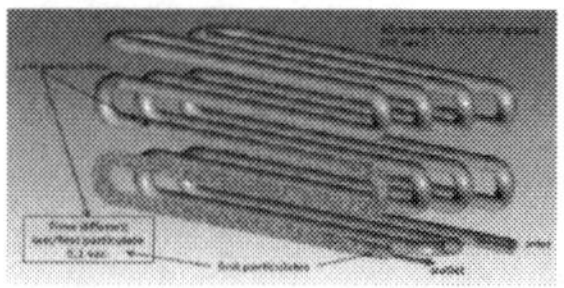

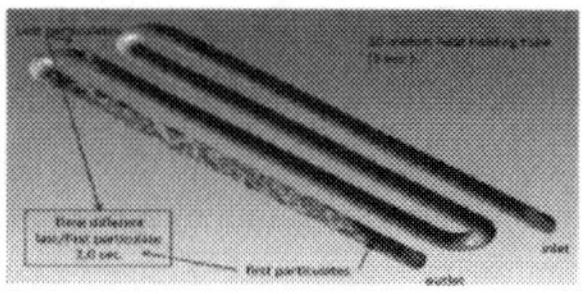

50 metres= 30 seconds; 10 metres= 6 seconds at 30 m3/h, DN 80; ∇= 1.66 m/s; Reynold - 300,000, dparticles= 10 μm; number of particles 65,000

With the short heat-holding sectoin, almost no under-pasteurisation of particles occurs. Fewer than 1 % of the particles entering leave the section before exiry of the 6-second heat-holding time. In the case of long heat-holding, by contrast, around 20 % of the particles entering leave before expiry of the 30-second heatholding time, and may thus not receive sufficient thermal treatment.

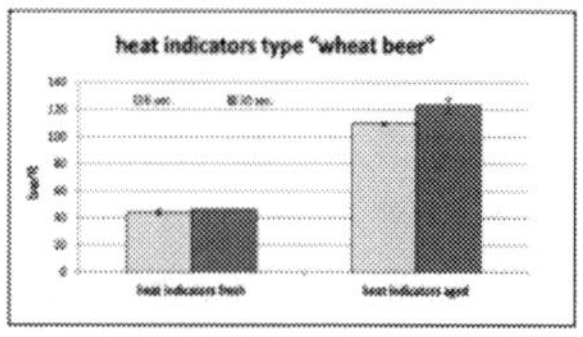

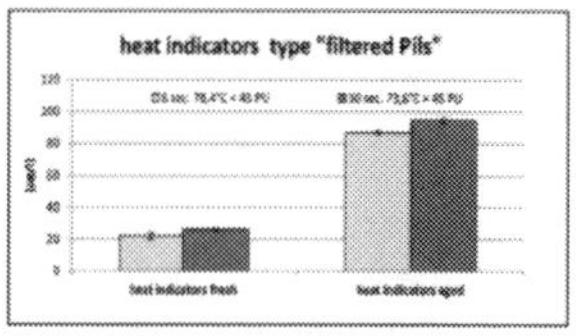

Short holding time with higher temperatures in comparison creates less heat indicators

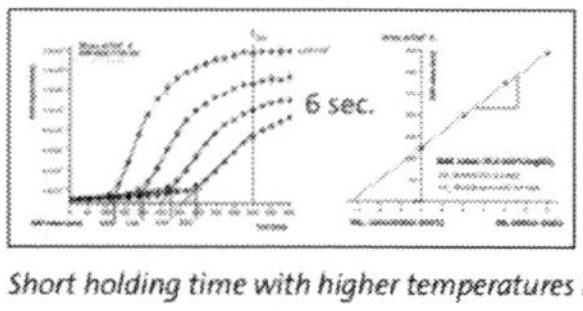

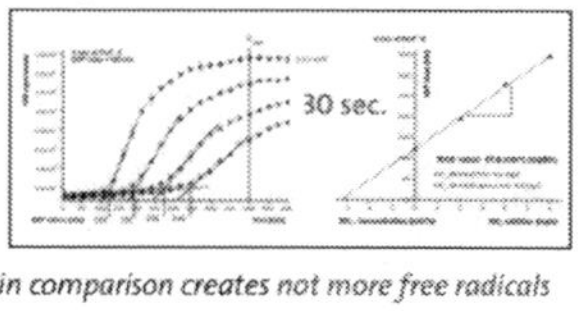

Short holding time with higher temperatures in comparison creates not more free radicals

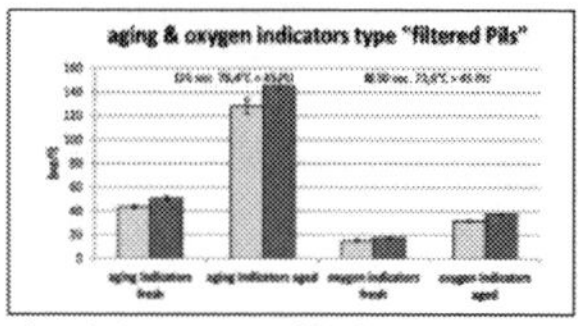

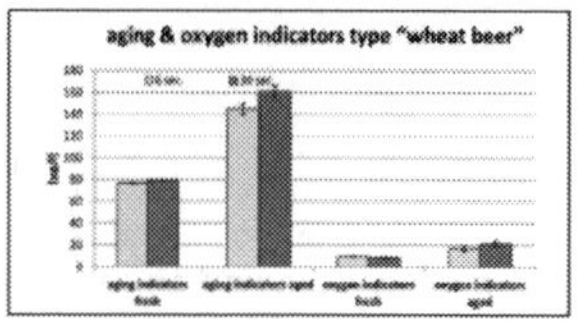

Short holding time with higher temperatures in comparison creates aging and oxygen indicators

When the filtered beers are examined individually, shortening the heart-holding time from the traditional 30 seconds to 6 seconds (beers containing alcohol) already produces advantages in terms of beer quality, and in conjunction with unfiltered beers, an analytical measurable reduction in thermal stress. The microbiological safety could be more safety by the 6 sec., because the dwell time distribution is more homogenous.

Shortening the time enables the critical temperature for a desired protein denaturation for haze stabilisation of the unfiltered beers to be achieved using one heat-holding section, without over-pasteurising the beer. Even an output adjustment is possible without letting the temperature fall below the critical value of 80°C with a target value of a constant 220 PUs.

All beers examined in the comparison showed lower key indicators for oxygen, heat and ageing with short heat-holding. The taste results shows no significant different.

So we can say that a shorter heat holding have benefits for the beer quality – but especially in the higher microbiological safety, material savings, pressure drops and foot print.

Conclusion

Recommended dimensioning parameters for beer flash pasteurisers

Beer types	Heat-holding time	Temperatures 100 % output*	Temperatures 50 % output*
Alcoholic, filtered beer	6 seconds	77°C (28 PU) 78.4°C (45 PU) Maximum	74.9°C (28 PU) 76.3°C (45 PU)
Non-alcoholic filtered beer's & "critical" beers**	10 seconds	78.2°C (70 PU) Maximum	76.1°C (70 PU)
Unfiltered beer	6 seconds	83.6°C (220 PU) 82.1°C (150 PU)	81.1°C (220 PU) 80.0°C (150 PU) Minimum
Non-alcoholic filtered beer or critical beer, in conjunction with alcoholic beer	10 seconds	**Unfiltered beer:** 81.5°C (220 PU) 80.5°C (150 PU) **Filtered alcoholic beer:** 75.5°C (28 PU) 76.9°C (45 PU) **Filtered non-alcoholic beer:** 78.2°C (70 PU) Maximum	**Unfiltered beer:** 81.1°C (220 PU) 80°C* (150 PU) *Minimum reached at 90 % of output! **Filtered alcoholic beer:** 73.4°C (28 PU) 74.8°C (45 PU) **Filtered non-alcoholic beer :** 76.1°C (70 PU)

KRONES

centec

Centec GmbH

Gesellschaft für

Labor- und Prozessmesstechnik

Martin Jörg[1], Lisa Riedl[1*], Dr. Robert Koukol[2]

Controlling of CO2-purity degree at the beer fermentation as a parameter for the quality control

Centec PureConStar: A STEP ACCORDING FDA-PAT TOWARDS THE FULL AUTOMATICAL CONTROLLED FERMENTATION PROCESS

Cost saving by the new and improved process and quality control Centec PureConStar: First of all the optimizing of the CO2 control processing for the CO2 recovery systems bears a potential economical return of invest for the Centec PureConStar system within a few years!

The poster focusses second on the new additional possibility of monitoring and controlling the fermentation process, which leads to a benefit of interest towards the full automatic fermentation process control, according the Food and Drug Administration (FDA), who has made a major contribution to this. The Process Analytical Technology Initiative (PAT Initiative) forces the industry towards the fastest full automatic real time quality control by in-line sensor analysis, to minimize the slow and expensive at-line lab analysis in the future.

Fermentation failures must not only occur through suboptimal yeast treatment, but may also result from technical deficits by processing. These errors can be recognized now more earlier by the measurement of the CO2 purity level than in the case of others like common pH or extract reduction. A non starting fermentation is one of the worst case scenario in the brewing process, which leads to several additional problems in the followed up production and beer quality.

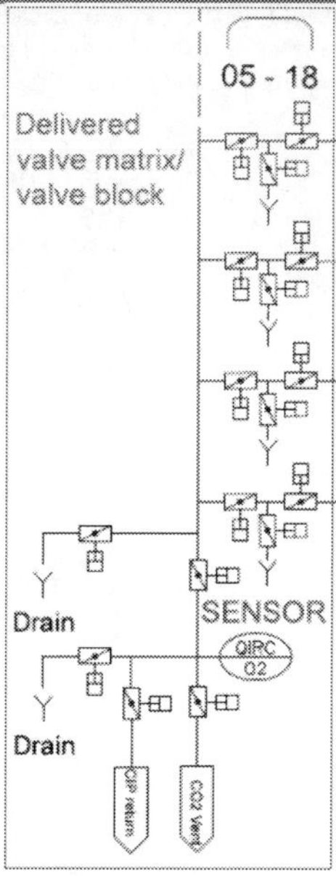

Figure 1: centec Oxytrans O2-monitor integration and installation into the fermation cellar tank farm.

Centec PureConStar shows the reproducibility of the measuring principle under constant conditions. That can be defined minimum via the standard deviation to a corridor of max./min. limit values for the control. Furthermore the reduction of O2 according the CO2 purity level can be measured much more earlier and significant than the common measured fermentation parameters.

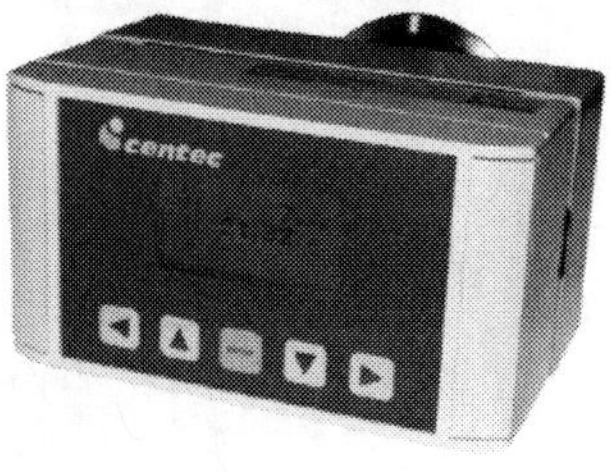

Figure 2: centec Oxytrans TR in-line O2 Sensor

In terms of an average market price of approximately 0,15 €/kg for CO2, the theoretical value of approximately 1 million € are produced of CO2 from 2 million hl beer according the Balling formula. Therefore each percentage more recovered CO2 by Centec PureConStar saves about 10.000 €.

O2 Reduction by Venting

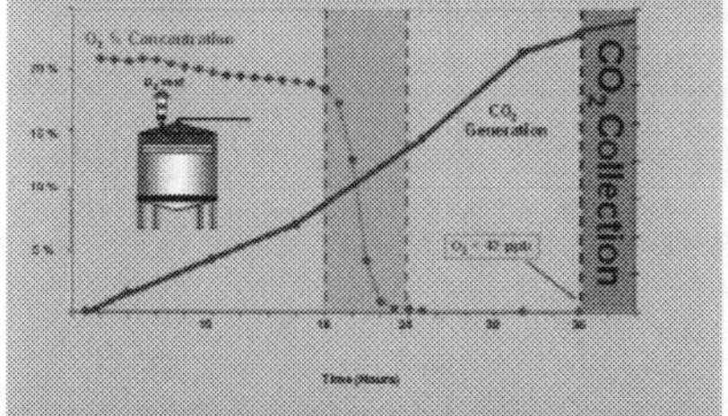

Figure 3: CO2/O2 within the fermentation process The figure 3 shows that the CO2 production during the fermentation is stoichiometrically almost linear, whereby the CO2 purity level development in relation to the O2 decrease starts after about 18 hours and reaches its final value within a few hours. This fact suggests that the measured O2 reduction within the CO2 purity development can be used as a parameter for the fermentation process control. Due CO2 can be only used for the CO2 recovery system by certain purity level, the measurement improves the recovery process from a empiric to a full measureable automatic process, where losses by failure are not present anymore and all usable CO2 will be used.

Fermentation QM monitoring parameters

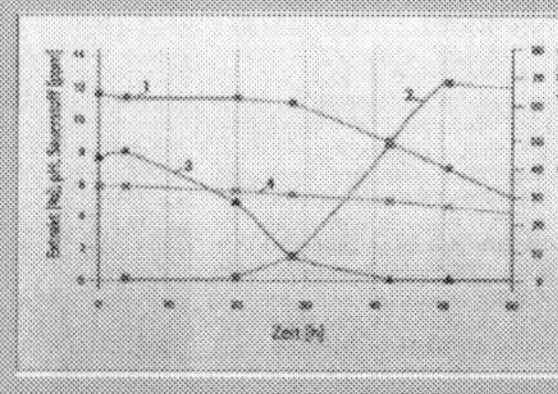

Optimum fermentation:
(1) Extract (2) Yeast cells (3) O2 (4) pH

Figure 4: Comparison of fermentation parameter measuremnents. The figure 4 shows that the CO2 purity increase with the decrease of the O2 can be measured much more earlier and significant compared to the usual existing fermentation parameters. It is also apparent that the change of the measured value drop is significantly more clearly detectable in the start than the pH or extract. As a result, deviations from the standard fermentation process can be detected much faster within 20 hours and the reaction time for process correction measures can be reduced according.

Process control

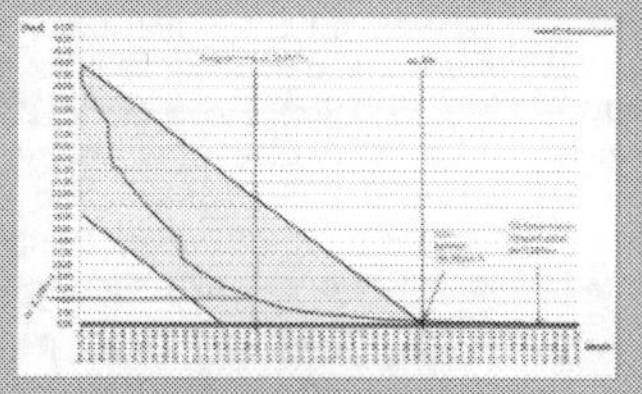

Corridor of max./min. limit values for the control.
Measurement start 23 hours after fermentation start by tank full

Figure 5: Min/max corrdiors. When constant process parameters are given and the fermentation process is referred to as relatively stable, a standard deviation to the measured value can be determined by the usual course of the CO2 purity degree development and thus constant deviation limits are calculated by the standard measurement value. If these minimum and maximum limits are interrupted, this can be detected by the Centec oxygen sensor and correspondingly reacted.
-> Extract and pH measurements show much later whether a fermentation or process failure is present!

→ Improvement of the fermentation in-line quality control
→ Improvement of the CO2 recovery within the advanced ecological "green foot print"
→ Acceptable Return of invest within the benefit of an improved fermentation process

centec

*Corresponding author

Dipl.Brmstr. Martin Jörg
E-Mail:
Tel. : +49 6181 71878 19
Fax: +49 6181 71878 50

[1]Contact

Centec - Gesellschaft für
Labor- und Prozessmesstechnik mbH
Wilhelm-Röntgen-Straße 10
D-63477 Maintal/Frankfurt

[2]Contact

Robert.Kuokol@CenteC.de
Centec GmbH
Wilhelm-Roentgen-Str. 10
D-63477 Frankfurt-Maintal / Germany

centec

Hop Citrus Character

Why are all the molecules floral?

R.J.Marriott ray.marriott@totallynaturalsolutions.com and C.Wilson colin.wilson@totallynaturalsolutions.com

Totally Natural Solutions Ltd, East Peckham, Kent, TN12 5LG, UK

Introduction

Citrus character from hop addition to beers is an important flavour characteristic that has been developed in new hop varieties that are now popular for dry hopping. There have been a number of studies that have shown a correlation between citrus character and some monoterpene alcohols such as linalool and citronellol but all the molecules identified that are either found in hops or metabolised from hop precursors are overwhelmingly floral in character [1,2]. Synergism between these is thought to develop citrus characters but this cannot be demonstrated with the pure compounds alone. In contrast almost all of the characteristic aroma molecules in citrus oils are aliphatic or terpene carbonyls for example geranial, citronellal, nootkatone and sinensal which have significant lower odour thresholds than the corresponding alcohols. This work explores the influence of terpene aldehydes that potentially contribute to citrus flavour and aroma in hops and beer.

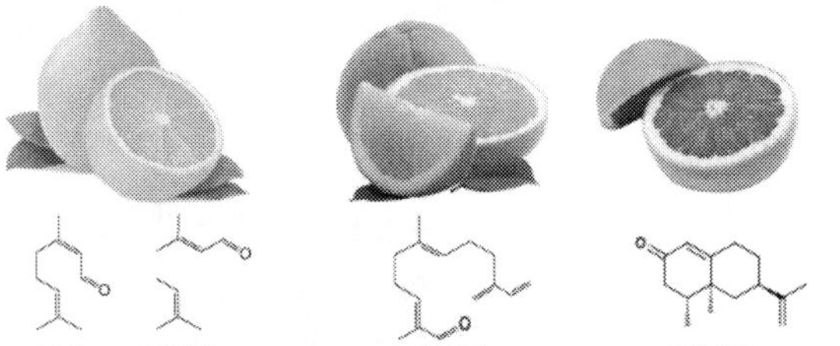

neral geranial sinensal nootkatone

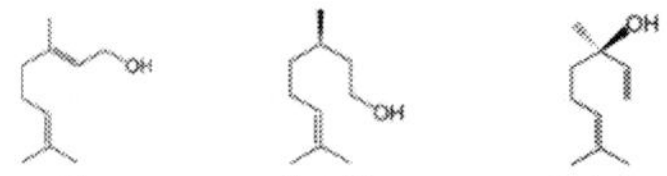

geraniol citronellol linalool

Influence of processing and brewing addition

Terpene aldehydes such as geranial are known to be lost during steam distillation both from hops [3] and other aromatic plants such as lemon balm. It is therefore to be expected that they will not survive wort boiling and we have not observed them in any wort samples that we have examined. Even if the aldehydes were to survive through to the wort the reductive environment of fermentation would almost certainly result in their conversion to the corresponding alcohol and then potentially metabolised to β-citronellol as proposed by King *et al* [4].

However dry hop addition does result in the transfer of the aldehydes into beer and the terpene aldehydes such as geranial can be successfully extracted from the hops using low temperature extraction using carbon dioxide as shown in figure 1.

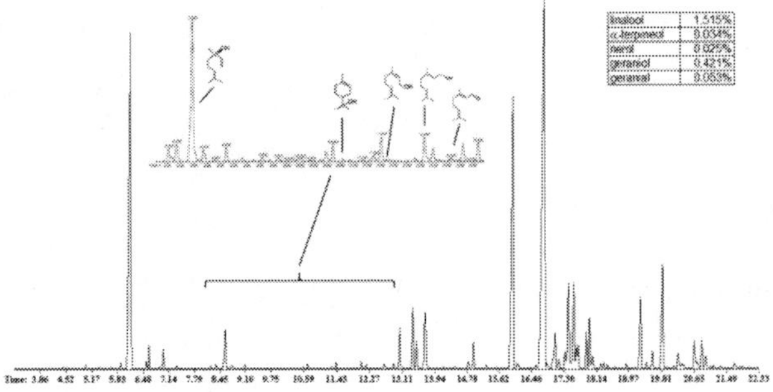

Figure 1. Monoterpene alcohols and aldehydes in hop oil from "Citra" hops extracted using low temperature carbon dioxide

Beer concentration and taste thresholds

The presence of terpene aldehydes such as geranial can be easily demonstrated in hops or hop oils but unless they are at a level that is above the taste threshold in beer contribution to citrus flavour is unlikely. However given the relatively high levels of dry hopping in craft beers and a transfer efficiency of 40-60% [5] linalool, geraniol and geranial are all above the reported taste threshold [6].

Compound	Citra hops (ppm)	Dry hopping (g/hl)	Transfer efficiency	Beer (ppb)
linalool	242.40	250	39%	236.3
α-terpineol	5.44	250	43%	5.8
nerol	4.00	250	44%	4.4
geraniol	67.36	250	41%	69.0
geranial	8.48	250	51%	10.8

Table 1. Concentration of monoterpene alcohols and aldehydes from "Citra" hops resulting from dry hopping as determined by headspace - SPME

Alternatively hop oil isolated using a benign method can be added directly to beer to provide the same levels of terpene alcohols and aldehydes.

Compound	Citra hop oil (%)	Oil addition (g/hl)	Transfer efficiency	Beer (ppb)
linalool	1.515	2	100%	303.0
α-terpineol	0.034	2	100%	6.8
nerol	0.025	2	100%	5.0
geraniol	0.421	2	100%	84.2
geranial	0.053	2	100%	10.6

Table 2. Concentration of monoterpene alcohols and aldehydes from the addition of "Citra" hop oil

Conclusions

This is only preliminary data but it appears that terpene aldehydes are unlikely to contribute to citrus flavour if the hops are added to the kettle or if the hop oil is isolated by steam distillation. However the level of geranial arising from traditional dry hopping or by addition of hop aroma isolated using a benign method is above the taste threshold in beer and is likely to contribute or modify citrus flavour. Further work to correlate the analytical levels with sensory profiles are ongoing.

References

1. Takoi, K., Itoga, Y., Koie, K., Kosugi, T., Shimase, M., Katayama, Y. & Watari, J. (2010). The Contribution of Geraniol Metabolism to the Citrus Flavour of Beer: Synergy of Geraniol and β-Citronellol Under Coexistence with Excess Linalool. *Journal of the Institute of Brewing*, *116*(3), 251-260.

2. Praet, T., Van Opstaele, F., Jaskula-Goiris, B., Aerts, G., & De Cooman, L. (2012). Biotransformations of hop-derived aroma compounds by Saccharomyces cerevisiae upon fermentation. *Cerevisia*, *36*(4), 125-132.

3. Nickerson, G. B., & Van Engel, L. (1992). Hop aroma component profile and the aroma unit. *J. Am. Soc. Brew. Chem*, *50*(3), 77-81.

4. King, A., Dickinson, J. R. (2003) Biotransformation of hop aroma terpenoids by ale and lager yeasts. *FEMS Yeast Res.*, 3, 53–62

5. Bailey, B., Schönberger, C., Drexler, G., Gahr, A., Newman, R., Pöschl, M., & Geiger, E. (2009). The influence of hop harvest date on hop aroma in dry-hopped beers. *Technical Quarterly of the Master Brewers Association of the Americas*, *46*(2), 1-7.

6. Takoi, K., Koie, K., Itoga, Y., Katayama, Y., Shimase, M., Nakayama, Y., & Watari, J. (2010). Biotransformation of hop-derived monoterpene alcohols by lager yeast and their contribution to the flavor of hopped beer. *Journal of agricultural and food chemistry*, *58*(8), 5050-5056.

PO80

Mini-oak chips: a promising tool for development of flavour enriched beer

UNIVERSITY OF CHEMISTRY AND TECHNOLOGY PRAGUE

Hanko V.*, Mačaj K., Nešpor J., Jelínek L., Dostálek P.

University of Chemistry and Technology, Prague, Department of Biotechnology, Technická 5, CZ-166 28 Prague 6, Czech Republic.

* Corresponding author: **Tel.:** +420 220 444 427, **Email adress:** hankovo@vscht.cz

KEYWORDS: maturation in wood, wooden chips, flavour enriched beer

INTRODUCTION: Maturation in wood is one of the key steps in the production of high-quality alcoholic beverages. Nowadays wooden barrels or small wooden fragments (chips, bars or cubes) from oak (*Quercus*), chestnut (*Castanea*), cherry tree (*Prunus*) and mulberry (*Morus*) are frequently used for production of wine and some distillates (whiskey, cognac, brandy etc.). Employment of these wooden materials in brewing industry is limited due to short time of beer maturation and also low content of ethanol in beer, leading only to minimum transfer of flavour active compounds (FACs) into beer. Extension of surface area and therefore improved transfer of FACs can be reached by milling or grinding wooden chips into smaller particles. Sensory profile of final beer produced by using wooden chips depends apart of kind of wood, rate of toasting, dosage, and fragments size also on conditions of technological process (temperature, pH and oxygen access), content of ethanol in beer and type of produced beer. In this work were tested leaching of selected FACs from different fractions of wooden chips in model beer. The most promising fractions were then dosed into green beer, which was subjected to sensory analysis after tree weeks.

MATERIAL AND METHODS

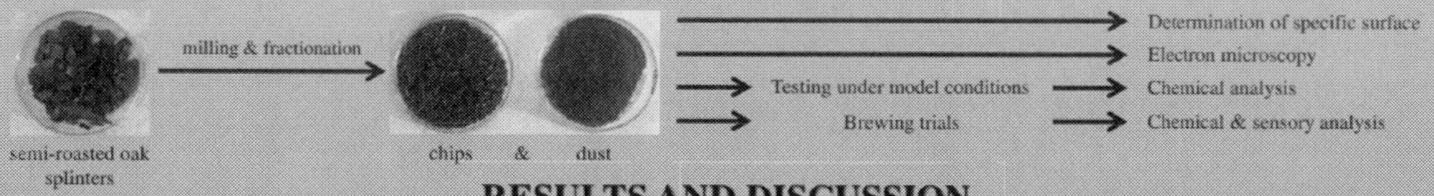

RESULTS AND DISCUSSION

Size and specific surface of wooden material

- Reduction of the particles size was successfully achieved by milling on cutting mill
- Specific surface of oak chips is ~ 20 rel.% larger then splinters
- Specific surface of oak dust is about **60 rel.%** larger then splinters
- Larger specific surface → higher transfer of substances into beer
- Oak dust could cause technological issues (extremelly small size)

Parameters	Oak splinters	Oak chips	Oak dust
Size (mm)	~ 10	> 5	< 0.5
Specific surface (m^2/g)	0.55	0.66	0.90

Electron microscopy of wooden fragments

Concentrations of FACs of oak wood (μg/l) in model solutions, depending on form and dose of fragment, time of maturation

Vanillin (vanilla aroma)

Taking samples after:	Dose of splinters (g/l)				Dose of chips (g/l)			
	2	4	6	12	2	4	6	12
1st week	334 ± 18	506 ± 9	837 ± 8	1497 ± 17	210 ± 6	367 ± 12	497 ± 10	886 ± 17
2nd week	343 ± 15	562 ± 11	937 ± 7	1712 ± 11	199 ± 17	386 ± 8	562 ± 15	1055 ± 11
3rd week	366 ± 10	624 ± 12	1071 ± 13	1887 ± 18	208 ± 8	399 ± 12	568 ± 9	1057 ± 10
4th week	374 ± 8	667 ± 11	1130 ± 11	1955 ± 15	203 ± 9	396 ± 7	578 ± 11	1041 ± 9

cis- oak lactone (oak, coconut aroma)

Taking samples after:	Dose of splinters (g/l)				Dose of chips (g/l)			
	2	4	6	12	2	4	6	12
1st week	17 ± 0.5	25 ± 0.9	35 ± 0.9	61 ± 0.5	16 ± 0.8	30 ± 0.9	37 ± 0.8	119 ± 0.6
2nd week	18 ± 0.5	30 ± 0.5	49 ± 0.8	92 ± 0.4	17 ± 0.4	35 ± 1.1	61 ± 0.6	135 ± 0.9
3rd week	12 ± 0.4	35 ± 0.7	60 ± 0.8	170 ± 1.5	17 ± 0.5	45 ± 1.1	99 ± 1.0	179 ± 1.0
4th week	96 ± 16.0	118 ± 1.2	145 ± 0.9	440 ± 5.9	67 ± 0.9	148 ± 1.3	216 ± 1.9	540 ± 10.0

Eugenol (clove aroma)

Taking samples after:	Dose of splinters (g/l)				Dose of chips (g/l)			
	2	4	6	12	2	4	6	12
1st week	-	2 ± 0.4	4 ± 0.8	6 ± 0.5	-	3 ± 0.3	4 ± 0.6	7 ± 0.5
2nd week	-	3 ± 0.2	5 ± 0.5	6 ± 0.4	-	4 ± 0.2	5 ± 0.5	9 ± 0.3
3rd week	-	3 ± 0.8	4 ± 0.5	6 ± 0.6	-	6 ± 0.4	6 ± 0.5	10 ± 0.9
4th week	-	6 ± 0.5	7 ± 0.9	16 ± 0.6	-	10 ± 0.5	12 ± 0.6	26 ± 1.0

Threshold value exceeded: < 5x < 10x < 20x < 30x < 40x

- The occurrence of individual substances is strongly influenced by the type of wood fragments
- The content of monitored substances depends on occurance in specific part (centre/surface) of fragment (vanillin), specific surface of fragment (oak lactone) or both (eugenol)

Concentrations of flavour active compounds of oak wood (μg/l) in beer

Addition of oak wood material during beer maturation	Vanilin	cis- oak lactone	Eugenol
None	-	-	-
Chips	826 ± 67.0	39 ± 8.1	4 ± 0.7
Splinters	447 ± 42.0	79 ± 9.6	8 ± 0.6

Threshold value exceeded: < 5x < 10x < 20x

- Lower content of monitored substances in beer compared with model solutions (~ 20 rel.%) was probably caused by their adsorption on yeast cells and degradation by yeast metabolism
- Practically all beers maturated with wood material exceeded threshold values several times

Comparison of sensory profile of beers maturated with oak fragments

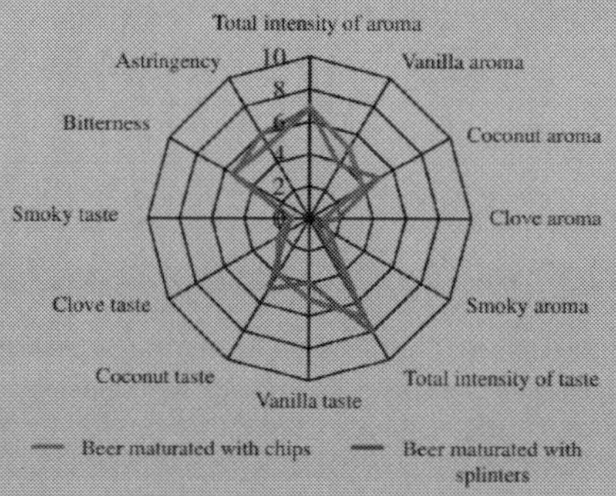

- Beer maturated with oak chips had strong vanilla flavour
- Vanilla flavour was noticed also in beer maturated with oak splinters, however, coconut and clove flavours were dominant

CONCLUSIONS

- Maturation in wood is appropriate especially for production of top-fermented, stronger, semi-dark or dark beers, like porter or stout, due to their coffee, chocolate and roasted aromas, which can be complemented with vanilla or coconut aroma from wood
- Even a small doses of wooden fragments during beer production can influence the sensory profile of beer
- The possibilities of reaching specific aroma/flavour:

  vanilla (vanillin) – oak chips, minimal dose 2 g/l, at least 2 weeks of maturation

  coconut (oak lactone) – oak splinters, minimal dose 6 g/l, at least 3 weeks of maturation

  clove (eugenol) – oak splinters, minimal dose 4 g/l, length of maturation does not matter

- Combination of wood splinters and chips is appropriate for reaching required aroma

Acknowledgement: This work was supported by TACR „Centre for Innovative Use and Strengthening of Competitiveness of Czech Brewery Raw Materials and Products" – TE02000177.

Can brewing with unmalted quinoa produce good beers?

Kordialik-Bogacka Edyta, Bogdan Paulina
edyta.kordialik-bogacka@p.lodz.pl

Lodz University of Technology
Faculty of Biotechnology and Food Sciences
Institute of Fermentation Technology & Microbiology

INTRODUCTION

The use of quinoa as a partial substitute for barley malt gives brewers the opportunity to develop final products with new sensory characteristics.
However, the properties of quinoa, also favourable from nutritional point of view, can cause technological problems and compromise the quality of the final product. Hitherto, there have not been detailed studies on processability of quinoa, which has induced in brewers doubts or even fear against its use.

AIM

The aim of this study was to evaluate the effect of partial malt replacement with unmalted quinoa on wort and beer quality. In this investigation the malt was substituted with up to 30% quinoa seeds or up to 40% quinoa flakes. The benefits of the application of commercial enzyme mixture for brewing with quinoa were also assessed.

MATERIALS AND METHODS

Unmalted quinoa (*Chenopodium quinoa*) in the form of seeds or flakes and barley malt were used to produce the wort and beer. During wort production the infusion-mashing procedure was applied: 30 min at 45°C, 60 min at 62°C, 30 min at 72°C and 10 min at 78°C. The mashing process with 30% quinoa flakes was performed without and with exogenous enzyme mixture (Brewers Compass® - an enzymatic complex consisting of β-glucanase, α-amylase and protease). After wort boiling with hop, the extract of all worts was adjusted to 12°P to compare the fermentability of worts produced with various quantities of quinoa. The hopped worts were analyzed according to the standard methods of Analytica EBC with regard to: colour, pH, viscosity, total soluble nitrogen (TSN), free amino nitrogen (FAN). Bottom-fermented beers were produced by fermentation of 400 ml worts (12°P) with *Saccharomyces pastorianus* W34/70 at 10°C for 14 days. After fermentation beers were matured for 14 days at 1°C. Experimental procedures were carried out in triplicate. Extract, attenuation, colour, pH, viscosity, TSN and ethanol content were measured in beers. The means of the results with standard deviations are presented. A student t-test was performed to determine the statistical significance. A p value below 0.05 was considered statistically significant. Sensory analysis of all-malt beer and beer produced with 30% quinoa flakes was carried out by a trained 42-member taste panel. Aroma, taste, mouthfeel, carbonation, bitterness and the overall mark were evaluated with a score from 1 (the worst) to 6 (the best).

RESULTS AND DISCUSSION

- The extract content of the worts obtained with quinoa was lower compared with the all-malt wort (Tab. 1).
  The extract content in the wort produced with 30% quinoa rose as a result of using enzyme preparation but it was still lower than that in all-malt wort.
  There were not statistically significant differences in an extract yield between worts produced with 30% quinoa seeds and flakes.
- The colour of worts and beers increased with increasing contribution of adjuncts (Tabs. 2 and 3).
  The use of exogenous enzymes during brewing with 30% quinoa flakes did not affect the colour of wort but had an impact on the colour of beer.
- The greater the amount of malt that was substituted with quinoa, the higher was the pH value of wort and beer.
- Wort and beer viscosity increased when malt was replaced with adjuncts. The use of exogenous enzymes during production of wort with 30% quinoa flakes resulted in decrease of wort and beer viscosity to the level corresponding to all-malt wort and beer.
- The use of adjuncts caused the TSN level in wort and beer to decrease.
- Replacing malt with unmalted quinoa resulted in the reduced level of FAN in wort. Enzyme preparation contributed to increasing both the TSN and FAN contents in wort. However, these increases, although statistically significant, were not substantial.
- The alcohol content and attenuation of beer decreased with increasing replacement of malt with adjuncts. The application of enzyme preparation improved attenuation of wort obtained with quinoa substantially.
- Beer produced with 30% quinoa received in sensory evaluation the overall mark of 4.2 while all-malt beer 3.4.

Tab. 1. The extract content of sweet worts produced with various quantities of quinoa

Type of wort		Extract content [°P]
all-malt		13.64 ± 0.54 a
10%	quinoa seeds	12.82 ± 0.44 b
20%		12.48 ± 0.34 c
30%		12.27 ± 0.41 d
10%	quinoa flakes	12.83 ± 0.46 b
20%		12.31 ± 0.32 d
30%		12.28 ± 0.41 d
40%		11.53 ± 0.21 f
30% quinoa flakes + enzyme		12.64 ± 0.37 e

Tab. 2. Characteristics of hopped worts (12°P) produced from barley malt and quinoa seeds or flakes

Type of wort		Colour [EBC unit]	pH	Viscosity [mPa·s]	TSN [mg/L]	FAN [mg/L]
all-malt		9.1 ± 0.1 a	5.68 ± 0.04 a	1.66 ± 0.04 a	937 ± 17 a	190.3 ± 5.2 a
10%	quinoa seeds	10.3 ± 0.2 b	5.77 ± 0.03 b	1.70 ± 0.03 b	900 ± 28 b	182.0 ± 4.1 b
20%		10.3 ± 0.2 b	5.80 ± 0.06 b	1.72 ± 0.02 b	860 ± 20 c	175.5 ± 6.9 c
30%		11.4 ± 0.2 c	5.91 ± 0.05 c	1.77 ± 0.02 c	845 ± 15 d	160.2 ± 5.9 d
10%	quinoa flakes	10.2 ± 0.3 b	5.78 ± 0.06 b	1.71 ± 0.02 b	895 ± 19 e	183.8 ± 6.9 b
20%		10.3 ± 0.2 b	5.82 ± 0.06 b	1.73 ± 0.02 b	870 ± 27 f	175.6 ± 4.5 e
30%		10.4 ± 0.2 b	5.90 ± 0.04 c	1.78 ± 0.03 c	860 ± 20 c	168.2 ± 5.2 f
40%		11.6 ± 0.1 c	5.96 ± 0.02 d	1.85 ± 0.03 d	820 ± 15 g	145.5 ± 4.2 g
30% quinoa flakes + enzyme		10.4 ± 0.3 b	5.85 ± 0.07 c	1.66 ± 0.03 a,b	877 ± 12 i	174.4 ± 7.4 e

Tab. 3. Characteristics of beers produced from barley malt and quinoa seeds or flakes

Type of beer		Extract [°P]	Alcohol [% vol]	Attenuation [%]	Colour [EBC unit]	pH	Viscosity [mPa·s]	TSN [mg/L]
all-malt		4.21 ± 0.51 a	5.40 ± 0.21 a	65	5.4 ± 0.1 a	4.55 ± 0.09 a	1.49 ± 0.02 a	695 ± 12 a
10%	quinoa seeds	4.19 ± 0.23 a	5.22 ± 0.29 b	66	6.8 ± 0.2 b	4.61 ± 0.08 a	1.49 ± 0.03 a	638 ± 17 b
20%		4.35 ± 0.45 b	4.94 ± 0.40 c	64	7.2 ± 0.2 c	4.69 ± 0.06 b	1.51 ± 0.04 a	617 ± 19 c
30%		4.69 ± 0.44 c	4.85 ± 0.31 d	62	8.9 ± 0.2 d	4.72 ± 0.08 b	1.56 ± 0.04 b	603 ± 20 d
10%	quinoa flakes	4.10 ± 0.33 d	5.29 ± 0.21 b	67	6.6 ± 0.2 b	4.59 ± 0.10 a	1.50 ± 0.01 a	631 ± 17 b
20%		4.28 ± 0.25 b	4.96 ± 0.28 c	65	7.1 ± 0.1 c	4.68 ± 0.05 b	1.51 ± 0.02 a	611 ± 19 e
30%		4.61 ± 0.51 c	4.92 ± 0.25 c	63	8.5 ± 0.2 e	4.70 ± 0.05 b	1.56 ± 0.02 b	601 ± 20 c
40%		5.24 ± 0.27 e	4.32 ± 0.20 e	57	9.0 ± 0.2 f	4.81 ± 0.07 c	1.58 ± 0.03 b	576 ± 11 f
30% quinoa flakes + enzyme		4.22 ± 0.34 a	5.26 ± 0.17 b	65	7.7 ± 0.1 g	4.65 ± 0.06 b	1.51 ± 0.01 a	610 ± 19 e

Acknowledgements The authors would like to thank Cezary Góral (Biochem-Art) for kindly providing enzymatic preparation.

36TH CONGRESS EBC 2017 EUROPEAN BREWERY CONVENTION 14-18 MAY LJUBLJANA SLOVENIA

Casa abierta al tiempo
UNIVERSIDAD AUTÓNOMA METROPOLITANA

LILLE ISA

"Sensory profile of a gluten free beer made 100% with corn malt"

M.A Romero-Medina[1], H.B Escalona-Buendia[1], J.R Verde-Calvo[1], M. Lelievre-Desmas[2].
Universidad Autónoma Metropolitana[1], México; Institut Supérieur d'Agriculture-Lille[2].
mangelika7@gmail.com

INTRODUCTION

Beer is one of the most popular fermented beverages all over the world. Traditionally brewed with barley and wheat malt, typical beers turns in an unsuitable beverage for people who suffer from celiac disease, the intolerance of gluten proteins contained in cereals like barley and wheat.
One appropriate alternative to produce gluten-free beers is to replace barley with gluten-free grains such as corn, which has been used to elaborate foodstuffs for celiac patients. However, the use of 100% corn malt in beer process is a challenge due to the different chemical composition of corn (proteins, sugars) that could impact their sensory characteristics and the quality of the beer.
Hence, using pigmented varieties of Mexican corn as the main ingredient in brewing allows to develop alternative and innovative styles for individuals with gluten intolerance and at the same time incorporate traditional flavors of Mexican beverages, without compromising the sensory quality of beer.

OBJECTIVE

The objective of this study was to determine the effect of two varieties of pigmented Mexican corn on the sensory profile of craft beers made with this gluten-free cereal.

MATERIAL

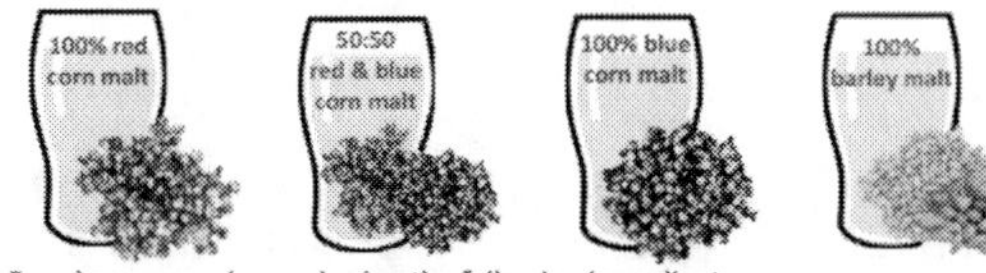

Four beers were brewed using the following ingredients:

Malts
- Two varieties of pigmented corn: red and blue Chalqueño breeds from Milpa Alta, Mexico City.
- Barley malt: 2-row American Brewers Malt and Roasted Caramel Malt 60L (Briess Malt & Ingredients Co., Chilton, WI, USA)

Hops
- Saaz (3-5 α-acids) for aroma purpose
- Magnum (13-15 α-acids) for bitter purpose

Yeast
- *Saccharomyces cerevisae* Safale US-05 (Fermentis Division of S. I. Lesaffre, Marcq-en-Baroeul, France)

Adjuncts
- Guajillo chili (*Capsicum annuum*) from Milpa Alta, Mexico city.

METHODOLOGY

Brewing process: All beers were produced using the typical brewing process and all steps remained constant, also all malting process was carried out according to the method described in Mexico Patent nº MX/a/2014/009140 (Verde Calvo et al., 2014)

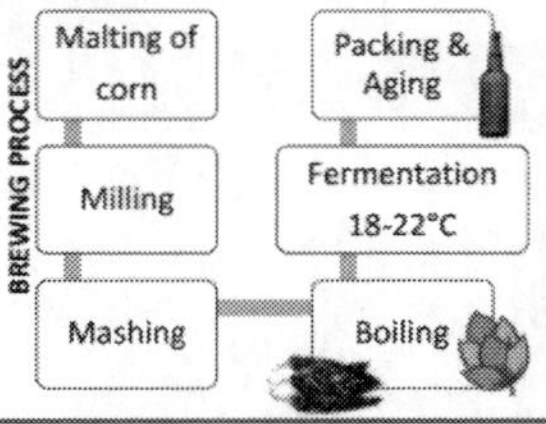

Sensory Evaluation
Beer prototypes were characterized by Quantitative Descriptive Analysis (QDA). Evaluations were conducted twice

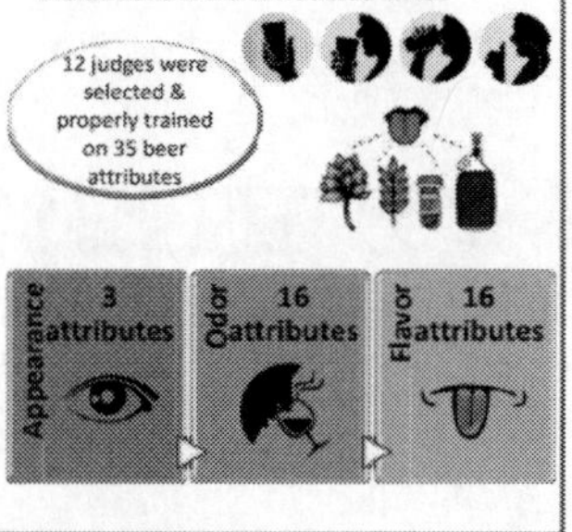

STATISTICAL ANALYSIS

Attributes were rated for intensity on non-structured scale (15cm) and the data were collected using FIZZ software (Biosystemes, FIZZ, Sensory Software version 2.5, Couternon, France).

The data were analyzed by principal component analysis (PCA) a multivariate method and analysis of variance (ANOVA) was applied to each attribute. Statistical analysis was conducted using R: A Language and Environment for Statistical Computing, Vienna, Austria, 2016.

RESULTS

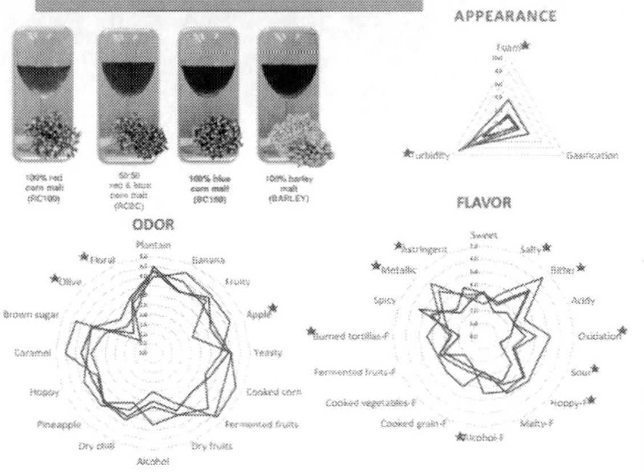

Fig. 1.Sensory profiles of beers made with different types of malts. The stars indicate a p-value <0.05 in the ANOVA.

The type of malt used in beers impacts their sensory profile. The 4 beers are significantly different ($p<0.05$) on 14 attributes like foam, turbidity, floral, olive and apple odor and salty, bitter, sour, burned tortillas, alcohol, hoppy flavors. Also astringent, metallic, oxidation sensation were significantly different.

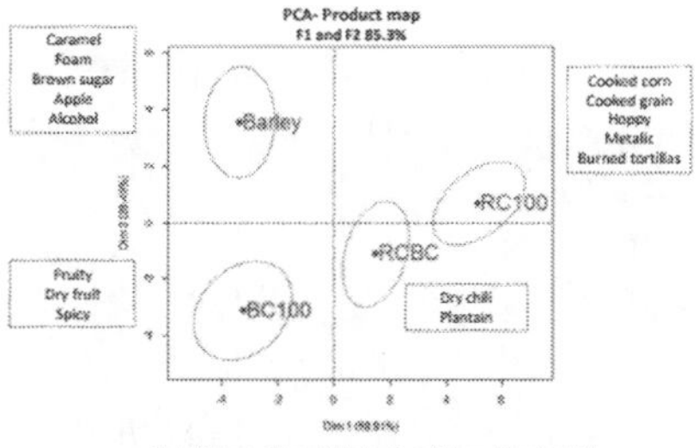

Fig. 2. PCA-map of the individuals with confidence ellipses ($p<0.05$).
The framed attributes describe the products surrounded by the same color

From the Fig. 2, it can be seen that:
- The 4 beers are well distinguished as the confidence ellipses do not overlap.
- Corn beers share attributes of typical barley beers: bitter, hoppy, floral and malty.
- Corn beers have also specific sensory characteristics: spicy, burned tortillas, olive, dry chili, cooked corn and fermented fruits.
- The blue corn beer is the most similar beer to the barley beer, sharing some characteristics: sweet, fruity and dry fruits.

CONCLUSION

Native varieties of Mexican corn were used for the first time as main ingredient of beer. Blue and red corn malt seem to be good gluten-free options for the brewing industry.
These new beers preserve the typical sensory characteristics of the traditional barley beers as well as interesting new flavors.
Thus, blue and red corn beers offer a modern way to preserve the typical Mexican flavors.

REFERENCES

Verde Calvo, José Ramón, Héctor Bernardo Escalona Buendía, Nicte Nataly Cruz Rodriguez, and María Angélica Romero Medina. 2014. Proceso de elaboración de cerveza a base de maíz. MX/a/2014/009140.

Comparative brewing performances of quinoa, amaranth, einkorn, millet and chia to produce beer

HAUTE ECOLE Lucia de Brouckère PÔLE UNIVERSITAIRE EUROPÉEN DE BRUXELLES WALLONIE

Anne Pietercelie, Charline Lepoivre, Laurence Van Nedervelde

Department of Brewing Sciences and Fermentation Technology
Institut Meurice – Haute Ecole Lucia De Brouckère
1 avenue Emile Gryzon (B10), B-1070 Brussels, Belgium
(e-mail: a.pietercelie@meurice.heldb.be)

INSTITUT MEURICE

INTRODUCTION

Research on alternative adjuncts as brewing substrates is still needed to overcome technical constraints and annual harvest variability. In this project, the use of pseudocereal quinoa (Belgian and Peruvian), millet, amaranth, einkorn and chia seeds as potential beer ingredients is reported. Various throw proportions of raw material were tested at lab scale. Usual parameters like yield extract, colour, viscosity or free amino nitrogen were analyzed allowing us to produce 5 innovative beers with good brewing characteristics and interesting organoleptic profiles.

Adjuncts characterization (Congress mash)

- Extract, colour, FAN, sugar content, wort viscosity and conversion time present satisfying values in terms of mashing performances for nearly all tested adjuncts and conditions (20 or 40% maximum of the recipe to fit belgian legal definition of a "beer") as summarized in table 1.
- Compared to 100% pure malt, good yield extracts with adjunct pre-gelatinization at 90°C are obtained as illustrated by figure 1, even at 40% for quinoa's and einkorn. However, amaranth and mostly chia gave lower values.
- Chia seeds are definitively not suitable for brewing purpose mainly due to high protein content producing a slimy gel layer on mash top and therefore extended filtration time and viscosity.

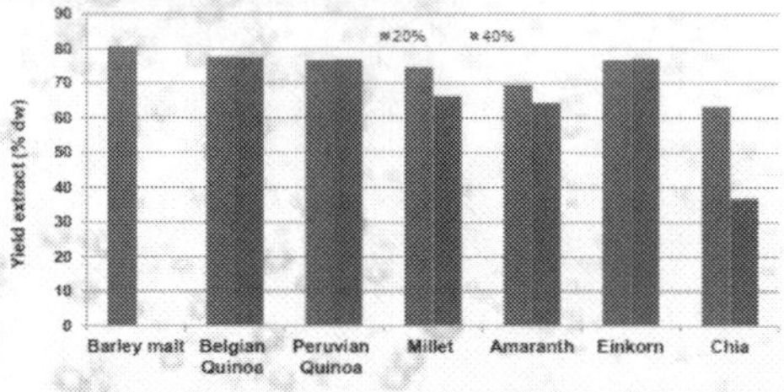

Fig. 1 : Yield extract with 20 or 40% of raw adjuncts (mean of duplicate)

Table 1: Brewing potential for Belgian and Peruvian Quinoa (BQ or PQ), Millet (M), Amaranth (A), Einkorn (E) and Chia (C) at 20 or 40% of 2RS pilsen malt (2RSP) throw (method EBC 6.3 with adjunct pre-gelatinization at 90°C) - Raw organic materials were purchased in local market - Harvest 2015

Wort Parameter	100% Pure malt	20% Belgian Quinoa	40% Belgian Quinoa	20% Peruvian Quinoa	40% Peruvian Quinoa	20% Millet	40% Millet	20% Amaranth	40% Amaranth	20% Einkorn	40% Einkorn	20% Chia	40% Chia
Extract (°P)	8.7	8.3	8.2	8.2	8.2	8.0	7.1	7.5	6.9	8.2	8.1	6.9	4.0
Colour (°EBC)	5.8	5.4	5.1	4.6	4.3	4.3	3.1	4.2	3.3	4.5	4.0	3.5	2.8
pH	5.7	5.9	6.1	6.0	6.1	5.8	5.8	5.9	6.0	5.8	5.6	6.0	6.2
Viscosity (mPa.s)	1.40	1.45	1.45	1.44	1.43	1.43	1.36	1.45	1.42	1.58	1.71	1.91	2.00
FAN (mg/L)	278	164	140	172	161	170	139	184	179	149	128	176	-
Fermentable sugars (g/L)	58	50	47	52	49	59	56	58	54	48	52	46	-
Conversion time (min)	7	10	21	10	15	12	19	10	10	12	17	10	12
Filtration time (min)	35	42	73	42	37	25	23	52	120	44	52	>120	>>120
Moisture (% dw)	5.0	9.6		7.4		10.3		9.4		10.0		8.6	
Total proteins (% dw)	10.1	12.5		11.6		12.2		16.0		11.8		17.0	

Lab-scale beer production (1L) with 20% of crude adjuncts

- Five beers (20% of adjunct) were produced in small amount (1L) with similar mashing programme and fermentation conditions in order to compare the raw cereal impact on finished product. Sugar profile of respective worts is presented in figure 2 and are in accordance with literature.
- Beer attributes are listed in table 2 and are globally within the range of usual values reported for a light blond ale beer.
- Good fermentation parameters (RDF) are mainly observed.
- High polyphenols content of quinoa and amaranth could be problematic for beer stability, final stabilization should be carefully optimized.
- Expert panel of 7 evaluators has briefly described beer sensory attributes. Despite a low alcohol content, a preference appeared for amaranth recipe (A20), perceived as well-balanced in terms of aroma and taste with fresh floral flavor and good drinkability whilst quinoa's, millet and einkorn seemed more neutral with a persistent astringency and/or roughness.

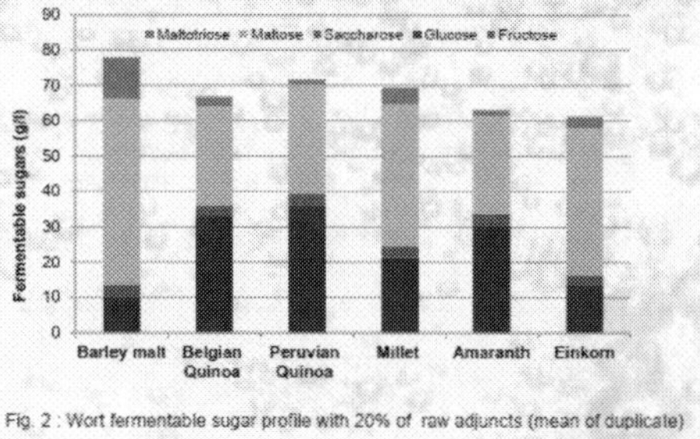

Fig. 2 : Wort fermentable sugar profile with 20% of raw adjuncts (mean of duplicate)

Table 2: Beer characteristics made with 20% of adjunct brewed with adapted mashing programme (10 min at 45°C - 25 min at 62°C - 30 min at 72°C - 2 min at 78°C), paper funnel filtration, boiling (60 min – hop Cluster addition to reach 30 EBU), cooling, pitching (1g/L) with Safbrew BE-256 from Fermentis, 7days of fermentation at 18°C, 7 days of maturation at 4°C.
Adjuncts were separately pre-gelatinizated at 90°C before mashing. Values are mean of duplicate.

Beer Parameter	20% Belgian Quinoa	20% Peruvian Quinoa	20% Millet	20% Amaranth	20% Einkorn
Original gravity (°P)	12.6	12.7	12.2	12.5	12.9
Alcohol (%v/v)	4.5	4.2	4.1	3.8	4.3
Real Degree Fermentation (%)	69.0	66.0	67.1	59.9	67.2
Colour (°EBC)	10.5	7.7	6.6	6.6	7.3
pH	4.5	4.4	4.3	4.4	4.4
Polyphenols (mg/L)	232	237	117	198	96
Bitterness (EBU)	27	28	26	30	27

CONCLUSIONS

Use of alternative ingredients as a part of the recipe is a natural way to produce new beers. In this work, the brewing abilities of raw quinoas (Belgian & Peruvian), proso millet amaranth or einkorn were clearly demonstrated. Five ale beers were produced in small volume and characterized. All present good values in terms of fermentability, colour or pH. The most appreciated recipe, with 20% of crude organic amaranth was scaled-up (20L) and a nice interesting gluten-free beer (light ale type) was produced, currently in refermentation, very promising to develop a fresh and tasty innovative beer.

"CO_2 FROM ANAEROBIC DIGESTION SOURCES"
A novel and safe source for the food and beverage industry?"

Author: Frank Verkoelen, Manager Strategic Project for QC - Pentair Haffmans - Venlo, The Netherlands – frank.verkoelen@pentair.com

INTRODUCTION

Climate changes are experiences worldwide. The temperature of the globe is rapidly increasing and this phenomenon is known as global warming.
At the climate summit in Paris, governments made agreements to bring the global CO_2 emissions to almost zero.
Brewing industry is already contributing using the green and sustainable CO_2 from fermentation. Biogas is another key renewable energy source that produces green and sustainable CO_2, with a carbon food print close to zero.

TECHNOLOGY/PROCESS

A biological material is broken down in an anaerobic fermentation process at which biogas is produced, containing primarily methane (CH_4) and carbon dioxide (CO_2). This CO_2 is an alternative CO_2 source and a second revenue stream, gaining popularity as a green and sustainable CO_2, for now and the future.

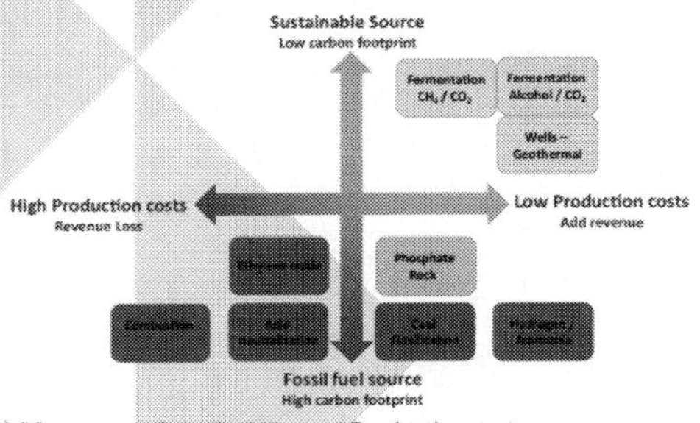

Figure 1) CO_2 sources, Sustainability and Production costs

Through legislation the feed process of an Anaerobic Digester is controlled. Various biogas sources are available and all have different contaminants that require different treatments. This also effects the treatment of the CO_2.

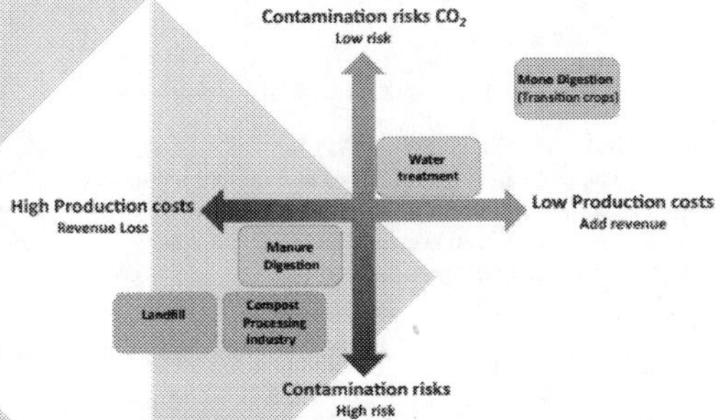

Figure 2) Biogas sources, CO_2 contaminations and Production costs

After collecting the Biogas, methane and CO_2 are separated. Following technologies are used to separate the methane and CO_2:
- Absorption
- Adsorption and
- Membranes

The CO_2 stream is further purified in a process with is similar to the purification of CO_2 from fermentation at breweries: the gas is washed with water, removing water solubles and ammonia, going through activated carbon treatment adsorbing components (incl. sulfur components and VOC), undergoes a heat treatment (during compression) where temperature rises to 110 °C as a sterilization step and is liquefied, removing all none condensable gasses including O_2.

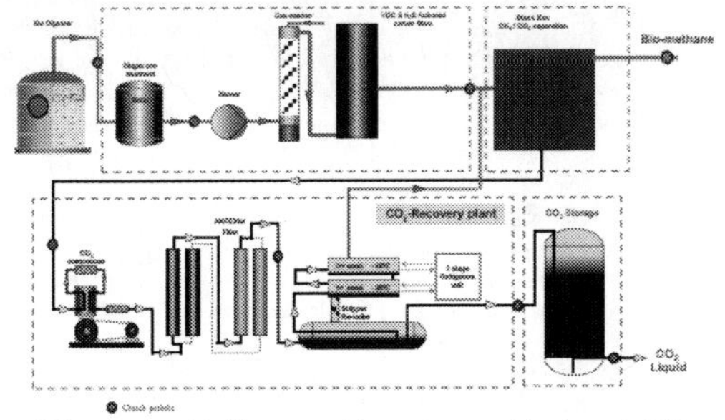

Figure 3) From Anaerobic Digester to bio-methane and food-graded CO_2

CO_2 Quality:

The complete process from fermentation, separation and CO_2 upgrading is highly monitored.
The recovered CO_2 meets the quality standards defined in the EIGA doc 70/17 "carbon dioxide food and beverage graded, source qualification, quality standards and verification" and has a purity > 99,998 %. A Food Safety Risk Assessment was completed with mayor gas companies with more than 700 CO_2 gas samples analyzed since June 2015, all within the contamination limits specified by EIGA.

CONCLUSION

The journey to a cleaner world has started with biogas producing bio-methane (CH_4) as a renewable energy sources. Recovering the CO_2 gas from the biogas production allows reducing use of fossil fuel for producing CO_2 only, that is still 70-75 % of the CO_2 produced and all with a carbon food print from almost zero. This green and sustainable CO_2 gas can be used for various applications and industries, including Food and Beverage, fulfilling the quality standards for food and beverage graded CO_2 specified by EIGA (Doc 70/17).

TECNOLÓGICO NACIONAL DE MÉXICO
INSTITUTO TECNOLÓGICO DE CELAYA

"Analysis of significant factors in the formation of sugars during the mash process in beer production and its modeling with neural networks"

Héctor Hugo Rivera-Yerena, José Francisco Louvier-Hernández, Teresa del Carmen Flores-Flores
Departament of Chemical Engineering, Tecnológico Nacional de México-Instituto Tecnológico de Celaya
Antonio García Cubas Pte #600 with Av. Tecnológico, Celaya, Guanajuato, 38090, México. **hector.rivera@iqcelaya.itc.mx**

Abstract
A statistical study based on a fractional experimental design with central points was carried out to simultaneously quantify the effects of temperature, pH, mash thickness, diastatic power and particle size distribution on starch hydrolysis during mash in beer production. The sugars, analyzed by HPLC, respond differently to each variable. The pH effect and the mash thickness are not included in previously published models. The results show that they are importants and should be included to achieve a better characterization of the system. The experimental data were used in the training of a neural network. The neural network presents a good fit to the experimental data, adequately representing the effects of pH and mash thickness on the formation of sugars.

Introduction

Brewing is a sequence of unit operations that produce a complex mixture of components wich determines beer physicochemical and sensory characteristics [1]. The most important operations are mash process and fermentation. In mashing the hydrolysis of different components occurs simultaneously, the most important process is the enzymatic conversion of the starch into simple sugars, since they represent more than 90% of the solubilized matter [2]. The carbohydrates have a direct influence on the yeast metabolism, influencing the profile of volatile compounds in beer [3] [4]. The sugar concentration is a function of several process variables.

Published models use mash temperature and malt enzymatic content as main variables in kinetic models [5]. It has been shown that pH and mash thickness ratio (kg water / kg malt) have important effects on the final sugar profile [6] [7], although no study has been published showing the magnitude of these factors simultaneously.

Artificial neural networks (ANN) have been shown to be useful in biochemical applications with reduced number of data [8] [9]. ANN have been implemented to make predictions about key compounds in the characteristics of beer [10] [11].

Method

Mashing, with water at 40 ppm of Ca^{2+}, was carried out in a laboratory batch reactor with temperature and agitation control. Fractional factorial design 2^{6-2}_{IV} with center points was carried out at the next levels under a typical infusion scheme: temperature 1 (T1, 50-60 ° C), temperature 2 (T2, 60-69 ° C), pH (pH 5.2-5.8), mash thickness (W, 2.33 -4), particle size distribution (Tpp, D1-D2) and diastatic power (DP, 120-140). The response was grams of carbohydrates per gram of malt for total sugars, maltose, maltotriose, glucose, fructose and sucrose. The concentration of sugars was performed by high performance liquid chromatography (HPLC) with a SUPELCOSIL LC-NH2 column and evaporative light scattering detector (ELSD). The mobile phase composition was 65% acetonitrile and 35% water at room temperature. Dilution 1:50 with deionized water and filtering with 0.2μm membranes was done before injection into HPLC system. The data were used to perform an analysis of main factors and interactions, as well as the regression and ANN models. The number of layers and nodes were determined by trial and error. Conventional feed forward ANN with Levenberg-Marquardt algorithm for training was used.

Results

Statistical analysis shows that the total sugar concentration is strongly influenced by the pH and mash thicknes ratio, in addition to the temperature. In all cases the particle size distribution was not significant. The results are presented in Figure 1.

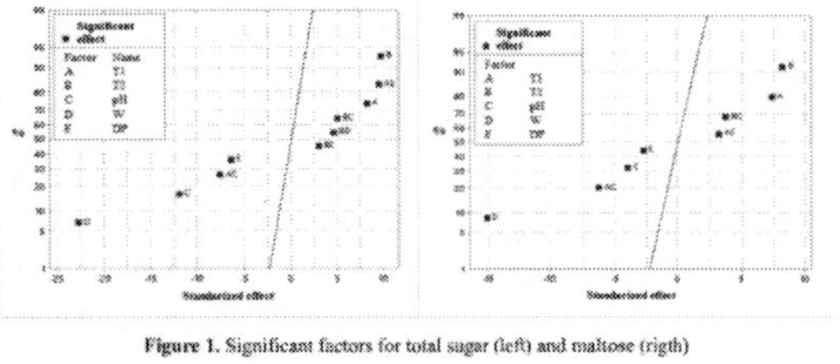

Figure 1. Significant factors for total sugar (left) and maltose (rigth)

Results of regression analysis that considers main effects and interactions shows the need to include second order factors due to lack of fit. The estructure of ANN model is presented in Figure 2.

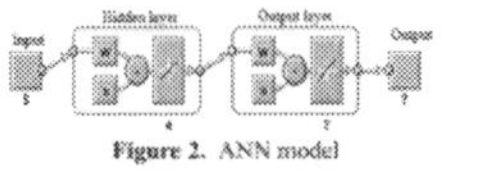

Figure 2. ANN model

Both regression and ANN models have good fits to experimental data for each carbohydrate, with more dispersion in regresión model, Figure 3.

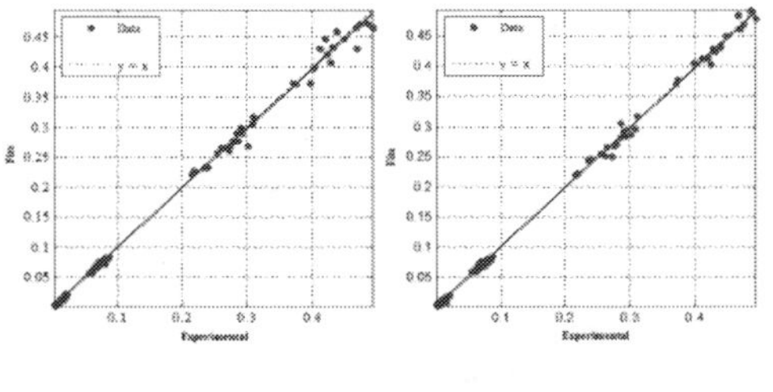

Figure 3. Regression model (left) and ANN model (rigth)

The Figure 4 shows the response of total sugars to the main factors, where curvature is evident.

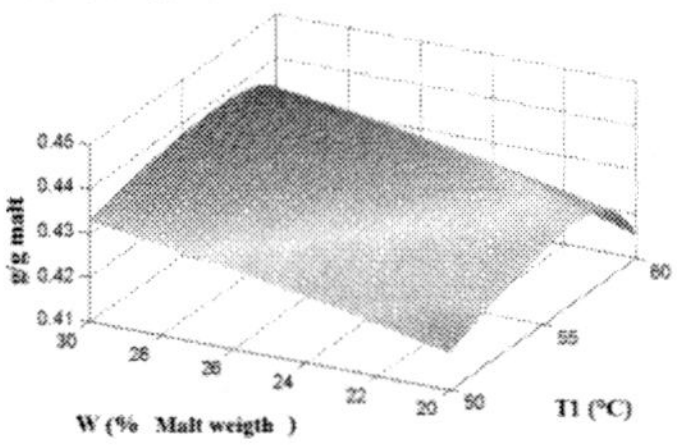

Figure 4. Total sugars in function of mash thickness and temeperature

Conclusions

The statistical evidence shows that pH and mash thickness ratio are as important as the temperature and enzymatic content in mashing. An should be included in prediction models. Regression and ANN models provides a good representation of mashing process.

References

[1] G. A. Da Silva, F. Augusto, and R. J. Poppi, "Exploratory analysis of the volatile profile of beers by HS-SPME-GC," Food Chem., vol. 111, pp. 1057–1063, 2008.
[2] D. Briggs, P. Brookes, R. Stevens, and C. Boulton, Brewing: science and practice. Woodhead Publishing, 2004.
[3] Y. He, J. Dong, H. Yin, Y. Zhao, R. Chen, X. Wan, P. Chen, X. Hou, J. Liu, and L. Chen, "Wort composition and its impact on the flavour-active higher alcohol and ester formation of beer – A review," J. Inst. Brew., vol. 120, no. 3, pp. 157–163, 2014.
[4] M. Riu Aumatell, P. Miró, A. Serra Cayuela, S. Buxaderas, and E. López Tamames, "Assessment of the aroma profiles of low-alcohol beers using HS-SPME-GC-MS," Food Res. Int., vol. 57, pp. 196–202, 2014.
[5] C. Brandam, X. M. Meyer, J. Proth, P. Strehaiano, and H. Pingaud, "An original kinetic model for the enzymatic hydrolysis of starch during mashing," Biochem. Eng. J., vol. 13, pp. 43–52, 2003.
[6] G. De Rouck, B. Jaskula, B. De Causmaecker, S. Malfliet, F. Van Opstaele, J. De Clippeleer, J. De Brabanter, L. De Cooman, and G. Aerts, "The influence of very thick and fast mashing conditions on wort composition," J. Am. Soc. Brew. Chem., vol. 71, pp. 1–14, 2013.
[7] D. E. Evans, M. Goldsmith, K. S. Redd, R. Nischwitz, and A. Lentini, "Impact of Mashing Conditions on Extract, its Fermentability, and the Levels of Wort Free Amino Nitrogen (FAN), beta-Glucan, and Lipids," J. Am. Soc. Brew. Chem., vol. 70, no. 1, pp. 39–49, 2012.
[8] D. Baş and I. H. Boyacı, "Modeling and optimization II: Comparison of estimation capabilities of response surface methodology with artificial neural networks in a biochemical reaction," J. Food Eng., vol. 78, pp. 846–854, 2007.
[9] R. Lanouette, J. Thibault, and J. L. Valade, "Process modeling with neural networks using small experimental datasets," vol. 23, pp. 1167–1176, 1999.
[10] C. Riverol and J. Cooney, "Estimation of the ester formation during beer fermentation using neural networks," J. Food Eng., vol. 82, pp. 585–588, 2007.
[11] Y. Zhang, S. Jia, and W. Zhang, "Predicting acetic acid content in the final beer using neural networks and support vector machine," J. Inst. Brew., vol. 118, no. October 2012, pp. 361–367, 2012.

Evaluation of poly phenols and bitter substances derived from hops and their yield by producing hopped wort concentrates for the production of beer and beer type malt beverages

Hochschule Anhalt

Anhalt University of Applied Sciences

Jean Titze*, Mandy Hirschmann and Renate Richter

Anhalt University of Applied Sciences • Department of Applied Biosciences and Process Engineering • Koethen

Introduction

In the last years, the use of hopped wort concentrates has increased significantly. Hopped wort concentrate is a concentrated brewer's wort with 60 to 70 °Plato [1]. The use of a falling film evaporator is commercially the most common method for the concentration step. Besides all the possibilities and benefits of these concentrates, the concentration process seems to have a negative effect on some valuable parameters. It has been shown that hop losses of 10 to 15% occurred during the concentration step [2]. Additionally, no answer has been given regarding the yield of poly phenols derived from hops in wort concentrates so far.

In this work, the influence of the concentration step on the yields of poly phenols and bitterness substances of three different hop products (pellets, hop extract, spent hops) were evaluated.

In a recent investigation antioxidative key substances during the brewing process were characterized, in particular to which extent the poly phenols were dissolved in wort [3]. Based on these results, the influence of the concentration step on the yields of poly phenols and bitterness substances were evaluated.

Materials & Methods

Preparation of the wort concentrate: An one hectoliter brewhouse equipped with a 2 roller mill and a lauter tun was used. The lab trials were done in a mash bath as well as a heating mantle and a flask. A liquor-to-grist ratio of 1:3.3 led to an extract value of around 20 °Plato (first wort). A typical high-short mashing regime (mashing-in at 62 °C, 1st rest at 62 °C for 30 min, 2nd rest at 72 °C for 30 min, mashing-off at 78 °C) and a boiling time of 90 minutes has been chosen [modified 3].

The concentration of the wort took place in a falling film evaporator (4-effect evaporator). The concentration for the lab trials were done in a vacuum rotary evaporator.

Hop products: Similar to the previous investigation [3], the hop variety Perle as pellets type 45, as spent hops (extraction residues from CO_2 extraction) and as CO_2 hop extract were used. The three hop products were added to the wort as follows: 1st hop dosage at the start of boiling and 2nd dosage at 10 minutes before the end of boiling [acc. to 3].

Analyzing methods: Bitterness substances were analyzed according to MEBAK 2.17.1 and total poly phenols according to MEBAK 2.16.1 [4].

Figure 1 shows, that hopped wort was analyzed at the start of boiling (A) and at the end of boiling (B). After separating the hot trub, the wort was concentrated from 20 to 60 °P. After storing it for one or two days, the wort concentrate was re-diluted back to its original gravity and (C) analyzed in the same way.

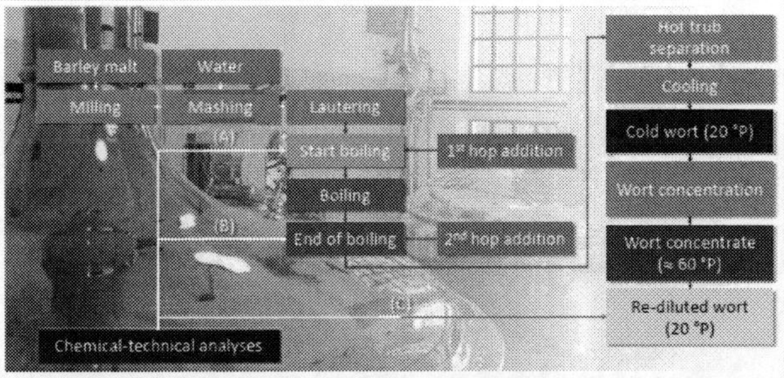

Fig. 1. Brewing trials to produce standard wort and re-diluted wort concentrate.

Acknowledgement

We thank Dr. Christina Schönberger, Technical Manager of Barth Innovations, for supplying all hop products that were necessary for the successful performance of the experiments, Ilona Jawinski and Jörg Wikert for their careful support of the comprehensive experiments and Joh. Barth & Sohn GmbH & Co. KG for its financial support.

References

1 TITZE, J., HUBER, H. W. and HAMEL, C., 2016: *The potential of wort concentrate used for beer production to enable cost optimisation in craft breweries.* – 12th International Trends in Brewing, Gent, Belgium, 03.-07.04.2016.

2 HUBER, H. W. and TITZE, J., 2016: *Chance für Craftbierbrauereien? Bierherstellung mit Würzekonzentrat.* – Brauindustrie 101, no. 11. pp. 30-33.

3 AIF 17474 N, 2015: *Aktivierungsorientierte Charakterisierung antioxidativer Schlüsselinhaltsstoffe und deren Wirkung auf die sensorische Qualität von Bier.* – Final report 2015.

4 JACOB, F. (Ed.), 2012: *Brautechnische Analysenmethoden. Würze – Bier – Biermischgetränke.* Methodensammlung der Mitteleuropäischen Analysenkommission (MEBAK), MEBAK, Freising.

* Corresponding author: jean.titze@hs-anhalt.de Prof. Dr.-Ing. Jean Titze
Anhalt University of Applied Sciences
Department of Applied Biosciences and Process Engineering
Bernburger Str. 55
06366 Koethen
Germany

Results & Discussion

To compare the results of the lab scale with the pilot scale trials, the thiobarbituric acid number showed, that the thermal load of the worts was increasing during the boiling and concentration step (from the start of boiling to the end of boiling till re-dilution after concentration took place, see Fig. 2). The increase during boiling in the lab trials was nearly 5 units higher than at the pilot scale, while the concentration influence was reduced by the use of the rotary evaporator (lab scale).

Figure 3 shows the losses of bittesness during the concentration step for the preboil wort and the cast wort. Dependent of the used hop product, hop peletts had 22-23% losses, spent hops 10-17% and hop extract 1-2% only. Even during the boiling step, losses in bitter units accured (e.g. for hop pellets 16-17%). This can be explained due to elimination during the protein coagulation.

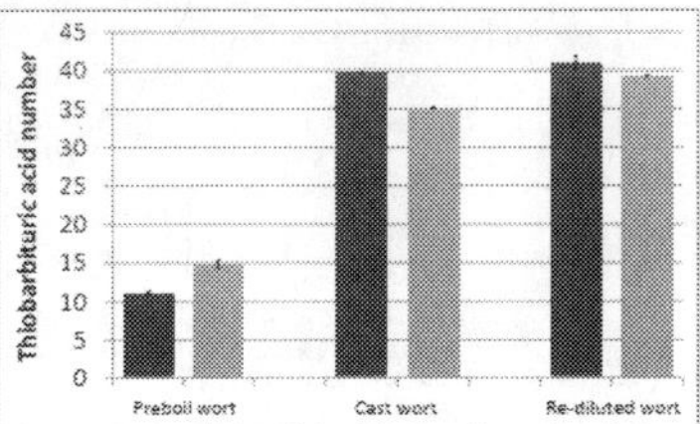

Fig. 2. Increase of the thiobarbituric acid number during the lab-scale (■) and the pilot-scale trials (■).

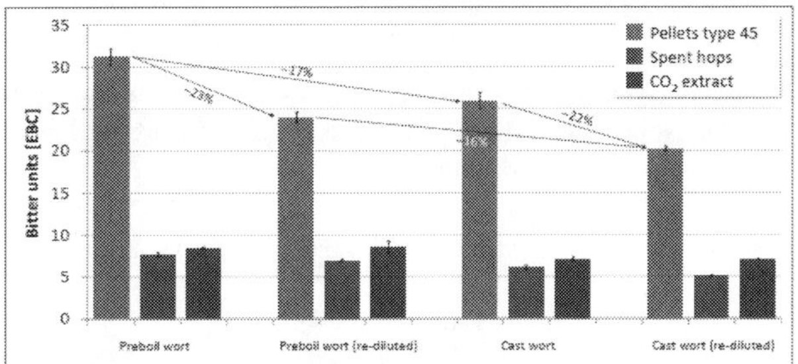

Fig. 3. Influence of the concentration step on the bitterness in wort.

In a normal wort, 80% of the poly phenols come from the malt and 20% from the hops. In relation to its addition, spent hops contributed to the highest poly phenol content (Fig. 4). Losses of poly phenols during the concentration step were in the same range as the losses of bitter units – approximately 30% for hop pellets, 20-30% for spent hops and only around 1% for the CO_2 hop extract. The very low losses in poly phenols can be explained because the CO_2 hop extract shows a very low poly phenol content. That means the analyzed poly phenols are derived from mainly malt.

The average loss of poly phenols during the concentration step was 44.9 mg/l. The results also showed that the losses in the pilot scale trails (using a falling film evaporator) were less than in the lab scale (use of a rotary evaporator).

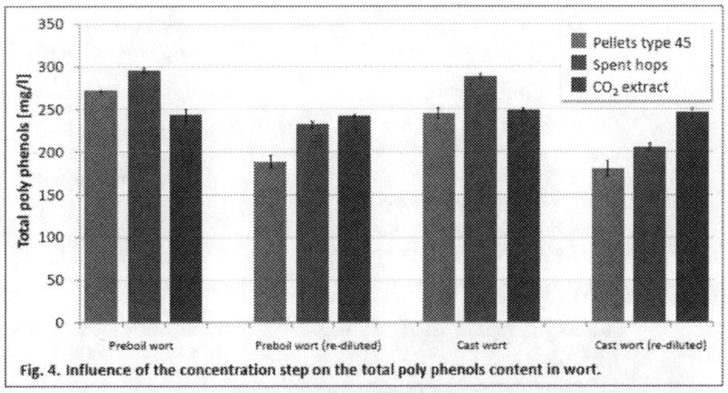

Fig. 4. Influence of the concentration step on the total poly phenols content in wort.

Conclusion

In this investigation the role of hop products and their yield in bitterness substances as well as the poly phenol content were evaluated for the use in hopped wort concentrates. Furthermore, the influence of the concentration step was evaluated. While good bitterness yields were achieved with CO_2 hop extract, it was also shown that approximately 45 mg/l of the poly phenols were lost.

The loss of bitterness of hopped wort concentrates during the concentration step reported in previous investigations could be confirmed [1,2]. The use of spent hops (extraction residues from CO_2 extraction) contributed to a significant amount of poly phenols in the wort concentrates while the bitterness contribution was very low. Maybe this could be a solution for a poly phenol enriched wort concentrate with less bitter taste for a trendy, healthy and refreshing malt beverage concept.

P090

Monitoring of an onion-like off-flavor component and its precursor in large-scale brewing

Taku Irie[1)], Shigekuni Noba[2)] and Minoru Kobayashi[2)]

1) Production Technology Center, Asahi Breweries, Ltd., Japan
2) Research Laboratories for Alcohol Beverages, Asahi Breweries, Ltd., Japan

Introduction

2-mercapto-3-methyl-butanol (2M3MB) and its precursor

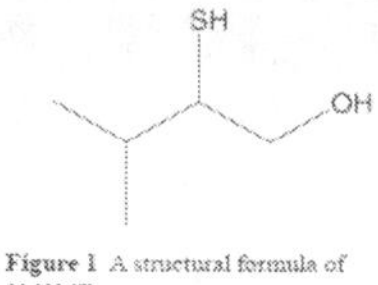

Figure 1 A structural formula of 2M3MB

-2M3MB imparts an "onion-like"[1)] or "sweaty" flavor to beer.
-Some factors that increase this component in beer have been reported previously[2) 3)], including "Hot aeration" in the production of wort.
-However, the mechanism of 2M3MB formation in brewing still remains unclear.

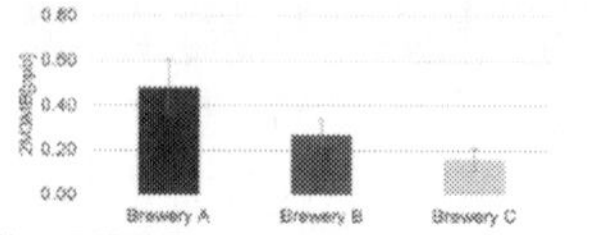

Figure 2 2M3MB levels in a certain brand of beer during production at different breweries (mean ± SD, n=4 each).

As shown in Figure 2, variations in 2M3MB levels have been observed among breweries, possibly leading to sensory variations.

The purpose of this study is to identify factors that control 2M3MB levels in large-scale brewing.

Recently, we purified the precursor of 2M3MB from hops and identified the compound as 2,3-epoxy-3-methyl-butanal (EMB)[4)].

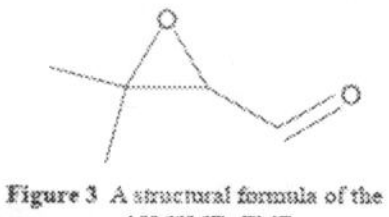

Figure 3 A structural formula of the precursor of 2M3MB, EMB.

By a series of lab-scale fermentation tests with this precursor, we showed that the formation of 2M3MB requires H_2S which generates as a yeast metabolite. We also observed that reduction by yeast enzyme was important for 2M3MB production.
(**For details, See oral presentation L13**)

In this poster presentation, we show the results of monitoring the levels of 2M3MB and its precursor during the large-scale brewing process, and propose a mechanism for how the precursor is formed and converted to 2M3MB.

Materials and Methods

Brewery	A	B	C
Brewhouse	700HL (Huppmann)	1,250HL (Ziemann)	1,250kL (Huppmann)
Fermentation tanks	2,800HL	5,000HL	5,000HL

Table 1 The characteristics of breweries where we monitored the 2M3MB and its precursor.

We monitored the levels of 2M3MB and its precursor during the brewing process for a certain brand of beer at three breweries with documented variability in 2M3MB production.

Results and Discussion

Precursor Contents in Wort

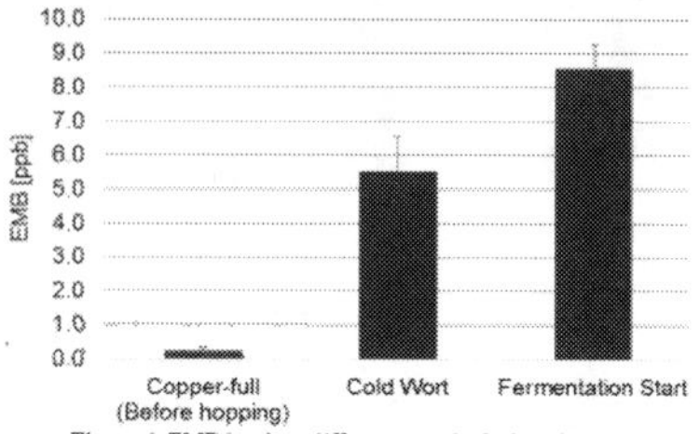

Figure 4 EMB levels at different steps in the brewing process (Brewery A) (mean ± SD).

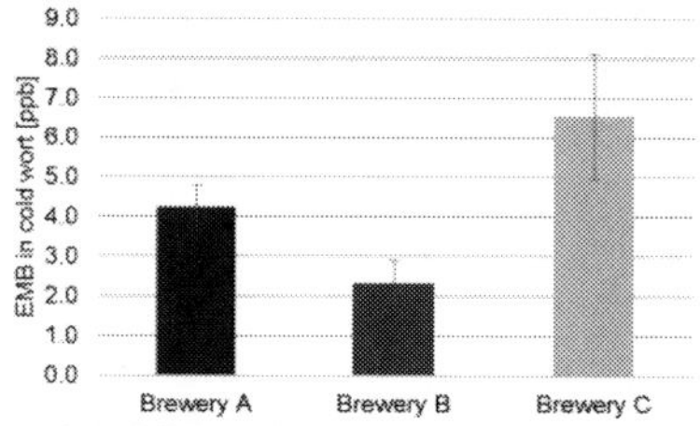

Figure 5 EMB levels in cold wort brewed at different breweries (mean ± SD, n=4 each).

EMB was generated primarily in the brewhouse (Figure 4), especially through the boiling process after adding hops (the source of the precursor). EMB levels in cold wort differed among the breweries. Notably, the highest EMB levels were seen at Brewery C, which brewed beer with the lowest 2M3MB levels (Figure 5).
Figure 4 also shows that precursor levels increased not only in the hot section, but also after wort cooling, when pitching and aeration occur. This result suggested that excessive aeration may also lead to increased onion-like flavor in the beer.

2M3MB and its precursor in Fermentation

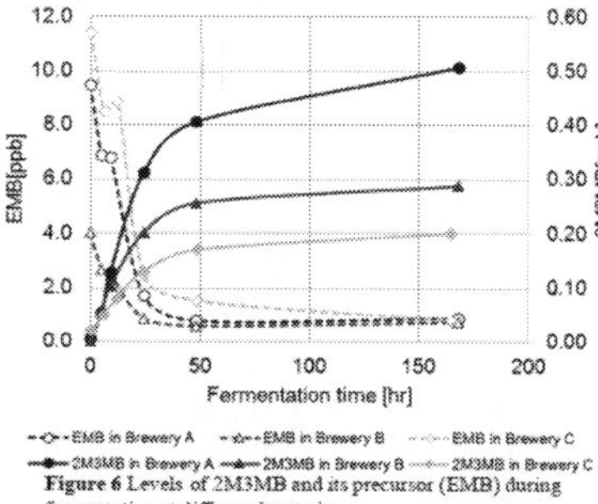

Figure 6 Levels of 2M3MB and its precursor (EMB) during fermentation at different breweries.

Table 2 Conversion ratio of the precursor to 2M3MB.

Brewery	A	B	C
EMB at Start [ppb]	9.5	4.0	11.4
2M3MB at End [ppb]	0.51	0.29	0.20
Conversion Ratio [mol/mol]	4.5%	5.9%	1.5%

EMB levels decreased immediately after the initiation of fermentation and fell below the limit of detection by 48h, concomitant with the accumulation of 2M3MB.
The conversion ratio of the precursor to 2M3MB was under 10%.
The difference in 2M3MB levels between Brewery A and B seemed to reflect the differences in EMB levels at the start of fermentation. On the other hand, Brewery C, with high precursor levels, showed a low conversion ratio, suggesting that 2M3MB levels were influenced not only by the amount of EMB in wort, but also by the fermentation process.

Proposed mechanism of the formation of 2M3MB

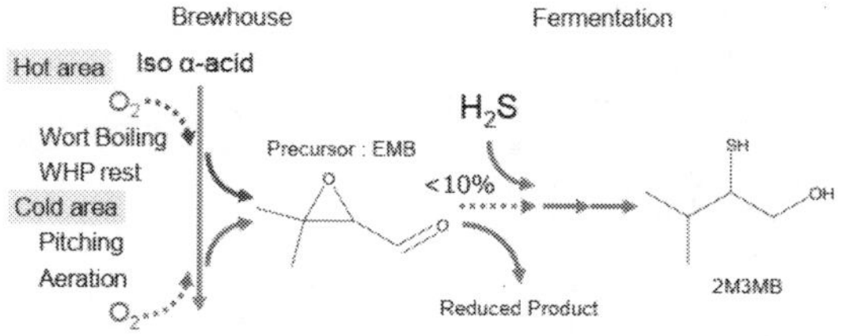

1. Formation of precursor
- Converted from iso α-acid in hop.
- Requires oxygen.
- Occurs in both hot and cold areas.

2. Conversion to 2M3MB
- Requires H2S.
- Mediated by yeast; specific mechanism remains unclear.

Figure 7 Proposed mechanism of the formation of EMB and 2M3MB.

Formation of both EMB and 2M3MB would depend on factors specific to each site's brewing conditions, leading to variability in 2M3MB production among breweries.

Summary & Future Work

-We monitored the levels of 2M3MB and EMB during the brewing process at breweries known to show variability in 2M3MB production.
-2M3MB levels were influenced both by the amount of EMB in wort and by the rate of conversion of EMB to 2M3MB in fermentation.
-This result would help us to understand the underlying reasons for variations in 2M3MB accumulation during production at different breweries.
-We are further investigating both the formation of precursor (**to be reported at ASBC2017**) and the conversion to 2M3MB during fermentation.

References

1) Olsen, A. *et al.* Carlsberg Res. Commun. 53:1-9 (1988)
2) Narziss, R. *Brauwelt 33* (1978)
3) Vermeulen, C. *et al. Proc. Eur. Brew. Conv. Congr.* (2005)
4) Noba et al. WBC (2016)

Contact

E-mail : taku.irie@asahibeer.co.jp

36TH CONGRESS EBC SLOVENIA 2017 14-18 MAY LJUBLJANA EUROPEAN BREWERY CONVENTION

Characterization of ale yeasts
- a study of technological, analytical, biological and sensory attributes

BRAUUNION ÖSTERREICH
University of Natural Resources and Life Sciences, Vienna

F. Webersinke [1,2], H. Klein [1], M. Flieher [1], H. Jäger [2], C. Forster [1]

[1] Brau Union Österreich AG, 4020 Linz, Austria;
[2] University of Natural Resources and Life Sciences, Vienna (BOKU)

Background:

Beer styles brewed with top fermenting yeast, in particular ale yeasts, have been increasingly successful in recent years powered by the craft beer trend. For product development and realizing a consistent product quality it is important to understand the specific flavor characteristics and the process performance of the different strains.

Methods:

Eleven ale yeasts and one wine yeast of different commercial suppliers have been compared in technological, analytical, biological and sensory aspects. The strains performed standardized fermentations in 60 L of plain worts in CCTs to obtain evaluation data.

In technological studies, a predictive model for the control of esters, acids and higher alcohols in dependence of pitching and fermenting conditions has been developed for the Scottish ale yeast WLP028. The experimental set-up has been developed with a central composite design, including all the wort aeration (0,3 mg/L – 5 mg/L dissolved oxygen) and fermentation temperature parameters (14 °C – 26 °C).

The fermentation process and the bottled beers were further analysed with GC (by-products, DMS, Diacetyl), HPLC (4-Vinylguaiacol/-phenol, Carbohydrates), Photometer (Total SO_2, FAN), Alcolyzer (Extracts, Alcohol), sensory descriptive profiling, NucleoCounter® (Cell numbers) and phase-contrast imaging.

Clustering:

A hierarchical clustering of all fermentation products, resulted in a sequenced arrangement of groups according to their origin. Yeasts from Belgium, the USA and UK could be separated by the two markers n-propanol and ethyl acetate only (P = 95 %).

The wine yeast 71B had the lowest homology, compared to the other ale yeasts. Also maltotriose was not affected by this strain.

Hierachical clustering based on fermentation by-products:

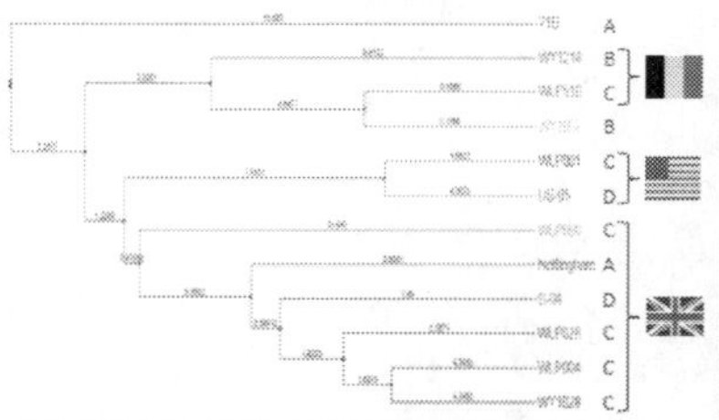

	A	B	C	D
Supplier	Lallemand	Wyeast	White Labs	Fermentis
Culture	Dry-yeast	Liquid	Liquid	Dry-yeast

Fig. 1: Clustering of the tested yeast strains

Yeast Comparison:

Marked differences were found in all analysed categories. For the interpretation of the fermentation by-products, a quartile clustering has been performed, as shown in table 1. The quartiles were calculated by a holistic approach of all sampled strains.

Tab. 1: Concentrations of fermentation by-products classified into quartiles

	71B	Nottingham	S-04	US-05	WLP001	WLP004	WLP028	WLP099	WLP510	WY1007	WY1028	WY1214	Cluster
Phenyl acetate	-			-	-			+			+	+	2
Ethyl butyrate	+	-		+	-	-	+		+	+		+	7
Ethyl caprate	+	-		-			+		-			+	2
Ethyl caproate	+	-					+	+	-	-	-		9
Ethyl caprylate	+	-	-					+		+	-		6
Furfuryl acetate	+	+	+	-	-	-	+	+	-	+	+	+	8
Hexyl acetate		-	+	-	-	-		+	-		+	+	6
Butyl acetate		+		-	-					-	+	+	4
Thiopropyl acetate		-	-	-	-			+	+	+		+	7
Ethyl acetate	-			-	-			+	+			+	4
Isoamyl acetate			-	-	-			+			+	+	4
Total esters	-			-	-			+	+			⊕	4
2-Methyl-1-butanol	-	+	-	-		+						+	3
3-Methyl-1-butanol	-	+	-	-		+						+	3
n-Phenylethanol	-		+	-		-			+	+		+	2
Furfuryl alcohol	-	-	-	-	+	-	-	+	+	+	-	+	5
Hexanol	-	-	-	+		+		-	+	+			4
i-Butanol		⊕		-		+	+			-		-	3
Linalool	-	-	-				+	-	+	+	-		5
n-Octanol	-						-		+	+	-	+	2
n-Propanol	-	+		+	+					-		-	3
Total higher alc.	-	+		-		+				-		+	3
Butyric acid		⊕	+	+	-				-			-	1
Valeric acid	-	+	+	+			-		-				1
Caproic acid	⊕	-	+	+		-	+	+			-		5
Caprylic acid	+	-				-			+	+	-		5
Capric acid	⊕	-		-			+	-				+	1
Lauric acid	⊕		+	-	-		+		-				2
Total acids	⊕		+		-	-	+				-		1
4-Vinylphenol	+			-		-		-				+	-
4-Vinylguaiacol	+							-		-	-	+	8
Acetaldehyde				-		-		+	-		+	+	2
Total SO_2	+	+		-		-		⊕	-				3
DMS	-	-			-			⊕		+			-
Diacetyl (free)	+	+						⊕	-	-	-		9

(+) yeast was classified into fourth quarter/ > $Q_{3\,(75\,\%)}$
(-) yeast was classified into first quarter / < $Q_{1\,(25\,\%)}$
(circled sign) yeast was verified as an outliner (P = 95 %)
Cluster: Reference to predictive modelling.

Predictive modelling:

The experimental data of the central composite design (CCD) was interpreted by principal component analysis (PCA) and hierarchical cluster analysis (HCA). Furthermore nine clusters of fermentation by-products and the sensorical profiling outputs have been applied into a variables factor map, demonstrating the effects of fermentation temperature and wort aeration rates. The constitutents of each cluster are listed in table 1.

Selected fermentation by-products are cited also in table 2, including linear correlations with the temperature (T), oxygen content in wort (O_2) and overall sensory preference (S.P.) at 95 % significance level (*), 99 % (**) and 99,9 % (***); (-) represents a negative correlation.

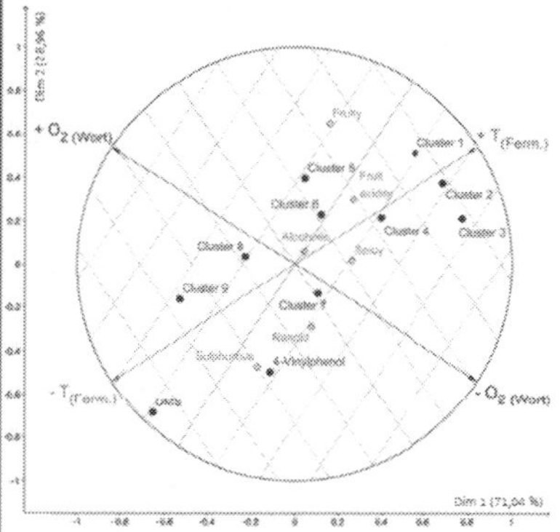

Fig. 2: Variables factor map of fermentation temperature and dissolved oxygen in wort – the constitutents of each cluster are listed in Tab. 1.

Tab. 2: Correlations of selected fermentation by-products

	Cluster	T	O_2	S. P.
Butyric acid	1	***		
Valeric acid	1	***		
n-Phenylethanol	2	**		
2-Methyl-1-butanol	3	***	* (-)	
3-Methyl-1-butanol	3	***	* (-)	
Phenyl acetate	2	***		
Acetaldehyde	2	***		* (-)
Ethyl acetate	4			*
Isoamyl acetate	4	*		
Ethyl caproate	9	* (-)		
Hexyl acetate	6			**

Conclusion:

Yeast strain, pitching and fermentation parameters affected beer composition and sensory attributes markedly. It was possible to enhance flavor capabilities by the investigation of the optimal conditions, which is especially relevant for future developments.

Genetic changes in brewing yeast *Saccharomyces pastorianus* during industrial and laboratory scale serial re-pitching

Neža Čadež[1], Jure Zupan[1], Sonja Smole Možina[1], Peter Raspor[1], Nataša Kočar[2]
[1]University of Ljubljana, Biotechnical faculty, Jamnikarjeva 101, Ljubljana, Slovenia;
[2]Pivovarna Laško Union d.o.o., Pivovarniška ulica 2, Ljubljana, Slovenia
neza.cadez@bf.uni-lj.si

Introduction

Successive batch fermentations with serially reused yeast biomass, i.e. re-pitching, has been a common practice in lager brewing. Re-pitching is usually performed six to twelve times, because of possible yeast aging effects on weakening of fermentation capacity, altered flocculation characteristics and changes in the chemical composition of the final product.

However, the rationale behind these aging effects is not understood completely. One of the reason for altered phenotype, mostly flocculation capacity are frequent recombinations between tandem repeats within intragenic region (1). On the other hand, there have been reports on extensive re-pitching at laboratory conditions, which did not show any significant phynotypic changes or changes in expression profile of lager yeast (2). Similar conclusions were drawn also for genetical stability in Ty and δ sequences (3) even though the brewery practice is different.

On this basis, the aim of our study was to monitor genetic changes, chromosomal length polymorphism, changes in gene copy number and in intergenetic microsatellite regions along with expression of genes which have impact on beer qualty during the re-pitching cycles repeated for 15 and 31-times at the industrial and laboratory-scale, respectively.

Materials and methods

1. Serial beer fermentations with *S. pastorianus* strain TUM34/70

a) Industrial scale - 15 re-pitchings b) Laboratory scale - 31 re-pitchings

2. Genetic changes

a) Electrophoretic karyotyping

Chromosomes were isolated from yeast biomass as desribed by Dobšak et al.(2014). The CHEF-DR III pulsed-field gel electrophoresis system(Bio-Rad) was used.

b) Gene copy number determination

The copy numbers of the transporter genes ScMAL31 and SeMAL31 (Nakao et al. 2004) were estimated with quantitative PCR of genomic DNA. ACT1 and TEF1 were used as reference gene. First re-pitching was used as positive control. Copy numbers were determined by Qbase+ computer program. Primers were constructed based genes annotated by Nakao et al.(2009).

c) Microsatellite analysis

Based on the genome of Saccharomyces pastorianus TUM34/70 1500 microsatellite DNA regions in proximity of genes relevant for brewing technological properties (20.000 bp up- and downstream) (5) were selected by computer program WebSat (http://wsmartins.net/websat/). 38 microsatellites were selected under criteria of least 6 times repeat units. PCR products were analysed on capilary electrophoresis by Macrogen Inc. The microsatellites were analysed by BioNumerics 7.6 computer program.

3. Gene expression

RNA was isolated using the commercial Qiagen RNeasy mini kits. The reverse transcription reaction was performed using commercial Superscript VILO cDNA synthesis kits (Invitrogen). The relative expression of the target genes ILV2, ILV6, BDH1 and BDH2 and the reference genes of ACT1, 18S rRNA, TEF1 was determined by ABI 7500 and analysed by QBase+ computer program.

Results

1. Genetic stability

a) Chromosomal stability during re-pitching

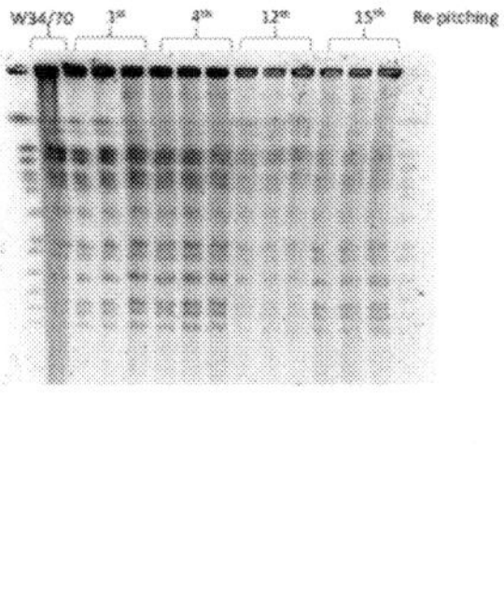

b) Determination of copy number of maltose transporter genes

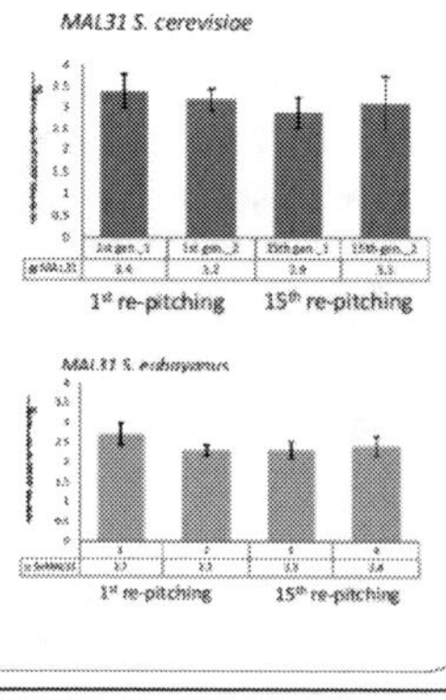

Results

c) Polymorphisms in microsatellite regions

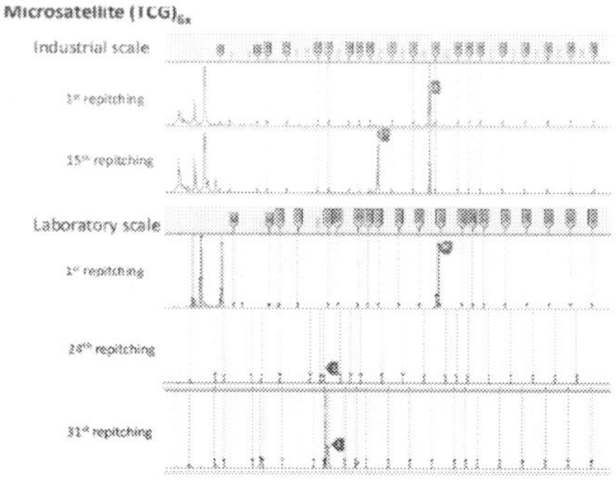

2. Gene expression of genes responsible for diacetyl production

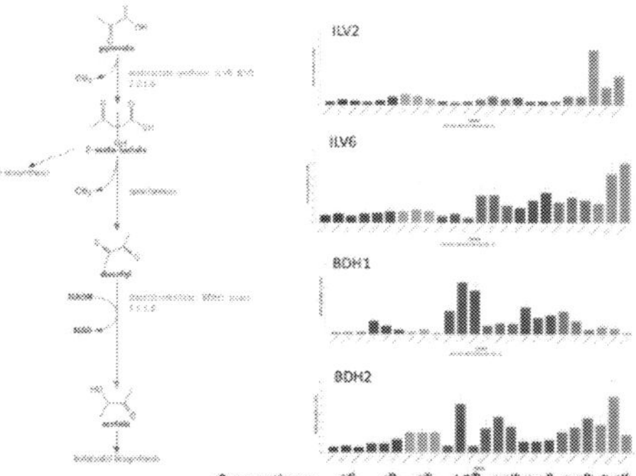

Discussion

The only genetic changes observed were on the level of changes in number of repeats in microsatellite regions where we found polymorphisms in three out of 37 microsatellite analysed.

The industrial and laboratory scale fermentations gave similar results regarding microsatellite polymorphism.

The reported changes in subtelomeric regions were not detected by elecrophoretic karyotyping or by changes in copy numbers of transporter genes for maltose on Saccharomyces cerevisiae or S. eubayanus chromosomes.

The first differences in metabolism were observed after the fifteenth re-pitching when variability in vitality, flocculation and expression of genes responsible for diacetyl production was significant.

1. Verstrepen, K. J., Jansen, A., Lewitter, F., & Fink, G. R. (2005). Intragenic tandem repeats generate functional variability. Nature Genetics, 37(9), 986–990.
2. Bühligen, F., Rüdinger, P., Fetzer, I., Stahl, F., Scheper, T., Harms, H., & Müller, S. (2013). Sustainability of industrial yeast serial repitching practice studied by gene expression and correlation analysis. Journal of Biotechnology, 168(4), 718–728.
3. Powell, C. D., Diacetis, A. N., & Brew, J. J. (2007). Long Term Serial Repitching and the Genetic and Phenotypic Stability of Brewer's Yeast, 113(1), 67–74.
4. Dobšak, M., Gebremariam, M. M., Čadež, N., Zupan, J., Raspor, P., Zarnkow, M., ... Košir, I. J. (2014). The influence of serial repitching of Saccharomyces pastorianus on its karyotype and protein profile during the fermentation of gluten-free buckwheat and quinoa wort. International Journal of Food Microbiology, 185, 93–102.
5. Nakao, Y., Kanamori, T., Itoh, T. et al. (2009). Genome sequence of the lager brewing yeast, and interspecific hybrid. DNA Res, 16:115–29.

Acknowledgement:
The project was financially supported by Ministry of Higher Education, Science and Technology by European Regional Development Fund and by Brewery Laško Union d.d. We acknowledge support of the project by Matej Oset. We thank for valuable assistance at laboratory to Aleš Ogrinc, Neja Jančič, Tadeja Tumpej, Tjaša Žlavs and Martina Podgoršek. Irena Lenarčič from Brewery Union is acknowledged for GC analyses of beer samples.

The Use of Selected Strains of *Lachancea thermotolerans* for Brewing Sour Beer

Prof. John Sheppard, North Carolina State University, Raleigh NC USA

Introduction:

Sour beers are currently experiencing the most rapid growth in demand of any beer style. However, producing a sour beer of consistent quality is difficult, time consuming and a risky venture for many brewers.

Since the vast majority of yeast are unable to produce significant quantities of organic acid, lactic bacteria must be a part of the process. This introduces considerable uncertainty and long delays in the fermentation.

Our research has focused on the isolation and domestication of specific strains of *Lachancea thermotolerans.* These yeast give good yields of ethanol from maltose, while also generating significant quantities of lactic acid, resulting in a pleasant tasting, sour beer in a fraction of the time needed for conventional methods.

Source of the yeast:

Isolated from the gut of wild bees, purified, selected and identified in the lab before commencing with brewing trials.

Fermentation data:

American Lambic Recipe:

65% malted 2-row barley + 35% malted pale wheat, fermented at 23°C in 2.5 BBL pilot-scale brewery.

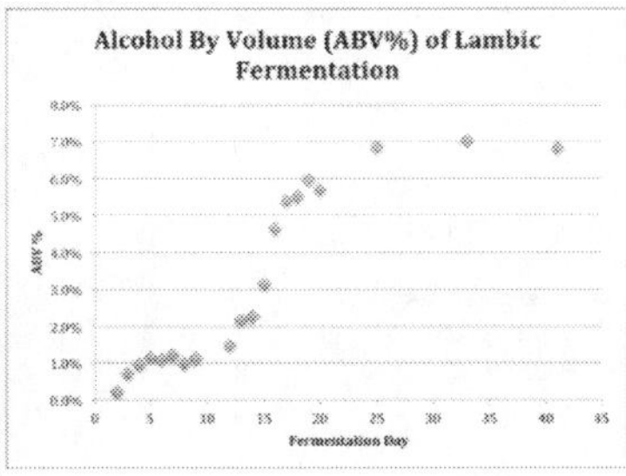

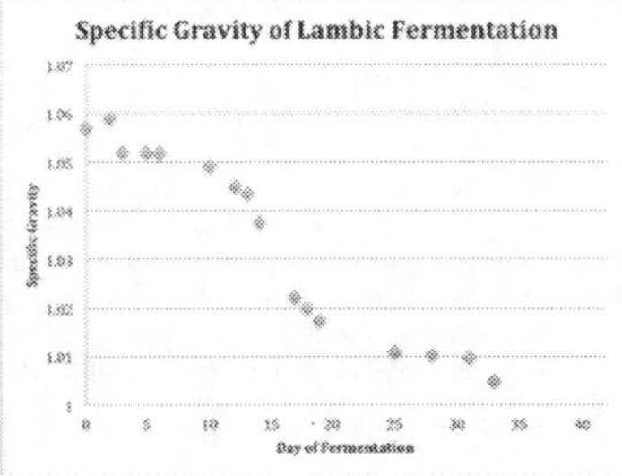

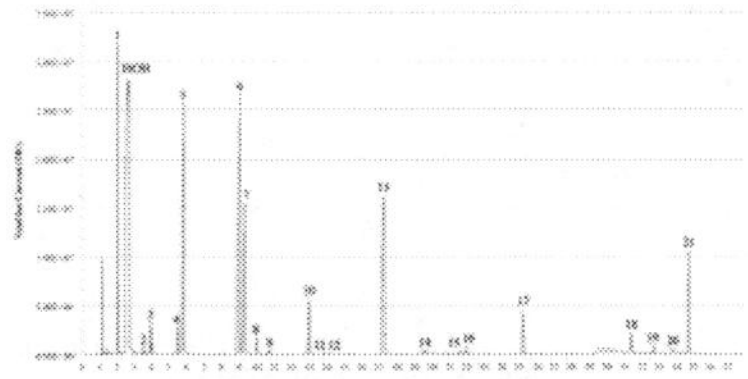

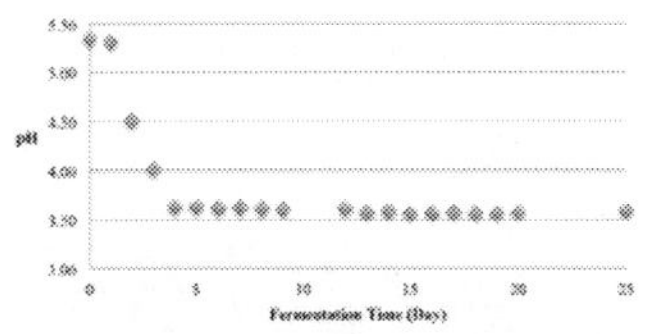

GC data of volatiles in Lambic Sour Beer

Peak #	Compound	Average Peak ratio	Standard Deviation
1	Ethyl acetate	3.56	0.59
2	Isobutyl acetate	0.08	0.01
3	Ethyl butyrate	0.41	0.03
4	Isobutanol	0.58	0.05
5	Isoamyl acetate	6.57	1.37
6	Isoamyl alcohol	9.77	0.79
7	Ethyl hexanoate	4.28	0.15
8	Styrene	0.35	0.03
9	Acetic acid, hexyl ester	0.08	0.01
10	2-Heptanol	1.00	0.00
11	Ethyl lactate	0.04	0.01
12	1-Hexanol	0.04	0.00
13	Ethyl octanoate	6.67	0.86
14	2-Ethylhexanol	0.07	0.00
15	2,3-Butanediol	0.06	0.03
16	Linalool	0.15	0.00
17	Ethyl decanoate	1.04	0.16
18	Phenylethyl acetate	0.47	0.04
19	Ethyl laurate	0.17	0.03
20	Caproic acid	0.38	0.11
21	Phenylethyl Alcohol	2.46	0.35

Summary Comparison

Parameter	Conventional sour	American sour
Culture	Mixed yeast + bacteria	*Lachancea sp.*
Fermentation time	6-24 months	**3 WEEKS**
All malt?	No	Yes
Fermentation temp	Ambient - cellar	Ambient
Flavors & aromas	++ --	++++
pH range	3.0 – 3.6	3.2 - 3.5
German purity law?	No	**YES**

Gratefully acknowledge Dr. Anne Madden and Dr. Robert Dunn of NCSU for their assistance in the isolation of the wild yeast, and Claire Svendsen for the beer analyses.

UNIVERSITY of GUELPH

CHANGING LIVES IMPROVING LIFE

Impact of *Dekkera/Brettanomyces bruxellensis* on Wort Primary, Secondary, and Co-Pitch/Mixed Fermentations

Caroline Tyrawa[1], Richard Preiss[1,2], and George van der Merwe[1]

[1] Department of Molecular and Cellular Biology, University of Guelph, Guelph, Canada

[2] Escarpment Laboratories, Guelph, Canada

MOLECULAR and Cellular Biology

INTRODUCTION

Brettanomyces bruxellensis (teleomorph *Dekkera*) has seen a recent surge in popularity amongst craft brewers due to its utility in wort fermentations. Previously it was predominantly studied as a spoilage organism in wine and other industrial fermentations; however, recently is has been found to positively contribute to the flavour profile of certain Belgian beers, producing unique products with characteristic 'fruity' or 'funky' flavour profiles[1]. Despite this increase in industry interest and use of *Brettanomyces*, little is known about the yeast and how it behaves in different types of fermentations. To this end, we investigated *B. bruxellensis* in primary, secondary, and co-pitch/mixed fermentation models (with *S. cerevisiae*) in order to determine the effect and extent of flavour compound production in different fermentations, whether primary *S. cerevisiae* strain selection is important, and whether there is a difference between secondary and co-pitch fermentations in terms of kinetics and flavour compound production.

MATERIALS & METHODS

Fermentations. Lab-scale fermentations were conducted at 22°C for 21 days using a standard wort (12.5° Plato). The fermentations were carried out in 70 mL of wort in 'spice jars' fitted with fermentation locks. Each yeast strain was tested individually in a primary fermentation and a combination of each *S. cerevisiae* strain with each *B. bruxellensis* strain was tested in a secondary and co-pitch/mixed fermentation. Table 1 summarizes yeasts strains used in this study and pitch day and pitch rate for each of the fermentations. Prior to pitching all yeasts were propagated in identical wort for 3 days.

Yeast Species	*B. bruxellensis*			*S. cerevisiae*		
Fermentation Type	Primary	Secondary	Co-Pitch/ Mixed	Primary	Secondary	Co-Pitch/Mixed
Pitch Day	0	7	0	0	0	0
Pitch Rate	$1.2x10^7$	$1.0x10^6$	$1.0x10^6$	$1.2x10^7$	$1.2x10^7$	$1.2x10^7$
Strain Name	BSI Drei	BBY011	PESTIV	US-05, American Ale	BBY002, English Ale	WY1762, Belgian Ale
Strain Characteristics	Beer isolate	Beer isolate	Wine isolate	Neutral; POF -	Fruity; POF -	Phenolic; POF +

Table 1. Yeast strains, pitch rate, and pitch day used in primary, secondary, and co-pitch/mixed fermentations in this study.

GC-MS Analysis. Following fermentation, samples were collected and filtered using a 0.45 µm syringe filter prior to metabolite analysis. Flavour metabolite analysis was performed using HS-SPME-GC-MS, with a method adapted from Rodriguez-Bencomo *et al.*, 2012[2]. 3-octanol and 3,4-dimethylphenol were used as internal standards.

HPLC Analysis. The ethanol concentration was measured using HPLC and a RI detector. The samples were run on an Aminex HPX-87H column, using 5 mM sulfuric acid as the mobile phase. The sample contained 50 µL of 6% (v/v) isopropanol as the internal standard.

RESULTS

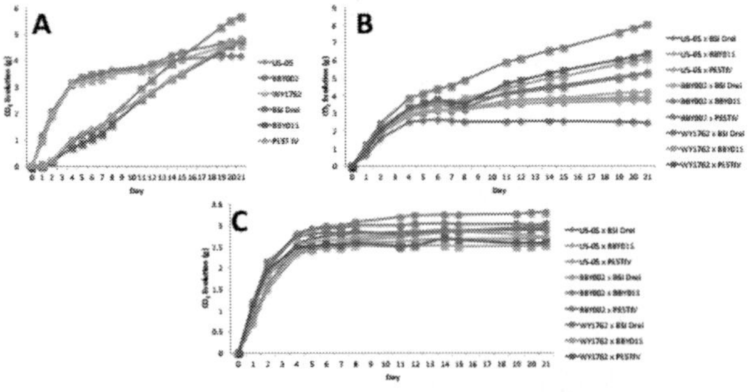

Figure 1. Fermentation profiles for primary (A), secondary (B), and co-pitch/mixed (C) fermentation models. Points represent the mean of biological replicates (n=3).

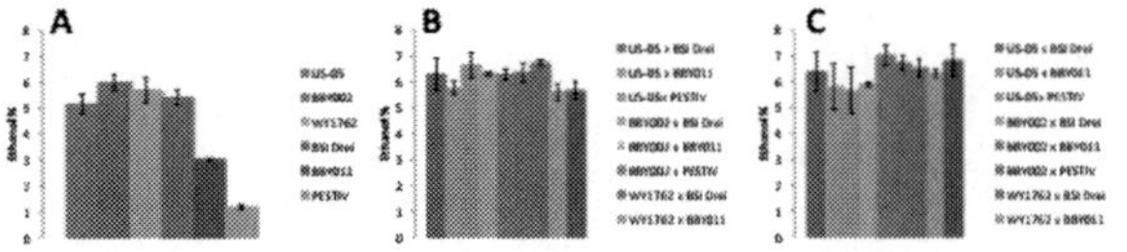

Figure 2. Final ethanol concentrations for primary (A), secondary (B), and co-pitch/mixed (C) fermentations.

RESULTS

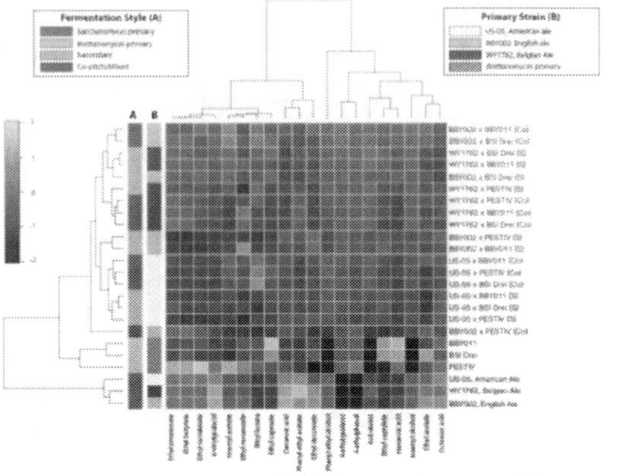

Figure 3. Heat map showing correlation of each fermentation type with flavour compound production. Each square represents the mean of biological replicates (n=3).

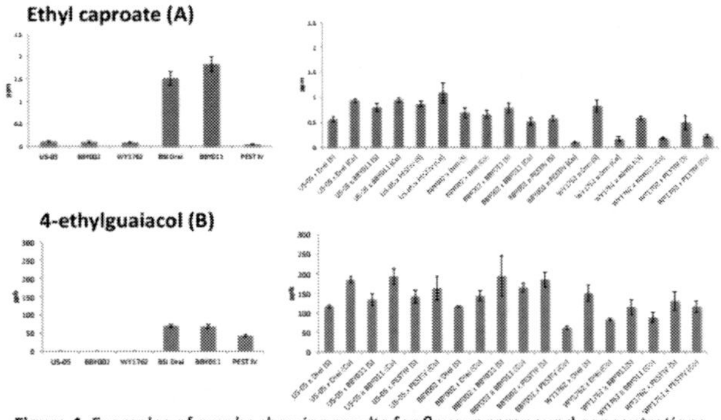

Figure 4. Examples of graphs showing results for flavour compound concentrations. These graphs showcase trends observed during the experiment, namely that secondary (S) and co-pitch/mixed (Co) fermentations are less fruity and more phenolic than primary fermentations.

DISCUSSION

- Typical fermentation profiles for *Saccharomyces* and *Brettanomyces* observed in 1° fermentation (Figure 1A).
- *Brettanomyces* was able to complete a 2° fermentation following the initial *Saccharomyces* fermentation. The highest fermentation rates were observed for samples fermented with the WY1762 *Saccharomyces* strain (Figure 1B).
- No significant differences between different strains were observed in co-pitch/ mixed fermentation profiles (Figure 1C).
- Higher ethanol content observed in 2° (Figure 2B)/co-pitch (Figure 2C) compared to 1° fermentation (Figure 2A), potentially due to *Brettanomyces*' ability to ferment additional sugars. This effect was not as evident in 1° *Brettanomyces* fermentations likely due to the fermentations not being fully completed at day 21.
- As shown by heatmap clustering, primary *S. cerevisiae* strain is distinguishable despite the influence of *Brettanomyces*.
- Primary strain selection does not appear to be a significant indicator of flavour profiles in 2° and co-pitch fermentations.
- 2° and co-pitch are less fruity than 1° fermentations, as evidenced by decreased levels of ethyl caproate and ethyl caprylate (Figure 3; Figure 4A).
- 2° and co-pitch are more funky than 1° fermentations , as evidenced by increased levels of 4-ethylguaiacol (Figure 3; Figure 4B).
- Taken together, these results show that mixed fermentation is an effective strategy to obtain *Brettanomyces* beers in an accelerated fashion.

Acknowledgments: This project was funded by the University of Guelph & OMAFRA and is carried out in collaboration with Escarpment Laboratories (Guelph, Canada). We thank Royal City Brewing Co. for their gracious donation of wort.

References

[1] Steensels, J., *et al.* (2015). *Brettanomyces* yeasts — From spoilage organisms to valuable contributors to industrial fermentations. *International Journal of Food Microbiology, 206*, 24–38.

[2] Rodriguez-Bencomo, J.J., *et al.* (2012). Optimization of a HS-SPME-GC-MS Procedure for Beer Volatile Profiling Using Response Surface Methodology: Application to Follow Aroma Stability of Beers Under Different Storage Conditions. *Food Analytical Methods*, 5(6), 1386–1397.

36TH CONGRESS EBC 2017 EUROPEAN BREWERY CONVENTION 14-18 MAY LJUBLJANA SLOVENIA

Recent notable microbiological contaminations of craft beer in the United States

Wade Begrow, M.S.
Microbiologist – Founders Brewing Company – Grand Rapids, MI, USA

Abstract

United States production of craft beer in the twenty-first century has increased significantly, and the US craft beer continues to increase in volume and market share. The majority of American craft beers are unpasteurized and are susceptible to microbial contamination and spoilage. Wild yeasts (e.g. *Saccharomyces cerevisiae* var. *diastaticus*) and lactic acid bacteria (e.g. *Lactobacillus acetotolerans*) are responsible for several major recalls, the US craft beer in the recent history of causing serious economic consequences for producers.

Povzetek

Združene države proizvodnja obrti piva v enaindvajsetem stoletju se je močno povečala, in ZDA obrtnih pivo še vedno povečanje obsega in tržnega deleža. Večina ameriških obrtnih piva so nepasteriziranega in so dovzetni za mikrobiološko kontaminacijo in kvarjenjem. Divje kvasovke (kot *Saccharomyces cerevisiae* var. *diastaticus*) in mlečnokislinske bakterije (kot *Lactobacillus acetotolerans*) so odgovorni za več večjih odpoklicev, ZDA obrt pivo v novejši zgodovini povzroča resne gospodarske posledice za proizvajalce.

Discussion

The United States brewing industry has greatly increased in complexity in the twenty-first century. From 2015 to 2016, the number of U.S. breweries increased from 4,548 to 5,301 (16.6% increase). A large number of breweries in the U.S. are classified as microbreweries or brewpubs (less than 17,600 hL production), yet there are several notable regional and national craft beer manufacturers[1]. Furthermore, the rapid growth of the U.S. craft beer segment has led to business partnerships with foreign entities and even complete buyouts.

Microbiological stability of U.S. craft beer is of great importance because the majority of the products are unpasteurized[7]. Furthermore, large craft breweries in the U.S. have a wide distribution network and it can be difficult to control inventory and storage conditions of packaged product. A major threat to the U.S. craft beer market is product degeneration due to microbiological contamination[6]. Wild yeast including *Saccharomyces cerevisiae* var. *diastaticus* was likely the causative organism in at least three major recalls of super attenuated beer[10,12,13], and *Lactobacillus acetotolerans* contamination resulted in a national recall of a prominent barrel aged imperial stout due to undesirable off flavors[12].

In contrast to traditional top-fermenting brewing yeast, *Saccharomyces cerevisiae var. diastaticus* secretes an extracellular glucoamylase that can increase the overall fermentable sugars in beer leading to over attenuation. It is capable of growth at 37 °C, but it can remain viable at refrigeration temperatures. This is particularly dangerous to consumers if beer contaminated with *S. cer.* var. *diastaticus* is packaged because it can cause package failure and explosion[3,11,15]. In addition, the increased fermentation can lead to alcohol labeling non-compliance.

Selective media including YM + $CuSO_4$ and lysine agar may be used in conjunction with 37 °C aerobic incubation to detect *diastaticus* in top-fermented beer, but some strains of *diastaticus* are inhibited. Detection of *S. cerevisiae* contamination including *diastaticus* in lager yeast can be achieved without the use of molecular diagnostics[16]. Detection of wild yeast and spoilage bacteria can be challenging for U.S. craft breweries because the use of advanced microbiological detection methods are not commonly adopted. This could be a result of a large focus on detecting contamination in lager beer in scientific literature. After all, lager beer production has dominated the U.S. beer market, but ale-focused breweries are gaining a foothold in the sector[1]. The American Society of Brewing Chemists (ASBC) recommends molecular diagnostic equipment for microbiological screening only for breweries that produce over 90,000bbls (~105,000hL), and traditional microbiology is recommended for breweries that produce at least 15,000bbls (17,600hL)[1]. Molecular diagnostics techniques including real-time PCR (qPCR) are currently the most widely accepted methods to detect and identify the aforementioned spoilage organisms with confidence[7]. Several commercially available molecular diagnostic kits are available to detect the spoilage organism with confidence. Additionally, reactive testing of packages can measure increased carbonation and alcohol content in beers that are suspected to be contaminated with *diastaticus*.

The U.S. Food and Drug Administration's (FDA) Food Safety Modernization Act (FSMA) of 2011 marked a major change in U.S. Food Safety by transferring more focus to preventing contaminants through utilization of Good Manufacturing Processes (GMPs). Craft breweries in the United States are subject to FDA oversight, and appropriate controls must be in place to mitigate risk to the consumer (21 C.F.R. § 1, 11, 16, 106, 110 2015). It is hypothesized that the FDA will continue to increase attention to producers of U.S. craft beer as the market expands.

Figures and Tables

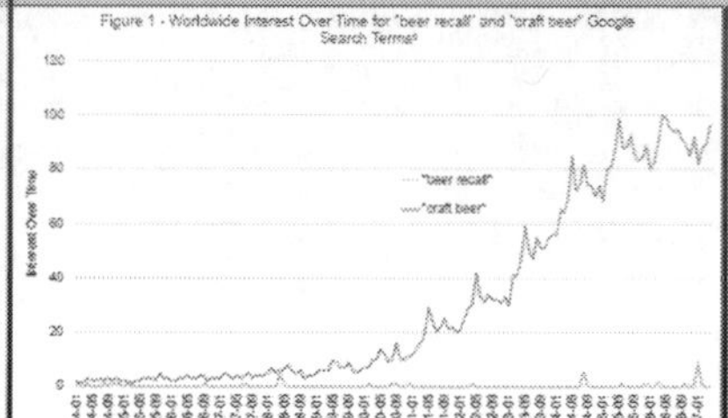

Figure 1 - Worldwide Interest Over Time for "beer recall" and "craft beer" Google Search Terms[5]

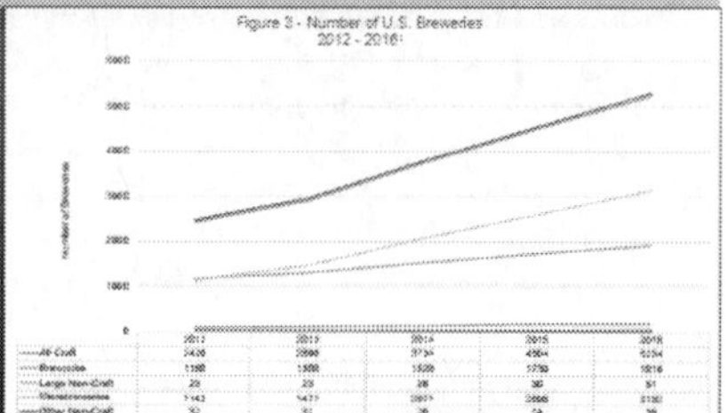

Figure 3 - Number of U.S. Breweries 2012 - 2016[1]

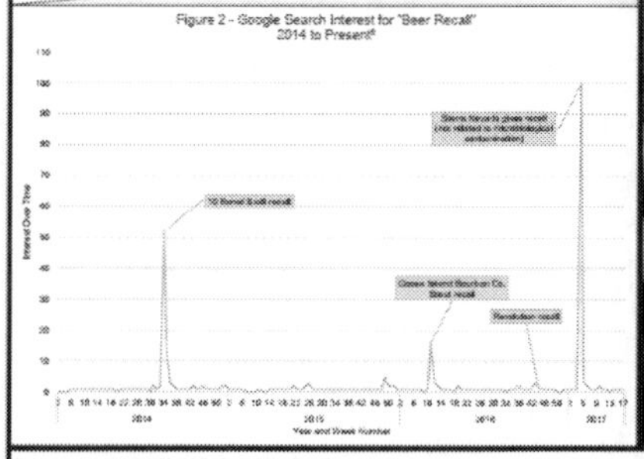

Figure 2 - Google Search Interest for "Beer Recall" 2014 to Present[5]

Figure – *Saccharomyces cerevisiae*[9]

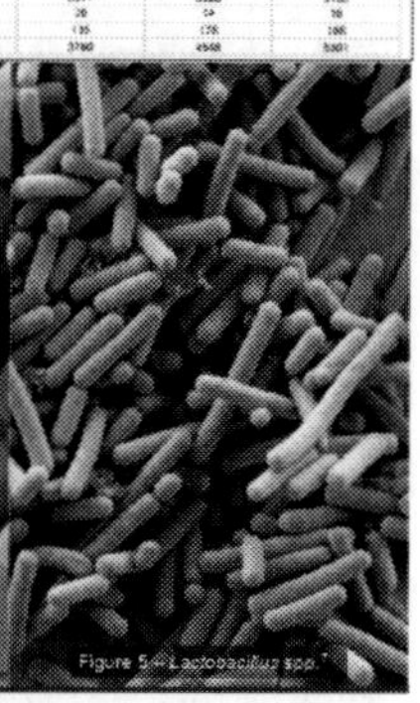

Figure 5 – *Lactobacillus spp.*

Table 1 – Recent notable U.S. Craft Beer Recalls

Date	Brewery	Style	ABV	Contaminant	Sources	Comments
7/30/2014	10 Barrel Brewing	Fruit Beer	4.50	wild yeast - likely *S. cer.* var. *diastaticus*	2	"[T]his product may be experiencing secondary fermentation in the bottle, causing over carbonation."
8/4/2015	Angry Orchard	Cider	5.00	wild yeast	11	"Angry Orchard is voluntarily recalling the cases due to ... signs of re-fermentation in the bottle, which could result in increased carbonation levels, and in some instances, bottles may overflow when opened or burst under increased pressure."
1/8/2016	Goose Island	Imperial Stout	14.20	*Lactobacillus acetotolerans*	8	"Goose Island launched the biggest recall in its 28-year history, which included refunds for all bottles of Bourbon County coffee stout and barley wine, whether opened or not."
9/9/2016	Left Hand Brewing	Sweet Stout	6.00	wild yeast - likely *S. cer.* var. *diastaticus*	9	"Left Hand slipped five spots because of lower production in 2016 after a recall of its number-one seller, Milk Stout Nitro, which was brewed with the wrong yeast."
10/17/2016	Revolution Brewing	Ale, IPA, Porter	5.00 - 10.00	wild yeast - likely *S. cer.* var. *diastaticus*	10	"Revolution Brewing is issuing a product withdrawal of batches of beers produced with our house ale yeast that do not meet our brand specifications for taste and flavor."

Conclusion

The craft beer segment of the U.S. market continues to grow in size and complexity[1], and microbiological spoilage is a major threat[6]. Widespread distribution of mostly unpasteurized products with rudimentary or nonexistent microbiological quality control and quality assurance poses risks to the consumer and manufacturer alike. Super attenuating yeast (e.g. *Saccharomyces cerevisiae* var. *diastaticus*) and spoilage bacteria such as *Lactobacillus acetotolerans* have caused major recalls in recent times[3,11,12,14,15]. As craft beer continues to mature and government oversight becomes more comprehensive, focus on preventing microbiological contamination must persevere in order to preserve the integrity of the industry[16]. Innovative diagnostic techniques like real-time PCR can provide reliable ways to detect and identify bacterial and fungal beer spoilers[7].

About Founders Brewing Company[4]

Founders Brewing Company is located in Grand Rapids, Michigan, USA. The company was founded in 1997 by Mike Stevens and Dave Engbers, and it is celebrating its 20th anniversary. Founders follows the philosophy of producing high quality beer with complex, bold flavors and aromatics. Founders is one of the highest recognized breweries in the U.S., and it is consistently ranked as a top brewery by organizations such as the Ratebeer.com and Beeradvocate.com. Mahou San Miguel (Spain) partnered with Founders and purchased a minority share in December 2014. Founders beer is distributed in 46 States in the U.S. and 26 countries including Slovenia.

References

1. Brewers Association. (2017). U.S. Craft Brewery Count by Category. Retrieved 2017, from Brewers Association Trends: https://www.brewersassociation.org/statistics/number-of-breweries/
2. DeJesus, E. (2014, July 30). 10 Barrel Recalls the Now-Ironically-Named 'Swill'. Eater Portland. Retrieved from http://pdx.eater.com/2014/7/30/6178779/10-barrel-recalls-the-now-ironically-named-swill
3. Elzinga, K. G., Tremblay, C. H., & Tremblay, V. J. (2015). Craft Beer in the United States: History, Numbers, and Geography. Journal of Wine Economics, 242–274.
4. Founders Brewing Company. (n.d.). History. Retrieved 2017, from Founders Brewing Company Website: https://foundersbrewing.com/history/
5. Google Inc. (2017). Google Trends. Retrieved 2017, from Google.com.
6. Hill, A. E. (2015). Brewing Microbiology: Managing Microbes, Ensuring Quality and Valorising Waste. Woodhead Publishing.
7. Horn, N., Wegmann, U., Dertli, E., Mulholland, F., Collins, S., & Waldron, K. (2013). Spontaneous Mutation Reveals Influence of Exopolysaccharide on Lactobacillus johnsonii Surface Characteristics. PLoS ONE. Retrieved from https://doi.org/10.1371/journal.pone.0059957
8. Howard, P. (2013). Too Big to Ale? Globalization and Consolidation in the Beer Industry. Michigan State University. Retrieved 2017
9. Murtey, & Ramasamy. (2016). Sample Preparations for Scanning Electron Microscopy – Life Sciences. In Janecek, & Kral, Modern Electron Microscopy in Physical and Life Sciences. InTech.
10. Nason, A. (2014, August 2). 10 Barrel Brewing Co. issues second recall notice this week. BeerPulse.com. Retrieved 2017, from http://beerpulse.com/2014/08/10-barrel-brewing-co-issues-second-recall-notice-this-week-3401/
11. Noel, J. (2016, May 26). Goose Island identifies bacteria that caused tainted beer, but not source. Chicago Tribune. Retrieved 2017, from http://www.chicagotribune.com/dining/ct-goose-island-tainted-beer-recall-bacteria-20160526-story.html
12. The Denver Post. (2016, September 12). Left Hand Brewing recalls Milk Stout Nitro, says foreign yeast put too much fizz in the bottle. The Denver Post. Retrieved 2017, from http://www.denverpost.com/2016/09/12/left-hand-brewing-recalls-milk-stout-nitro/
13. Trotter, G. (2016, October 18). Revolution Brewing founder on huge beer recall: 'It's like a death in the family'. Chicago Tribune. Retrieved 2017, from http://www.chicagotribune.com/business/ct-revolution-beer-recall-1019-biz-20161018-story.html
14. United States Food and Drug Administration. (2015, August 5). Angry Orchard Cider Company Announces Voluntary Recall of Select Cases of Angry Orchard Crisp Apple Hard Cider. Retrieved from Recalls, Market Withdrawals, & Safety Alerts: https://www.fda.gov/Safety/Recalls/ucm457397.htm
15. University of California, Davis. (2017). Retrieved from Viticulture & Enology: http://wineserver.ucdavis.edu/eaimg/EA2/images/stories/Photos/s.c.1000.jpg
16. Werner, B. (2005). Colour Atlas and Handbook of Beverage Biology. Nuremberg, Germany: Fachverlag Hans Carl.

A rapid high throughput quantitative polymerase chain reaction (qPCR) based diagnostic test for premature yeast flocculation (PYF) in malts

Mandeep Kaur[1], D Evan Evans[2], John P Bowman[1], Alex Speers[3], Doug Stewart[4] and Anthony Koutoulis[5]

[1]Tasmanian Institute of Agriculture, University of Tasmania, Hobart, Australia, [2]Tassie Beer Dr, Hobart, Australia, [3]Canadian Institute of Fermentation Technology, Dalhousie University, Halifax, Canada, [4]Coopers Brewery Ltd, Adelaide, Australia, [5]School of Biological Sciences, University of Tasmania, Hobart, Australia

Introduction

PYF is an intermittent issue in the brewing industry that results in incomplete yeast utilisation of fermentable sugars in the wort, resulting in out of specification beer and disrupted brewing production schedules. Its occurrence leads to significant economic losses for the affected brewer. Previous investigations have identified that PYF is associated with certain batches of malt and due to unfavourable but ill-defined interactions with barley microbes during malting. The standard procedure for the detection of barley malt that causes PYF is by fermentation assays. While there exists a standard ASBC assay (Yeast-14) it is desirable to have a more rapid less laborious method to compliment this assay. Despite PYF often being related to fungal infection of grain, little effort has been directed to identity PYF-responsible fungal taxa and further developing a diagnostic test for these taxa. This study focused on the identification of microbes that cause PYF using DNA fingerprinting techniques, such as pyrosequencing and the development of a qPCR based test for PYF malt batches that is a reliable, efficient and economic screening tool for this intermittent brewing fermentation problem of PYF.

Methodology and Outcomes

Tag encoded pyrosequencing of the conserved fungal LSU and ITS regions of rRNA gene was applied to comprehensively assess the fungal population diversity of malts whose PYF status had been determined by a variety of fermentation tests and subsequently identify putative PYF casual microbes (Figure 1). From this analysis along with review of the literature fungal taxa were selected to design 16 unique primer-probe pairs for qPCR analysis. After optimizing the qPCR conditions, 85 samples were analysed by qPCR (Figure 2). The rRNA gene copy numbers of the specific taxa were calculated for each malt sample. Statistical analysis of qPCR data identified 6 primer-probe pairs (5 genera) that were of most practical value in determining malt PYF status. These primer-probe pairs were further validated with 200 barley malt samples from commercial maltsters and PYF sensitive breweries.

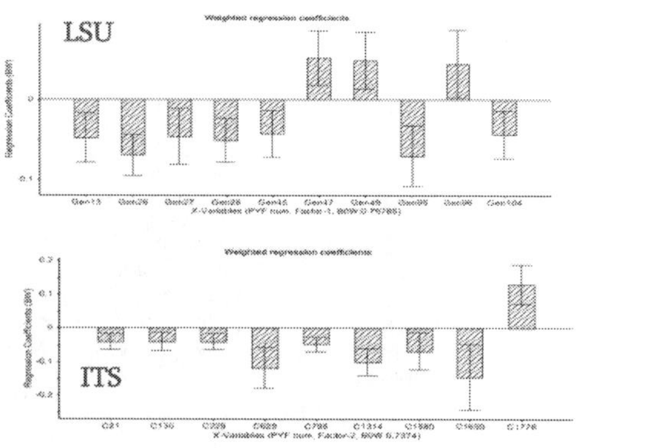

Figure 1: Partial least square (PLS) analysis of pyrosequencing data. Each bar represents individual fungal taxon identified by sequence analysis of LSU/ITS region of rRNA gene. The bars above the X axis are positively correlated with PYF and the height of the bar represents the strength of correlation between specific fungal taxon and PYF.

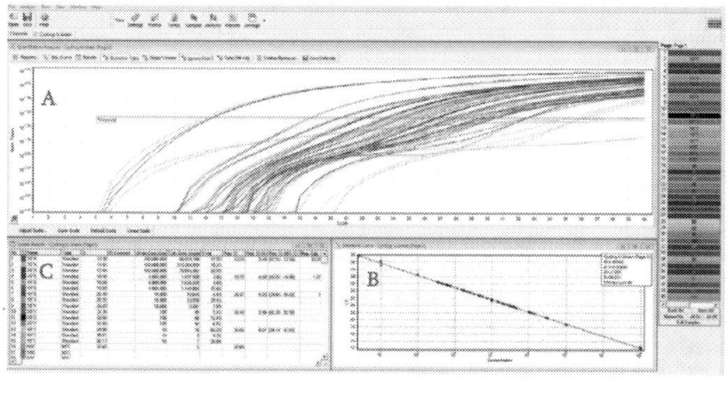

Figure 2: (A) qPCR amplification plots (relative fluorescence vs. cycle number) representing the accumulation of PCR product over time, (B) standard curve in a typical qPCR run to calculate the rRNA gene copy numbers of unknown malt samples (C) resultant fungal gene copy number enumeration of studied malt samples in a typical qPCR run with single primer-probe pair specific for one fungal taxon.

A comprehensive statistical analysis on qPCR data showed two fungal genera were positively correlated with the occurrence of PYF. For statistical analysis studied malt samples were assigned into PYF +ve and PYF –ve malts in two different ways; based on the PYF status provided by the supplier and based on visual observation of the qPCR results of two key primer-probe pairs at University of Tasmania (UTAS). Chi-square analysis of the two status outcomes indicated a significant association using Fisher's test (Fisher's exact test, P=0.0142).

A logistic regression model was fitted (eq. (i)) to both sets of data (supplier's PYF status and UTAS PYF status, separately) to determine the probability of PYF occurrence along with 95% confidence intervals for the predicted probabilities:

$$\text{Log}(p/1-p) = a+b^*x \qquad \text{(i)}$$

Where p is the probability of the outcome (i.e. positive flocculation), x is the predictor (i.e. qPCR result of a primer-probe pair), and a and b are parameters to be estimated. The a is the y-intercept and corresponds to the state in which x is zero. If b is positive then the probability of the outcome increases with x, and it decreases if b is negative.

The probability of flocculation for a given value of x can be calculated using eq. (ii):

$$\text{Predicted probability} = 1/1+\exp(-(a+b^*x)) \qquad \text{(ii)}$$

Supplier PYF status

The PYF outcome was significantly associated with primer-probe pair 1. There is 50% probability of PYF occurrence if the value of primer-probe pair 1 is 108 gene copy numbers/ng genomic DNA at a 95% confidence interval. The outcome was also significantly associated with primer-probe pair 2 as well. There is 50% probability of PYF occurrence if the value of primer-probe pair 2 is 207 gene copy numbers/ng genomic DNA at a 95% confidence interval.

UTAS PYF status

The outcome was significantly associated with primer-probe pair 1. There is 50% probability of PYF occurrence if the value of primer-probe pair 1 is 74 gene copy numbers/ng genomic DNA at a 95% confidence interval. The probability of PYF occurrence increases to 90% if the value of primer-probe pair 1 is 1573 copy numbers/ng DNA at a 95% confidence interval.

For primer-probe pair 2 the logistic regression on continuous value failed since there was a cut point that equals 74 gene copy numbers/ng genomic DNA. For primer-probe pair 2 ≤74 gene copy numbers/ng genomic DNA PYF did not occur while for values >74 PYF always occurred.

Commercialisation of the Test and Future Research Opportunities

Research has resulted in a timely, reproducible and cost efficient qPCR based test for routine malt PYF status determination. The test has the potential to be commercialised. Successfully adopted the test will greatly benefit the malting and brewing industry. Authors welcome malting and brewing industry/researchers' interest for using this test and/or partnership in commercialisation of this test and future research opportunities.

We wish to thank Grain Research and Development Corporation (GRDC), Joe White Maltings Pty Ltd and Cargill for their financial support.

Combined effect of CO_2 saturation and alcohol content on yeast performance during beer refermentation

A.-C. Vandeville[1], J. Depierreux[1], Y. Gosselin[2] and L. Van Nedervelde[1]
[1]Department of Brewing Sciences and Fermentation Technology - Meurice R&D
Avenue Emile Gryzon 1, B-1070 Brussels, Belgium (lvannedervelde@meurice.heldb.be)
[2]Fermentis, Division of S.I. Lesaffre, 137 Rue Gabriel Péri, Fr-59703 Marcq en Baroeul (ygosselin.fermentis@lesaffre.fr)

INTRODUCTION

Bottle refermentation is well known by Belgian brewers and is growingly used worldwide so as to protect beer from oxidation and to keep a "craft style". Unfortunately, brewers are confronted with yeast performance troubles when refermenting beers that present simultaneously substantial alcohol content and CO_2 pre-saturation. Note that even if the effect of alcohol content on fermentation rate has been known for a long time, the influence of CO_2 is less studied. Therefore, we investigated the combined effect of ethanol and CO_2 on bottle refermentation speed.

MATERIALS and METHODS

After rehydration (10% w/v sterile physiological water), the dried yeast Safbrew F2 from Fermentis was tested on 3 different industrial top fermented beers (A, B and C) with respectively 5.2%, 8.0% and 12.0% v/v of ethanol and a CO_2 content adjusted from 0.5 to 6.0 g/L. The concentration of added glucose was fixed in order to achieve the same final CO_2 content (7.0 g/L). Refermentations were performed in 33 cl bottles at 24°C during three weeks. Alcohol and extract contents were monitored with a DMA-Alcolyzer (Anton Paar). Volatile compounds were quantified by headspace GC, residual fermentable sugars by HPLC and CO_2 content was measured with a CarboQC (Anton Paar). Yeast cell concentration was determined with a Vi-CELL (Beckman Coulter).

RESULTS

The yeast strain F2 is known, among other things, to be resistant to alcohol by reaching and undergoing 12,0% v/v (data not shown). However, when pitched into a beer containing concurrently 12.0% v/v and 6.0 g/L of CO_2, this yeast shows refermentation performance disorders.

In greater detail, if we compare the 3 beers A, B and C starting from 1.5 and 6.0 g/L initial saturation (figures 1 and 2, respectively), we observe the negative and well-known effect of ethanol on refermentation rate. Indeed the higher ethanol content, the lower CO_2 production rate. Moreover if we take a closer look at beer C in figure 2, we can see the damaging combined effect of a substantial alcohol content and a high pre-saturation.

■ Beer A – 5.2% v/v ♦ Beer B – 8.0% v/v ▲ Beer C – 12.0% v/v

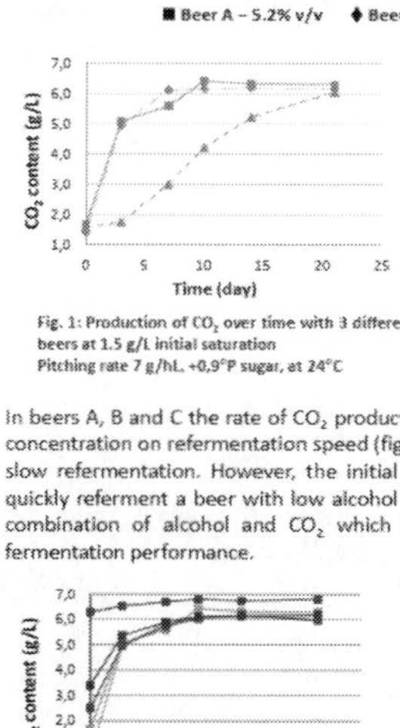

Fig. 1: Production of CO_2 over time with 3 different beers at 1.5 g/L initial saturation
Pitching rate 7 g/hL, +0,9°P sugar, at 24°C

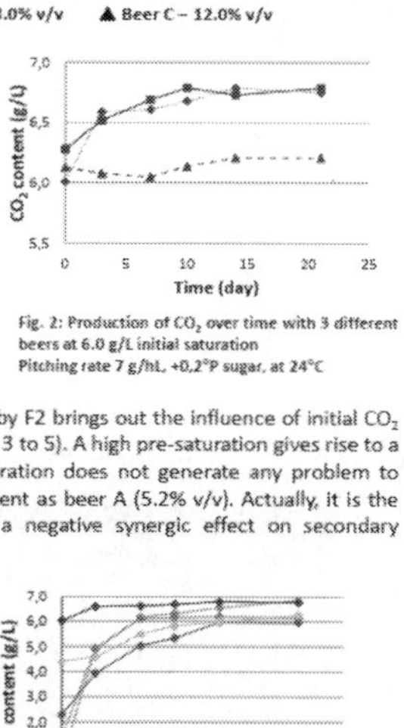

Fig. 2: Production of CO_2 over time with 3 different beers at 6.0 g/L initial saturation
Pitching rate 7 g/hL, +0,2°P sugar, at 24°C

In beers A, B and C the rate of CO_2 production by F2 brings out the influence of initial CO_2 concentration on refermentation speed (figures 3 to 5). A high pre-saturation gives rise to a slow refermentation. However, the initial saturation does not generate any problem to quickly referment a beer with low alcohol content as beer A (5.2% v/v). Actually, it is the combination of alcohol and CO_2 which has a negative synergic effect on secondary fermentation performance.

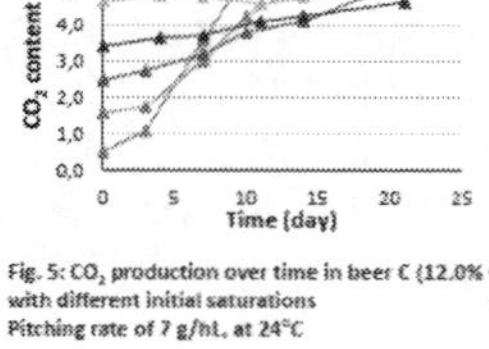

Fig. 3: CO_2 production over time in beer A (5,2% v/v) with different initial saturations
Pitching rate of 7 g/hL, at 24°C

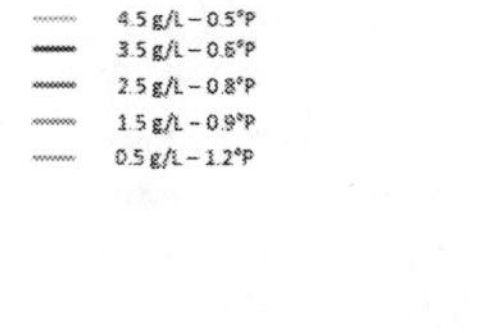

Fig. 4: CO_2 production over time in beer B (8.0% v/v) with different initial saturations
Pitching rate of 7 g/hL, at 24°C

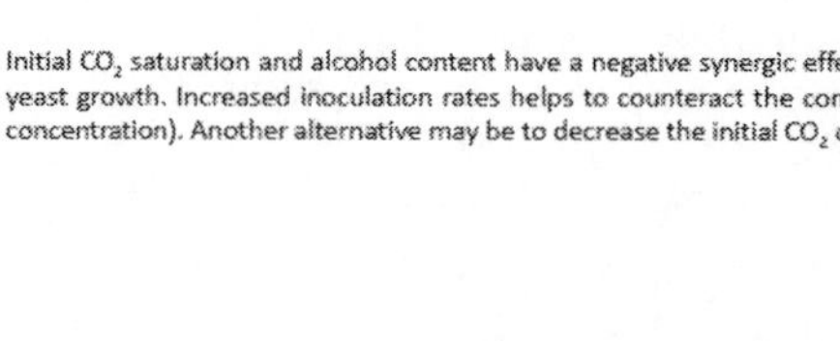

Fig. 5: CO_2 production over time in beer C (12.0% v/v) with different initial saturations
Pitching rate of 7 g/hL, at 24°C

Figure 6 shows the percentage of glucose taken up by F2 at the end of refermentation (21 days) for the 3 beers pre-saturated at 1.5 and 6.0 g/L. It clearly appears that the negative combined effect of ethanol and CO_2 causes refermentation performance troubles. Indeed only 16% of glucose were consumed in beer presenting simultaneously 12.0% v/v of ethanol and 6.0 g/L of CO_2.

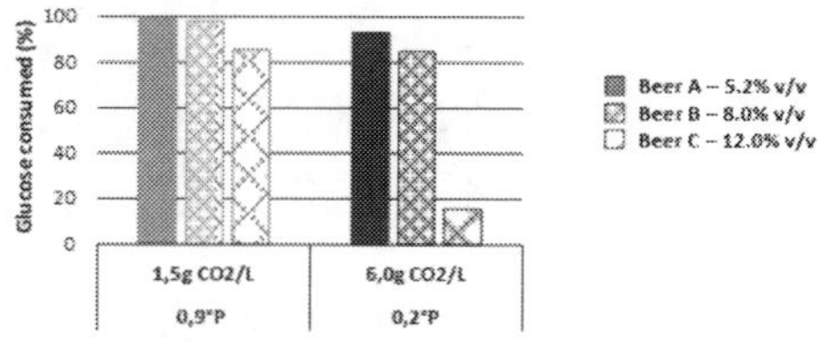

Fig. 6: Percentage of glucose consumed after 21 days of refermentation
Pitching rate of 7g/hL, at 24°C

Note that the decrease of the refermentation rate can be correlated with the reduction of the observed yeast growth when the initial pre-saturation and ethanol content increase (figure 7). One might think that it is related to the amount of glucose added but this is not the case. For example, the yield of dry biomass produced per gram of glucose consumed for beers A, B and C with 1.5 g/L initial saturation are respectively 0.137, 0.067 and 0.020 g dry yeast/g glucose.

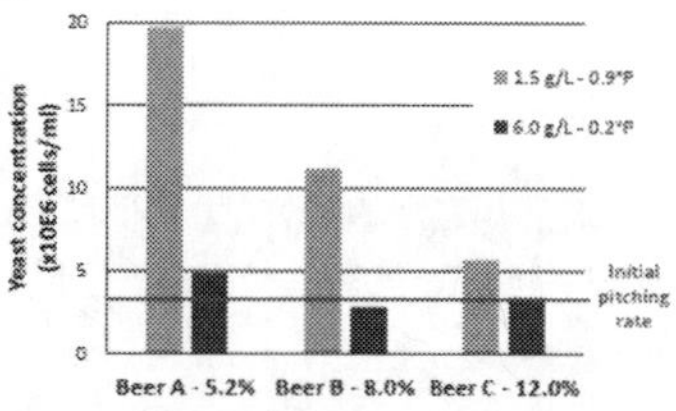

Fig. 7: Yeast cell concentration at the end of 21 days of refermentation
Pitching rate of 7g/hL, at 24°C

As mentioned above the combination of high initial saturation of CO_2 and alcohol content affects yeast growth and consequently refermentation rate. Fortunately, by increasing the pitching rate, the speed of fermentation is improved without yeast growth (figures 8A and 8B). Moreover beer flavour profiles remain unchanged after complete refermentation with F2 (data not shown).

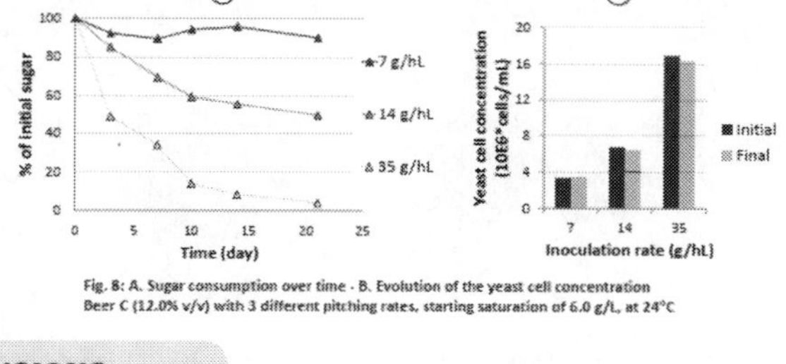

Fig. 8: A. Sugar consumption over time - B. Evolution of the yeast cell concentration
Beer C (12.0% v/v) with 3 different pitching rates, starting saturation of 6.0 g/L, at 24°C

CONCLUSIONS

Initial CO_2 saturation and alcohol content have a negative synergic effect on refermentation rate. Our results indicate that this decrease of fermentation rate can be attributed to inhibition of yeast growth. Increased inoculation rates helps to counteract the combined deleterious effect of CO_2 and ethanol without affecting the quality of the beer and the yeast deposit (final cell concentration). Another alternative may be to decrease the initial CO_2 content for high-alcohol beers.

Use of probiotic yeast in production of beer

B. Šenkárčinová, I. A. Graça Dias, M. Karabín, T. Brányik

University of Chemistry and Technology, Prague, Department of Biotechnology, Technická 5, CZ-166 28 Prague 6, Czech Republic

UNIVERSITY OF CHEMISTRY AND TECHNOLOGY PRAGUE

AIM OF THE WORK: To use non-traditional microorganisms, such as probiotic yeast in beer fermentation process, as one of the possibilities for new product development.

INTRODUCTION

Probiotics are beneficial to human health, typically in improving immunity and maintaining the balance of intestinal microflora. Yeast *Saccharomyces cerevisiae* (*boulardii*) has been subject of study due to its probiotic effects. Many clinical studies showed the beneficial consequences of this yeast in improving immunity and maintaining the balance of intestinal microflora. It would be a visionary idea to combine these probiotic benefits of yeast *Saccharomyces cerevisiae* (*boulardii*) to produce (alcohol-free) beer. However, so far little is known about the ability of this yeast to ferment brewery wort and produce flavor active compounds.

MATERIAL AND METHODS

The growth of probiotic yeast *Saccharomyces cerevisiae* (*boulardii*) was tested at different medium compositions, temperature, ethanol and iso-α-acid content. Subsequently three important technological parameters (original extract, fermentation temperature and pitching rate) were optimized (central composite design of experiments, statistical analysis of data) from the standpoint of flavor active volatile compounds formation (higher alcohols, esters). The experiments were carried out at laboratory scale (300 mL fermenting wort). Esters and higher alcohols were analysed by HS-SPME/GC-MS (Agilent Technologies, USA).

RESULTS & DISCUSSION

Optimization of flavor formation by varying independent variables:
E, Original extract (2 – 12 % wt.)
T, Temperature (2 – 14 °C)
P, Pitching rate (2 – 12×10^{6}cells/mL)

Flavor active by-products formed under different conditions were unified by the legally admitted maximum ethanol content ≤0.5 % alcohol by volume (3.945 g/L).

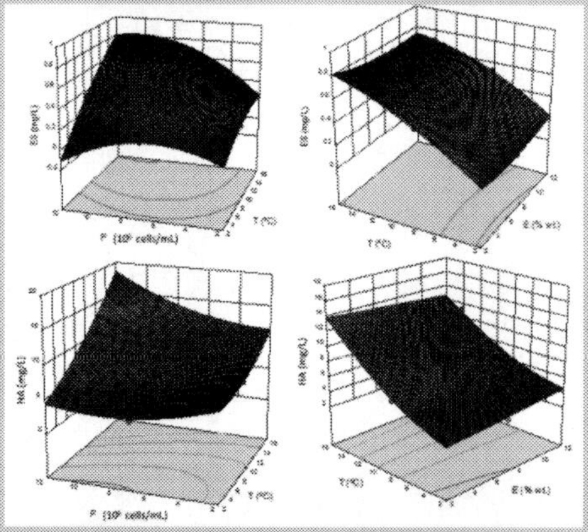

Response surface plots of the interactive effects of two independent linear variables on formation of total esters (ES) and higher alcohols (HA) by probiotic yeast *Saccharomyces cerevisiae* (*boulardii*).

Optimal variable values to maximize ES and HA formation during alcohol-free beer production by limited fermentation.

Parameter	Optimal Value
E – original extract (% wt.)	6.9
T – temperature (°C)	14
P – pitching rate (10^6 cells/mL)	11.8

Sensorial analysis: beer produced by fermentation with probiotic yeast *S. cerevisiae* (*boulardii*) was characterized by slightly acidic flavor with tones of plums, cherries and bitter almonds.

CONCLUSIONS

Saccharomyces cerevisiae (*boulardii*)

- growth was detected even at 0°C
- growth was unaffected by iso-α acids (60 mg/L)
- growth was unaffected by ethanol (4 g/L), at 32 g/L the growth rate was reduced by 25%

The results showed that probiotic yeast were able to ferment brewery wort and that temperature was statistically the most important parameter for formation of volatile compounds during fermentation.

Acknowledgement

The authors thank the project **NAZV QJ1610202** (Ministry of Agriculture of the Czech Republic) for financial support.

Novel spoilage yeast detection method for bottom fermented beer and pitching yeast

P099

36th EBC Congress 2017 Ljubljana, Slovenia

Technische Universität München

Michel M.*[1], Meier-Dörnberg T.[1], Wagner S.R.[2], Jacob F.[1], Hutzler M.[1]
[1]Research Center Weihenstephan for Brewing and Food Quality, TU München, Germany,
[2]Central Washington University, Brewing Program, Washington, United States
E-mail: *m.michel@wzw.tum.de

Introduction: *Saccharomyces cerevisiae* var. *diastaticus* as well as wild types of *S. cerevisiae* and some non-*Saccharomyces* yeasts are frequently found as spoilage yeast in breweries. Most notably a contamination of the pitching yeast can have great influence of the finished product concerning over attenuation and flavor. *S. cerevisiae* and most non-*Saccharomyces* spoilage yeasts can grow at 37°C whereas bottom fermenting brewing yeast *S. pastorianus* strains can not. A detection of the spoiler especially in pitching yeast can be performed on agar plates, taking about 3-5 days. We used a sterile 50 ml disposable mini fermenter supplied by Bactest (Cambridge, U.K.), witch was incubated (and stirred) by the Speedy Breedy device also supplied by Bactest. The CO_2 produced by the spoilers growing at 37°C was detected by the device due to pressure measurement. A faster detection and a low detection limit could be obtained by the new method.

Method:

- **50 ml of YM broth** was autoclaved an filled into the mini fermenters **(Fig. 1.)**
- The broth was incubated by spiked samples of bottom fermenting pitching yeast.
- Four strains of *S. pastorianus* (**TUM 34/70, 34/78, 44, 66**) and five strains of *S. cerevisiae* var. *diastaticus* (**TUM SY 1, DSMZ 70487, TUM PIBB 105, TUM PIBA 109, TUM BLQ 27**) were used to investigate influence of different strains.
- Pitching yeast was measured at varying vitalities (Acidification potential) to investigate false positive results at higher temperature resistance due to high or low vitality.
- Mini fermenters were placed in Speedy Breedy device (Two incubations possible at once, **all trials performed in triplicates**)
- Results were verified by **Real-Time PCR.**
- Samples were spiked with **contamination rates of 0.001 % (10 cells/ml) to 10 % (100.000 cells/ml)** at a pitching yeast cell count of **1 x 10⁶ cells/ml.**
- Speedy Breedy protocol was set to: pressure stability target: **1.5 mBar/min, rotor speed: 60 rpm, temperature: 37°C**

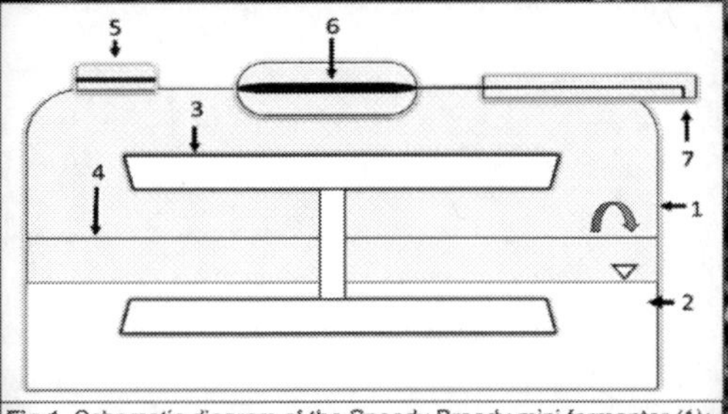

Fig 1. Schematic diagram of the Speedy Breedy mini fermenter (1): YM broth (2), magnetic stirrer (3), stirrer axis (4), incubation rubber septum (5), pressure measure membrane (6) and pressure-measuring connection to the Speedy Breedy device (7)

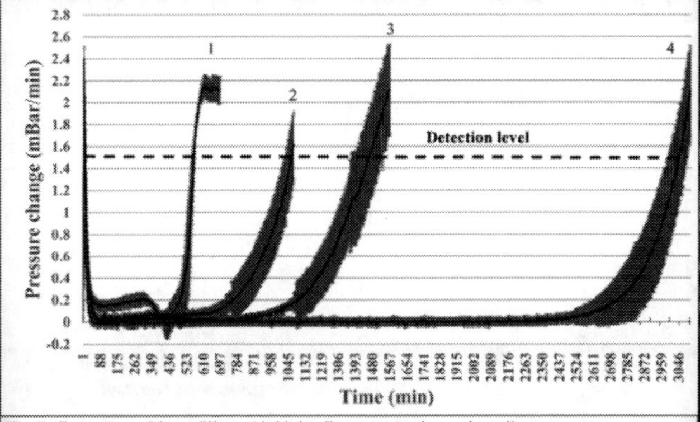

Fig 2. Detection of four different initial cell concentrations of spoilage yeast *S. cerevisiae* var. *diastaticus* **TUM SY1** in pitching yeast of **TUM 34/70** performed in **triplicates with standard deviation.**
(initial cell concentrations: 1 = 100.000 cells/ml, 2 = 1.000 cels/ml, 3 = 100 cel/ml, 4 = 10 cells/ml)

Results:

- **All spiked samples could be detected** in 9 to 50 hours depending on the spoilage yeast concentration **(Fig. 2)**.
- **Vitality** of the pitching yeast showed **no influence** of the test
- This **test can easily be performed** as part of brewery's daily routine with minimum effort.
- **Contaminations in unfiltered beer and finished beer** could also **be detected** reliably.
- All results successfully verified by Real-Time PCR **(Fig 3)**.

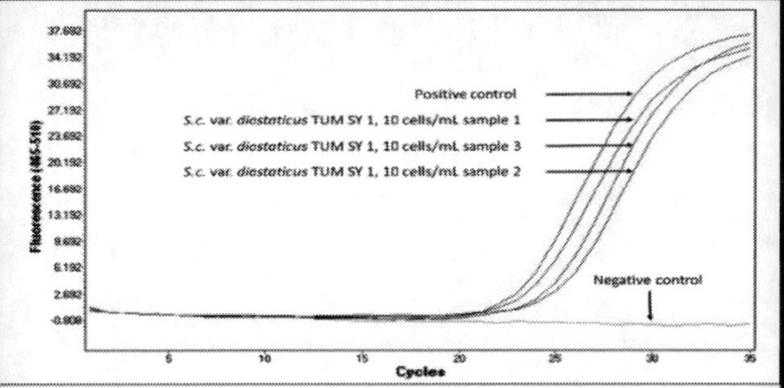

Fig 3. Amplification curves of the three detected samples of *S. cerevisiae* var *diastaticus* TUM SY1 at 10 cells/ml after being detected by the Speedy Breedy device. Furthermore, one curve shows the positive control and another the negative control of the Real-Time PCR system.

Literature:

According publication can be found on the ASBC website:
http://www.asbcnet.org/publications/Pages/default.aspx

Title: A New Approach for Detecting Spoilage Yeast in Pure Bottom-Fermenting and Pure *Torulaspora delbrueckii* Pitching Yeast, Propagation Yeast, and Finished Beer

Authors: Maximilian Michel, Tim Meier-Dörnberg, Anna Kleucker, Fritz Jacob and Mathias Hutzler

J. Am. Soc. Brew. Chem. 74 (3): Pages 200 - 205, 2016

Acknowledgments: Thanks to Dr. D. Walrafen, I&L Biosystems GmbH for supply of material and Dr. S.R. Wagner for support

Research Center Weihenstephan for Brewing and Food Quality
www.blq-weihenstephan.de

Development of surface-engineered brewing yeast for acceleration of maturation and improvement of colloidal stability of beer

UNIVERSITY OF CHEMISTRY AND TECHNOLOGY PRAGUE

Dostálek Pavel[1*], Cejnar Rudolf[1], Hložková Kateřina[2], Jelínek Lukáš[1], Kotrba Pavel[2]

University of Chemistry and Technology, Prague, [1]Department of Biotechnology, [2]Department of Biochemistry and Microbiology, Technická 5, CZ-166 28 Prague 6, Czech Republic [*]Corresponding author. Tel.: +420 220444037; e-mail: Pavel.Dostalek@vscht.cz

KEYWORDS: *Saccharomyces cerevisiae*, beer, QPF peptide, procyanidin removal, α-acetolactate decarboxylase

INTRODUCTION:

Beer production is one of the oldest known traditional biotechnological processes, but is nowadays facing increasing demands not only for enhanced product quality, but also for improved production economics. In beer production the most important parameter is the strain of brewing yeast, which can be classified as top-fermenting or bottom-fermenting strains. Among the top-fermenting yeasts are the many species of the genus *Saccharomyces*, most of them closely related to *Saccharomyces cerevisiae*, thus creating a heterogeneous group of polyploid and sometimes hybrid yeast strains. Bottom-fermenting strains belong to the species *Saccharomyces pastorianus* and are genetic hybrids of *Saccharomyces cerevisiae* and *Saccharomyces bayanus*. Bottom brewed beer today contributes to 90% of the total beer market. Targeted genetic modification of a yeast strain is one way to increase beer quality and to improve the economics of beer production, and various transformed brewer's yeast strains have been prepared.

Cell surface display is a powerful tool for endowing novel functions on the host cell by displaying functional proteins on its surface. Proteins covalently bounded on the yeast surface are simultaneously more active than free proteins in solution. Here we show that display of *A.aceti ssp xylinum* ALDC fusions with α-aglutinin (AGa1Cp) produces a whole-cell biocatalyst capable of eliminating the accumulation of diacetyl in fermented wort. Another application was done with a proline-rich peptide QPF attached to its cell wall [1,2].

YEAST SURFACE DISPLAY STRATEGY

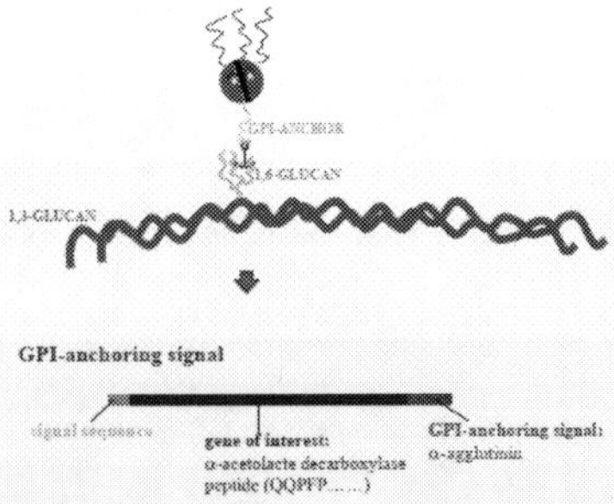

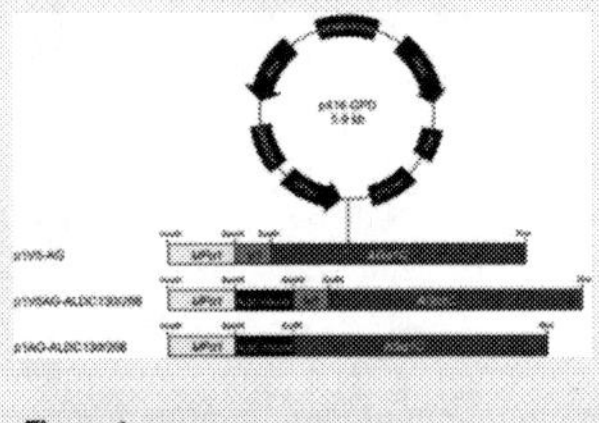

Figure 1 Organization of expression vectors for the surface display of αα-acetolactate decarboxylase (ALDC) fusions to V5-AGα1Cp

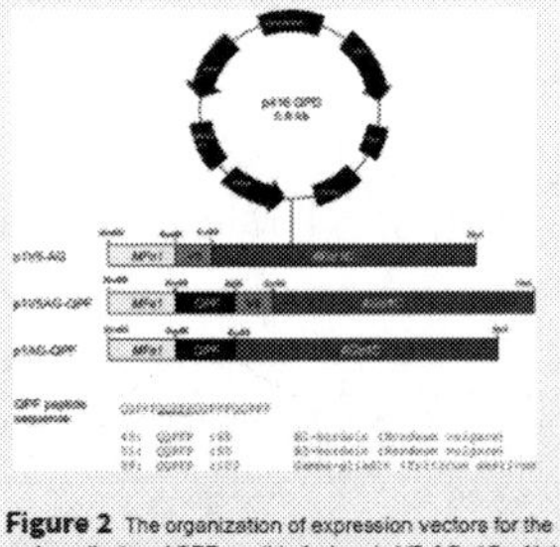

Figure 2 The organization of expression vectors for the surface display of QPF peptide fusions to V5-AGα1Cp. Note that the 21-amino acid sequence of MFα1 leader peptide and C-terminal 33-amino acid fragment of AGα1Cp are removed in S. cerevisiae during the recombinant protein targeting and C-terminal GPI anchor attachment, respectively . The occurrence of the QQPFP sequence in hordeins and gamma-gliadin is shown at the bottom

RESULTS

Figure 3 Growth curves and diacetyl concentrations during wort fermentation

a Fermentation with *S. cerevisiae* W303-1A (p1V5-AG).
b Fermentation with *S. cerevisiae* W303-1A (p1V5AG-ALDC208).
c Fermentation with *S. cerevisiae* W303-1A (p1AG-ALDC208)

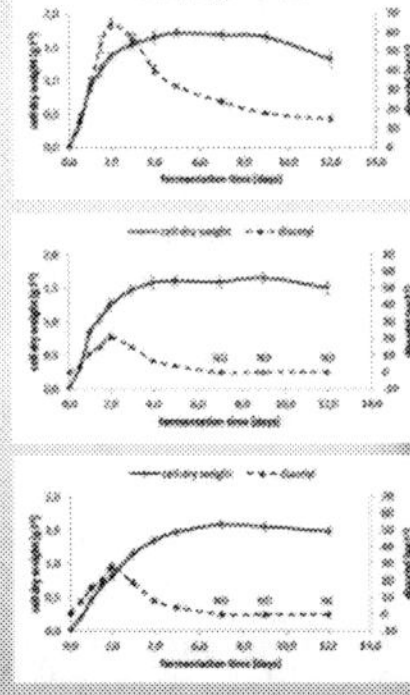

RESULTS

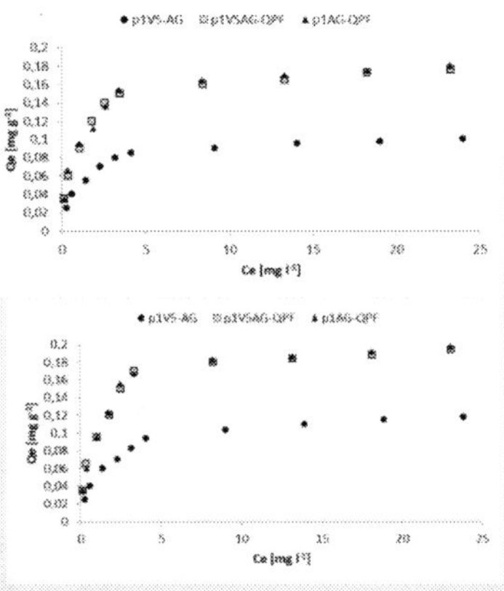

Figure 4 Isotherm equilibrium data for catechin adsorption (A) and epicatechin adsorption (B) onto yeast cells

CONCLUSIONS

Two yeast strains were prepared by surface display technique:

1) with ALDC activity,

2) with repetetive QQPFP peptides

- ✓ Display of ALDC activity on the cell wall of wort-fermenting yeasts was expected to rapidly decrease the concentration of the undesirable diacetyl by-product.
- ✓ Our data demonstrate that expression of ALDC activity on yeast cell walls prevented accumulation of diacetyl in the wort and the remaining low concentration was readily removed by yeast metabolism. This approach can, therefore, be regarded as an effective way to decrease the accumulation of diacetyl during wort fermentation and thus accelerate the costly process of beer maturation.
- ✓ The QPF peptide served to bind polyphenols during fermentation and, thus, to decrease their concentration.
- ✓ Strains displaying QPF were able to bind about twice as much catechin and epicatechin as a control strain displaying only the anchoring domain.
- ✓ Depending on the concentration of yeast, uptake of polyphenols was 1.7–2.5 times higher.

REFERENCES

[1] Rudolf Cejnar, Kateřina Hložková, Pavel Kotrba, Pavel Dostálek (2016) Surface-engineered Saccharomyces cerevisiae displaying α-acetolactate decarboxylase from *Acetobacter aceti ssp xylinum*. Biotechnol. Lett. 38: 12. 2145-2151.

[2] Rudolf Cejnar, Kateřina Hložková, Lukáš Jelínek, Pavel Kotrba, Pavel Dostálek (2017) Development of engineered yeast for biosorption of beer haze-active polyphenols. Appl. Microbiol. Biotechnol. 101: 4. 1477-1485.

Acknowledgement: This work was supported by TA CR "Centre for Innovative Use and Strengthening of Competitiveness of Czech Brewery Raw Materials and Products" – TE02000177.

UCC
Coláiste na hOllscoile Corcaigh, Éire
University College Cork, Ireland

36th CONGRESS EBC 2017 EUROPEAN BREWERY CONVENTION LJUBLJANA 14-18 MAY

36th Congress of the European Brewery Convention

Sour brewing: A comparison of acidification methods in the brewing process on wort and beer qualities

Lorenzo Peyer[1], Martin Zarnkow[2], Fritz Jacob[2], David De Schutter[3], and Elke Arendt[1]

[1] *Cereal and Beverage Science, School of Food and Nutritional Sciences, University College Cork, Cork, Ireland*
[2] *Forschungszentrum Weihenstephan für Brau- und Lebensmittelqualität, Technische Universität München, Freising-Weihenstephan, Germany*
[3] *Fund Baillet Latour, Leuven, Belgium*

Introduction

Sour beers are experiencing a renaissance. Along with the traditional method used in the production of Belgian *Lambics* or German *Berliner Weisse*, involving acidification and ageing taking place over several months, increasing demand for quicker production of acidic beers has called for alternative approaches that circumvent the long fermentation and maturation times. Commonly, this involves an acidification of mash and/or wort before or during alcoholic fermentation, and lactic acid bacteria (LAB) has been the preferred acidifying agents. The choice on when and how to acidify along the brewing process, however, could have implications on the final beer properties. To date, very little research has been conducted to understand the impact that LAB acidification methods cause on the processing and product quality of sour beers. This study aims to compare the fermentative profiles as well as the technological and final attributes of three beers acidified differently throughout the brewing process.

Material and Methods

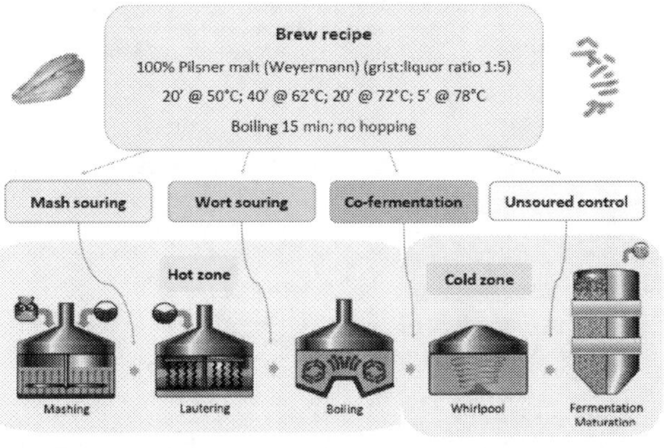

Fig. 1. Souring trials done at different stages along the brewing process.

Lactic acid fermentation
Lb. amylovorus FST2.11
7 log CFU/ml
40°C for 18 h
→ **5-6 g/L lactic acid**

Alcoholic fermentation
Saccharomyces cerevisiae (Safale US-05)
7 log CFU/ml
20°C for 2 weeks (primary)
0°C for 2 weeks (maturation)

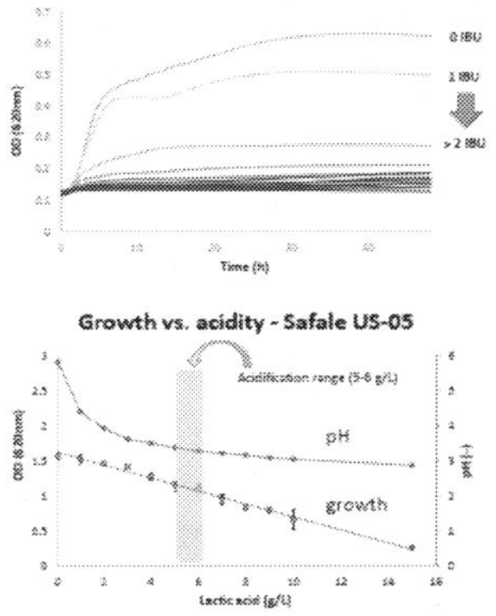

Fig. 2. Growth of *Lb. amylovorus* FST2.11 and Safale US-05 in relation to the addition of iso-α-acids (above) and lactic acid (below) in wort, respectively.

Results

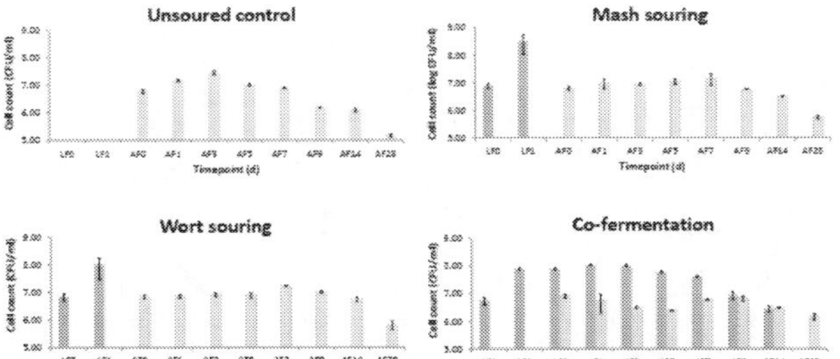

Fig. 3. Growth of *Lb. amylovorus* FST2.11 (green) and Safale US-05 (yellow) (above) and fermentation profiles (below) during lactic (LF) and alcoholic (AF) fermentation.

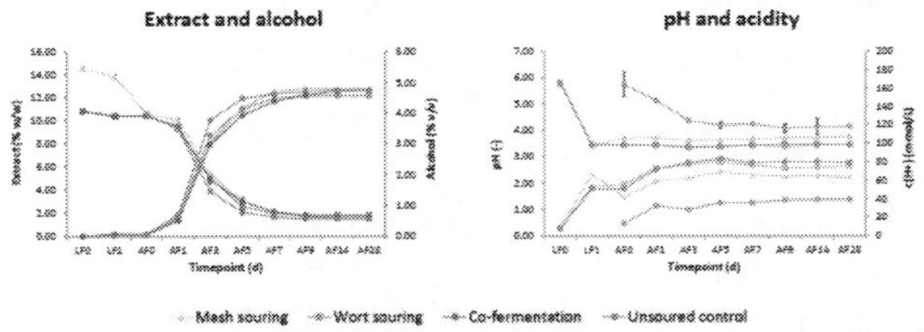

Table 1. Metabolic output, qualitative attributes and flavour compounds analysed for the different treatments

	Lactic acid production	LAB nutrients consumption	Lautering times	Haze	Colour	Foam
Unsoured control	-	-	-	-	-	-
Mash souring	↑(↑)	↓	↓/↑	↓	↓(↓)	↑
Wort souring	↑	↓(↓)	-	↓	↓(↓)	↓
Co-fermentation	↑	↓(↓)	-	↑	↓	-

	Isoamyl acetate	Diacetyl	Acetoin	Acetaldehyde	Isovaleric acid	Sensory
Unsoured control	-	-	-	-	-	typical
Mash souring	↑	-	-	↑↑	-	not clean
Wort souring	(↑)	-	-	↑	-	clean
Co-fermentation	-	↑	↑↑	↑	↑	astringent

Lb. amylovorus FST2.11 was selected for its fast acidification in wort, moderate resistance to alcohol, as well as hop sensitivity. The lactic culture consumed maltose preferentially over monosaccharides, and uptake of free amino nitrogen was limited (up to 13 mg/L). Yeast growth was delayed by 2 to 4 days in the soured trials, leading to a faster attenuation in the unsoured control. During co-fermentation, lactobacilli outnumbered yeast cells (30:1 at day 5) until day 9.

Among the soured beers, lowest levels of off-flavours were found in wort-soured beers, while co-fermented beers led to opalescent beers containing high levels of diacetyl and acetoin. The low pH and the high level of staling compounds such as acetaldehyde suggest that sour beers suffered from low oxidative stability.

Conclusions

Final beer qualities varied considerably depending on the souring method applied. Souring of wort by *Lb. amylovorus* FST2.11 before the boiling step (= *wort souring*) was found to produce bright, tart beer with minor organoleptic failures while limiting the risk of bacterial cross-contamination within the brewing facility.

Fonds Baillet Latour

This study was funded by the Baillet Latour Fund within the framework of a scholarship for doctoral students.

Technische Universität München

Microbial Beer Stabilization

Dipl. Ing. Reiner Gaub, Pall GmbH - Germany
Dipl. Ing. Ursula Brendel-Thimmel, Pall Filtersystems GmbH Germany

PALL Pall Corporation

Introduction

Consumers expect a microbial safe product when they have a beer. Due to the natural preservative effects of beer components, pathogenic risks are low and easily mitigated by the production process. However, there is a group of beer spoilage microorganisms which, if present, can negatively affect the taste, odor and clarity. The microbial quality of beer is therefore the most important quality attribute in order for brewers to sustain their business.

Qualification and measurement of microbial reduction processes and their barrier effectiveness

The microbial quality of a product and measurement of microbial treatment effectiveness is well defined in the pharmaceutical industry (1,2). The Food and Beverage industry has no overall standard to qualify effectiveness of microorganisms reducing process steps . Depending on the applied technology and the supplier, different ways of qualifying microbial effectiveness are in use. In general, the amount of microorganism before and after treatment is measured and put into relation to each other. This relation is known as Titer Reduction (TR) and defined as:

$$TR = \frac{\text{Total colony count upstream CFU*}}{\text{Total colony count downstream CFU}}$$

(*CFU = colony forming units)

As the ratio is usually a very high number and inconvenient to handle, microbiologists often use the term Log Reduction Value (LRV), which is the Log 10 of the titer reduction; also percentage removal efficiency is used (Table 1).

Challenge tests are established to assess microbiological reduction capabilities under controlled conditions with standard test organisms. Those have been well studied and thus guaranty reliable results, e.g. *Brevundimonas diminuta* or *Serratia marcescens*.
Challenge tests can also get done with typical beer spoilers in beer for further characterization under conditions which are closer to reality. In such tests the filters are challenged with a very high number of microorganisms in order to mimic worst case. The results from the colony counts in unfiltrate and filtrate are calculated into Titer Reduction or LRV(3).

LRV	Titer Reduction	% Removal Efficiency
1	10:1	90
2	100:1	99
3	1.000:1	99,9
4	10.000:1	99,99
5	100.000:1	99,999
6	1.000.000:1	99,9999
7	10.000.000:1	99,99999
8	100.000.000:1	99,999999
9	1.000.000.000:1	99,9999999

Table 1: TR – Various ways of representing

In addition to laboratory based microbial qualification full scale challenge tests can also be applied on installed equipment in the brewery using harmless variants of beer spoiling microorganisms (7).

Filter selection by removal characteristics

Against common understanding pore size of a filter is determined by the removal characteristics of microorganisms and not by a physical measurement of pore size. Different filter manufacturers may all offer 0,45 µm pore sized filters but with different microbial reduction capabilities.

	Product A	Product B	Product C	Product D
Poresize	0,45 µm	0,45 µm	0,45 µm	0,45 µm
LRV *Lactobacillus brevis*	>8	7,4	--	>10
LRV *Serratia marcescence*	11	7,4	>9	>8

Table 2: Log reduction values for 0,45 µm membrane cartridges from different brands

Thus the selection of the best suited final filter for beer should always be based on microbiological removal data (3). It allows to choose the filter in direct relation to the expected or known contamination with spoilers and the probability for a bottle to be contaminated. Published LRV data should be based on proper statistics in terms of the number of samples and confidence intervals.

Thermal reduction of microorganism

With thermal treatment, Pasteurization Units (PU) are used to express the measure of treatment effectiveness. For pasteurization a few published papers are available, which put pasteurization units into relation with a direct measured titer reduction.

Zufall et.al. and Feilner et.al. published data for the inactivation of specific beer spoiler considering the impact from high temperature/short time pasteurization(4,5). Results show a less effective inactivation at higher temperature/shorter exposure time especially for *Lactobacillus brevis* and *Saccharomyces diastaticus*.

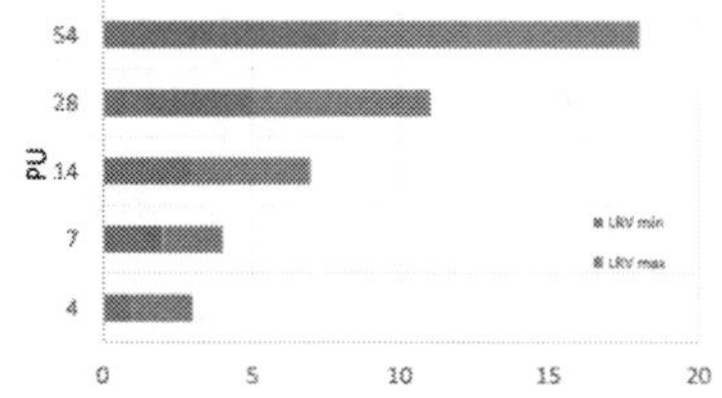

Graph 1 Log reduction values at different Pasteurization Unit levels

Both papers report a none linear relationship between microbial load and inactivation efficiency. Thus the correlation factor of 1.39 used for the calculation of the PU units in breweries is differing. Deviation results from system design, type of microorganisms and temperature level. This forces brewers to apply a higher amount of pasteurization units to the product to compensate for variability in the PU factor.

Microbial removal steps – impacts in practice

Modern cleaning regimes combined with microbial control concepts reduce contamination risk significantly. However, the size and complexity of brewery installations, the increased number of processed brands and the ability of microorganisms to adapt to changing conditions are increasing the contamination risk, often in unexpected locations and situations. Cell counts in process easily reach 10^7 to 10^9 CFU/ml in hidden pockets of an installation and the .
In final packaged product, beer processing accounts for 40% and packaging accounts for 60% of spoiled product (6). Main contaminants in the production process are still *Lactobacillus spec.* and *Pediococcus spec.*, wild yeast is minor. While the species stays unchanged, the ability to adjust to various conditions is increasing. This microbial flexibility in combination with variations in beer type and mix provides new challenges to brewers and requires a different view on safety.
Key is to assess microbial separation efficiency for each individual production step and have a risk profile on hand for the various scenarios.

LRV by process step	*Saccharomyces sp.*		*Lactobacillus brevis*		*Pediococcus damnosus*		Data Source
	min	max	min	max	min	max	
Centrifuge	2	6	0	1	0	1	Lit 7
DE Filter	2	7	0	3	0	3	Field data
PROFi System	7	9	2	5	2	4	Lit 7
Particle Filter	0	1	0	1	0	1	Field Data
Fine filter	2	9	1	5	1	5	Supplier data sheet
Membrane Filter	>12	>12	6	9	6	9	Supplier data sheet
Flash Pasteurizer	4	7	4	7	4	7	Lit 4;5
Tunnel Pasteurizer	4	>6	4	>6	4	>6	Supplier data sheet

Table 3 Overview typical log reduction values for different treatment steps

Depending on the installed system arrangement and way of operation, the microbial removal effectiveness of the installed steps varies in a wide range. To ensure sufficient safety, each brewer needs to define the brewery specific removal level according to the actual situation. The following graphs gives variations in microbial safety for different arrangements expressed as total LRV

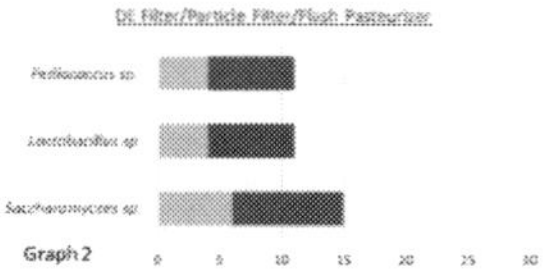

Graph 2

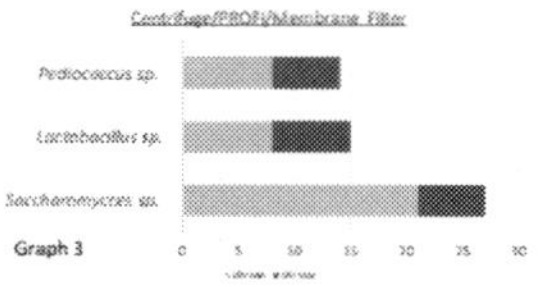

Graph 3

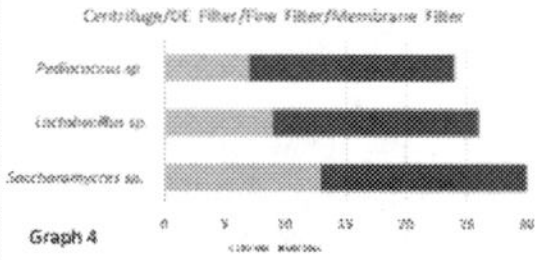

Graph 4

Graphs 2-4
Variations in LRV at three installation examples:
2. DE Filter/Particle Filter/Flash pasteurizer
3. Centrifuge/PROFi/Membrane filter
4. Centrifuge/DE Filter/Fine Filter/Membrane Filter

Microbial removal steps – effectiveness on product safety

The risk of likelihood for contamination in a packed container depends on:

- The sum of separation/inactivation steps with the different titer reductions reductions
- Upstream contamination (type of microorganism and quantity)

A *Lactobacillus* contamination in unfiltrate at the level of 10^6/ml (= 10^9/l) is reduced to different contamination levels in packaged product for the above shown combinations as follows

	Inlet CFU/L	LRV	LRV	Outlet CFU/l		Potentially contaminated bottles at max LRV	
Lactobacillus brevis		min	max	max	min	%	min
DE/Particle/Flash Pasteurizer	10^9	4	11	10^5	10^{-2}	1	1 out of 100
Cent./PROFi/Membrane	10^9	8	15	10^1	10^{-6}	0,000001	1 out of 1.000.000
Cent./DE/Fine/Membrane	10^9	7	17	10^0	10^{-8}	0,00000001	1 out of 100.000.000

Table 4 Log reduction values and theoretical contamination levels for three different combinations

The example shows that with heavy upstream contamination even a multi-step filtration and separation concept might fail if conditions are not well controlled.
Selecting the correct filter materials or applying the relevant Pasteurization Units, significantly lowers the risk of contamination in the final product.

Conclusion

Microbial stabilization of beer depends on the hygienic status in process, and the effectiveness in microorganism reduction for the different treatment steps. Each individual process step contributes to the overall microbial reduction in a certain range. Thus safety of the final product is determined by the optimum combination of the different treatment steps and how they are operated.
Membrane filtration gives best product safety without compromising taste. Fine filtration is an important contributor to microbial safety but not as effective as membranes. Thermal treatments show in general lower effectiveness compared to membranes.

Literature

1. FDA Guidance for Industry - Sterile Drug Products Produced by Aseptic Processing Current Good Manufacturing Practice (Section IX. Validation of Aseptic Processing and Sterilization), 2004
2. ASTM No. F838-05, rev. 2005
3. Brendel-Thimmel U.et al,: Sterilfiltration von Flüssigkeiten und Gasen, November 2006, 1655, CITEAH78/11 1579-1750 ISSN 0009-286X, 2006
4. C.Zufall , K.Wackerbauer,: ,: The biological impact of flash pasteurization over a wide temperature interval; Journal of The Institute of Brewing 106 (2000), 3, 166 ff.
5. Feilner,R. Oehmichen,T.: .: Flash pasteurization of filtrated and wheat beer with one heat holding tube; Brewing Science 69 (2016), Sept/Oct
6. Back, W.: Die biologische Situation durch Bierschädlinge im Jahr 2008, 42.Technologisches Seminar 2009, Freising Weihenstephan, Januar 2009
7. Back W.; Bohak I.; Krottenthaler M.: Biologische Abnahme PROFi Anlage.; Gea/Pall Studie 02/2005

DAUSCH Technologies

Deeper Insight into Results of a Round Robin Test by Multivariate Data Analysis

B. Kessler [1], W. Kessler [2], M. Dausch [1], N. Rettberg [3]

[1] Dausch Technologies GmbH, D-76829 Landau
[2] Steinbeis Transferzentrum für Prozesskontrolle und Datenanalyse, D-72764 Reutlingen
[3] Versuchs- und Lehranstalt für Brauerei in Berlin e.V., FI Instrumentelle Bier- und Getränkeanalytik, D-13353 Berlin

Introduction and Method

Between 2008 and 2016, under the guidance of VLB, 40 breweries took part in a round robin test to characterize 7 beer types from 6 different German breweries. Numerous independent scientists performed the standard tests by using similar methods and equipment. These data were measured as described in DIN ISO 5725 Part 1 and 2.

In many data structures there is a high redundancy of the measured parameters. Thus, a principal component analysis (PCA) is performed to extract the most important information. PCA is a chemometric method that splits a two-dimensional data table – in this case comprising 233 different samples and 11 different parameters – into a bilinear model of latent variables, the so-called principal components (PC). The principal components are calculated so that they explain as much variance of the data as possible. The objects, in our case the different beers, are then described in a lower dimensional PC space. A PCA allows a linearly independent clustering and structuring of the samples and facilitates the interpretation of the data.

Results and Discussion

The common way to describe the correlations between quality parameters is to calculate the correlation coefficients r. Table 1 shows the correlation coefficients for the 11 measured parameters. The intensity of the colour indicates the correlation strength. 'Red' means a positive correlation, 'blue' a negative one, 'no colour' indicates that there is no correlation.

	E_{app} W-%	E_r W-%	Alc. W-%	Alc. V-%	E_o W-%	pH-value	CO_2	Bitterness	Colour	Foam	Turbidity
E_{app} W-%	1,00	0,97	0,74	0,74	0,95	0,66	0,36	-0,15	0,28	-0,08	-0,28
E_r W-%	0,97	1,00	0,88	0,88	0,99	0,76	0,17	-0,00	0,08	-0,05	-0,32
Alc. W-%	0,74	0,88	1,00	1,00	0,88	0,77	-0,20	0,22	-0,33	0,03	-0,43
Alc. V-%	0,74	0,88	1,00	1,00	0,88	0,76	-0,20	0,24	-0,33	0,04	-0,40
E_o W-%	0,95	0,99	0,88	0,88	1,00	0,75	-0,08	0,14	-0,18	0,01	-0,37
pH-value	0,66	0,76	0,77	0,76	0,75	1,00	-0,01	0,29	-0,29	0,04	-0,26
CO_2	0,36	0,17	-0,20	-0,20	-0,08	-0,01	1,00	[illegible]	[illegible]	-0,20	0,20
Bitterness	-0,15	-0,00	0,22	0,24	0,14	0,29	[illegible]	1,00	[illegible]	0,14	0,67
Colour	0,28	0,08	-0,33	-0,33	-0,18	-0,22	[illegible]	[illegible]	1,00	-0,11	0,06
Foam	-0,08	-0,05	0,03	0,04	0,01	0,04	-0,20	0,14	-0,11	1,00	-0,05
Turbidity	-0,28	-0,32	-0,43	-0,40	-0,37	-0,26	0,20	0,67	0,06	-0,03	1,00
Legend	1,00	0,80	0,60	0,40	0,20	0,00	-0,20	-0,40	-0,60	-0,80	-1,00

Tab. 1: Correlation matrix of the different quality parameters

Extract (real and apparent), alcohol and the pH-value all show a correlation coefficient higher than 0.8. All the other parameters don't show a clear correlation. This representation summarizes the measured parameters but does not classify the beers.

With a PCA it is possible to get a summary on how the correlations between the measured parameters are as well as a clustering according to the similarities of the beers. The beer clusters are shown in the scores plot (Fig. 1), and the correlations of the measured parameters are shown in the loadings plot (Fig. 2).

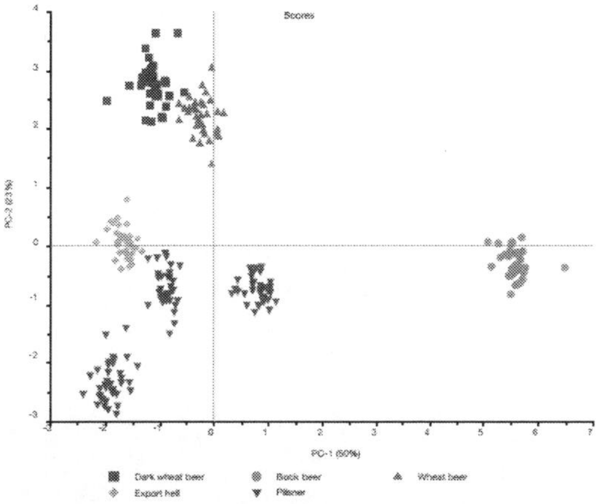

Fig. 1: Scores plot with different clusters of various beer types

The cluster of Bock beer is very different to the others. Bock beer has the highest score in PC 1 and Export hell the lowest. Surprisingly both are neutral to PC 2. The three different Pilsner beers can be differentiated mainly by PC 1 but also to a certain extend by PC 2. The results of the two different Wheat beers are very close to each other on PC 1 and PC 2.

Fig. 2 shows the loadings plot of PC 1 and 2. Alcohol (weight- and volume-%), extract (real and apparent) and pH-value represent PC 1 (see high correlations in table 1, mainly composition). PC 2 is orthogonal to PC 1 and represents mainly colour, CO_2 and also bitterness and turbidity which have different signs in the loadings (mainly appearance and taste). With regard to these data this means that beer with a higher CO_2 level has lower bitterness and turbidity on average. Neither PC 1 nor PC 2 inclose foam.

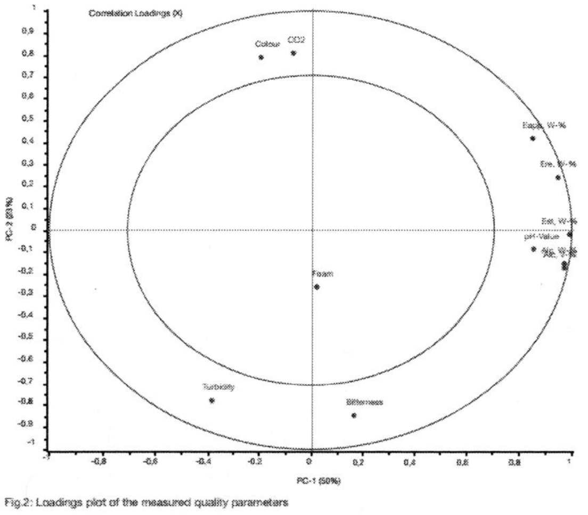

Fig.2: Loadings plot of the measured quality parameters

Conclusion and Perspectives

The PCA allows to characterize the measured quality parameters according to their intercorrelations. Despite the assumption that the same beer is produced all the time the PCA shows that the beers within one type (e.g. Pilsner) are different to each other. Approaches like the optical spectroscopy used as inline analysis could offer a chance to qualify the beers during production. This gives the opportunity for adapting the brewing process to a defined quality. These approaches have already been verified for different quality parameters in the beverage industry and will soon be expanded to more parameters.

We acknowledge the support of VLB Berlin for providing the data of the Round Robin test.

European Brewery Convention
c/o The Brewers of Europe
Rue Caroly 23-25
1050 Brussels
Belgium

ISBN 978-1-5108-5391-1

9 781510 853911

CURRAN ASSOCIATES INC.
proceedings
.com